Lecture Notes in Electrical Engineering

Volume 1427

The book series *Lecture Notes in Electrical Engineering* (LNEE) publishes the latest developments in Electrical Engineering—quickly, informally and in high quality. While original research reported in proceedings and monographs has traditionally formed the core of LNEE, we also encourage authors to submit books devoted to supporting student education and professional training in the various fields and applications areas of electrical engineering. The series cover classical and emerging topics concerning:

- Communication Engineering, Information Theory and Networks
- Electronics Engineering and Microelectronics
- Signal, Image and Speech Processing
- Wireless and Mobile Communication
- Circuits and Systems
- Energy Systems, Power Electronics and Electrical Machines
- Electro-optical Engineering
- Instrumentation Engineering
- Avionics Engineering
- Control Systems
- Internet-of-Things and Cybersecurity
- Biomedical Devices, MEMS and NEMS

For general information about this book series, comments or suggestions, please contact leontina.dicecco@springer.com.

To submit a proposal or request further information, please contact the Publishing Editor in your country:

China

Jasmine Dou, Editor (jasmine.dou@springer.com)

India, Japan, Rest of Asia

Swati Meherishi, Editorial Director (Swati.Meherishi@springer.com)

Southeast Asia, Australia, New Zealand

Ramesh Nath Premnath, Editor (ramesh.premnath@springernature.com)

USA, Canada

Michael Luby, Senior Editor (michael.luby@springer.com)

All other Countries

Leontina Di Cecco, Senior Editor (leontina.dicecco@springer.com)

**** This series is indexed by EI Compendex and Scopus databases. ****

Surender Reddy Salkuti
Editor

Artificial Intelligence for Integrated Smart Energy Systems in Electric Vehicles

Springer

Editor
Surender Reddy Salkuti
Department of Railroad and Electrical Engineering
Woosong University
Daejeon, Korea (Republic of)

ISSN 1876-1100 ISSN 1876-1119 (electronic)
Lecture Notes in Electrical Engineering
ISBN 978-3-031-94275-4 ISBN 978-3-031-94276-1 (eBook)
https://doi.org/10.1007/978-3-031-94276-1

This Springer imprint is published by the registered company Springer Nature Switzerland AG
The registered company address is: Gewerbestrasse 11, 6330 Cham, Switzerland

Preface

In today's ever-evolving energy landscape, renewable energy sources (RESs) have emerged as a vital component of modern power systems and it has presented profound challenges and opportunities to the power industry. The extensive use of electric vehicles (EVs) gaining importance due to low maintenance cost, zero emissions, and dependency on fossil fuels. These RESs and EVs have gained immense popularity for their environmental benefits, helping nations reduce their carbon emissions and combat climate change. The inherent intermittency and variability of RESs have posed significant challenges to system operators and changed many aspects of power system control, operations, and planning. The growing popularity of EVs poses unique challenges for grid operators and infrastructure, which requires effectively managing these vehicles' integration into the grid. While EVs offer a greener alternative to traditional internal combustion engine vehicles, their charging patterns introduce additional load fluctuations into the grid. EVs are typically charged during peak demand hours, which can strain the grid and further exacerbate frequency deviations.

The uncertainties related to wind turbines, solar photovoltaics (PVs), and other uncertain factors increase the uncertainty of the grid. The incorporation of EVs into the utility grid further adds complexity, as their charging patterns introduce additional load fluctuations. One of the main barriers to accepting EVs is the range anxiety that is the limitation of charging stations, which consists of high-charge batteries and high-energy charging infrastructure. Electric vehicle charging stations (EVCSs) must be integrated with the distribution system to charge the EVs. However, the integration of EVCSs may cause new problems in smart energy systems such as voltage instability, harmonics, power imbalance, and low power factor. These problems may not be solved by analytical methods, and need to be solved using learning algorithms within the field of Artificial intelligence (AI). In recent years, AI has seen fast growth and evolution in the realization of modern hybrid grids. AI-based approaches can be employed in modern hybrid grids for load forecasting, renewable energy forecasting, planning, operation, fault diagnosis and protection of various equipment in electrical networks, transient stability, and control, analysis of the behavior of customer electricity consumption, etc. The book consists of 33 chapters.

This book presents the emerging technologies that are required for sustainable and resilient future smart grids.

Chapter "Artificial Intelligence in Electric Vehicles for Sustainable Driving" examines the advantages of eco-driving EVs and several AI-powered methods to improve eco-driving behaviors. These methods include traffic prediction, route planning, intelligent speed adaptation, adaptive cruise control, and more. This chapter also assesses how well various AI-based techniques work in terms of user acceptability, security, comfort, and energy efficiency. The practical consequences and problems of deploying AI-driven eco-driving solutions are clarified by analyzing user experiences and performance indicators. The chapter "Modernization of Electric Grids for Charging of Electric Vehicles" outlines the practical applications of more recent methods to advance the study of efficient grid-connected EV charging. By utilizing modern analytical and optimization skills, researchers can look into several aspects of grid charging efficiency. This enables them to offer perspectives and solutions that are critical for sustaining grid sustainability and stability while assisting the growing EV market. Chapter "Mitigating Impacts of Electric Vehicle Charging Stations to the Distribution Systems by Optimal Operation of Soft Open Point" presents a method to limit the impact of EVCSs on voltage deviation and power loss of the distribution system based on optimizing Soft Open Points (SOP) operation installed in the distribution system. The optimal operation power for SOP at each hour of the day is determined by the Archimedes Optimization Algorithm (AOA). The evaluation results of the proposed solution are performed on a 33-bus system integrating a Photovoltaic System (PVS) system and an EVCS for twenty-four hours. The results show that optimizing SOP operation can limit the impact of EVCSs on the distribution system by improving the voltage profile and reducing power loss of the system.

Chapter "Performance Evaluation of Artificial Neural Networks for Electric Vehicle State of Charge Estimation Across Different Driving Cycles" provides a detailed analysis of SOC estimation in EVs, focusing on the application of Artificial Neural Networks (ANNs) and comparing their performance with the Extended Kalman Filter (EKF) and coulomb counting methods. The study is conducted using the Tesla Model S across two distinct driving conditions: the Highway (HW) driving cycle and the City (CY) driving cycle, providing a comprehensive evaluation across varied operational environments. The results show that the ANN method consistently outperforms the EKF, achieving up to 27.4% lower Root Mean Square Error (RMSE) in the highway driving cycle, indicating superior accuracy in SOC estimation. Chapter "GJO-Pattern Search Algorithm Based DG and Capacitor Placement in Distribution Network with Zone Based Installation of EVCSs" presents the adaptation of EV charging stations (EVCSs) in radial distribution networks (RDNs) raises particular issues. Proper coordination of optimally placed distributed generations (DGs) and capacitors integrated with EVCSs can mitigate the effects of increased active power loss, a decline in voltage stability margin, deterioration in voltage profile, and a worsening in network dependability caused by EVCS integration. The distribution test systems are divided into different zones. The location-based installations of the EVCS technique ensure effective coverage of EVs and reachable

EVCS infrastructure across the networks. Chapter "Optimizing Renewable Energy Integration and Electric Vehicle Charging Stations in Distribution Networks Using Hybrid Whale Optimization Algorithm" introduces a methodology for integrating renewable energy sources (RESs) and electric vehicle charging stations (EVCSs) into radial distribution networks, leveraging a novel Hybrid Whale Optimization Algorithm (HWOA). The efficacy of this algorithm is thoroughly examined through the analysis of IEEE 33 and IEEE 69 bus radial distribution systems (RDSs), which serve as test cases to evaluate its performance and suitability for real-world applications. The HWOA is specifically tailored to tackle the intricate optimization tasks associated with RESs and EVCSs integration, with the primary objectives of minimizing power losses, improving voltage stability, and guaranteeing adequate coverage of EVCSs across the distribution network. The chapter "An Introduction to Energy Management in Smart Energy Systems" presents the integration of energy storage systems and their role in stabilizing the grid amid variations in renewable generation. Advanced forecasting techniques for renewable energy and load demand are also introduced to alleviate the effects of variability. Furthermore, hybrid energy management systems combining rule-based approaches with optimization algorithms are explored for improved efficiency. The potential of demand-side management in reducing peak loads and enhancing grid reliability is emphasized, highlighting its critical role in future energy systems.

The chapter "A Review of AI-Driven Energy Optimization Strategies for Electric Vehicles" explores the pivotal role of Artificial Intelligence (AI) in advancing smart energy management for EVs. Covering various aspects of AI applications, such as predictive modeling for energy consumption, adaptive control strategies, genetic algorithms for charging station optimization, and integration of AI techniques for synergistic benefits. Beginning with predictive modeling, we explore how AI augments predictive analytics in EVs, delving into data sources, feature selection, and applications in forecasting energy consumption and scheduling maintenance tasks. This chapter further delves into AI-driven adaptive control strategies, detailing the techniques involved in managing EV energy efficiently, including dynamic energy distribution algorithms, optimal charging strategies, and real-time adaptation to grid conditions and user preferences. The chapter "AI-Powered Strategies for Efficient EV Energy Management" provides a comprehensive overview of the methodologies and technologies driving AI applications in EV energy management. By harnessing the power of AI algorithms, sensor integration, and edge computing, we unlock new possibilities for optimizing efficiency, enhancing safety, and reducing environmental impact in the transportation sector. The journey towards a cleaner, greener, and more sustainable transportation ecosystem is ongoing, and AI remains a crucial enabler in this endeavor. V2G systems facilitate the bidirectional flow of electric current between automobiles and the power grid, offering benefits such as grid stabilization, demand response, and the integration of RE. Chapter "CdS-LDR Optimization in Dual-Axis Solar Tracking System with NPC Multilevel Inverter for Electric Vehicle Application" gives an idea to implementation and design a dual-axis solar tracker using light dependent resistor, 3-phase Neutral Point Clamped multilevel inverter,

IR2110 switch gate driver, Arduino Uno, motor driver, PIC microcontroller generating pulse width modulation signals, ULN2003 motor driver, 12 V Permanent Magnet DC motors, and a buck-boost converter. In addition, the project utilized the tracker's design, Lambert's cosine law, and the inverse square law. The dual-axis solar tracker was designed to improve the efficiency of solar panels by tracking the sun's movement.

Chapter "Development of Artificial Intelligence Based Advanced Battery Management System with Enhanced Charging Monitoring and Fire Protection for Electric Vehicles" presents the design and development of an advanced battery management system (ABMS) leveraging artificial intelligence (AI) which mainly concentrates on enhanced charging monitoring and fire protection to the battery pack in EV. The developed ABMS integrates with efficient sensing technologies and AI-based computational algorithms to provide a real-time battery pack module with cell balancing mechanisms to monitor the charging process which ensures the optimal charging conditions and in turn, maximizes the battery lifespan in EVs. In the current research work, within the hardware model, a fire protection system is implemented by utilizing AI to detect early signs of thermal runaway which ensures fire safety in BMS. The experimental validation is done on the proposed ABMS through comprehensive testing in the simulation model and real-time operating conditions. Chapter "A Comparative Analysis of A* and Ant Colony Algorithms for EV Routing" delves into the Electrical Vehicle Routing Problem (EVRP), by considering the unique requirements of electric vehicles, such as battery range limitations and the availability of charging stations. A significant addition to this work is the inclusion of different driving modes such as Eco, Ride, Sport, and Warp which influence energy consumption rates and travel times. Also, it aims to optimize routes not only for efficiency but also by factoring in these variable driving modes to adapt to drivers' preferences or situational needs. The A* algorithm is employed for its effectiveness in quickly determining the shortest and least costly path by estimating future costs. Chapter "AI-Driven Optimization Techniques for Electric Vehicle Charging Infrastructure" explores various AI-driven optimization techniques tailored specifically for EV charging infrastructure. Reinforcement learning algorithms enable charging stations to learn and adapt their behavior based on real-time feedback, resulting in improved scheduling and resource allocation. Genetic algorithms provide an evolutionary approach to optimizing charging infrastructure layout and operation, considering factors such as charging demand, grid capacity, and geographical constraints. Neural networks offer predictive capabilities for forecasting charging demand, optimizing charging schedules, and mitigating peak load demands on the grid. Real-world implementations and case studies are examined to showcase the effectiveness of AI-driven optimization techniques in EV charging infrastructure.

Chapter "AI-Driven Innovations in Electric Vehicle Energy Management: Challenges and Opportunities" explores how AI technologies optimize integrated smart energy systems within EVs with a view of their transformative potential and the key challenges involved. Energy management algorithms driven by AI allow EVs to adaptively regulate their power use and charging patterns based on real-time data from sources such as traffic conditions, customer preferences, and network dynamics.

By applying predictive analytics, EVs can predict energy demand, optimize routing options, and minimize battery charging costs leading to maximum energy efficiency while reducing environmental pollution. Chapter "Dead-Beat Predictive Control of a Grid-Tied PV-Hybrid Energy Storage System" presents a three-phase, four-wire grid-integrated PV-hybrid energy storage system (HESS) controlled by a predictive dead-beat control. With faster dynamics and enhanced steady-state response, the predictive dead-beat control emphasizes quick error reduction between reference and grid currents for generating inverter switching signals. The proposed control enables the system to perform active and reactive power control, harmonics elimination, neutral current compensation, and load balancing at the point of common coupling (PCC) to maintain DC bus and AC bus voltage stability, even during dynamic conditions. Chapter "Automated Open Circuit Fault Detection in Cascaded H-Bridge Multilevel Inverters Using Artificial Intelligence Technique for EV Application" proposes an automated fault detection method for identifying open circuit (OC) and source faults in CHBMLIs using artificial neural networks (ANN). The ANN is trained using a dataset comprising voltage and current measurements under normal operating conditions and various fault scenarios. Once trained, the ANN is capable of accurately detecting OC faults in real time, thus enabling proactive maintenance and preventing potential system downtime. Simulation results using MATLAB/Simulink demonstrate the effectiveness and reliability of the proposed fault detection approach, highlighting its potential for enhancing the reliability and performance of CHBMLI systems in practical applications.

Chapter "Introduction to Smart Grids and Its Key Components" addresses the need for smart grids (SGs), key terminologies and advantages of SG technologies, converters, and maximum power point tracking methodologies available in SG, charging infrastructure, and artificial intelligence techniques. Customers may take an active role in energy management through demand response, which allows them to react to pricing signals and modify their energy use appropriately. Integration of electric cars with the infrastructure needed for charging them, enabling grid balancing and environmentally friendly mobility. Chapter "A Resilient Passivity-Based Control Approach for a Two-Input DC-DC Buck Converter with Constant Power Loads in Distributed Generation Systems" presents a resilient nonlinear control strategy for a DC-DC buck power converter with a CPL in a distributed generation system. In constant power load, the input voltage decreases/increases concerning increases/decreases of input current, and at its terminals, it exhibits the attributes of a negative impedance. Due to this behavior, it will impact the dynamics, stability, and power quality of the converter system. Moreover, the DC-bus voltage changes will happen when the system experiences disturbances like changes in load and change in input voltage. The passivity-based control (PBC which is a simple form of PD) strategy has been utilized to address the instability issues caused by the CPLs. Chapter "A Two-Stage Scheduling Based Energy Management in a Microgrid" addresses the need for energy management in a microgrid (MG). Outliers in MG's, representing anomalies or unexpected variations in data, can significantly disrupt the performance and reliability of energy management systems (EMS). The proposed work uses an unsupervised K-means clustering technique which is a prominent one for outlier detection and imputation. In addition to outliers, there is a need to model the uncertainties for effective EMS. This paper addresses the uncertainties based on Monte Carlo simulation for scenario generation and K-means clustering for scenario reduction. Furthermore, a two-stage scheduling approach is proposed to minimize day-ahead operational costs and real-time imbalance costs.

Chapter "A Novel Self-balanced Switched Capacitor Seven-Level Inverter" presents a novel, one-DC source, seven-level inverter that is suitable for wind and solar power applications. Floating capacitors (FC) combined with a switched-capacitor design yields the desired outcome. This configuration was selected to minimize control problems when generating a tripolar staircase pattern because of its sensor-less capacitor voltage balancing potential. Inadvertent overvoltage of the capacitor is prevented over time, resulting in a further benefit. The phase disposition pulse-width modulation (PD-PWM) technique has been used to regulate the operation of the topology. Moreover, fewer semiconductor devices are utilized in the suggested topology. Chapter "RST-Based Robust Digital PID Controller Design for Dual-Input SEPIC Converters" presents a robust multivariable control technique for the design of a controller for integrated DC-DC converters. The digital decentralized controllers are designed for two input-integrated converters to achieve stable and reliable output voltage. A robust PID controller is designed with RST control architecture for the regulation of TI-BS and TI-MBS converters using the pole placement technique. The detailed procedural steps to design a robust controller using this technique are presented. A sixth-order closed-loop characteristic equation is formulated for Voltage and current loops and controllers are designed based on this. The desired pole location for both voltage and current control loops is assumed to be the same. Chapter "Deep Learning Approaches to V2G Strategy and Its Effects on Microgrid Dynamics" explains the scheme, functionalities, and possible advantages of V2G while also illustrating how it might be used on a microgrid. Explain how electric cars can act as a turning replacement by holding additional grid power stored in their batteries and reintroducing it to the system when needed. This aids in controlling the frequency of the grid within its nominal specifications. EVs are essential for achieving amazing features; however, integrating a large number of EVs also causes grid congestion. Uncontrolled charging could therefore result in under voltage and intricate grid congestion.

Chapter "Frequency Stability of Highly Ranged Renewable Energy Penetrated Multi-area Interconnected Power System Under a 3DOFPID Controller" presents the coordinated control approach of capacitive energy storage (CES) and a thyristor-controlled series compensator (TCSC) is employed for the investigative model in this work under the OOA-tuned 3DOFPID controller. The improvement in the dynamic behavior of the multi-area IPS is achieved with the deployment of the additional CES-TCSC control strategy. Finally, the multi-area IPS is targeted with a wide range of loads and parametric uncertainties to validate the robustness of the implemented control techniques. Chapter "A Cascaded (1 + PI)-PID Controller for the Improvement in Load Frequency Control of the Diversified Interconnected Power System" develop a robust controller design for the optimal operation and stability of the diversified interconnected power system (IPS). With regards to this connection, a cascade one plus proportional, integral (1 + PI), and proportional-integral-derivative (PID) ((1 + PI)-PID) is presented to regulate the deviations in the behavior of the diversified IPS. However, a recent soft computing technique called the butterfly optimization algorithm (BFOA) is utilized for the parameter tuning of the cascade (1 + PI)-PID controller to obtain the most optimal performance. Chapter "FOTIDF Controller for Stability of Renewable Integrated Multi-area Realistic Power System with AC-DC Links" focuses on the design of a fractional order tilt-integral-derivative plus filter controller (FOTIDF) using the water cycle optimization algorithm (WCOA) as load frequency controller (LFC) for the dynamical stability of multi-area renewable integrated realistic power system (MARIRPS). The analysis on MARIRPS is carried out by injecting the step load disturbance (SLD) of 10% in area 1. Moreover, the investigative model is perpetuated with realistic constraints for obtaining the analysis near the real-time environment.

Chapter "Simultaneous Optimal Allocation and Sizing of Distributed Generations and Shunt Capacitors in Distribution Network for Voltage Profile Improvement and Loss Reduction Using IMDE Algorithm" proposes an IMDE optimization method to optimize the integration of renewable DGs and SCs in the distribution networks under consideration, to reduce power loss and voltage deviation, and to enhance the voltage profile of the network while meeting a variety of operational restrictions. Comparing the simulation results of the proposed algorithm on two IEEE 33-bus and IEEE 69-bus test networks with recently published methods in this field. Chapter "Application of Advanced Photovoltaic Control Algorithms for Stability of the Power Grid" presents a new algorithm of model-based (MB) tracking for the regulation of power networks. The algorithm can now change the curtailment of power based on the factor of reduction provided by the operators of the power system, thanks to the improved formulation. The new algorithm has outstanding performance and precision, as demonstrated by the results. It can regulate power between 10% and 100% of the maximum power available. Chapter "Performance Evaluation of Quasi Z-Source Inverter Utilize for Grid Connected with Wind-PV Co-generation Systems" provides a thorough performance evaluation of a Quasi Z-Source Inverter (qZSI) used in grid-connected wind-photovoltaic (PV) co-generation systems. Reducing dependency on fossil fuels and attaining sustainable energy objectives need the integration of clean energy sources, especially solar and wind. The qZSI topology offers several advantages, including improved voltage-boosting capabilities, reduced component count, and enhanced system reliability. This study examines the operational characteristics of the qZSI in combination with wind and PV systems, focusing on efficiency, power quality, and dynamic response under varying load conditions and environmental factors.

Chapter "Application of Swarm Intelligence Techniques for Quasi-Z-Source Inverters in Renewable Energy Systems" explores how swarm intelligence techniques, namely Particle Swarm Optimization (PSO) and Artificial Bee Colony (ABC) algorithms, might improve quasi-Z-source inverters' (QZSI) efficiency for grid-connected renewable energy systems. We found that using these optimization algorithms to adjust the parameters of PI controllers leads to significant improvements in system efficiency, voltage regulation, and stability under various operational conditions. Our comparison of PSO and ABC shows that PSO has faster convergence, while ABC offers robustness in avoiding local optima. This demonstrates the effectiveness of swarm intelligence approaches in addressing the challenges of dynamic and nonlinear renewable energy systems. Chapter "Grey Wolf Optimization Based Optimal Charging Scheduling for Electric Vehicles with Real-Time Traffic Congestion Monitoring" presents an optimization approach for the scheduling of electric vehicle (EV) charging and the management of traffic congestion, utilizing the Grey Wolf Optimization (GWO) procedure. The recommended method includes several modifications designed to enhance performance and minimize operational costs for EVs. Comprehensive simulation results, alongside theoretical calculations, have been rigorously validated, demonstrating the algorithm's effectiveness. Our findings reveal that the proposed approach can significantly improve the efficiency of EV usage by identifying optimal routes with available charging stations, thus saving time and reducing costs. The simulation results highlight the algorithm's ability to optimize the Time of Use (TOU) of grid power and alleviate congestion at charging stations, underscoring its practical benefits. Chapter "Unit Template-Based Control Algorithm for Control of Voltage and Frequency in Solar-Hydro System" introduces a unit template control algorithm that uses less number of sensors as compared to various control algorithms of variable frequency control (VFC) off-grid asynchronous generators (IAG). The IAG is run by an uncontrolled hydro tur-bine for power generation integrated with a solar photovoltaic PV system with consumer loads. Solar PV system consists of a DC-DC converter, and solar panels with a maximum power point tracking (MPPT) controller. The VFC with a four-leg IGBT converter and storage battery on a DC link with a connected point of common coupling (PCC). The four-leg IGBT converter is utilized for suppressing neutral current.

Chapter "Overview of Electric Vehicle Management Using Artificial Intelligence in Smart Energy Systems" discusses new trends and emerging technologies of EVs integrated with smart energy systems and optimized by artificial intelligence (AI) solutions. This work looks into the issues of integrating EVs with smart energy systems and provides the enhanced applications of integrated into EVs along with smart energy systems. AI also plays a transformative role in EVs significantly enhancing various aspects of development in autonomous driving, battery optimization, route optimization, predictive maintenance, market analysis, etc. Chapter "Design of D-STATCOM for Mitigation of Current Harmonics Using Class Topper Algorithm" focuses on the design and implementation of a D-STATCOM (Distributed Static Compensator) to mitigate current harmonics in non-linear systems. These harmonics, which are integral multiples of the supply frequency, cause issues such as increased heat generation, reduced power factor, voltage distortion, and decreased system reliability. A novel Class Topper Algorithm is implemented for real-time estimation and control of the current harmonics. This optimization algorithm regulates the DC link capacitor voltage fed to the inverter in the D-STATCOM, minimizing harmonics in the source current and improving system performance compared to traditional methods.

The emphasis of this book is to present a computational platform for the performance analysis and optimization of future hybrid energy systems forging a path towards a greener, more equitable energy future. This will provide alternatives to current static models used in legacy systems. It will help to address the variability of RESs and uncertainties of EV loads, which will help to improve system security, reliability, efficiency and reduce dependence on fossil fuels. This book also presents an introduction to the Internet of Things (IoT) in smart spaces including smart homes, smart offices, smart cities, etc. as well as IoT applications in energy management and conversion, smart grids, and electric vehicles. This book also discusses various challenges and considerations, future trends, and advancements in implementing the IoT in smart spaces. Apart from the existing operational challenges, cybersecurity also has become an essential component in modern smart energy systems. Mitigating the threats related to cybersecurity is critical to maintaining the stability and security of utility grids.

Happy Reading!

Daejeon, Korea (Republic of) Surender Reddy Salkuti

Contents

About the Editor

Surender Reddy Salkuti has been working with Woosong University, Daejeon, Republic of Korea as an Associate Professor in the Department of Railroad and Electrical Engineering since April 2014. He received a Ph.D. degree in Electrical Engineering from the Indian Institute of Technology Delhi (IITD), New Delhi, India, in 2013. He was a Postdoctoral Researcher at Howard University, Washington, DC, USA, from 2013 to 2014. His research interests include power system restructuring issues, smart grid development with the integration of wind and solar photovoltaic energy sources, battery storage, and electric vehicles, demand response, power system analysis and optimization, artificial intelligence and soft computing techniques application in power systems, and renewable energy. He has published three edited volumes with Springer (LNEE and GREEN) and over 300 research articles in peer-reviewed international journals and conference proceedings. He served as Guest Editor for various international journals. He is also an editorial board member for many journals. He is the recipient of the 2016 Distinguished Researcher Award from Woosong University Educational Foundation, Republic of Korea, and the POSOCO Power System Award (PPSA) 2013, India. He is listed in the top 2% of scientists published in a study conducted by researchers of ICSR Lab, Elsevier BV and Stanford University, USA. He has Google Scholar citations of 6839 with an h-index of 41 and an i10-index of 150. He is a Member of IEEE and IEEE Power and Energy Society.

Artificial Intelligence in Electric Vehicles for Sustainable Driving

Debani Prasad Mishra, Aashutosh Rath, and Surender Reddy Salkuti

Abstract The adoption of electric vehicles (EVs) has great potential to cut greenhouse gas emissions and lessen reliance on fossil fuels. However, an EV operator's driving style significantly impacts both the vehicle's energy consumption and environmental impact. One of the most important tactics for maximizing EV efficiency is eco-driving, which modifies acceleration, deceleration, speed, and other driving behaviors to reduce emissions and fuel consumption. Artificial intelligence (AI) plays a key role in enabling eco-driving by offering drivers knowledgeable direction, constructive criticism, and assistance. This chapter examines the advantages of eco-driving EVs and several AI-powered methods to improve eco-driving behaviors. These methods include traffic prediction, route planning, intelligent speed adaptation, adaptive cruise control, and more. This chapter also assesses how well various AI-based techniques work in terms of user acceptability, security, comfort, and energy efficiency. The practical consequences and problems of deploying AI-driven eco-driving solutions are clarified by analyzing user experiences and performance indicators. The conversation also touches on the difficulties and possible uses of AI in improving eco-driving for EVs. The chapter emphasizes the revolutionary role that AI can play in encouraging sustainable transportation behaviors and lowering the environmental impact of EVs by addressing obstacles and examining opportunities.

Keywords Energy-efficient driving · Eco-driving · Intelligent speed adaptation · Adaptive cruise control · Route planning · Traffic forecasting

D. P. Mishra · A. Rath
Department of Electrical and Electronics Engineering, IIIT Bhubaneswar, Bhubaneswar, Odisha, India
e-mail: debani@iiit-bh.ac.in

S. R. Salkuti (✉)
Department of Railroad and Electrical Engineering, Woosong University, Daejeon, Republic of Korea
e-mail: surender@wsu.ac.kr

S. R. Salkuti (ed.), *Artificial Intelligence for Integrated Smart Energy Systems in Electric Vehicles*, Lecture Notes in Electrical Engineering 1427,
https://doi.org/10.1007/978-3-031-94276-1_1

Nomenclature

EV	Electric Vehicle
EM	Energy Management
BDP	Battery Degradation Policy
MP	Motion Planning
MSC	Multi-Scale Control
RA	Regression Analysis
CA	Classification Analysis
HEV	Hybrid Electric Vehicle
PHEV	Plug-in Hybrid Electric Vehicle
FCV	Fuel Cell Vehicle
IoV	Internet of Vehicles
RCS	Resilient Control Schemes
ESB	Energy Saving Barriers
HMEDM	Human Machine Environment Data Mining
GO	Global Optimization
VPTC	Vehicle Power Terrain Control
AD	Automatic Driving
CV	Connected Vehicles
EA	Energy Automation

1 Introduction

With the potential to significantly contribute to the worldwide shift towards Net Zero and the decarbonization of road transportation, electric vehicles (EVs) are quickly gaining traction in the market. Nonetheless, there are difficulties in incorporating EVs into the current supply chains, refueling infrastructure, and electrical networks. AI is a game-changer that turns EVs from passive energy users into proactive members of a cleaner and smarter grid. Artificial Intelligence (AI) algorithms provide EV intelligence, allowing them to anticipate battery depletion, optimize energy use, and navigate on their own [1]. With the help of AI's invisible hand, these quiet companions on our journeys are now helping to create a more sustainable future. We are opening up new opportunities for eco-driving, energy management, and seamless connection in our changing transportation scene as we investigate the uses of AI in EV integration. With their greener alternative to conventional fossil fuel-powered cars, EVs have the potential to completely transform the transportation industry. AI which endows EVs with intelligence and efficiency, is at the core of this revolution. AI algorithms improve energy use through eco-driving, and predictive models forecast battery degeneration. Internal combustion engines (ICEs) and electric motors are combined in hybrid electric vehicles (HEVs) and plug-in hybrid EVs (PHEVs), while fuel cell vehicles (FCVs) that run on hydrogen provide further environmentally friendly choices [2]. Motion planning algorithms guarantee safe trajectories as EVs progress toward autonomous driving, while the Internet of Vehicles (IoVs) improves connection. Obstacles continue, but robust control strategies adjust to a variety of circumstances. This symbiotic relationship between AI and EVs is creating a smarter, greener mobility ecology.

1.1 AI-Based Control System

Machine learning (ML) algorithms are used by AI-controlled industrial systems to streamline production procedures. These systems estimate maintenance requirements, monitor quality, and make autonomous parameter adjustments. AI systems, for example, may dynamically adjust robot motions on an assembly line based on real-time sensor data, guaranteeing accuracy and efficiency. AI-driven control systems examine enormous volumes of data in complicated networks (like telecoms or data centers). They anticipate network overload, spot possible problems, and distribute resources as efficiently as possible [3]. Through the process of historical pattern recognition, these technologies improve network performance and dependability. AI-based control plays a major role in autonomous cars. These systems, which interpret sensor data (camera, radar, and lidar) to make snap choices, range from self-driving automobiles to drones. Lane maintenance, accident avoidance, and route planning are all handled by AI algorithms. An autonomous car's brain constantly absorbs information from its surroundings and adjusts to a variety of situations. In smart grids, power distribution is managed by AI-driven control systems. They forecast load variations, optimize energy flow, and maintain supply and demand equilibrium. These technologies decrease energy waste and improve grid stability by studying past use trends. During production, AI-based controls keep an eye on the quality of the output. They make sure that the output is consistent, identify flaws, and modify process settings. AI algorithms, for instance, improve etching procedures in semiconductor production to produce exact designs. Chemical factories managed by AI can optimize processes, keep an eye on safety criteria, and avert dangerous situations [4]. Figure 1 depicts the general AI-based control system in a vehicle to keep things at their ideal, these systems modify flow rates, pressure, and temperature. The capacity of AI to learn from anomalous circumstances improves safety procedures.

1.2 Hybrid Electric Vehicles (HEV)

The greatest features of both an electric propulsion system and a traditional internal combustion engine (ICE) are combined in a hybrid electric vehicle (HEV). These are the salient features of HEVs: Two Sources of Power Two power sources power an HEV: an electric motor and an internal combustion engine, which runs on gasoline most of the time. Together, these power sources enable smooth switching between

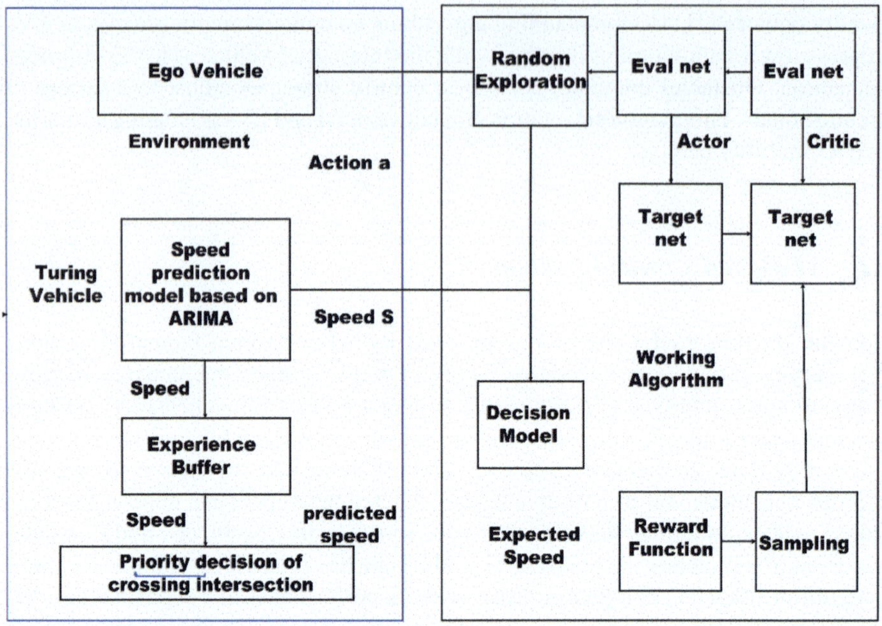

Fig. 1 General AI-based control system in a vehicle

them in response to varying driving circumstances and energy requirements. Fuel economy and efficiency: Higher fuel efficiency is the goal of HEVs as opposed to conventional cars. By helping with acceleration and low-speed driving, the electric motor lessens the need for the gasoline engine [5]. When braking, regenerative braking gathers energy and transforms it into electrical power to replenish the battery. Hybrid Parallel: The wheels can be driven separately or in tandem by the engine and electric motor. Series Hybrid: The electric motor powers the wheels while the engine functions as a generator to charge the batteries. Power-Dividing Hybrid combines the best qualities of series and parallel hybrids. Benefits: Lower emissions: Compared to conventional cars, HEVs produce fewer emissions. HEVs may run entirely on electricity in silent electric mode while traveling at slow speeds. Regenerative braking prolongs battery life by capturing energy while braking. Negative aspects Cost: Because HEVs have two powertrain components, they are usually more expensive upfront. Limited electric range: HEVs need fuel to travel farther than plug-in hybrids. Complex drivetrain: Specialist knowledge may be needed for maintenance and repairs.

1.3 Plug-In Hybrid Electric Vehicle (PHEV)

Dual Power Sources: PHEVs, in contrast to conventional hybrids, have a bigger battery pack that enables them to operate at slower speeds solely on electricity. PHEVs may run entirely on electricity for a certain amount of time (usually between 30 and 50 km, though this varies by model) when they are fully charged. When the battery runs out of power or when traveling at greater speeds is necessary, the internal combustion engine (ICE) starts up (such as highway driving). Having two powertrains allows for flexibility: gasoline is used for longer travels and electricity for shorter ones. Charging and Regenerative Braking: Plug-in hybrid electric vehicles (PHEVs) can be charged by connecting them to an external power source, hence the term. Depending on the type of charger (Level 1, Level 2, or DC fast charging), different charging times apply [6]. Owners of PHEVs have the option of charging at home, at work, or in public charging stations. Regenerative braking recovers energy lost during deceleration and feeds it back into the battery, therefore increasing electric range. The fuel efficiency and electric range of a PHEV are influenced by the battery's capacity. While some versions have a longer range (perfect for driving in cities), others have more comprehensive electric-only features. PHEVs have exceptional fuel efficiency, particularly on short daily commutes. Electric power is frequently sufficient for commuters to finish their entire trip; on longer journeys, the gasoline engine guarantees that drivers won't run out of fuel. Environmental Impact: Compared to traditional cars, PHEVs emit fewer greenhouse gases. They emit no tailpipe emissions when powered by electricity. However, the type of electricity utilized for charging determines the overall environmental impact. The advantages increase if renewable energy sources are used to generate electricity. Cost Considerations: Because PHEVs require larger batteries and additional components than standard hybrids, they are typically more expensive upfront. Nevertheless, over time, fuel savings and government subsidies can balance the initial cost. Maintenance expenses are cheaper for automobiles that run exclusively on gasoline.

1.4 Fuel Cell Vehicle (FCV)

An EV that uses a fuel cell to produce power for its onboard electric motor is referred to as a fuel cell vehicle (FCV) or fuel cell electric vehicle (FCEV). The following are the main details of FCVs: How FCVs Operate: The electric motors in FCVs are powered by electricity, just like in all-electric cars. But FCVs use a fuel cell to generate electricity rather than just a battery. Through chemical reactions, the fuel cell produces energy. Typically, compressed hydrogen and oxygen from the air are used. As a result, the car produces solely heat and water vapor as emissions. Fundamentals of Fuel Cells: An electrolyte separates the anode and cathode in a fuel cell. The anode receives hydrogen gas, which divides into protons and electrons there. While the electrons flow through an external circuit to produce an electric current, the protons go through the electrolyte to the cathode [7]. Water is a result of the combination of protons and electrons at the cathode with airborne oxygen. Benefits of FCEVs: Zero Emissions: Since FCVs only emit water vapor, they are categorized as zero-emission cars. Extended Range: FCVs frequently provide a larger driving range when compared to battery-only electric vehicles. Fast Refuelling: Filling up a hydrogen-powered vehicle takes minutes, just like filling up a traditional gasoline-powered vehicle. Obstacles and Facilities: Hydrogen Infrastructure: It's still difficult to build a comprehensive network of hydrogen filling stations. Production of Hydrogen: Presently, the majority of hydrogen is produced from reformed natural gas, which has environmental issues. Cost: Because fuel cell technology is so sophisticated, FCVs can be costly.

2 Multi-scale Energy Management

Multi-scale Management system in a vehicle refers to the optimization and coordination of energy flow across different scales within a system. It involves managing energy at various levels, from individual components (microscale) to entire systems (macroscale). Figure 2 depicts the key considerations in MSEM design including selecting appropriate energy storage technologies (such as batteries and supercapacitors), efficient energy conversion methods, intelligent control strategies, and demand-side management. Applications of MSEM include EVs, smart grids, smart buildings, and industrial processes. As we look to the future, integrating AI and decentralizing energy management within microgrids are promising directions.

2.1 Energy Storage System

Let's explore the exciting field of distribution systems' multi-time-scale energy storage optimization. Maintaining a steady power balance over a range of time periods

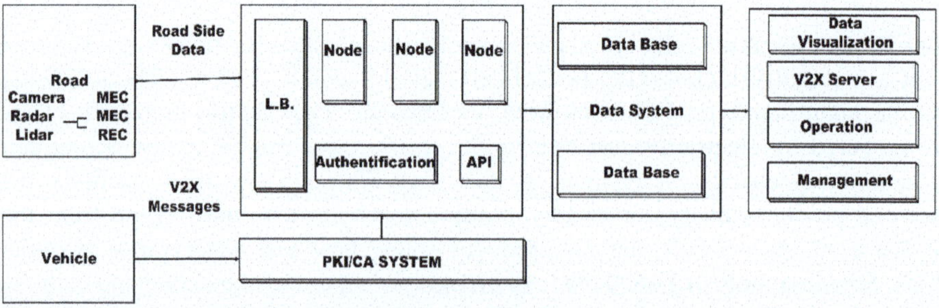

Fig. 2 Multi-scale management system in a vehicle

has become a major problem for contemporary power systems as the use of renewable energy sources increases. Let's examine some important facets of this subject: Goal and Difficulties: The main goal is to allocate energy storage capacity as efficiently as possible while taking power imbalance hazards into account over various time frames. The erratic and unpredictable nature of renewable energy sources like solar and wind power as a result of weather variations presents difficulties. Methodology: There are multiple steps in the suggested procedure [8]. Temporal Decoupling: Real operational data is temporally decoupled utilizing the Seasonal and Trend decomposition using the Loess (STL) decomposition method. Power Balance Computation: Power balance calculations are carried out at different time scales using the decoupling data. Energy Storage Allocation: To maximize power balance, various energy storage capacities such as electrochemical and hydrogen storage are distributed. Risk Assessment: The chosen allocation's imbalance risks are evaluated. Types of Energy Storage: Two energy storage systems are taken into consideration in this context: Storage of hydrogen: A means of reducing long-term uncertainty. Utilized in the renewable energy system to manage short-term uncertainty is electrochemical storage. Validation in Real-World Settings: The efficacy of this methodology is confirmed through the use of recent real-world data from a southern Chinese region. Hourly applications show how the suggested strategy is practically applicable.

2.2 *Battery State Estimation*

Lithium-ion battery status estimation on multiple scales. To improve battery utilization efficiency and maintain safety, EVs must estimate battery parameters like state of charge (SoC) and state of power (SoP) with accuracy. The following are some salient observations: Difficulties with SoP Estimation: Most of the previous research concentrates on estimating SoP while taking into account a single restriction, frequently ignoring the estimation of battery state over a variety of time scales. Ignoring differences over many time scales can result in inaccurate SoP estimates and possible safety risks. SoP Estimation with Multiple Time Scales Proposed: The suggested approach uses an H∞ filter to integrate multi-time scale and multi-constraint SOP estimation. This is how it operates: Battery Model: A ternary lithium-ion monolithic battery is modeled using a second-order RC equivalent circuit. Parameter Identification: To identify parameters, recursive least squares (RLS) with a forgetting factor are employed. H∞ Filtering: After estimating SOC, a joint multi-time scale, multi-constrained SOC-SOP estimation is carried out using the H∞ filtering algorithm [9]. Validation: Under UDDS (Urban Dynamometer Driving Schedule) conditions, the joint estimation technique is verified. The estimation of SOC's mean absolute value relative error (MARE) is 1.17%, while the estimation of SOP's MARE at various time scales is less than 1.6% Other Strategies: Scholars have investigated several techniques for estimating SOC, such as: For online parameter measurement and SOC estimate, combine an extended Kalman filter (EKF) with an unscented Kalman filter (UKF). For SOC estimation, the Composite Multiscale Wavelet Transform is utilized, which examines input data at many scales. Accurate battery status estimation by multi-dimensional, multi-level, and multi-scale signal information mining.

2.3 Battery Degradation Prediction

Predicting battery degradation with an emphasis on lithium-ion batteries in particular. Ensuring long-term stability and optimizing battery consumption require accurate prediction of battery cycle life. Here are a few observations: Battery Degradation Complexity: Degradation of batteries is a difficult nonlinear issue. The difficulty is exacerbated by a variety of chemistries, designs, and degradation mechanisms. Variations in charge/discharge rates and other dynamic cycle variables make the forecast even more difficult. Researchers have looked into several methods, including feature engineering and machine learning, to forecast battery lifetime: Empirical Models: These models are based on empirical relationships and historical data. Equivalent circuit models are commonly used for estimating the state of health since they use electrical components (such as resistors, capacitors, etc.) to reflect battery behavior. Data-Driven Models: Based on discharge voltage curves, battery cycle life can be predicted using machine learning techniques like linear regression and Gaussian process regression [10]. Feature extraction: From the discharge voltage curves in this case, 53 features including 18 newly generated features are produced. Algorithm Performance: In the worst scenario, Robust Linear Regression (RLR) produced an RMSE of 6.90%. The RMSE for Gaussian Process Regression (GPR) was 6.33%. Real-Time Applications: Efficient assessment of battery cycle life is made possible by the combination of feature engineering and machine learning based only on discharge voltage curves. For shipboard applications, where real-time monitoring and optimization are crucial, such forecasts may be useful. Predicting battery degradation accurately is essential to ensuring consistent battery performance in applications such as energy storage systems and electric cars.

2.4 Power-Trains

Powertrains in systems of several scales. Power trains are essential to attaining efficiency, sustainability, and peak performance in the transportation and energy sectors. The following are some significant facets of this field: Train Energy Simulation and Drivetrain Power Structures: Electric trains are powered by the railway traction power network. Comprehending the energy flow inside the system is essential for energy-efficient train operations [11]. Railway power network models and train movement models are integrated into an energy evaluation simulation. This simulator computes energy flow while taking several train driving controls and schedules into account. Substation energy, power transmission networks, monitoring trains, and regenerative braking trains are all included in the analysis. AC and DC Traction Networks: 3000–5000 DC, 15 kV, 16 2/3 Hz AC, and 25 kV 50 Hz AC are the different voltages at which modern railway traction power networks operate. By transferring energy from braking to moving trains, regenerative braking systems increase overall efficiency. A large percentage of the world's electric railway lines are covered by DC electric railway networks. Energy-efficient optimization: Train paths can be optimized analytically to maximize the use of external power flow [12]. Flexible control techniques are needed to integrate renewable energy sources into power networks. Examples of these tactics include storing energy as thermal energy in district heating systems or using surplus electricity for thermal systems. Large-scale wind power integration's strain on peak regulation can be reduced by using multi-type flexible resources, such as demand response, energy storage, and deep peak regulation of thermal units. Multi-Scale Modelling and Simulation: Creating multi-scale models enables thorough simulation at all scales, from micro to macro. These models characterize the geometry, physics, behavior, and other aspects of the system. Such multi-scale modeling is advantageous for urban train traction power supply systems.

2.5 *Vehicle Dynamics*

Vehicle dynamics and multi-scale energy systems intersect. We'll examine the complex interactions that arise between energy management, system optimization, and vehicle behavior in this setting. The following are some salient observations: Multiple-Body Dynamics in Automobile Design: Evolution in History: The origins of multi-body dynamics (MBD) can be traced back to the Sumerian invention of wheels for chariots and carts. Galileo Galilei, who focused on ideas like force and moment, established the scientific basis for mechanical viewpoints. Use in Vehicle Engineering: MBD is crucial for examining the noise, vibration, and harshness (NVH) of a vehicle [13]. It includes a wide range of phenomena, such as infinitesimal amplitude electrodynamics and huge displacement rigid body dynamics. Strategies for Dynamic Energy Management: Switching to Electrical Power Systems: Electrical power systems are replacing traditional ones in multi-energy driving systems.

These systems are essential to reducing pollutants and fuel consumption. System for Dynamic Energy Management (EMS): An EMS distributes power among a powertrain's constituents (such as auxiliary batteries and diesel engines) as efficiently as possible. Making decisions in real-time guarantees effective energy use. Dynamics of Hybrid Powertrains with Multiple Frequency Excitations: Modelling at Multiple Time Scales: Different time scales are subject to periodic excitations in hybrid powertrains. Multi-time scale models have been constructed by researchers to better comprehend their dynamic behavior. Examining the Dynamic Properties: We can learn more about the performance and stability of hybrid powertrains by examining them under multi-frequency excitations. Modeling Battery Systems and Actual Driving Behaviour: Enormous Battery Systems: Accurate modeling of battery systems is essential for developing electric-drive automobiles. User-Determined Driving Style: Designing systems and energy management techniques are informed by an understanding of how real-world users operate their cars [14]. Vehicle Energy Dynamics Simplified: Science-Based Models: Simpler energy models may be more useful for investigations of collective energy consumption (such as multi-vehicle traffic flow) than more sophisticated instruments that necessitate an in-depth understanding of vehicle dynamics.

3 AI in Smart Batteries

3.1 Optimization of Energy Usage with AI

Energy optimization in the context of AI and smart batteries. The combination of AI technology and smart batteries is propelling innovation in sustainable transportation as the market for EVs grows. Here are several ways AI can improve battery lifespan, performance, and energy consumption: Energy Efficiency Improvement: Adaptive Battery Optimization: By proactively evaluating consumption patterns and modifying system settings to conserve energy without sacrificing performance, AI-powered algorithms transform power management. Load Forecasting: Artificial Intelligence methods are essential for forecasting energy use. Machine learning and deep learning models can forecast load requirements by examining past data, which enables smart batteries to optimize energy use. Anomaly Detection: AI assists in locating unusual battery system activity. Smart batteries are capable of identifying irregularities and taking appropriate action by utilizing a range of learning techniques, such as optimization tactics [14]. Demand Response: AI makes it possible to make dynamic changes in response to current demand. AI approaches can be utilized to optimize demand response schemes that are either price-based or incentive-based.

3.2 Enhance Performance

Enhanced Energy Efficiency: Customized Battery Management: AI systems evaluate variables like braking patterns, travel distances, and ambient circumstances. AI optimizes energy distribution and charging cycles by adjusting battery utilization based on individual driving patterns. This method extends the battery's overall life and increases the vehicle's range. Real-Time Process Optimization: AI can evaluate large volumes of data produced by robotic systems in the manufacturing and testing of batteries. Artificial Intelligence improves battery performance by spotting patterns and real-time process optimization. Increased Battery Life: Patterns of Charging and Discharging: AI algorithms maximize cycles of charging and discharging. They prolong battery life by strategically managing the utilization of stored power through data analysis. Accurate Battery Health Diagnostics: By identifying problems, controlling temperature, and assessing battery health, artificial intelligence techniques improve the safety, dependability, and effective operation of EVs [15, 16]. Energy Efficiency and Sustainability: Skilful Energy Administration AI guarantees effective energy use by evaluating enormous volumes of data. Intelligent power management using smart batteries promotes sustainability and energy conservation.

3.3 Extend the Life-Span of Smart Batteries

Optimal Energy Usage: Dynamic Battery Management: AI-enabled smart batteries can dynamically modify their management systems in response to various real-time events, including traffic congestion, weather, and driving habits. They guarantee effective energy consumption while increasing driving range by optimizing energy distribution and charging cycles. AI-powered predictive maintenance is made possible by its ability to analyze real-time battery health parameters [17]. It foresees possible problems before they arise, guaranteeing vehicle safety and saving EV owners money on maintenance. Personalized Energy Solutions: Adaptive Learning: Over time, smart batteries can adjust to user preferences and habits thanks to machine learning algorithms. These batteries offer individualized energy management strategies by analyzing each driver's unique driving style. User-Centric Efficiency: Smart batteries modify their behavior to improve driving while preserving energy, whether it's via maximizing regenerative braking or modifying power output during acceleration. Longevity Extension: Deep Learning Lifespan Models A Deep Learning (DL) lifespan model is the central component of smart battery technology. Gradient optimization is used to update this model, which is created using gathered data [18]. It prolongs the life of the battery and maximizes performance. AI regularly checks the status of batteries, anticipating problems and streamlining the charging and discharging procedures [19]. Smart batteries protect the battery and increase its lifespan by preventing overcharging or undercharging. Savings and Effect on the Environment: Lower Upkeep Expenses: For EV owners, proactive maintenance and efficient energy use result in financial benefits. Environmental Benefits: Smart batteries help create a more sustainable transportation ecology by reducing waste and prolonging the life of batteries.

4 AI in Smart Batteries

The term vehicle-to-everything (V2X) refers to all of a vehicle's connected communications. Imagine a world in which automobiles form dynamic links with pedestrians, other cars, traffic lights, road signs, and even entire cities, in addition to communicating with their drivers. What you should know is as follows: V2X components include: Vehicle-to-Vehicle (V2V): Direct communication exists between cars [20]. They exchange vital details like position, speed, and possible dangers. For example, your automobile will tell you if a car ahead brakes suddenly so you may respond quickly. Vehicles and roadside infrastructure, such as traffic lights, signs, toll booths, and smart intersections, are connected via V2I (Vehicle-to-Infrastructure). Smarter traffic management and safety features are made possible by this interaction. Cars and pedestrians engage in vehicle-to-pedestrian (V2P) interactions. Consider a situation in which a person enters a crosswalk and your vehicle is signaled to halt or slow down. Vehicles and wider networks exchange data through V2N (Vehicle-to-Network). Access to real-time information on accidents, road closures, and traffic conditions improves navigation and the entire driving experience. Safety and Collision Avoidance: The safety aspect of V2X technology is greatly enhanced. Vehicles can prevent crashes by exchanging real-time information [21]. For example, a car can notify other surrounding vehicles to change lanes or reduce speed if it detects an obstruction or sudden braking ahead.

Emergency services also gain from this. V2X-enabled cars provide vital information to emergency personnel in an accident, accelerating help. Efficiency and Optimization of Traffic: V2X enhances traffic movement. Envision an intelligent crossroads where automobiles convey their plans to one another. By dynamically adjusting their timing, traffic lights minimize waiting times and reduce congestion. Real-time traffic information facilitates route optimization, fuel economy, and increased productivity. V2X even allows EVs to return extra energy to the grid during periods of high demand, promoting sustainability and stability in the grid. Impact on the environment: Less emissions result from smarter traffic movement. V2X technology reduces stop-and-go traffic, which helps to purify the air [22]. By taking part in demand response programs, EVs with V2X capability can balance the supply and demand for electricity. They transform into movable energy storage devices that are advantageous to the environment and the grid. Difficulties and Prospects for the Future: Although the idea behind V2X is intriguing, it will take time for it to be widely adopted. Connectivity: Dependable Communication is crucial [23, 24]. To close the gap between automobiles, infrastructure, and pedestrians, the auto sector must use large-scale networks like LTE and 5G more widely. Standardization: Creating uniform guidelines and standards guarantees smooth communication across various car makes and models. Privacy and Security: Guarding user information and averting online dangers are important factors to take into account. Infrastructure Investment: A significant amount of money is needed to upgrade traffic patterns and roadways to accommodate V2X.

5 AI-Based Eco-driving Assistant Systems

The Human Interactive Driving (HID) team at Toyota Research Institute (TRI) is steadfastly dedicated to active safety systems that increase human drivers' engagement, even though many are aiming toward driverless cars. The creation of autonomous vehicles has dominated recent developments in the field of automotive technology. Instead of replacing human drivers, the HID team at TRI is focusing on active safety technologies that empower people [25]. By utilizing the advantages of artificial and human intelligence, this approach seeks to make driving safer and more pleasurable. A media event was arranged by TRI to introduce the public to its new Accelerated Concepts program, a targeted effort that aims to develop and deploy cutting-edge safety technology quickly. To address some of the most important issues facing vehicle safety and expedite the release of life-saving technologies, the program will bring together a group of elite researchers and engineers enabling people to drive Using a variety of sensors and algorithms, active safety systems keep an eye on the environment around the car and take appropriate action to avoid collisions. Lane departure warning (LDW), which notifies the driver when the car inadvertently strays from its lane, automated emergency braking (AEB), which applies the brakes automatically to prevent or lessen a collision, and adaptive cruise control (ACC), which keeps a safe distance from the car in front of it and modifies speed as necessary, are a few examples of active safety features [26].

The HID team at TRI is working on developing active safety solutions that go beyond these fundamental features. Their objective is to develop AI-powered helpers who work in unison with human drivers to offer real-time direction and assistance. By working together, we hope to improve driving as a whole and make it safer, more effective, and more pleasurable. Increasing the level of happiness and engagement among drivers. The HID team at TRI thinks that active safety systems can promote a more enjoyable driving experience by keeping the driver involved and in charge [27]. Drivers will continue to be active participants in the driving process rather than feeling like passengers in their cars, gaining access to AI's insights and capabilities without giving up control. This strategy is in line with TRI's vision of Symbiosis Autonomy, in which machines and people collaborate to provide the best results. Symbiosis Autonomy in the driving context refers to a smooth cooperation between the driver and the AI-powered components of the vehicle [28, 29]. A wide variety of technologies developed under the three main research pillars of the HID team were demonstrated by each research vehicle and simulator: The first was human-focused learning, which builds behavioral models of people using data-driven ML approaches. These models guarantee that the technology will comprehend drivers and assist them in the most natural manner conceivable, from driver awareness to action and intent [30]. The second was safety and driver/vehicle performance, where the team used AI to develop expert-level driving abilities. Developing autonomous technology that can help drivers avoid accidents in difficult situations requires imbuing AI with advanced driving abilities. The last element is shared autonomy, in which drivers and AI collaborate to make driving safer and more pleasurable [31].

The Global Research Innovation Platform (GRIP), a state-of-the-art research vehicle with in-wheel electric motors and four-wheel steering constructed specifically for TRI, was made available for participants to drive [32]. The purpose of GRIP is to advance driving research. Because of GRIP's special features, such as its ability to simulate in-car dynamics, novel scenarios can be created for driver instruction in a secure setting. The future of interactive driving with humans. The HID team at TRI is leading the way in creating cutting-edge active safety technologies that will completely change driving. Their study has a great deal of potential for developing more efficient, fun, and safe transportation systems [33]. We may anticipate seeing increasingly more advanced human-interactive driving systems that improve the interaction between drivers and their cars as AI technology develops.

6 Conclusions

Enhanced Driving Range, Artificial intelligence (AI) algorithms reduce energy use and lessen range anxiety in electric vehicle drivers. AI-powered battery health prediction algorithms predict battery degradation and increase battery life. In situations involving autonomous driving, motion planning powered by AI improves productivity and security. Energy Efficiency, AI regulates regenerative braking, power distribution, and energy consumption in electric vehicles. AI optimizes efficiency and reduces waste by dynamically modifying power flow. EVs participate in load balancing by communicating with the electrical grid through Vehicle-to-Grid (V2G) Integration. AI algorithms regulate cycles of charging and discharging to assist in maintaining energy homeostasis. AI facilitates predictive maintenance by estimating the amount of repair that will be needed. In addition to ensuring optimum vehicle performance, this cuts downtime. Customized Driving Settings: AI adjusts driving settings based on user preferences. Customization improves user experience and promotes environmentally friendly driving practices.

All things considered, AI research and practical commercialization in EVs and associated infrastructure are still in their infancy. A greater number of EV applications utilizing AI are currently being commercialized than others. The techniques' technical maturity is insufficient to meet the requirements for their practical deployment. Some future orientations are listed below to help overcome current obstacles like high-throughput data generation, carrying out in situ calculations, and the scalability of data from battery-level to battery-pack scale, and to further facilitate the development of AI techniques used in EV and related infrastructure. Motion Planning for Autonomous Driving: AI-powered methods for motion planning improve EVs' capacity for autonomous driving. To generate safe and effective driving routes, these algorithms take into account current traffic conditions, roadblocks, and safety restrictions.

Acknowledgements This research work was supported by "Woosong University's Academic Research Funding—2025".

References

1. Diao X (2023) Research on electric vehicle charging safety warning based on A-LSTM algorithm. IEEE Access 11:55081–55093. https://doi.org/10.1109/ACCESS.2023.3281552
2. Hu Z, Mehrjardi TR, Ehsani M (2024) On the lifetime emissions of conventional, hybrid, plug-in hybrid and electric vehicles. IEEE Trans Ind Appl 60(2):3502–3511. https://doi.org/10.1109/TIA.2023.3330950
3. Timilsina L, Badr PH, Hoang PH, Ozkan G, Papari B, Edrington CS (2023) Battery degradation in electric and hybrid electric vehicles: a survey study. IEEE Access 11:42431–42462. https://doi.org/10.1109/ACCESS.2023.3271287
4. Abul Masrur M (2021) Hybrid and electric vehicle (HEV/EV) technologies for off-road applications. Proc IEEE 109(6):1077–1093. https://doi.org/10.1109/JPROC.2020.3045721
5. Payarou T, Pillay P (2023) Integrated multipurpose power electronics interface for electric vehicles. IEEE Trans Transp Electrific 9(2):2429–2443. https://doi.org/10.1109/TTE.2022.3209098
6. Ehsani M, Singh KV, Bansal HO, Mehrjardi RT (2021) State of the art and trends in electric and hybrid electric vehicles. Proc IEEE 109(6):967–984. https://doi.org/10.1109/JPROC.2021.3072788
7. Phan D, Bab-Hadiashar A, Fayyazi M, Hoseinnezhad R, Jazar RN, Khayyam H (2021) Interval type 2 fuzzy logic control for energy management of hybrid electric autonomous vehicles. IEEE Trans Intell Veh 6(2):210–220. https://doi.org/10.1109/TIV.2020.3011954
8. Gautam AK, Tariq M, Pandey JP, Verma KS, Urooj S (2022) Hybrid sources powered electric vehicle configuration and integrated optimal power management strategy. IEEE Access 10:121684–121711. https://doi.org/10.1109/ACCESS.2022.3217771
9. Liu J, Wang Z, Zhang L (2022) Integrated vehicle-following control for four-wheel-independent-drive electric vehicles against non-ideal V2X communication. IEEE Trans Veh Technol 71(4):3648–3659. https://doi.org/10.1109/TVT.2022.3141732
10. Yuvaraj T, Devabalaji KR, Kumar JA, Thanikanti SB, Nwulu NI (2024) A comprehensive review and analysis of the allocation of electric vehicle charging stations in distribution networks. IEEE Access 12:5404–5461. https://doi.org/10.1109/ACCESS.2023.3349274
11. Agrawal A (2023) Mathematical modeling of driving forces of an electric vehicle for sustainable operation. IEEE Access 11:95278–95294. https://doi.org/10.1109/ACCESS.2023.3309728
12. Hosseini SM, Soleymani M, Kelouwani S, Amamou AA (2023) Energy recovery and energy harvesting in electric and fuel cell vehicles, a review of recent advances. IEEE Access 11:83107–83135. https://doi.org/10.1109/ACCESS.2023.3301329
13. Nezamuddin ON, Nicholas CL, Santos CE (2022) The problem of electric vehicle charging: state-of-the-art and an innovative solution. IEEE Trans Intell Transp Syst 23(5):4663–4673. https://doi.org/10.1109/TITS.2020.3048728
14. Cui D, Wang Z, Zhang Z, Liu P, Wang S, Dorrell DG (2022) Driving event recognition of battery electric taxi based on big data analysis. IEEE Trans Intell Transp Syst 23(7):9200–9209. https://doi.org/10.1109/TITS.2021.3092756
15. Jeong SI, Choi DH (2022) Electric vehicle user data-induced cyber attack on electric vehicle charging station. IEEE Access 10:55856–55867. https://doi.org/10.1109/ACCESS.2022.3177842
16. Wang X, Sun C, Wang R, Wei T (2020) Two-stage optimal scheduling strategy for large-scale electric vehicles. IEEE Access 8:13821–13832. https://doi.org/10.1109/ACCESS.2020.2966825
17. Schwenk K, Meisenbacher S, Briegel B, Harr T, Hagenmeyer V, Mikut R (2021) Integrating battery aging in the optimization for bidirectional charging of electric vehicles. IEEE Trans Smart Grid 12(6):5135–5145. https://doi.org/10.1109/TSG.2021.3099206
18. Hafdaoui HE, Khallaayoun A (2023) Mathematical modeling of social assessment for alternative fuel vehicles. IEEE Access 11:59108–59132. https://doi.org/10.1109/ACCESS.2023.3284844

19. Cao W, Zhu Z, Nan J, Yang Q, Gu G, He H (2022) An improved motion control with cyber-physical uncertainty tolerance for distributed drive electric vehicle. IEEE Access 10:770–778. https://doi.org/10.1109/ACCESS.2021.3136573
20. Husain I (2021) Electric drive technology trends, challenges, and opportunities for future electric vehicles. Proc IEEE 109(6):1039–1059. https://doi.org/10.1109/JPROC.2020.3046112
21. Li Y (2024) Key technologies and prospects for electric vehicles within emerging power systems: insights from five aspects. CSEE J Power Energy Syst 10(2):439–447. https://doi.org/10.17775/CSEEJPES.2024.00190
22. Salkuti SR (2019) Optimal operation of microgrid considering renewable energy sources, electric vehicles and demand response. In: E3S web conference, vol 87, p 01007. https://doi.org/10.1051/e3sconf/20198701007
23. Chen H, EL-Refaie AM, Zuo Y, Cai S, Cao L, Lee CHT (2022) Comparative study and design optimization of a dual-mechanical-port electric machine for hybrid electric vehicle applications. IEEE Trans Veh Technol 71(8):8341–8353. https://doi.org/10.1109/TVT.2022.3175476
24. Chen H, EL-Refaie AH, Zuo Y, Cai S, Cao L, Lee CHT (2022) Comparative study and design optimization of a dual-mechanical-port electric machine for hybrid electric vehicle applications. IEEE Trans Veh Technol 71(8):8341–8353, https://doi.org/10.1109/TVT.2022.3175476
25. Ding X, Wang Z, Zhang L, Wang C (2020) Longitudinal vehicle speed estimation for four-wheel-independently-actuated electric vehicles based on multi-sensor fusion. IEEE Trans Veh Technol 69(11):12797–12806. https://doi.org/10.1109/TVT.2020.3026106
26. Jung CM, Pagidipala S, Salkuti SR (2024) Emerging technologies for the integration of renewable energy, energy storage and electric vehicles. In: Salkuti SR (eds) Energy and environmental aspects of emerging technologies for smart grid. green energy and technology. Springer, Cham. https://doi.org/10.1007/978-3-031-18389-8_5
27. Arandhakar S, Jayaram N, Shankar YR, Gaurav KPSV, Halder S (2022) Emerging intelligent bidirectional charging strategy based on recurrent neural network accosting EMI and temperature effects for electric vehicle. IEEE Access 10:121741–121761. https://doi.org/10.1109/ACCESS.2022.3223443
28. Salkuti SR (2023) Advanced technologies for energy storage and electric vehicles. Energies 16(5):2312. https://doi.org/10.3390/en16052312
29. Salkuti SR (2022) Emerging and advanced green energy technologies for sustainable and resilient future grid. Energies 15(18):6667. https://doi.org/10.3390/en15186667
30. Zhu X, Yan R, Huang Z, Wei W, Yang J, Kudratova S (2020) Logistic optimization for multi depots loading capacitated electric vehicle routing problem from low carbon perspective. IEEE Access 8:31934–31947. https://doi.org/10.1109/ACCESS.2020.2971220
31. Liu Y (2023) Study on the design and speed ratio control strategy of continuously variable transmission for electric vehicle. IEEE Access 11:107880–107891. https://doi.org/10.1109/ACCESS.2023.3318323
32. Teng F, Ding Z, Hu Z, Sarikprueck P (2020) Technical review on advanced approaches for electric vehicle charging demand management, Part I: Applications in electric power market and renewable energy integration. IEEE Trans Ind Appl 56(5):5684–5694. https://doi.org/10.1109/TIA.2020.2993991
33. Ahmed I, Rehan M, Basit A, Tufail M, Hong KS (2023) Neuro-fuzzy and networks-based data driven model for multi-charging scenarios of plug-in-electric vehicles. IEEE Access 11:87150–87165. https://doi.org/10.1109/ACCESS.2023.3303963

Modernization of Electric Grids for Charging of Electric Vehicles

Debani Prasad Mishra, Aashutosh Rath, Goutam Kumar, Nischay Jha, and Surender Reddy Salkuti

Abstract It is critical to increase grid charging efficiency given the growing demand for electric vehicles (EVs). This presentation looks at innovative approaches to analysis and optimization. Demand-side management, integrating renewable energy sources, smart grid technology, the infrastructure for current charging, and system resilience are all covered. This chapter outlines the practical applications of more recent methods to advance the study of efficient grid-connected EV charging. By utilizing modern analytical and optimization skills, researchers can look into several aspects of grid charging efficiency. This enables them to offer perspectives and solutions that are critical for sustaining grid sustainability and stability while assisting the growing EV market. This multidisciplinary approach eases the transition to a future that is more environmentally friendly and supportive of the usage of electric cars by addressing present problems and fostering innovation in energy management systems. Collaboration between researchers, businesses, and legislators is still essential to putting these advancements into practice and managing the dynamic landscape of energy and transportation infrastructure.

Keywords Vehicle to grid · Battery · Grid to vehicle · Electric vehicle · Efficiency management · Sustainability

D. P. Mishra · A. Rath · G. Kumar · N. Jha
Department of Electrical and Electronics Engineering, IIIT Bhubaneswar, Bhubaneswar, Odisha, India
e-mail: debani@iiit-bh.ac.in

G. Kumar
e-mail: b321050@iiit-bh.ac.in

N. Jha
e-mail: b321023@iiit-bh.ac.in

S. R. Salkuti (✉)
Department of Railroad and Electrical Engineering, Woosong University, Daejeon, Republic of Korea
e-mail: surender@wsu.ac.kr

S. R. Salkuti (ed.), *Artificial Intelligence for Integrated Smart Energy Systems in Electric Vehicles*, Lecture Notes in Electrical Engineering 1427,
https://doi.org/10.1007/978-3-031-94276-1_2

Nomenclature

PV	Photovoltaic
EV	Electric Vehicle
EM	Energy Management
BM	Battery Management
DL	Deep Learning
CA	Control Algorithms
DLM	Dynamic Load Management
CI	Charging Infrastructure
HEV	Hybrid Electric Vehicle
PHEV	Plug-in Hybrid Electric Vehicle
FS	Feature Selection
GNN	Graph Neural Network
MG	Modwave Graph
ESB	Energy Saving Barriers
OCS	Optimal Charging Strategies
G2V	Grid to Vehicle
V2G	Vehicle to Grid
AD	Automatic Driving
CV	Connected Vehicles
EA	Energy Automation

1 Introduction

A paradigm shift in the transportation sector has been sparked by the global shift toward sustainable energy solutions, with electric vehicles (EVs) emerging as key participants in the race for lower carbon emissions and a greener future. The infrastructure that supports EVs is becoming more and more important as EV adoption soars, especially when it comes to the infrastructure for charging EVs and how well it integrates into the current power grid. The global commitment to reducing environmental degradation and minimizing the effects of climate change is demonstrated by the rise in the ownership of EVs [1]. EVs are a key component of this revolutionary journey that governments, businesses, and consumers are all on toward sustainable transportation options [2]. However, the reliability and effectiveness of the infrastructure that supports their charging is crucial to the successful integration of EVs into the mainstream of transportation.

To maximize the efficiency of the infrastructure used for charging EVs, smart grid technologies are essential. To enable real-time monitoring, data analytics, and dynamic adjustments to meet variable energy demands, the traditional power grid is integrated with advanced communication and control technologies, a concept known as a smart grid. The need for intelligent charging systems is becoming more and more apparent as EV integration advances. Smart charging includes a dynamic and responsive interaction between the car, the charging station, and the grid in addition to just providing electricity to an EV [3]. By adjusting charging schedules in response to changes in electricity pricing, grid load, and the availability of renewable energy (RE), these systems can improve the sustainability and overall efficiency of the grid. Another crucial aspect of the evolving landscape is the seamless integration of renewable energy sources (RESs) into the EV charging infrastructure [4]. The proliferation of solar and wind power presents an opportunity to align EV charging with clean energy generation. By strategically deploying charging stations powered by RESs, the environmental footprint of EVs can be further reduced, marking a significant step towards a truly sustainable transportation ecosystem. However, there are obstacles to this life-changing trip. To avoid grid congestion and potential interruptions, careful control of the increased load on the grid is necessary, particularly during peak charging hours. Furthermore, to guarantee a smooth and user-friendly charging experience, standardized protocols for compatibility between various charging networks and EV models must be developed. In this environment, grid resilience is the ability of the electricity grid to tolerate disruptions and bounce back emerges as a crucial factor. It becomes crucial to consider elements like redundancy, grid stability, and cyber-physical security to ensure a safe and dependable infrastructure for charging EVs. Technological advancements in energy storage (ES) also contribute significantly to increased grid resilience by reducing the effects of variable RESs and supplying backup power during times of high demand [5]. The deployment of the vehicle-to-grid (V2G) technology enhances the stability and dependability of the grid. EVs can function as dynamic grid assets thanks to their bidirectional power flow, which enables them to absorb excess energy during off-peak hours and inject electricity during peak demand. This helps to improve overall grid resilience, balance the grid, and lessen the chance of blackouts. The adoption of V2G power flow encourages security and independence in energy use. Reliance on centralized power plants and outside energy sources is reduced by decentralizing power generation and storage [6]. This change ensures a more resilient and self-sustaining energy ecosystem, improving energy security by lowering susceptibility to disturbances. The deployment of V2G encourages technological development and innovation in the energy and electric car industries.

Research and development are driven by the requirements for bidirectional power flow, sophisticated energy management systems, and smart grid integration [7]. This in turn encourages creativity, which accelerates the development of energy solutions that are more sustainable and efficient. V2G technology helps cut down on carbon emissions by encouraging the use of stored energy during peak hours. V2G contributes to reducing environmental effects and supporting international efforts to tackle climate change by maximizing the use of RESs and minimizing dependency on conventional power plants. The extensive use of V2G technology can promote economic expansion and job creation. The need for qualified experts in manufacturing, technology development, and maintenance is rising as the electric car and RE sectors grow to support V2G integration. Employment prospects and economic success are subsequently enhanced by this. By adding another level of flexibility to the grid, V2G deployment improves grid flexibility in response to fluctuating energy demands. When coupled to the grid, EVs provide an adaptable energy storage system (ESS) that may be used tactically to control variations in energy generation and consumption. This adaptability makes the energy grid nimbler and more responsive [8]. The possibility of taking part in V2G initiatives adds even more motivation for people to switch to electric cars. EVs become more economically appealing when one considers the possibility of making money from the energy stored in the vehicle as well as the possible earnings from engaging in grid services [9]. This promotes the wider adoption of environmentally friendly transport options. By actively integrating electric vehicles into the energy ecosystem, V2G technology facilitates more efficient use of energy resources [10]. Excess RE can be stored in car batteries to minimize waste during times of low energy demand. This stored energy can be smoothly reintroduced back into the grid at times of high demand, making the most use of the resources at hand and fostering a more sustainable energy environment [11].

2 Methodology

In the heart of a microgrid (MG), there stands a synchronous generator and a duo of EV battery sets, forming the backbone of sustainable energy dynamics. Each battery cluster symbolizes a community of twenty EVs seamlessly connected to the grid. Within the microgrid's orchestration, the operational frequency gracefully sways between 0.999 and 1.001 units. Here, the battery (G2V) block embodies the essence of daily-charged EVs, adorned with charge levels below 80%, while its counterpart, the battery (V2G) block, mirrors the prowess of EVs boasting a state of charge (SOC) exceeding 90%. These high SOC vehicles become heroes during peak grid demand, actively participating in Vehicle-to-Grid (V2G) operations [12].

As the MG saga unfolds, envision two batteries intertwining with the grid, embarking on a rhythmic charging ballet. With starting SOCs of 70 and 98% respectively, their journey gracefully ascends to higher SOC levels. A modest 1 MW load connecting to the grid triggers a poetic dip in the system frequency below 0.999 per unit after 10 s [13]. Wisely choreographing stability, the battery (V2G) block takes the lead, discharging its power to bring the network frequency back into the comforting range of 0.999–1.001 units.

Following this, at the 30-s mark, a 2 MW load says farewell to the grid, causing a brief incident in which the network frequency briefly surpasses 1.001 per unit. With lightning speed, the battery (V2G) block resumes its charging schedule and expertly maneuvers the network frequency back into the designated range. The battery (G2V) block subtly integrates into the charging ensemble during this melodic charging performance, adding its distinct charm to the developing energy plot [14, 15].

3 Microgrid with Electric Vehicle Circuit

Optimizing energy distribution and usage is achieved by integrating an MG with an EV charging circuit. It controls power fluctuations in the grid and enables effective EV charging. This system offers a dependable and sustainable energy supply by combining conventional energy sources with RESs like wind and solar power [16]. Dynamic load management is made possible by smart grid technology, guaranteeing the best possible use of the resources at hand. Carbon emissions are decreased and clean transportation is encouraged when EV charging is integrated into the MG. All things considered, this creative circuit supports environmental sustainability, improves energy resilience, and eases the shift to a greener future. First, photovoltaic (PV) panels: these solar panels provide clean, RESs by turning sunlight into electricity. They are necessary to use solar energy. Wind energy is captured and transformed into electrical power using wind turbines. They aid in the production of renewable power and are especially useful in windy areas. Diesel generators are backup power sources; however, they are not as environmentally friendly as other solutions [17, 18]. In times of emergency or insufficient production of renewable energy, they can be helpful. Maintaining a balance between supply and demand requires energy storage. Excess energy from wind turbines and solar panels is stored in batteries and released when needed. Converters change the direct current (DC) produced by wind turbines and solar panels into alternating current (AC) that is appropriate for use in homes. Inverters carry out the opposite task of charging batteries by transforming grid or generator AC power into DC electricity. Compared to conventional batteries, supercapacitors store energy more effectively [19, 20]. They can improve system performance overall and provide powerful bursts quickly. All components are connected by appropriately sized and insulated cables, which provide effective energy transfer throughout the system.

3.1 Types of Loads

The function here represents the electrical load connected to the grid and the role in the circuit introduces variations in the grid frequency based on its connection or disconnection from the grid for which regulation to load changes. Stack association (10 s)—upon accepting a 1 MW stack, the battery (V2G) piece recognizes the recurrence diminish and responds by sending power to the network. The recurrence won't drop underneath the assigned lower restrain much appreciated for this activity [21, 22]. The battery (V2G) square returns to charge status when a 2 MW stack disengages from the network (30 s). By doing this, additional control is retained and the recurrence is kept underneath the upper restrain that has been set.

3.2 *Types of Loads*

For the system to remain stable and reliable, power regulation is necessary. It entails matching the supply and consumption of electricity in real-time by balancing the generation of power with demand. Frequency control, voltage regulation, load shedding, and other techniques are used to control variations and stop blackouts. By maximizing energy flow and reducing waste, cutting-edge technologies like demand response programs, ESSs, and smart grids allow for more effective regulation. Good power regulation contributes to a sustainable and dependable energy infrastructure that is necessary for modern society's operation and economic growth. It also strengthens grid resilience, facilitates the integration of RESs, and guarantees a steady supply of electricity [23]. Figure 1 depicts the actual real-life charging system real-world charging systems in microgrids integrate multiple components to effectively manage EV charging while preserving grid stability. Dynamic scheduling depending on grid circumstances and user preferences is possible with smart charging stations that are outfitted with communication protocols. The energy management system (EMS) of the MG, which optimizes charging schedules to reduce peak demand and balance load throughout the grid, communicates with these stations. Furthermore, during times of high generation, EVs can be directly charged by RESs like solar or wind power, which lessens dependency on the main grid. Within the MG ecosystem, this real-world charging method increases efficiency, reduces expenses, and fosters sustainability.

Figure 2 depicts the flow chart of power regulation successful control is fundamental to keeping up the microgrid's soundness and constancy [24, 25]. This involves overseeing control employing an assortment of components and administration procedures, especially when the stack changes and the EV batteries are at changing states of charge (SOC).

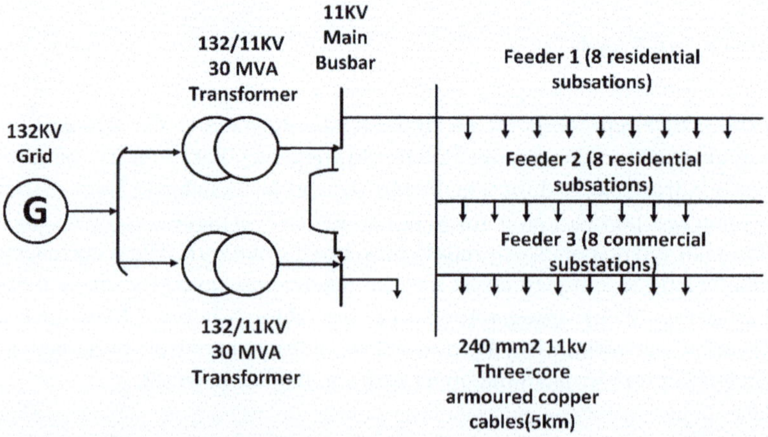

Fig. 1 Actual real-life charging system

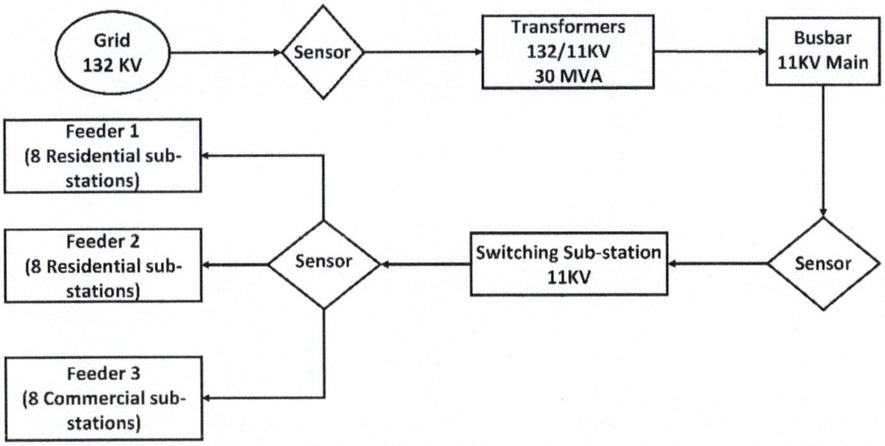

Fig. 2 Flow chart of power regulation

In arrange to help the network as required, the MG uses an energetic and versatile control methodology that utilizes the battery (V2G) block's capabilities [25]. MG outlines how V2G bolster may effectively keep up lattice soundness by observing framework recurrence and adjusting the charging or discharging status of the EV batteries [26], guaranteeing that the control supply remains inside the correct recurrence extend.

3.3 State of Charge

The term state of charge (SoC) describes how much energy a battery has left over after reaching its maximum capacity. It's an essential variable for controlling battery longevity and performance in a variety of applications, such as RESs and electric cars. Users can prevent deep draining or overcharging, which can deteriorate battery health, by optimizing charging and discharging patterns with the help of SoC monitoring [29, 30]. To precisely estimate SoC, sophisticated algorithms and sensors are used, guaranteeing effective use and increasing battery life [27, 28]. Predicting remaining operating time and organizing recharging activities are two other ways that SoC information improves overall system reliability and makes the most use of ESSs.

When EVs acquire energy from the grid, the SOC graph for G2V systems shows the battery charge level of the vehicles over time, highlighting charging patterns and efficiency [31, 32]. Figure 3 depicts the SOC graph (G2V) might reveal information about how well the charge and discharge procedures work. In the SOC curve, steep slopes could represent quick charging or discharging, whereas gentle slopes might represent slower energy transfer rates.

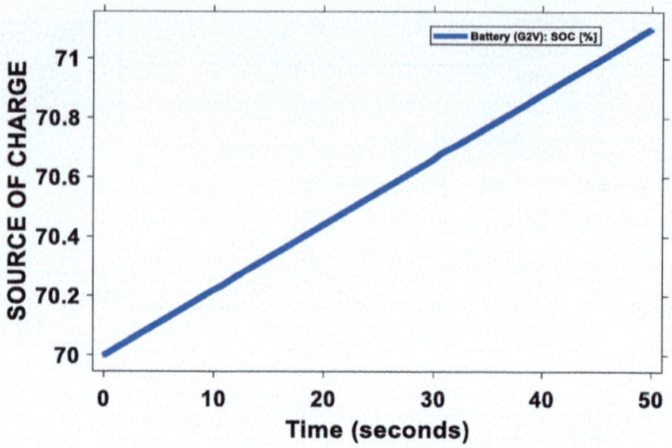

Fig. 3 SOC (G2V) graph

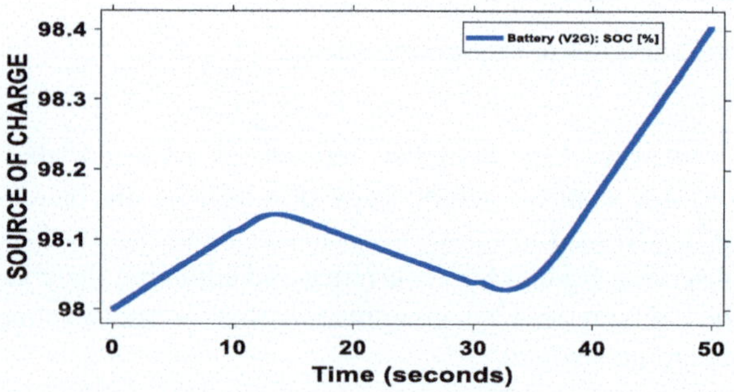

Fig. 4 SOC (V2G) graph

The link between an EV's battery charge level and its ability to release electricity back into the grid is depicted in the SOC versus V2G graph. It illustrates how V2G capabilities for energy management and grid assistance are impacted by SOC [33, 34]. Figure 4 depicts the SOC (V2G) graph trends in battery charging and discharging over time are displayed on the SOC graph (V2G). The SOC rises when the batteries are being charged since they are getting power from renewable or grid-supplied sources [35, 36]. On the other hand, when the batteries are discharging and supplying energy to the grid or other loads, the SOC drops.

4 Simulation Results

The plot underneath shows the AC framework recurrence, the dynamic control that streams to two battery branches, and the state-of-charge of batteries in moo level of devotion.

4.1 Average Voltage Source Converter Output

In control hardware and electrical buildings, the normal voltage source converter (VSC) yield could be a vital parameter. It appears the normal voltage created by the VSC over a given sum of time. Planning and overseeing control transformation frameworks [37], such as voltage source converters utilized in renewable vitality frameworks, engine drives, and high-voltage direct current (HVDC) transmission, require an understanding of the normal VSC yield. To ensure dependable electrical framework execution, unfaltering framework operation, and successful vitality transformation, engineers look at and optimize the normal VSC yield [38]. Controlling the normal VSC yield too makes it conceivable to control voltage levels, receptive control trade, and control stream in interconnected networks, all of which improve the soundness and steadfastness of the framework as an entirety [39, 40].

Figure 5 depicts the voltage output (voltage versus time) graph modulation waveform generated by VSCs over time displayed on the average VSC output graph. It shows how the modulated waveform's amplitude and frequency vary, which is crucial for examining VSC behavior and improving power conversion procedures.

Figure 6 depicts the voltage output (current output versus time) graph voltage waveform produced by VSCs over a certain time period is seen in the Average VSC output voltage versus time graph. It provides insights into VSC performance, stability, and grid compatibility by displaying voltage fluctuations over time.

Figure 7 depicts the voltage output (power dissipated versus time) graph power variations produced by VSCs during a certain interval are depicted in the average VSC output power versus time graph. It displays power output variations, which are important for assessing VSC effectiveness, system performance, and grid integration compatibility.

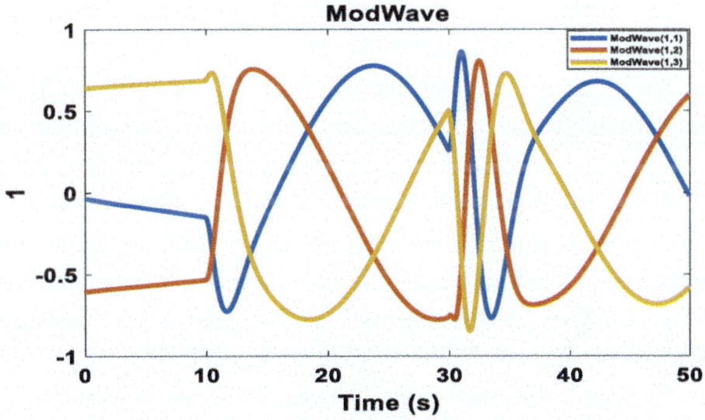

Fig. 5 Voltage output (voltage versus time) graph

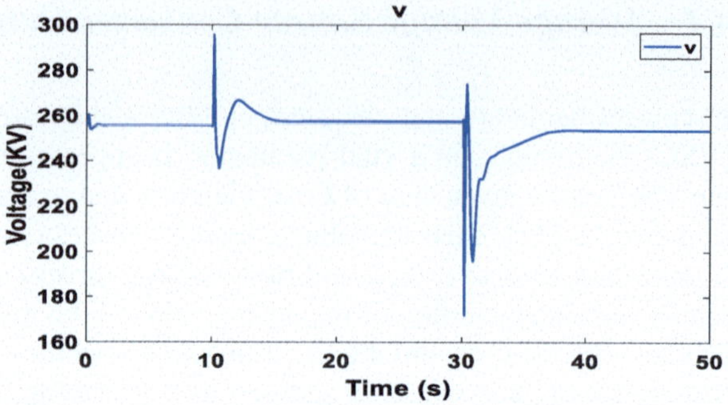

Fig. 6 Voltage output (current output versus time) graph

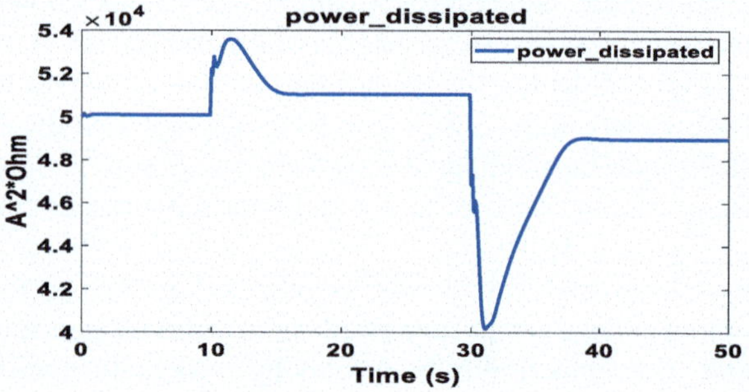

Fig. 7 Voltage output (power dissipated versus time) graph

4.2 Battery (V2G)

A basic portion of V2G innovation is the battery (V2G) component, which permits control to stream both ways between EVs and the electrical framework. Battery (V2G) module in V2G frameworks permits EV batteries to release control back into the network as fundamental, in expansion to charging from it. Because of this highlight, EVs can act as portable vitality capacity gadgets and offer useful framework administrations counting recurrence direction, stack adjusting, and top shaving. In arrange to maximize vitality productivity, move forward lattice soundness, and energize the integration of renewable vitality sources into the electrical lattice, the battery (V2G) is fundamental. Figure 8 depicts the V2G battery power capacity used for grid services is tracked by the G2 versus time graph in a battery V2G system. It helps with grid stability, renewable integration, and V2G operation optimization by displaying variations in power absorption or injection.

Figure 9 depicts the N (vehicle-to-grid) state of charge (SOC) of batteries in vehicle-to-grid (V2G) systems over time shown by the battery V2G N versus time graph. Visualizing variations in battery charge levels helps optimize grid interactions and charging schedules for better energy management. Figure 10 depicts the V_{bat} battery vehicle-to-grid (V2G) systems, the battery voltage (V_{bat}) versus time graph shows the voltage levels of the battery over a certain period of time. It monitors voltage

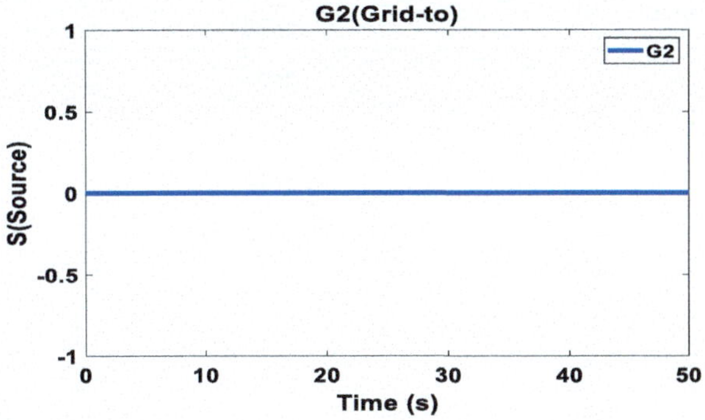

Fig. 8 G2G (vehicle-to-grid)

variations to offer information on system performance, charging/discharging trends, and battery health.

Figure 11 depicts the battery charge levels over a certain period of time in the battery V2G charge versus time graph. Highlighting charging trends makes it possible to analyze charging rates, energy storage capacity, and V2G system performance optimization.

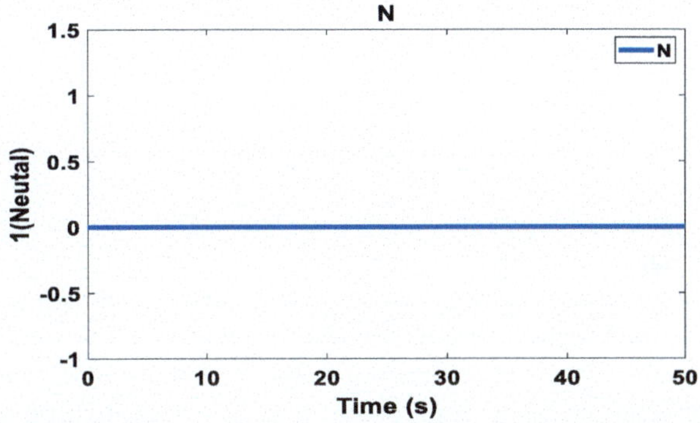

Fig. 9 N (vehicle-to-grid)

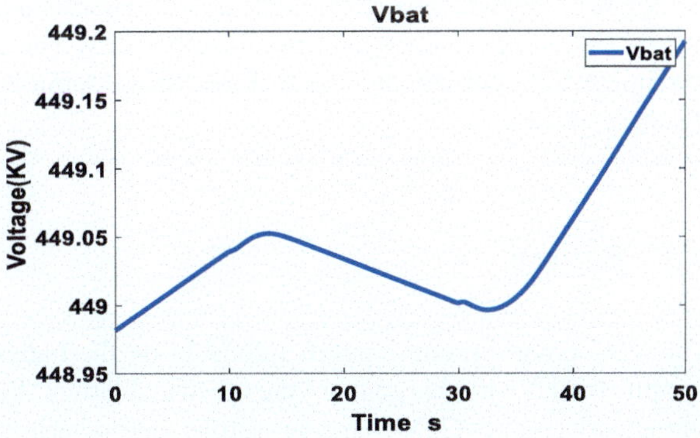

Fig. 10 V_{bat} (vehicle-to-grid)

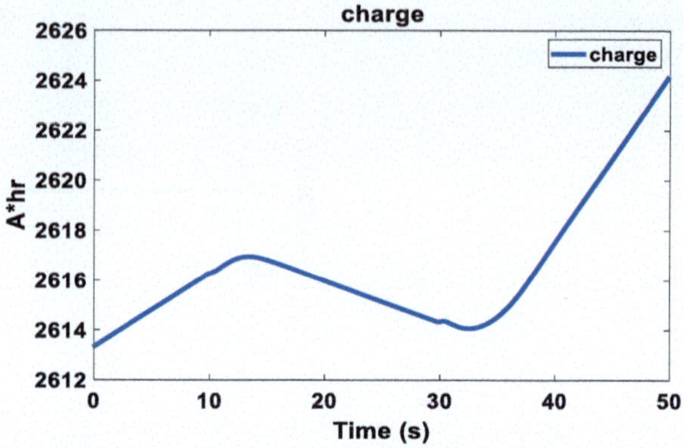

Fig. 11 Charge (V2G)

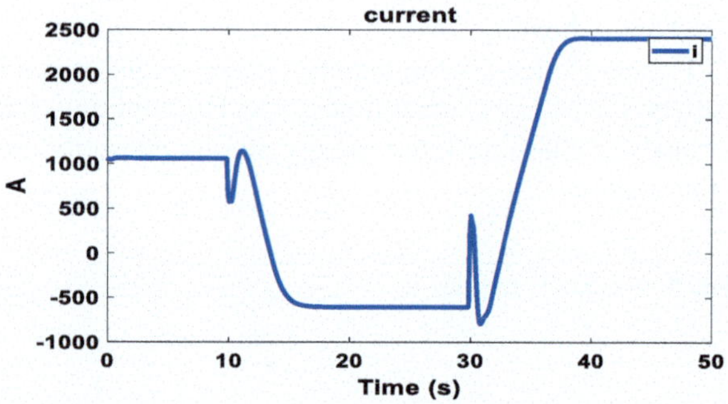

Fig. 12 Current versus time variations

Figure 12 depicts the current dynamic variations in the electric current delivered or taken by batteries over time in V2G systems shown in the battery V2G versus time graph. For energy management analysis, it displays charging, discharging, and grid interaction patterns. Figure 13 depicts the power dissipated rate at which energy is lost in the battery during V2G operations is shown by the battery V2G power dissipation versus time graph. It assists in evaluating battery efficiency and system optimization by tracking dissipation over time.

Figure 14 depicts the battery capacity (AH) fluctuates with time during the vehicle-to-grid (V2G) operations and time. It is crucial to keep an eye on battery health and maximize V2G performance as it shows the dynamic variations in battery charge or discharge levels.

4.3 Battery

The V2G environment depends intensely on the battery (G2V) component, which permits the EVs to charge from the electrical lattice. Battery (G2V) module in V2G frameworks makes G2V charging simpler, empowering EVs to beat off their batteries

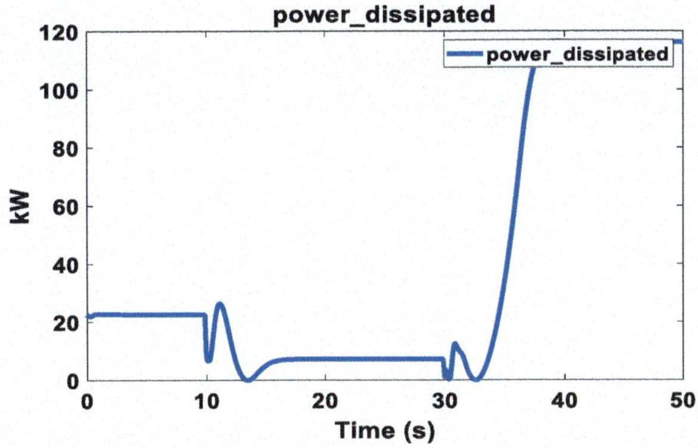

Fig. 13 Power dissipated versus time

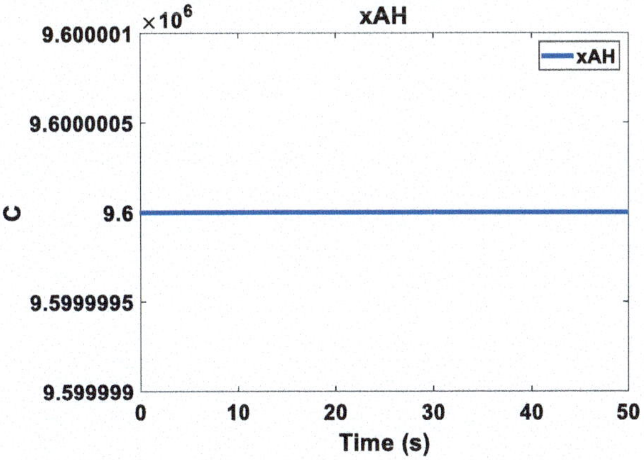

Fig. 14 Battery capacity versus time variations

at off-peak times or at whatever point lattice circumstances permit. With the assistance of this highlight, EV proprietors can utilize the network as a commonsense and promptly accessible source of vitality for their cars. Inside the V2G system, the battery (G2V) is basic for maximizing framework utilization, supporting successful vitality administration, and optimizing charging procedures.

Figure 15 depicts the G2 versus time graph which shows the dynamics of energy transfer from the grid to electric cars over time in the context of G2V systems. It helps to understand charging patterns, load balancing, and the best way to integrate EVs into the grid infrastructure by displaying fluctuations in the amount of energy delivered from the grid to vehicles.

Figure 16 depicts the N versus time graph which is commonly used to illustrate the pace at which EV batteries charge or discharge over time in G2V systems. The variable N may stand for the EV batteries' SOC fluctuations over a certain period of time, charging power level, or charging/discharging pace. This graph facilitates understanding charging patterns and grid management by providing a visual representation of the dynamic charging behavior of EVs.

Fig. 15 G2 versus time graph

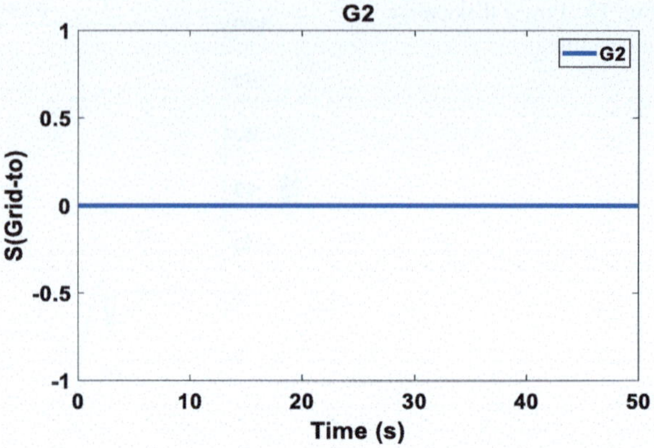

Fig. 16 N versus time graph

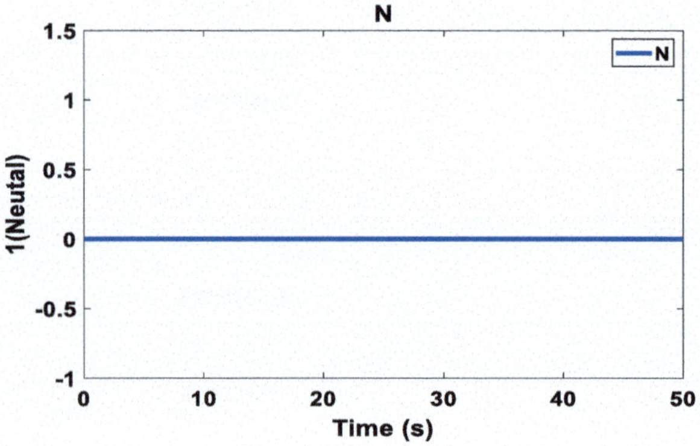

Figure 17 depicts the battery voltage change in electric cars over time during charging in the G2V context by the V_{bat} (battery voltage) versus time graph. It offers perceptions of the dynamics of battery charging and aids in the optimization of charging plans for effective energy management. Figure 18 depicts the charge versus time graph that shows how much energy accumulates in the batteries of EVs over a given period of time. It makes grid connection and energy management easier by providing a visual depiction of the charging process and showing how car batteries use energy over time. Figure 19 shows the dynamic flow of electric current over time from the grid to electric cars. It draws attention to differences in charging rates, which helps with grid management and electric car charging strategy optimization.

Figure 20 depicts the power dissipated rate of energy consumption by electric cars from the grid over time is shown by the power dissipated versus time graph in a G2V setting. It draws attention to fluctuations in power use, which helps with grid management and infrastructure resource allocation for charging. Figure 21 depicts the cumulative energy consumption by EVs from the grid over time (Ampere-Hours) versus the time graph. It helps with grid management and understanding car charging behaviors by disclosing charging patterns and total energy consumption.

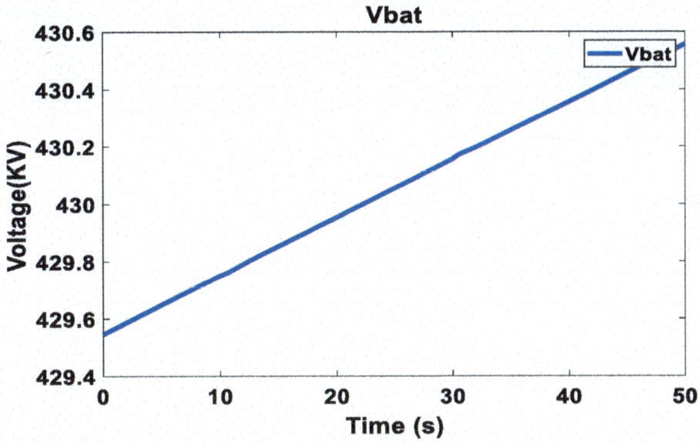

Fig. 17 Battery voltage versus time in G2V mode of operation

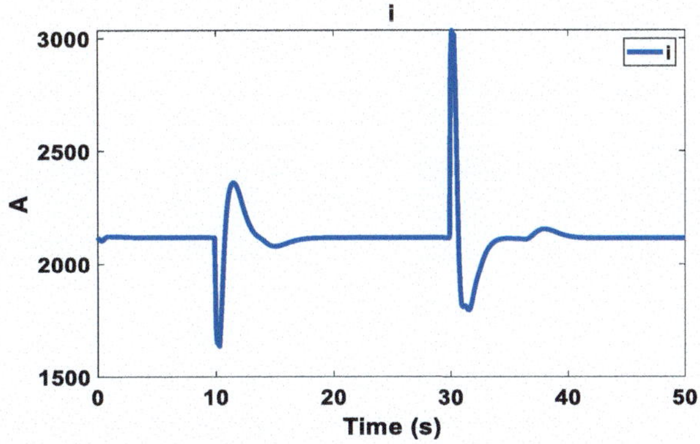

Fig. 18 Charge versus time in G2V mode of operation

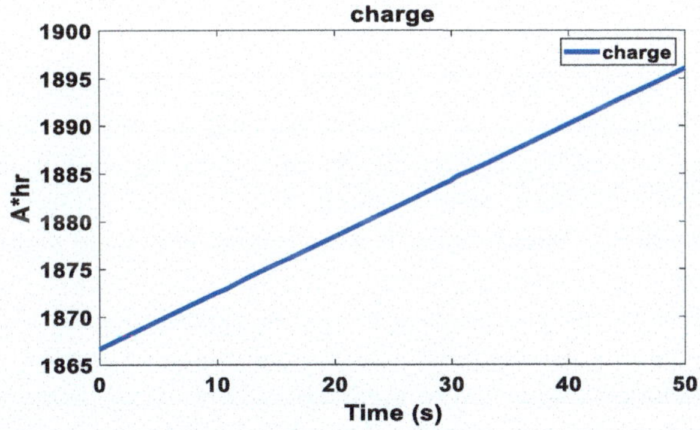

Fig. 19 Current versus time graph in G2V mode of operation

5 Conclusions

By using a prediction-based model, it is possible to extrapolate from the available data points and minimize differences in power consumption and utilization, which greatly increases efficiency. Through the examination of data trends and patterns,

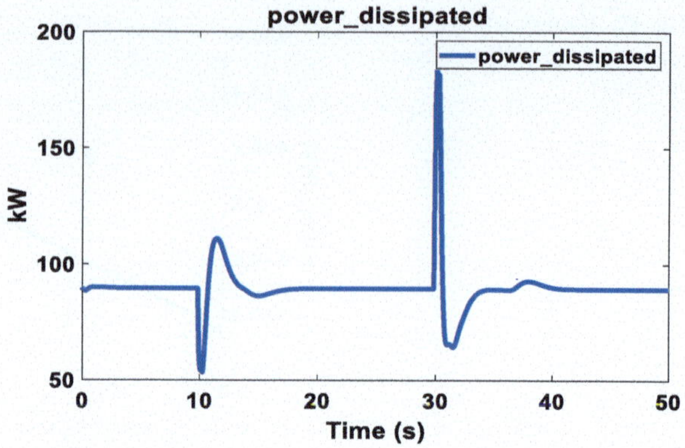

Fig. 20 Power dissipated versus time graph in G2V mode of operation

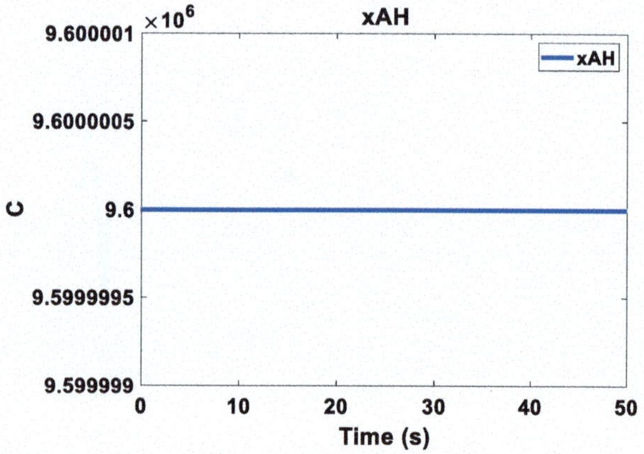

Fig. 21 Energy consumption versus time in G2V mode of operation

the model can precisely predict future consumption, enabling pre-emptive actions to maximize resource distribution and utilization. This method not only helps to spot possible inefficiencies but also gives decision-makers the ability to implement focused improvement methods. Using ongoing improvement and modification, the prediction-based model has the potential to yield significant improvements in efficacy, guaranteeing the most efficient use of resources while minimizing wastage. In the end, integrating these predictive capabilities could result in significant improvements to sustainability and energy efficiency across a variety of industries. V2G includes a wide run of applications and offers plenty of choices for tried-and-true control era and capacity. Also, V2G offers a more determined technique. Another vital issue is the environment. It still gets a parcel of feedback, even though. The essential causes of this feedback are the tall introduction taking a toll, the nonattendance of government endowments, and the producer's and people's unwillingness to alter. Most people have a restricted and careless point of view since they as it were assessing the current circumstances disregard the potential benefits of V2G in the future. Promising outcomes from the foremost later V2G arrangement ventures have provoked more ponder within the range. V2G will spread widely as and when more vigorous batteries and sensibly estimated lattice lines get to be standard. The conclusions of people and producers are to be observed until that point. At long last, the way to V2G usage will be made significantly simpler by the governments of affluent and creating countries broadly spreading the concept, as well as its conceivable outcomes and openings. Besides, V2G will be upheld by keen lattice advances, which can meet future control requests. Future V2G microgrids with electric vehicles have the potential to change transportation foundations and vitality frameworks.

Acknowledgements This research work was supported by "Woosong University's Academic Research Funding –2025".

References

1. Lam HS, Yuan H, Tan SC, Mi CC, Pou J, Hui SYR (2022) Bidirectional AC–DC modular multilevel converter with electric spring functions for stabilizing renewable AC power grid at the distribution voltage level. IEEE J Emerg Select Top Power Electron 10(6):7589–7600. https://doi.org/10.1109/JESTPE.2022.3173809
2. Acharige SSG, Haque ME, Arif MT, Hosseinzadeh N, Hasan KN, Oo AMT (2023) Review of electric vehicle charging technologies, standards, architectures, and converter configurations. IEEE Access 11:41218–41255. https://doi.org/10.1109/ACCESS.2023.3267164
3. Rivera S (2023) Charging infrastructure and grid integration for electromobility. Proc IEEE 111(4):371–396. https://doi.org/10.1109/JPROC.2022.3216362
4. Ahmed I, Rehan M, Basit A, Tufail M, Hong KS (2023) A dynamic optimal scheduling strategy for multi-charging scenarios of plug-in-electric vehicles over a smart grid. IEEE Access 11:28992–29008. https://doi.org/10.1109/ACCESS.2023.3258859
5. Wu S, Wang L, Song B, Gao J (2024) Live-line detection of deteriorated insulators on overhead transmission lines based on electric field distribution. IEEE Sens J 24(2):1687–1695. https://doi.org/10.1109/JSEN.2023.3332849
6. Xi L, Li H, Zhu J, Li Y, Wang S (2024) A novel automatic generation control method based on the large-scale electric vehicles and wind power integration into the grid. IEEE Trans Neural Netw Learn Syst 35(5):5824–5834. https://doi.org/10.1109/TNNLS.2022.3194247
7. Wang Q (2023) A nested decomposition method for the AC optimal power flow of hierarchical electrical power grids. IEEE Trans Power Syst 38(3):2594–2609. https://doi.org/10.1109/TPWRS.2022.3194077
8. Abdelsattar M, Ismeil MA, Aly MM, Saber Abu-Elwfa S (2024) Analysis of renewable energy sources and electrical vehicles integration into microgrid. IEEE Access 12:66822–66832. https://doi.org/10.1109/ACCESS.2024.3399124
9. Baharom R (2022) A view of wireless charging system for electric vehicle: technologies, power management and impacts. In: IEEE industrial electronics and applications conference (IEACon), Kuala Lumpur, Malaysia, pp 146–151. https://doi.org/10.1109/IEACon55029.2022.9951794
10. Suseela PB, Bhanu CVK (2023) Design of solar-powered electric vehicle charging system. In: International conference for advancement in technology (ICONAT), Goa, India, pp 1–8. https://doi.org/10.1109/ICONAT57137.2023.10080468
11. Sivakumar N, Raja SC, Saravanan M, Xavier SAE (2023) Analysis the performance of energy transfer system (ETS) for EV battery charging method using non-isolated SEPIC converter-based CCCV control. In: 9th international conference on electrical energy systems (ICEES), Chennai, India, pp 377–381. https://doi.org/10.1109/ICEES57979.2023.10110262

12. Chehab MH, Khelil M, Salah CB (2023) Enhancing efficiency in wireless solar charging stations through IoT control and optimization strategies. In: IEEE international conference on artificial intelligence & green energy (ICAIGE), Sousse, Tunisia, pp 1–6. https://doi.org/10.1109/ICAIGE58321.2023.10346409
13. Diao X (2023) Research on electric vehicle charging safety warning based on A-LSTM algorithm. IEEE Access 11:55081–55093. https://doi.org/10.1109/ACCESS.2023.3281552
14. Yousuf AKM, Wang Z, Paranjape R, Tang Y (2023) Electric vehicle charging station infrastructure: a comprehensive review of technologies, challenges, and mitigation strategies. In: 2023 IEEE Canadian conference on electrical and computer engineering (CCECE), Regina, SK, Canada, pp 588–592. https://doi.org/10.1109/CCECE58730.2023.10289005
15. ElGhanam E, Sharf H, Odeh Y, Hassan MS, Osman AH (2022) On the coordination of charging demand of electric vehicles in a network of dynamic wireless charging systems. IEEE Access 10:62879–62892. https://doi.org/10.1109/ACCESS.2022.3182700
16. Jeong SI, Choi D-H (2022) Electric vehicle user data-induced cyber attack on electric vehicle charging station. IEEE Access 10:55856–55867. https://doi.org/10.1109/ACCESS.2022.3177842
17. Dubey A, Santoso S (2015) Electric vehicle charging on residential distribution systems: impacts and mitigations. IEEE Access 3:1871–1893. https://doi.org/10.1109/ACCESS.2015.2476996
18. Jung H, Jang H, Han H, Cho K, Kang B, Park S (2022) "Electric Vehicle charging system to reduce carbon emissions using photovoltaic power generation and ESS. In: 2022 IEEE 5th student conference on electric machines and systems (SCEMS), Busan, Republic of Korea, pp 1–4. https://doi.org/10.1109/SCEMS56272.2022.9990052
19. Xiaotao H (2021) Operation and maintenance system of electric vehicles' charging and discharging facilities based on repository. In: IEEE 3rd international conference on civil aviation safety and information technology (ICCASIT), Changsha, China, pp 894–897. https://doi.org/10.1109/ICCASIT53235.2021.9633471
20. Ray PK, Bharatee A, Panda S, Satiawan INW (2021) Modeling and power management of electric vehicle charging system. In: International conference on smart-green technology in electrical and information systems (ICSGTEIS), Sanur, Bali, Indonesia, pp 100–105. https://doi.org/10.1109/ICSGTEIS53426.2021.9650377
21. Zhang J (2021) An optimal dispatching strategy for charging and discharging of electric vehicles based on cloud-edge collaboration. In: 3rd Asia energy and electrical engineering symposium (AEEES), Chengdu, China, pp 827–832. https://doi.org/10.1109/AEEES51875.2021.9403163
22. Bai S (2022) Overview of electric vehicle charging standards. IEEE/IAS industrial and commercial power system Asia (I&CPS Asia), Shanghai, China, pp 700–705. https://doi.org/10.1109/ICPSAsia55496.2022.9949844
23. Pasetti M, Iacono SD, Zaninelli D (2023) Real-time state of charge estimation of light electric vehicles based on active power consumption. IEEE Access 11:110995–111010. https://doi.org/10.1109/ACCESS.2023.3322651
24. Bose B, Kumar Tayal V, Moulik B (2020) Multi-loop multi-objective control of solar hybrid EV charging infrastructure for workplace. In: 2nd international conference on advances in computing, communication control and networking (ICACCCN), Greater Noida, India, pp 491–496. https://doi.org/10.1109/ICACCCN51052.2020.9362768
25. Mazumdar D, Biswas PK, Sain C, Ustun TS (2024) GAO optimized sliding mode based reconfigurable step size Pb&O MPPT controller with grid integrated EV charging station. IEEE Access 12:10608–10620. https://doi.org/10.1109/ACCESS.2023.3344275
26. Jie B, Baba J, Kumada A (2024) Contribution to V2G system frequency regulation by charging/discharging control of aggregated EV group. IEEE Trans Ind Appl 60(1):1129–1140. https://doi.org/10.1109/TIA.2023.3292814
27. Wang X (2021) Tri-level scheduling model considering residential demand flexibility of aggregated HVACs and EVs under distribution LMP. IEEE Trans Smart Grid 12(5):3990–4002. https://doi.org/10.1109/TSG.2021.3075386

28. Liu Z, Wang L (2022) A robust strategy for leveraging soft open points to mitigate load altering attacks. IEEE Trans Smart Grid 13(2):1555–1569. https://doi.org/10.1109/TSG.2021.3134176
29. Kurukuru VSB, Haque A, Padmanaban S, Khan MA (2022) Rule-based inferential system for microgrid energy management system. IEEE Syst J 16(1):1582–1591. https://doi.org/10.1109/JSYST.2021.3094403
30. Parise G, Parise L, Malerba A, Pepe FM, Honorati A, Chavdarian PB (2017) Comprehensive peak-shaving solutions for port cranes. IEEE Trans Ind Appl 53(3):1799–1806. https://doi.org/10.1109/TIA.2016.2645514
31. Paaso A, Kushner D, Bahramirad S, Khodaei A (2018) Grid modernization is paving the way for building smarter cities [technology leaders]. IEEE Electrif Mag 6(2):6–108. https://doi.org/10.1109/MELE.2018.2816848
32. Bhadane K (2021) A comprising study on modernization of electric vehicle subsystems, challenges, opportunities and strategies for its further development. In: 4th biennial international conference on nascent technologies in engineering (ICNTE), NaviMumbai, India, pp 1–9. https://doi.org/10.1109/ICNTE51185.2021.9487757
33. Das HS, Li S, Chowdhury MFF, Ghosh T (2020) Game theoretical energy management of EV fast charging station with V2G capability. In: IEEE international conference on power and energy (PECon), Penang, Malaysia, pp 344–349. https://doi.org/10.1109/PECon48942.2020.9314460
34. Begum R, Rathi M (2022) Design of PFC with buck boost converter for plug-in electric vehicles and battery charging applications. In: IEEE 2nd international conference on sustainable energy and future electric transportation (SeFeT), Hyderabad, India, pp 1–6. https://doi.org/10.1109/SeFeT55524.2022.9909283
35. Salkuti SR (2023) Advanced technologies for energy storage and electric vehicles. Energies 16(5):2312. https://doi.org/10.3390/en16052312
36. Salkuti SR (2019) Optimal operation of microgrid considering renewable energy sources, electric vehicles and demand response. E3S Web Conf 87:01007. https://doi.org/10.1051/e3sconf/20198701007
37. Jiang L, Liu H, Wang C (2021) Summary of EMC test standards for wireless power transfer systems of electric vehicles. In: Asia-pacific international symposium on electromagnetic compatibility (APEMC), Nusa Dua - Bali, Indonesia, pp 1–4. https://doi.org/10.1109/APEMC49932.2021.9597175
38. Kaleybar HJ, Brenna M, Castelli-Dezza F, Carmeli MS (2023) Smart hybrid electric railway grids: a comparative study of architectures. In: IEEE international conference on electrical systems for aircraft, railway, ship propulsion and road vehicles & international transportation electrification conference (ESARS-ITEC), Venice, Italy, pp 1–6. https://doi.org/10.1109/ESARS-ITEC57127.2023.10114844
39. Salkuti SR (2022) Emerging and advanced green energy technologies for sustainable and resilient future grid. Energies 15(18):6667. https://doi.org/10.3390/en15186667
40. Rubio F, Pereda J, Rojas F, Poblete P (2022) Hybrid sorting strategy for modular multilevel converters with partially integrated 2nd life battery energy storage systems for fast EV charging. In: IEEE 7th southern power electronics conference (SPEC), Nadi, Fiji, pp 1–7. https://doi.org/10.1109/SPEC55080.2022.10058403

Mitigating Impacts of Electric Vehicle Charging Stations to the Distribution Systems by Optimal Operation of Soft Open Point

Thuan Thanh Nguyen, Viet Anh Truong, Trung Nhan Nguyen, and Hoai Phong Nguyen

Abstract Connecting electric vehicle charging stations (EVCSs) to the distribution system can cause negative impacts on the system such as increasing voltage drop and power loss. This work presents a method to limit the impact of EVCSs on voltage deviation and power loss of the distribution system based on optimizing Soft Open Points (SOP) operation installed in the distribution system. The optimal operation power for SOP at each hour of the day is determined by the Archimedes Optimization Algorithm (AOA). The evaluation results of the proposed solution are performed on a 33-bus system integrating a Photovoltaic System (PVS) system and an EVCS for twenty-four hours. The results show that optimizing SOP operation can limit the impact of EVCSs on the distribution system by improving the voltage profile and reducing power loss of the system. The compared results of AOA and Particle Swarm Optimization (PSO) show that AOA reaches a better performance than PSO in finding the optimal power of SOP in each interval. In addition, the impact of uncertainty of industrial, residential, and commercial loads, EVCS, and PVS on SOP operation results is also evaluated based on the Monte Carlo simulation method. The results show that the SOP operation solution also limits the impact of uncertainty factors on voltage deviation and power loss indicators of the distribution system.

T. T. Nguyen (✉) · H. P. Nguyen
Faculty of Electrical Engineering Technology, Industrial University of Ho Chi Minh City, Ho Chi Minh City, Viet Nam
e-mail: nguyenthanhthuan@iuh.edu.vn

H. P. Nguyen
e-mail: nguyenhoaiphong@iuh.edu.vn

V. A. Truong
Ho Chi Minh City University of Technology and Education, Ho Chi Minh City, Viet Nam
e-mail: anhtv@hcmute.edu.vn

T. N. Nguyen
Office of Academic Affairs, Industrial University of Ho Chi Minh City, Ho Chi Minh City, Viet Nam
e-mail: nguyentrungnhan@iuh.edu.vn

S. R. Salkuti (ed.), *Artificial Intelligence for Integrated Smart Energy Systems in Electric Vehicles*, Lecture Notes in Electrical Engineering 1427,
https://doi.org/10.1007/978-3-031-94276-1_3

Keywords Electric vehicle charging stations · Soft open points · Photovoltaic system · Distribution system · Archimedes optimization algorithm

Nomenclature

EVCS	Electric Vehicle Charging Stations
SOP	Soft-Open-Points
AOA	Archimedes Optimization Algorithm
PVS	Photovoltaic System
PSO	Particle Swarm Optimization
EV	Electric Vehicles
DG	Distributed Generations
MCS	Monte Carlo Simulation
IRC	Industrial, Residential, and Commercial
STD	Standard Deviation

1 Introduction

Electric vehicles (EVs) are an effective solution to reduce human dependence on fossil energy sources as well as reduce air pollution. However, in many countries, due to the rapid increase in the number of electric vehicles, electric vehicle charging stations (EVCSs) are built and connected to the existing distribution systems. Therefore, EVCS causes some negative impacts on the technical indicators of the distribution systems such as voltage reduction and increasing power loss [1, 2]. In addition, the stochastic characteristics of electric vehicles, other types of loads, and some distributed generations (DG) in the distribution systems make it even more difficult to overcome the above limitations.

To overcome the aforementioned limitations of EVCSs on the distribution systems, many studies have optimized EVCS placement to reduce its impact on the power systems. In [3], the location of EVCS and capacitors in the distribution system is optimized to maintain the voltage profile, increase the net gain, and minimize power loss. In [4], the location of EVCS and DG is optimized to reduce power loss in the distribution system. In [5], the problem of optimal placement of EVCS in the distribution system to minimize power loss, voltage deviation, and voltage stability index is proposed. In [6], the location of EVCS, DG, DSTATCOM, and battery energy storage systems is optimized to mitigate the impact of EVCS on the power loss of the distribution system. Similarly, in [7], the location of EVCS, DG, and DSTATCOM is optimized to reduce power loss, voltage stability, and reliability of the distribution system. However, the optimal location of EVCS depends not only on the technical factors of the distribution system but also on many other factors such as the installation area and the convenience for users. Therefore, the actual locations of the EVCS may not be the most optimal locations to achieve positive technical indicators for the distribution system. Therefore, after installing EVCS, technical solutions are needed to reduce the impact of EVCS such as installing DG, compensating reactive power, and installing electronic devices to control the power flow of the distribution system. Soft Open Points (SOPs) are power devices used to connect feeders in the distribution system [8]. The SOP configuration may include back-to-back voltage-source converters (VSCs) [9], static series synchronous compensators [10], or a unified-power-flow-controller [11] to allow power control between feeders. Therefore, using SOP in the distribution system with DG and EVCS promises to improve the system's indicators, contributing to reducing the impact of EVCSs on the system.

In this work, on a distribution system integrating Photovoltaic System (PVS) and EVCS, the SOP power connected between two feeders is optimally controlled to improve the voltage profile and reduce the power loss of the distribution system. The Archimedes Optimization Algorithm (AOA) is adapted to find the optimal operating power of the SOP at different intervals of the day. AOA is inspired by an interesting law of physics Archimedes' principle [12] wherein, the positions of all objects in the fluid are considered as the solutions to the optimization problems. In [12], AOA has demonstrated better performance than many metaheuristic algorithms such as genetic algorithms, Particle Swarm Optimization (PSO), whale optimization algorithm, sine–cosine algorithm, Harris' hawk optimization, and equilibrium optimizer. In the field of electrical engineering, AOA has been used for many problems such as the placement of EVCSs [5], photovoltaic model parameters identification [13, 14], maximum power point tracking of photovoltaic module [15], and wind energy generation system [16] and optimal power flow [17]. This is the foundation for this study to use AOA for the SOP power optimization problem to reduce the impact of EVCSs on the voltage profile and power loss of the distribution system. In addition, after the optimal power value of SOP at each interval of the day is determined, the uncertainty of EVCSs and some parts of the distribution system such as load demand and PVS are also evaluated using the Monte Carlo Simulation (MCS) method. The novelties of this work are summarized as follows:

- Optimize SOP power to reduce voltage deviation and power loss on the distribution system with PVS and EVCSs.
- Apply AOA to optimize SOP power at each interval of the day.
- Investigate the uncertainty of EVCS, load demands, and PVS using the MCS method.

The main contributions of this work are as follows:

- A model of EVCS, load demands, and PVS with their uncertainty is presented.
- Details of the steps to apply AOA to the SOP power optimization problem to reduce voltage deviation and power loss of the distribution system are presented.
- Details of the MCS method to investigate the uncertainty of the EVCS, PVS, industrial, residential, and commercial (IRC) loads are presented.

- Simulation results show that optimal operation of SOP can reduce the impact of EVCS on voltage deviation and power loss of the distribution system and AOA is a potential method for optimizing the power of SOP.

The entire work is structured as follows: the model of EVCS, load demands, and PVS, and the objective function are described in the next section. The application of AOA for finding the optimal power of SOP in each interval of the day is shown in Sect. 3. Section 4 provides the numerical results of the 33-bus distribution system. Finally, the conclusions are presented.

2 Problem Formulation

2.1 Model of EVCS

The load of an EVCS depends on many factors such as the time of EV connects to the EVCS, the travel distance of each EV, and the SOC of each EV when entering the EVCS [18]. However, in this study, to evaluate the impact of EVCS on the parameters of the distribution system, the EVCS is described as a real load that consumes active power from the grid when the EV is charged at EVCS. Thus, when EVCS is connected to a bus of the distribution system, the load at that bus is determined as follows [19]:

$$P_{j,new}^{t} = P_{j}^{t} + P_{EV}^{t} \tag{1}$$

where $P_{j,new}^{t}$, and P_{j}^{t} are the power at bus j after and before EVCS is installed at that bus at interval t, respectively. P_{EV} is the EVCS power at interval t.

Due to the uncertainty of EVCS load in different time periods. It is necessary to consider the uncertainty of EVCS load. The uncertainty of EVCS can be described mathematically using the Normal PDF [20] or Truncated distribution [21]. In this study, the uncertainty of EVCS is modeled using the Normal distribution function as follows:

$$PDF(P_{EV}) = \frac{1}{\delta_{EV}\sqrt{2\pi}} \cdot \exp\left(\frac{-(P_{EV} - \mu_{EV})^2}{2\delta_{EV}{}^2}\right) \tag{2}$$

where μ_{EV} and δ_{EV} are mean and standard deviation (STD) of the load demand of EVCS. Based on the load changes of EVCS at different times through mean and STD values, EVCS load samples are generated using Normal distribution to assess the uncertainty of EVCS.

2.2 Model of Other Loads in the Distribution System

The load at other load buses in the distribution system is assumed to include three types industrial, residential, and commercial (IRC). Each type accounts for a certain proportion. Then, the load at each bus is determined as follows [22]:

$$\begin{cases} P_j^t = \sum_{k=1}^{3} P_{j,max} \,.\, A_{j,k} \,.\, C_k^t \\ Q_j^t = \sum_{k=1}^{3} Q_{j,max} \,.\, A_{j,k} \,.\, C_k^t \end{cases} \tag{3}$$

where $P_{j,max}$ is the maximum load at node j. $A_{j,k}$ is the ratio of each type of load at bus j. C_k^t 1 is the ratio of each type of load at interval t. Similar to EVCS, the I.R.C. load uncertainty in the system is also described by the Normal PDF [23, 24].

2.3 Model of PVS

The power of PVS which depends on solar irradiance and ambient temperature is determined by the following equation.

$$P_{PV}^t = N_{PV} \cdot \mathrm{FF} \cdot \mathrm{V}_{ock}^t \cdot I_{ock}^t \tag{4}$$

where N_{PV} is the number of PV modules. FF is the fill factor. V_{ock} and I_{sck} are the open voltage and short circuit current at the irradiance s and temperature T_{amb}. FF, $V_{ock,}$ and I_{sck} are determined as follows:

$$\begin{cases} FF = \frac{I_{MPP} \cdot V_{MPP}}{V_{OC} \cdot I_{SC}} \\ \mathrm{V}_{ock}^t = V_{OC} \cdot \left[1 - K_V \cdot \left[T_{amb} + s \cdot \left(\frac{NOCT-20}{0.8}\right) - 25\right]\right] \\ I_{ock}^t = s \cdot I_{SC} \cdot \left[1 + K_I \cdot \left[T_A + s \cdot \left(\frac{NOCT-20}{0.8}\right) - 25\right]\right] \end{cases} \tag{5}$$

Here I_{MPP}, $V_{MPP,}$ and I_{SC}, V_{OC} are the PV current and voltage at the maximum point and sort circuit current as well as open circuit voltage. K_I and K_V are respectively temperature coefficients of I_{SC} and V_{OC}. $NOCT$ is the nominal operating cell temperature. The PV power depends strongly on temperature. The uncertainty of irradiance can be modeled by the Beta PDF distribution as follows [25–27].

$$PDF(s) = \begin{cases} \frac{\Gamma(\alpha+\beta)}{\Gamma(\alpha)\Gamma(\beta)} \cdot s^{(\alpha-1)} \cdot (1-s)^{(\beta-1)}; \alpha \geq 0, \beta \geq 0; 0 \leq s \leq 1 \\ 0; \, otherwise \end{cases} \tag{6}$$

where s is solar irradiance (kw/m^2). α and β are factors determined based on the mean (γ_s) and STD (δ_s) of irradiance as follows:

$$\begin{cases} \alpha = \frac{\gamma_s \cdot \beta}{1-\gamma} \\ \beta = (1-\gamma_s) \cdot \left(\frac{\gamma_s \cdot (1+\gamma_s)}{\delta_s^2} - 1 \right) \end{cases} \tag{7}$$

Based on the irradiance changes at different times through mean and STD values, irradiance samples are generated using Beta distribution to assess the uncertainty of solar irradiance.

2.4 *Model of Soft Open Point Device*

SOP is a power electronic device that is used to connect two feeders in the distribution system. The details of SOP can be from reference [9]. In the distribution system, SOP can be used to adjust active power between its two terminals and supply reactive power at terminals. The powers of two terminals of SOP can be determined as follows:

$$\begin{cases} P_I^t + P_{II}^t = 0 \\ \sqrt{(P_I^t)^2 + (Q_I^t)^2} \le S_I \\ \sqrt{(P_{II}^t)^2 + (Q_{II}^t)^2} \le S_{II} \end{cases} \tag{8}$$

where $P_I^t + jQ_I^t$ and $P_{II}^t + jQ_{II}^t$ are the complex power of the terminal I and II at the interval t, respectively. S_I and S_{II} are the allowed apparent power of terminals I and II, respectively.

2.5 *The Objective of Optimal Operation of SOP*

When EVCSs are connected to the distribution system, the system's parameters such as voltage profile and power loss are adversely affected by the sudden increase in load at the buses connected to EVCSs. In active distribution systems that involve DGs, this negative situation can be mitigated if the power from these DGs as well as from other feeders can be transferred to the EVCSs-connected feeder. In this study, the SOP device is used to connect the feeders to control the desired amount of active power between the feeders. In addition, the SOP also has the control capability to provide the amount of reactive power at the connected nodes to perform reactive power compensation for the system. Therefore, the objective of the SOP's active and reactive power control in this study to improve the voltage profile and reduce the power loss of the system is mathematically described as follows:

$$f_{obj}^{t} = \mathrm{w}_1 \cdot \frac{\sum_{j=1}^{N_{bus}} \left|1 - V_j^t\right|}{\sum_{j=1}^{N_{bus}} \left|1 - V_{0j}^t\right|} + \mathrm{w}_2 \cdot \frac{\Delta P^t}{\Delta P^{0t}} \tag{9}$$

where V_j^t and ΔP^t are the voltage amplitude of bus j and power loss of the system at the interval t. V_{0j}^t and ΔP^{0t} are the voltage amplitude of bus j and power loss of the system in the initial condition at interval t. N_{bus} is the number of buses in the system. w_1 and w_2 are weight factors. The operation of SOP does not negatively affect the current and voltage profiles of the system. This is demonstrated by the following constraints:

$$0.95 \leq V_j^t \leq 1.05; \quad j = 1, \ldots, N_{bus} \tag{10}$$

$$KI_i^t \leq 1; \quad i = 1, \ldots, N_{branch} \tag{11}$$

where V_j^t and KI_i^t are the voltage amplitude of bus j and the current-carrying factor of branch i at interval t. N_{branch} is the number of branches of the system.

3 Finding the Optimal Powers of SOP Using the AOA Algorithm

In this section, the details of AOA for finding the optimal power of SOP at each interval are presented.

3.1 Initialization

For solving the optimal problem based on AOA, each object is considered as a solution. Thus, each object contains information on SOP powers consisting of active and reactive powers at its terminals. Each object (O_i) in the population with the size of N and the volume (vo), density (de), and acceleration (ac) of each are generated randomly by Eqs. (12) and (13), respectively.

$$O_i = L_i + r_1 \cdot (U_i - L_i); i = 1, 2, \ldots, N \tag{12}$$

$$\begin{cases} vo_i = r_2 \\ de_i = r_3 \\ ac_i = L_i + r_1 \cdot (U_i - L_i) \end{cases} \tag{13}$$

where L_i and U_i are the lower and upper bounds of objects. r_1, r_2 and r_3 are random vectors in [0, 1]. The generated solutions will be updated to the parameter of the distribution system. Then, the load flow is executed, and the fitness value of each solution is determined as follows:

$$F_i^t = f_{obj}^t + k_p.\left[\begin{array}{l}\sum_{j=1}^{N_{bus}} \max\left(0.95 - V_j^t, 0\right) + \sum_{j=1}^{N_{bus}} \max\left(V_j^t - 1.05, 0\right) \\ \qquad + \sum_{i=1}^{N_{branch}} \max\left(KI_i^t - 1, 0\right)\end{array}\right] \quad (14)$$

where F_i^t is the fitness value of solution i at interval t. k_p is the penalty coefficient. Based on the fitness values, the best position (O_{best}) and the corresponding volume (vo_{best}), density (de_{best}) and acceleration (ac_{best}) are determined.

3.2 Update the Population

In this section, the volume, density, acceleration, and position of objects are updated. The details of the updated process are as follows:

Objects tend to return to an equilibrium state after colliding with each other. In AOA, this state transition is described by the transfer factor TF defined as follows:

$$TF = \exp\left(\frac{t - t_{max}}{t_{max}}\right) \quad (15)$$

where t_{max} and t are the maximum and current number of iterations respectively. Similarly, the density reduction coefficient (dr) is also determined as follows:

$$dr = \exp\left(\frac{t_{max} - t}{t_{max}}\right) - \frac{t}{t_{max}} \quad (16)$$

The volume and density of objects are updated based on their current values and the best one by the following equation:

$$\begin{cases} de_{i,new} = de_i + r_4 \cdot (de_{best} - de_i) \\ vo_{i,new} = vo_i + r_4 \cdot (vo_{best} - vo_i) \end{cases} \quad (17)$$

where r_4 is a random number in [0, 1]. From Eq. (15), the coefficient TF increases from 0 to 1. When this value is less than 0.5, it is considered that an interaction occurs between objects. On the contrary, objects tend to reach a state of equilibrium. Based on this idea, the C is updated as follows:

$$ac_{i,new} = \begin{cases} \frac{de_{mr}+vo_{mr}\cdot ac_{mr}}{de_{i,new}\cdot vo_{i,new}}; \ if \quad TF < 0.5 \\ \frac{de_{best}+vo_{best}\cdot ac_{best}}{de_{i,new}\cdot vo_{i,new}}; \ otherwise \end{cases} \tag{18}$$

where de_{mr}, $vo_{mr,}$ and ac_{mr} are the density, volume, and acceleration of a random object in the population. Then, the acceleration of objects is normalized as follows:

$$ac_{i,norm} = 0.9 \cdot \frac{ac_{i,new} - \min(ac)}{\max(ac) - \min(ac)} + 0.1 \tag{19}$$

Finally, the position of objects is updated as follows:

$$O_{i,new} = \begin{cases} O_i + c_1 \cdot r_5 \cdot ac_{i,norm} \cdot dr \cdot (O_r - O_i); \ if \ TF < 0.5 \\ O_{best} + f \cdot c_2 \cdot r_6 \cdot ac_{i,norm} \cdot dr \cdot (c_3 \cdot TF \cdot O_{best} - O_i); \ otherwise \end{cases} \tag{20}$$

where c_1, c_2 and c_3 are constants that are selected to 2, 6, and 2, respectively. r_5 and r_6 are random numbers in [0, 1]. f is the motion direction factor that is defined as follows:

$$f = \begin{cases} 1; \ if 2 \cdot r_7 - 0.5 < 0.5 \\ -1; \ otherwise \end{cases} \tag{21}$$

where r_7 is a random number in [0, 1]. Then, the fitness value of the new solution is determined by Eq. (15). For each object, based on the fitness value, if the new position is better than the corresponding current one, the new position will replace the current one. Otherwise, the current position of the object is maintained for the next iteration as shown in Eq. (22). In addition, the best solution is also updated at the final step of the updated process.

$$O_i = \begin{cases} O_{i,new}; \ if F_{i,new} < F_i \\ O_i; \ otherwise \end{cases} \tag{22}$$

where $F_{i,new}$ is the fitness value of the new solution i.

3.3 Stopping Condition of Searching for New Solutions

The process of updating the positions of objects is performed continuously until the number of iterations reaches the maximum value for the previous. The entire implementation steps of AOA for searching optimal powers of SOP at each interval are described as follows:

- Step 1: Generate the current position of objects by Eq. (12) and volume, density, and acceleration by Eq. (13)
- Step 2: Evaluate the fitness of the current position of objects by Eq. (14).

- Step 3: Determine O_{best}, vo_{best}, and de_{best}.
- Step 4: Set the current iteration equal to 1.
- Step 5: Determine *TF* and *dr* by Eqs. (15) and (16).
- Step 6: Update the volume and density of objects by Eq. (17).
- Step 7: Update the volume and density of objects by Eqs. (18) and (19).
- Step 8: Update the position of objects by Eq. (20).
- Step 9: Evaluate the fitness of the new position of objects by Eq. (21).
- Step 10: Update the current position of objects by Eq. (22), and O_{best}, vo_{best} and de_{best}.
- Step 11: Increase the current iteration.
- Step 12: Return to Step 5 until the current iteration reaches to maximum value.
- Step 13: Output the best solution O_{best}.

4 Results and Discussion

4.1 Parameters of the System, IRC Load, PVS, and EVCS

The 33-node distribution system has thirty-two I.R.C. loads. There are 5 tie-switches on the system including {33-34-35-36-37} connecting between feeders. On the system, there is a PVS installed and connected to bus 8. The parameters of the PVS including N_{PV}, T_{amb}, $NOCT$, I_{MPP}, V_{MPP}, I_{SC}, V_{OC}, K_I and K_V have values respectively 2000, 31 °C, 45 °C, 2.44 A, 188.8 V, 2.59 A, 222.9 V, 0.0004 and 0.0028. The EVCS consists of 12 power poles with a power of 60 kW of each connected at node 18.

Based on this system connection, tie-switch 34 will be replaced by an SOP to control the power-sharing from PVs to EVCS. The single-line diagram of the entire system is shown in Fig. 1 and the maximum load parameters at buses and the line impedances are referenced from [28]. The ratio of I.R.C. loads on each bus is given in Table 1. The ratio of load types in each hour is given in Table 2. At each interval of the day, the STD of the load of the buses is assumed to be the same. Therefore, the STD value of the load at the intervals of the day is calculated as a percentage compared to the average load at the corresponding interval given in Table 2. The irradiance data at Go Vap District, Ho Chi Minh City in February 2024 is used for PVS [29].

Based on the load data in Tables 1 and 2 and the average irradiance and average EVCS load in Table 3, the average power at the buses and the total power of the IRC load demand, PVS, and EVCS at different intervals of the day are shown in Fig. 2. The power changes at different intervals of the day at buses are shown in Fig. 2a and the total power of the load, PVS, and EVCS are shown in Fig. 2b.

Based on the data of IRC load demand, EVCS, and PVS, the AOA algorithm is used to determine the active and reactive power of the SOP to aim at improving the voltage profile and reducing power loss on the system when EVCS is connected.

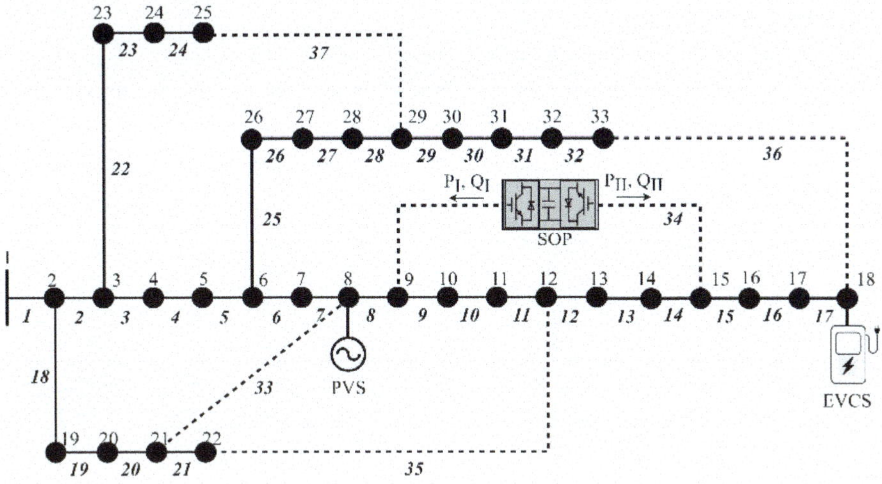

Fig. 1 The considered distribution system

Table 1 Ratio of IRC load types at buses

Bus	A_I	A_R	A_C	Bus	A_I	A_R	A_C	Bus	A_I	A_R	A_C
2	0.3	0.5	0.2	13	0.0	0.4	0.6	24	0.1	0.5	0.4
3	0.2	0.3	0.5	14	0.2	0.4	0.4	25	0.0	0.4	0.6
4	0.3	0.2	0.5	15	0.4	0.1	0.5	26	0.0	0.3	0.7
5	0.3	0.1	0.6	16	0.1	0.7	0.2	27	0.9	0.0	0.1
6	0.2	0.4	0.4	17	0.4	0.3	0.3	28	0.1	0.3	0.6
7	0.4	0.0	0.6	18	0.1	0.5	0.4	29	0.5	0.1	0.4
8	0.4	0.3	0.3	19	0.3	0.2	0.5	30	0.1	0.2	0.7
9	0.0	0.6	0.4	20	0.0	0.3	0.7	31	0.2	0.4	0.4
10	0.7	0.0	0.3	21	0.2	0.3	0.5	32	0.7	0.1	0.2
11	0.2	0.2	0.6	22	0.7	0.0	0.3	33	0.0	0.7	0.3
12	0.0	0.5	0.5	23	0.1	0.4	0.5				

To evaluate the effectiveness of the proposed solution, the following cases are considered:

- Case 1: The system without SOP operation.
- Case 2: The system with optimal SOP operation by AOA.
- Case 3: Investigating the uncertainty of I.R.C. load, PVS, and EVCS parameters.

In this case, the results of case 2 will be compared with those of case 1 to show the effectiveness of SOP optimization in reducing the impact of EVCS in terms of voltage deviation reduction and power loss reduction. Based on the results determined in case

Table 2 Ratio of IRC load types at hours

Hour	C_I	C_R	C_C	δ_{IRC} (%)	Hour	C_I	C_R	C_C	δ_{IRC} (%)
1	0.3	0.1	0.1	12.09	13	0.5	0.7	0.6	12.51
2	0.3	0.1	0.1	12.55	14	0.6	0.6	0.7	7.55
3	0.4	0.1	0.0	7.76	15	0.8	0.5	0.8	10.06
4	0.2	0.1	0.0	11.8	16	0.8	0.6	1.0	11.99
5	0.2	0.4	0.0	11.55	17	0.7	0.7	1.0	13.91
6	0.3	0.3	0.0	6.63	18	0.7	0.8	0.8	14.59
7	0.3	0.4	0.1	6.19	19	0.8	0.9	0.6	10.47
8	0.4	0.4	0.2	9.98	20	0.9	1.0	0.7	6.39
9	0.8	0.3	0.7	14.6	21	1.0	1.0	0.6	6.49
10	1.0	0.3	0.8	8.4	22	0.8	0.6	0.2	7.58
11	0.9	0.5	0.8	10.85	23	0.5	0.5	0.1	13.41
12	0.9	0.6	0.7	7.24	24	0.4	0.2	0.1	7.54

Table 3 Mean irradiance and EVCSs load in the day

Hour	γ_s $\left(\frac{kw}{m^2}\right)$	δ_s $\left(\frac{kw}{m^2}\right)$	P_{EV} (MW)	δ_{EV} (MW)	Hour	γ_s $\left(\frac{kw}{m^2}\right)$	δ_s $\left(\frac{kw}{m^2}\right)$	P_{EV} (MW)	δ_{EV} (MW)
1	0	0	0.380	0.041	13	0.547	0.282	0.250	0.030
2	0	0	0.235	0.023	14	0.492	0.235	0.320	0.040
3	0	0	0.190	0.010	15	0.418	0.237	0.275	0.026
4	0	0	0.385	0.032	16	0.344	0.204	0.285	0.017
5	0	0	0.475	0.031	17	0.264	0.163	0.345	0.025
6	0	0	0.240	0.031	18	0.088	0.041	0.325	0.046
7	0.041	0.022	0.445	0.036	19	0	0	0.505	0.033
8	0.291	0.182	0.330	0.034	20	0	0	0.350	0.046
9	0.430	0.217	0.315	0.021	21	0	0	0.410	0.043
10	0.533	0.253	0.290	0.032	22	0	0	0.225	0.034
11	0.589	0.273	0.165	0.013	23	0	0	0.520	0.030
12	0.584	0.284	0.190	0.022	24	0	0	0.595	0.056

2, and in case 3, the MCS method is used to evaluate the impact of uncertainty of I.R.C. load, PVS, and EVCSs on the system.

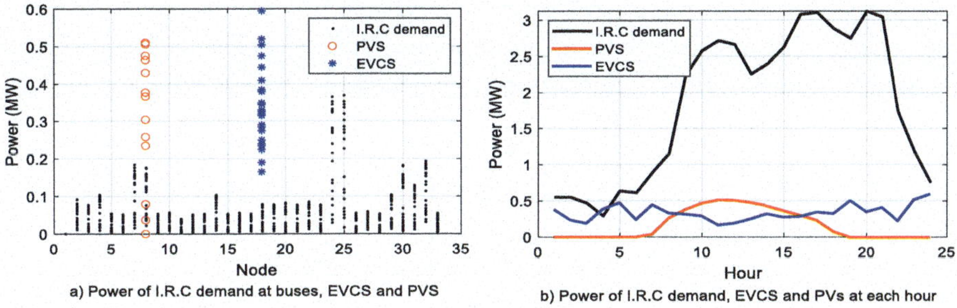

Fig. 2 The power of each bus and the power of each part at each hour

4.2 System Without SOP Operation

As the SOP is not operating, the total voltage deviation of the buses and the total power loss are given in Table 4. In which, the total voltage deviation of the buses $\sum_{j=1}^{N_{bus}}\left|1-V_j^t\right|$ called obj_1 at times during the day is 23.9913 p.u. and the total power loss ΔP^t called obj_2 is 2036.3288 kW. The voltage profile of many buses at intervals during the day has values outside the allowable limits. Figure 3 shows that the voltage at buses 8–18 and buses 28–33 is less than 0.95 pu at many intervals of the day. At bus 18, the boxplot values for 24 h including minimum, low adjacent, 25th percentile, median, 75th percentile, upper adjacent, maximum, and number finite outliers are respectively 0.89552, 0.89552, 0.92001, 0.93567, 0.95265, 0.97599, 0.97599 and 0 pu. This shows that for more than half of the day, the voltage at bus 18 is below the allowable value.

Table 4 The sum of voltage deviation and power loss of the initial system

Hour	obj_1	obj_2	Hour	obj_1	obj_2	Hour	obj_1	obj_2
1	0.5020	21.8973	9	1.0807	82.3157	17	1.5720	171.2678
2	0.3996	12.3380	10	1.1944	100.5852	18	1.5098	156.6012
3	0.3351	8.6138	11	1.1404	93.5168	19	1.6235	183.4373
4	0.3890	15.9730	12	1.1320	91.5215	20	1.6800	192.6992
5	0.6001	33.1208	13	0.9858	70.4939	21	1.6916	195.7967
6	0.4244	13.9928	14	1.1291	90.9070	22	0.9254	58.3557
7	0.6818	38.1597	15	1.2376	107.2220	23	0.8887	62.3566
8	0.6285	30.0665	16	1.4874	153.8339	24	0.7525	51.2562

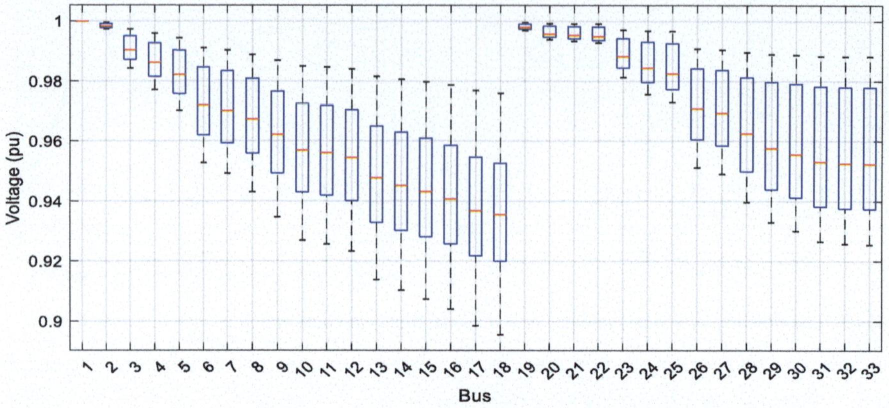

Fig. 3 The boxplot of voltage at each bus during the day of the initial system

4.3 System with Optimal SOP Operation by AOA

To operate the SOP to improve the voltage profile and reduce the loss by connecting EVCS to the system, the optimal power value of the SOP at each time of the day needs to be determined. Then, the AOA algorithm is applied to search the optimal power value of the SOP with the average power data of the IRC load, PVS, and EVCS load at each interval of the day. For the AOA algorithm, the number of control variables in each solution is set to 3 to represent the active power P_I, reactive power Q_{I} and Q_{II}. The maximum population size and number of iterations are chosen to be 30 and 200. The weight values w_1 and w_2 are chosen to be $\{0.2, 0.8\}$, $\{0.5, 0.5\}$ and $\{0.8, 0.2\}$, respectively. Where the first pair of values represents the priority for reducing power loss, the second pair of values represents the balance between the two objective functions, and the last pair of values gives priority to improving the voltage profile of the system.

The optimal SOP operation results with values of w_1 and w_2 selected as $\{0.2, 0.8\}$ are presented in Fig. 4 and Table 5. Figure 4a shows that at all times of the day, the voltage profile is always between 0.95 and 1 pu. Buses 31, 32, and 33 have the largest voltage changes during the day, but at all times of the day, their values are all greater than 0.95. Figure 4b shows that the power loss at many times of the day is smaller than the initial one. However, at 20:00 and 21:00, the power loss is larger than the initial one, but on the contrary, the voltage deviation is significantly improved. The total voltage deviation obtained when operating the SOP optimally for 24 h is 14.106 pu while the total power loss is 1955.238 kW. Both values are smaller than that of the original configuration shown in Case 1.

The observed results when operating the SOP optimally when the values of w_1 and w_2 are selected as $\{0.5, 0.5\}$ are shown in Table 6 and Fig. 5 wherein voltage profile is shown in Fig. 5a and power loss at each interval is shown in Fig. 5b. The total voltage deviation and power loss are 11.9821 pu and 2049.249 kW, respectively.

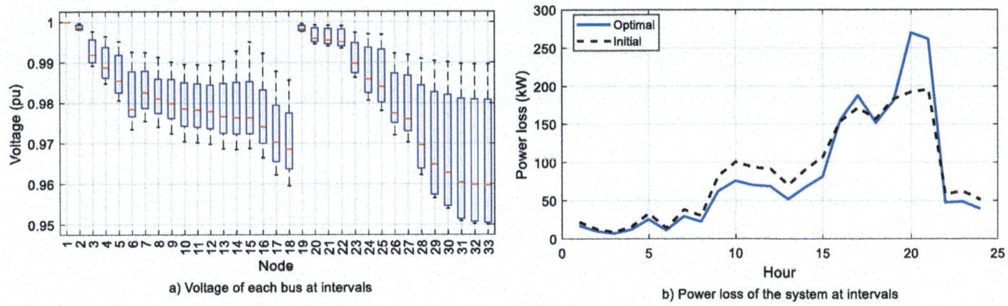

Fig. 4 Voltage profile and power loss of the system during the day with w_1 and w_2 set to $\{0.2, 0.8\}$

Table 5 The SOP power, voltage deviation, and power loss in the day with a value of $\{w_1, w_2\}$ set to $\{0.2, 0.8\}$

Hour	SOP power $[P_I, Q_I, Q_{II}]$ (MW)	obj_1	obj_2
1	0.3348, – 0.1864, – 0.0416	0.3615	17.1593
2	0.2249, – 0.1692, – 0.0339	0.2905	9.9273
3	0.1842, – 0.1361, – 0.0277	0.2471	7.0417
4	0.3167, – 0.1150, – 0.0319	0.2794	12.2583
5	0.4288, – 0.2054, – 0.0584	0.4256	25.4258
6	0.2422, – 0.1718, – 0.0409	0.3079	11.2189
7	0.4279, – 0.2739, – 0.0667	0.4816	29.4837
8	0.3601, – 0.3292, – 0.0695	0.4248	22.3645
9	0.4465, – 0.6618, – 0.1157	0.7272	62.2439
10	0.4600, – 0.7615, – 0.1298	0.7978	75.7405
11	0.3880, – 0.7836, – 0.1364	0.7507	69.9795
12	0.4050, – 0.7584, – 0.1363	0.7466	68.4208
13	0.4150, – 0.6395, – 0.1233	0.6443	51.9485
14	0.4780, – 0.6968, – 0.1310	0.7494	67.7732
15	0.4632, – 0.7699, – 0.1376	0.8331	81.5174
16	0.5283, – 1.9288, – 0.1878	0.6943	154.8743
17	0.5408, – 1.9255, – 0.4353	0.6704	188.0629
18	0.5458, – 1.7429, – 0.1649	0.7756	151.7404
19	0.6888, – 1.8431, – 0.1671	0.8213	180.9400
20	0.4826, – 1.9409, – 0.8855	0.6071	269.8984
21	0.5022, – 1.9359, – 0.8027	0.6449	261.8972
22	0.3478, – 0.4784, – 0.0963	0.6621	47.4268
23	0.5183, – 0.3597, – 0.0862	0.6303	48.6217
24	0.5251, – 0.2662, – 0.0664	0.5325	39.2731

The results show that the voltage is significantly improved at all times of the day. The large voltage fluctuations still occur at buses 31, 32, and 33, but the voltage amplitude of these nodes has been significantly improved compared to those of case 1 and case {0.2, 0.8}. However, the level of power loss reduction does not reach as in case 1.

The improvement of voltage and power loss as the values of w_1 and w_2 selected as {0.8, 0.2} are shown in Table 7 and Fig. 6 wherein voltage profile and power loss at each interval is presented in Fig. 6a, b, respectively. The total voltage deviation and power loss are 7.3897 pu and 2716.988 kW, respectively. The voltage is significantly improved at all times of the day and the voltage fluctuations of the nodes at different times of the day are also smaller than the above cases. However, the power loss of the system is larger than in the above cases.

Table 6 The gained results with w_1 and w_2 set to {0.5, 0.5}

Hour	SOP power $[P_I, Q_I, Q_{II}]$ (MW)	obj_1	obj_2
1	0.4056, – 0.3253, – 0.0936	0.2837	19.2756
2	0.2752, – 0.2680, – 0.0711	0.2351	10.9940
3	0.2261, – 0.2186, – 0.0589	0.2008	7.7835
4	0.3838, – 0.2464, – 0.0813	0.2058	14.1421
5	0.5180, – 0.3801, – 0.1237	0.3276	28.7944
6	0.2958, – 0.2771, – 0.0806	0.2489	12.4339
7	0.5180, – 0.4505, – 0.1327	0.3825	32.9371
8	0.4377, – 0.4809, – 0.1266	0.3397	24.8996
9	0.5671, – 0.8978, – 0.2044	0.5953	68.4905
10	0.5926, – 1.0208, – 0.2274	0.6530	83.3168
11	0.5174, – 1.0366, – 0.2322	0.6097	77.1653
12	0.5328, – 1.0081, – 0.2308	0.6074	75.4191
13	0.5292, – 0.8621, – 0.2075	0.5198	57.4849
14	0.6054, – 0.9456, – 0.2246	0.6103	74.7359
15	0.5990, – 1.0357, – 0.2376	0.6847	89.5092
16	0.6814, – 1.8396, – 0.2765	0.6560	157.4291
17	0.6826, – 1.8799, – 0.4801	0.6398	190.3304
18	0.7032, – 1.6324, – 0.2745	0.7335	154.4311
19	0.8506, – 1.7156, – 0.2932	0.7760	184.1546
20	0.5927, – 1.9101, – 0.9155	0.5842	271.5853
21	0.6401, – 1.8948, – 0.8423	0.6158	263.8536
22	0.4473, – 0.6744, – 0.1699	0.5523	51.7388
23	0.6299, – 0.5785, – 0.1675	0.5074	53.9857
24	0.6344, – 0.48 01, – 0.1457	0.4125	44.3586

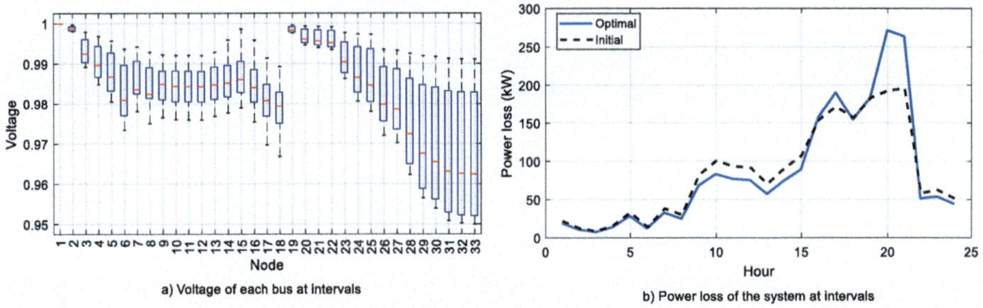

Fig. 5 Voltage profile and power loss of the system during the day with w_1 and w_2 set to $\{0.5, 0.5\}$

Table 7 The gained results with w_1 and w_2 set to $\{0.8, 0.2\}$

Hour	SOP power $[P_I, Q_I, Q_{II}]$ (MW)	obj_1	obj_2
1	0.4740, – 0.7874, – 0.1364	0.1175	33.3768
2	0.3369, – 0.6060, – 0.1124	0.1067	18.8968
3	0.2807, – 0.5030, – 0.0955	0.0912	13.4344
4	0.4211, – 0.6630, – 0.1030	0.0681	24.8792
5	0.6039, – 0.9572, – 0.1766	0.1241	50.8484
6	0.3637, – 0.6381, – 0.1262	0.1109	21.5249
7	0.6169, – 1.0394, – 0.1954	0.1673	56.5641
8	0.5114, – 0.9843, – 0.1728	0.1630	41.5976
9	0.6745, – 1.6675, – 0.2703	0.3326	107.6057
10	0.6905, – 1.8581, – 0.2844	0.3774	128.1907
11	0.5596, – 1.8400, – 0.2487	0.3681	114.7353
12	0.5900, – 1.8064, – 0.2580	0.3605	113.4874
13	0.5911, – 1.5808, – 0.2385	0.2932	88.7222
14	0.7097, – 1.7524, – 0.2873	0.3401	117.1051
15	0.7148, – 1.8679, – 0.3230	0.3957	137.6560
16	0.8098, – 1.8287, – 0.5846	0.5196	188.1357
17	0.9029, – 1.7846, – 0.6912	0.5521	211.5187
18	0.8894, – 1.7914, – 0.6267	0.5225	202.6465
19	1.0764, – 1.6856, – 0.7983	0.5536	242.6158
20	0.7154, – 1.8677, – 0.9608	0.5629	275.7574
21	0.8597, – 1.8058, – 0.9389	0.5717	273.2667
22	0.5919, – 1.3431, – 0.2692	0.2939	84.7619
23	0.7763, – 1.3129, – 0.2622	0.2310	92.5950
24	0.7395, – 1.1809, – 0.2094	0.1659	77.0659

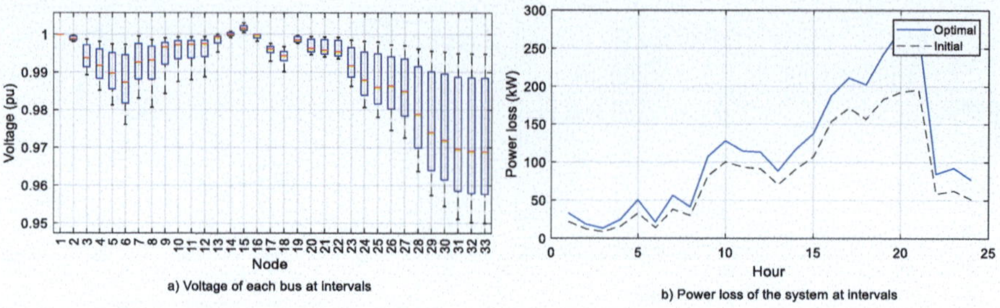

Fig. 6 Voltage profile and power loss of the system during the day with w_1 and w_2 set to $\{0.8, 0.2\}$

Thus, the operation of SOP can significantly improve the voltage profile of the system and reduce the influence of EVCSs on the voltage profile of the system. In addition, by setting the weight values of w_1 and w_2 with the priority level for w_2, we can obtain the power values of SOP to reduce the power loss as in the case of w_1 and w_2 set to $\{0.2, 0.8\}$. To evaluate the effectiveness of AOA for the considered problem, the PSO algorithm is also applied to find the optimal power of SOP in intervals of the day. The parameters of PSO including population size and number of iterations are chosen like AOA. The velocity scaling factors C_1, C_2 are chosen to be 2 [30, 31]. The compared results between AOA and PSO are shown in Fig. 7 wherein, the result of comparing the total voltage deviation is displayed in Fig. 7a and total power loss is presented in Fig. 7b. The results show that the total voltage deviation obtained by AOA is smaller than that of PSO for different values of w_1 and w_2 while the total power loss obtained by AOA is smaller than that of PSO for the cases with w_1 and w_2 set to $\{0.2, 0.8\}$ and $\{0.5, 0.5\}$. The average convergence characteristics of the two algorithms for the cases are presented in Fig. 8, wherein the convergence of two algorithms for cases $\{w_1, w_2\}$ of $\{0.2, 0.8\}$,$\{0.5, 0.5\}$ and $\{0.8, 0.2\}$ are shown in Fig. 8a–c, respectively. The figure show that AOA converges to a smaller value than PSO. This result shows that AOA is a potential method for this problem.

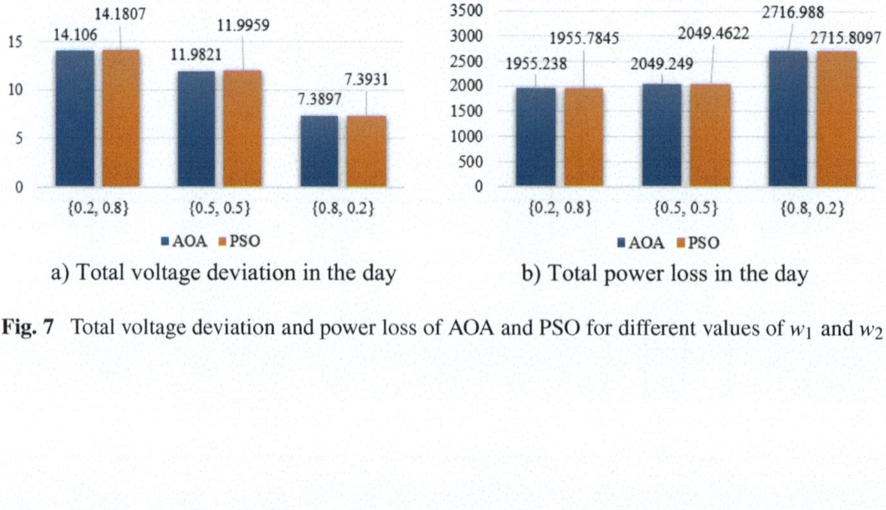

Fig. 7 Total voltage deviation and power loss of AOA and PSO for different values of w_1 and w_2

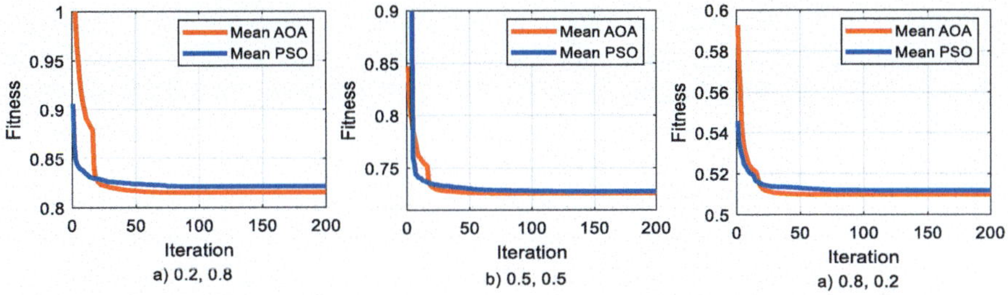

Fig. 8 Convergence curve of AOA and PSO for different values of w_1 and w_2

4.4 Investigating the Uncertainty of IRC Load, PVS, and EVCS Parameters

In this study, the MCS method is used to assess the uncertainty of input variables such as I.R.C. load, EVCS, and irradiance. The MCS method is based on random numbers of distribution functions to generate input variables for the models. The steps to implement the MCS method to assess the uncertainty of I.R.C. load, EVCS and irradiance on the voltage profile and power loss of the system are as follows:

- Step 1: Select the survey interval of the day from 1:00 h to 24:00.
- Step 2: Determine the average value and STD of I.R.C. load, EVCS, irradiance, and SOP power during the survey interval.
- Step 3: Randomly generate n input variables for I.R.C. and EVCS loads based on normal distribution, irradiance based on Beta distribution, and determine the corresponding PVS power.
- Step 4: Set the current number of iterations to $i = 1$.
- Step 5: Update the ith data in n data of I.R.C, EVCS loads, and PVS power as well as the optimal SOP power at survey interval into the system.
- Step 6: Perform power flow and determine the total voltage deviation and power loss of the ith input variables.
- Step 7: If i is less than n go back to step 5. Otherwise, go to the next step
- Step 8: Display the voltage and loss configuration results in the cases as a boxplot.

Although it is possible to investigate the system uncertainty at all intervals of the day, since the uncertainty assessment process is the same at each interval, in this study, the uncertainty is investigated at some intervals of the day when the voltage profiles and power loss of the system are most susceptible to influence, including 11:00 am and 19:00 with the coefficients w_1 and w_2 set to $\{0.5, 0.5\}$. At 11:00, the PVS generates the largest power and at 19:00 the EVCS and I.R.C. loads are high, but the PVS does not generate power.

The samples are generated from the mean and STD from Tables 2 and 3, where the I.R.C. and EVCS loads are generated from the normal distribution while the random irradiance is generated from the Beta distribution. The generated samples are depicted in Fig. 9 while the voltage profile and power loss results at 11:00 are presented in Fig. 10.

Figure 10 shows that the fluctuations of IRC, EVCS loads, and PVS power at 11:00 do not significantly affect the voltage profile of the system. The total average voltage deviation and average power loss are about 0.6124 pu and 77.9817 kW. At the EVCS bus, the boxplot values including minimum, low adjacent, 25th percentile, median, 75th percentile, upper adjacent, maximum, and finite outliers are all in the range of 0.97–0.99 pu. Meanwhile, without SOP, the total average voltage deviation and average power loss are about 1.1424 pu and 94.2889 kW. Figure 11 shows that without SOP, the voltage at many buses in the system fluctuates in the range of less than 0.95 pu even when PVS operates at high power during this time. In addition, the power loss is also larger than that in the case of SOP.

Similarly, at 19:00 with 1000 samples generated in Fig. 12, the results obtained in Fig. 13 are an average voltage deviation of 0.7763 pu and an average power loss of about 184.5063 kW. The figure shows that the voltage amplitude at buses 32 and 33

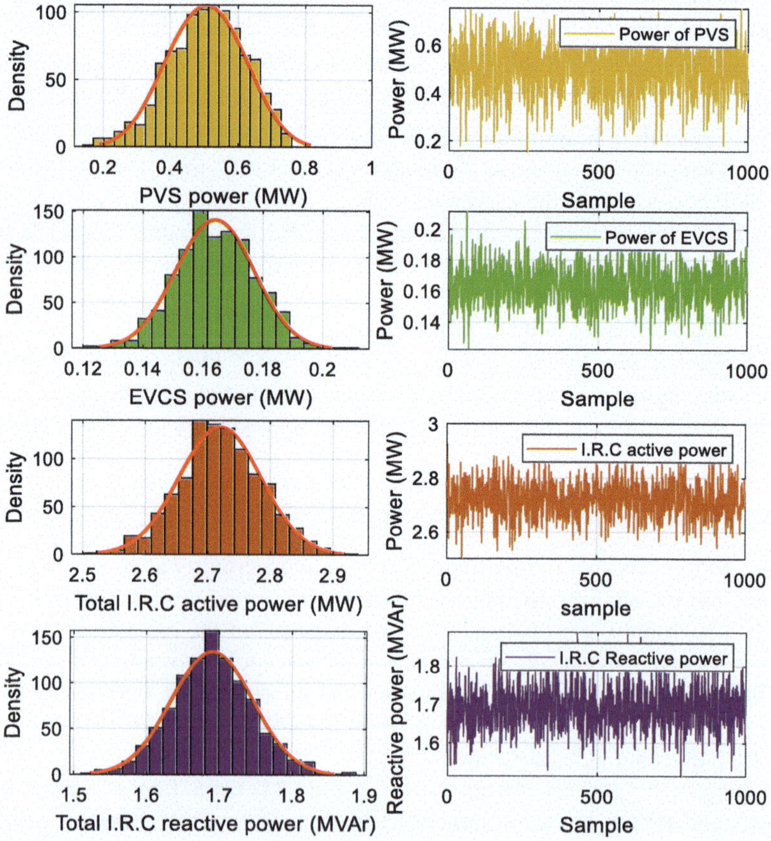

Fig. 9 Samples of PVS power, EVCS power, and IRC power at the interval 11

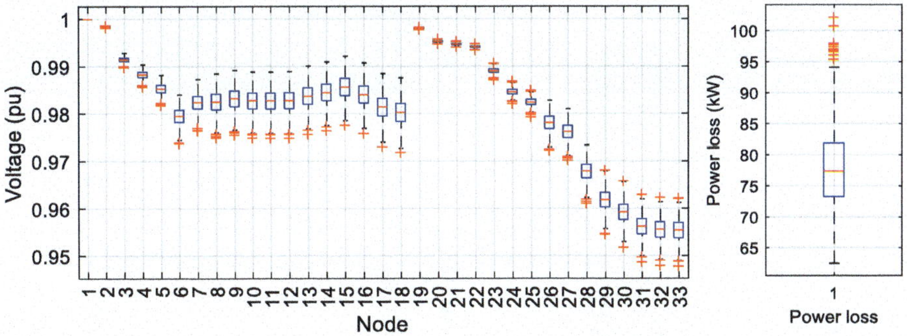

Fig. 10 Voltage profile and power loss over 1000 samples at the interval 11 with SOP

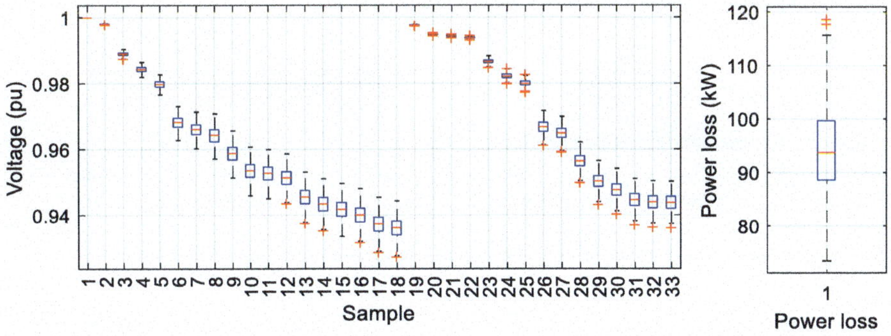

Fig. 11 Voltage profile and power loss over 1000 samples at the interval 11 without SOP

has about 50% of the samples having a voltage level lower than 0.95 pu at these two buses. On the EVCS bus, the voltage value is always in the range of 0.95–0.98 pu. Also, in this interval, if the SOP is not operated, the voltage profile obtained in Fig. 14 shows that many buses in the system such as buses 8–18 and 28–33 have a voltage fluctuation range below 0.95 pu. The total average voltage deviation and average power loss are about 1.6237 pu and 183.7845 kW. Thus, optimal SOP operation has contributed to improving the voltage profile to reduce the impact of EVCS load on these system indicators.

5 Conclusions

This study aims to limit the load of EVCS on the voltage profile and power loss of the distribution system by optimizing the operation of SOP power. On the radial distribution system with PVS, based on the installed SOP to connect between two feeders, the power of SOP is optimized during each hour by the AOA optimization algorithm. The results show that the optimal operation of SOP can improve the

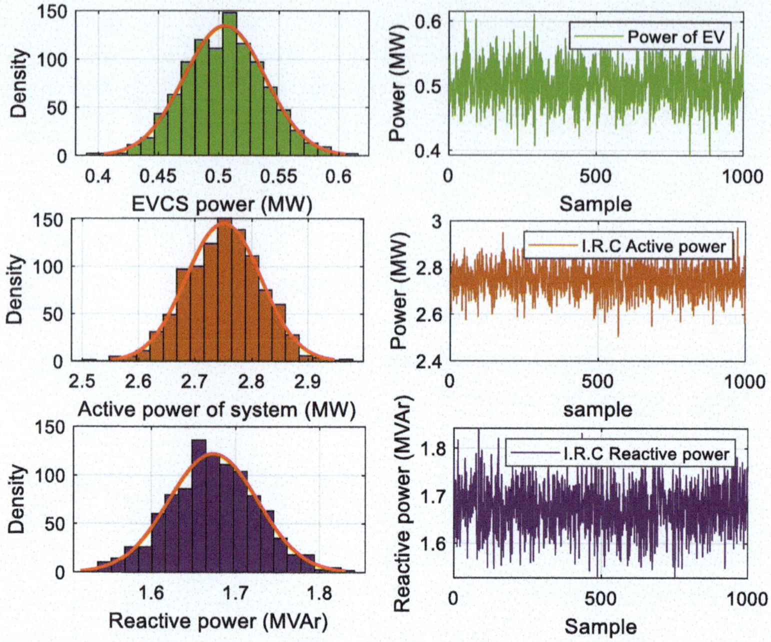

Fig. 12 Samples of PVS power, EVCS power and IRC power at the interval 19

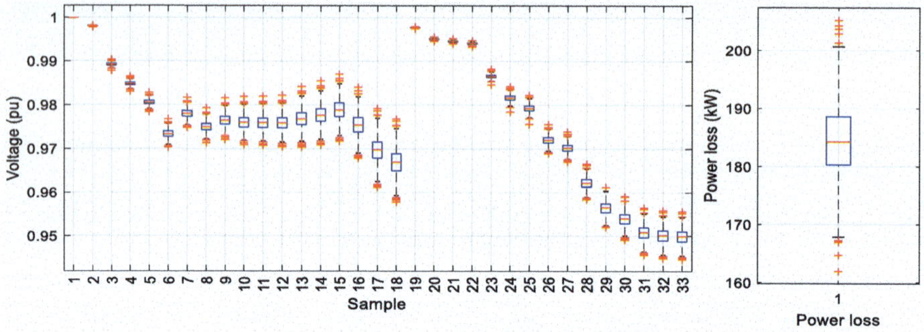

Fig. 13 Voltage profile and power loss over 1000 samples at the interval 19 with SOP

voltage profile and reduce the power loss of the distribution system depending on the priority level for the component objectives. Specifically, without operating SOP, the total voltage deviation obtained for 24 h is 23.9913 p.u while the total power loss is 2036.3288 kW. In the case where the objective function weight is chosen as {0.2, 0.8}, the total voltage deviation and total power loss obtained by the optimal operation of SOP for 24 h are {14.106 p.u, 1955.238 kW}, respectively, corresponding to a reduction of {41.20%, 3.98%} compared to those of the original case. When the objective function weight is selected as {0.5, 0.5}, the total voltage deviation

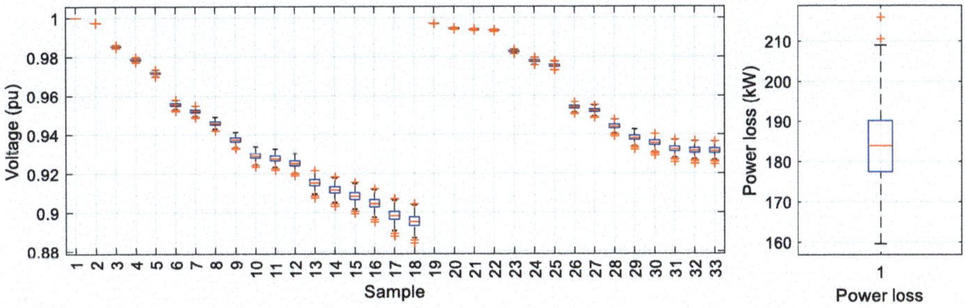

Fig. 14 Voltage profile and power loss over 1000 samples at the interval 19 without SOP

and total power loss obtained by operating optimally SOP for 24 h are {11.9821 pu, 2049.249 kW} respectively, corresponding to a reduction of {50.06%, – 0.63%} compared to those of the original case. As the objective function weight is selected as {0.8, 0.2}, the total voltage deviation and total power loss obtained for 24 h are {7.3897 p.u, 2716.988 kW} respectively, corresponding to a reduction of {69.20%, – 33.43%} compared to the original configuration. In addition, in this study, the uncertainty of input factors including IRC load, EVCSs, and irradiance is also evaluated to examine the influence of these parameters on the voltage profile and power loss of the system as well as the voltage profile at the EVCS bus. The survey results of the uncertainty of IRC load, EVCSs, and PVS also show that the optimal power of SOP helps that the voltage profile is within the allowable limit as there is fluctuation of IRC load, EVCSs, and PVS. Furthermore, the comparison with PSO shows that AOA is a potential method for the problem of optimizing the power of SOP in each interval of the day.

References

1. Chadha S, Jain V, Singh HR (2022) A review on Smart charging impacts of electric vehicles on grid. Mater Today Proc 63:751–755. https://doi.org/10.1016/j.matpr.2022.05.122
2. Deb S, Kalita K, Mahanta P (2018) Review of impact of electric vehicle charging station on the power grid. In: Proceedings of 2017 IEEE international conference on technological advancements in power and energy: exploring energy solutions for an intelligent power grid. TAP Energy 2017, vol 1, pp 1–6. https://doi.org/10.1109/TAPENERGY.2017.8397215
3. Geetha BT, Prakash A, Jeyasudha S, Dinakaran KP (2023) Hybrid approach based combined allocation of electric vehicle charging stations and capacitors in distribution systems. J Energy Storage 72:108273. https://doi.org/10.1016/j.est.2023.108273
4. Kathiravan K, Rajnarayanan PN (2023) Application of AOA algorithm for optimal placement of electric vehicle charging station to minimize line losses. Electr Power Syst Res 214:108868. https://doi.org/10.1016/j.epsr.2022.108868
5. Nurmuhammed M, Akdağ O, Karadağ T (2023) A novel modified Archimedes optimization algorithm for optimal placement of electric vehicle charging stations in distribution networks. Alex Eng J 84:81–92. https://doi.org/10.1016/j.aej.2023.10.055
6. Yuvaraj T, Suresh TD, Meyyappan U, Aljafari B, Thanikanti SB (2023) Optimizing the allocation of renewable DGs, DSTATCOM, and BESS to mitigate the impact of electric vehicle

charging stations on radial distribution systems. Heliyon 9:e23017. https://doi.org/10.1016/j.heliyon.2023.e23017
7. Yuvaraj T, Arun S, Suresh TD, Thirumalai M (2024) Minimizing the impact of electric vehicle charging station with distributed generation and distribution static synchronous compensator using PSR index and spotted hyena optimizer algorithm on the radial distribution system. e-Prime Adv Electr Eng Electron Energy 8:100587. https://doi.org/10.1016/j.prime.2024.100587
8. Saboori H, Heshmati Y (2023) Novel load partitioning and prioritization for distribution restoration with storage integrated-soft open point and energy-constrained resources. Electr Power Syst Res 214:108947. https://doi.org/10.1016/j.epsr.2022.108947
9. Cao W, Wu J, Jenkins N, Wang C, Green T (2016) Operating principle of soft open points for electrical distribution network operation. Appl Energy 164:245–257. https://doi.org/10.1016/j.apenergy.2015.12.005
10. Bloemink JM, Green TC (2013) Benefits of distribution-level power electronics for sup-porting distributed generation growth. IEEE Trans Power Deliv 28:911–919. https://doi.org/10.1109/TPWRD.2012.2232313
11. Cao W, Wu J, Jenkins N (2015) Feeder load balancing in MV distribution networks using soft normally-open points. In: IEEE PES innovative smart grid technologies conference, Europe, pp 1–6. https://doi.org/10.1109/ISGTEurope.2014.7028874
12. Hashim FA, Houssein EH, Hussain K, Mabrouk MS, Al-Atabany W (2022) Archimedes optimization algorithm: a new metaheuristic algorithm for solving optimization problems. Appl Intell 192:84–110. https://doi.org/10.1016/j.matcom.2021.08.013
13. Mahmoud MF, Mohamed AT, Swief RA, Said LA, Radwan AG (2022) Arithmetic optimization approach for parameters identification of different PV diode models with FOPI-MPPT. Ain Shams Eng J 13:101612. https://doi.org/10.1016/j.asej.2021.10.007
14. Krishnan H, Islam MS, Ahmad MA, Rashid MIM (2023) Parameter identification of solar cells using improved Archimedes optimization algorithm. Optik 295:171465. https://doi.org/10.1016/j.ijleo.2023.171465
15. Sajid I, Sarwar A, Tariq M, Bakhsh FI, Ahmad S, Shah Noor Mohamed A (2023) Archimedes optimization algorithm (AOA)-based global maximum power point tracking for a photovoltaic system under partial and complex shading conditions. Energy 283. https://doi.org/10.1016/j.energy.2023.129169
16. Fathy A, Alharbi AG, Alshammari S, Hasanien HM (2022) Archimedes optimization algorithm based maximum power point tracker for wind energy generation system. Ain Shams Eng J 13:101548. https://doi.org/10.1016/j.asej.2021.06.032
17. Akdag O (2022) A improved Archimedes optimization algorithm for multi/single-objective optimal power flow. Electr Power Syst Res 206:107796. https://doi.org/10.1016/j.epsr.2022.107796
18. Hadian E, Akbari H, Farzinfar M, Saeed S (2020) Optimal allocation of electric vehicle charging stations with adopted smart charging/discharging schedule. IEEE Access 8:196908–196919. https://doi.org/10.1109/ACCESS.2020.3033662
19. Aggarwal S, Kumar Singh A, Singh Rathore R, Bajaj M, Gupta D (2024) Revolutionizing load management: a novel technique to diminish the impact of electric vehicle charging stations on the electricity grid. Sustain Energy Technol Assess 65:103784. https://doi.org/10.1016/j.seta.2024.103784
20. Gholami K, Karimi S, Rastgou A, Nazari A, Moghaddam V (2024) Voltage stability improvement of distribution networks using reactive power capability of electric vehicle charging stations. Comput Electr Eng 116:109160. https://doi.org/10.1016/j.compeleceng.2024.109160
21. Polisetty SS, Jayanthi R, Veerraju MS (2023) An intelligent optimal charging stations placement on the grid system for the electric vehicle application. Energy 285:129500. https://doi.org/10.1016/j.energy.2023.129500
22. Milani AE, Haghifam MR (2013) An evolutionary approach for optimal time interval determination in distribution network reconfiguration under variable load. Math Comput Model 57:68–77. https://doi.org/10.1016/j.mcm.2011.05.047

23. Fakharian A, Sedighizadeh M, Khajehvand M (2021) Optimal operation of unbalanced microgrid utilizing copula-based stochastic simultaneous unit commitment and distribution feeder reconfiguration approach. Arab J Sci Eng 46:1287–1311. https://doi.org/10.1007/s13369-020-04965-x
24. Sedighizadeh M, Esmaili M, Jamshidi A, Ghaderi MH (2019) Stochastic multi-objective economic-environmental energy and reserve scheduling of microgrids considering battery energy storage system. Int J Electr Power Energy Syst 106:1–16. https://doi.org/10.1016/j.ijepes.2018.09.037
25. Dong XJ, Shen JN, Liu CW, Ma ZF, He YJ (2024) Simultaneous capacity configuration and scheduling optimization of an integrated electrical vehicle charging station with photo-voltaic and battery energy storage system. Energy 289:129991. https://doi.org/10.1016/j.energy.2023.129991
26. Nandini KK, Jayalakshmi NS, Jadoun VK (2024) A combined approach to evaluate power quality and grid dependency by solar photovoltaic based electric vehicle charging station using hybrid optimization. J Energy Storage 84:110967. https://doi.org/10.1016/j.est.2024.110967
27. Himabindu N, Hampannavar S, Deepa B, Longe OM, Mansani S, Komanapalli V (2023) Assessment of microgrid integrated biogas–photovoltaic powered electric vehicle charging station (EVCS) for sustainable future. Energy Rep 9:139–143. https://doi.org/10.1016/j.egyr.2023.09.098
28. Baran ME, Wu FF (1989). Network reconfiguration in distribution systems for loss reduction and load balancing. https://doi.org/10.1109/61.25627
29. The World Bank Group Global Solar Atlas. https://globalsolaratlas.info/detail?c=10.788117,106.316071,10&s=10.834563,106.67396&m=site&pv=medium,180,12,100. Last accessed 4 Aug 2024
30. Eberhart R, Kennedy J (1995) A new optimizer using particle swarm theory. In: MHS'95. Proceedings of the sixth international symposium on micro machine and human science, pp 39–43. https://doi.org/10.1109/MHS.1995.494215
31. Rezaee Jordehi A (2018) Enhanced leader particle swarm optimisation (ELPSO): an efficient algorithm for parameter estimation of photovoltaic (PV) cells and modules. Sol Energy 159:78–87. https://doi.org/10.1016/j.solener.2017.10.063

Performance Evaluation of Artificial Neural Networks for Electric Vehicle State of Charge Estimation Across Different Driving Cycles

Battula Amrutha Raju, Dushyant Sharma, and Surender Reddy Salkuti

Abstract With the growing adoption of electric vehicles (EVs), precise State of Charge (SOC) estimation is essential for optimizing battery management and ensuring reliable vehicle performance. This chapter provides a detailed analysis of SOC estimation in EVs, focusing on the application of Artificial Neural Networks (ANNs) and comparing their performance with the Extended Kalman Filter (EKF) and coulomb counting methods. The study is conducted using the Tesla Model S across two distinct driving conditions: the Highway (HW) driving cycle and the City (CY) driving cycle, providing a comprehensive evaluation across varied operational environments. The results show that the ANN method consistently outperforms the EKF, achieving up to 27.4% lower Root Mean Square Error (RMSE) in the highway driving cycle, indicating superior accuracy in SOC estimation. Even in the city driving cycle, the ANN reduces RMSE by up to 3.8%, demonstrating its robustness across various temperatures and driving conditions. These findings underscore the potential of ANN for SOC estimation, making it a viable and promising alternative for EVs.

Keywords Artificial Neural Networks (ANN) · Battery management systems · Electric Vehicles (EVs) · Extended Kalman Filter (EKF) · State of Charge (SOC)

B. A. Raju · D. Sharma
Department of Electrical Engineering, Indian Institute of Technology, Dhanbad, Jharkhand, India
e-mail: dushyant@iitism.ac.in

S. R. Salkuti (✉)
Department of Railroad and Electrical Engineering, Woosong University, Daejeon, Republic of Korea
e-mail: surender@wsu.ac.kr

S. R. Salkuti (ed.), *Artificial Intelligence for Integrated Smart Energy Systems in Electric Vehicles*, Lecture Notes in Electrical Engineering 1427,
https://doi.org/10.1007/978-3-031-94276-1_4

1 Introduction

The worldwide shift toward electric vehicles (EVs) marks a crucial step forward in combating climate change and reducing reliance on fossil fuels. As stricter emissions regulations are enforced and consumers increasingly favor sustainable transportation, EVs have become a vital solution for decreasing greenhouse gas emissions and improving air quality [1]. This shift towards electric mobility is motivated not only by environmental concerns but also by the promise of reduced operational costs, technological innovation, and improved driving experiences. At the epicenter of EV performance and reliability is the battery system. The battery is crucial, powering the vehicle and determining its range, efficiency, and overall performance. Lithium-ion (Li-ion) batteries, including variants like Lithium Iron Phosphate ($LiFePO_4$) and Lithium Nickel Cobalt Manganese Oxide (NCM), dominate the battery landscape thanks to their high energy density, extended cycle life, and comparatively lower cost [2–4]. Emerging technologies, such as solid-state batteries, promise further advancements in safety, energy density, and charging speeds. Despite these technological advancements, managing battery performance accurately remains a complex challenge.

A crucial element of battery management is the State of Charge (SoC) [5], which indicates the remaining energy in the battery. Accurately determining SoC is vital for optimizing performance and ensuring the vehicle operates within safe parameters, including battery health, temperature, power output, voltage, and charging rates [6]. Accurate SoC estimation is vital for energy management, range prediction, and battery protection. It helps prevent unexpected vehicle stoppages, aids in planning charging schedules, and extends battery lifespan by avoiding overcharging or deep discharging [7].

Several methods have been developed for SoC estimation, each offering distinct advantages and facing unique limitations. Traditional techniques include Coulomb Counting (CC), which tracks charge and discharge cycles to estimate SoC, and the Kalman Filter (KF), an algorithm that integrates noisy measurements with a predictive model to provide accurate estimates [8]. The Kalman Filter, introduced by Rudolf Kalman, has become a widely utilized method across various fields, including aerospace, robotics, and battery management systems (BMS) for EVs [9]. Its ability to provide accurate SoC estimates under uncertain conditions and its capability to fuse noisy sensor data with predictive models are among its notable strengths. However, the Kalman Filter and its variants, including the Extended Kalman Filter (EKF) and Unscented Kalman Filter (UKF), encounter difficulties in handling scenarios with highly non-linear dynamics or non-Gaussian noise characteristics [9].

The advent of machine learning and artificial intelligence has introduced new methodologies for SoC estimation, offering the potential for more accurate, adaptive, and robust solutions [10]. Machine learning techniques, especially deep learning approaches like Long Short-Term Memory (LSTM) networks and Hybrid Artificial Neural Networks, have proven effective in modeling the intricate, non-linear interactions found in battery systems [11]. Unlike traditional model-based approaches, machine learning methods learn directly from data, allowing them to adapt to varying operational conditions without the need for detailed system modeling [12–14].

Machine learning approaches also benefit from their ability to handle large datasets, making them suitable for real-time applications in EVs where continuous data generation is common [15, 16]. Furthermore, these models can improve over time as more data becomes available, enhancing their predictive capabilities. This adaptability and potential for continuous learning make machine learning a compelling alternative to traditional estimation methods like the Kalman Filter [17–19].

The objective of this chapter is to present a comprehensive comparative study of SoC estimation techniques, focusing on the performance of traditional methods like the Kalman Filter and modern machine learning approaches. By evaluating these techniques under various operating conditions, the chapter aims to highlight the advantages of machine learning methods in terms of accuracy, adaptability, and real-time applicability. The contributions of this chapter are twofold:

- It provides a detailed review of various SoC estimation methods used in EVs.
- Explores different driving cycles in Tesla Model S at various environmental conditions for SoC estimation, with a focus on comparison with ANN.

This chapter is structured as follows: Sect. 2 introduces the battery management system, outlining its main components and functions. Section 3 delves into a comprehensive analysis of the different SOC estimation techniques. Section 4 features a case study on SOC estimation for the Tesla Model S, assessing its performance across various driving cycles. The chapter concludes with a summary of the key findings.

2 System Modelling

The battery is often considered the heart of an EV, playing a crucial role in the overall functioning of the system. The battery's condition significantly impacts the vehicle's performance, efficiency, and longevity [20]. To effectively monitor the battery's status, the key parameter typically assessed is the SOC. This metric is particularly vital for Lithium-ion batteries, which are the preferred choice in modern electric vehicles. A Li-ion battery can be modeled using specific parameters that offer insights into its behavior under various conditions [21]. A common model aims to establish a direct correlation between battery consumption and open circuit voltage (OCV) [22]. However, this method often falls short due to the complexity of the battery's internal dynamics and external influences [22]. To accurately model a battery and estimate its SOC, it is essential to understand several key definitions and parameters that play a pivotal role in this process as given below. The battery management operation to assess the SOC estimation is provided in Fig. 1.

State of Charge (SOC): SOC indicates the battery's current energy level relative to its full capacity and is usually represented as a percentage. Estimating SOC is essential

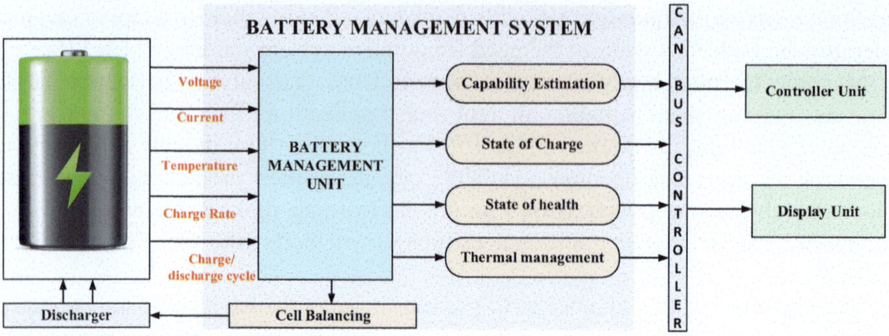

Fig. 1 Battery system management

for understanding the remaining charge in the battery and predicting the vehicle's available range. The United States Advanced Battery Consortium (USABC) [23] has established a standard convention for SOC, defining it in terms of the battery's current charge (Q), measured in ampere-hours which Q_{Remain} and Q_{Rated} as the remaining and rated capacity of the charge in the battery.

$$SOC = \frac{Q_{Remain}}{Q_{Rated}} \times 100\% \tag{1}$$

State of Health (SOH): SOH measures the overall health and efficiency of the battery relative to its ideal or original condition [23]. It provides an indication of the battery's aging process and its ability to hold charge, which directly impacts performance over time.

$$SOH\% = 1 - SOC\% \tag{2}$$

Charge Rate (C): The charge rate to the speed at which the battery is charged. It is usually expressed in terms of the battery's capacity (e.g., C-rate), where a 1C rate means charging the battery in one hour. The charge rate affects both the SOC and SOH, influencing the battery's lifespan and efficiency [24]. Here in Eq. (3), I_{charge} is the charging current (in Amperes, A). $C_{nominal}$ is the nominal capacity of the battery (in Ampere-hours, Ah).

$$C_{rate} = \frac{I_{charge}}{C_{nominal}} \tag{3}$$

Charging Cycle: A charging cycle is defined as one complete charge and discharge process. The number of charging cycles a battery can undergo before its capacity significantly degrades is a key factor in determining its SOH [24]. The relationship between charging cycles and SOC is important for understanding the battery's long-term performance.

Several variables are essential for accurate SOC estimation, including temperature, current, voltage, charge rate, and charging cycles [25]. Temperature, for example, can significantly affect the battery's performance and the accuracy of SOC estimation. Higher temperatures may accelerate battery degradation, while lower temperatures can reduce the battery's capacity and efficiency. The current drawn from or supplied to the battery is another critical factor, as it affects the internal resistance and the accuracy of SOC estimation algorithms. The rate of current flow influences the battery's internal resistance and directly impacts the precision of SOC estimation [26]. Voltage, particularly the voltage across the battery terminals, is fundamental in determining SOC, though relying solely on open circuit voltage can be misleading due to variations in load conditions and temperature. As mentioned earlier, the rate at which the battery is charged and the number of completed charging cycles provide important data for SOC estimation. These factors influence the battery's chemical processes, capacity retention, and overall health [27].

A comprehensive understanding of key concepts and parameters is essential for accurately modeling a Lithium-ion battery for SOC estimation in electric vehicles [28]. This section lays the groundwork for the subsequent discussion, which will explore SOC estimation methods in greater detail, building on the principles introduced here to develop a reliable and precise model for EV batteries [29].

3 SOC Estimation Methods

This section delves into the methods used for SOC estimation, building upon the principles discussed here to develop a reliable and precise model for electric vehicle batteries. SOC estimation methods can be broadly categorized into two key approaches: direct and indirect methods. A direct method relies on real-time measurements of actual battery parameters such as current, voltage, and impedance, making it typically easier to implement for immediate SOC estimation. On the other hand, indirect methods use models or data analysis to estimate SOC, offering more comprehensive insights but often requiring more complex calculations and assumptions about battery behavior. SOC estimation methods are classified in Fig. 2.

3.1 Direct Methods

3.1.1 Open Circuit Voltage (OCV) Method

This method estimates the SOC by measuring the battery's OCV after allowing the battery to rest. The relationship between OCV and SOC is generally stored in a look-up table [16]. While this approach offers reliable estimates when the s is idle,

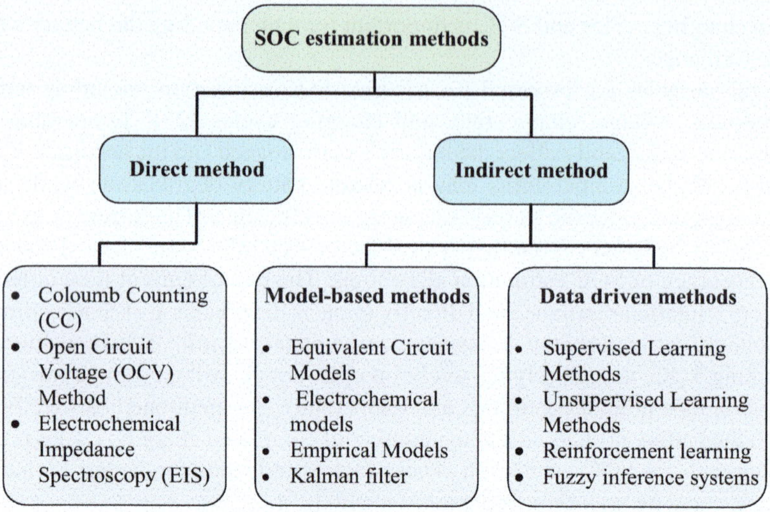

Fig. 2 SOC estimation method classification

its requirement for the battery to be disconnected from any load restricts its use in real-time applications, as it cannot continuously monitor SOC during operation.

3.1.2 Electrochemical Impedance Spectroscopy (EIS)

EIS is a method that involves applying a low AC voltage (5–50 mV) to the battery and measuring the resulting current to analyze its impedance characteristics [14]. SOC is then inferred from the impedance data. The EIS method offers valuable insights into the internal state of the battery and can be used to characterize battery health. Despite its advantages, EIS is complex and requires specialized equipment, making it less suitable for real-time applications and more appropriate for diagnostic purposes in controlled environments.

3.1.3 Coulomb Counting (CC)

The CC technique estimates the SOC by tracking the charge flowing into and out of the battery over time. It calculates the available charge by integrating the measured current. This method is particularly accurate for short-term assessments and is relatively simple to implement [18]. However, one significant drawback is the potential for cumulative error over time, especially if the initial SOC is not accurately known, which can lead to inaccuracies in long-term SOC estimation.

$$SOC_k = SOC_{k-1} + \left(\frac{1}{Q_{nominal}} \int_{k-1}^{k} I_{battery}(d\tau) \right) \times 100\% \tag{4}$$

3.2 *Indirect Methods*

3.2.1 Model-Based Methods

Model-based methods for SOC estimation in batteries employ mathematical models to simulate battery behavior under various conditions. These methods are categorized based on the complexity and type of model used, each with distinct advantages and limitations.

3.2.2 Equivalent Circuit Models (ECM)

ECM represents the battery using a combination of electrical components such as resistors, capacitors, and voltage sources. These components are adjusted to replicate the battery's behavior accurately. The primary advantage of ECMs lies in their simplicity and computational efficiency, allowing for quick SOC estimates [9]. However, the drawback is that they may not capture the intricate electrochemical processes occurring within the battery, particularly under varying operational conditions. The commonly used equivalent circuit models in electric vehicle applications include the resistive internal model (R_{int} model), the Thevenin model, the Partnership for a New Generation of Vehicles (PNGV model), and the Generalized Nonlinear Model (GNL model) [31]. Thevenin's (RC-based model) model of ECM is shown in Fig. 3.

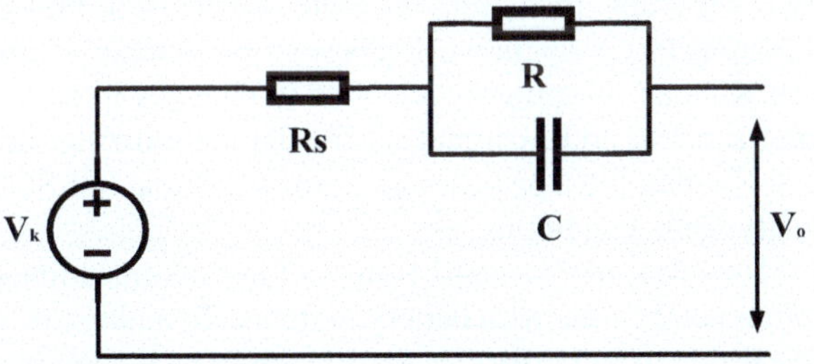

Fig. 3 Thevenin model used in ECM method

3.2.3 Electrochemical Models

Electrochemical models simulate the internal chemical and physical processes of a battery by using differential equations that consider factors such as ion transport, charge transfer, and concentration gradients. These models are primarily categorized into two types. The first is the Pseudo-Two-Dimensional (P2D) model, which offers a detailed and accurate representation by accounting for the distribution of lithium ions in both the solid and liquid phases of the electrodes [12]. The second is the Single-Particle (SP) model, which simplifies the P2D approach by focusing on a single particle of active material, making it less complex and more suitable for real-time applications. While these electrochemical models provide high accuracy in predicting battery behavior and SOC under various conditions, they are also computationally intensive and may require considerable time for parameter identification and calibration [31].

3.2.4 Empirical Models

Empirical models are based on experimental data and use statistical methods to establish relationships between measurable parameters, such as voltage and current, and SOC. Unlike electrochemical models, empirical models do not rely on understanding the underlying physical processes. Their main advantage is the ease of development and implementation, as they are derived from observed data. However, their ability to generalize beyond the conditions under which they were developed is limited, making them less reliable in diverse operational scenarios [17].

3.2.5 Kalman Filter-Based Models

The Kalman filter is a model-based method used for estimating the SOC in batteries by combining system predictions with real-time measurements. It operates through two key steps: predicting the state and its uncertainty based on previous data, and updating this prediction using new measurements. This continuous refinement makes the Kalman filter highly adaptive to changes in battery behavior, providing robust and accurate SOC estimates [26]. By minimizing the estimation error covariance, the Kalman filter provides an optimal estimate of SOC, making it an effective tool in dynamic conditions.

However, the standard Kalman filter assumes linear system dynamics, which is often not the case with battery systems. To address this, EKF is used, which extends the Kalman filter to handle non-linearities by linearizing the system around the current estimate [8]. This makes the EKF particularly suitable for the complex, non-linear nature of battery SOC estimation, ensuring more accurate and reliable results.

3.3 Data-Driven Methods

Data-driven methods for SOC estimation in batteries utilize machine learning and statistical techniques to analyze data and make predictions. These methods are categorized based on their underlying algorithms and approaches, each offering unique advantages and challenges.

3.3.1 Supervised Learning Methods

Supervised learning methods rely on labeled datasets where input features such as current, voltage, and temperature are associated with known SOC values. These models are trained to map these inputs to the corresponding SOC outputs using the training data. Several algorithms are commonly used in supervised learning for SOC estimation. Regression models, including linear regression, polynomial regression, and support vector regression, predict SOC based on continuous input features. Decision trees and random forests are effective at capturing non-linear relationships and interactions between different features, making them suitable for more complex SOC estimations [11]. ANNs, particularly deep learning models like feedforward neural networks, are also widely used in supervised learning. ANNs are capable of learning intricate patterns from large datasets, which enables them to deliver highly accurate SOC estimations, especially in scenarios where complex relationships exist among the input features.

3.3.2 Unsupervised Learning Methods

Unsupervised learning methods operate without the need for labeled data, focusing instead on uncovering patterns or clusters within the data without predefined SOC values. These methods are valuable for identifying underlying structures in the data that can inform SOC estimation. Clustering algorithms like K-means and hierarchical clustering are commonly used to group similar operating conditions [23]. By analyzing these clusters, insights into SOC can be inferred, even without direct labels. Principal Component Analysis (PCA) is another key technique, which reduces the dimensionality of the data [10]. PCA helps to identify the most significant features that influence SOC estimation, making it easier to analyze and interpret complex datasets.

3.3.3 Reinforcement Learning

Reinforcement learning involves training an agent to make decisions based on feedback from the environment. The agent learns to optimize SOC estimation by maximizing a reward signal. This approach is advantageous as it can adapt to changing conditions and learn optimal strategies over time [29]. However, it requires a well-defined reward structure and can be complex to implement effectively.

3.3.4 Fuzzy Logic Systems

Fuzzy inference systems (FIS) utilize fuzzy logic to manage uncertainty and imprecision in input data. These systems apply rules based on expert knowledge to estimate SOC. Fuzzy logic systems are particularly effective in dealing with noisy data and can incorporate human expertise into the model. However, they require careful design of fuzzy rules and membership functions to ensure accurate SOC estimation [32].

4 SOC Estimation of Tesla S Model: A Case Study Approach

This section provides a detailed analysis of State of Charge (SOC) estimation using two approaches: the EKF and ANN, both developed based on the CC method. The study evaluates these methods under two distinct driving cycles' city and highway using the Tesla Model S. The comparison aims to highlight the performance and accuracy of each method in various driving conditions.

4.1 Methodological Approaches: EKF and ANN

The analysis provides a comparative perspective on the effectiveness of EKF an indirect model-based method and ANN an indirect data-driven method in accurately estimating SOC under real-world scenarios.

4.1.1 Extended Kalman Filter Method (EKF)

The EKF is employed for the SOC of batteries in EVs by predicting voltage and updating SOC estimates with real-time measurements. Key aspects to understand include state variables, control variables, uncertainties, linearization, and SOC calculation. The EKF adapts to non-linear systems by linearizing non-linear models around the current estimate using a first-order Taylor expansion [33]. This adaptation makes the EKF effective for SOC estimation, given the non-linear nature of battery dynamics. The initial step in implementing the EKF involves defining the state-space model, which encompasses state equations and measurement equations.

State equation: These represent the internal states of the battery system that need to be estimated. In the context of SOC estimation, the primary state variable is the SOC itself. As per the model considered i.e., Thevenin's model additional state variables included are voltage drops across resistive and capacitive (V_{RC}) elements in an equivalent circuit model [27]. The state equation can be represented as the following as shown in Eqs. (5) and (6).

$$x_k = f(x_{k-1}, u_{k-1}) + w_{k-1} \tag{5}$$

$$x_k = \begin{bmatrix} SOC_k \\ V_{RC,k} \end{bmatrix} \tag{6}$$

where x_k is the state vector at time step *k*, representing the SOC; u_{k-1} is the input current (I); w_{k-1} is the process noise, assumed to be normally distributed with covariance (Q_c).

The battery model used in SOC estimation is non-linear and is typically represented by an equivalent circuit model as in Fig. 3. The most common model used is Thevenin's, which includes an OCV, a series resistance R_s, and a parallel RC network representing the dynamic behavior of the battery. The SOC update equation can be derived from coulomb counting [3].

$$SOC_k = SOC_{k-1} - \frac{\eta . I_{k-1} . \Delta t}{C} + w_{k-1} \tag{7}$$

$$V_{RC.k} = e^{-\frac{\Delta t}{RC}} V_{RC,k-1} + (1 - e^{-\frac{\Delta t}{RC}}) R I_{k-1} + w_{k-1} \tag{8}$$

Measurement Equation: It relates the SOC to measurable quantities like battery voltage and current (I). It is represented as z_k. The OCV-SOC relationship is often non-linear and is determined experimentally for the specific battery type, in this case, the lithium-ion battery which is presented in Eq. (9),

$$z_k = h(V_k) \tag{9}$$

where the function $h(V_k)$ is defined as,

$$V_k = OCV(SOC_k) - I_k R_s - V_{RC,k} \tag{10}$$

where V_k is the measured terminal voltage, $OCV(SOC_k)$ is the open-circuit voltage as a non-linear function of SOC and R_s is the series resistance [4]. Since the system is non-linear, EKF linearizes the state and measurement equations using a first-order Taylor series expansion around the current estimate. This linearization is achieved through Jacobian matrices.

The EKF algorithm proceeds in two main steps: prediction and update. During the prediction step, the EKF utilizes the RC battery model to estimate the SOC and RC Voltage based on the current state and input conditions. This step begins with the calculation of the predicted SOC by applying the state transition equations of the *RC* model, which account for the dynamics of the battery, including resistive and capacitive effects. The prediction integrates the influence of the input current and the inherent characteristics of the RC circuit, providing a forecast of how the SOC evolves over time. Additionally, the prediction includes an estimation of the associated uncertainty. This involves propagating the uncertainty from the previous state through the RC model, considering factors such as process noise and potential inaccuracies in the model [34]. By doing so, the EKF prepares an updated estimate of the SOC and its uncertainty for the subsequent correction phase.

The update step focuses on refining the predicted SOC using real-time measurements. This process begins by calculating the predicted terminal voltage based on the forecasted SOC and the RC model. The EKF then compares this predicted voltage with the actual measured voltage, determining the measurement residual, which represents the deviation between the predicted and observed values. Using this measurement residual, the EKF adjusts the predicted SOC to improve its accuracy. This adjustment involves correcting the SOC estimate to minimize the prediction error. Concurrently, the EKF updates the uncertainty (covariance) associated with the SOC estimate, reflecting the increased confidence in the adjusted estimate. Finally, the covariance matrix is revised to incorporate the new measurement, representing the reduced uncertainty and providing a refined estimate of the SOC [34]. This comprehensive update ensures that the SOC estimation remains accurate and responsive to the real-time behavior of the battery. The EKF model structure for SOC estimation is shown in Fig. 4.

For SOC estimation using the EKF, key parameters include SOC and V_{RC} as the primary state variables. The process noise covariance Q_c is typically set between 0.01 and 0.1. Initial state estimates are often around 80% SOC and the initial error covariance P might be diag(0.5, 0.5, 0.5). The sampling time Δt is usually 1 min. These parameters help the EKF accurately estimate SOC by adjusting for both model and measurement uncertainties.

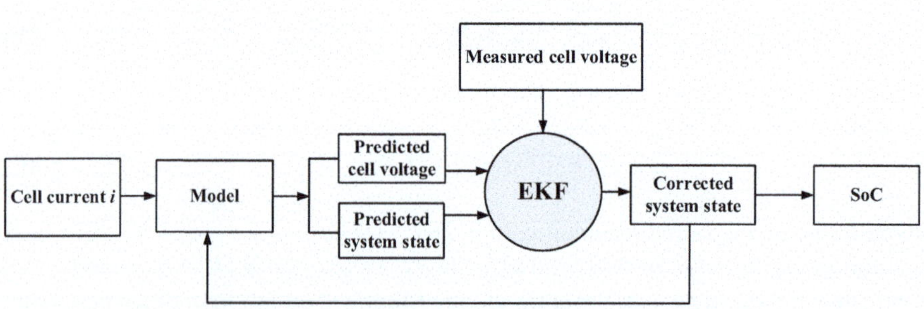

Fig. 4 EKF model structure for SOC estimation

4.1.2 Artificial Neural Network Model (ANN)

Artificial Neural Networks (ANNs) are advanced computational models that excel in estimating the State of Charge (SOC) of batteries in electric vehicles due to their ability to capture complex, non-linear relationships from data. This detailed explanation covers how an ANN model is used for SOC estimation, including the necessary equations and a step-by-step approach [11].

Data Preparation: The first step in using an ANN for SOC estimation involves preparing a comprehensive dataset that includes various features and corresponding SOC values. Essential input features typically include current (I), voltage (V), temperature (T), and charge/discharge status to indicate whether the battery is being charged or discharged. For illustrative purposes, let us assume the dataset consists of N samples, each with input features $v_i = [I_i, V_i, T_i]$ and corresponding SOC values y_i. This dataset is divided into training, validation, and test sets to build and evaluate the ANN model.

Network Architecture: The ANN used for SOC estimation comprises several layers:

Input Layer: Receives the input features (I, V, T). If there are n features, this layer has n neurons.

Hidden Layers: Consists of multiple neurons that apply activation functions to learn complex patterns [35]. The sigmoid function is used in this study as in Eq. (11).

$$f(v) = \frac{1}{1 + e^{-v}} \tag{11}$$

Output Layer: Produces the estimated SOC, usually with a single neuron for continuous output. The general form of the ANN model can be expressed as,

$$y = f(W_L \cdot f(W_{L-1} \ldots f(W_1 \cdot v + b_1) b_{L-1} + b_L)) \tag{12}$$

where x is the input vector W_i and b_i are the weights and biases of the ith layer and y represents the output SOC estimate.

Training the ANN: Training the ANN involves two key steps:

Forward Propagation: Compute the predicted SOC by propagating the input features through the network:

$$y_i = f(W_L \cdot f(W_{L-1} \ldots f(W_1 \cdot v_i + b_1) b_{L-1} + b_L)) \tag{13}$$

Backpropagation: Adjust the weights and biases to minimize the loss function using Gradient Descent optimization algorithms. This involves computing gradients of the loss function with respect to the weights and biases and updating them accordingly as shown in Eq. (14).

$$W_i \leftarrow W_i - \lambda \frac{\partial Er}{\partial W_i} \tag{14}$$

where λ is the learning rate and Er is the error calculated.

SOC Estimation: After training, the ANN is used to estimate SOC for new input data. Given a new set of inputs v_{new}, the trained network computes the SOC estimate as in Eq. (15).

$$y_{new} = f(W_L \cdot f(W_{L-1} \ldots f(W_1 \cdot v_{new} + b_1)b_{L-1} + b_L)) \tag{15}$$

The performance of the ANN model is evaluated using, RMSE [35]. The actual values are considered from the test data to analyze the RMSE as shown in Eq. (16).

$$RMSE = \sqrt{\frac{1}{N}\sum_{i=1}^{N}(y_i - \hat{y}_i)^2} \tag{16}$$

The ANN model for SOC estimation effectively learns from historical data to predict SOC based on current, voltage, and temperature inputs. Through a structured process of data preparation, network training, and evaluation, the ANN captures complex relationships and provides accurate SOC estimates, crucial for managing battery performance in electric vehicles (EVs) [36]. The ANN model of SOC estimation is shown in Fig. 5.

For SoC estimation using an Artificial Neural Network (ANN), the data requirements and parameters are as follows: The dataset consists of 7864 samples, with 70% (5506 samples) allocated for training. This training data should include diverse conditions such as current, voltage, temperature, and charge/discharge cycles. The remaining 30% (2359 samples) is used for testing the model's performance. The ANN model features an architecture with three hidden layers, each containing 128 neurons, with Rectified Linear Unit (ReLU) activation functions. Initial weights are set using the He initialization method (Kaiming initialization), and biases are initialized to zero. The model employs an Adam optimizer with a learning rate of 0.001 and uses RMSE as the loss function. Data preprocessing includes normalization,

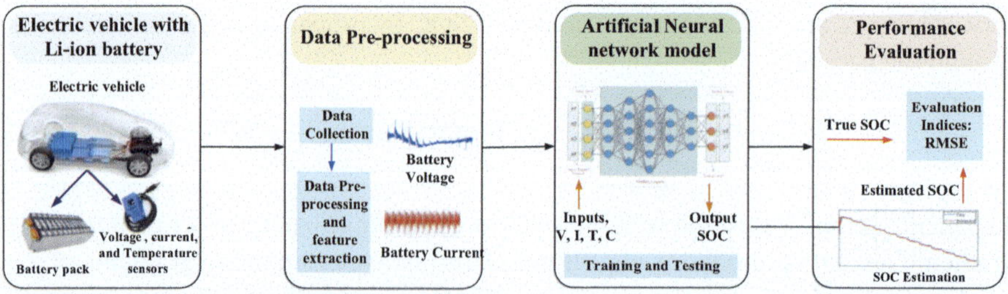

Fig. 5 ANN model-based SOC estimation method

handling missing values, and optional augmentation. An additional validation set, making up 15% of the total dataset (1179 samples), was utilized for hyperparameter tuning and to prevent overfitting.

4.2 *Tesla Model S: Modeling and Simulation Parameters*

The test system in this study involves the Tesla Model S, specifically focusing on its battery under varied operating conditions. Key parameters such as current, voltage, and temperature are considered, as they significantly impact SOC estimation. These parameters are crucial in modeling the Tesla Model S battery's performance across the city and highway driving cycles. The Tesla Model S is a leading electric vehicle known for its high performance and efficiency [37]. A comprehensive understanding of its operational characteristics involves examining both the battery system and the vehicle's physical attributes. The data for the ANN to perform is obtained from the open-source GitHub platform for the battery and driving cycle [38].

4.2.1 Physical Characteristics

The Tesla Model S features advanced engineering with key physical attributes that influence its performance. The vehicle's transmission system has a gear ratio of 9.3 and an efficiency of 80%, translating to effective power delivery. The electric motor provides a peak torque of 660 Nm and a maximum power output of 592 kW (794 hp), with regenerative braking contributing a 70% efficiency in energy recovery [37]. The vehicle's wheels have a radius of 0.23 m, while the brake discs measure 0.18 m in radius. The distance from the center of gravity to both the front and rear axles is 2.5 m, with the center of gravity positioned 0.7 m above the axles. Aerodynamic efficiency is characterized by a drag coefficient of 0.24 and a frontal area of 2.34 square meters, with the vehicle's total mass at 2250 kg.

4.2.2 Battery Characteristics

The Tesla Model S is equipped with a lithium-ion battery pack designed to enhance its driving range and performance. The battery configuration includes 86 cells connected in parallel within each module and 96 cells arranged in series across all modules. This configuration results in a total energy capacity of approximately 87.68 kWh. Each cell has a nominal voltage of 3.7 V and a nominal capacity of 2.862 A-h with a diameter of 18 mm and a length of 65 mm. The total voltage of the battery pack is therefore 355.2 V, and the capacity of each parallel module is 246.852 A-h. This configuration significantly contributes to the vehicle's overall energy storage and performance capabilities. The nominal capacity of 100 kWh considers the potential degradation and aging of the battery over time, ensuring that the battery performance remains reliable and meets the expected driving range despite the natural reduction in capacity [38]. The cell volume is 1.658×10^{-5} m^3. Thermal management is supported by a radial thermal conductivity of 3.819 W/(m K) and a heat capacity of 37.72 J/K, alongside a convective heat transfer coefficient of 10 W/m^2 K and a thermal resistance of 0.000216 m^2 K/W. These characteristics are modeled using MATLAB 2022b to analyze and optimize the Tesla Model S's performance, providing insights into energy consumption, thermal management, and overall vehicle efficiency.

4.3 *Results and Implications for SOC Estimation*

This section presents a comprehensive comparison of the EKF and ANN methods based on their performance in SOC estimation for the Tesla Model S. The analysis uses RMSE, to evaluate the accuracy of each approach in different driving environments.

4.3.1 City Driving Cycle

For the analysis of the city (CY) driving profile, the FTP-75 (Federal Test Procedure) has been employed, a standard used for testing light-duty vehicles in the United States. This procedure encompasses specific parameters, including a distance of 12.07 km (7.5 miles) covered over a duration of 1369 s, and repeats the pattern. The average speed during the test is 31.5 km/h (19.6 mi/h), with a maximum speed reaching 91.2 km/h (56.7 mi/h) [40], as illustrated in Fig. 6.

The SOC for the city driving profile was tested on the Tesla Model S, where the battery's SOC is monitored, taking approximately 10 h to deplete. SOC estimation

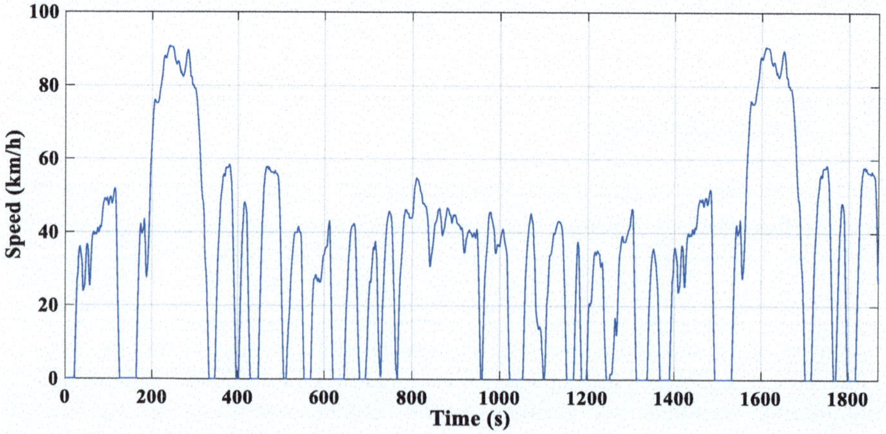

Fig. 6 City driving cycle of an electric vehicle

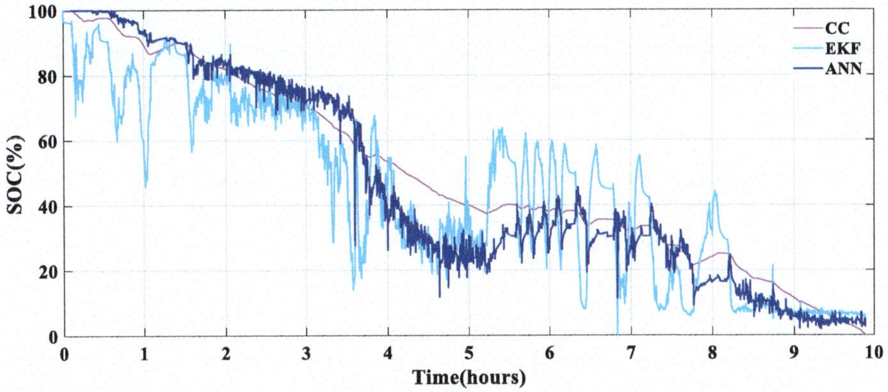

Fig. 7 Comparison of SOC of CC, EKF, and ANN for CY driving cycle

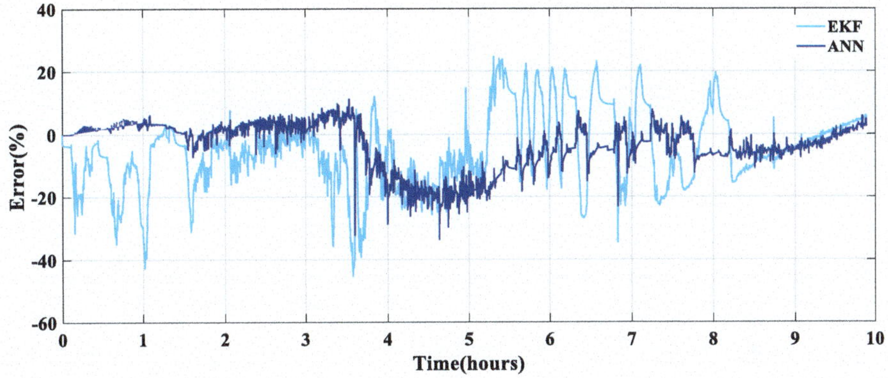

Fig. 8 Error percentage value of the EKF and ANN for the CY driving cycle

during this driving cycle was carried out using three different methods: CC, the EKF, and ANN, as depicted in Fig. 7. Additionally, the error analysis between the EKF and ANN methods is presented in Fig. 8, highlighting the comparative accuracy of these SOC estimation techniques.

4.3.2 Highway Driving Cycle

The study of the Tesla S Model is tested under a highway driving profile using the Highway Fuel Economy Test (HWFET) cycle. The HWFET, developed by the U.S. Environmental Protection Agency (EPA), is a chassis dynamometer driving schedule specifically designed to determine the highway fuel economy of light-duty vehicles. In this test, the vehicle underwent two runs, with a brief break between them. The actual test, which included emission measurements, lasted for 765 s, covering a total distance of 16.45 km (10.26 miles) at an average speed of 79.7 km/h (49.3 mi/h) [41]. The driving profile is shown in Fig. 9.

The SOC for the highway driving profile was tested on the Tesla Model S, with the battery's SOC monitored over an approximate duration of 4 h. During this highway driving cycle, SOC estimation was performed using three different methods: CC, EKF, and ANN. The results of this SOC estimation are illustrated in Fig. 10. Additionally, Fig. 11 provides an error analysis between the EKF and ANN methods, demonstrating the comparative accuracy of these SOC estimation techniques.

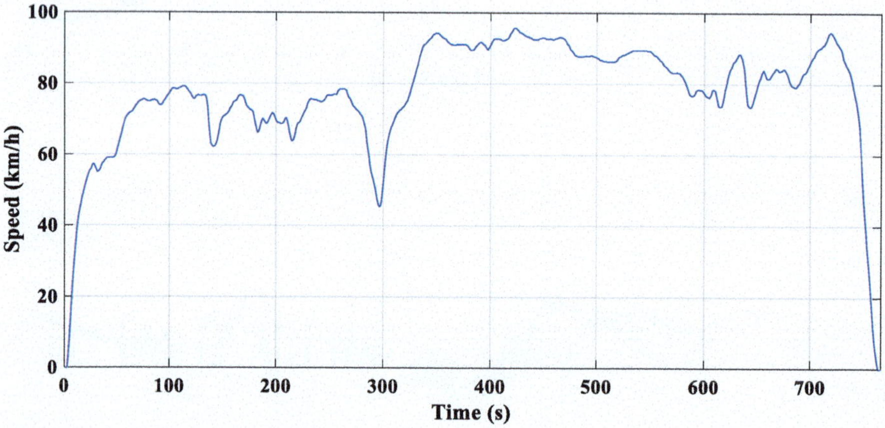

Fig. 9 Highway driving cycle of an electric vehicle

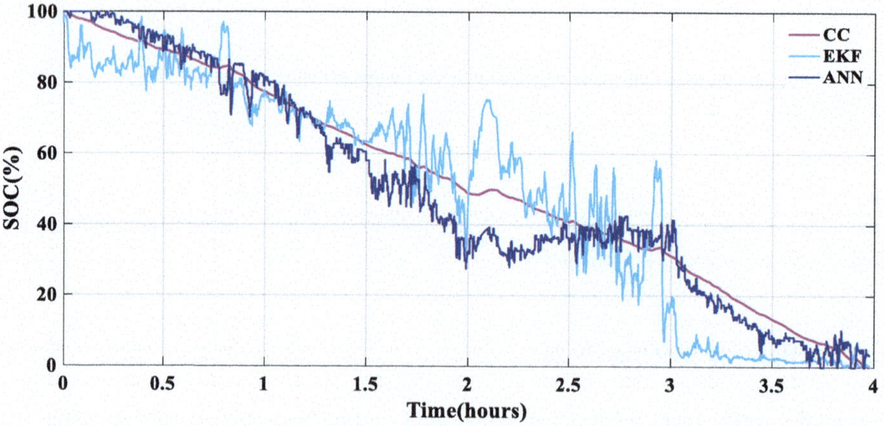

Fig. 10 Comparison for SOC of CC, EKF and ANN for HW driving cycle

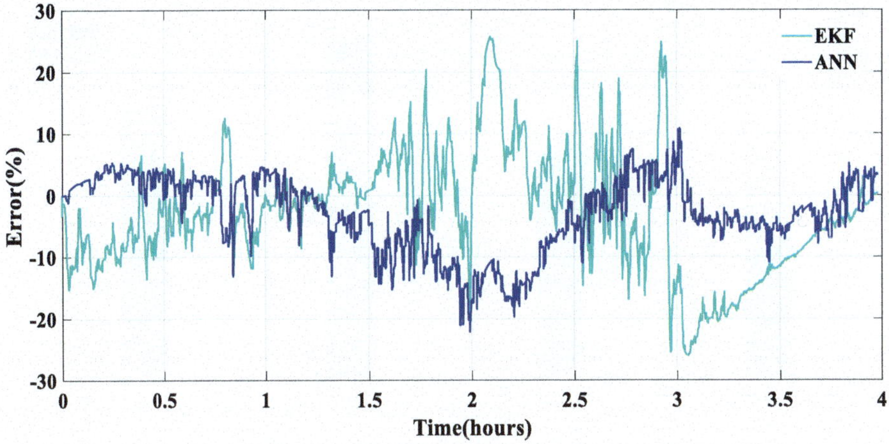

Fig. 11 Error percentage value of the EKF and ANN for the HW driving cycle

4.3.3 Analysis

The comparative analysis of RMSE values for SOC estimation reveals that the ANN consistently outperforms the EKF across different driving cycles. Additionally, the analysis incorporates RMSE values at various temperatures, as temperature significantly impacts SOC estimation accuracy, making its inclusion essential. Specifically, in the city driving cycle, the ANN demonstrates a reduction in RMSE of approximately 7.7% at 25 °C and 3.8% at 40°C compared to the EKF. In the highway driving cycle, the ANN shows a substantial improvement, achieving an RMSE that is 27.4% lower at 25 °C and 19.6% lower at 40 °C than the EKF. These results underscore the ANN's advantage over accuracy and robustness in SOC estimation, particularly under challenging driving conditions and temperature variations. Figure 12 represents the following analysis and two different temperatures and driving cycles.

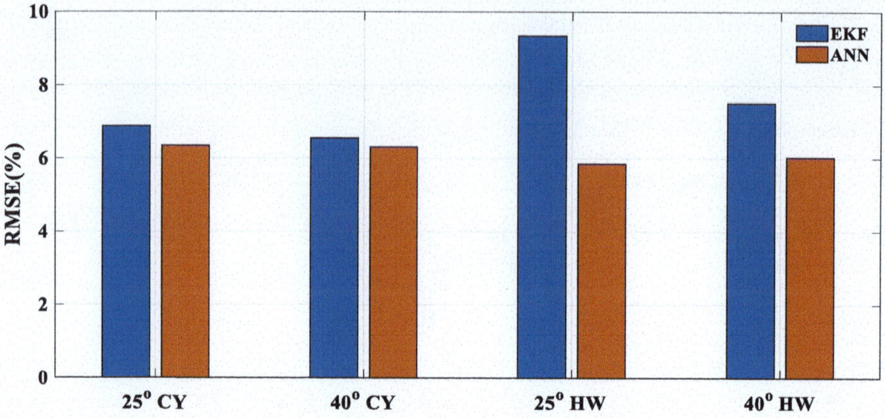

Fig. 12 Analysis of the highway and city driving cycles at 25 and 40 °C

5 Conclusion

This work has demonstrated the effectiveness of Artificial Neural Networks (ANN) for State of Charge (SOC) estimation in electric vehicles (EVs), particularly in comparison to the Extended Kalman Filter (EKF) and Coulomb Counting methods. The ANN approach consistently provided more accurate SOC estimates, especially in the Highway driving cycle, where it significantly reduced the Root Mean Square Error (RMSE) compared to the EKF. The results highlight ANN's robustness across different driving conditions and temperatures, making it a compelling choice for enhancing battery management systems in EVs. This study underscores the potential of ANN as a reliable and efficient tool for SOC estimation, contributing to the development of more advanced and accurate EV technologies.

Acknowledgements This research work was supported by "Woosong University's Academic Research Funding—2025".

References

1. Jian L, Yongqiang Z, Larsen GNS, Snartum A (2020) A comprehensive study on electric transportation challenges. eTransportation 6:100072. https://doi.org/10.1016/j.etran.2020.100072
2. Wang Y, Tian J, Sun Z, Wang L, Xu R, Li M, Chen Z (2020) A comprehensive review of battery modeling and state estimation approaches for advanced battery management systems. Renew Sustain Energy Rev 131:110015. https://doi.org/10.1016/j.rser.2020.110015
3. Raju BA, Reddy PKM, Sandeep V (2024) Leveraging hybrid energy storage for distributed secondary frequency regulation with consensus-driven event-triggered approach. Electr Eng. https://doi.org/10.1007/s00202-024-02747-4

4. Zhao L, Lin M, Chen Y (2016) Analysis on energy efficiency in renewable systems. Int J Energy Res 40(10):1389–1399. https://doi.org/10.1002/er.3530
5. Zhang C, Jiang J, Zhang L, Liu S, Wang L, Loh PC (2016) Review of battery system designs and performance. Energies 9(11):900. https://doi.org/10.3390/en9110900
6. Westerhoff U, Kroker T, Kurbach K, Kurrat M (2016) Energy storage system applications in grid support. J. Energy Storage 8:244–256. https://doi.org/10.1016/j.est.2016.09.001
7. Raju BA, Reddy PKM, Sandeep V (2023) A day ahead demand aggregation in islanded microgrid considering uncertainty. In: 3rd international conference on artificial intelligence and signal processing (AISP), India. https://doi.org/10.1109/AISP57993.2023.10134788
8. Ramadan HS, Becherif M, Claude F (2017) Hydrogen energy system developments for transportation. Int J Hydrog Energy 42(48):29033–29046. https://doi.org/10.1016/j.ijhydene.2017.07.219
9. Wang L, Lu D, Liu Q, Liu L, Zhao X (2019) New developments in electrochemical energy storage systems. Electrochim Acta 296:1009–1017. https://doi.org/10.1016/j.electacta.2018.11.156
10. Sepasi S, Ghorbani R (2014) A novel on-board state-of-charge estimation method for aged li-ion batteries based on model adaptive extended Kalman filter. J Power Sources 245:337–344. https://doi.org/10.1016/j.jpowsour.2013.06.040
11. Bai W, Zhang X, Gao Z, Xie S, Chen Y, He Y, Zhang J (2022) Innovations in electric power system research. Electr Power Syst Res 213:108751. https://doi.org/10.1016/j.epsr.2022.108751
12. Chen Z, Sun H, Dong G, Wei J, Wu J (2019) State estimation and management of power sources. J Power Sources 414:158–166. https://doi.org/10.1016/j.jpowsour.2019.01.012
13. Xia B, Zhang Z, Lao Z, Wang W, Sun W, Lai Y, Wang M (2018) Energy management and power generation innovations. Energies 11(6):1481. https://doi.org/10.3390/en11061481
14. How DNT, Hannan MA, Lipu MSH, Ker PJ (2019) A review of electric vehicle energy management systems. IEEE Access 7:136116–136136. https://doi.org/10.1109/ACCESS.2019.2942213
15. Pothireddy KMR, Battula AR, Vuddanti S (2023) Demand response: What is the optimal curtailment for economical benefit? In: 9th international conference on electrical energy systems (ICEES), Chennai, India, pp 235–239. https://doi.org/10.1109/ICEES57979.2023.10110210
16. Vaishnav V, Sharma D, Jain A (2023) A study on power systems engineering applications. IEEE Syst J 17(3):3401–3412. https://doi.org/10.1109/JSYST.2023.3240171
17. He T, Li D, Wu Z, Xue Y, Yang Y (2017) A modified Luenberger observer for SOC estimation of lithium-ion battery. In: Chinese Control Conference (CCC), China. https://doi.org/10.23919/ChiCC.2017.8028442
18. Tang X, Liu B, Gao F, Lv Z (2016) Energy-efficient battery management systems. Energies 9(5):336. https://doi.org/10.3390/en9050336
19. Tang X, Wang Y, Chen Z (2015) Advanced techniques in power management for electric vehicles. J Power Sources 282:452–459. https://doi.org/10.1016/j.jpowsour.2015.02.059
20. Hannan MA, Lipu MSH, Hussain A, Mohamed A (2017) Energy efficiency and optimization techniques in electric vehicles. Renew Sustain Energy Rev 78:834–854. https://doi.org/10.1016/j.rser.2017.05.001
21. Salkuti SR (2021) Challenges and innovations in electric grid management. J Electr Eng Technol 16(1):239–248. https://doi.org/10.1007/s42835-020-00608-1
22. Chen Z, Qiu S, Masrur MA, Murphey YL (2011) Neural networks applications in energy systems. In: Proceedings of international joint conference on neural networks. https://doi.org/10.1109/IJCNN.2011.6033495
23. Yang S, Liu X, Li S, Zhang C (2023) Advanced battery management system for electric vehicles. Springer, Huazhong University of Science and Technology Press. https://doi.org/10.1007/978-981-19-3490-2
24. Salkuti SR (2022) Emerging and advanced green energy technologies for sustainable and resilient future grid. Energies 15(18):6667. https://doi.org/10.3390/en15186667

25. Li Z, Huang J, Liaw BY, Zhang J (2017) Advanced research in power storage systems. J Power Sources 348:281–301. https://doi.org/10.1016/j.jpowsour.2017.03.001
26. Raju BA, Mohan Reddy PKM, Vuddanti S (2024) Decentralized coordination of electric vehicle charging in microgrids using game theory. In: IEEE 3rd international conference on electrical power and energy systems (ICEPES), India. https://doi.org/10.1109/ICEPES60647.2024.10653600
27. Chen H, Wang X, Li Z, Chen W, Cai Y (2019) Power system protection strategies. Prot Control Mod Power Syst 4(1):1–8
28. Sajid S, Salkuti SR et al (2022) Renewable energy resources and challenges. Energy Sour A 44(4):9861–9881. https://doi.org/10.1080/15567036.2022.2143948
29. Vuddanti S, Battula AR (2021) Electrical power system analysis for energy efficiency. Int J Emerg Electr Power Syst 23(3):317–328. https://doi.org/10.1515/ijeeps-2021-0093
30. Coleman M, Lee CK, Zhu C, Hurley WG (2007) Advanced power electronics for energy systems. IEEE Trans Ind Electron 54(5):2550–2557. https://doi.org/10.1109/TIE.2007.899926
31. Abu-Sharkh S, Doerffel D (2004) Power source advancements for electric vehicles. J Power Sour 130(1–2):266–274. https://doi.org/10.1016/j.jpowsour.2003.12.001
32. Sadeque F, Gursoy M, Sharma D, Mirafzal B (2024) Applications of power electronics in energy systems. IEEE Trans Ind Appl 60(3):4313–4323. https://doi.org/10.1109/TIA.2024.3357043
33. Lu L, Han X, Li J, Hua J, Ouyang M (2013) Power storage systems for electric vehicles. J Power Sourc 226:272–288. https://doi.org/10.1016/j.jpowsour.2012.10.060
34. Khan K, Jafari M, Gauchia L (2023) Battery management system innovations in electric vehicles. Appl Energy 341:121014. https://doi.org/10.1016/j.apenergy.2023.121014
35. Zhang K, Li L, Tao Y (2021) Intelligent energy management in electric vehicles. Energies 14(4):1065. https://doi.org/10.3390/en14041065
36. Bezha M, Nagaoka N (2019) Online learning ANN model for SoC estimation of the Lithium-Ion battery in case of small amount of data for practical applications. In: 10th international conference on power electronics and ECCE Asia (ICPE 2019—ECCE Asia), Busan, Korea, pp 1–5. https://doi.org/10.23919/ICPE2019-ECCEAsia42246.2019.8797079
37. Ragone M, Yurkiv V, Ramasubramanian A, Kashir B, Mashayek F (2021) Data driven estimation of electric vehicle battery state-of-charge informed by automotive simulations and multi-physics modeling. J Power Sources 483:229108. https://doi.org/10.1016/j.jpowsour.2020.229108
38. pyBMS (n.d.) Tesla model s battery pack documentation. Available at: https://github.com/bjpirt/pyBMS/blob/main/battery/tesla_model_s/README.md
39. Chademunhu BF, Gill A, Singh M (2023) Tesla model S battery pack modelling for voltage and current monitoring. In: 3rd Asian conference on innovation in technology (ASIANCON), Ravet, India, pp 1–5. https://doi.org/10.1109/ASIANCON58793.2023.10270222
40. National Renewable Energy Laboratory (n.d.) DriveCAT: drive cycle analysis tool. Available [Online]. www.nrel.gov/transportation/drive-cycle-tool
41. Transport Policy (n.d.) US light-duty highway fuel economy cycle. Available at: https://www.transportpolicy.net/standard/us-light-duty-highway-fuel-economy-cycle-hwfet/

GJO-Pattern Search Algorithm Based DG and Capacitor Placement in Distribution Network with Zone Based Installation of EVCSs

Rampreet Manjhi, Deepak Kumar Lal, Sandeep Biswal, and Surender Reddy Salkuti

Abstract Electric vehicles (EVs) serve a significant role in supporting sustainable development. Their adoption is consistent with numerous fundamental aims of sustainable development, including lowering greenhouse gas emissions, reducing dependency on finite fossil fuels and imported oil, and improving energy security. EV charging stations (EVCSs) in radial distribution networks (RDNs) raise particular issues. Proper coordination of optimally placed distributed generations (DGs) and capacitors integrated with EVCSs can mitigate the effects of increased active power loss, a decline in voltage stability margin, deterioration in voltage profile, and a worsening in network dependability caused by EVCSs integration. The distribution test systems are divided into different zones. The location-based installations of EVCSs technique ensure effective coverage of EVs and reachable EVCSs infrastructure across the networks. Subsequently, the Golden Jackal Optimization-Pattern Search (GJO-PS) algorithm, a novel meta-heuristic optimization method, is applied to identify the optimal locations for reactive power compensation devices and DGs. This approach is tested on 33-bus and 69-bus RDNs, with simulation results analyzed under various test scenarios.

R. Manjhi · D. K. Lal (✉)
Department of Electrical Engineering, Veer Surendra Sai University of Technology, Burla, Odisha, India
e-mail: dklal_ee@vssut.ac.in

R. Manjhi
e-mail: rampreetmanjhiphd21@vssut.ac.in

S. Biswal
Department of Electrical Engineering, O. P. Jindal University, Raigarh, Chhattisgarh, India
e-mail: sandeep.biswal@opju.ac.in

S. R. Salkuti
Department of Railroad and Electrical Engineering, Woosong University, Daejeon, Republic of Korea
e-mail: surender@wsu.ac.kr

S. R. Salkuti (ed.), *Artificial Intelligence for Integrated Smart Energy Systems in Electric Vehicles*, Lecture Notes in Electrical Engineering 1427,
https://doi.org/10.1007/978-3-031-94276-1_5

Keywords Electric vehicle charging station · Radial distribution network · Distributed generation · Shunt capacitors · Golden Jackal optimization algorithm · Pattern search technique

1 Introduction

Concern over the environment, rising petroleum prices, and desire to reduce dependency on fossil fuels combined with advancements in battery technology, government incentives, and the expansion of charging infrastructure, are propelling the rise of the electric vehicle (EV) market globally. Reducing dependency on fossil fuels is a strategic goal for many countries, aiming to enhance energy security and reduce geopolitical risks associated with oil imports. EVs contribute to this goal by utilizing electricity, which can be generated from a variety of sources, including renewable energy. EVs in radial distribution networks (RDNs) present unique challenges [1]. RDNs characterized by their tree-like structure with branches radiating out from a single substation, are common in many power distribution networks. The integration of EV charging stations (EVCSs) into these systems can have significant implications. EVCS is treated as an additional load to the network. The widespread adoption of EVs can lead to a substantial increase in electricity demand, particularly during peak charging times. This additional load can stress the existing infrastructure, potentially leading to overloading of transformers and feeders. It may also lead to voltage regulation challenges and power quality issues [2, 3]. EV charging patterns can be unpredictable and vary significantly, making it difficult for utilities to plan and manage the distribution network effectively [4, 5]. But, EVs in RDNs present bigger opportunities also. EVs can be integrated into demand response programs, where charging can be shifted to off-peak times or controlled dynamically for load balancing [6, 7]. This can help in smoothing the load curve and reducing peak demand. EVs can role as distributed energy storage units, providing services back to the grid such as peak shaving, frequency regulation, and voltage support. V2G technology permits EVs to discharge electricity back into the grid when needed [8–10]. Optimal placement of EVCS and real-time coordination for its charging can minimize the impact of its integration and increase reliability [11, 12].

The trend toward distribution automation underscores the importance of meticulous planning in electric power distribution systems. This planning aims to optimize operations, reduce costs, minimize power losses, and enhance the quality of power delivered to consumers. To ensure that distribution automation systems are economically viable, it is essential to identify the most efficient operating scenarios. This involves optimizing the distribution network to reduce capital investments while still achieving desired performance levels. Effective planning and the implementation of automation can help identify and mitigate power losses [13, 14]. Conventional power distribution systems are one-way communication, with no feedback-control applications from the consumers to send signals or data for line operators. An active electrical distribution network not only delivers electricity but also controls the flow of electric power from the generation site to reach its consumers. Distributed generation (DG) resources, like solar PV panels, wind turbines, energy storage units, etc., can be easily combined on active networks. This integration enables the network to handle variable generation sources effectively [15–18]. Active networks rely on robust communication systems that facilitate two-way communication between utilities and consumers. This communication enables better demand-side management and allows consumers to participate in energy markets [19]. An active electrical distribution network represents a modernized approach to electrical power distribution, leveraging advanced technologies and real-time management to enhance efficiency, reliability, and sustainability. Shunt capacitor banks, which provide reactive power, are installed on electrical distribution network feeders to lower power losses, optimize the voltage characteristics of feeders, and boost feeder loadability [20–23]. DGs are highly effective in minimizing system real power losses, and their performance can be further enhanced when used in conjunction with shunt capacitors. Combining DGs with shunt capacitors enables greater reductions in power losses and improved system efficiency. However, the selection of installation sites, capacities, and types of DG technologies must be carefully planned to ensure compatibility with the existing distribution system and seamless grid integration. Several heuristic approaches have been suggested for identifying the optimal placement and sizing of DGs and reactive power compensation devices [24–29].

DGs and distribution network reactive power compensators can mitigate the consequences of increased power loss, a decline in voltage stability margin, a deterioration in voltage profile, and a worsening in network dependability that comes with EVCSs integration. EVs when combined with distributed generation sources like solar PV can enhance the local generation capacity. A genetic algorithm has been applied to discover the best optimal locations of EVCS-PV DG-battery energy storage systems in 33-bus RDN. The network is divided into different zones for EVCSs placements [30]. This can reduce the dependency on the central grid and improve the reliability of the supply system. Bi-directional fast-charging EVCSs with a distinctive reactive power compensation method are illustrated for voltage regulation [31]. Reactive power compensation at the optimal location of EVs reduces its impact and improves the voltage profile [32, 33]. Spotted Hyena Optimizer Algorithm is used for the distribution system planning with DGs, capacitors, and EVCSs [31]. AI-based approach is used for reliability analysis of RDNs in the presence of EVs [34]. A noble double-level framework is developed for a regulating energy system with optimal locations of DGs and EVCS in 33-bus and 69-bus RDNs [35]. A strategy based on the whale optimization algorithm is used to solve the multi-objective problem of allocating EVCS along with DGs and capacitors simultaneously to improve voltage stability [36]. Optimal allocation of EVCS and renewable DGs with energy storage in RDNs by taking into account time sequence features of DG and power demand is considered [37].

Motivated by the present literature distribution system planning with EVCS is performed. MATLAB software was used for the distribution system load flow (DSLF) calculations. Simultaneous optimization is performed to determine the optimum locations and sizes of capacitors and DGs, i.e., bus-wise choices as well as power output ratings to make a positive contribution towards voltage profile enhancement with reduced network losses. The RDNs are divided into several zones. Each "zone" should provide as many EV chargers as the supply grid can bear. A sufficient number of EV charging stations are placed. Each of these most likely reflects a specific region with a variety of characteristics, such as population density, transportation infrastructure, and EV adoption rates. This location-focused strategy ensures reliable EV services while paving the way for a nationwide, easily accessible charging infrastructure. By looking at traffic patterns, charging demands, and existing EVCS infrastructure surrounding each zone, agencies can select the best areas for an EVCS in every zone. This move would help lower range anxiety, encourage greener transport solutions across the city, and also support increasing EV adoption. In this work, the Golden Jackal Optimization (GJO) approach is employed to determine the best positions and sizes of capacitors and DGs in RDNs. In 2022, Chopra and Ansari introduced the GJO algorithm [38]. This is a sophisticated algorithm based on the socially organized hunting behavior of two golden jackals. The GJO method has a high ability to investigate the global search space, but it is prone to premature convergence and can become trapped in local optima. To address this issue and improve the efficiency of this algorithm, this work proposes a hybrid GJO algorithm that incorporates the pattern search (PS) algorithm. The GJO-PS will ensure that the DGs and capacitors perform to their utmost ability. The increased voltage profile and incorporation of EVCS will be the project's most significant contributions. Figure 1 shows a visual illustration of the technique's proposed layout, demonstrating its potential for optimizing RDN performance. The arrangement displayed in the picture represents the strategic placement and scaling of components such as EVCS, DGs, and capacitors. The importance of the research is summarized as follows:

- The distribution test systems are divided into different zones. The location-based installations of the EVCS technique are used to ensure the accessibility of EV charging infrastructure across the networks.

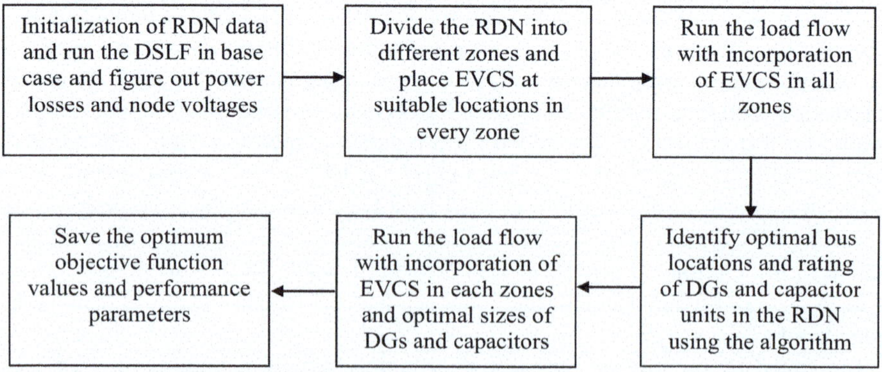

Fig. 1 Layout of the proposed approach

- A hybrid meta-heuristic optimization algorithm GJO-PS is employed to establish suitable locations for reactive power compensation devices and DGs. The method is implemented on a 33-bus and a 69-bus test RDNs.
- The success of the proposed method is evaluated using different test scenarios.

After the introduction in Sects. 1 and 2 elaborates on the DSLF method and problem formulation. Section 3 offers a summary of GJO-PS and details the procedure for addressing the optimization problem. Simulation results and discussions on the test systems are presented in Sect. 4, while Sect. 5 concludes with key inferences.

2 Methodology

2.1 Distribution System Load Flow (DSLF) Solution

The DSLF analysis evaluates the impact of integrating EVCS on voltage profiles, line currents, and power losses. It ensures that the additional load from EVCS does not exceed the network's capacity. The analysis employs the direct approach outlined in reference [39], utilizing two key matrices: the branch-current to bus voltage (BCBV) matrix and the bus-injection to branch-current (BIBC) matrix, which form the foundation of the method. The apparent power flow from the bus m is calculated using Eq. (1) as,

$$S_m = P_m + jQ_m \tag{1}$$

The injected current at each bus m is calculated as,

$$I_m = \left(\frac{P_m + jQ_m}{V_m}\right)^* \tag{2}$$

The line currents matrix is,

$$[J] = [BIBC][I] \tag{3}$$

where I refers to the matrix of branch current. The bus voltages can be calculated as,

$$[\Delta V] = [DSLF][I] \tag{4}$$

where $[DSLF] = [BCBV][BIBC]$. The results for DSLF are obtained iteratively during the kth iteration.

$$\left[\Delta V^{k+1}\right] = [DSLF]\left[I^k\right] \tag{5}$$

$$[\Delta V^{k+1}] = [\Delta V^{0}][\Delta V^{k+1}] \tag{6}$$

The power flow equation is written as,

$$P_n = P_m - P_{L,n} - R_{m,n}\left(\frac{P_m^2 + Q_m^2}{|V_m|^2}\right) \tag{7}$$

$$Q_n = Q_m - Q_{L,n} - X_{m,n}\left(\frac{P_m^2 + Q_m^2}{|V_m|^2}\right) \tag{8}$$

$P_{L,n} + jQ_{L,n}$ is the connected load at bus n. The active and reactive power losses are determined as,

$$\mathrm{P_{Loss}(m, n)} = \mathrm{R_{m,n}}\left(\frac{\mathrm{P_m^2 + Q_m^2}}{|\mathrm{V_m}|^2}\right) = \mathrm{BIBC.P_{RLPM}} \tag{9}$$

$$\mathrm{Q_{Loss}(m, n)} = \mathrm{X_{m,n}}\left(\frac{\mathrm{P_m^2 + Q_m^2}}{|\mathrm{V_m}|^2}\right) = \mathrm{BIBC.Q_{REPM}} \tag{10}$$

where P_{RLPM} and Q_{REPM} represent the active and reactive power matrices respectively. The sum of network power losses are,

$$P_{TLoss} = \sum_{l=1}^{NB-1} P_{Loss}(l) \tag{11}$$

$$Q_{TLoss} = \sum_{l=1}^{NB-1} Q_{Loss}(l) \tag{12}$$

where RDN has NB number of buses. The number of branches is $NB - 1$. The branch number is $l = 1, 2, 3, \ldots, NB - 1$.

2.2 Problem Formulation

The goal function of the DG and capacitor allocation is as follows:

$$Minimize\ F = \min P_{TLoss} \tag{13}$$

Subject to the constraints,

$$|V_m|^{min} \leq |V_m| \leq |V_m|^{max} \tag{14}$$

$$|I_l| \leq |I_l|^{max} \tag{15}$$

The total real power provided by the DGs should not exceed the combined real power demand and the total real power losses in the lines. Likewise, to prevent over-compensation, the reactive power supplied by the capacitors should remain below the required amount.

$$\sum_{m=1}^{ndg} P_{DG,m} \leq \sum_{m=2}^{NB} P_{L(m)} + \sum_{l=1}^{NB-1} P_{Loss}(l) \tag{16}$$

$$\sum_{m=1}^{nc} Q_{c,m} \leq \sum_{m=2}^{NB} Q_{L(m)} \tag{17}$$

3 Golden Jackal Optimization-Pattern Search Algorithm

The GJO algorithm was motivated by golden jackals' communal hunting activities. The three fundamental steps of GJO are probing the prey, surrounding, and attacking. Throughout the iterations, the shifting postures of the golden jackal pair in relation to the fitness and location of their prey lead to the most exciting possibilities thus far. In contrast to existing meta-heuristic algorithms, the Pattern Search (PS) with meta-heuristic algorithm provides an adaptable and balanced operator for improving local search capability [40, 41]. Hence, to improve the performance of GJO, PS is combined and the GJO-PS hybrid algorithm is proposed to solve the optimization problem.

3.1 GJO Algorithm

The GJO algorithm operates according to the steps outlined below [37].

3.1.1 Search Area Design

The beginning points of the GJO algorithm are randomly distributed across the search space, as follows:

$$\alpha_{initial} = \alpha_{min} + R_n(\alpha_{max} - \alpha_{min}) \tag{18}$$

where "R_n" is a random value from 0 to 1, $\alpha_{initial}$ refers to the position of the initial golden jackals, α_{min} and α_{max} are the boundary solutions. The initial *Prey* matrix of size $[p \times q]$ is given in Eq. (19).

$$Prey = \begin{bmatrix} \alpha_{1,1} & \alpha_{1,2} \cdots & \alpha_{1,q} \\ \vdots & \ddots & \vdots \\ \alpha_{p,1} & \alpha_{p,2} \cdots & \alpha_{p,q} \end{bmatrix} \tag{19}$$

There is "p" number of preys with "q" variables. From *Prey* matrix two fittest are opted as the jackal pair. In the procedure of optimization, an objective function is employed to determine the fitness of each prey. Every prey's (*FPrey*) fitness is established by,

$$FPrey = \begin{bmatrix} f(\alpha_{1,1}; \alpha_{1,2}; \ldots, \alpha_{1,q}) \\ f(\alpha_{2,1}; \alpha_{2,2}; \ldots, \alpha_{2,q}) \\ \vdots \\ f(\alpha_{p,1}; \alpha_{p,2}; \ldots, \alpha_{p,q}) \end{bmatrix} \tag{20}$$

In which, "f" is the objective function.

3.1.2 Phase of Exploration

Jackals are adept at monitoring their prey. The mathematical equations for updated positions of the jackal pair are as follows.

$$\alpha_1(t) = \alpha_{MJ}(t) - E_n|\alpha_{MJ}(t) - r_s.Prey(t)| \tag{21}$$

$$\alpha_2(t) = \alpha_{FJ}(t) - E_n|\alpha_{FJ}(t) - r_s.Prey(t)| \tag{22}$$

$Prey(t)$ is the location of the prey in t iterations, and $\alpha_{MJ}(t)$ and $\alpha_{FJ}(t)$ signify the jackal pair's positions, $\alpha_1(t)$ and $\alpha_2(t)$ are new positions of the jackal pair, respectively. E_n is decaying prey energy ascertained as,

$$E_n = E_{n1}.E_{n0} \tag{23}$$

E_{n0} and E_{n1} represent the prey's beginning and declining energy, respectively. E_{n0} ranges from $[-1 \ to \ 1]$ and calculated as,

$$E_{n0} = 2.r - 1 \tag{24}$$

where "r" is a random parameter. E_{n1} is found as,

$$E_{n1} = c_1.(1 - (t/T)) \tag{25}$$

Here T refers to the total number of iterations and c_1 is a constant. Gradually E_{n1} declined from 1.5 to 0 during the iterations. The value of $|\alpha_{MJ}(t) - r_s.Prey(t)|$ and $|\alpha_{FJ}(t) - r_s.Prey(t)|$ represent the distance that the golden jackal has to go to reach its monitored prey in Eqs. (21) and (22) and is added or deducted from the jackal's present location based on E_n. "r_s" is evaluated as follows.

$$r_s = 0.05LF(\alpha) \tag{26}$$

The function of fitness of levy flight LF is computed as follows:

$$LF(\alpha) = 0.01 \times (\mu \times \sigma)/\left(\left|v^{(1/\beta)}\right|\right) \tag{27}$$

$$\sigma = \left(\frac{\Gamma(1+\beta) \times \sin(\pi\beta/2)}{\Gamma\left(\frac{1+\beta}{2}\right) \times \beta \times \left(2^{\frac{\beta-1}{2}}\right)}\right) \tag{28}$$

where μ and v are random values and β is a constant. The positions of the jackals have been altered as follows:

$$\alpha(t+1) = \frac{\alpha_1(t) + \alpha_2(t)}{2} \tag{29}$$

3.1.3 Phase of Exploitation

The jackals follow, tackle, and harass the prey and kill. The updated position of the jackal's pair is articulated as,

$$\alpha_1(t) = \alpha_{MJ}(t) - E_n|r_s.\alpha_{MJ}(t) - Prey(t)| \tag{30}$$

$$\alpha_2(t) = \alpha_{FJ}(t) - E_n|r_s.\alpha_{FJ}(t) - Prey(t)| \tag{31}$$

3.1.4 Shift from Phase of Exploration to Exploitation

E_n Value is used in the method to transition from the phase of exploration to exploitation. Whenever the prey is involved in evasive activity, its level of energy plummets dramatically. When it falls from 0 to − 1, the prey weakens; when it rises from 0 to 1, it strengthens. If $|E_n|$ is greater than 1, the jackal couple searches for prey, and if $|E_n|$ is less than 1, jackals assault the prey before involving in exploitation.

3.2 Pattern Search (PS) Technique

This section outlines the PS algorithm's fundamental mechanism [40, 41]. The search direction for ith coordinate for gth dimension of the ′n_p′ variables search area can be stated as,

$$S_g^h = \begin{cases} 1; g = h \\ 0; g \neq h \end{cases} \quad (32)$$

where, $g = 1, 2, \ldots, n_p$ and $h = 1, 2, \ldots, n_p$. The randomly generated step length is given in Eq. (33).

$$\lambda_g^* = \lambda_g^{min} + rand \times \left(\lambda_g^{max} - \lambda_g^{min}\right) \quad (33)$$

Here λ_g^{min} and λ_g^{max} are the boundaries of step length for g^{th} decision variable. The decision parameter (x_g) is updated once along the coordinate direction (h) as,

$$x_g = x_g + \lambda_g^* \times S_g^h \quad (34)$$

The control variables vector is updated based on the minimum objective function value. This technique is repeated for each ′n′ coordinate. The search direction for the subsequent optimization cycle is determined as follows:

$$S_g^h = x_g - Z_g \quad \left(g = 1, 2, \ldots, n_p; h = n_p + 1\right) \quad (35)$$

In which, Z_g is the base value of the choice parameter x_g. Moreover, in the pattern ′m′ direction, one of the coordinate directions will be disregarded, as in this case:

$$S_g^m = S_g^h\left(g = 1, 2, \ldots, n_p; h = n_p + 1\right) \quad (36)$$

The procedure is repeated until the complete direction of the vector is eliminated, at which point all operations return to one of the vector directions. In the end, until the operation has reached its maximum number of iterations, the updating process continues.

3.3 Implementation of GJO-PS Algorithm for Optimization Problem

The goal of incorporating PS with GJO is to maximize the benefits of both algorithms. The PS algorithm uses the beginning point determined by the GJO method. Each iteration involved the consecutive application of the GJO and PS algorithms. The ideal control variable setting supplied by the GJO algorithm serves as the starting point for the PS approach to obtain a superior outcome. This improved answer given by the PS technique substitutes the previous one in the population of the GJO algorithm and serves as the basis for further refinement in the following iteration by the GJO algorithm. The technique is continued for a predetermined iteration number. The optimized parameters of the hybrid GJO-PS algorithm are given in Appendix. The flowchart for the hybrid GJO-PS algorithm is provided in Fig. 2.

4 Results and Discussions

Simulating the optimal location of DGs and capacitors, along with the incorporation of EVCSs in the 33-bus and 69-bus test RDNs are involved using MATLAB power system simulation software. This process facilitates the analysis of the distribution networks.

4.1 A 33-Bus Test RDN

The 12.66 kV, 33-bus RDN is depicted in Fig. 3 [42]. After DSLF, a real power loss without DGs, capacitors, and EVCS at rated load is 210.9884 kW. The 33-bus RDNs are divided into five numbers of zones as depicted in Table 1. The EV chargers of capacity 200 kW each are placed in each zone [29]. This location-based approach ensured effective EV coverage and readily available EV charging facilities throughout the region [43]. The ideal positions and capacities of shunt capacitors and DGs are then determined using the GJO-PS algorithm. The simulation results, based on 30 independent trials and various test scenarios, are summarized in Table 2. The integration of DGs and capacitors has notably reduced network power losses and enhanced the node voltages. A large amount of loss reduction is obtained with the simultaneous allocation of capacitors and DGs incorporation with EVCSs. Figure 4 illustrates the voltage profile under various placement scenarios, while Fig. 5 depicts the convergence characteristics of the hybrid GJO-PS algorithm. The fast convergence rate proves the achievement of optimal solutions quickly.

4.2 A 69-Bus Test RDN

To confirm the validity of the established approach, the technique is extended to a hypothetical 12.66 kV, 69-bus RDN as displayed in Fig. 6 [43]. The 69-bus RDNs are divided into five numbers of zones as depicted in Table 3. The EV chargers of capacity 200 kW each are placed in each zone. The middle buses of each zone are considered suitable locations for EVCS installations as presented in Table 3. The

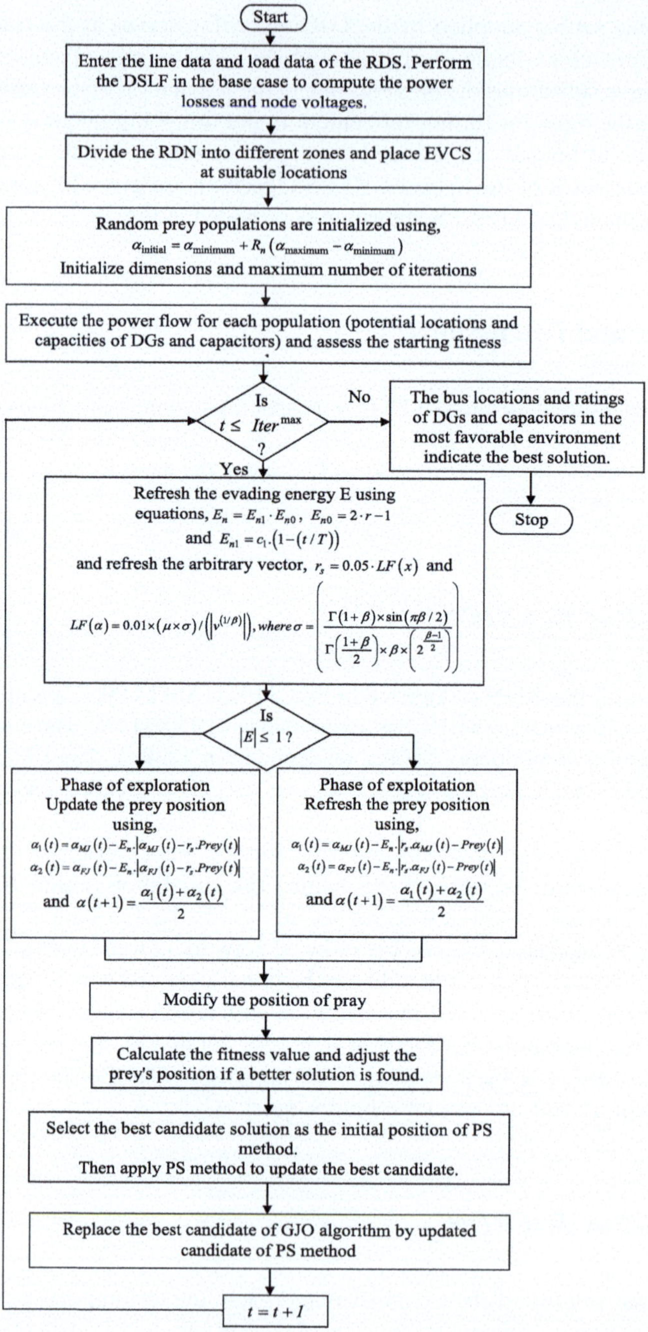

Fig. 2 Flowchart of the GJO-PS algorithm for the optimization problem incorporation of EVCSs

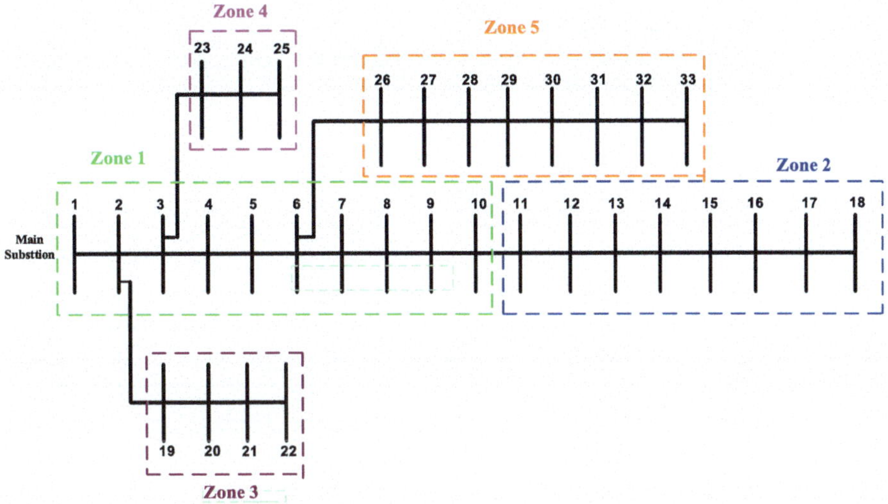

Fig. 3 Single line diagram of the 33 bus RDN

Table 1 Locations of EVCS in different zones of the 33 bus RDN

Zone (bus to bus)	Location of EVCS
1 (2–10)	8
2 (11–18)	17
3 (19–22)	21
4 (23–25)	25
5 (26–33)	30

simulation results, detailed in Table 4, are based on different test cases conducted over independent 30 trials. The incorporation of DGs and capacitors has notably decreased network power losses and enhanced the node voltages, demonstrating a reduced impact from the integration of EVCS. Figure 7 presents the voltage profile across various placement scenarios.

5 Conclusions

Integration of EVCSs into radial distribution systems presents several challenges; it also offers opportunities for grid modernization, improved load management, and enhanced reliability. GJO-PS, a newly developed meta-heuristic optimization technique, is employed to identify the optimal sites for reactive power compensation devices and DGs. The method is implemented on the 33-bus and 69-bus test RDNs. The computation results with different test scenarios are studied. The successful integration of DGs and shunt capacitor banks with the incorporation of EVCSs into

Table 2 Simulation results for the 33 bus RDN

System	EVCS locations (bus number)	kVAR rating of Capacitor/ located at bus number	DG capacity (kW)/ located at bus number	P_{TLoss}(kW)	Q_{TLoss} (kVAr)	V_{min}(p.u.)
Base case	–	–		210.9884	143.1376	0.90377 at bus 18
5 EVCS	8, 17, 21, 25, 30	–	–	319.5092	220.1313	0.87589, at bus 18
5 EV + 2 Capacitors	8, 17, 21, 25, 30	495/12 1085/30	–	242.3817	167.8660	0.90467 at bus 18
5 EV + 3 Capacitors	8, 17, 21, 25, 30	467/12 556/24 1039/30	–	238.5083	165.4121	0.90446 at bus 18
5 EV + 2 DGs	8, 17, 21, 25, 30	–	1114/14 1447/30	103.3757	72.5216	0.96634 at bus 18
5 EV + 3 DGs	8, 17, 21, 25, 30	–	1067/14 1096/25 1340/30	81.5665	58.2278	0.96653 at bus 18
5 EV + 1 Capacitor + 1 DG	8, 17, 21, 25, 30	1264/30	2347/8	96.2603	72.1051	0.95541 at bus 18
5 EV + 2 Capacitors + 2 DGs	8, 17, 21, 25, 30	450/12 1045/30	1145/14 1325/30	44.0575	32.3825	0.97611 at bus 25

RDNs solved the technical and economic issues due to the unplanned placement of EVCS. The zone-specific EVCS installation techniques are suggested. Effective EV coverage and readily available EV charging infrastructure were ensured throughout the region by this location-based approach. Utilities can dramatically improve power quality, efficiency, and reliability, as well as expand the use of green technologies, lessen reliance on fossil fuels, and mitigate environmental problems, by addressing critical aspects and executing optimal techniques.

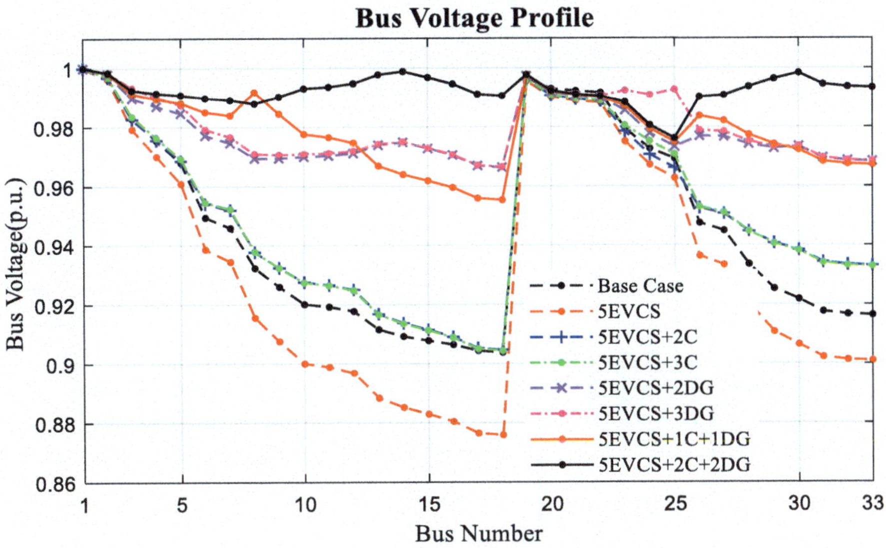

Fig. 4 Bus voltage profile of the 33 bus RDN included with EVCS, shunt capacitors, and DGs

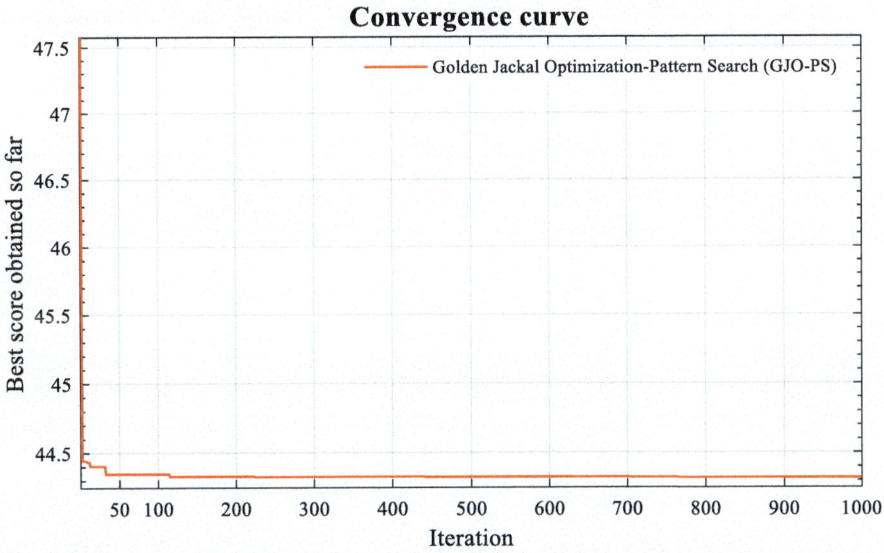

Fig. 5 Objective value variation across iterations for the optimal placement and sizing of 5 EV + 2 Capacitors + 2 DGs in the RDN

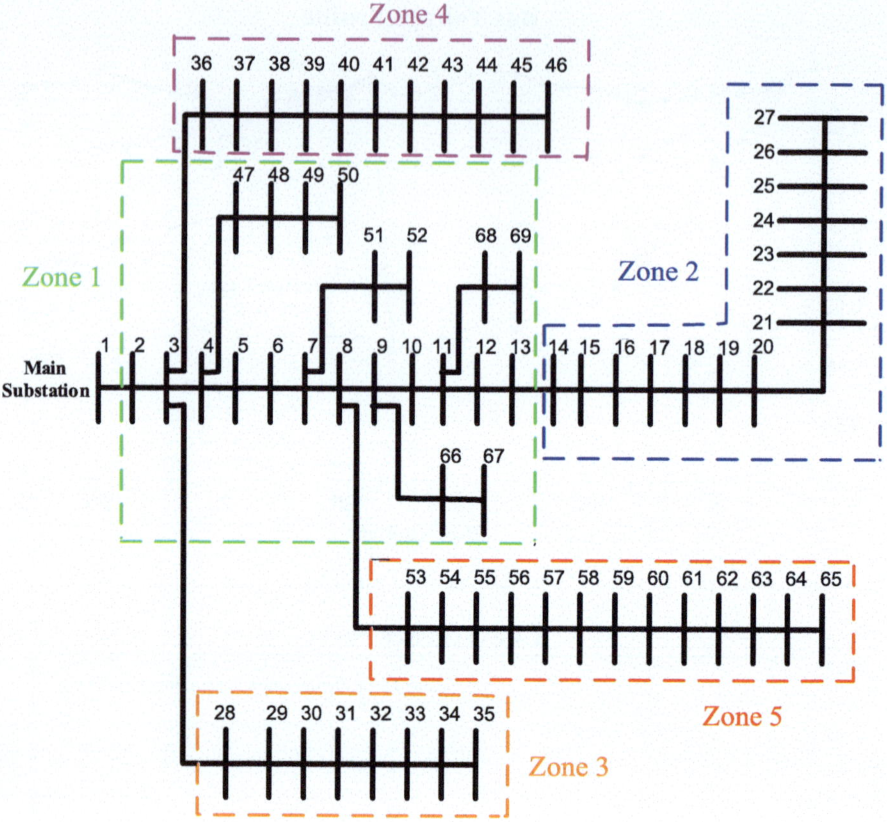

Fig. 6 Single line diagram of the 69 bus RDN

Table 3 Locations of EVCS in different zones of the 69 bus RDN

Zone (bus to bus)	Location of EVCS
1 (2–13, 47 to 52, 66–69)	7
2 (14–27)	21
3 (28–35)	32
4 (36–46)	41
5 (53–65)	59

Table 4 Simulation results for the 69 bus RDN

System	EVCS locations (bus number)	Capacitor size (kVAR)/ located at bus number	DG capacity (kW)/ located at bus number	P_{TLoss}(kW)	Q_{TLoss} (kVAr)	V_{min}(p.u.)
Base case	–	–	–	224.9827	102.1584	0.90918 at bus 65
5 EVCS	7, 21, 32, 41, 58	–	–	278.6660	126.3653	0.90042 at bus 65
5 EV + 1 Capacitor + 1 DG	7, 21, 32, 41, 58	1303/61	2053/61	38.4836	20.9741	0.9621 at bus 27
5 EV + 2 Capacitors + 2 DGs	7, 21, 32, 41, 58	350/17 1239/61	704/21 1924/61	10.2296	9.8913	0.99293 at bus 69
5 EV + 3 Capacitors + 3 DGs	7, 21, 32, 41, 58	350/11 250/21 1200/61	475/11 610/21 1858/61	6.1926	8.0663	0.99426 at bus 50

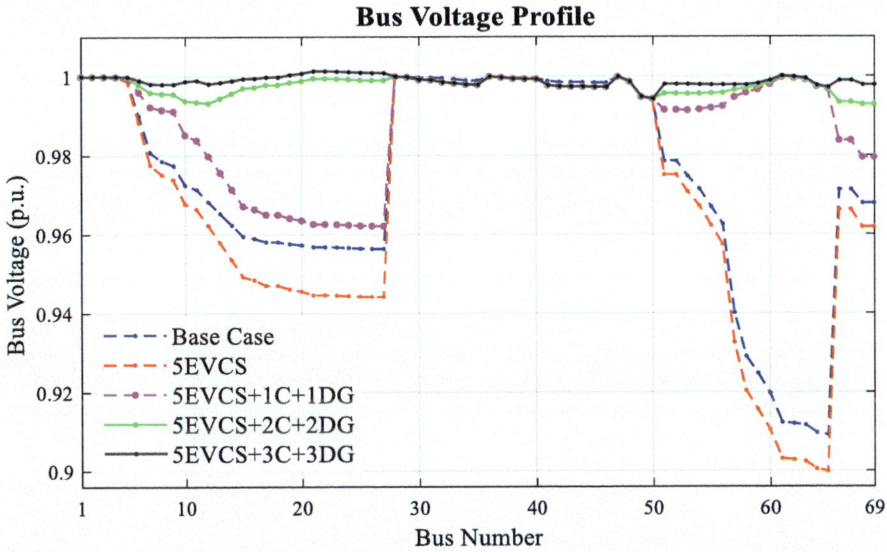

Fig. 7 Bus voltage profile of the 69 bus RDN included with EVCS, shunt capacitors, and DGs

Appendix

The optimum controlling parameters of the GJO algorithm are given below:

Number of search agents = 30, maximum number of iterations = 100, Rn is an arbitrary values in (0,1), r is in [0, 1], $c_1 = 0.5$, $\beta = 1.5$, values of μ and v are in [0, 1]. PS is executed with the following algorithm controlling parameters: mesh size = 1, mesh expansion factor = 2, mesh contraction factor = 0.5.

References

1. Fernández LP, Román TGS, Cossent R, Domingo CM, Frías P (2011) Assessment of the impact of plug-in electric vehicles on distribution networks. IEEE Trans Power Syst 26(1):206–213. https://doi.org/10.1109/TPWRS.2010.2049133
2. Habib S, Khan MM, Abbas F, Numan M, Ali Y, Tang H, Yan X (2020) A framework for stochastic estimation of electric vehicle charging behavior for risk assessment of distribution networks. Front Energy 14:298–317. https://doi.org/10.1007/s11708-019-0648-5
3. Ponnavaikko M, Prakasa Rao KS (1982) An approach to optimal distribution system planning through conductor gradation. IEEE Trans Power Appar Syst 6:1735–1742. https://doi.org/10.1109/TPAS.1982.317227
4. Božič D, Pantoš M (2015) Impact of electric-drive vehicles on power system reliability. Energy 83:511–520. https://doi.org/10.1016/j.energy.2015.02.055
5. Qian K, Zhou C, Allan M, Yuan Y (2011) Modeling of load demand due to EV battery charging in distribution systems. IEEE Trans Power Syst 26(2):802–810. https://doi.org/10.1109/TPWRS.2010.205745
6. Shahbazi A, Moradi Cheshmeh Beigi H, Abdi H, Shahbazitabar M (2023) Probabilistic optimal allocation of electric vehicle charging stations considering the uncertain loads by using the Monte Carlo simulation method. J Oper Autom Power Eng 11(4):277–284. https://doi.org/10.22098/joape.2023.10427.1738
7. Abid MS, Apon HJ, Alavi A, Hossain MA, Abid F (2023) Mitigating the effect of electric vehicle integration in distribution grid using Slime Mouldalgorithm. Alex Eng J 64:785–800. https://doi.org/10.1016/j.aej.2022.09.022
8. Etemesi AB, Megahed TF, Kanaya H, Mansour DEA (2024) Design of a bidirectional EV charger using unit template-based current-controlled synchronous reference frame hysteresis. J Electr Eng 75(1):29–40. https://doi.org/10.2478/jee-2024-0005
9. Hadian E, Akbari H, Farzinfar M, Saeed S (2020) Optimal allocation of electric vehicle charging stations with adopted smart charging/discharging schedule. IEEE Access 8:196908–196919. https://doi.org/10.1109/ACCESS.2020.3033662
10. Bilal M, Rizwan M (2021) Integration of electric vehicle charging stations and capacitors in distribution systems with vehicle-to-grid facility. Energy Sour Part A: Recov Util Environ Effects 1–30. https://doi.org/10.1080/15567036.2021.1923870
11. Salkuti SR (2022) Emerging and advanced green energy technologies for sustainable and resilient future grid. Energies 15(18):6667. https://doi.org/10.3390/en15186667
12. Deilami S, Masoum AS, Moses PS, Masoum MAS (2011) Real time coordination of plug-in electric vehicle charging in smart grids to minimize power losses and improve voltage profile. IEEE Trans Smart Grid 2(3):456–467. https://doi.org/10.1109/TSG.2011.2159816
13. Das D (2004) Maximum loading and cost of energy loss of radial distribution feeders. Int J Electr Power Energy Syst 26(4):307–314. https://doi.org/10.1016/j.ijepes.2003.09.00
14. Miu KN, Chiang HD (2000) Electric distribution system load capability: problem formulation, solution algorithm, and numerical results. IEEE Trans Power Deliv 15(1):436–442. https://doi.org/10.1109/61.847286

15. Hung D, Mithulananthan N (2013) Multiple distributed generators placement in primary distribution networks for loss reduction. IEEE Trans Industr Electron 60(4):1700–1708. https://doi.org/10.1109/TIE.2011.2112316
16. Aman MM, Jasmon GB, Mokhlis H, Bakar AHA (2012) Optimal placement and sizing of a DG based on a new power stability index and line losses. Int J Electr Power Energy Syst 43(1):1296–1304. https://doi.org/10.1016/j.ijepes.2012.05.053
17. Kaur S, Kumbhar G, Sharma J (2014) A MINLP technique for optimal placement of multiple DG units in distribution systems. Int J Electr Power Energy Syst 63:609–617. https://doi.org/10.1016/j.ijepes.2014.06.023
18. Manjhi R, Lal DK, Biswal S (2022) Optimal allocation of DGs in radial distribution network for power loss minimization based on LSF and GJO algorithm, In: 2022 IEEE 2nd international symposium on sustainable energy, signal processing and cyber security (iSSSC), pp 1–6. https://doi.org/10.1109/iSSSC56467.2022.10051616
19. Davarzani S, Granell R, Taylor GA, Pisica I (2019) Implementation of a novel multi-agent system for demand response management in low-voltage distribution networks. Appl Energy 253:113516. https://doi.org/10.1016/j.apenergy.2019.113516
20. Sravanthi P, Sandeep V (2022) Solving realistic reactive power market clearing problem of wind-thermal power system with system security. Int J Emerg Electr Power Syst 23(2):125–144. https://doi.org/10.1515/ijeeps-2021-0060
21. Grainger JJ, Civanlar S (1985) Volt/var control on distribution systems with lateral branches using shunt capacitors as Voltage regulators-part I, II and III. IEEE Trans Power Apparat Syst 104(11):3278–3297. https://doi.org/10.1109/TPAS.1985.318842
22. Jung CM, Pagidipala S, Salkuti SR (2024) Emerging technologies for the integration of renewable energy, energy storage and electric vehicles. In: Salkuti SR (eds) Energy and environmental aspects of emerging technologies for smart grid. Green energy and technology. Springer, Cham. https://doi.org/10.1007/978-3-031-18389-8_5
23. Manjhi R, Lal DK, Biswal S (2022) LSF and Golden Jackal optimization algorithm based optimal placement and sizing of capacitors in distribution system. In: 2022 IEEE 2nd international symposium on sustainable energy, signal processing and cyber security (iSSSC), pp 1–6. https://doi.org/10.1109/iSSSC56467.2022.10051298
24. Manjhi R, Lal DK, Biswal S, Abdelaziz AY (2024) Technical indices evaluation of radial distribution network with optimally allocated DGs and capacitors using golden jackal optimization algorithm. ECTI Trans Electr Eng Electron Commun 22(2). https://doi.org/10.37936/ecti-eec.2024222.250677
25. Abdelaziz AY, Ali ES, AbdElazim SM (2016) Flower pollination algorithm and loss sensitivity factors for optimal sizing and placement of capacitors in radial distribution systems. Int J Electr Power Energy Syst 78:207–214. https://doi.org/10.1016/j.ijepes.2015.11.059
26. Salkuti SR (2023) Advanced technologies for energy storage and electric vehicles. Energies 16(5):2312. https://doi.org/10.3390/en16052312
27. Sambaiah KS, Jayabarathi T (2019) Optimal allocation of renewable distributed generation and capacitor banks in distribution systems using Salp Swarm algorithm. Int J Renew Energy Res 9(1):96–107. https://doi.org/10.20508/ijrer.v9i1.8581.g7567
28. Mohanty PK, Lal DK (2021) Voltage stability index and butterfly optimization algorithm-based DG placement and sizing in electrical power distribution system. In: Green technology for smart city and society. Springer, Berlin. pp 311–326. https://doi.org/10.1007/978-981-15-8218-9_27
29. Lakshmi GVN, Laxmi AJ, Veeramsetty V, Salkuti SR (2022) Optimal placement of distributed generation based on power quality improvement using self-adaptive Lévy Flight Jaya algorithm. Clean Technol 4(4):1242–1254. https://doi.org/10.3390/cleantechnol4040076
30. Deeum S, Charoenchan T, Janjamraj N, Romphochai S, Baum S, Ohgaki H, Mithulananthan N, Bhumkittipich K (2023) Optimal placement of electric vehicle charging stations in an active distribution grid with photovoltaic and battery energy storage system integration. Energies 16(22):7628. https://doi.org/10.3390/en16227628

31. Yong JY, Ramachandaramurthy VK, Tan KM, Mithulananthan N (2015) Bi-directional electric vehicle fast charging station with novel reactive power compensation for voltage regulation. Int J Electr Power Energy Syst 64:300–310. https://doi.org/10.1016/j.ijepes.2014.07.025
32. Manjhi R, Lal DK, Biswal S, Salkuti SR (2024) Optimal location selection of electric vehicle charging stations and capacitors in radial distribution networks using GJO algorithm. In: Energy and environmental aspects of emerging technologies for smart grid, pp 711–731. Springer International Publishing, Cham. https://doi.org/10.1007/978-3-031-18389-8_31
33. Yuvaraj T, Arun S, Suresh TD, Thirumalai M (2024) Minimizing the impact of electric vehicle charging station with distributed generation and distribution static synchronous compensator using PSR index and Spotted Hyena Optimizer algorithm on the radial distribution system. e-Prime-Adv Electr Eng Electron Energy 8:100587. https://doi.org/10.1016/j.prime.2024.100587
34. Bilal M, Rizwan M, Alsaidan I, Almasoudi FM (2021) AI-based approach for optimal placement of EVCS and DG with reliability analysis. IEEE Access 9:154204–154224. https://doi.org/10.1109/ACCESS.2021.3125135
35. Liu L, Zhang Y, Da C, Huang Z, Wang M (2020) Optimal allocation of distributed generation and electric vehicle charging stations based on intelligent algorithm and bi-level programming. Int Trans Electr Energy Syst 30(6):12366. https://doi.org/10.1002/2050-7038.12366
36. Mehroliya S, Arya A (2024) Optimal planning of power distribution system employing electric vehicle charging stations and distributed generators using metaheuristic algorithm. Electr Eng 106:1–17. https://doi.org/10.1007/s00202-023-02198-3
37. Balu K, Mukherjee V (2023) Optimal allocation of electric vehicle charging stations and renewable distributed generation with battery energy storage in radial distribution system considering time sequence characteristics of generation and load demand. J Energy Storage 59:106533. https://doi.org/10.1016/j.est.2022.106533
38. Chopra N, Ansari MM (2022) Golden jackal optimization: a novel nature-inspired optimizer for engineering applications. Expert Syst Appl 198:116924. https://doi.org/10.1016/j.eswa.2022.116924
39. Teng JH (2003) A direct approach for distribution system load flow solutions. IEEE Trans Power Delivery 18(3):882–887. https://doi.org/10.1109/TPWRD.2003.813818
40. Gupta S, Kumar N, Srivastava L (2021) An efficient Jaya algorithm with Powell's Pattern Search for optimal power flow incorporating distributed generation. Energy Sour Part B 16(8):759–786. https://doi.org/10.1080/15567249.2021.1942595
41. Narang N, Dhillon JS, Kothari DP (2012) Multiobjective fixed head hydrothermal scheduling using integrated predator-prey optimization and Powell search method. Energy 47(1):237–252. https://doi.org/10.1016/j.energy.2012.09.004
42. Kashem MA, Ganapathy V, Jasmon GB, Buhari MI (2000) A novel method for loss minimization in distribution networks. In: DRPT 2000 proceedings (Cat. No. 00EX382) IEEE international conference on electric utility deregulation and restructuring and power technologies, pp 251–256. https://doi.org/10.1109/DRPT.2000.855672
43. Das D (2008) Optimal placement of capacitors in radial distribution system using Fuzzy-GA method. Int J Electr Power Energy Syst 30:361–367. https://doi.org/10.1016/j.ijepes.2007.08.004

Optimizing Renewable Energy Integration and Electric Vehicle Charging Stations in Distribution Networks Using Hybrid Whale Optimization Algorithm

Varun Krishna Paravasthu, Balasubbareddy Mallala, B. Mangu, and Surender Reddy Salkuti

Abstract This work introduces a methodology for integrating renewable energy sources (RESs) and electric vehicle charging stations (EVCSs) into radial distribution networks, leveraging a novel Hybrid Whale Optimization Algorithm (HWOA). The efficacy of this algorithm is thoroughly examined through the analysis of IEEE 33 and IEEE 69 bus radial distribution systems (RDSs), which serve as test cases to evaluate its performance and suitability for real-world applications. The HWOA is specifically tailored to tackle the intricate optimization tasks associated with RESs and EVCSs integration, with the primary objectives of minimizing power losses, improving voltage stability, and guaranteeing adequate coverage of EVCSs across the distribution network. In pursuit of these objectives, the algorithm synergistically integrates the Genetic Whale Optimization Algorithm (GWOA) and Thermal Exchange Optimization (TEO) techniques, harnessing their capabilities to optimize the placement and sizing of RESs and EVCSs components. Extensive analysis validates the proposed algorithm's impressive efficacy in achieving its intended outcomes.

V. K. Paravasthu
Department of Electrical and Electronics Engineering, ACE Engineering College (Autonomous), Ghatkesar, Hyderabad, Telangana, India

B. Mallala
Department of Electrical and Electronics Engineering, CBIT, Hyderabad, Telangana, India
e-mail: balasubbareddy_eee@cbit.ac.in

B. Mangu
Department of Electrical Engineering, University College of Engineering, Osmania University, Hyderabad, Telangana, India
e-mail: mangu.b@uceou.edu

S. R. Salkuti (✉)
Department of Railroad and Electrical Engineering, Woosong University, Daejeon, Republic of Korea
e-mail: surender@wsu.ac.kr

S. R. Salkuti (ed.), *Artificial Intelligence for Integrated Smart Energy Systems in Electric Vehicles*, Lecture Notes in Electrical Engineering 1427,
https://doi.org/10.1007/978-3-031-94276-1_6

Through strategic placement of RESs at optimal sites and determination of appropriate capacities, the HWOA efficiently mitigates power losses while concurrently enhancing the voltage profile of the RDSs. The algorithm identifies optimal positions for EVCS units, guaranteeing comprehensive coverage to address the growing demand for electric vehicle charging stations. Results obtained through applying the HWOA underscore its capacity to substantially improve the overall performance and dependability of RDSs. By incorporating multiple optimization objectives and leveraging sophisticated optimization techniques, the proposed algorithm presents a promising solution for seamlessly integrating RESs and EVCSs, thereby facilitating the evolution of more sustainable and resilient energy systems. This research represents a significant contribution to the advancement of distribution system optimization, laying the foundation for the practical implementation of renewable energy and electric vehicle technologies within modern power grids.

Keywords Renewable energy sources · Electric vehicle fast charging stations · Hybrid whale optimization algorithm · Radial distribution network · Optimization techniques · Power loss reduction

Nomenclature

EVs	Electric vehicles
RES	Renewable energy sources
EVCS	Electric vehicle fast charging stations
HWOA	Hybrid whale optimization algorithm
TLBO	Teaching–learning-based optimization
ILP	Integer linear programming
GA	Genetic algorithm
GA-PSO	Genetic algorithm-particle swarm optimization
PDF	Probability density function
BWPO	Binary waterwheel plant optimization
AGWO	Adaptive grey wolf optimizer

1 Introduction

Traditional road transportation systems significantly contribute to greenhouse gas emissions [1]. Electric vehicles (EVs) have emerged as a promising alternative to Internal Combustion Engine (ICE) vehicles in addressing this environmental challenge.

However, the increase in EVs necessitates the establishment of adequate charging infrastructure along travel routes to alleviate concerns about driving range [2].

To address this need, numerous mathematical models have been proposed for the optimal placement of charging infrastructure [3–5]. These models typically incorporate multiple objectives, including minimizing investment costs, reducing energy losses within the distribution network, and improving voltage profiles to ensure efficient operation. The objective of these models is to strategically position charging stations along road networks in a manner that maximizes their utility and effectiveness. By minimizing investment costs, the models aim to optimize the allocation of resources while ensuring widespread accessibility to charging facilities. Additionally, by reducing energy losses within the distribution network, the models seek to improve the efficiency of the charging infrastructure [6]. Enhancing voltage profiles is vital for maintaining the stability and reliability of the distribution system. These models prioritize voltage profile improvement as an objective to minimize fluctuations and ensure a steady power supply to EV charging stations [7].

The Netherlands currently leads globally in charging station density, with approximately 19–20 stations per 100 km of road network [8]. China closely follows as the second-best performer, providing around 3–4 charging stations per 100 km.

Predictions suggest that the UK will reach around 3 EVCSs per 100 km by 2030. Low or near-zero deployment levels of charging stations hinder the widespread assumption of EVs. To encourage greater uptake of EVs among both private and public vehicle users, there is a critical need to expand the charging infrastructure network [9]. Achieving this expansion is vital to addressing the service distribution effectively. The service distribution revolves around optimizing the number and placement of charging stations within the network while adhering to specific constraints. Strategically identifying the best locations and quantities of charging stations encourages EV adoption by providing convenient access to charging facilities for EV drivers [10].

The challenge of charging station deployment is approached within a multi-objective framework, considering various factors such as distribution network constraints, and operational considerations. These constraints encompass critical aspects like voltage strength, network reliability, power loss minimization, charging station accessibility, and waiting times at EVCSs. To tackle this complex problem, two optimization algorithms, Teaching–Learning-Based Optimization (TLBO) and Chicken Swarm Optimization (CSO), are applied in [11]. These algorithms optimize the best locations for EVCSs in the distribution network. They handle the sizing and placement of both RESs and EVCSs simultaneously. The utilization of TLBO and CSO enables the optimization process to efficiently handle the complexities involved in determining the optimal configuration of charging infrastructure. By combining these algorithms, the placement of charging stations can be optimized to ensure effective network operation while minimizing costs and meeting the diverse needs of EV drivers. However, TLBO and CSO algorithms typically require careful parameter tuning to achieve optimal performance. Choosing suitable parameter values can be difficult and may demand extensive experimentation, rendering these algorithms less user-friendly and more resource-intensive than other optimization techniques.

In [12], a multi-objective optimization problem is framed to achieve several goals, including reducing power losses, minimizing voltage fluctuations, and optimizing demand. To address this complex problem, a novel optimization approach called the genetic algorithm-particle swarm optimization (GA-PSO) technique is proposed. The efficacy of the GA-PSO technique is evaluated through a case study conducted on an IEEE 33-bus system. This study aims to validate the efficiency of the GA-PSO method in concurrently optimizing the placement and capacity of RESs and EVCSs. GA-PSO optimization may face scalability issues when applied to large-scale optimization problems with numerous finding variables and constraints. As the problem size increases, the computational complexity of the algorithm grows, resulting in longer convergence times and increased memory requirements. In [13], a method based on integer linear programming (ILP) for both placement and sizing of EVCSs is proposed, leveraging a grid partitioning approach. This technique involves dividing the distribution network into grid cells and then selecting the most suitable grid cells for EVCSs placement based on predefined criteria such as proximity to demand centers, accessibility, and available infrastructure. The sizing of EVCSs is optimized to ensure adequate charging capacity to meet the demand for EVs in each selected grid cell. The grid partitioning approach may lack flexibility in adapting to dynamic changes in EV demand patterns or network conditions.

In [14] presented a new method for optimizing the positioning of charging stations and allocating charging piles within an EV charging infrastructure network. The optimization process involves multiple criteria, aiming to find the best locations for charging stations while considering various constraints like recharging demand and cruising range limits. To tackle this optimization problem, the study utilizes two different algorithms: Lazy Greedy with Direct Gain (LGDG) and Lazy Greedy with Effective Gain (LGEG). Both algorithms are rooted in the greedy method, which entails iteratively selecting the best available option at each step based on certain criteria. The performance of greedy algorithms like LGDG and LGEG can be highly sensitive to the initial conditions and the order in which decisions are made. Small variations in the input parameters or starting configurations may lead to significantly different outcomes, making these algorithms less robust and reliable in certain contexts. The methodology in [15] addresses the EV charging station allocation problem by developing a strategy to identify optimal charging station locations and determine the necessary number of stations to meet charging demands effectively. This comprehensive approach aims to optimize both the spatial distribution and quantity of charging stations within a given area. To achieve this, the method relies on Genetic Algorithms (GAs), a heuristic optimization technique learned by the process of biological adaptation and inheritance. GA is employed to estimate the optimal longitudes and latitudes for the placement of EVCS. By leveraging the evolutionary principles of selection, crossover, and mutation, GAs iteratively refine candidate solutions to converge toward an optimal or near-optimal solution. Genetic Algorithms may struggle to explore the entire solution space thoroughly, particularly in high-dimensional or constrained optimization problems. This limitation can result in suboptimal solutions, especially if the algorithm becomes trapped in local optima.

Sustainable sources, such as solar panels or wind turbines, can add EV charging infrastructure by providing renewable energy to power charging stations. Improving the placement and position of both DG units and charging stations ensures efficient energy management and reduces reliance on centralized power sources. RES units can offer grid support services, such as voltage regulation and peak shaving, which can enhance the solidity and consistency of the electrical grid [16]. Co-optimizing DG placement with EV charging station locations enables the strategic deployment of DG units to mitigate grid congestion and improve overall grid performance. By incorporating RES optimization into EV charging station planning, renewable energy sources can be leveraged to charge electric vehicles, reducing greenhouse gas emissions, and promoting sustainable transportation. This integration aligns with the transition towards clean energy and supports environmental sustainability goals [17].

This chapter addresses the existing gaps in the optimization of EVCSs and RESs placement by developing a novel optimization model using a hybrid whale optimization algorithm. The model aims to obtain the best locations and capacities for RESs and EVCSs within a specified area to effectively meet the dispersed demand for EV charging. It uses a facility-to-site approach, offering versatility in EVCSs placement. The optimization problem is framed as a tri-objective model with three main goals: reducing overall losses, enhancing voltage stability, and maximizing EV charging station coverage. The model considers voltage and capacity constraints imposed on the charging stations. The objectives of the proposed work in this chapter are as follows:

- Develop an innovative Hybrid Whale Optimization Algorithm (HWOA) for the efficient combination of RESs and EVCSs into RDSs.
- Combine Genetic Whale Optimization Algorithm (GWOA) and Thermal Exchange Optimization (TEO) techniques within the HWOA framework to address the complex optimization challenges associated with RESs and EVCSs integration.
- Evaluate the use of the proposed HWOA through extensive analysis and experimentation on IEEE 33 and 69 bus systems.
- Optimize the placement and sizing of RESs and EVCSs elements using the HWOA to minimize power losses, increase voltage stability, and confirm sufficient coverage of EVCSs infrastructure.

2 The System Model

This study concentrates on integrating RESs and EVCSs into the distribution network.

2.1 Solar Energy Model

The solar system transfers sunlight into electrical energy by various approaches. It consists of multiple cells arranged in series to generate the necessary voltage. The electricity generation is influenced by irradiance and temperature, necessitating consideration of both factors when assessing the solar generation's performance. The Normal Operating Cell Temperature (NOCT) can be defined as follows [18].

$$T_c = T_a + \frac{NOCT - 20^\circ}{0.8} G \quad (1)$$

The power generated by the cells can be expressed as follows:

$$P_c = P \times [\eta \times (T_c - 25^\circ)] \quad (2)$$

The power generated by the PV panel can be given as,

$$P_{panel} = P_c \times n_{cell} \quad (3)$$

The daily outlines of temperature and solar radiation used in the evaluation are in Fig. 1 [19].

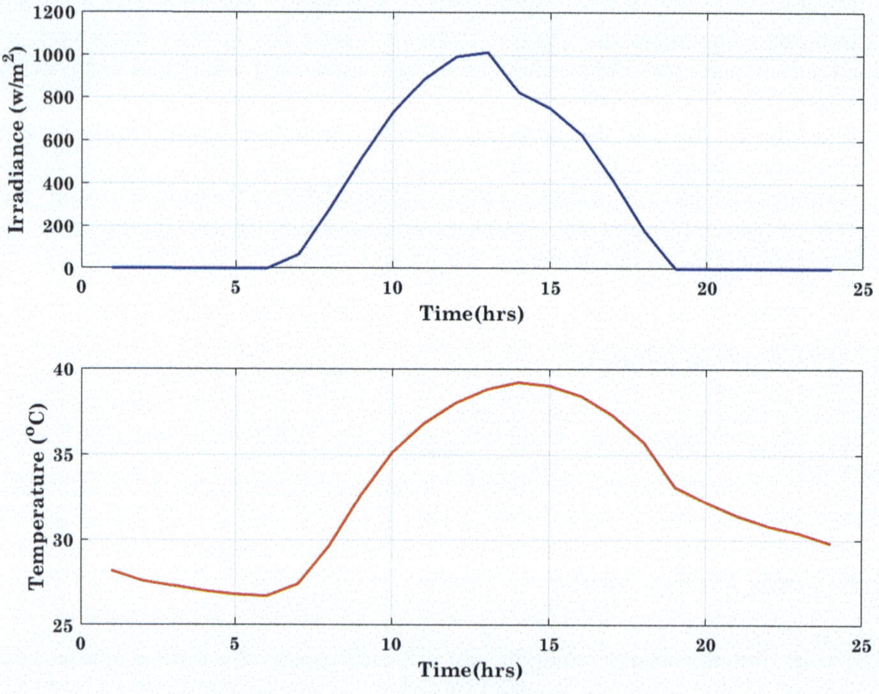

Fig. 1 Irradiance in w/m^2 and temperature in °C for 24 h of a day

2.2 Wind Turbine (WT) Model

The power output of the WT varies on factors such as direction and speed of the wind, topographical site of the place, and wind density is given as,

$$P_{WT}(t) = \begin{cases} 0 & V > V_{cut-off} \text{ or } V < V_{cut-in} \\ p_{WT}^{r}\left(\frac{p_{WTO}-p_{WT}^{r}}{V_{cut-off}-V_{r}}\right)(V(t)-V_{r}) & V_{r} < V \leq V_{cut-off} \\ p_{WT}^{r}\left(\frac{-V_{cut-in}+V(t)}{-V_{cut-in}+V_{r}}\right)^{3} & V_{cut-in} \leq V \leq V_{cut-off} \end{cases} \quad (4)$$

where p_{WT}^{r} is the rated power of the WT, V_{cut-in}, V_{r}, $V_{cut-off}$ are the cut-in, rated, and cut-off speeds of the turbine, V is the wind speed, and p_{WTO} is the WT power at the cut-off speed. Power generated from wind turbines is given as,

$$P_{Windtotal} = n_{WT} \times P_{WT} \quad (5)$$

where n_{WT} denotes the number of wind turbines. The daily wind speed pattern is depicted in Fig. 2 [19].

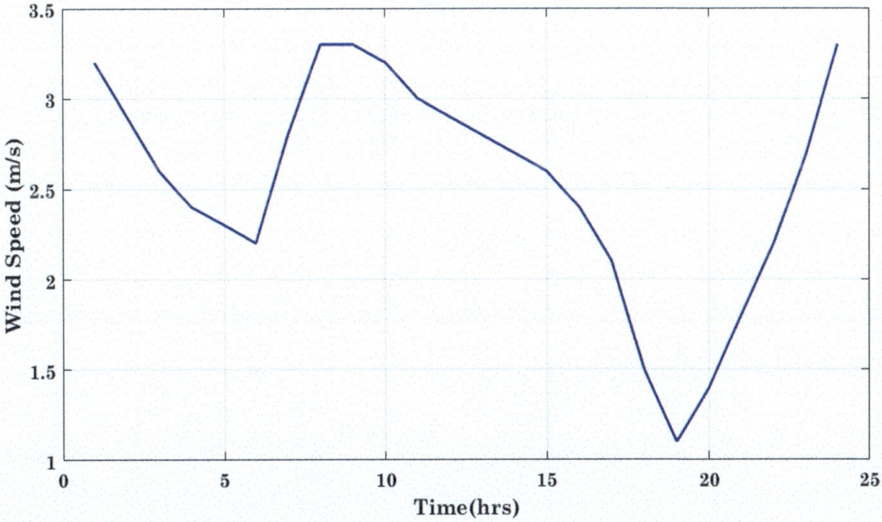

Fig. 2 Variation of wind speed for 24 h of a day

2.3 Model of an Electric Vehicle

Energy consumption per kilometer, mileage per day, and charging and waiting time are the factors that should be considered while modeling EVCSs. The expected mileage per day can be modeled using a lognormal distribution [20]. The probability density function (PDF) of the lognormal distribution can be calculated as follows:

$$f(x) = \frac{1}{\sqrt{2\pi\sigma x}} \exp\left(-\frac{(\ln(x) - \mu)^2}{2\sigma^2}\right), \ x > 0 \tag{6}$$

In this context, 'x' represents a random number with a zero mean and a unit variance. 'μ' is a locative parameter, and 'σ' is the scaling parameter which is determined as,

$$\mu = \ln\left(\frac{m}{\sqrt{1+\frac{v}{m^2}}}\right)\sigma = \sqrt{\ln\left(1 + \frac{v}{m^2}\right)} \tag{7}$$

where 'm' represents the standard deviation, and 'v' denotes the mean derived from historical data. The anticipated daily is given as,

$$M_d = em^{\left(\mu_m+\sigma_m\times\sqrt{-2\times\ln c_1}\times\cos(2\pi c_2)\right)} \tag{8}$$

where μ_m is the lognormal probability and σ_m is scaling probability, and c_1 and c_2 are random variables that follow the normal distribution; μ_m and σ_m are calculated as,

$$\mu_m = \ln\left(\frac{\mu_{md}^2}{\sqrt{\mu_{md}^2+\sigma_{md}^2}}\right)\sigma_m = \ln\left(1 + \frac{\sigma_{md}^2}{\mu_{md}^2}\right) \tag{9}$$

μ_m is standard deviation and σ_m is the mean of EV mileage data. The consumed energy per mile, a crucial parameter in EV modeling, can be computed as [21],

$$E_m = \propto \times K_{EV}^b \tag{10}$$

The maximum mileage of a fully charged EV is given as,

$$M_{dMax} = \frac{B_{Cap}}{E_m} \tag{11}$$

Demand for charging by an EV is given as,

$$E_d = \begin{cases} B_{Cap} & M_d \geq M_{dMax} \\ M_d \times E_m & M_d < M_{dMax} \end{cases} \tag{12}$$

The waiting period by EV is estimated by a Gaussian distribution as follows:

$$t_a = \mu_a + \sigma_a.x_1,\ t_d = \mu_d + \sigma_d.x_2 \tag{13}$$

$$t_{dur} = t_d - t_a \tag{14}$$

Arrival time is t_a, departure time is t_d and time for charging is t_{dur}, μ_a is standard deviation x_1 and x_2 are random values. The desired state of charge $SOC^{desired}$ of the EV battery is given as,

$$SOC^{desired} = min\left\{\left(SOC^{init} + \frac{E_d}{B_{Cap}}\right), \left(SOC^{init} + \frac{t_{dur}}{B_{Cap}}.r_{ch}\right)\right\} \tag{15}$$

3 Objective Function

This section aims to define the objective function and constraints necessary for the allocation and management of RES and EVCS. The objectives encompass minimizing overall power loss in the network while mitigating voltage violations. Related constraints involve maintaining supply–demand equilibrium, adhering to bus voltage limits, respecting generation limits, and addressing EV-related constraints.

3.1 Electrical Distribution Power Loss

The first part of the objective function is given by,

$$f_1 = Minimize\ P_{loss} \tag{16}$$

$$P_{loss} = \sum_{t=1}^{24} \sum_{i=1}^{n_d} \sum_{j>1}^{n_d} y_{ij}\left(V_{i,t}^2 + V_{i,t}^2 + 2V_{i,t}V_{j,t}\cos\left(\delta_{i,t} - \delta_{j,t}\right)\right) \tag{17}$$

The number of branches is n_d, magnitude and angle of the voltage at ith bus tth time are $V_{i,t}$ and $\delta_{i,t}$, respectively, and at jth bus tth time are $V_{j,t}$ and $\delta_{j,t}$. Reducing network losses is a key objective from the perspective of the distribution system operator.

3.2 Voltage Irregularity Index

The second objective is to reduce network voltage violations. RESs and EVCSs in the grid can raise the supply capacity, thereby reducing losses and improving voltage stability. This objective can be formulated as follows:

$$f_2 = Minimize \sum_{t=1}^{24} \sum_{i=1}^{n_d} \left|1 - V_{i,t}\right| \tag{18}$$

3.3 Maximizing EV Charging Station Accessibility

Maximizing coverage ensures that EV drivers have an approach to the charging station within a specified maximum distance, regardless of their location in the traffic network. The problem of positioning EV charging stations is akin to a facility location problem. In the traditional facility location problem, service demand is assumed to be fixed and located at specific points within a traffic network. The primary factors influencing placement decisions include the weighted distance or time required for travel to these facilities and the distance or time it takes for the farthest user to reach them. These parameters are essential control measures for determining optimal placement strategies. The mathematical model is formulated to determine the optimal locations of facilities to maximize coverage:

$$f_3 = \sum_{i \in I} c_i s_i \tag{19}$$

Constrained by the criteria,

$$\sum_{j \in D_i} \alpha_j \geq \beta_i \forall i \in I \tag{20}$$

$$\sum_{j \in J} \alpha_j = \gamma \tag{21}$$

$$\alpha_j = \begin{cases} 1, & if\ a\ charging\ station\ is\ assigned\ to\ area\ j \\ 0, & else \end{cases} \tag{22}$$

$$\beta_i = \begin{cases} 1, & if\ one\ or\ more\ charging\ stations\ are\ assigned\ at\ areas\ in\ the\ set\ N_i \\ 0, & otherwise \end{cases} \tag{23}$$

3.4 Constraints

The formulated problem considers five constraints: supply–demand balance, limits on bus voltage, thermal limits, generation limits, and constraints related to EVs.

3.4.1 Supply–Demand Balance

This constraint ensures that the supplied power at each bus equals the demand power plus the power losses of the branches connected to that bus. It can be expressed as follows:

$$P_{gi,t} - P_{di,t} = P_{chi,t} + |V_{i,t}| \sum_{j=1}^{n_d} |y_{i,j}||V_{j,t}| \cos(\delta_{i,t} - \delta_{j,t} - \theta_{i,j}) \quad (24)$$

$$Q_{gi,t} - Q_{di,t} = |V_{i,t}| \sum_{j=1}^{n_d} |y_{i,j}||V_{j,t}| \sin(\delta_{i,t} - \delta_{j,t} - \theta_{i,j}) \quad (25)$$

where $P_{gi,t}$, $P_{di,t}$, $P_{chi,t}$ are the generated, demand, and EV-charged active powers at bus i during time t, individually, $Q_{gi,t}$ and $Q_{di,t}$ are caused and claim reactive powers at bus i during time t, separately, and $\theta_{i,j}$ is the angle of $y_{i,j}$.

3.4.2 Bus Voltage Constraint

When integrating charging stations and RESs, it's imperative to maintain the bus voltage within its normal operating limits, which can be expressed as follows:

$$V^{min} \leq V_{i,t} \leq V^{min} \quad (26)$$

$$\delta^{min} \leq \delta_{i,t} \leq \delta^{min} \quad (27)$$

where min and max indicate minimum and maximum values.

3.4.3 Thermal Constraint

The introduction of EVs into the grid elevates power flow through transmission lines, consequently leading to heightened temperatures in these lines. It's crucial to ensure that the power flow remains within acceptable limits, which can be articulated as follows:

$$|S_{i,t}| \leq |S_i^{max}|, i = 1, 2, \ldots, n_d \quad (28)$$

$S_{i,t}$ is the power flow in line i at time t, while S_i^{max} is the maximum suitable flow in line i.

3.4.4 Generation Limit

The generated power from RESs must be within its normal limits, stated as follows:

$$P_{RES}^{min} \leq P_{RESi,t} \leq P_{RES}^{max} \tag{29}$$

3.4.5 EV Constraint

The power demanded by the EVs must fall within predefined minimum and maximum thresholds, which can be expressed as follows:

$$P_{EV}^{min} \leq P_{EVi,t} \leq P_{EV}^{max} \tag{30}$$

4 Optimization Algorithm

4.1 *Whale Optimization Algorithm*

The WOA imitates the chasing behavior of humpback whales, comprising two primary stages: exploitation and exploration.

During the first stage, the process simulates the behavior of whales such as surrounding prey, and employs the bubble-net intrusion technique. In the second stage, it involves searching for a target.

4.1.1 Exploitation Phases

Equations (31) and (32) in the algorithm imitate the behavior of humpback whales surrounding the prey. In this stage, the best solution serves as the target prey, and the remaining solutions aim to converge towards it.

$$\vec{u} = \left| \vec{\beta} . \overrightarrow{x_b(t)} - \overrightarrow{x(t)} \right| \tag{31}$$

$$\vec{x}(t+1) = \left| \overrightarrow{x_b(t)} - \vec{\alpha} . \vec{u} \right| \tag{32}$$

t is the present iteration, $\overrightarrow{x_b}$ is the current best solution, and $\overrightarrow{\alpha}$ and $\overrightarrow{\beta}$ are determined as,

$$\overrightarrow{\alpha} = \overrightarrow{2a}.\overrightarrow{r} - \overrightarrow{a} \tag{33}$$

$$\overrightarrow{\beta} = 2\overrightarrow{r} \tag{34}$$

The algorithm involves a linear decrease of the coefficient. $\overrightarrow{a}$, which goes from 2 to 0 as the iterations progress. $\overrightarrow{r}$ is a random vector with components uniformly distributed between 0 and 1. The adjustment of the present position concerning the current best is regulated by modifying the values of the $\overrightarrow{\alpha}$ and $\overrightarrow{\beta}$.

To replicate the bubble-net attacking behavior of humpback whales, two mathematical models are proposed:

- Reducing the area of the surrounding: In this model, the value of the coefficient $\overrightarrow{a}$ gradually decreases linearly from 2 to 0 using $\overrightarrow{r}$ and the $\overrightarrow{\alpha}$ throughout the iterations.
- Circular position update: In this process, the algorithm initially computes the distance relating the whale and its prey. Subsequently, the humpback whale performs a surrounding maneuver around the prey in a logarithmic spiral motion. The numerical representation of this motion is as follows:

$$\overrightarrow{x}(t+1) = \overrightarrow{u'} \times e^{\gamma\psi} \times \cos(2\pi\psi) + \overrightarrow{x_b(t)} \tag{35}$$

where $\overrightarrow{u'} = \overrightarrow{x_b(t)} - \overrightarrow{x(t)}$ is the distance between the prey and the humpback whale, γ is a constant determining the shape of the logarithmic spiral, and ψ is a random value chosen from the range (− 1, 1). In the operation phase, once the location of the prey is identified, the humpback whale dives and initiates the formation of spiral bubbles around the prey while ascending towards the surface. During this process, the humpback whale progressively narrows its movement into a circular path while simultaneously pursuing the prey along a spiral trajectory. This hunting behavior indicates that both the reducing circle and the circular path have an equal likelihood of being implemented when updating the position of the humpback whale during the iteration process.

4.1.2 Exploration Phase

During this phase, the WOA takes to push the solution away from the current best solution and explores the search space randomly. To achieve this, the algorithm utilizes a random vector ($\overrightarrow{\alpha}$) with a range of [+ 1 and − 1] and randomly picks a reference solution. This process, combined with the condition $|\overrightarrow{\alpha}| > 1$, facilitates global exploration and is given as,

$$\overrightarrow{u} = \left| \overrightarrow{\beta} .\overrightarrow{x_{rand}} - \overrightarrow{x} \right| \tag{36}$$

$$\overrightarrow{x}(t+1) = \overrightarrow{x_{rand}} - \overrightarrow{\propto}.\overrightarrow{u} \tag{37}$$

where $\overrightarrow{x}_{rand}$ denotes the position vector of the solution randomly selected from the current population.

4.1.3 Overview of WOA

In the WOA algorithm, the adaptive change of the coefficient $\overrightarrow{\propto}$ enables smooth move among exploration and exploitation phases: exploration occurs when the magnitude of $|\overrightarrow{\propto}| > 1$, while exploitation occurs when the magnitude of $|\overrightarrow{\propto}| < 1$. In the exploitation phase, WOA can switch relating the shrinking circle and spiral-shaped path strategies. The WOA algorithm primarily relies on two internal parameters $\overrightarrow{\propto}$ and $\overrightarrow{\beta}$, which require adjustment.

4.2 *Thermal Exchange Optimization (TEO)*

The TEO algorithm is inspired by Newton's law of cooling, which asserts that the rate of heat loss of an object is proportional to the temperature difference between the object and its surroundings. In the TEO algorithm, specific solutions are assigned as cooling objects, while others denote the environment. The updated environmental temperature, adjusted according to the previous temperature of the solution, is defined as follows:

$$T_i^{env} = \left(1 - (c_1 + c_2 \times (1-t) \times Rand) \times T_i^{'env}\right) \tag{38}$$

$$t = \frac{Iter}{MaxIter} \tag{39}$$

In Eq. (39), c_1 *and* c_2 are the control variables, $T_i^{'env}$ represents The temperature update equation for each solution is expressed as follows, based on the previous temperature of the solution T_i^{env}, the current iteration (Iter), and the maximum number of iterations (*MaxIter*):

$$T_i^{new} = T_i^{env} + \left(T_i^{old} - T_i^{env}\right) \exp^{(-\varepsilon t)} \tag{40}$$

$$\varepsilon = \frac{\cos(object)}{\cos(worstobject)} \tag{41}$$

In the equation above, when an object has a lower ε, to alter the capability to escape local optima, a parameter Pro is introduced. This parameter decides whether a component of each cooling object should undergo alteration or remain unchanged. If $Ran(i)$ $(i = 1, 2, \ldots, n) < Pro$, one dimension of the ith object is randomly selected, and its value is revived as follows:

$$T_{i,j} = T_{j,min} + Rand\left(T_{j,max} - T_{j,min}\right) \tag{42}$$

In Eq. (42), $T_{i,j}$ denotes the jth variable of the ith solution. $T_{j,max}$ *and* $T_{j,min}$ refer to the upper and lower bounds of the jth variable, individually.

4.3 Crossover Operator

Genetic Algorithm (GA) is a popular evolutionary algorithm used for optimization. It employs various genetic operators such as selection, crossover, mutation, and replacement. Initially, two parent chromosomes are selected based on their fitness values. Then, the crossover operator is applied to these parents to produce offspring chromosomes. The mutation operator is applied to introduce localized changes in the offspring chromosomes. Finally, the replacement operator removes chromosomes with insufficient fitness values from the population. This mechanism ensures that the fittest chromosomes produce more offspring, aiding in the convergence towards the global optimum.

5 Genetic Whale Optimization Algorithm–Thermal Exchange Optimization for Global Optimization Problem

This study introduces five mechanisms to improve the WOA. By integrating aspects of GA, solutions with high potential are identified through crossover operations, resulting in more effective solutions, and enhancing the algorithm's global exploration capability. Additionally, the Thermal Exchange Optimization (TEO) algorithm's position update equation replaces the original equation, while its TM is integrated with the proposed algorithm to store historical-best-so-far solutions. Through these mechanisms, the TEO algorithm enhances the local exploitation capability of the WOA algorithm.

5.1 Generating a High-Quality Initial Population

In global optimization problems, having solutions in proximity to the region containing the global optimum can significantly enhance the performance of optimization algorithms. To achieve this, a mechanism is proposed that integrates a local search algorithm to generate high-quality initial populations. Initially, a larger initial population is generated to cover the search space comprehensively, followed by the selection of the best-performing solutions. This step aims to pinpoint potential areas containing the global optimum. Then, the position update equation of the TEO algorithm is used to explore nearby regions, increasing the chances of the population approaching the global optimum.

5.2 Improvement of the Exploitation Phase

In the basic WOA, two mathematical formulas are used in the local exploitation phase: the encircling prey formula and the spiral position updating formula. The circling prey formula in the basic WOA has a drawback: it only follows the best solution in the population. This means that if the best solution gets trapped in a local optimum, other solutions may also get stuck. To address this limitation and improve the algorithm's exploitation capabilities, enhancements are needed.

The proposed algorithm incorporates the Thermal Exchange Optimization (TEO) algorithm, known for its strong exploitation abilities, to address this issue. In this modified approach, the position update equation of the TEO algorithm replaces the prey-around formula used in the basic WOA.

5.3 The Proposed Algorithm GWOA-TEO

The algorithm includes Thermal Memory (TM) along with four improvements: generating a high-quality initial population, enhancing the exploitation phase, implementing crossover operations, and updating the best solution found so far. Algorithm 1 and Fig. 3 outline the pseudocode and flowchart, respectively. To prevent potential infinite loops and excessive computational costs, an Implementation Time (IT) parameter is introduced. In scenarios with low-dimensional problems, where the population range is limited, implementing the TM mechanism may take considerable time. Additionally, the size of the thermal memory TM impacts computational costs, as the algorithm checks whether solutions match the memory one by one during the process of escaping local optima. Hence, selecting appropriate values for IT and TM is crucial for optimizing computational efficiency. For all case studies in this research, we have set IT (Iterations) to 20 and the size of thermal memory (TM) to 10.

Pseudo code: the process of the HWOA

```
Initialize the population X_i(i = 1, 2, ..., n) of position and rating of RES and EVCS
Evaluate the fitness function for each one of the population using load flow analysis
Construct thermal memory TM_i(i = 1, 2, ..., L), L is the size of TM
Set the optimal solution X* in the population which provides minimum fitness value
while (t < T), where T is a maximum number of iterations
Set the leading solution LX* based on TM
for every solution
    Revise a, α and β
p is the probability of a reduction in the search area
if (p < 0.5)
  if (|α| < 1)
    Eq. (7) to update the current solution
  else if (|α| ≥ 1)
    Eq. (8) to update the current solution
  end if
else if (p ≥ 0.5)
      Eq. (6) to update the current solution
end if
for i < L
    if the current solution same with TM_i
      Implement the crossover operator mechanism
      times = times + 1
    end if
    if times > IT
  times = 0;
end if
end for
end for
Evaluate the fitness value of each solution
Revise TM if obtain the better solution
Revise LX* based on TM
Revise X* if X* is the better solution
t = t + 1
end while
return X*
```

6 Results and Discussion

The algorithm is subjected to analysis on both the IEEE 33 and IEEE 69 bus networks to optimize the placement of RES and EVCS, in addition to determining their respective sizes. This comprehensive examination delves into identifying the most effective locations for RES and EVCS within the power distribution networks, while also evaluating the optimal capacities required for their efficient operation.

HWOA is adopted for minimization of the fitness function with 100 iterations and a particle size of 50. MATLAB/Simulink tools are used for optimization and load

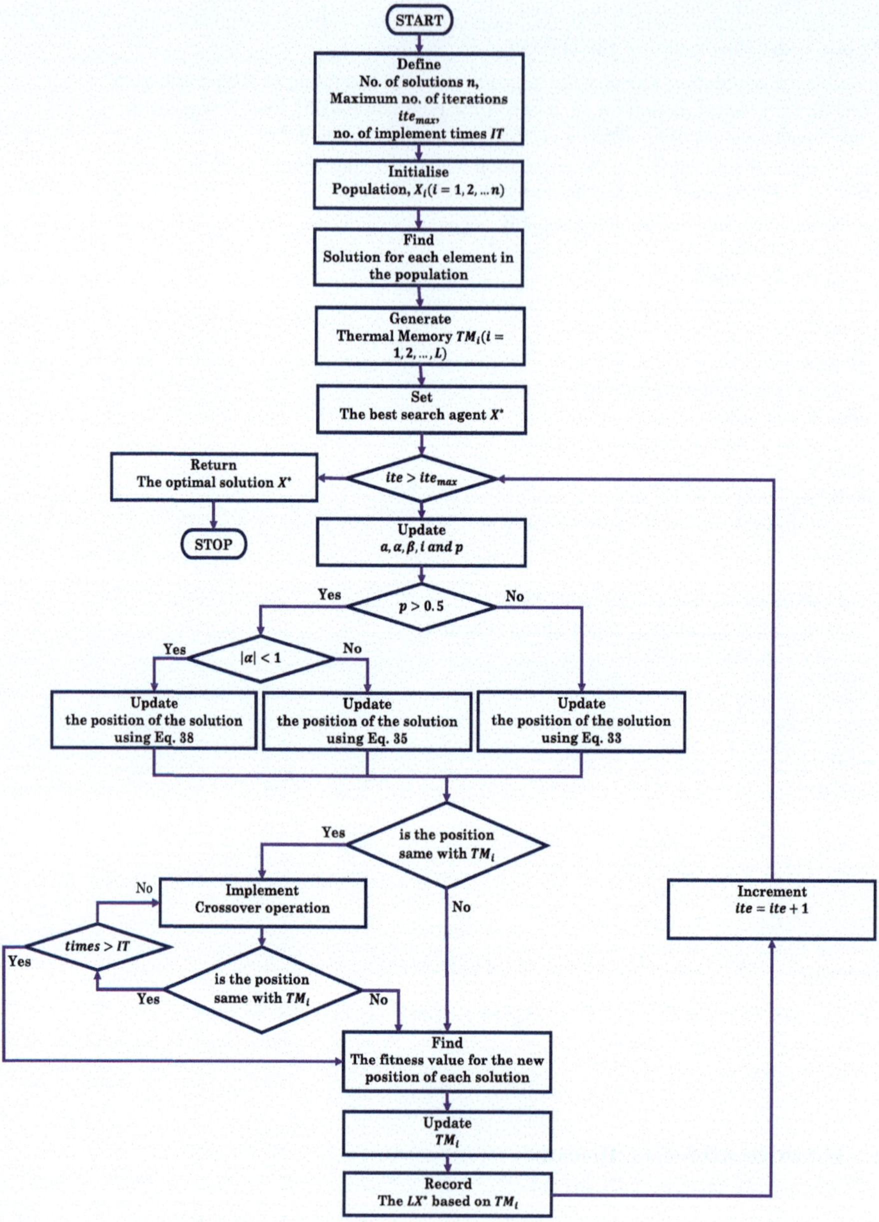

Fig. 3 Flow chart of HWOA

flow analysis using flow charts and pseudo code detailed in Sect. 3. Total network power loss of 24 h, voltage irregularity index, and traffic coverage of charging stations are considered in fitness function while optimizing the position and size of RES and EVCS. The load curve presented in Fig. 5 is considered to calculate the total power loss and voltage irregularity index for 24 h of 33 and 69 bus systems. The maximum possibility of generation for RES was set to 1800kW and the maximum rating of EVCS is assigned to 2000 kW. Three cases are considered to check the effectiveness of the proposed algorithm which are given as,

- Power loss minimization and coverage of EVCSs.
- Voltage irregularity index minimization and coverage of EVCSs.
- Power loss minimization, voltage irregularity index, and minimization and coverage of EVCSs.

6.1 IEEE 33 Bus Radial Distribution System

Figure 4 shows the single-line diagram of the IEEE 33-bus RDS comprising 33 nodes and 32 branches. The system uses a voltage of 12.66 kV with a total rated power of 100 MVA. The aggregate base load connected to the system amounts to 3.715 MW and 2.3 MVAR. Load demand over a 24-h period is graphically represented in Fig. 5. Details regarding bus-connected loads, branch resistance, and inductance values were obtained from [22]. Through load flow analysis of the 33-bus distribution system, the total active and reactive power losses for the base case were determined to be 3865.782 and 2584.639, respectively. The proposed Hybrid Whale Optimization Algorithm (HWOA) was implemented and subjected to comparative analysis alongside several other optimization methodologies. These methodologies include the Binary Waterwheel Plant Optimization (BWPO) [23], the Skill Optimization Algorithm (SOA) [24], the Adaptive Grey Wolf Optimizer (AGWO) [25], the Improved Whale Optimization Algorithm (IWOA) [26], the Harris Hawk Optimization (HHO) [27], Cuckoo search algorithm (CSA) [28] and Flower pollination algorithm (FPA) [29] all of which were fully programmed and executed using MATLAB/SIMULINK for comparison.

6.1.1 Case 1: Optimization Using the Proposed Algorithm Incorporates Both Power Loss Reduction and EVCSs Coverage Within the Fitness Function

A summary of the results obtained in these circumstances is presented in Table 1. With the proposed algorithm, the optimized positions for two RESs are identified as buses 13 and 30, with respective capacities of 935.68 and 922.67 kW. Additionally, for the two EVCS units, the best locations are determined to be buses 29 and 3, with ratings of 80.74 and 95.21 kW, respectively. The active and reactive power losses, over a 24-h period, associated with these optimized configurations are measured at

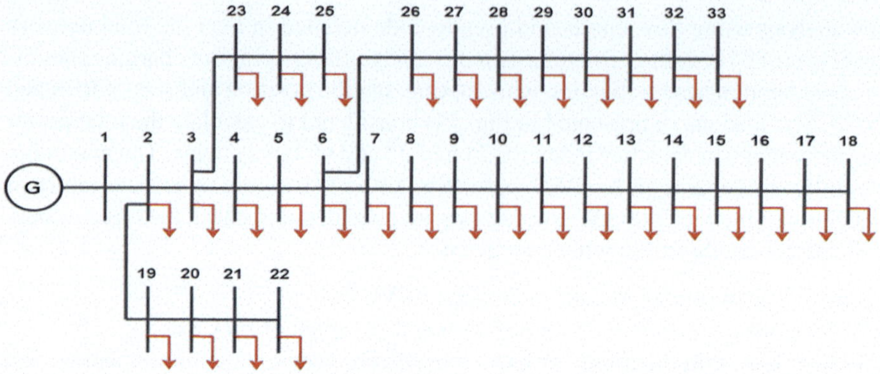

Fig. 4 IEEE-33 bus network single line diagram

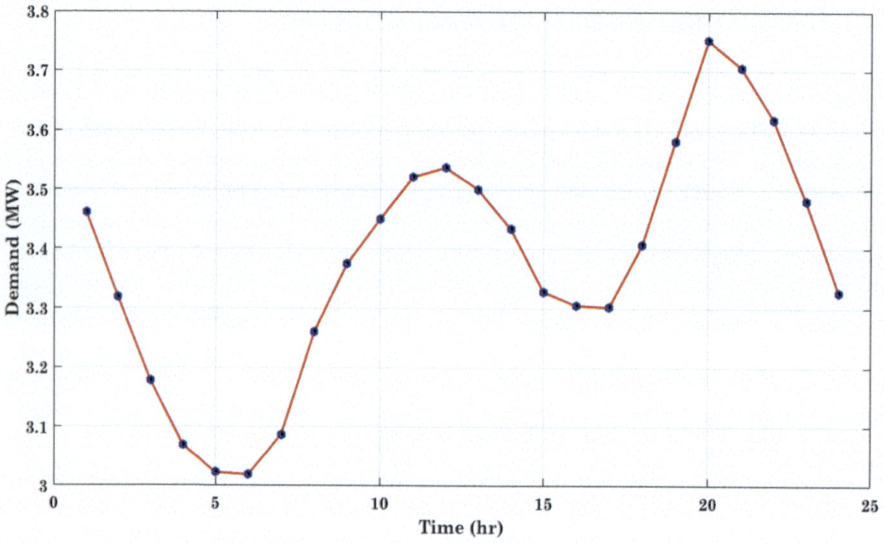

Fig. 5 The daily load demand of the IEEE-33 bus network

1615.015 kW and 908.6 KVAR, respectively. Particularly, in comparison to the base case without RES and EVCS, there is a significant drop in active power losses by 58% and reactive power losses by 64%. Also, when compared with the optimization algorithms explored in this study, the proposed algorithm returns lower levels of active and reactive power losses, as presented in Table 1.

In this scenario, while optimizing the system with the proposed algorithm, the focus remains primarily on minimizing power losses and ensuring adequate coverage of EVCS. However, it's worth noting that the optimization process does not explicitly consider voltage irregularity as a component within the fitness function. Consequently, the voltage irregularity index stands at 12.97 after the installation of RES

Table 1 Results obtained for the IEEE 33 bus system after improving the fitness function with power loss reduction and EVCSs coverage

	CSA	FPA	AGWO	IWOA	HHO	Proposed HWOA
Real power loss (kW)	1656.108	1631.219	1710.760	1633.728	1630.05	1615.015
Reactive power loss (kVAR)	1023.5	996.84	1056.08	995.61	919.73	908.61
Voltage Irregularity index (pu)	16.21	16.80	18.06	16.19	16.01	12.97
Voltage minimum and location	0.970, 33	0.9712, 33	0.9661, 18	0.9710, 18	0.9731, 33	0.9743, 33
Voltage maximum and location	1.00, 1	1.00, 1	1.00, 1	1.00, 1	1.00, 1	1.00, 1
Size and position of RES-1	1083.65, 30	1274.34, 10	1140.69, 13	1245.08, 10	1208.74, 8	935.68, 13
Size and position of RES-2	935.2, 8	1045.67, 30	1162.44, 30	1010.54, 30	1008.65, 30	922.67, 30
Size and position of EVCSs	85.32, 29; 110.5, 3	102.65, 30; 98.46, 3	114.32, 29; 102.47, 2	81.24, 30; 95.32, 3	80.36, 29; 98.46, 3	80.74, 29; 95.21, 3

and EVCS at their optimized positions and capacities determined by the proposed HWOA. When compared to alternative optimization algorithms, the HWOA exhibits a notable improvement in the system's voltage profile. This enhancement in voltage stability is crucial for granting the consistency and efficiency of the distribution network. By analyzing Fig. 6, which illustrates the relationship between active power losses and the number of optimization iterations, it becomes clear that the HWOA consistently reduces power losses throughout the iterative process. Figure 7 provides insights into the behaviour of the voltage irregularity index over the optimization iterations, highlighting the effectiveness of the HWOA in keeping a more stable voltage profile compared to other algorithms. Figure 8 visually represents the voltage magnitudes across all 33 buses in the distribution system, showcasing the substantial improvements achieved by the HWOA in enhancing voltage levels post-optimization. These findings underscore the aptitude of the future Hybrid Whale Optimization Algorithm not only in reducing power losses but also in improving the voltage profile of the distribution network, thereby enhancing its overall performance and reliability.

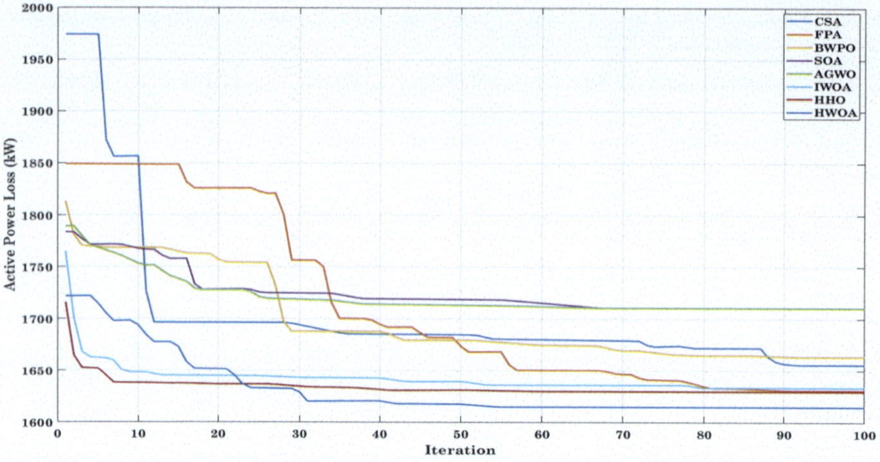

Fig. 6 Active power losses of IEEE 33 bus system for case 1

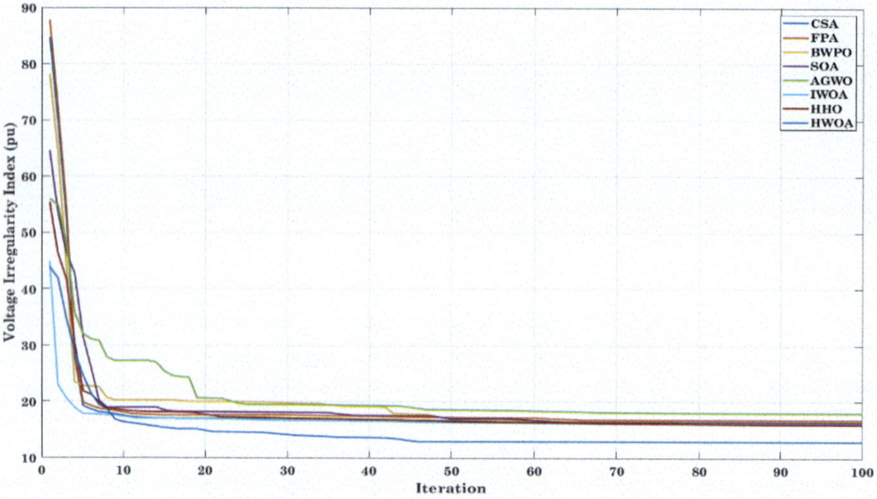

Fig. 7 Voltage irregulating index of IEEE 33 bus system for case 1

6.1.2 Case 2: Optimization Process Employing the Proposed Algorithm Incorporates Two Critical Factors Into Its Fitness Function: The Voltage Irregularity Index and the Coverage of EVCS

Unlike the first case, the optimization does not explicitly consider active power losses within this situation. A comprehensive summary of the results obtained from this approach is provided in Table 2. Utilizing the proposed algorithm, the optimization identifies the optimal positions for two RESs as buses 10 and 31, with respective

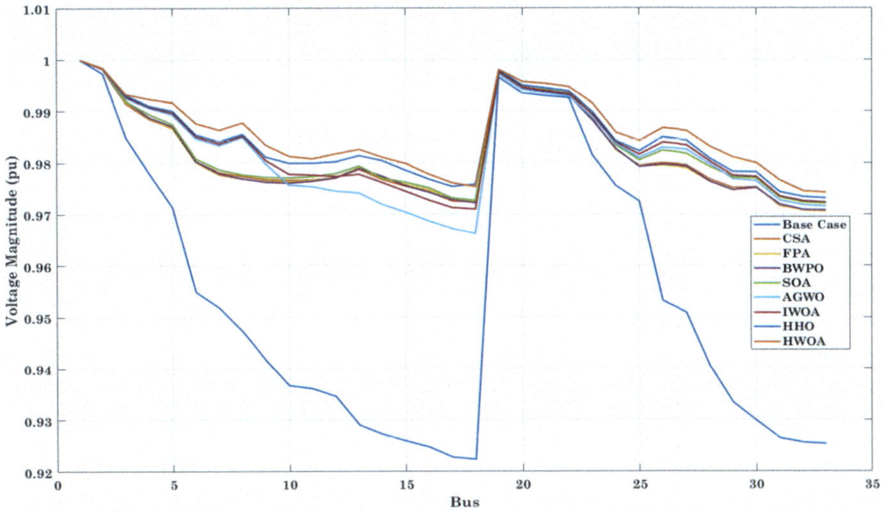

Fig. 8 Voltage magnitude of IEEE 33 bus system for case 1

capacities of 1254.37 and 1293.76 kW. For the two EVCS units, the most suitable locations are determined to be at buses 2 and 33, each with ratings of 52.61 and 110.84 kW, respectively. Over a 24-h period, the optimized configurations yield active and reactive power losses of 1623.92 kW and 913.62 KVAR, separately. The addition of the voltage irregularity index in the optimization process results in a significant reduction, with the index value lowered to 2.9494. This reduction underscores the effectiveness of the proposed Hybrid Whale Optimization Algorithm (HWOA) in improving the stability and efficiency of the distribution network, especially in terms of voltage management.

Observing Fig. 9, depicting the correlation between active power losses and the number of optimization iterations, reveals a consistent trend of reduction in power losses facilitated by the HWOA throughout the iterative process. Figure 10 offers valuable insights into the behaviour of the voltage irregularity index across optimization iterations, illustrating the HWOA's effectiveness in maintaining a stable voltage profile compared to alternative algorithms. Figure 11 provides a visual representation of the voltage magnitudes across all 33 buses in the distribution system.

6.1.3 Case 3: Multi-Objective Problem Incorporating Power Loss, Voltage Irregularity, and the Coverage of EVCSs

Table 3 outlines the best solutions obtained, mentioning the installation of RESs at buses 10 and 31, with ratings of 1254.37 and 1293.76 kW, respectively. Additionally, EVCSs installations are suggested at buses 2 and 33, with ratings of 52.61 and 110.87 kW, respectively. Following the installation of RESs and EVCSs according to these

Table 2 Results obtained for the IEEE 33 bus system once optimizing the fitness function with the voltage irregularity index and the coverage of EVCSs

	CSA	FPA	AGWO	IWOA	HHO	Proposed HWOA
Real power loss (kW)	1668.29	1659.22	1735.94	1652.76	1646.04	1623.92
Reactive power loss (kVAR)	1032.025	1023.95	1071.62	1007.20	928.75	913.62
Voltage irregularity index (pu)	3.475	3.468	3.727	3.418	3.429	2.9494
Voltage minimum and location	0.9826, 25	0.9824, 25	0.9841, 25	0.9844, 25	0.9847, 25	0.9855, 25
Voltage maximum and location	1.00, 1	1.00, 1	1.0029, 10	1.0032, 10	1.0035, 10	1.0038, 10
Size and position of RES-1	1385.26, 10	1346.85, 10	1410.52, 10	1342.50, 10	1340.29, 10	1254.37, 10
Size and position of RES-2	1402.85, 31	1398.04, 30	1450.87, 31	1382.74, 30	1378.54, 30	1293.76, 31
Size and position of EVCS	63.25, 2; 175.28, 33	75.64, 3; 125.86, 33	82.65, 2; 131.65, 33	74.86, 3; 122.45, 33	74.02, 3; 123.68, 33	52.61, 2; 110.87, 33

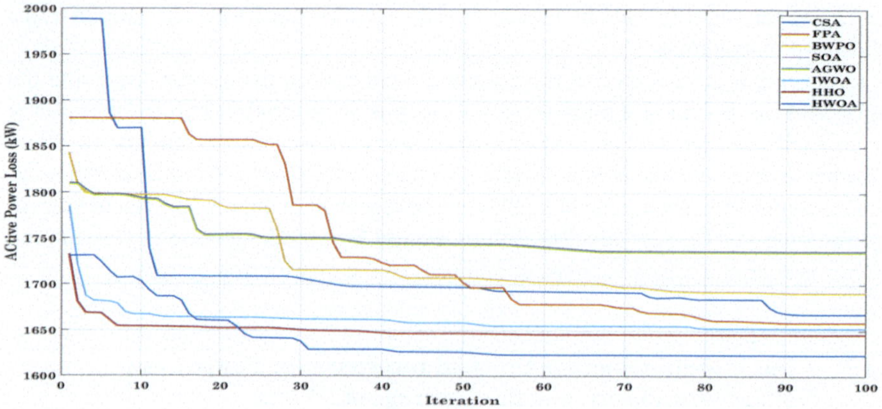

Fig. 9 Active power loss of IEEE 33 bus system for case 2

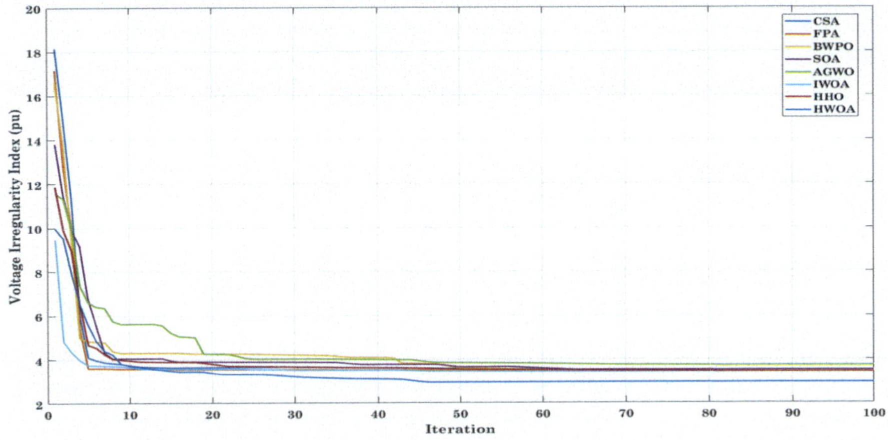

Fig. 10 Voltage irregularity index of IEEE 33 bus system for case 2

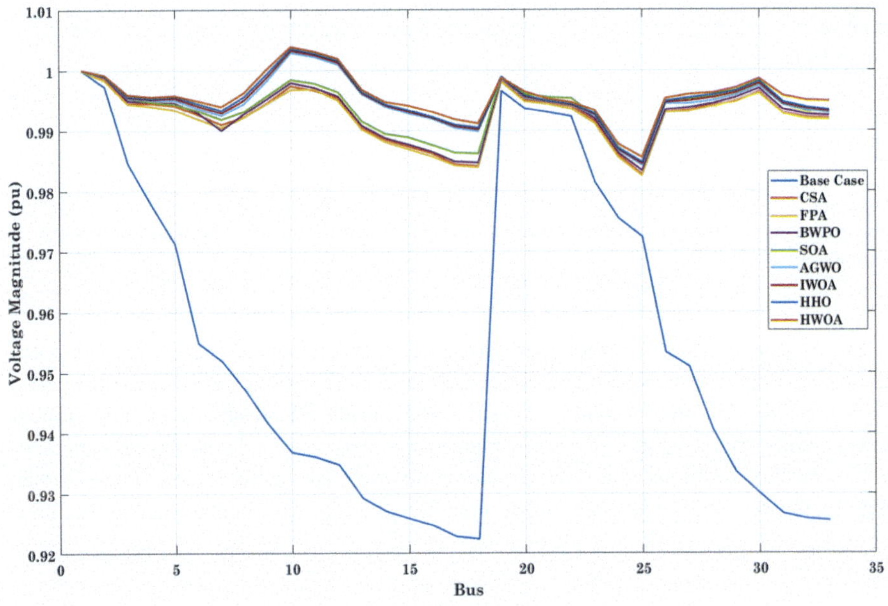

Fig. 11 Voltage magnitude of IEEE 33 bus system for case 2

recommendations, the system exhibits an active power loss of 1624.97 kW, a reactive power loss of 913.62 kVAR, and a voltage irregularity index of 2.9494 p.u. Comparatively, against the next best solutions provided by the remaining considered algorithms, the proposed HWOA demonstrates superior performance by reducing active power losses by 16.95 kW, reactive power losses by 15.13 kVAR, and the voltage irregularity index by 0.4796 p.u. These results underscore the use and supremacy

Table 3 Results obtained for IEEE 33 bus system after increasing the fitness function with the power loss, the voltage irregularity index and the coverage of EVCSs

	CSA	FPA	AGWO	IWOA	HHO	Proposed HWOA
Real power loss (kW)	1695.80	1653.20	1734.09	1652.60	1641.92	1624.97
Reactive power loss (kVAR)	1032.025	1023.95	1071.62	1007.20	928.75	913.62
Voltage irregularity index (pu)	3.475	3.468	3.727	3.418	3.429	2.9494
Voltage minimum and location	0.9826, 25	0.9824, 25	0.9841, 25	0.9844, 25	0.9847, 25	0.9855, 25
Voltage maximum and location	1.00, 1	1.00, 1	1.0029, 10	1.0032, 10	1.0035, 10	1.0038, 10
Size and position of RES-1	1385.26, 10	1346.85, 10	1410.52, 10	1342.50, 10	1340.29, 10	1254.37, 10
Size and position of RES-2	1402.85, 31	1398.04, 30	1450.87, 31	1382.74, 30	1378.54, 30	1293.76, 31
Size and position of EVCS	63.25, 2; 175.28, 33	75.64, 3; 125.86, 33	82.65, 2; 131.65, 33	74.86, 3; 122.45, 33	74.02, 3; 123.68, 33	52.61, 2; 110.87, 33

of the proposed HWOA over other optimization approaches. Figure 12 illustrates the variation in active power losses throughout each iteration for both the proposed algorithm and the other considered algorithms. Figure 13 depicts the reactive power losses, while Fig. 14 presents the voltage irregularity index. These visual representations further emphasize the significant improvements achieved by the proposed HWOA in optimizing the distribution system, affirming its effectiveness in simultaneously addressing multiple optimization objectives while outperforming alternative algorithms.

6.2 *IEEE-69 Bus Network*

The proposed algorithm was tested on the IEEE-69 bus system, comprising 68 branches and 69 nodes. Operating at a standard voltage of 12.66 kV and a base power of 100 MVA, this network handles significant loads.

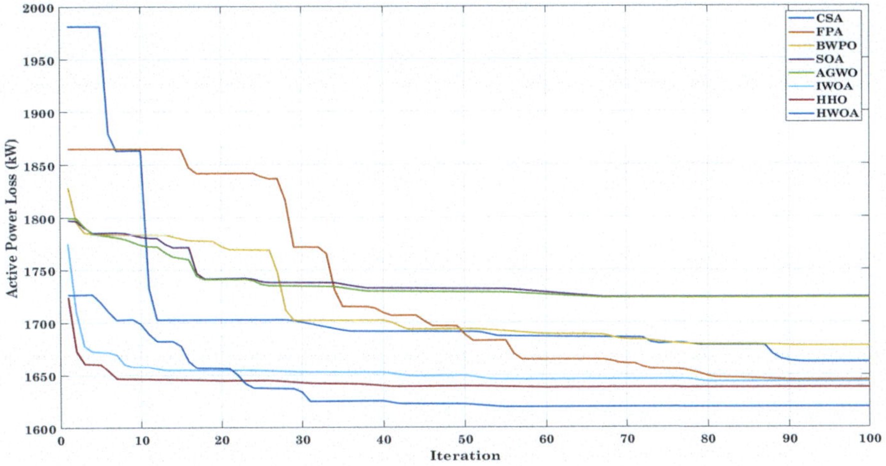

Fig. 12 Active power loss of IEEE 33 bus system for case 3

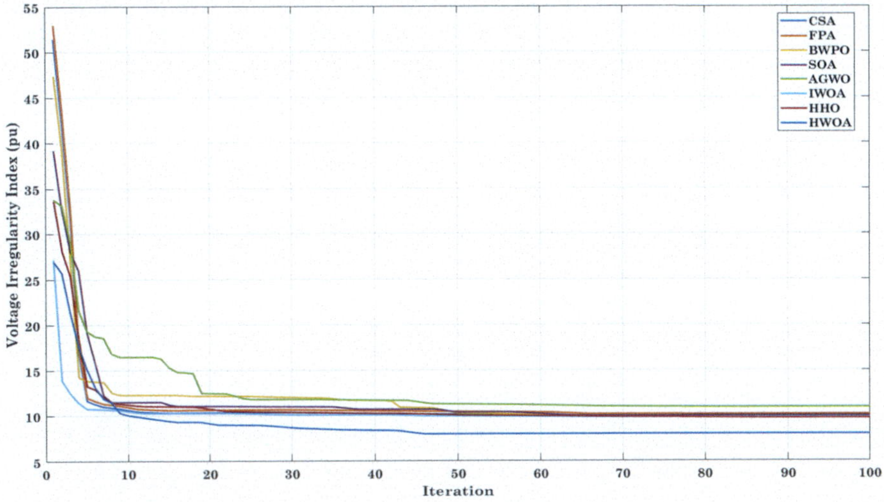

Fig. 13 Voltage irregularity index of IEEE 33 bus system for case 3

The rated active power load is related to 3801.89 kW, while the rated reactive power load connected is 2694.10 kVAR, indicating the significant energy demands of the system. A comprehensive view of the network's structure and components is provided through the single-line diagram presented in Fig. 15. This diagram offers insights into the network's topology, essential for devising effective optimization strategies. The network's demand and branch data, crucial for the optimization process, were obtained from [30, 31]. The network's load profile, reflecting the 24-h demand pattern as a percentage of the rated load, is depicted in Fig. 5. This profile

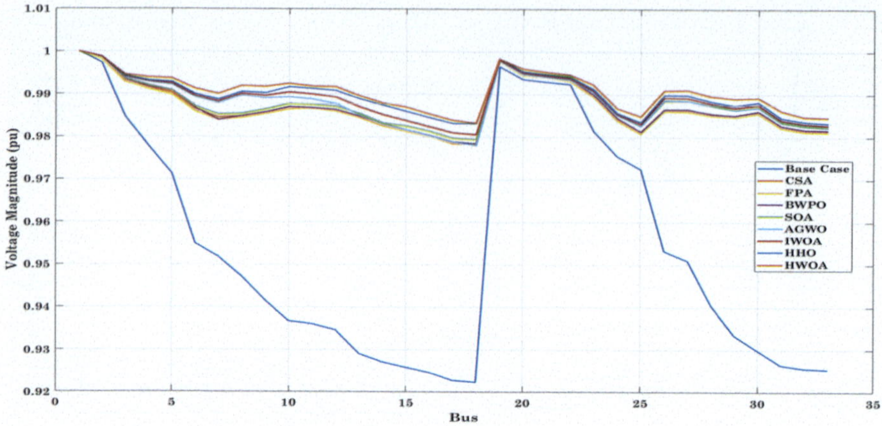

Fig. 14 Voltage magnitude of IEEE 33 bus system for case 3

highlights the dynamic nature of energy consumption throughout the day, posing assignments for optimization algorithms [32]. In the absence of RESs and EVCSs installations, the base case records an active power loss of 8598.623 kW and a reactive power loss of 3896.745 kVAR. These values serve as benchmarks for assessing the efficacy of optimization strategies in minimizing losses and improving network efficiency.

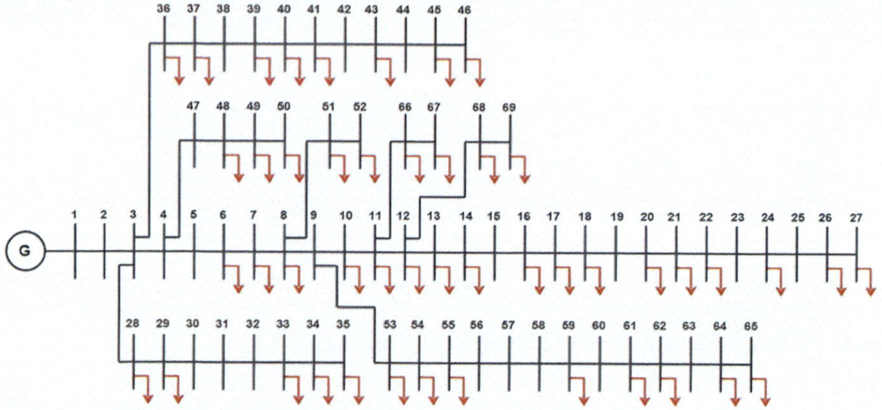

Fig. 15 IEEE-69 bus network single line diagram

6.2.1 Case 1: Optimization Process Utilizing the Proposed Algorithm Integrates Both Power Loss Reduction and EVCSs

A comprehensive summary of the results obtained in this context is presented in Table 4. Applying the proposed algorithm, the optimization identifies the optimal positions for three RESs at buses 60, 17, and 12, with respective capacities of 1315.75, 745.347, and 43.0875 kW. Additionally, for the EVCS units, optimal locations are determined at buses 5, 11, and 17, each with ratings of 151.24, 54.95, and 143.57 kW, respectively. The active and reactive power losses seen with these optimized configurations, measured over a 24-h period, amount to 2670.25 kW and 1152.031 kVAR, respectively. Compared to the base case without RES and EVCS installations, there is a real drop in active power losses by 69% and reactive power losses by 71%. When associated with the optimization algorithms explored in this study, the proposed algorithm consistently yields lower levels of both active and reactive power losses, as depicted in Table 4. These results emphasize how effective the proposed algorithm is in optimizing the 69-bus system., achieving significant reductions in power losses while enhancing EVCS coverage. Such improvements contribute to the total efficacy and reliability of the distribution network, proving the superiority of the proposed algorithm over alternative optimization approaches.

In this case, the optimization process utilizing the proposed algorithm primarily focuses on minimizing power losses and ensuring sufficient coverage of EVCSs. Significantly, the optimization process does not explicitly incorporate voltage irregularity as a parameter within the fitness function. The voltage irregularity index remains at 12.179 following the installation of RES and EVCS at their optimized positions and capacities determined by the proposed HWOA. Despite the absence of voltage irregularity consideration in the fitness function, the HWOA demonstrates a notable improvement in the system's voltage profile compared to alternative optimization algorithms. This enhancement in voltage stability is vital for improving the consistency and efficiency of the distribution network. Figure 16 illustrates the relationship between active power losses and the number of optimization iterations, it becomes evident that the HWOA consistently reduces power losses throughout the iterative process. This demonstrates the algorithm's effectiveness in achieving its primary objective of minimizing power losses. Figure 17 provides insights into the behaviour of the voltage irregularity index over the optimization iterations, highlighting the HWOA's effectiveness in maintaining a more stable voltage profile compared to other algorithms. This further underscores the algorithm's capability to indirectly address voltage stability despite the absence of explicit consideration within the fitness function. Figure 18 represents the voltage magnitudes across all 69 buses in the distribution system. This presents the substantial improvements achieved by the HWOA in enhancing voltage levels post-optimization, further affirming its efficacy in improving the overall performance and reliability of the distribution network.

Table 4 Results attained for the IEEE 69 bus system after optimizing the fitness function with the power loss and the coverage of EVCSs

	CSA	FPA	AGWO	IWOA	HHO	Proposed HWOA
Real power loss (kW)	2876.026	2831.3231	2719.919	2699.027	2708.562	2670.25
Reactive power loss (kVAR)	1376.07	1354.36	1258.56	1220.875	1238.24	1152.031
Voltage irregularity index (pu)	23.707	19.000	20.799	16.966	14.576	12.179
Voltage minimum and location	0.9797, 66	0.9811, 66	0.9824, 66	0.9827, 66	0.9829, 66	0.9846, 66
Voltage maximum and location	1.00, 1	1.00, 1	1.00, 1	1.00, 1	1.00, 1	1.00, 1
Size and position of RES-1	1536.25, 61	1504.78, 61	1410.87, 60	1398.40, 61	1402.77, 61	1315.75, 60
Size and position of RES-2	785.64, 17	763.55, 17	746.85, 17	763.874, 17	753.69, 17	745.347, 17
Size and position of RES-3	46.0097, 14	45.2167, 10	43.7721, 10	44.8217, 14	44.7350, 14	43.0875, 12
Size and position of EVCSs	75.64, 5; 48.52, 11; 89.65, 21	86.32, 5; 49.63, 11; 90.32, 20	116.42, 3; 55.78, 13; 116.84, 21	130.65, 4; 46.12, 12; 122.34, 17	140.25, 4; 52.08, 11; 120.65, 17	151.24, 5; 54.95, 11; 143.57, 17

6.2.2 Case 2: Optimization Process Employing the Proposed Algorithm Integrates Two Critical Factors: Voltage Irregularity Index and the Coverage of EVCSs

Unlike the first case, active power losses are not explicitly considered within this scenario. A comprehensive summary of the results obtained from this approach is provided in Table 5. Utilizing the implied algorithm, the optimization identifies the optimal positions for three at buses 69, 17, and 12, with respective capacities of 1340.62, 736.84, and 44.0241 kW. For the EVCS units, the most suitable locations are determined to be buses 5, 11, and 17, each with ratings of 147.12, 53.465, and 141.28 kW, respectively.

Over a 24-h period, the optimized configurations yield active and reactive power losses of 3201.20 kW and 1381.099 kVAR, individually. Importantly, the inclusion of the voltage irregularity index in the optimization process results in a significant

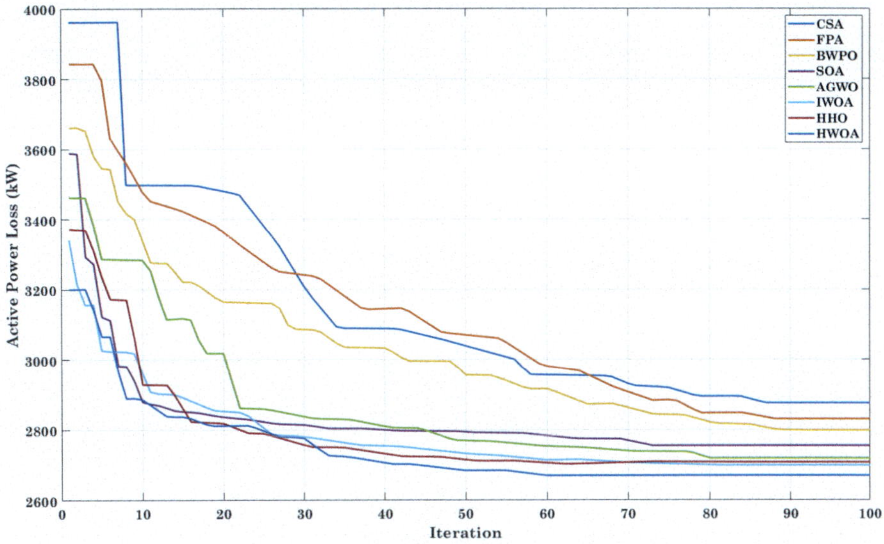

Fig. 16 Active power loss of IEEE 69 bus system for case 1

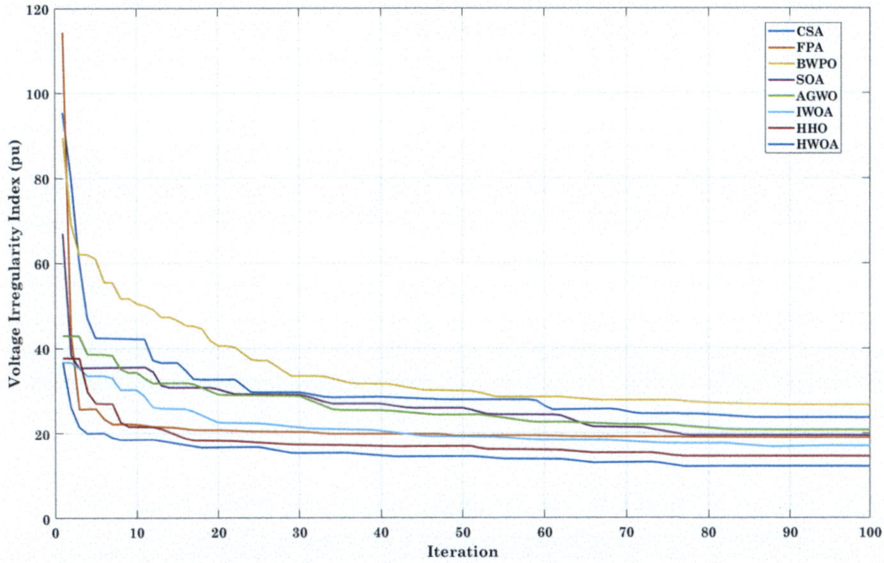

Fig. 17 Voltage irregularity index of IEEE 69 bus system for case 1

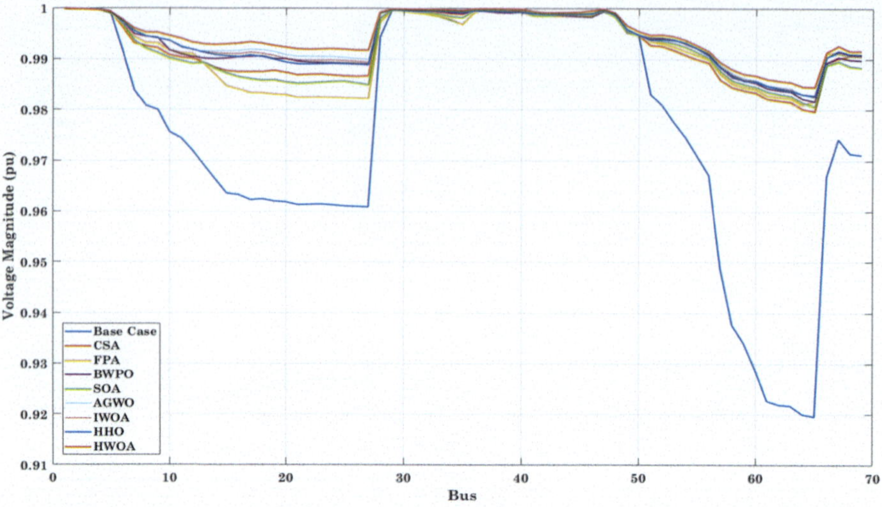

Fig. 18 Voltage magnitude of IEEE 69 bus system for case 1

reduction, with the index value lowered to 2.484. This reduction underscores the effectiveness of the proposed HWOA in enhancing the stability and efficiency of the distribution network, particularly concerning voltage management. Figure 19 depicts the correlation between active power losses and the number of optimization iterations, revealing a consistent trend of reduction in power losses facilitated by the HWOA throughout the iterative process. Figure 20 illustrates the behaviour of the voltage irregularity index across optimization iterations, demonstrating the HWOA's effectiveness in maintaining a stable voltage profile compared to alternative algorithms. Figure 21 provides a visualization of the voltage magnitudes across all 69 buses in the distribution system, showcasing the substantial improvements achieved by the HWOA in enhancing voltage levels post-optimization. These findings highlight the effectiveness of the HWOA in optimizing the distribution network. It shows its ability to handle multiple objectives, enhancing overall network performance and reliability.

6.2.3 Case 3: A Multi-Objective Problem Integrating Power Loss, Voltage Irregularity Index, and the Coverage of EVCSs

Table 6 outlines the best solutions obtained, recommending the installation of RES at buses 69, 17 and 12, with ratings of 1264.21, 652.318 and 35.681 kW respectively. EVCS installations are suggested at buses 5, 11 and 17, with ratings of 158.37 kW, 59.34 kW and 155.34 kW respectively. Following the installation of RES and EVCS according to these recommendations, the system exhibits an active power loss of 3201.20 kW, automatic power loss of 1266.56 kVAR, and a voltage irregularity index of 2.484 p.u. Figure 22 illustrates the variation in active power losses

Table 5 Results attained for the IEEE 69 bus system after optimizing the fitness function with the voltage irregularity index and the coverage of EVCSs

	CSA	FPA	AGWO	IWOA	HHO	Proposed HWOA
Real power loss (kW)	3160.829	3019.612	3003.947	2970.60	2988.28	2935.72
Reactive power loss (kVAR)	1658.604	1544.495	1521.408	1466.92	1493.99	1381.099
Voltage irregularity index (pu)	14.459	11.469	12.561	10.419	8.763	7.331
Voltage minimum and location	0.9846, 66	0.9868, 62	0.9898, 66	0.9911, 66	0.9925, 66	0.9957, 66
Voltage maximum and location	1.00, 1	1.00, 1	1.0029, 13	1.0020, 13	1.0014, 69	1.006, 69
Size and position of RES-1	1554.24, 61	1496.58, 61	1436.14, 60	1422.82, 61	1411.57, 69	1340.62, 69
Size and position of RES-2	790.24, 17	785.70, 17	759.84, 17	771.672, 17	762.84, 17	736.84, 17
Size and position of RES-3	47.89, 14	44.315, 10	42.574, 13	43.5746, 13	45.6742, 14	44.0241, 12
Size and position of EVCS	74.32, 5; 46.14, 11; 88.79, 21	84.31, 5; 46.71, 11; 89.92, 20	112.15, 3 54.91, 13 110.57, 21	122.91, 4 45.73, 12 116.728, 17	135.47, 4 50.47, 11 119.54, 17	147.12, 5 53.465, 11 141.28, 17

throughout each iteration for both the proposed algorithm and the other considered algorithms. Additionally, Fig. 23 depicts the reactive power losses, while Fig. 24 presents the voltage irregularity index. These visual representations further emphasize the significant improvements achieved by the proposed HWOA in optimizing the distribution system, affirming its effectiveness in simultaneously addressing multiple optimization objectives while outperforming alternative algorithms.

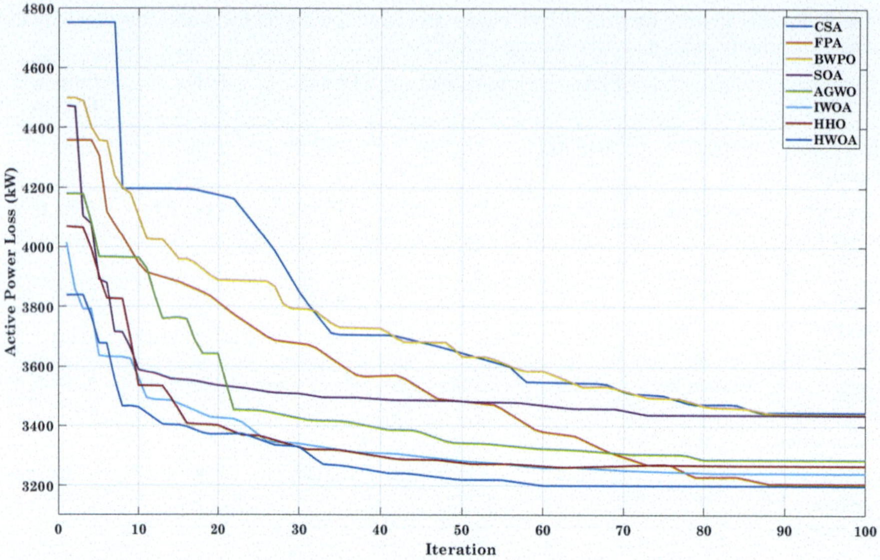

Fig. 19 Active power loss of IEEE 69 bus system for case 2

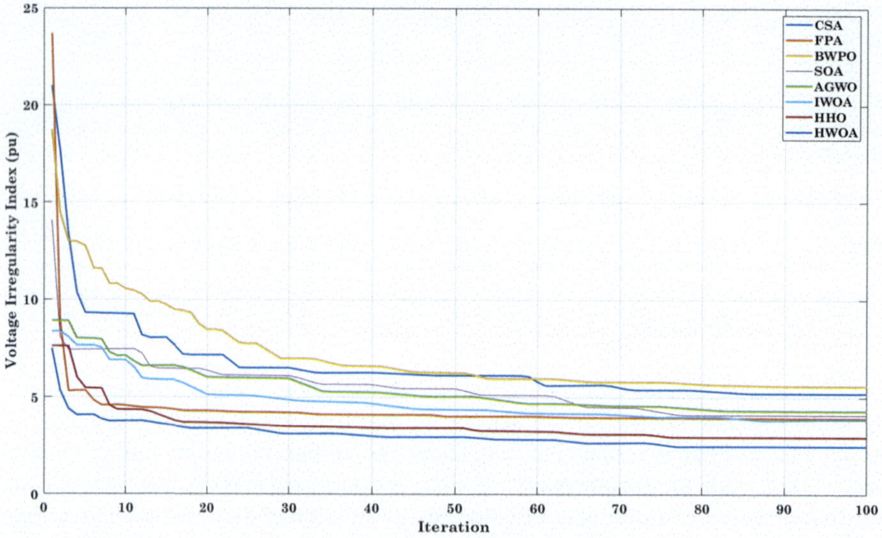

Fig. 20 Voltage irregularity index of IEEE 69 bus system for case 2

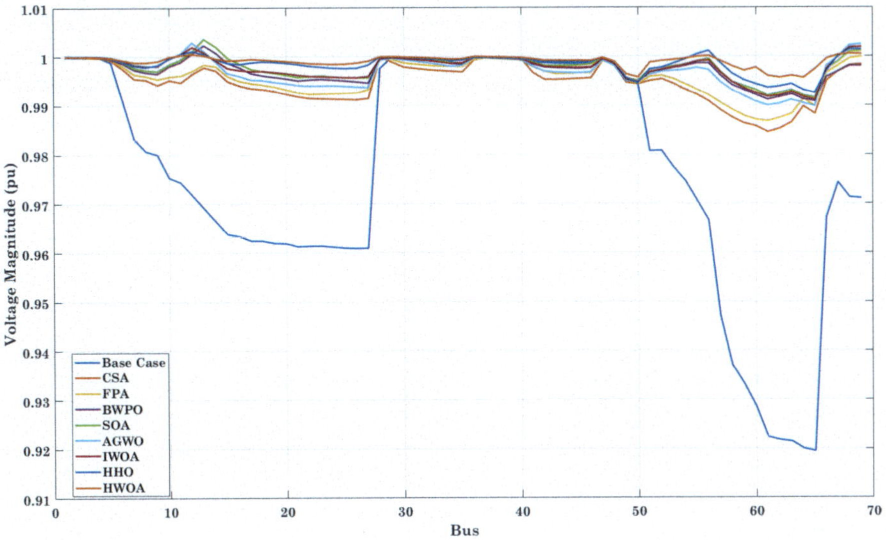

Fig. 21 Voltage magnitude of IEEE 69 bus system for case 2

7 Conclusions

The research presented in this chapter highlights the efficacy of the Hybrid Whale Optimization Algorithm (HWOA) in addressing the tasks related to integrating RES EVCS interested in RDSs. Through comprehensive analysis and investigation conducted on both the IEEE 33 and 69 bus systems, the algorithm has demonstrated its proficiency in optimizing the placement and sizing of RES and EVCS. This optimization process results in the minimization of power losses, improvement of voltage stability, and the provision of adequate coverage for EVCS infrastructure. By leveraging a combination of Genetic Whale Optimization Algorithm (GWOA) and Thermal Exchange Optimization (TEO) techniques, the HWOA offers a versatile and efficient approach to tackle the complex optimization problems inherent in RES and EVCSs integration. The algorithm's ability to consider multiple objectives, including power loss reduction, voltage stability improvement, and EVCSs coverage enhancement, underscores its suitability for practical application in real-world distribution networks. The results obtained from the application of the HWOA illustrate significant improvements in the performance and reliability of radial distribution networks. By strategically placing RESs at optimal locations and determining suitable capacities, the algorithm successfully mitigates power losses while improving the overall voltage profile of the network. The optimized placement of EVCS units confirms

Table 6 Results obtained for IEEE 69 bus system after adjusting the fitness function with the power loss, the voltage irregularity index and the coverage of EVCSs

	CSA	FPA	AGWO	IWOA	HHO	Proposed HWOA
Real power loss (kW)	3445.631	3207.90	3287.97	3242.97	3268.00	3201.20
Reactive power loss (kVAR)	1521.51	1453.840	1389.98	1343.71	1366.11	1266.56
Voltage irregularity index (pu)	5.212	3.937	4.322	3.872	2.950	2.484
Voltage minimum and location	0.9834, 66	0.9849, 62	0.9861, 66	0.9869, 66	0.9877, 66	0.9902, 65
Voltage maximum and location	1.00, 1	1.00, 1	1.00, 1	1.00, 1	1.00, 1	1.00, 1
Size and position of RES-1	1435.62, 61	1412.85, 61	1365.74, 60	1355.42, 61	1359.73, 69	1264.21, 69
Size and position of RES-2	705.49, 17	695.67, 17	685.49, 17	692.75, 17	683.73, 17	652.31, 17
Size and Position of RES-3	41.53, 14	38.65, 10	36.95, 13	37.95, 13	38.37, 14	35.68, 12
Size and position of EVCS	82.65, 5; 56.34, 11; 95.64, 21	91.37, 5; 55.33, 11; 98.57, 20	118.24, 3; 60.25, 13; 115.74, 21	128.95, 4; 55.34, 12; 121.674, 17	138.64, 455.64, 11 120.35, 17	158.37, 5; 259.34, 17; 111.55, 34

complete coverage to meet the growing need for an EV charging structure. The findings of this research aid in advancing the field of distribution system optimization and pave the way for the widespread adoption of sustainable energy solutions in modern power grids. The proposed HWOA offers a promising avenue for well-integrating RESs and EVCSs, thereby facilitating the transition towards more resilient, efficient, and environmentally sustainable energy systems.

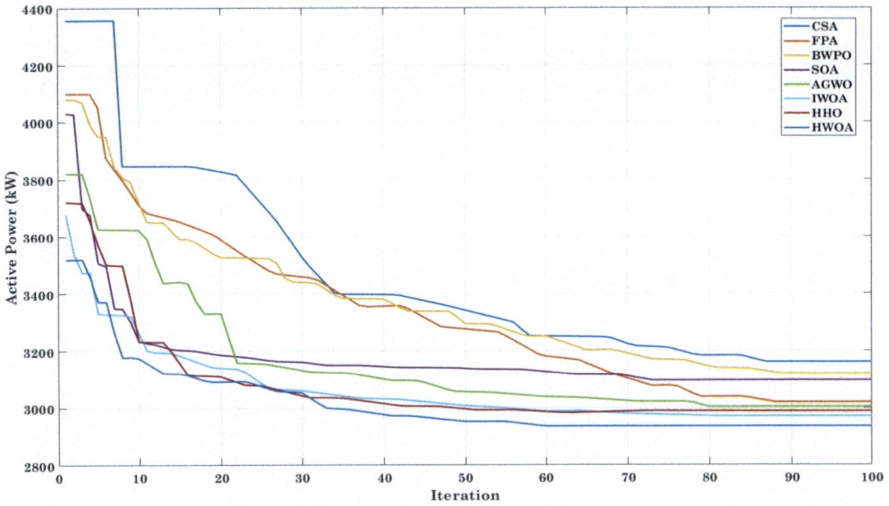

Fig. 22 Active power loss of IEEE 69 bus system for case 3

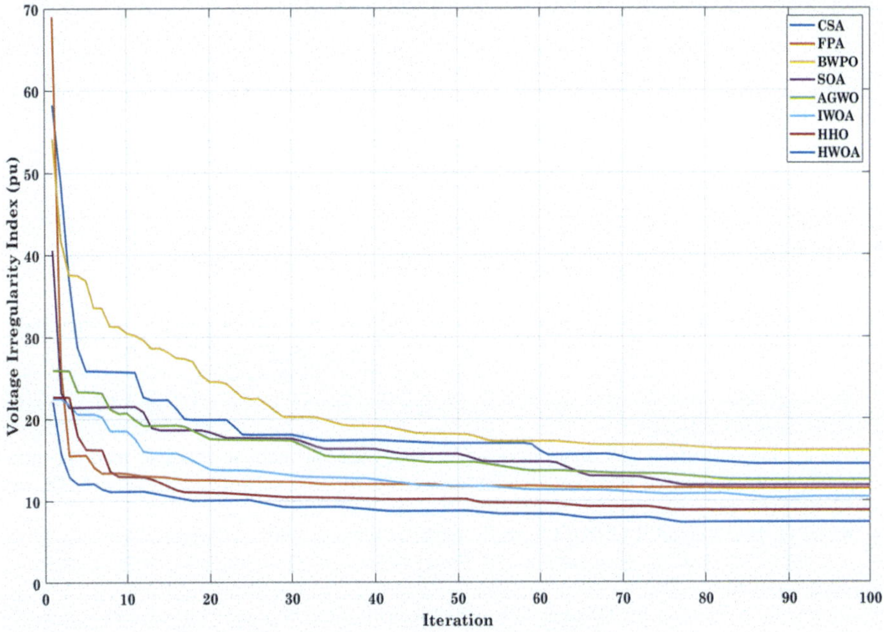

Fig. 23 Voltage irregularity index of IEEE 69 bus system for case 3

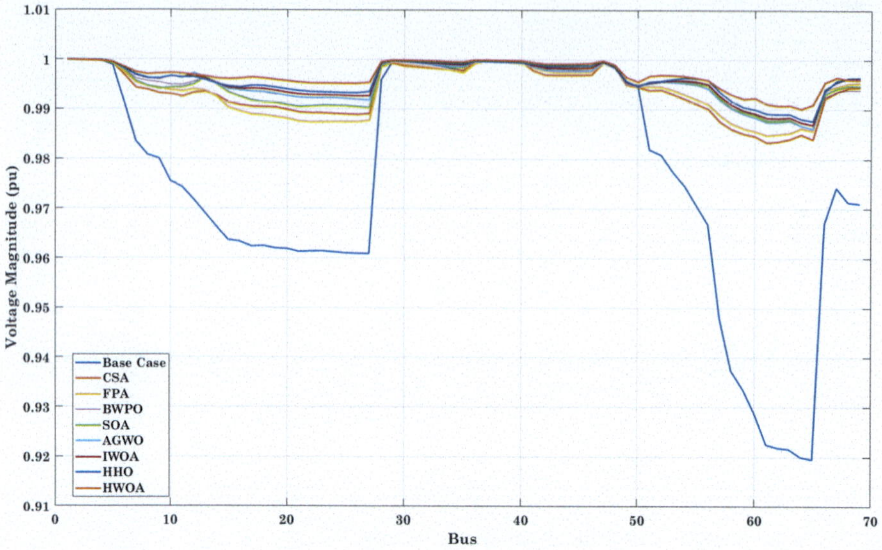

Fig. 24 Voltage magnitude of IEEE 69 bus system for case 3

Acknowledgements This research work was supported by "Woosong University's Academic Research Funding—2025".

References

1. Dalia S, Tomas B, Ligita B (2013) Comparative assessment of road transport technologies. Renew Sustain Energy Rev 20:611–618. https://doi.org/10.1016/j.rser.2012.12.021
2. Lance N, Gerardo ZDR, Benjamin KS, Johannes K (2019) Fear and loathing of electric vehicles: The reactionary rhetoric of range anxiety. Energy Res Soc Sci 48:96–107. https://doi.org/10.1016/j.erss.2018.10.001
3. Tim U, Jeppe R, Peter BA, Seyedmostafa H (2022) Electric vehicle charging infrastructure planning for integrated transportation and power distribution networks. A review. E Transp 12:1001163. https://doi.org/10.1016/j.etran.2022.100163
4. Mohd B, Mohammad R (2020) Electric vehicles in a smart grid: a comprehensive survey on optimal location of charging station. IET Smart Grid 3:267–279. https://doi.org/10.1049/iet-stg.2019.0220
5. Alexander K, Roman M, Dietmar G (2017) Electrification of a city bus network—an optimization model for cost-effective placing of charging infrastructure and battery sizing of fast-charging electric bus systems. Int J Sustain Transp 11(10):707–720. https://doi.org/10.1080/15568318.2017.1310962
6. Salkuti SR (2023) Advanced technologies for energy storage and electric vehicles. Energies 16(5):2312. https://doi.org/10.3390/en16052312
7. Mohammad HH, Mohammad HRK, Pooria D, Payman D (2022) Voltage and energy control in distribution systems in the presence of flexible loads considering coordinated charging of electric vehicles. Energy 239:121880. https://doi.org/10.1016/j.energy.2021.121880

8. Emilia MS (2023) Electric vehicle charging infrastructure along highways in the EU. Energies 16(2):895. https://doi.org/10.3390/en16020895
9. Salkuti SR (2022) Emerging and advanced green energy technologies for sustainable and resilient future grid. Energies 15(18):6667. https://doi.org/10.3390/en15186667
10. Miguel FA, Bernard G, Martim JM (2020) Increasing electric vehicle adoption through the optimal deployment of fast-charging stations for local and long-distance travel. Eur J Oper Res 285(1):263–278. https://doi.org/10.1016/j.ejor.2020.01.055
11. Sanchari D, Kari T, Xiao-Zhi G, Karuna K, Pinakeswar M (2020) A hybrid multi-objective chicken swarm optimization and teaching learning based algorithm for charging station placement problem. IEEE Access 8:92573–92590. https://doi.org/10.1109/ACCESS.2020.2994298
12. Ahmad FI, Atif A, Imtiaz M, Mousa K, Irfan (2022) Placement of electric vehicle fast charging stations in distribution network considering power loss, land cost and electric vehicle population. Energy Sour Part A: Recov Util Environ Effects 44(1):1693–1709. https://doi.org/10.1080/15567036.2022.2055233
13. Mehul B, Bivas RD, Nivedita S, Smruti RS (2023) PC-ILP: a fast and intuitive method to place electric vehicle charging stations in smart cities. Smart Cities 6(6):3060–3092. https://doi.org/10.3390/smartcities6060137
14. Ying Z, Yanhao W, Fanyu L, Bin W, Yao-Yi C, Xin Z (2020) Efficient deployment of electric vehicle charging infrastructure: Simultaneous optimization of charging station placement and charging pile assignment. IEEE Trans Intell Transp Syst 22(10):6654–6659. https://doi.org/10.1109/TITS.2020.2990694
15. Guangyou Z, Zhiwei Z, Sumei L (2022) Location optimization of electric vehicle charging stations: based on cost model and genetic algorithm. Energy 247:123437. https://doi.org/10.1016/j.energy.2022.123437
16. Ibrahim A, Abido MA, Muhammad K, Andrey VS (2020) A comprehensive review of recent advances in smart grids: a sustainable future with renewable energy resources. Energies 13(23):6269. https://doi.org/10.3390/en13236269
17. Shimaa B, Ahmed IO, Elsayed TE, Ahmad AT, Hala MAM, Samy MM (2023) Achieving green mobility: Multi-objective optimization for sustainable electric vehicle charging. Energ Strat Rev 53:101351. https://doi.org/10.1016/j.esr.2024.101351
18. Ramanan CJ, King HL, Jundika CK, Sukanta R, Bhaskor JB, Bhaskar JM (2024) Design study on the parameters influencing the performance of floating solar PV. Renew Energy 223:120064. https://doi.org/10.1016/j.renene.2024.120064
19. Available. [Online]: https://nsrdb.nrel.gov/
20. Weiying W, Toshihiro O (2024) Lognormal distribution of daily travel time and a utility model for its emergence. Transp Res Part A: Policy Pract 183:104058. https://doi.org/10.1016/j.tra.2024.104058
21. Maksymilian M, Tiziana C (2023) Energy consumption of electric vehicles: analysis of selected parameters based on created database. Energies 16(3):1437. https://doi.org/10.3390/en16031437
22. Sunday AS, Gafari AA, Isaiah GA, Harun ORH, Samson OA, Oludamilare BA (2023) Impact of distributed generators penetration level on the power loss and voltage profile of radial distribution networks. Energies 16(4):1943. https://doi.org/10.3390/en16041943
23. Amel AA, Abdelaziz AA, El-Sayed MEK, Abdelhameed I, Marwa ME, Doaa SK, Ayman EA (2023) A binary waterwheel plant optimization algorithm for feature selection. IEEE Access 11:94227–94251. https://doi.org/10.1109/ACCESS.2023.3312022
24. Givi H, Hubalovska M (2023) Skill optimization algorithm: a new human-based metaheuristic technique. Comput Mater Contin 74(1). https://doi.org/10.32604/cmc.2023.030379
25. Kazem M, Amir PH, Seyedali M, Amir BF (2022) Adaptive grey wolf optimizer. Neural Comput Appl 34(10):7711–7731. https://doi.org/10.1007/s00521-021-06885-9
26. Shen Y, Chen Z, Farhad SG, Seyedali M (2023) An improved whale optimization algorithm based on multi-population evolution for global optimization and engineering design problems. Expert Syst Appl 215:119269. https://doi.org/10.1016/j.eswa.2022.119269

27. Tansel D, Yavuz SO (2023) A new robust Harris Hawk optimization algorithm for large quadratic assignment problems. Neural Comput Appl 35(17):12531–12544. https://doi.org/10.1007/s00521-023-08387-2
28. Amir HG, Xin-She Y, Amir HA (2013) Cuckoo search algorithm: a metaheuristic approach to solve structural optimization problems. Eng Comput 29:17–35. https://doi.org/10.1007/s00366-012-0308-4
29. Xin She Yang (2012) Flower pollination algorithm for global optimization. In: International conference on unconventional computing and natural computation. Springer, Berlin, pp 240–249. https://doi.org/10.1007/978-3-642-32894-7_27
30. Prakash DB, Lakshminarayana C (2016) Multiple DG placements in distribution system for power loss reduction using PSO algorithm. Procedia Technol 25:785–792. https://doi.org/10.1016/j.protcy.2016.08.173
31. Jung CM, Pagidipala S, Salkuti SR (2024) Emerging technologies for the integration of renewable energy, energy storage and electric vehicles. In: Salkuti SR (eds) Energy and environmental aspects of emerging technologies for smart grid. Green Energy and Technology. Springer, Cham. https://doi.org/10.1007/978-3-031-18389-8_5
32. Sravanthi P, Sandeep V (2022) Solving realistic reactive power market clearing problem of wind-thermal power system with system security. Int J Emerg Electr Power Syst 23(2):125–144. https://doi.org/10.1515/ijeeps-2021-0060

An Introduction to Energy Management in Smart Energy Systems

Krishna Mohan Reddy Pothireddy, Gummala Babu, Kampa Narendra Babu, Sandeep Vuddanti, and Surender Reddy Salkuti

Abstract The electrical industry is entering a new era where the conventional power system is being transformed into a smart grid with two-way power flow and information flow. The main drivers of this evolution are the increased demand for electricity, an increasing lack of interest in traditional energy sources, and the unpredictable and stochastic outputs of distribution energy resources (DERs). Comprehensive energy management for DC, AC, and hybrid microgrids operating in both off-grid or isolated modes as well as grid-connected modes is addressed in this chapter. Additionally, this chapter discusses energy management strategies for addressing the uncertainties associated with the generation of renewable electricity. In the end, an energy management strategy based on cost comparison is considered a case study to minimize operational costs and maximize the use of renewable energy. Besides, the chapter concluded with an introduction to demand response, demand response programs, and price elasticity.

Keywords Distribution energy resources · AC microgrid · DC microgrid · Hybrid microgrid · Energy management · Operational cost

Abbreviations

MG	Microgrid
PV	Photovoltaic
WT	Wind turbine

K. M. R. Pothireddy · G. Babu · K. N. Babu · S. Vuddanti
Department of Electrical Engineering, National Institute of Technology Andhra Pradesh, Tadepalligudem, Andhra Pradesh, India
e-mail: sandeep@nitandhra.ac.in

S. R. Salkuti (✉)
Department of Railroad and Electrical Engineering, Woosong University, Daejeon 34606, Republic of Korea
e-mail: surender@wsu.ac.kr

S. R. Salkuti (ed.), *Artificial Intelligence for Integrated Smart Energy Systems in Electric Vehicles*, Lecture Notes in Electrical Engineering 1427,
https://doi.org/10.1007/978-3-031-94276-1_7

DER's	Distributed energy resources
SG	Smart grid
EG	Electrical grid
ICT	Information and communication technologies
RES	Renewable energy sources
HES	Hybrid energy systems
HV	High voltage
AC	Alternating current
DC	Direct current
MPPT	Maximum power point tracking
MPC	Model predictive control
FC	Fuel cell

1 Introduction

To cope with the occasional peak demand, the inflexible, inefficient, and expensive peak load generators need to be turned on. This solution creates economic, environmental, and technical impossibilities. The problem is alleviated by deploying distributed energy resources (DERs) like wind turbines (WT) and photovoltaic (PV), however, the volatile nature of DERs is one of their inherent behaviors. To meet the abrupt imbalance of power, a system with a high number of stochastic and variable sources needs a huge spinning reserve [1]. Additionally, the energy scheduling method must handle the uncertainty in the decision variables [2], the inclusion of these sources makes it more difficult. Moreover, it is evident that supply-side generation adjustments [3] will be made in response to any imbalance in electrical power, to avoid these drawbacks, flexible and effective energy management strategies are to be deployed into the power systems [4].

A smart grid (SG) is a technologically advanced and modernized electrical grid (EG) that incorporates a range of digital and communication technologies to enhance the reliability of energy generation, transmission, distribution, and utilization [5]. In contrast to conventional electrical grids, an SG integrates information and communication technologies (ICT) to monitor, control, and optimize various components of the power system. Figure 1 illustrates the key features of an SG. Traditional power systems have enormous complex issues related to centralized and decentralized conditions. Indeed, not only mitigation of these and modernization, but smart grids and microgrids have come to live light. Microgrids have several lucrative futures related to the power sector. Renewable energy sources (RESs) play a promising role with respective grid connection and integration of systems to empower quality and efficiency [6].

Huge and incredible facilities are underway in the energy industry due to the availability, understandability, and usability of solar and wind-based energy sources [7]. Previous periods indicate that these sources are not given as much attention due

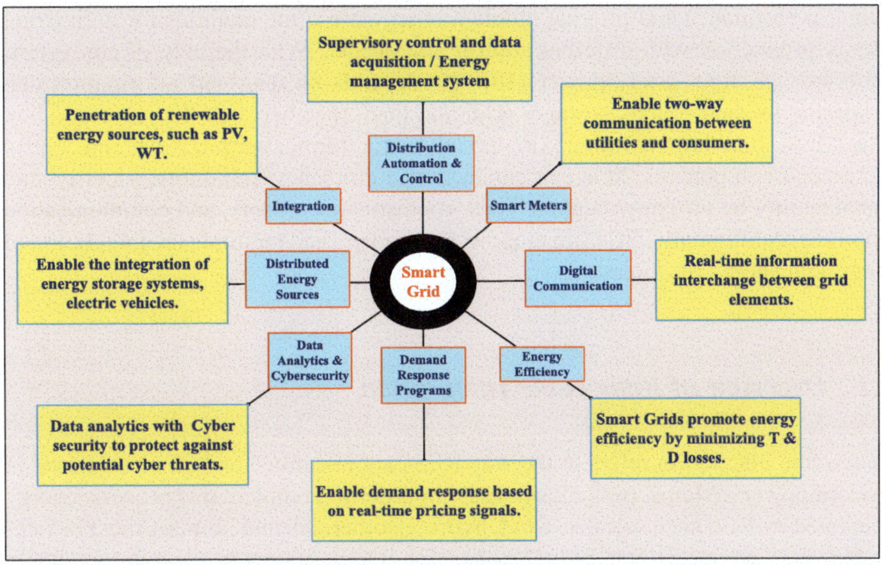

Fig. 1 Key futures and components of the smart grid

to a number of limitations, including cost, availability of data specific to a place, and technological hazards. As a result of the installation of advanced technology, limitations, and obstacles are changing with the times, and the potential for taking risks has grown. India's installation of solar and wind-powered energy increased from 0.5 to 5.6% and 1.1 to 1.8%, respectively, between 2012 and 2022.

There are several maximum power point tracking techniques (MPPT) available since the output from these sources is erratic and fluctuating. The system's operational point is continually perturbed, for instance, only when conditions change quickly using the perturb and observe approach [8]. In order to match the derivative in conductance, the incremental conductance approach modifies the functional point, which improves dynamic operation precision. The fractional open circuit voltage approach is appropriate for low-cost implementation as it controls the duty cycle of the converter to keep voltage around fractional V_{OC} levels. Furthermore, the short circuit current I_{SC} in panels is the subject of fractional short circuit current. Model predictive control adjusts control inputs and may require additional processing resources based on how effectively the MPPT is doing [9]. The above-mentioned techniques are conventional methods of MPPT and also there are metaheuristic, optimization, and adaptive strategies-based algorithms for tracking. Besides, solar, wind-based MPPT methods adjust operating parameters under intermittent conditions. Tip speed ratio control modifies the rotor speed; pitch angle control regulates the aerodynamic forces acting on the rotor and is employed in variable speed wind turbines; generator torque control uses sensor feedback to measure wind speed; rotor speed and generator torque to calculate control signal and handed-down for variable-speed wind turbines with direct-drive or geared generators and MPC to improve RES output performance and provide efficient energy are some methods for optimizing energy production with different control approaches. With the help of converters, batteries can store steady power [10]. SG depends on data and information that is specific to each source, storage, grid, and load, together with the corresponding power flow. Enabling real-time grid operation monitoring, optimization, and control are crucial components. SGs can enhance grid efficiency, reliability, security, and sustainability by utilizing data analytics, sophisticated sensors, and communication networks. Additionally, SGs can empower customers and facilitate the integration of renewable energy resources.

1.1 Overview of Renewable Integration

Renewable integration refers to the way RESs are integrated and efficiently used in modern power systems, particularly EGs. In order to complete this process, energy generated by RES such as solar, wind, hydro, geothermal, and biomass must be integrated into the conventional EG. The objective is to develop a more environmentally friendly and sustainable energy balance [11].

1.2 Integration of RESs

SG enables to integration of RESs like solar and wind by efficiently managing their variable and intermittent nature. This integration minimizes dependency on conventional fossil fuels (e.g. petroleum and coal) and improves sustainability. The integration of RESs into SGs is an important step in modernizing and optimizing the power infrastructure. SGs use advanced technologies to effectively handle the unpredictable and intermittent characteristics of RESs like solar and wind. Energy generation methods are classified into centralized and decentralized generation [12].

1.2.1 Wind Turbine Generating Power Plant

The system produces extremely low voltage at certain periods due to unpredictable circumstances, hence, MPPT may assist in maintaining that variable power response to fixed power response [13].

$$V = \frac{d}{t} \tag{1}$$

$$P_{wdp} = \frac{1}{2}\rho_{kg}A_{rsa}V_{mps} \tag{2}$$

$$P_{cap} = \eta_{wd} A_{sa} P_{wdp} \tag{3}$$

$$P_{otp} = \frac{1}{2} \rho_{kg} A_{sa} V_{mps}^{3} C_{pow} \tag{4}$$

'V' indicates the speed of the wind, 'd' is the distance, time period 't', wind speed V_{mps}, P_{wdp} is represented as wind turbine output power, ρ_{kg} is the air density in kg/m^3, A_{rsa} is the area swept by the wind turbine rotor in m^2. P_{otp} is the total output power of a wind turbine. C_{pow} is coefficients of wind power [14].

1.2.2 Solar PV System

$$G_i = G_{sct} * EOAC \tag{5}$$

$$P_{ot} = G_i * A * \eta_{ml} \tag{6}$$

$$\eta_{ml} = \frac{P_{ot}}{G_i * A} * 100\% \tag{7}$$

$$E_y = \int_{t1}^{t2} P_{ot}(t)dt \tag{8}$$

G_i indicates solar irradiance, units are W/m^2, G_{sct} referred as solar constant and atmospheric effects due to clouds, dust, and shades. P_{ot} represents output power in watts and A is the total area of the solar panel in m^2. η_{ml} indicates the efficiency of PV modules. E_y indicates the energy yield in kWh calibrated according to the given time period and t_1 and t_2.

1.2.3 Battery Storage System

Storage system plays a crucial role in energy storage and grid stabilization [15].

$$SOC = \frac{Q_{sto}}{Q_{max}} * 100 \tag{9}$$

$$\eta_{cge} = \frac{E_{ot}}{E_{it}} \tag{10}$$

$$\eta_{dge} = \frac{E_{ot}}{E_{it}} \tag{11}$$

$$E = Q * V \tag{12}$$

$$\eta_{Eng} = \frac{E_{ot}}{E_{it}} \tag{13}$$

where SOC indicates the state of charge and Q_{sto} is the actual charge stored and Q_{max} is the maximum capacity of the battery. η_{cge} and η_{dge} represent the charging and discharging efficiencies and E represents the total amount of electrical energy.

1.2.4 Centralized Generation

Centralized generation is a power generation concept in which electrical energy is generated at large, centralized facilities and then transmitted over long distances through the power grid to end users [16]. In this case, electrical energy can be generated by generating stations that use various kinds of fuels, such as coal, natural gas, nuclear, and renewable energy sources. The centralized generation technique is separate from decentralized or distributed energy generation, which generates electrical energy near the place of utilization, commonly utilizing smaller-scale, local facilities. Centralized generation has been a traditional and predominant model for supplying electricity to communities and industries. Figure 2 shows an easily understood illustration of centralized generation.

1.2.5 Decentralized Generation

Decentralized energy generation means generating electrical energy regionally, therefore reducing the need for significant transmission infrastructure. Solar panels and wind turbines improve grid efficiency and stability [17]. In smart grids, this technique minimizes distribution losses, which leads to a more efficient and adaptable energy system. The integration of distributed energy resources, such as rooftop solar panels, encourages develop a more sustainable and resilient energy system. Figure 3 shows a practical representation of decentralized generation.

1.3 Renewable Energy in Smart Grids

The intermittent supply and irregularities of RESs like WT and PV present significant challenges to RESs integration [18]. These sources' availability varies depending on conditions such as weather and time of day. To efficiently incorporate RE, robust SGs with improved communication to adopt variations in supply and demand. These EGs enable real-time monitoring and efficient management of renewables. Hybrid energy systems, such as solar-wind combinations, provide an uninterrupted supply

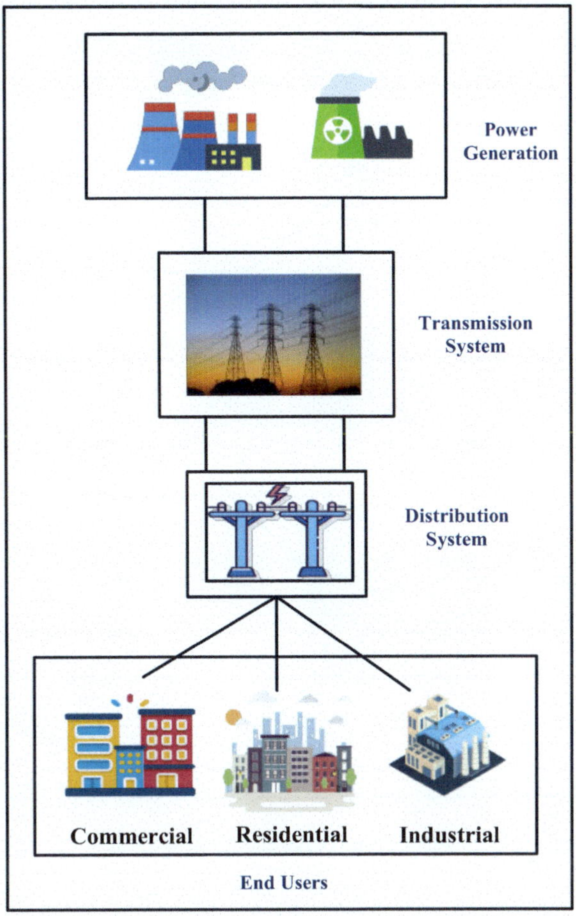

Fig. 2 Schematic diagram of centralized generation

of electricity by balancing with each source's intermittent nature. This strategy maximizes the utilization of resources and improves the stability of the grid, contributing to a more reliable and sustainable energy system [19].

1.3.1 Distributed Energy Resources in the Smart Grid

DERs include dispatchable sources, non-dispatchable sources, and energy storage systems.

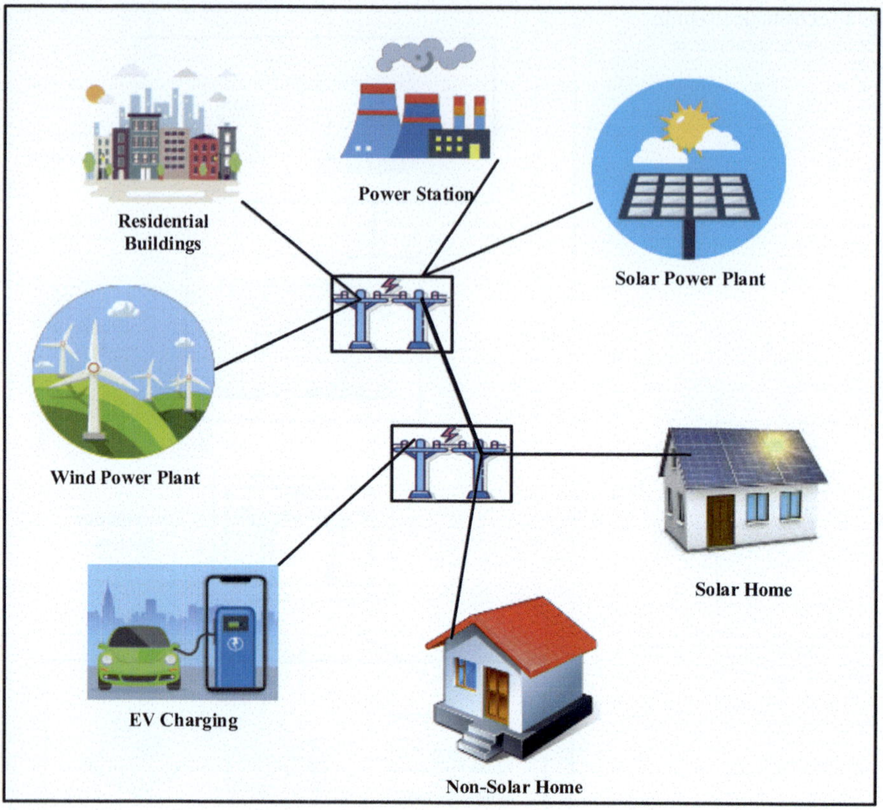

Fig. 3 Schematic diagram of decentralized generation

1.3.2 Renewable Energy Sources

RESs are listed below and a brief introduction about its generation of electrical energy. Figure 4 shows the schematic diagram of RESs.

- Solar Power: Photovoltaic cells use sunlight to generate electrical energy [20].
- Wind Power: To generate electrical energy from the wind, it is harnessed using wind turbines [21].
- Hydro Power: The flow of water, frequently through dams, electrical energy is generated.
- Geothermal Power: The process of generating electrical energy harnesses heat from the Earth's interior.
- Biomass: Organic materials such as garbage, wood, and agricultural wastes are transformed into electrical energy.

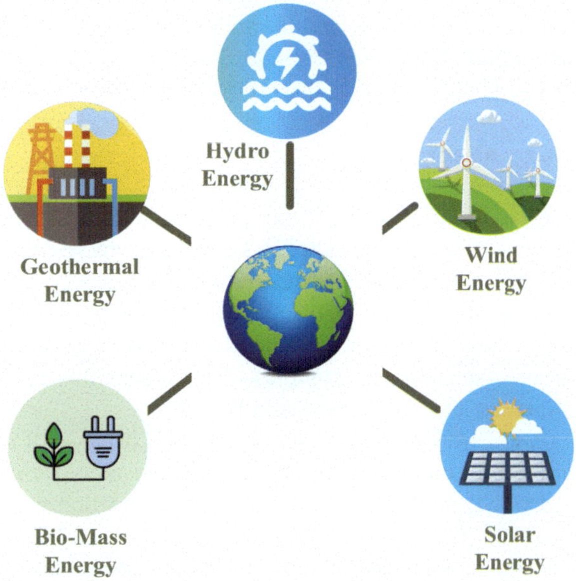

Fig. 4 Types of renewable energy sources

1.4 Electric Grid Operation

An electric grid is an interconnected network that distributes electricity from generation to the consumer.

1.4.1 Electric Grid

Grid typically refers to a network, an electric grid is an interconnected network that distributes electricity from producers/suppliers to the end user. It is made up of generating stations and high voltage transmission lines that transmit power from distant sources to demand centers, and distribution lines that connect individual customers [22].

1.4.2 Super Grid

Super grid also known as Mega grid, is a wide-area transmission network that transmits huge amounts of power over long distances. A smart grid is a modernized electric grid that uses digital or analog information and communication technology to automatically gather and act on information such as supplier and customer preferences to improve the reliability and sustainability of energy production and distribution. An MG is an integrated regional grid made up of DERs such as FC, PV, WT, power electronic converters, energy storage devices (such as batteries, supercapacitors, and flywheels), load, and it can be disconnected from the main grid, and operate autonomously in off-grid mode.

1.4.3 Microgrid

The electrical grid was facing the diffusion of distributed energy resources and the advance of power clusters [9], known as a microgrid. Alternating current (AC) MG and direct current (DC) MG are the two types of MG. DERs, such as PV and FC, often operate at a DC. To accommodate DC loads, a DC link can be established via power electronics circuits. PV nodes, batteries, DC terminals of power electronics circuits, and FC are examples of DC nodes that support obtaining the anticipated voltage level of the power supply [23].

A DC-MG is chosen over an AC-MG because of its sophisticated reliability and uninterruptible power supply [24], nonexistence of reactive power ensuing in lower losses [25], higher efficiency, simpler connection with the DC bus, lack of synchronization [26], frequency aspect [27], and future DC homes-DC loads such as LED, TV, laptop, dryer, and washer. MG has the potential to outperform AC MGs in terms of control parameters (DC: voltage and current, AC: voltage, current, frequency, phase, and reactive power), as well as DC being more efficient by removing the last rectification stage (AC/DC), as DC power generates the majority of the device end utilization. The distributed nature allows for the exclusion or optimization of the connection to the centralized grid, improving power quality to sensitive loads. In an islanded mode, an MG functions independently, utilizing DERs to meet load demands whereas in the grid-connected mode, the MG connects to the main grid to meet load demands [28].

1.5 Necessity of a DC MG?

Figure 5 shows the structure of DC MG. DC MGs have an advantage over AC MGs in terms of system efficiency, cost, and size. Using fewer power electronic converters leads to higher efficiency. Furthermore, AC/DC converters do not require a transformer, reducing the size of the DC MG significantly. DC MGs, like AC MGs, also require an energy management system.

However, no more than voltage stabilization is required. DC MGs do not require frequency stabilization support, as AC MGs do. The major components in a DC MG are solar photovoltaic systems, wind energy systems, battery energy storage systems, utility grids, power electronics converters, and electrical loads [29].

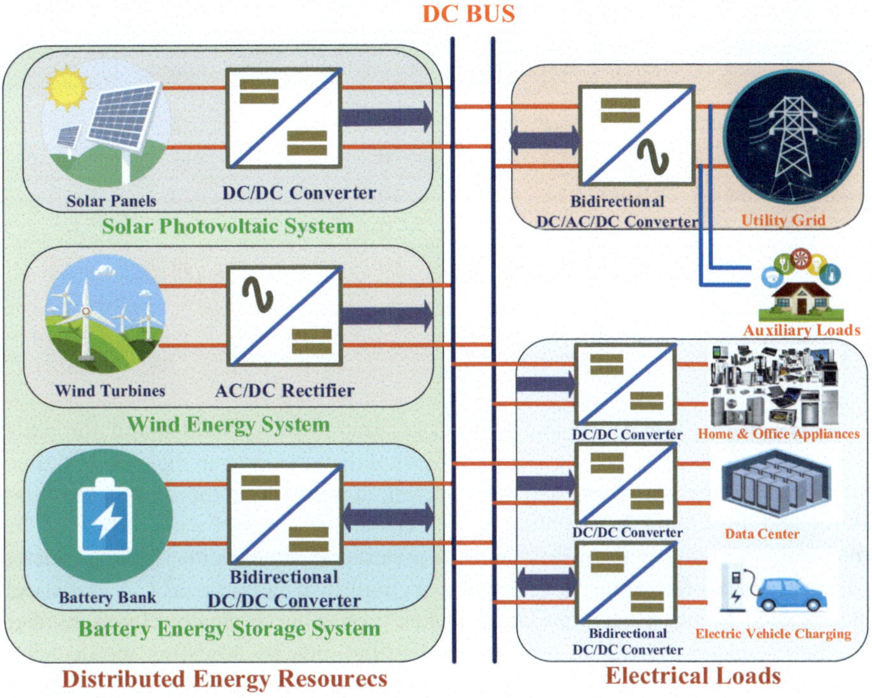

Fig. 5 Structure of DC MG

2 Problem Modelling

The primary objective is to minimize the total operational costs (TOC), which include the generation costs, no-load costs, and start-up/shutdown costs of generators. Table 1 indicates the generator cost coefficient and minimum and maximum power limits.

$$OC^t_{G1} = b_o\left(G^t_1\right)^2 + b_1 G^t_1 + b_2 \tag{14}$$

$$OC^t_{G2} = b_o\left(G^t_2\right)^2 + b_1 G^t_2 + b_2 \tag{15}$$

$$OC^t_{G,} = b_o\left(G^t_3\right)^2 + b_1 G^t_3 + b_2 \tag{16}$$

where OC^t_{G1} is the operational cost of generator 1; b_0, b_1 and b_2 indicates the cost coefficients of diesel generators.

Table 1 Generator cost coefficients

S.No	Parameters	Cost coefficient			Minimum power in kW	Maximum power in kW
		b_0	b_1	b_2		
1	Generator 1	0.01	2	10	0	12
2	Generator 2	0.02	3	8	0	15
3	Generator 3	0.015	1	12	0	20
4	PV	–	–	–	0	20
5	WT	–	–	–	0	20

3 Case Study

An MG consisting of WT, PV, and 3 diesel generators with load demand and electricity price is considered for the study. A Weibull distribution may be used to define wind speeds at a specific height and location, which is useful for monitoring the wind resource and improving the structure as well as the efficiency of wind turbines. The shape parameter 'k' affects the distribution curve shape, whereas the scale parameter 'c' determines the frequency at which the Weibull distribution peaks. A statistical method called Monte Carlo simulation makes it possible to create a large number of random scenarios based on probability distributions. By adjusting the decision variables, the objective is to reduce the microgrid's TOC [30]. The active power supplied by each generator, start-up, and shutdown times, and binary decision variables resulting from unit commitment are the decision variables in the OC minimization issue under re-distribution. Figures 6 and 7 indicate the Weibull distribution of wind turbines and the Beta distribution of PV power. Beta distribution parameters are mean and standard deviation and are considered as 13.45 and 0.42, respectively. Figure 8 indicates the monthly wind distribution. Deterministic optimization assumes that every variable and parameter in the optimization problem is known with certain and searches for the most optimal solution provided the fixed values of the parameters. It works well in situations when there are either very few or no uncertainties [31]. It could end in undesirable solutions if there are plenty of uncertainties or if the parameters alter in real-world situations.

Uncertainties in PV and wind can be addressed using stochastic optimization and robust optimization. Stochastic optimization is ideal for issues including inherent uncertainty in energy systems with intermittent renewable generation. Stochastic optimization issues are frequently tackled using scenario-based optimization and Monte Carlo simulation [32], whereas, robust optimization focuses on finding out with solutions that can handle a certain amount of uncertainty. Figure 9 indicates the scenario generation using MCS. It is evident from Fig. 9 that in total 100 scenarios have been generated and were reduced to 10 as shown in Fig. 10.

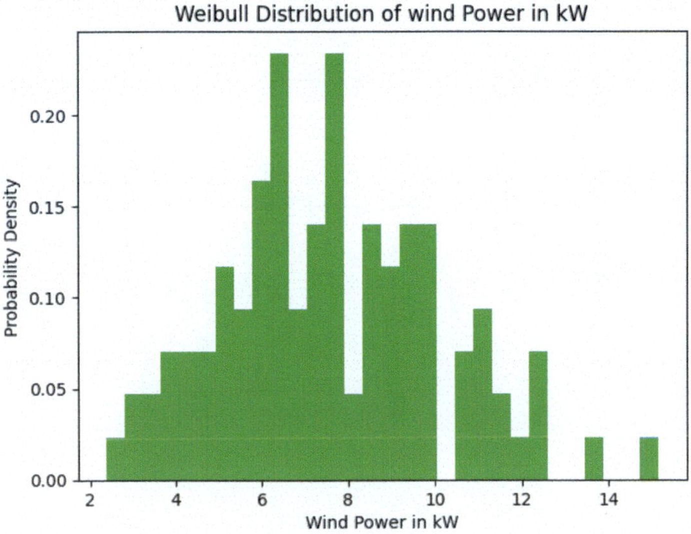

Fig. 6 Probability distribution of wind turbine

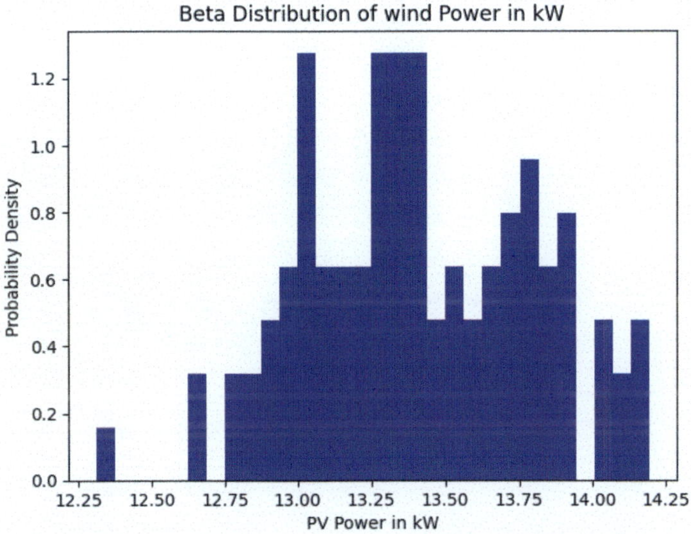

Fig. 7 Probability distribution of PV

Figure 11 indicates the hourly electricity price of the grid for 24-h scheduling instant. The energy management system compares the electricity price and the incremental operational cost of the MG and decides the commitment status of the generator. If the grid electricity price is greater than the incremental operational cost of the diesel generator then the generator is turned on and vice versa. The objective of this

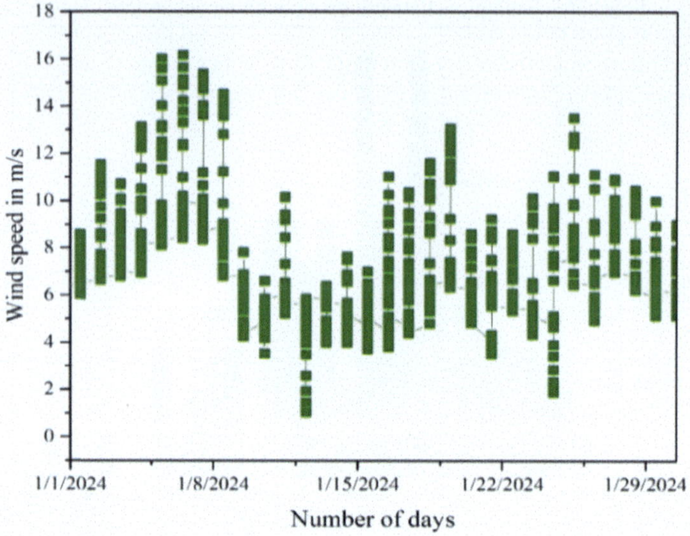

Fig. 8 Wind speed variation in a month

Fig. 9 Scenario generation using Monte Carlo simulation

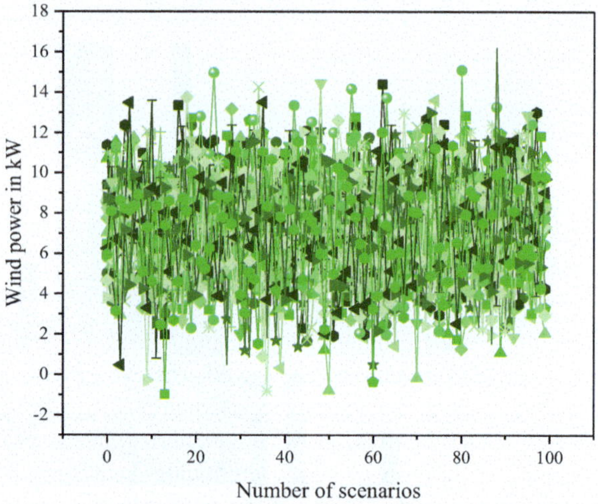

paper is to minimize the environmental impact by maximizing the renewable use of energy. Figure 12 indicates the hourly load, wind, and PV power.

Determine both the probability distributions and the unknown parameters.

Using random numbers, sample from these distributions to generate numerous scenarios. Every scenario depicts a potential way in which the unknown parameters may come into existence. K-means clustering is used to combine related situations after an extensive set of scenarios is generated. This reduces the number of scenarios

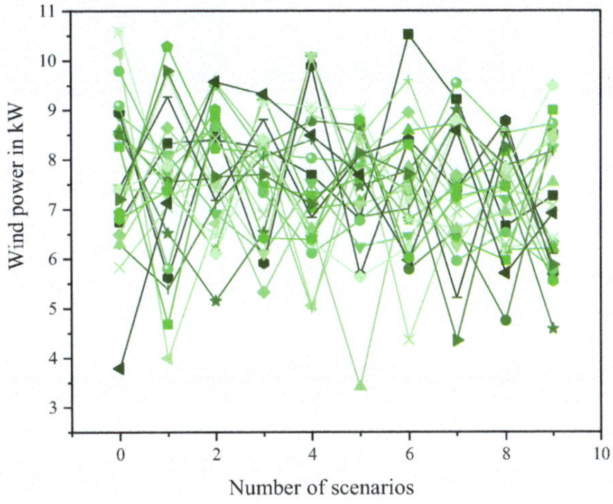

Fig. 10 Scenario reduction using K-means clustering

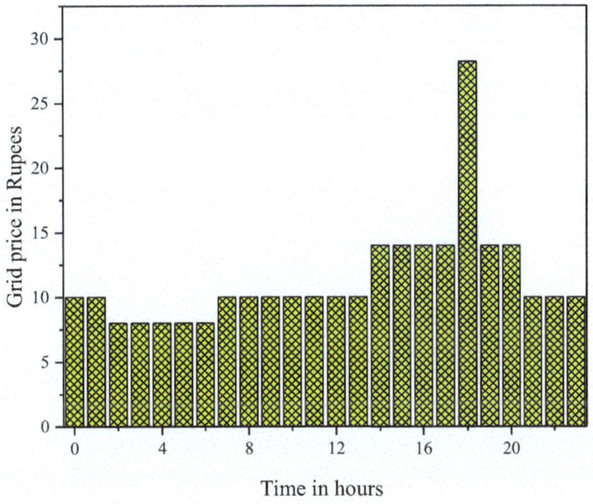

Fig. 11 Grid electricity price

yet preserves important features [33]. Table 2 indicates the unit commitment status of the sources along with the energy exchange with the grid.

A key aspect of DSM called demand response (DR) involves altering the way customers use power in response to signals from utilities or grid managers. It attempts to lessen grid stress, control demand during peak hours, and prevent the need for expansion. Demand response is the term used to describe how patterns of energy use alter in response to variations in the cost of electricity or incentives under regular or abnormal circumstances. Following fluctuations in the price of electricity, users

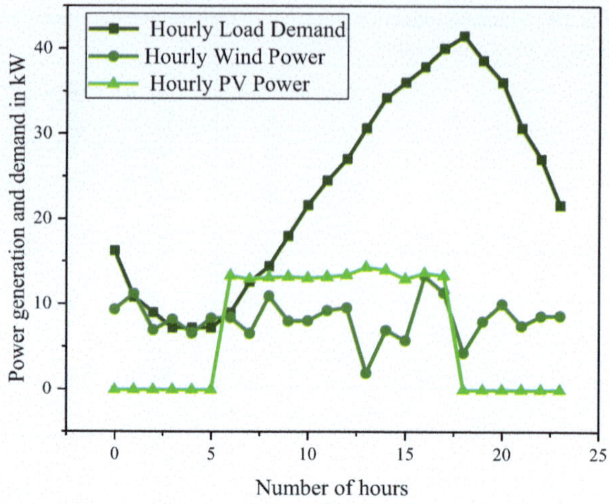

Fig. 12 Hourly load demand, PV power generation, and wind power generation

plan their loads according to their preferences. Customers reduce their loads in high-priced power regions and redistribute their loads in low-priced electricity regions. Figure 13 indicates the demand response programs available for energy management.

It Increases grid stability and dependability when demand is at its peak. It minimizes the requirement for infrastructure and expansion of power plants. Offers a flexible and adaptable way to address changes in the demand for power. It reduces the demand for fossil fuel-based power during peak hours, which promotes environmental sustainability. Figure 14 indicates the price elasticity of demand.

4 Conclusions

An introduction to conventional sources of energy generation, renewable energy sources, and their demerits have been presented in the beginning, and an introduction to smart grid technologies and various methods of power generation and various methods of energy management have been presented in detail. Further, the issues with uncertainties in renewable power generation are addressed by using MCS and K-means clustering. Demand response, demand response programs, and their advantages have been presented at last. Additionally, the integration of energy storage systems and their role in stabilizing the grid amid variations in renewable generation is discussed. Advanced forecasting techniques for renewable energy and load demand are also introduced to alleviate the effects of variability. Furthermore, hybrid energy management systems combining rule-based approaches with optimization algorithms are explored for improved efficiency. Lastly, the potential of demand-side management in reducing peak loads and enhancing grid reliability is emphasized, highlighting its critical role in future energy systems.

Table 2 Unit commitment status, energy exchange, price for energy exchange, and total generation

Hours	Generator 1	Generator 2	Generator 3	Energy exchange with Grid (kW)	Energy exchange with cost	Total generation in (kW)
1	OFF	OFF	OFF	6.8434	68.4348	9.4015
2	OFF	OFF	OFF	– 0.4141	– 4.1417	11.244
3	OFF	OFF	OFF	2.0092	16.0737	7.0157
4	OFF	OFF	OFF	– 1.028	– 8.2304	8.2488
5	OFF	OFF	OFF	0.5887	4.7101	6.6312
6	OFF	OFF	OFF	– 1.1617	– 9.2939	8.3817
7	OFF	OFF	OFF	– 12.824	– 102.59	21.849
8	OFF	OFF	OFF	– 6.9407	– 69.407	19.575
9	OFF	OFF	OFF	– 9.7349	– 97.349	24.174
10	OFF	OFF	OFF	– 3.2832	– 32.832	21.333
11	OFF	OFF	OFF	0.45926	4.5926	21.200
12	OFF	OFF	OFF	1.99403	19.940	22.553
13	OFF	OFF	OFF	3.96119	39.611	23.113
14	OFF	OFF	OFF	14.3328	143.32	16.352
15	ON	ON	ON	10.2441	143.41	24.050
16	ON	ON	ON	14.3061	200.28	200.28
17	ON	ON	ON	7.83063	109.62	30.074
18	ON	ON	ON	12.2754	171.85	27.795
19	ON	ON	ON	34.1586	963.20	7.3563
20	ON	ON	ON	27.6126	386.57	11.014
21	ON	ON	ON	23.0416	322.58	13.058
22	OFF	OFF	OFF	23.1822	231.82	7.5027
23	OFF	OFF	OFF	18.4286	184.28	8.6463
24	OFF	OFF	OFF	12.9688	129.68	8.6911

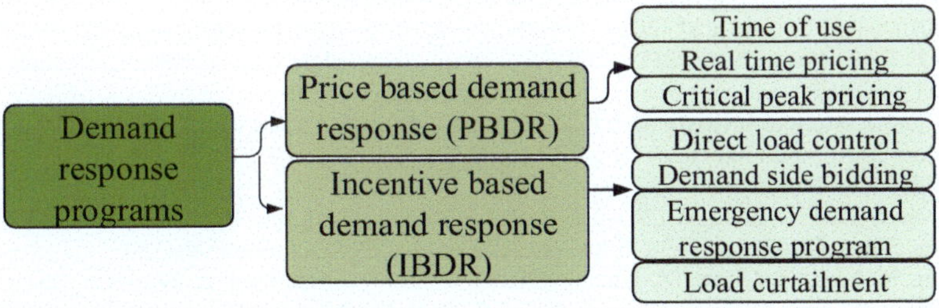

Fig. 13 Types of demand response programs

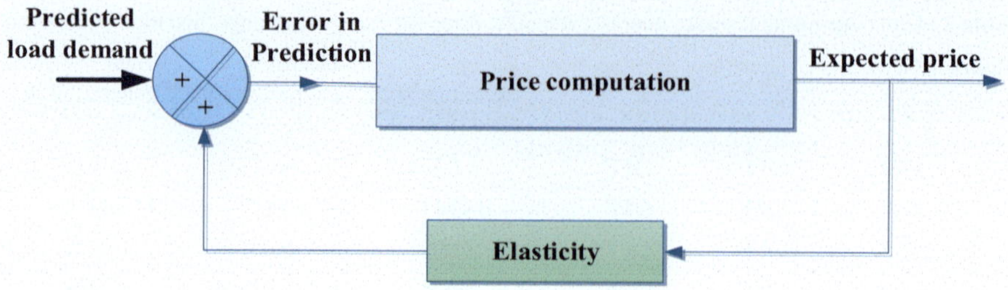

Fig. 14 Price elasticity of demand response

Acknowledgements This research work was supported by "Woosong University's Academic Research Funding—2025".

References

1. Talaq J, El-Hawary F, El-Hawary E (1994) A summary of environmental/economic dispatch algorithms. IEEE Trans Power Syst 9(3):1508–1516. https://ieeexplore.ieee.org/document/336110
2. Lamont J, Obessis E (1995) Emission dispatch models and algorithms for the 1990s. IEEE Trans Power Syst 10(2):941–947. https://ieeexplore.ieee.org/abstract/document/387937
3. Delson J (1974) Controlled Emission Dispatch. IEEE Trans Power Appar Syst 93(5):1359–1366. https://ieeexplore.ieee.org/document/4075500
4. Dibavar A, Ivatloo B, Zare K, Khalili T, Bidram A (2021) Economic-emission dispatch problem in power systems with carbon capture power plants. IEEE Trans Ind Appl 57(4):3341–3351. https://ieeexplore.ieee.org/document/9428528
5. Uddin M, Romlie M, Abdullah M, Halim S, Bakar A, Kwang T (2018) A review on peak load shaving strategies. RSER 82(3):3323–3332. https://doi.org/10.1016/j.rser.2017.10.056
6. Kirschen D, Strbac G, Cumperayot P, Mendes D (2000) Factoring the elasticity of demand in electricity prices. IEEE Trans Power Syst 15(2):612–617. https://ieeexplore.ieee.org/document/867149
7. Conejo A, Morales M, Baringo L (2010) Real-time demand response model. IEEE Trans Smart Grid 1(3):236–242. https://ieeexplore.ieee.org/document/5607339
8. Pourmousavi S, Nehrir M (2014) Introducing dynamic demand response in the LFC model. IEEE Trans Power Syst 29(4):1562–1572. https://ieeexplore.ieee.org/abstract/document/7285784
9. Medina J, Muller N, Roytelman I (2010) Demand response and distribution grid operations: opportunities and challenges. IEEE Trans Smart Grid 1(2):193–198. https://ieeexplore.ieee.org/document/5545489
10. Parvania M, Firuzabad M (2010) Demand response scheduling by stochastic SCUC. IEEE Trans Smart Grid 1(1):89–98. https://ieeexplore.ieee.org/document/5464384
11. Deng R, Yang Z, Chow M, Chen J (2015) A survey on demand response in smart grids: mathematical models and approaches. IEEE Trans Ind Inform 11(3):570–582. https://ieeexplore.ieee.org/document/7063260
12. Rahimi F, Ipakchi A (2010) Demand response as a market resource under the smart grid paradigm. IEEE Trans Smart Grid 1(1):82–88. https://ieeexplore.ieee.org/document/5454394
13. Mathieu J, Price P, Kiliccote S, Piette M (2011) Quantifying changes in building electricity use, with application to demand response. IEEE Trans Smart Grid 2(3):507–518. https://ieeexplore.ieee.org/document/5772947

14. Palensky P, Dietrich D (2011) Demand side management: demand response, intelligent energy systems, and smart loads. IEEE Trans Ind Inform 7(3):381–388. https://ieeexplore.ieee.org/document/5930335
15. Zhou X (2022) Exploiting integrated demand response for operating reserve provision considering rebound effects. IEEE Access 10(2):15151–15162. https://ieeexplore.ieee.org/document/9701332
16. Xiaoyang Z, Weixin Y, Johnson R (2016) Forecast load impact from demand response resources. In: 2016 IEEE power and energy society general meeting 2(3):1–5. https://ieeexplore.ieee.org/document/7741992
17. Bitaraf H, Rahman S (2018) Reducing curtailed wind energy through energy storage and demand response. IEEE Trans Sustain Energy 9(1):228–236. https://ieeexplore.ieee.org/document/7972993
18. Shi W, Li N, Xie X, Chu C, Gadh R (2014) Optimal residential demand response in distribution networks. IEEE J Selected Areas Commun 32(7):1441–1450. https://ieeexplore.ieee.org/abstract/document/6840288
19. Sibo N, Zhou M, Li G (2017) Optimal residential community demand response scheduling in smart grid. Appl Energy 210(2):1280–1289. https://doi.org/10.1016/j.apenergy.2017.06.066
20. Tarish H, See O, Elmenreich W (2016) A review of residential demand response of smart grid. RSER 59(2):166–178. https://doi.org/10.1016/j.rser.2016.01.016
21. Niamh C, Pierre P, Madsen M, Mark O (2014) Benefits and challenges of electrical demand response: a critical review. RSER 39(1):686–699. https://doi.org/10.1016/j.rser.2014.07.098
22. Rezaee A (2019) Optimisation of demand response in electric power systems, a review. RSER 103(1):308–319. https://doi.org/10.1016/j.rser.2018.12.054
23. Hussain M, Yan G (2018) A review of demand response in an efficient smart grid environment. Electr J 31(5):55–63. https://doi.org/10.1016/j.tej.2018.06.003
24. Siano P (2014) Demand response and smart grids—a survey. RSER 30(4):461–478. https://doi.org/10.1016/j.rser.2013.10.022
25. Jung CM, Pagidipala S, Salkuti SR (2024) Emerging technologies for the integration of renewable energy, energy storage and electric vehicles. In: Salkuti SR (eds) Energy and environmental aspects of emerging technologies for smart grid. Green energy and technology. Springer, Cham. https://doi.org/10.1007/978-3-031-18389-8_5
26. Salkuti SR (2022) Modeling of various renewable energy resources for smart electrical power systems. In: Salkuti SR, Ray P (eds) Next generation smart grids: modeling, control and optimization. Lecture notes in electrical engineering, vol 824. Springer, Singapore. https://doi.org/10.1007/978-981-16-7794-6_2
27. Faria S, Spínola VZ (2018) Reschedule of distributed energy resources by an aggregator for market participation. Energies 11(713):1–20. https://doi.org/10.3390/en11040713
28. Zhang G, Li Y, Deng X (2020) K-means clustering-based electrical equipment identification for smart building application. Information 11(27):25–35. https://doi.org/10.3390/info11010027
29. Faria S, Spínola J, Vale Z (2018) Distributed energy resources scheduling and aggregation in the context of demand response programs. Energies 11(1):987–992. https://doi.org/10.3390/en11081987
30. Abrishambaf O, Faria S, Vale Z (2020) Ramping of demand response event with deploying distinct programs by an aggregator. Energies 13(1):138–145. https://doi.org/10.3390/en13061389
31. Lezama F, Faria S, Vale Z (2020) Demand response of residential houses equipped with PV-battery systems: an application study using evolutionary algorithms. Energies 13(1):24–36. https://doi.org/10.3390/en13102466
32. Salkuti SR, Ray P, Pagidipala S (2022) Overview of next generation smart grids. In: Salkuti SR, Ray P (eds) Next generation smart grids: modeling, control and optimization. Lecture notes in electrical engineering, vol 824. Springer, Singapore. https://doi.org/10.1007/978-981-16-7794-6_1
33. Sravanthi P, Sandeep V (2022) Solving realistic reactive power market clearing problem of wind-thermal power system with system security. Int J Emerg Electr Power Syst 23(2):125–144. https://doi.org/10.1515/ijeeps-2021-0060

A Review of AI-Driven Energy Optimization Strategies for Electric Vehicles

Debani Prasad Mishra, Priyanshu Yadav, Arul Kumar Dash, and Surender Reddy Salkuti

Abstract The integration of electric vehicles (EVs) into modern transportation systems necessitates sophisticated energy management strategies to optimize their performance and sustainability. This chapter explores the pivotal role of Artificial Intelligence (AI) in advancing smart energy management for EVs. Covering various aspects of AI applications, such as predictive modeling for energy consumption, adaptive control strategies, genetic algorithms for charging station optimization, and integration of AI techniques for synergistic benefits. Beginning with predictive modeling, we explore how AI augments predictive analytics in EVs, delving into data sources, feature selection, and applications in forecasting energy consumption and scheduling maintenance tasks. This chapter further delves into AI-driven adaptive control strategies, detailing the techniques involved in managing EV energy efficiently, including dynamic energy distribution algorithms, optimal charging strategies, and real-time adaptation to grid conditions and user preferences. The chapter also explores genetic algorithms for optimizing the locations of charging stations, covering an overview of genetic algorithms, the factors influencing optimization decisions, and the objectives and constraints involved. Additionally, it investigates the integration and synergies among various AI techniques in EV energy optimization, discussing potential combinations, and synergies between battery management.

Keywords Electric vehicle · Artificial intelligence · Genetic algorithm · Predictive modelling · Battery · Energy system

---

D. P. Mishra · P. Yadav · A. K. Dash
Department of Electrical and Electronics Engineering, IIIT Bhubaneswar, Bhubaneswar, Odisha, India
e-mail: debani@iiit-bh.ac.in

S. R. Salkuti (✉)
Department of Railroad and Electrical Engineering, Woosong University, Daejeon, Republic of Korea
e-mail: surender@wsu.ac.kr

S. R. Salkuti (ed.), *Artificial Intelligence for Integrated Smart Energy Systems in Electric Vehicles*, Lecture Notes in Electrical Engineering 1427,
https://doi.org/10.1007/978-3-031-94276-1_8

Abbreviations

EV	Electric vehicle
AI	Artificial intelligence
SoH	State of health
SoC	State of charge
RUL	Remaining useful life
ML	Machine learning
ANN	Artificial neural network
MLR	Multiple linear regression
SVR	Support vector regression
KNN	K-nearest neighbors
RF	Random forest
GA	Genetic algorithm
SI	Swarm intelligence
RL	Reinforcement learning
NLP	Natural language processing
GNN	Graph neural network
DL	Deep learning
FL	Fuzzy logic

1 Introduction

Electric vehicles (EVs) have swiftly evolved from a niche market into a mainstream transportation solution, poised to revolutionize the automotive industry. This transformation is driven by rising concerns over climate change, advancements in battery technology, enhanced charging infrastructure, and supportive government policies. EVs have emerged as a sustainable alternative to traditional internal combustion engine vehicles. Central to this EV revolution is the concept of smart energy management, an advanced ecosystem comprising algorithms, sensors, and control systems, all designed to maximize the efficiency of the energy stored in the vehicle's battery [1]. Unlike vehicles powered by fossil fuels, EVs depend entirely on electricity for propulsion, necessitating a fundamental shift in energy management and utilization practices. Smart energy management involves various strategies aimed at optimizing every aspect of energy consumption within an EV. This includes monitoring the state of charge and the health of the battery pack, dynamically adjusting powertrain operations, and managing the usage of ancillary systems. Every aspect of energy use is meticulously orchestrated to ensure optimal performance, extended range, and a superior user experience. The effectiveness of these smart energy management systems is crucial for the success of EV technology. With consumers demanding vehicles that offer longer ranges, faster charging times, and lower operating costs, the pressure to innovate and enhance energy usage is immense [2]. In this context, the integration of artificial intelligence (AI) has emerged as a transformative force, providing exceptional capabilities in data analysis, pattern recognition, and real-time decision-making. AI-driven energy optimization represents a significant shift in how EVs are managed and operated. Utilizing the power of machine learning and data analytics, AI algorithms can analyze vast amounts of data, including driving patterns, traffic conditions, battery health, and environmental factors, to make informed decisions regarding energy usage [3]. This predictive and adaptive approach allows EVs to anticipate and respond to changing conditions, maximizing efficiency and range in real-time. AI-driven energy optimization offers the potential for personalized user experiences, where the vehicle learns and adapts to individual driving habits, preferences, and environmental conditions. Through continuous learning and optimization, EVs equipped with AI-driven energy management systems can tailor their performance to meet the unique needs of each driver, thereby enhancing satisfaction and adoption rates. In summary, smart energy management powered by AI is foundational to the EV revolution, unlocking unprecedented levels of efficiency, performance, and sustainability. As the automotive industry continues its shift towards electrification, the role of AI in optimizing energy usage will become increasingly important, shaping the future of transportation for generations to come.

1.1 Overview of Smart Energy Management in EVs

EVs represent a revolutionary shift in transportation, offering a sustainable solution to the environmental challenges posed by traditional internal combustion engine (ICE) vehicles. Central to this transformation is smart energy management, a dynamic and intricate system designed to maximize the utilization of energy stored within the vehicle's battery. This multifaceted approach includes a range of advanced technologies and strategies meticulously engineered to extract the utmost efficiency from electric propulsion. Battery management is a cornerstone of smart energy management in EVs. The battery pack, as the primary energy reservoir, must be utilized efficiently to optimize the vehicle's range and performance [4]. This involves sophisticated algorithms and sensors that monitor the battery's state of charge (SoC) and state of health (SoH), ensuring optimal charging and discharging cycles to prolong its lifespan and maintain peak performance. Powertrain control is another critical component of smart energy management, managing the distribution of power from the battery to the electric motor(s) and other propulsion components. Through real-time adjustments of torque, speed, and regenerative braking, powertrain control systems optimize energy usage while meeting the driver's performance demands. This dynamic optimization ensures a seamless driving experience while maximizing energy efficiency. Thermal management systems play a vital role in maintaining the optimal operating temperature of the battery pack and other critical components. By regulating temperature fluctuations, these systems prevent overheating or excessive cooling, thereby optimizing energy efficiency and extending the lifespan of the battery and other components. Additionally, ancillary systems optimization involves strategies aimed at minimizing overall power consumption. From intelligent climate control to efficient lighting systems, every aspect of ancillary energy usage is scrutinized and optimized to reduce energy wastage and extend the vehicle's range. Regenerative braking is a hallmark feature of EVs, enabling the vehicle to recover kinetic energy during deceleration and braking. This energy is then stored back in the battery, further enhancing overall efficiency and extending the vehicle's range [5]. Smart energy management in EVs represents a convergence of advanced technologies and innovative strategies aimed at maximizing efficiency, performance, and sustainability. By leveraging the capabilities of battery management, powertrain control, thermal management, ancillary systems optimization, and regenerative braking, EVs are set to revolutionize the future of transportation, offering a cleaner, greener, and more efficient mode of mobility.

1.2 Importance of AI in Energy Optimization

In the relentless pursuit of enhancing the range and performance of EVs, the pivotal role of AI in energy optimization emerges as a linchpin of innovation. AI algorithms, leveraging the prowess of machine learning and data analytics, serve as the backbone of real-time energy optimization strategies, scrutinizing an extensive array of parameters to fine-tune energy usage dynamically. The unparalleled ability of AI to analyze complex datasets, encompassing driving patterns, traffic dynamics, weather forecasts, and battery health metrics, enables EVs to adapt swiftly to the ever-changing environment, thereby maximizing efficiency and extending range. The significance of AI in energy optimization lies in its capacity to orchestrate a symphony of control across the vehicle's systems, ranging from powertrain dynamics to thermal management and ancillary components. By harnessing AI-driven algorithms, EVs can achieve a level of precision and adaptability that was hitherto unattainable, thereby unlocking new thresholds of performance and efficiency [6]. Through intelligent modulation of powertrain operations, AI optimizes torque delivery, throttle response, and regenerative braking, ensuring an optimal balance between power demand and energy consumption, even in the face of fluctuating driving conditions.

Furthermore, AI empowers EVs to navigate the complexities of thermal management with finesse, maintaining optimal operating temperatures for the battery pack and critical components. By predicting and pre-emptively mitigating thermal fluctuations, AI-driven thermal management systems not only enhance energy efficiency but also safeguard the longevity and reliability of key components. Moreover, auxiliary systems optimization, facilitated by AI algorithms, ensures that energy-intensive functionalities such as climate control and infotainment operate with maximal efficiency, minimizing unnecessary energy drain and maximizing the vehicle's driving range. In essence, the integration of AI in energy optimization heralds a new era of intelligent mobility, where EVs transcend their conventional limitations to deliver unprecedented levels of performance, efficiency, and user satisfaction. As AI continues to evolve and permeate every facet of automotive technology, the trajectory toward a sustainable and electrified future becomes increasingly promising, propelled by the transformative potential of intelligent energy optimization in EVs.

2 Battery Life Cycle Management

2.1 *Battery Degradation*

Battery degradation in EVs is a multifaceted process characterized by the gradual loss of battery capacity and performance over time. This degradation stems from various factors intrinsic to battery chemistry, usage patterns, and environmental conditions. EV batteries commonly employ lithium-ion chemistry, subject to chemical reactions during charging and discharging cycles. These reactions induce physical changes within battery cells, contributing to degradation. Additionally, each charge–discharge cycle contributes to degradation, as the active materials within battery electrodes undergo structural changes, resulting in reduced capacity and efficiency. Even during periods of inactivity, batteries experience calendar aging, where chemical reactions within cells lead to the breakdown of electrolytes and electrode materials, further diminishing performance. Temperature plays a pivotal role in battery degradation [7]. High temperatures accelerate chemical reactions and material breakdown within battery cells, exacerbating degradation processes. Conversely, extremely low temperatures can also negatively impact battery performance and longevity. Depth of discharge is another critical factor; deeper discharge cycles, where batteries are drained to lower levels, can accelerate degradation compared to shallow discharge cycles. Charging practices, such as fast charging and frequent rapid charging, also contribute to degradation by generating heat within battery cells. This heat can accelerate degradation processes and lead to premature battery aging. Overcharging or over-discharging the battery beyond recommended limits can cause irreversible damage to battery cells, hastening degradation and shortening battery life. Battery degradation is often quantified in terms of SoH, representing the battery's capacity relative to its original capacity. A decline in SoH over time indicates degradation and reduced performance [8]. Environmental factors, including humidity, exposure to pollutants, and vibrations, also influence battery degradation. Proper storage and maintenance practices are essential to mitigate the effects of these factors and prolong battery life.

2.2 State of Charge (SoC) and State of Health Monitoring (SoH)

In EVs, the SoC and the SoH of the battery serve as indispensable metrics, offering profound insights into the battery's present state and functionality. These parameters play a pivotal role in determining the operational capabilities, longevity, and overall efficiency of the battery system within EVs. Understanding and effectively managing SoC and SoH are paramount for optimizing driving range, ensuring reliability, and prolonging the lifespan of EV batteries. SoC reflects the instantaneous energy level stored within the battery, informing drivers of available range and facilitating efficient energy management. Meanwhile, SoH provides a comprehensive assessment of the battery's health and remaining capacity, enabling proactive maintenance and informed decisions regarding battery replacement or refurbishment [9]. In this discussion, we'll explore the significance, measurement, and implications of SoC and SoH in the context of EV battery management, highlighting their indispensable role in driving the transition toward sustainable transportation solutions.

The SoC signifies the current level of charge stored within the battery, expressed as a percentage of its total capacity. Essentially, it serves as a direct indicator of the amount of energy accessible in the battery at any given moment. This dynamic parameter fluctuates as the battery undergoes charging or discharging cycles during regular operation. SoC is commonly depicted as a percentage, ranging from 0% when the battery is completely discharged to 100% when it's fully charged. This representation offers users real-time visibility into the available energy for driving purposes. Whether displayed on the vehicle's dashboard or accessed through the Battery Management System (BMS), SoC enables drivers to accurately assess the remaining driving range before necessitating a recharge. The maintenance of an optimal SoC range is imperative for maximizing both battery lifespan and performance. Extreme SoC levels, whether fully charged or fully discharged, can accelerate degradation and significantly diminish the overall longevity of the battery [10]. Therefore, strategic SoC management is crucial for EVs to ensure the efficient utilization of available energy, optimize driving range, and mitigate the risk of premature battery wear and deterioration. In summary, SoC serves as a vital parameter in EV battery management, offering real-time insights into energy availability and facilitating informed decision-making regarding charging and driving practices. By maintaining an optimal SoC range, EV owners can enhance battery durability, prolong operational life, and optimize overall vehicle performance.

The SoH of the battery provides a comprehensive evaluation of its overall health and condition in relation to its original capacity. This parameter serves as a crucial indicator of the battery's degradation over time and its remaining usable capacity compared to when it was new. SoH is typically expressed as a percentage, representing the ratio of the battery's current capacity to its original capacity. A SoH of 100% signifies that the battery is operating at its full capacity, while a lower SoH indicates degradation and reduced capacity. Several factors influence SoH, including battery chemistry, usage patterns, operating conditions, and environmental factors.

Over time, these variables contribute to gradual degradation and a decline in SoH [11]. Monitoring SoH is essential for evaluating battery performance, predicting the remaining lifespan, and determining the need for maintenance or replacement. It enables proactive maintenance strategies and facilitates optimization of battery usage to extend its operational lifespan. To assess SoH accurately, advanced diagnostic techniques such as predictive modeling and battery health monitoring systems are employed. These techniques analyze data from the battery management system, including voltage, current, temperature, and cycling behavior, to estimate SoH and identify potential degradation trends.

2.3 Predicting Battery Health

Lithium-ion (Li-ion) batteries are widely praised for their numerous advantages, such as their impressive capacity to store energy and their ability to deliver high power. Consequently, they are extensively utilized in a range of sectors, including renewable energy storage, electric transportation, telecommunications infrastructure, and various commercial uses. However, Li-ion batteries face performance degradation over time due to the irreversible deterioration of their internal chemical components, a phenomenon known as battery aging [12]. This degradation poses significant challenges to the safety, reliability, and performance of batteries, potentially leading to system failures, fires, or explosions.

To address these issues and ensure the safe and reliable operation of lithium-ion batteries, the field of battery health management has made significant strides. Key metrics for evaluating battery health during use include SoH and Remaining Useful Lifetime (RUL). Researchers have investigated numerous methods for estimating SoH and predicting RUL, with current strategies primarily falling into two categories: physical model-based approaches and data-driven techniques. With the advent of advanced data analytics and artificial intelligence technologies, data-driven methods have become the leading approach in battery health analysis. These techniques involve collecting extensive data on battery usage, extracting features related to degradation through feature engineering, and using data-driven algorithms to develop models that reflect battery aging characteristics [13]. Unlike traditional physical model-based methods, data-driven approaches do not require precise modeling of complex electrochemical processes or detailed battery parameter identification, offering greater flexibility, robustness, and adaptability [4].

In Fig. 1, the prediction of battery life using data is illustrated. The field of predicting the Remaining Useful Lifetime (RUL) of lithium-ion batteries primarily utilizes three methodologies: model-based, data-driven, and hybrid approaches. Model-based methods rely on the Butler-Volmer kinetic equation to evaluate side reaction rates and the increase in membrane resistance. Data-driven techniques, such as support vector machines and neural networks, have gained popularity due to their simplicity and ease of implementation. Hybrid methods aim to enhance predictive

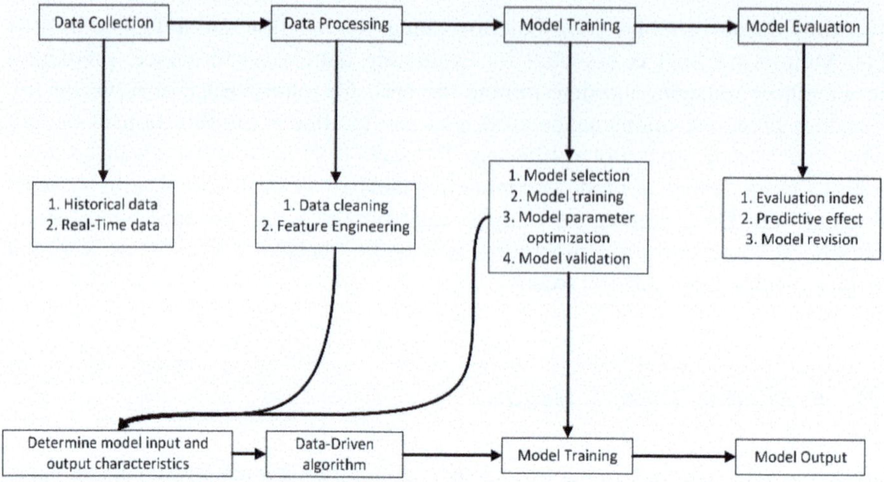

Fig. 1 Battery life prediction through data-driven model

accuracy but often involve complex parameter identification processes. While model-based methods offer valuable insights into electrochemical processes, they are often cumbersome and impractical for real-world applications. In contrast, data-driven approaches, including neural networks like backpropagation (BP), are preferred for their broad applicability in estimating battery SoH and predicting RUL. However, traditional BP neural networks may encounter convergence issues during training, underscoring the need for improved optimization techniques [14]. The integration of artificial intelligence and data-driven methodologies holds significant promise for advancing battery health management. These approaches ensure the longevity, reliability, and safety of lithium-ion batteries in various applications, making them a vital component in the ongoing development of battery technology.

2.3.1 Effect of Voltage on the Lifespan of Lithium Batteries

Theoretical analysis highlights the significant impact of voltage on the lifespan of lithium-ion batteries. Overcharging or over-discharging can lead to material damage and safety risks. Deep discharges below thresholds such as 2.49 V or 1.9 V can rapidly accelerate battery degradation, resulting in irreversible damage like internal metal short circuits that make the battery unusable [15]. While BMSs have become standard solutions to mitigate overcharge and deep discharge risks, determining the optimal charging voltage remains a complex challenge.

Lithium-ion batteries typically use materials such as NMC (Nickel Manganese Cobalt) and NCA (Nickel Cobalt Aluminium), whose structural stability greatly influences their capacity. During charging cycles, lithium ions are released from NMC materials, which increase capacity. However, empirical data from leading research institutions indicate a proportional relationship between the charging voltage and the extent of lithium extraction in NMC materials [16]. Higher voltages increase material instability, thereby reducing cycling life expectancy. While a higher voltage enhances battery capacity, it also accelerates damage to material cycle stability. Therefore, selecting a charging voltage that aligns with the battery's design life is crucial. The optimal charging voltage is closely tied to the characteristics of the battery materials. Studies show that NMC811 material reaches peak discharge energy density at a charging voltage of 4.31 V. Additionally, NMC111 material attains its highest energy density at a 4.51 V charging voltage [5]. However, due to material degradation at higher voltages, sustained cycling reveals that lower voltages result in higher discharge energy densities. Furthermore, materials with lower lithium content have poorer structural stability, highlighting the complex interplay between charging voltage, material characteristics, and battery performance.

2.3.2 Effect of Current on the Lifespan of Lithium Batteries

The lifespan of lithium-ion batteries is significantly influenced by the currents used during charging and discharging due to their impact on the durability of battery materials. At a fixed voltage limit, lower currents result in greater lithium extraction, leading to increased material instability. During the battery formation process, smaller currents are typically used to reach the voltage limit, resulting in relatively higher capacities. Data on the charging capacity of lithium-ion materials at various current levels indicate that smaller currents yield higher capacities, while larger currents result in lower capacities [6]. Interestingly, despite the higher initial capacities achieved with smaller currents, extensive formation tests show that materials formed with higher currents exhibit greater capacities over time. This suggests that higher currents cause less structural damage to the materials, improving stability and cycle performance. However, determining the optimal current for battery formation is complex and necessitates ongoing research within battery management systems [17]. Balancing the goal of high initial capacity achieved with smaller currents and improved cycle life attained through increased currents requires careful consideration and continuous refinement.

2.3.3 Evaluating the Life of Lithium Batteries Based on Data-Driven Methods

Assessing the lifespan of lithium batteries using data-driven methods involves applying regression analysis techniques or machine learning algorithms to identify relationships between battery health indicators and degradation patterns. By utilizing data extracted from battery performance, such as degradation state and maximum capacity or internal resistance, predictive models can be developed to estimate the battery's SOH and predict its Remaining Useful Life (RUL). Unlike traditional approaches, data-driven methods circumvent the complexities of electrochemical reactions and nonlinear degradation mechanisms within batteries, relying solely on battery data accumulation to construct mathematical models for analysis and prediction [18]. Consequently, these methods offer robustness and adaptability, making them widely utilized for SOH estimation and RUL prediction. Typically, after lithium battery production, the lifespan ranges between 300 and 500 charge–discharge cycles, with testing data primarily confined to 300 cycles. Despite variations in attenuation curves among different lithium batteries, a common exponential decay pattern emerges with increasing cycle numbers, showcasing a key characteristic of battery capacity attenuation. Although current data-driven technologies for Li-ion battery SOH estimation and RUL prediction are relatively mature, challenges persist.

Primarily, data-driven models often lack physical interpretability, relegating them to black-box models. Integrating physical models with data-driven approaches has been attempted to improve interpretability, but accurately describing battery electrochemical characteristics remains elusive. Additionally, data-driven models rely heavily on the quality and relevance of the data used, often resulting in poor prediction performance for untrained operating modes. Future trends suggest leveraging advanced technologies like industrial big data and cloud computing to analyze extensive battery data from practical scenarios, enhancing SOH analysis and evaluation.

Lastly, current battery data-driven models predominantly utilize easily measurable parameters such as voltage, current, and temperature. However, non-invasive detection technologies offer the potential for more comprehensive battery health monitoring through direct detection of internal structural changes, temperature distribution, and electrolyte degradation [19]. These technologies could significantly enhance battery management systems, improving SOH monitoring and evaluation capabilities.

3 Predictive Modelling of Energy Consumption and Maintenance

The predictive modeling of energy consumption and maintenance stands at the forefront of innovation in optimizing the performance and sustainability of EVs. As the integration of EVs into modern transportation systems accelerates, the need for sophisticated energy management strategies becomes increasingly paramount. Predictive modeling offers a proactive approach to address challenges related to energy consumption forecasting and maintenance scheduling, ensuring efficient utilization of resources while enhancing vehicle reliability and longevity.

In this section, we delve into the significance of predictive modeling in the context of EV energy management, exploring its role in forecasting energy consumption patterns and scheduling maintenance tasks. From analyzing data sources to selecting

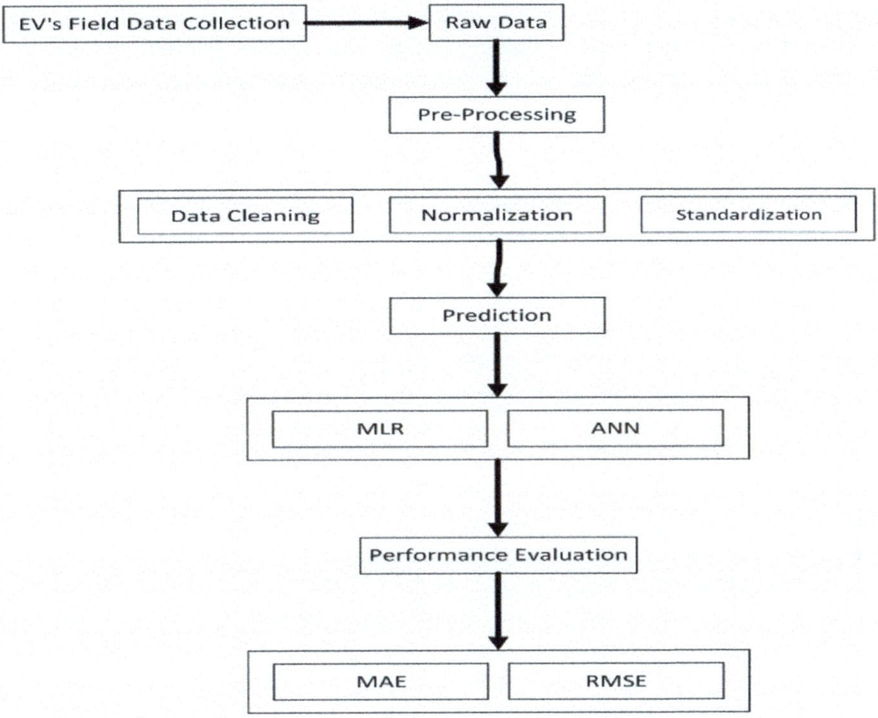

Fig. 2 Framework for predicting energy consumption in the proposed EVs

appropriate machine learning (ML) algorithms, we navigate through the intricacies of predictive modeling techniques tailored specifically for EVs [20]. By harnessing the power of predictive analytics, EV manufacturers and operators can gain valuable insights into energy usage trends, enabling them to optimize charging schedules, enhance battery performance, and ultimately extend vehicle lifespan. Throughout this exploration, we not only uncover the technical intricacies of predictive modeling but also highlight its practical implications for the EV industry. From improving operational efficiency to mitigating maintenance costs, predictive modeling emerges as a transformative tool in shaping the future of sustainable transportation. Join us as we embark on a journey to unlock the full potential of predictive modeling in revolutionizing energy management for EVs [7]. Figure 2 shows the outline of the energy consumption method.

3.1 Importance of Predictive Maintenance

The importance of predictive maintenance in the predictive modeling of energy consumption cannot be overstated, particularly in the context of addressing the urgent need for energy conservation and sustainability. As energy conservation has become a top priority in recent years, organizations are increasingly focused on reducing energy wastage in large energy-consuming entities such as buildings, factories, and office spaces. One of the major challenges in this effort is the lack of awareness about the sources of energy wastage and effective strategies to reduce it. Studies have shown that well-maintained machines and assets can potentially use up to 20% less energy, demonstrating the significant energy-saving potential of maintenance practices [8]. Achieving such energy savings requires the adoption of proactive maintenance strategies, such as Energy-Centred Maintenance (ECM). ECM is a continuous improvement maintenance regime that integrates physical preventive and predictive maintenance tasks with energy-related maintenance tasks aimed at preserving equipment operational parameters and efficiency. By incorporating energy-related maintenance tasks into their maintenance regimes, organizations can effectively maintain equipment efficiency, reduce energy waste, and prevent energy-intensive equipment from consuming more energy than necessary [21]. This not only results in significant cost savings but also contributes to reducing the organization's carbon footprint, aligning with broader sustainability objectives. Additionally, automating maintenance tasks, particularly in identifying maintenance requirements using modern digital technologies, further enhances the effectiveness of energy-centered maintenance. Predictive maintenance plays a crucial role in the predictive modeling of energy consumption by enabling organizations to anticipate maintenance needs and optimize resource allocation. By leveraging predictive models to forecast equipment failures or maintenance requirements, organizations can proactively address energy-related issues, minimize downtime, and maximize operational efficiency [22]. Ultimately, predictive maintenance is essential for achieving energy-saving targets and promoting sustainable practices within organizations, making it an indispensable component of predictive modeling of energy consumption.

3.2 ML Algorithms for Predictive Maintenance

Recent research indicates that Artificial Neural Networks (ANNs) and Multiple Linear Regression (MLR) could be highly effective for the predictive modeling of energy consumption and maintenance in EVs. An ANN model could be utilized to forecast driving behavior and EV energy usage by considering variables such as vehicle speed, acceleration, and jerk. Additionally, a neural network might be employed for residual range estimation of EVs, taking into account internal vehicle factors. While these models are likely to demonstrate promising accuracy, they might not capture all critical internal influences. On the other hand, traditional machine learning models like K-Nearest Neighbours (KNN) and Random Forest (RF) have been used to predict electric bus energy consumption, with MLR and Support Vector Regression (SVR) often showing superior performance. Advanced machine learning algorithms, such as those developed for gradient boosting, could offer even greater predictive capabilities. These advanced algorithms might significantly improve the accuracy of energy consumption predictions in EVs, paving the way for more efficient energy management in the future [9].

3.2.1 Regression Model

Linear regression (LR) stands as a fundamental model utilized in both statistics and machine learning, tracing its roots back to statistical analysis of numerical predictor-response relationships. This model hinges on the notion of a linear link between input variables (x) and a single output variable (y). Simple linear regression entails a solitary input variable (x), whereas multiple linear regression (MLR) encompasses several inputs. Through the least squares method, the MLR model is estimated by minimizing the sum of squared residuals to identify a regression line optimizing data ft. In this context, the MLR approach integrates multiple input variables, making it ideal for analyzing the impact of various predictors (x) on the outcome variable (y) [10]. In an MLR model, various predictor variables X_1, X_2, X_3, ..., X_n are used, provided they satisfy the assumptions of the regression model. The regression model is represented in the equation.

$$Y = \beta_0 + \beta_1 X_1 + \beta_2 \beta X_2 + \beta_3 X_3 + \cdots + \beta_n X_n + \varepsilon \tag{1}$$

where X denotes the predictor variables, Y denotes the response variable, β_n is the regression coefficient, and ε is the random error. Figure 3 depicts the simple linear regression model.

Equation (2) shows the estimated regression model, and it is expressed by,

$$\widehat{Y} = \widehat{\beta_0} + \widehat{\beta_1} X_1 + \widehat{\beta_2} X_2 + \cdots + \widehat{\beta_k} X_k \tag{2}$$

where $\widehat{Y}$ represents the predicted value and $\widehat{\beta_k}$ denotes the estimated of regression coefficients. MLR is a widely used method for predicting a dependent variable Y_i,

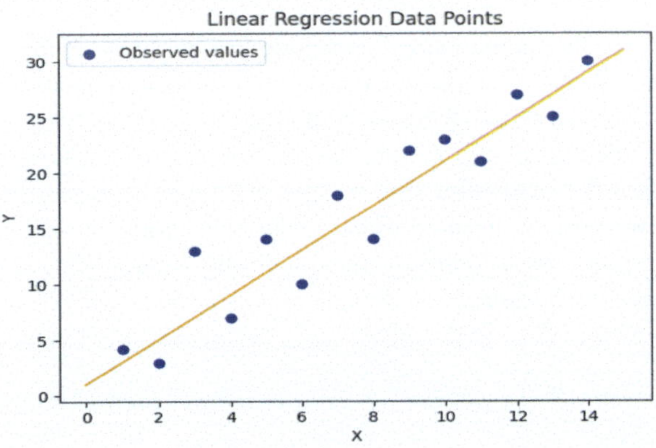

Fig. 3 Linear regression model

based on a set of independent variables X_{ij}. The MLR model can be expressed by using Eq. (3).

$$Y_i = \sum_{j=1}^{k} \beta_j X_{ik} + \varepsilon_i \tag{3}$$

where k represents the number of independent variables, β_j signifies the linear regression coefficient, X_{ij} indicates the value of j for observation i, and ε_i represents the residual error.

3.2.2 Neural Networks

ANNs also known as multilayer perceptron (MLP), are versatile ML algorithms commonly applied in regression and classification tasks. Inspired by the nervous system of living organisms, ANN models consist of interconnected nodes or neurons, akin to biological neurons and synapses. Typically, an ANN comprises input, hidden, and output layers. The input layer has nodes representing attributes, while the output layer nodes correspond to classes. Each layer contains multiple nodes, with information passing from one layer to the next. Neurons within layers are connected by numerical weights, determining the strength of signal transmission between them [23]. The configuration of neurons and hidden layers varies based on the task and available training data. Each neuron's output is calculated based on inputs from the preceding layer, with associated weights influencing signal propagation [11]. The architecture of an ANN, including its layers and connections, is depicted in Fig. 4. The mathematical expression of an ANN is represented by an equation, defining how inputs are processed through the network to produce outputs.

$$n_k^h = \sum_{j=1}^{X} w_{kj}^h t_j + b_k^h \quad k = 1 to z \tag{4}$$

In the mathematical expression of an ANN, X denotes the number of input variables, while z indicates the number of hidden neurons. Each input variable is represented by t, with b representing the bias of the hidden layer and w denoting its weight. These calculated factors, along with the bias, are inputted into the activation function. The activation function, crucial for determining the output of each neuron, often employs the sigmoid function due to its widespread usage. Outputs are derived from the sum of weighted values, with the sigmoid function transforming this sum into the final output.

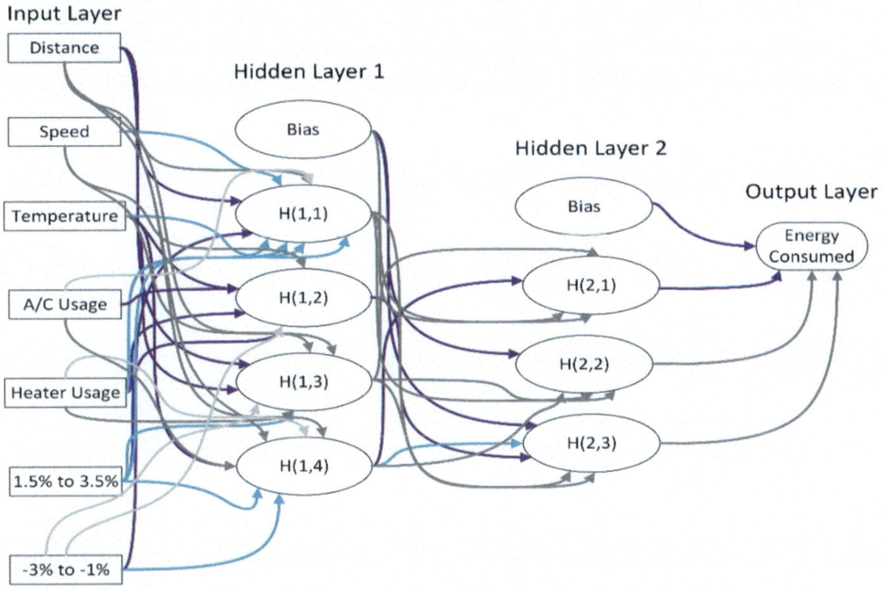

Fig. 4 ANN architecture

4 Adaptive Control Strategies for Energy Distribution and Charging

4.1 Challenges

As EVs continue to gain traction in the automotive market, the demand for efficient energy distribution and charging infrastructure presents formidable challenges. Chief among these obstacles is the inherent variability in energy demand and availability, stemming from factors such as fluctuating driving patterns, diverse charging preferences, and the unpredictable nature of renewable energy sources. This variability often leads to congestion and inefficiencies in the charging network, as charging stations struggle to meet the diverse needs of EV owners while maintaining optimal grid balance. Furthermore, the limited capacity of existing charging infrastructure exacerbates these challenges, resulting in long wait times and inconvenience for EV owners, particularly during peak demand periods. This capacity constraint not only impedes the widespread adoption of electric vehicles but also hampers the transition to a sustainable transportation ecosystem [24]. Moreover, the lack of coordination between charging stations and the electrical grid further compounds the strain on the grid, leading to potential reliability and stability issues.

The integration of renewable energy sources into the grid adds another layer of complexity to the challenges of energy distribution and charging. While renewable energy offers a clean and sustainable alternative to fossil fuels, its intermittent nature poses significant challenges for grid operators and charging infrastructure providers. Without effective coordination and management strategies, the integration of renewable energy sources (RESs) can lead to grid instability, voltage fluctuations, and potential disruptions in energy supply [12]. Addressing these multifaceted challenges requires innovative adaptive control strategies that can dynamically manage energy distribution and charging to optimize efficiency and reliability. By leveraging advanced technologies such as AI, data analytics, and smart grid solutions, these adaptive control strategies aim to balance supply and demand, minimize congestion, and enhance grid stability. Additionally, investments in infrastructure upgrades, grid modernization, and interoperability standards are crucial to building a robust and resilient charging network capable of meeting the evolving needs of EV owners and supporting the widespread adoption of EVs [13].

4.2 Adaptive Control Strategies

In response to the challenges posed by the dynamic nature of energy distribution and charging infrastructure, adaptive control strategies offer innovative solutions to optimize efficiency, reliability, and sustainability. These strategies leverage advanced technologies and real-time data analytics to dynamically adjust energy distribution and charging parameters based on evolving conditions. Here, we delve into two key adaptive control strategies: dynamic pricing and load balancing.

4.2.1 Dynamic Pricing

Dynamic pricing mechanisms play a pivotal role in optimizing energy distribution and charging in EVs. By adjusting electricity prices in real time based on supply and demand dynamics, dynamic pricing incentivizes EV owners to charge their vehicles during off-peak hours when electricity is abundant and cheaper. This not only helps alleviate congestion and strain on the grid during peak demand periods but also promotes the use of RESs, which are typically more available during off-peak hours [25]. Dynamic pricing encourages EV owners to adopt smart charging practices, such as scheduling charging sessions during times of low demand, further optimizing energy usage, and reducing overall costs.

Moreover, dynamic pricing enhances grid flexibility by enabling demand response programs, where EV owners can voluntarily reduce or shift their charging load in response to grid conditions or price signals. By participating in demand response programs, EV owners can benefit from lower electricity prices while contributing to grid stability and reliability [26]. This symbiotic relationship between dynamic pricing and demand response (DR) facilitates a more efficient and resilient energy ecosystem, where supply and demand are harmonized in real time to optimize resource utilization and minimize environmental impact.

4.2.2 Load Balancing

Load balancing strategies are vital for optimizing energy distribution and managing charging across the network. By dynamically assigning charging resources based on factors like the battery's SoC, charging station availability, and grid capacity, load balancing algorithms ensure fair access to charging infrastructure while reducing congestion and wait times. For example, these algorithms can prioritize EVs with lower SoC to minimize downtime and maximize charging station use. Additionally, load-balancing strategies enable bidirectional energy flow between EVs and the grid, allowing EVs to act as distributed energy resources that enhance grid stability and resilience [26]. Through vehicle-to-grid (V2G) technology, EVs can store excess energy during low-demand periods and feed it back into the grid during peak demand times, helping to balance supply and demand and support grid stability. This bidirectional energy flow not only aids grid operators but also creates extra revenue streams for EV owners, further promoting the adoption of EVs.

5 Genetic Algorithm for Optimizing Charging Station Locations

5.1 Overview of Genetic Algorithm

Genetic Algorithm (GA) is a heuristic optimization technique modeled after the principles of natural selection and genetic processes. It falls under the category of evolutionary algorithms and is extensively employed for addressing optimization and search issues. The fundamental concept of a genetic algorithm is to simulate natural evolutionary processes, wherein a group of possible solutions evolves over multiple generations through selection, crossover, and mutation [27]. Here's a brief overview of how a GA works:

- Initialization: The algorithm begins by generating an initial population of potential solutions, often referred to as individuals or chromosomes. Each of these solutions corresponds to a possible answer to the optimization problem.
- Evaluation: Each individual in the population is assessed using a fitness function, which measures the quality or effectiveness of the solution in relation to the optimization objective.
- Selection: Individuals from the current population are chosen to act as parents for the subsequent generation. The likelihood of selection is based on the individual's fitness, meaning that better-performing solutions have a greater probability of being selected.
- Crossover: The selected individuals, or parents, undergo crossover or recombination, during which their genetic information is exchanged to produce new offspring solutions. This process emulates reproduction in natural evolution.

- Mutation: Random alterations, known as mutations, are applied to the offspring solutions. These mutations help maintain genetic diversity within the population and prevent the algorithm from becoming trapped in local optima.
- Replacement: The fresh offspring solutions supplant certain individuals in the existing population. This guarantees that the population size remains consistent across generations.
- Termination: The algorithm progresses through the stages of selection, crossover, and mutation for a predetermined number of generations or until a termination condition is satisfied (such as achieving a satisfactory solution).

Now, let's discuss how a GA can be used for optimizing charging station locations. In the context of EVs, optimizing charging station locations involves determining the optimal placement of charging stations to maximize accessibility and coverage while minimizing costs and congestion. This is a complex optimization problem with multiple objectives and constraints. A GA can be applied to this problem as follows:

- Representation: Each potential solution in the GA represents a set of candidates' charging station locations. This could be represented as a binary string, where each bit corresponds to a potential location.
- Fitness Function: The fitness function evaluates each candidate solution based on criteria such as proximity to major roads, population density, existing charging infrastructure, and expected demand for charging services. The goal is to maximize coverage and accessibility while minimizing costs and congestion.
- Genetic Operators: In the GA selection, crossover, and mutation operators are employed to advance the population of potential solutions across successive generations. Selection prioritizes superior solutions for parent selection, crossover generates fresh offspring solutions by amalgamating genetic information from chosen parents, and mutation introduces random modifications to uphold diversity.
- Optimization: The GA iterates through multiple generations, gradually improving the quality of candidate solutions. Through the process of natural selection and genetic operators, the algorithm converges towards a set of charging station locations that optimize the defined objectives and constraints.
- Evaluation and Deployment: Once the GA terminates, the resulting set of charging station locations is evaluated for feasibility and effectiveness. The selected locations can then be deployed in the real world, subject to further validation and refinement based on practical considerations and stakeholder feedback.

5.2 *Application of GA*

The application of GA offers a versatile approach to solving complex optimization problems across various domains. In this context, we explore its utility by considering a hypothetical area characterized by specific demographics, traffic patterns, and transportation infrastructure. By collecting and processing essential data such as traffic simulation models and Origin–Destination matrices tailored to this area, the power of GA to optimize solutions can be used. This entails iteratively refining potential solutions to address challenges like efficient placement of charging stations for EVs or optimal facility locations for urban services. Through this application, we navigate the intricacies of problem-solving within a defined context, harnessing the capabilities of GA to arrive at robust solutions.

5.2.1 Study Area

In a specific region within a country, approximately 320,000 individuals reside, while its metropolitan area accommodates around 1.09 million inhabitants. Each day witnesses approximately 1.29 million journeys, with half of them made by private vehicles, 38% via public transportation, and the remainder by bicycle or foot. The public transportation system predominantly relies on buses, although plans for a metro network are underway. The road network spans about 2180 km, supported by an 11-km bicycle infrastructure and a bike-sharing initiative [28]. This area serves as the economic, cultural, and administrative hub of the region, attracting visitors from nearby locales and neighboring countries.

Considering a slight increase in traffic congestion, energy consumption, and air pollution, authorities have shifted their focus toward sustainable transportation initiatives. These initiatives encompass a Sustainable Urban Mobility Plan (SUMP), an adaptable traffic management system, and the provision of information services to promote environmentally friendly travel options. With the goal of transforming this area into a sustainable city, local leaders are actively promoting electromobility through incentives and public charging facilities [29]. Current research is focused on identifying suitable locations for fast-charging stations within the municipality, particularly in areas experiencing a significant influx of daily trips within the metropolitan region.

5.2.2 Data

Assuming data is collected and processed corresponding to the hypothetical area. To forecast the forthcoming demand for EV charging within the Municipality, we rely on data gleaned from the latest trip Origin–Destination analysis and the traffic simulation model tailored to the area's specifics. These invaluable resources are often developed by research institutions as part of projects aimed at managing urban mobility. The Origin–Destination matrix for the region stems from extensive phone surveys conducted among residents, with participation from around 4800 individuals, whose daily travel patterns were meticulously recorded [30]. Leveraging these survey findings, trip generation models were crafted, and utilizing a gravity model, attractors and generators were strategically assigned to shape the initial OD matrix. Further fine-tuning was accomplished by calibrating the matrix with real-world data, resulting in 24 hourly matrices for an average day, totaling approximately 884,000 vehicles.

The region is geographically segmented into 338 traffic zones, with 123 zones falling within the Municipality's jurisdiction. Considering prevailing economic conditions, it is hypothesized that by the projected year 2021, passenger vehicle numbers and trip patterns will closely mirror those of 2010. Drawing insights from governmental reports, the anticipated penetration rate of EVs is estimated to hover between 3 and 8%, with a conservative projection of 6% EVs by 2021. A uniform penetration rate across all regions is assumed, with EV adoption expected to have negligible effects on commuter behaviors. Charging demand estimation hinges on the distances vehicles traverse from their origins to their destinations. Vehicles that cover a certain distance, depleting most of their stored energy, are presumed to require charging upon reaching their destinations. It is assumed that each charging station can cater to a single vehicle at a time. Charging demand per zone during peak hours (8:00–9:00) is computed based on these assumptions and serves as the foundation for our analysis in this section [14].

5.2.3 Genetic Algorithm and Tool

GA emerges as a potent artificial optimization instrument tailored for tackling assorted combinatorial optimization challenges. Devised by Holland, GA functions as a stochastic search heuristic grounded in the principles of natural selection and evaluation. Here, chromosomes embody potential solutions to the quandary, typically manifested as binary bit strings. GA undergoes successive iterations, refining solutions based on the evaluation of a fitness function. To deploy GA in resolving the conundrum of EV charging infrastructure placement, researchers can devise a bespoke tool utilizing R and pertinent packages like GA, ggplot2, ggmap, and gWidgets. The tool necessitates input data in CSV format, encompassing candidate charging point locations, their coordinates, anticipated charging demand, and inter-location distances [31]. Researchers can set parameters like the preferred number of stations, coverage radius per station, charging velocity, and generalized cost per kilometer.

Before commencing the optimization algorithm, users can fine-tune supplementary parameters including the initial population size, crossover probability, and mutation probability. By iteratively assessing fitness and selecting parent chromosomes for reproduction, GA refines solutions to pinpoint optimal charging station locales. A sensitivity analysis can be conducted across various scenarios, adjusting the number of charging stations. The outcomes of the optimization endeavor can be depicted through tables and visual aids, facilitating a comparison of GA's efficacy against prior studies. Through scrutinizing fitness values and iteration patterns, researchers can gauge the performance of diverse charging station arrangements. Fine-tuning parameters such as service distance can further enhance the optimization process, ensuring enhanced coverage of charging demand [15].

6 Integrating AI Techniques for Energy Optimization

6.1 AI Combinations for Enhanced Optimization

In the quest for sustainable transportation solutions, EVs have surfaced as a promising alternative to conventional internal combustion engine vehicles. However, optimizing the energy efficiency and performance of EVs remains a critical challenge. Leveraging the capabilities of AI offers a transformative approach to tackle this challenge. By combining various AI techniques such as Machine Learning, Reinforcement Learning, Genetic Algorithms, Swarm Intelligence, Natural Language Processing, and Graph Neural Networks, innovative solutions can be developed to optimize energy usage, enhance driving range, and improve overall efficiency in EVs. These combinations of AI empower EVs with intelligent features like predictive maintenance, adaptive energy management, dynamic route planning, and personalized user experiences, ultimately driving the transition towards a greener and more sustainable future of mobility [16].

6.1.1 Machine Learning (ML) and Reinforcement Learning (RL)

By combining ML and RL techniques, EVs can benefit from both predictive capabilities and adaptive decision-making. ML models can provide insights into long-term trends and patterns, while RL algorithms can respond to real-time changes and uncertainties in the environment. For example, ML models can predict energy consumption based on historical data, while RL agents can dynamically adjust charging schedules and driving behavior to optimize energy usage in response to changing conditions [17]. This combination enables EVs to adapt to dynamic environments, optimize energy management strategies, and improve overall performance and efficiency. It also lays the foundation for autonomous and self-optimizing EV systems that continuously learn and improve over time.

6.1.2 Genetic Algorithms and Swarm Intelligence (SI)

Combining GA with SI allows for the exploration of solution spaces using evolutionary principles while leveraging the collective intelligence and decentralized decision-making capabilities of swarm-based algorithms. For example, GA can be used to evolve populations of charging station placement strategies, while SI algorithms can coordinate the charging behavior of EV swarms at these stations in real-time [18]. This combination enables EVs to adaptively optimize their behavior and interactions with the environment, leading to improved efficiency, scalability, and robustness in complex transportation and energy systems.

6.1.3 Natural Language Processing (NLP) and Graph Neural Networks (GNN)

Combining NLP with GNN enables the integration of textual data and graph-structured data to enhance decision-making and optimization in EV systems. For example, NLP techniques can be used to analyze textual data from user reviews and social media posts about EVs, extracting insights that inform the construction and analysis of transportation networks modeled by GNNs. Similarly, GNNs can learn representations of nodes and edges in transportation networks, incorporating textual information from NLP models to enrich the graph-based representations and improve the accuracy of predictions and recommendations related to EV route planning, charging infrastructure deployment, and energy management [19].

6.1.4 Deep Learning (DL) and Fuzzy Logic (FL)

Combining DL with FL enables the development of hybrid intelligent systems that leverage the strengths of both approaches to tackle complex optimization problems in EVs. DL models can learn from large-scale data sources to extract meaningful features and patterns, while FL controllers can interpret and apply this knowledge to make context-aware decisions in real-time [32]. For example, DL models can analyze sensor data to detect anomalies or predict future states of EV components, while FL controllers can adjust control parameters based on these predictions and adapt to changing operating conditions. By integrating DL and FL, EVs can benefit from robust, adaptive, and efficient optimization strategies that enhance performance, safety, and user satisfaction.

6.2 Battery, Predictions, and Control Synergies

The predictive maintenance, driven by ML algorithms, serves as a proactive means to identify potential battery faults or degradation through thorough data analysis. By forecasting battery health and performance trends, predictive maintenance algorithms enable timely servicing, minimizing downtime, and pre-empting unexpected failures. This proactive stance ensures that batteries operate at peak efficiency and reliability, thereby maximizing their operational lifespan. Battery state estimation, facilitated by ML models, provides precise real-time assessments of the SOC and SOH of EV batteries. These estimations empower meticulous monitoring of battery performance and degradation, allowing for the implementation of adaptive control strategies to optimize energy utilization and uphold battery health effectively [20].

Integrated control strategies leverage insights from predictive maintenance and battery state estimations to dynamically manage energy in EVs. Machine learning algorithms analyze diverse factors like driving conditions, traffic patterns, and user preferences to optimize energy consumption and enhance driving range. By continually adjusting parameters like throttle response, regenerative braking, and power distribution, AI-driven control systems ensure that EV batteries operate within their optimal parameters, thus maximizing efficiency and overall performance. Furthermore, control synergies enable the deployment of adaptive charging strategies, where ML algorithms learn from historical charging data and user behavior to predict future charging requirements [33]. These systems take into account factors such as time-of-use electricity rates, grid demand, and battery health to schedule charging sessions during off-peak hours, thereby minimizing grid impact. Additionally, they dynamically adjust charging rates and patterns to prevent overcharging and minimize degradation, thereby prolonging battery lifespan [17].

6.3 Cross-Disciplinary Collaboration Opportunities

This concept underscores the synergy between various fields and domains, including engineering, data science, policy, and industry, to harness the power of AI in optimizing energy systems. By leveraging expertise from diverse disciplines, collaborative efforts can unlock new insights, develop robust methodologies, and address complex challenges in energy optimization. This introduction sets the stage for exploring how collaborative endeavors can propel the integration of AI technologies towards creating more sustainable, efficient, and resilient energy ecosystems [21].

- Engineering and Data Science Integration: Collaboration between engineers and data scientists can facilitate the development of AI-driven energy optimization solutions. Engineers bring domain expertise in energy systems, while data scientists contribute advanced analytics and machine learning techniques to analyze and optimize energy consumption patterns.
- Policy and Industry Partnerships: Collaboration between policymakers, industry stakeholders, and AI experts can drive the adoption of AI techniques for energy optimization. Policymakers provide regulatory frameworks and incentives to encourage industry adoption, while industry partners offer real-world data and insights to inform AI model development and deployment.
- Academic and Industry Research Collaborations: Collaborative research efforts between academia and industry can lead to the development of innovative AI-based energy optimization technologies. Academic institutions contribute cutting-edge research and expertise in AI algorithms, while industry partners provide access to data, infrastructure, and practical insights to validate and implement AI solutions.
- Interdisciplinary Teams for System Integration: Interdisciplinary teams comprising experts from diverse fields such as electrical engineering, computer science, economics, and environmental science can collaborate to address complex challenges in energy optimization. By integrating insights from multiple

disciplines, these teams can develop holistic solutions that consider technical, economic, and environmental factors.

- Cross-sector Collaboration for Data Sharing: Collaboration between different sectors, such as transportation, manufacturing, and utilities, can enable data-sharing initiatives to support AI-driven energy optimization efforts. By sharing anonymized data on energy consumption, production, and distribution, stakeholders can collectively develop more accurate and robust AI models for optimizing energy usage across various sectors.
- International Collaboration for Global Impact: International collaboration and knowledge exchange can accelerate the adoption of AI techniques for energy optimization on a global scale. By sharing research findings, methodologies, and resources across borders, countries can collectively address shared energy challenges and work towards sustainable and resilient energy systems worldwide [22].

6.4 Future Trends and Challenges

In the future, integrating AI techniques for energy optimization will see significant advancements across various fronts. Firstly, AI algorithms, such as deep learning and reinforcement learning, will evolve to handle larger datasets, leading to more accurate predictions and optimization strategies. Secondly, the integration of IoT and sensor technologies will enable real-time data collection, allowing AI algorithms to optimize energy usage and predict maintenance needs more accurately [23]. Additionally, the rise of edge computing and decentralized systems will drive the need for localized AI solutions, enhancing decision-making and system reliability. Transparency and explainable AI will gain importance, ensuring trust and acceptance of AI-driven optimization strategies. Collaboration between experts from diverse disciplines will lead to the development of holistic energy optimization solutions [24]. Resilience and sustainability will be prioritized, with AI techniques optimizing energy systems for resilience against extreme weather events and maximizing the use of renewable energy sources. Autonomous and adaptive energy optimization systems will become more common, continuously learning and improving their performance over time. Finally, global collaboration and knowledge sharing will accelerate the adoption of AI-driven energy optimization technologies worldwide, revolutionizing the management of energy systems for greater efficiency and sustainability [25].

Despite the promising advancements in AI-driven energy optimization strategies for EVs, several challenges remain to be addressed. One significant challenge is the interoperability and standardization of AI algorithms and systems across different EV models and charging infrastructure. The lack of standardized protocols and data formats hinders seamless integration and interoperability, limiting the scalability and effectiveness of AI-driven energy optimization solutions. Another challenge lies in the availability and quality of data required for training and deploying AI algorithms. While EVs generate vast amounts of data related to energy consumption, driving patterns, and battery health, accessing and processing this data in real time poses technical and logistical challenges. Moreover, ensuring the accuracy, reliability, and privacy of data collected from EVs raises concerns regarding data security and user privacy, necessitating robust data governance frameworks and privacy-preserving mechanisms.

Furthermore, the complex and dynamic nature of energy systems and grid infrastructure poses challenges for AI-driven energy optimization. Factors such as grid congestion, renewable energy integration, and fluctuating energy prices introduce uncertainties and complexities that AI algorithms must navigate to optimize energy distribution and charging effectively. Addressing these challenges requires advanced AI techniques capable of adapting to dynamic and uncertain environments while ensuring grid stability and reliability.

7 Conclusions

The integration of artificial intelligence (AI) into energy optimization strategies for electric vehicles (EVs) represents a significant advancement towards enhancing the efficiency, reliability, and sustainability of transportation systems. Through AI-driven predictive modeling, adaptive control strategies, and genetic algorithms, EVs can achieve optimal energy consumption, manage charging infrastructure effectively, and extend battery life. However, the realization of these benefits hinges upon overcoming challenges such as interoperability, data availability, and grid complexity. Tackling these challenges demands collaborative endeavors from industry stakeholders, policymakers, and researchers to establish standardized protocols, enhance data accessibility and quality, and advance AI techniques tailored to the distinctive demands of EV energy optimization. Nevertheless, with the rapid advancements in AI technology and the increasing adoption of EVs, the future holds immense promise for AI-driven energy optimization to pave the way toward cleaner, more sustainable transportation.

Acknowledgements This research work was supported by "Woosong University's Academic Research Funding—2025".

References

1. Hou ZY, Lou PY, Wang CC (2017) State of charge, state of health, and state of function monitoring for EV BMS. In: 2017 IEEE international conference on consumer electronics (ICCE), Las Vegas, NV, USA, pp 310–311. https://doi.org/10.1109/ICCE.2017.7889332
2. Feng Y, Duan J, Xue C, Han F (2021) State-of-charge and state-of-health estimation algorithms of lithium-ion batteries using SMO. In: 2021 IEEE 5th conference on energy internet and energy system integration (EI2), Taiyuan, China, pp 2104–2108. https://doi.org/10.1109/EI252483.2021.9713303

3. How DNT, Hannan MA, Lipu MSH, Sahari KSM, Ker PJ, Muttaqi KM (2020) State-of-charge estimation of li-ion battery in electric vehicles: a deep neural network approach. IEEE Trans Ind Appl 56(5):5565–5574. https://doi.org/10.1109/tia.2020.3004294
4. Chen C, Chen B, Wang T, Zheng J, Lin Z (2022) Battery health assessment and life prediction in battery management system. In: 2022 3rd international conference on electronic communication and artificial intelligence (IWECAI), Zhuhai, China, pp 87–90. https://doi.org/10.1109/IWECAI55315.2022.00025
5. Liu W, Placke T, Chau K (2022) Overview of batteries and battery management for electric vehicles. Energy Rep 8:4058–4084. https://doi.org/10.1016/j.egyr.2022.03.016
6. Zhang F, Hu X, Langari R, Cao D (2019) Energy management strategies of connected HEVs and PHEVs: recent progress and outlook. Prog Energy Combust Sci 73:235–256. https://doi.org/10.1016/j.pecs.2019.04.002
7. Ullah I, Liu K, Yamamoto T, Mamlook REA, Jamal A (2021) A comparative performance of machine learning algorithm to predict electric vehicles energy consumption: a path towards sustainability. Energy Environ 33(8):1583–1612. https://doi.org/10.1177/0958305x211044998
8. Ragone M, Yurkiv V, Ramasubramanian A, Kashir B, Mashayek F (2021) Data driven estimation of electric vehicle battery state-of-charge informed by automotive simulations and multi-physics modeling. J Power Sour 483:229108. https://doi.org/10.1016/j.jpowsour.2020.229108
9. Thomas S, Philip AO, Vishwanath N (2023) ML based data driven energy centered predictive maintenance. In: 2023 2nd international conference on edge computing and applications (ICECAA), Namakkal, India, pp 994–1001. https://doi.org/10.1109/ICECAA58104.2023.10212246
10. Wu J, Zeng W, Chen Z, Tang XF (2016) Hierarchical temporal memory method for time-series-based anomaly detection. In: 2016 IEEE 16th international conference on data mining workshops (ICDMW), Barcelona, Spain, pp 1167–1172. https://doi.org/10.1109/ICDMW.2016.0168
11. Farbiz F, Miaolong Y, Yu Z (2020) A cognitive analytics based approach for machine health monitoring, anomaly detection, and predictive maintenance. In: 2020 15th IEEE conference on industrial electronics and applications (ICIEA), Kristiansand, Norway, pp 1104–1109. https://doi.org/10.1109/ICIEA48937.2020.9248409
12. Ullah I, Liu K, Yamamoto T, Zahid M, Jamal A (2021) Electric vehicle energy consumption prediction using stacked generalization: an ensemble learning approach. Int J Green Energy 18(9):896–909. https://doi.org/10.1080/15435075.2021.1881902
13. López FC, Fernández RÁ (2020) Predictive model for energy consumption of battery electric vehicle with consideration of self uncertainty route factors. J Clean Prod 276:124188. https://doi.org/10.1016/j.jclepro.2020.124188
14. Chen S, Shi Y, Chen X, Qi F (2015) Optimal location of electric vehicle charging stations using genetic algorithm. In: 2015 17th Asia-Pacific network operations and management symposium (APNOMS), Busan, Korea (South), pp 372–375. https://doi.org/10.1109/APNOMS.2015.7275344
15. Zhou G, Zhu Z, Luo S (2022) Location optimization of electric vehicle charging stations: based on cost model and genetic algorithm. Energy 247:123437. https://doi.org/10.1016/j.energy.2022.123437
16. Nigro M, Ferrara M, Vincentis RD, Liberto C, Valenti G (2021) Data driven approaches for sustainable development of e-mobility in urban areas. Energies 14(13):3949. https://doi.org/10.3390/en14133949
17. Yang H et al (2020) Lithium-ion battery life cycle prediction with deep learning regression model. In: 2020 IEEE applied power electronics conference and exposition (APEC), New Orleans, LA, USA, pp 3346–3351. https://doi.org/10.1109/APEC39645.2020.9124049
18. Efthymiou D, Chrysostomou K, Morfoulaki M et al (2017) Electric vehicles charging infrastructure location: a genetic algorithm approach. Eur Transp Res Rev 9:27. https://doi.org/10.1007/s12544-017-0239-7

19. Salmasi FR (2007) Control strategies for hybrid electric vehicles: evolution, classification, comparison, and future trends. IEEE Trans Veh Technol 56(5):2393–2404. https://doi.org/10.1109/tvt.2007.899933
20. Salkuti SR (2023) Advanced technologies for energy storage and electric vehicles. Energies 16(5):2312. https://doi.org/10.3390/en16052312
21. Rauf M, Kumar L, Zulkifli SA, Jamil A (2024) Aspects of artificial intelligence in future electric vehicle technology for sustainable environmental impact. Environ Challenges 14:2024. https://doi.org/10.1016/j.envc.2024.100854
22. Aysan AF, Yüksel S, Eti S, Dinçer H, Akin MS, Kalkavan H, Mikhaylov A (2024) A unified theory of acceptance and use of technology and fuzzy artificial intelligence model for electric vehicle demand analysis. Decis Anal J 11:100455. https://doi.org/10.1016/j.dajour.2024.100455
23. Salkuti SR (2019) Optimal operation of microgrid considering renewable energy sources, electric vehicles and demand response. E3S Web Conf 87:01007. https://doi.org/10.1051/e3sconf/20198701007
24. Salkuti SR (2022) Emerging and advanced green energy technologies for sustainable and resilient future grid. Energies 15(18):6667. https://doi.org/10.3390/en15186667
25. Saleem S, Ahmad I, Ahmed SH, Rehman A (2024) Artificial intelligence based robust nonlinear controllers optimized by improved gray wolf optimization algorithm for plug-in hybrid electric vehicles in grid to vehicle applications. J Energy Storage 75:109332. https://doi.org/10.1016/j.est.2023.109332
26. Ahmed M, Zheng Y, Amine A, Fathiannasab H, Chen Z (2021) The role of artificial intelligence in the mass adoption of electric vehicles. Joule 5(9):2296–2322. https://doi.org/10.1016/j.joule.2021.07.012
27. Oladosu TL, Pasupuleti J, Kiong TS, Koh SPJ, Yusaf T (2024) Energy management strategies, control systems, and artificial intelligence-based algorithms development for hydrogen fuel cell-powered vehicles: a review. Int J Hydrogen Energy 61:1380–1404. https://doi.org/10.1016/j.ijhydene.2024.02.284
28. Salkuti SR (2021) Energy storage and electric vehicles: technology, operation, challenges, and cost-benefit analysis. Int J Adv Comput Sci Appl 12(4). https://doi.org/10.14569/IJACSA.2021.0120406
29. Mishra DP, Dash AK, Patro KA, Salkuti SR (2024) Integration of electric vehicles with smart grid. In: Salkuti SR (eds) Energy and environmental aspects of emerging technologies for smart grid. Green energy and technology. Springer, Cham. https://doi.org/10.1007/978-3-031-18389-8_12
30. Jung CM, Pagidipala S, Salkuti SR (2024) Emerging technologies for the integration of renewable energy, energy storage and electric vehicles. In: Salkuti SR (eds) Energy and environmental aspects of emerging technologies for smart grid. Green energy and technology. Springer, Cham. https://doi.org/10.1007/978-3-031-18389-8_5
31. Viswa Teja A, Razia Sultana W, Salkuti SR (2023) Analysis of various speed control methods for PMSM drive-based electric vehicles. In: Salkuti SR, Ray P, Singh AR (eds) Power quality in microgrids: issues, challenges and mitigation techniques. Lecture notes in electrical engineering, vol 1039. Springer, Singapore. https://doi.org/10.1007/978-981-99-2066-2_14
32. Viswa Teja A, Razia Sultana W, Salkuti SR (2023) Performance explorations of a PMS motor drive using an ANN-Based MPPT controller for solar-battery powered electric vehicles. Designs 7(3):79. https://doi.org/10.3390/designs7030079
33. Jauhar SK, Sethi S, Kamble SS, Mathew S, Belhadi A (2024) Artificial intelligence and machine learning-based decision support system for forecasting electric vehicles' power requirement. Technol Forecast Soc Chang 204:123396. https://doi.org/10.1016/j.techfore.2024.123396

AI-Powered Strategies for Efficient EV Energy Management

Debani Prasad Mishra, Abhipsha Dash, Ayush Kumar, and Surender Reddy Salkuti

Abstract AI-driven smart charging, vehicle-to-grid (V2G) systems, and decentralized energy trading enable the effective utilization of renewable energy (RE), supporting grid stability and reducing dependency on fossil fuels. This chapter provides a comprehensive overview of the methodologies and technologies driving AI applications in EV energy management. By harnessing the power of AI algorithms, sensor integration, and edge computing, we unlock new possibilities for optimizing efficiency, enhancing safety, and reducing environmental impact in the transportation sector. The journey towards a cleaner, greener, and more sustainable transportation ecosystem is ongoing, and AI remains a crucial enabler in this endeavor. V2G systems facilitate the bidirectional flow of electric current between automobiles and the power grid, offering benefits such as grid stabilization, demand response, and the integration of RE. The connection between EVs and the power grid allows for efficient use of resources, making the grid more stable and helping us move towards a future that uses RE and is less centralized. AI improves the user experience by giving personalized advice on charging stations, routes, and how to drive to save energy. As AI becomes more widely used in smart energy systems for EVs, strong cybersecurity measures are needed to protect against possible threats and keep data private. To fully unlock the potential of AI in the development of sustainable transportation solutions, we need to work together across disciplines, set up rules and regulations, and get stakeholders involved. In short, merging AI with smart energy systems in EVs can potentially transform the future of transportation. With continued research, innovation, and teamwork, AI-powered smart energy systems will be crucial in creating a transportation system that is cleaner, greener, and more adaptable.

D. P. Mishra · A. Dash · A. Kumar
Department of Electrical and Electronics Engineering, IIIT Bhubaneswar, Bhubaneswar, Odisha, India
e-mail: debani@iiit-bh.ac.in

S. R. Salkuti (✉)
Department of Railroad and Electrical Engineering, Woosong University, Daejeon, Republic of Korea
e-mail: surender@wsu.ac.kr

S. R. Salkuti (ed.), *Artificial Intelligence for Integrated Smart Energy Systems in Electric Vehicles*, Lecture Notes in Electrical Engineering 1427,
https://doi.org/10.1007/978-3-031-94276-1_9

Keywords Artificial intelligence · Electric vehicles · Smart energy systems · Battery optimization · Grid integration · Vehicle-to-grid

Abbreviations

EV	Electric vehicle
AI	Artificial intelligence
V2G	Vehicle-to-grid
IoT	Internet of things
CNN	Convolutional neural network
RNN	Recurrent neural network
ANN	Artificial neural network
ML	Machine learning
BMS	Battery management system
MPC	Model predictive control
SIL	Software-in-the-loop
ACC	Adaptive cruise control
SOC	State of charge
HVAC	Heating, ventilation, and air conditioning
HMI	Human machine interface
NLP	Natural language processing
GNN	Graph neural network
DL	Deep learning

1 Introduction

The integration of artificial intelligence (AI) into electric vehicle (EV) energy management systems heralds a new era of sustainable transportation. This section delves deeper into the methodological aspects of AI applications in EV energy management. From AI algorithms and models to sensor integration and edge computing, this introduction sets the stage for exploring the practical implementation and real-world applications of AI in enhancing the efficiency, sustainability, and user experience of EVs [1]. As the global shift towards electrification gains momentum, the role of AI in optimizing EV energy consumption becomes increasingly pivotal. By leveraging real-time data processing and machine learning algorithms, AI enables dynamic route planning, adaptive cruise control, and predictive eco-driving assistance, enhancing both the efficiency and safety of EVs on the road. In addition, AI plays a key role in optimizing energy consumption and extending the operational life of EV batteries.

The methodology section of this chapter provides a comprehensive overview of the tools and techniques underlying AI-driven power management strategies. From predictive maintenance and advanced control systems to simulation techniques and energy consumption optimization, AI serves as the backbone of innovative solutions to maximize available resources while minimizing environmental impact [2]. One of the key areas of focus in this section is the integration of renewable energy sources (RESs) into EV energy management. AI-driven smart charging, vehicle-to-grid (V2G) systems, and decentralized energy trading enable effective utilization of RE, supporting grid stability and reducing dependency on fossil fuels.

By harnessing the power of AI algorithms, sensor integration, and edge computing, we unlock new possibilities for optimizing efficiency, enhancing safety, and reducing environmental impact in the transportation sector. The adventure toward a cleaner, greener, and more sustainable transportation environment is ongoing, and AI remains a crucial enabler in this enterprise. Continued research, innovation, and collaboration across disciplines will be essential in realizing the full potential of AI-driven EV energy management, propelling us toward a future where sustainable transportation is not just a possibility but a reality [3]. In the following sections, we will delve into the methodologies and technologies driving AI applications in EV energy management, exploring the practical implementation of AI algorithms, sensor integration, and edge computing. Through case studies, examples, and real-world applications, we will illustrate how AI is revolutionizing the way we manage energy in EVs, paving the way for a more efficient, sustainable, and user-friendly transportation future.

2 Methodology

The integration of AI into EV energy management signifies a revolutionary approach poised to transform the efficiency, reliability, and sustainability of EVs, paving the way for a shift in the transportation landscape. A systematic process and a suite of technological innovations are aimed at achieving these objectives. This approach harnesses the power of advanced AI and machine learning (ML) techniques to analyze vast amounts of data and extract actionable insights that maximize energy consumption, charging patterns, and overall vehicle performance [4]. By integrating sensors and Internet of Things (IoT) devices into EVs, this methodology enables real-time monitoring of key parameters such as battery health, energy usage, and driving patterns, facilitating proactive decision-making and ensuring optimal operation under varying conditions. Moreover, the adoption of edge computing architectures enables AI-driven energy management systems to process data locally within the vehicle, reducing latency and enhancing responsiveness while preserving privacy and security [5].

Furthermore, the deployment of sophisticated ML training strategies, such as reinforcement learning and transfer learning, enables AI models to continuously adapt and improve over time, refining their predictive capabilities and enhancing their ability to optimize EV energy systems in dynamic and uncertain environments. Additionally, predictive maintenance models leverage AI-driven analytics to anticipate and mitigate potential failures before they occur, maximizing vehicle uptime and minimizing maintenance costs while ensuring the reliability and safety of EV operations [6]. Advanced control strategies, informed by AI insights and real-time data, enable precise and adaptive control of energy flows within the vehicle, optimizing efficiency and extending range while enhancing the driving experience. Moreover, rigorous simulation and validation techniques, including virtual testing environments and real-world validation scenarios, ensure that AI applications in EV energy management meet stringent performance and safety requirements, instilling confidence in their efficacy and reliability [7]. This comprehensive methodology provides a roadmap for the successful deployment of AI in optimizing EV energy systems, offering a transformative path toward a more sustainable transportation future. By harnessing the power of AI-driven energy management solutions, stakeholders can unlock new opportunities for efficiency, reliability, and sustainability in the realm of electric transportation, driving positive societal, environmental, and economic impacts on a global scale.

2.1 Overview of AI Algorithms and Models

The methodology for integrating AI into EV energy management represents a multifaceted approach aimed at harnessing the power of advanced algorithms and models to optimize energy consumption, enhance vehicle performance, and promote sustainability. This methodology begins with the identification and utilization of appropriate AI models tailored to the specific requirements of EV energy management [8]. Neural networks, decision trees, and support vector machines are some of the tools that serve as the foundation of machine learning. Historical data on energy usage, driving patterns, and environmental conditions can be used to develop models that anticipate future energy demand. Deep learning models, such as CNNs and recurrent neural networks, play a pivotal role in handling complex data analysis and decision-making processes within EV energy management systems [9]. CNNs are particularly well-suited for tasks involving image recognition and signal processing, making them invaluable for analyzing sensor data and identifying patterns related to vehicle performance and energy usage. Similarly, RNNs excel at processing sequential data and time-series information, enabling them to model temporal dependencies and make informed decisions based on past experiences. These deep learning models are adept at handling large datasets and extracting meaningful insights that drive efficient energy management in EVs, thereby optimizing battery usage, extending vehicle range, and enhancing overall performance [10].

Moreover, the methodology emphasizes the importance of continuous refinement and optimization of AI algorithms and models through iterative training and validation processes. By leveraging feedback from real-world data and user interactions, AI-driven energy management systems can continuously adapt and improve over time, ensuring that they remain effective and reliable in dynamic and uncertain environments. Additionally, the methodology emphasizes the integration of AI algorithms and models with sensors and Internet of Things (IoT) devices embedded within EVs. These sensors capture real-time data on various parameters such as battery health, energy consumption, and driving behavior, providing valuable inputs for AI-driven decision-making processes [11]. By integrating sensor data with AI algorithms, EV energy management systems can dynamically adjust charging schedules, optimize energy usage, and anticipate maintenance needs, thereby maximizing vehicle efficiency and reliability.

Furthermore, the methodology underscores the importance of interoperability and compatibility between AI-driven energy management systems and existing EV infrastructure. By adhering to common standards and protocols, AI algorithms can seamlessly integrate with charging stations, grid networks, and other components of the EV ecosystem, enabling holistic and interoperable solutions that enhance user experience and promote wider adoption of electric transportation [12]. The methodology for integrating AI into EV energy management represents a comprehensive approach aimed at leveraging advanced algorithms, models, and data analytics techniques to optimize energy usage, enhance vehicle performance, and promote sustainability. By harnessing the power of machine learning and deep learning models, integrating sensor data, and ensuring interoperability with existing EV infrastructure, stakeholders can unlock new opportunities for efficiency, reliability, and innovation in the realm of electric transportation, driving positive societal and environmental impacts on a global scale [13].

2.2 Integration of Sensors and IoT Devices

A pivotal aspect of AI-driven energy management lies in the seamless integration of sensors and IoT devices, catalyzing a transformative shift in the efficiency, reliability, and sustainability of EVs and their associated energy systems. At the forefront of this integration are sensors embedded within EVs, acting as vigilant sentinels that capture and transmit real-time data on a multitude of critical parameters [14]. These sensors, strategically positioned throughout the vehicle, continuously monitor various aspects such as battery health, energy consumption, driving patterns, and environmental conditions, providing a comprehensive and dynamic view of the vehicle's operational state. By harnessing this wealth of real-time data, AI algorithms gain invaluable insights into the performance and behavior of EVs, enabling them to make informed decisions and optimize energy management strategies with unprecedented precision and efficiency.

Moreover, IoT devices play a pivotal role in facilitating seamless communication and coordination between EVs, charging stations, and the power grid, forming a cohesive and interconnected ecosystem that fosters synergy and collaboration. These IoT devices serve as conduits for data exchange and command transmission, enabling

EVs to communicate their energy needs, charging status, and operational requirements to charging stations and the grid in real-time. Similarly, charging stations and grid infrastructure use the internet of things to convey relevant information such as energy prices, grid demand, and availability of RESs to EVs, enabling them to make intelligent decisions regarding charging schedules and energy sources [15]. This bidirectional flow of information ensures optimal resource allocation, grid stability, and user experience, ultimately enhancing the overall efficiency and effectiveness of EV energy management systems.

Furthermore, the real-time data collected by sensors and IoT devices serves as the lifeblood of AI-driven energy management systems, empowering them to perform a myriad of critical functions ranging from energy optimization to predictive maintenance [16]. Similarly, charging stations and grid infrastructure use the IoT to convey relevant information such as energy prices, grid demand, and availability of renewable energy sources (RESs) to EVs, enabling them to make intelligent decisions regarding charging schedules and energy sources. This enables proactive interventions such as adaptive charging strategies, predictive maintenance scheduling, and fault detection, minimizing downtime, reducing operational costs, and ensuring the reliability and safety of EV operations. Additionally, the integration of sensors and IoT devices enables AI-driven energy management systems to adapt and respond to dynamic and evolving conditions in real-time, optimizing energy usage, and grid interaction based on changing environmental, operational, and market factors. This agility and responsiveness are crucial for maximizing the efficiency and resilience of EV energy systems, especially in the face of fluctuating energy prices, grid congestion, and RE variability.

Moreover, the seamless integration of sensors and IoT devices fosters innovation and collaboration across the electric mobility ecosystem, driving the development of new services, business models, and value-added solutions that enhance user experience, promote sustainability, and accelerate the adoption of electric transportation. By harnessing the power of real-time data collection, communication, and analytics, stakeholders can unlock new opportunities for efficiency, reliability, and innovation in EV energy management, paving the way for cleaner, smarter, and more sustainable future transportation. The integration of IoT sensors and devices is a critical component of AI-driven energy management in EVs, enabling real-time data collection, communication, and analysis that under pins energy consumption optimization, predictive maintenance, and network interaction [17]. By leveraging this integration, stakeholders can unlock new opportunities for efficiency, reliability, and innovation, driving positive societal, environmental, and economic impacts on a global scale.

2.3 Role of Edge Computing in AI Applications

Edge computing emerges as a cornerstone in revolutionizing the efficiency and responsiveness of AI applications within EV energy management systems, heralding a new era of real-time decision-making and dynamic optimization. At the heart of this paradigm shift lies the decentralized processing of data, with edge computing facilitating the localization of computational tasks either directly on the vehicle or on nearby edge devices. This localized approach circumvents the traditional reliance on centralized cloud-based processing, offering unparalleled benefits in terms of reduced latency, minimized bandwidth usage, and enhanced scalability.

By harnessing the computational power of edge devices, such as onboard processors, edge servers, and IoT gateways, AI algorithms can analyze and act upon data in real time, ensuring swift and seamless execution of critical functions without the constraints of network latency or connectivity. This real-time responsiveness is particularly crucial in the context of EV energy management, where split-second decisions regarding energy distribution, battery management, and vehicle performance can have significant implications for efficiency, reliability, and safety.

Moreover, edge computing enables AI applications to adapt and respond to changing conditions on the fly, dynamically adjusting energy management strategies based on evolving environmental, operational, and user-specific factors. This agility and adaptability are paramount for optimizing energy usage, maximizing vehicle range, and ensuring a smooth and uninterrupted driving experience for EV owners. Furthermore, the localized processing capabilities afforded by edge computing empower AI-driven energy management systems to operate autonomously and independently of centralized infrastructure, even in remote or bandwidth-constrained environments. This resilience and self-sufficiency are essential for ensuring continuous operation and reliability, particularly in scenarios where network connectivity may be intermittent or unreliable [18].

Additionally, edge computing facilitates the seamless integration of AI applications with onboard sensors and IoT devices, enabling real-time data acquisition, analysis, and decision-making at the source. By processing sensor data locally on the vehicle or surrounding devices, AI algorithms can extract useful insights and derive meaningful conclusions without the need to transmit raw data to centralized servers for processing. This not only reduces latency and bandwidth usage but also increases privacy and security by minimizing the exposure of sensitive data to external networks. Moreover, edge computing enables AI applications to leverage contextual information and situational awareness, enhancing the accuracy and relevance of decision-making in dynamic and uncertain environments. By combining real-time sensor data with localized processing power, edge computing empowers AI-driven energy management systems to make informed decisions that optimize energy usage, enhance vehicle performance, and improve overall efficiency.

Furthermore, the adoption of edge computing in EV energy management holds significant implications for scalability, enabling AI applications to scale seamlessly across a distributed network of edge devices without overburdening centralized infrastructure [19]. This distributed approach ensures that computational tasks are distributed efficiently and effectively, maximizing resource utilization and minimizing bottlenecks. Moreover, edge computing facilitates the deployment of AI applications in resource-constrained environments, such as remote or mobile settings, where centralized cloud-based processing may be impractical or infeasible. By bringing intelligence closer to the source of data generation, edge computing unlocks new opportunities for innovation and efficiency in EV energy management, driving positive societal, environmental, and economic impacts on a global scale.

In conclusion, edge computing plays a pivotal role in enhancing the efficiency, responsiveness, and scalability of AI applications within electric vehicle energy management systems, enabling real-time decision-making, dynamic optimization, and autonomous operation at the edge. By processing data locally on the vehicle or nearby edge devices, edge computing reduces latency, minimizes bandwidth usage, and ensures continuous operation in diverse and challenging environments. This decentralized approach not only enhances the efficiency and reliability of EV energy management but also unlocks new opportunities for innovation and scalability, driving positive impacts on sustainability, mobility, and society as a whole.

2.4 *Strategies for Machine Learning Training*

Effective ML training strategies stand as pivotal pillars in the successful deployment of AI within EV energy management systems, laying the groundwork for predictive analytics, pattern recognition, and dynamic optimization. At the heart of these strategies lies the judicious utilization of both historical and real-time data, harnessing the collective insights gleaned from past experiences and current observations to train models capable of discerning patterns and extrapolating future outcomes with precision and accuracy [20]. This iterative process includes a spectrum of techniques, from supervised and unsupervised learning to reinforcement learning, each tailored to specific EV energy management challenges and goals. In supervised learning, models are trained on labeled data sets, with each data point paired with corresponding ground truth labels, allowing the model to learn to map input features to target outputs by iteratively adjusting model parameters.

Conversely, unsupervised learning techniques aim to uncover underlying patterns and structures within unlabeled data, clustering similar data points together or identifying latent factors that capture the inherent structure of the data. Reinforcement learning, on the other hand, revolves around the concept of learning through interaction with the environment, whereby the model receives feedback in the form of rewards or punishments based on its actions, allowing it to learn the optimal decision policy through trial and error [21]. These diverse ML training approaches enable the development of robust and adaptive models capable of capturing the complexity and variability inherent in EV energy management, including factors such as driving patterns, environmental conditions, and energy consumption dynamics.

Moreover, the iterative nature of ML training ensures that models continually evolve and improve over time, as training datasets are augmented with new data and model parameters are fine-tuned to enhance performance and generalization capabilities. Continuous data updates enable models to adapt to evolving conditions and capture emerging trends, ensuring that they remain effective and relevant in dynamic and uncertain environments. In addition, techniques such as cross-validation and hyperparameter tuning play a key role in optimizing model performance and preventing overfitting, thus increasing model robustness and generalizability. Cross-validation techniques such as k-fold cross-validation divide the dataset into multiple subsets, with each subset used alternately for training and validation, allowing for a more robust estimation of model performance and generalization error. Hyper-parameter tuning involves systematically searching for the optimal combination of hyperparameters such as learning rates, regularization powers, and model architectures using techniques such as grid search or random search, ensuring that models are tuned to achieve peak performance [22].

By leveraging these sophisticated ML training strategies, stakeholders can develop AI-driven energy management systems that are not only accurate and reliable but also adaptive and resilient, capable of navigating the complexities and uncertainties inherent in the domain of EV energy management. In conclusion, effective machine learning training strategies form the bedrock of successful AI deployment in EV energy management, enabling the development of robust models capable of predictive analytics, pattern recognition, and dynamic optimization. By harnessing historical and real-time data, employing diverse ML techniques, and leveraging iterative training processes, stakeholders can develop AI-driven energy management systems that are accurate, reliable, and adaptive, driving positive impacts on efficiency, reliability, and sustainability in the realm of electric transportation.

2.5 Models for Predictive Maintenance

AI models for predictive maintenance represent a cornerstone in the arsenal of tools aimed at ensuring the reliability, longevity, and performance of EV components, particularly batteries. Carefully crafted to anticipate and mitigate potential problems before they manifest as critical failures, these models use advanced data analytics and machine learning algorithms to extract actionable insights from the vast data streams coming from battery management systems (BMSs) [23]. The BMS serves as a treasure trove of information, capturing critical parameters such as temperature, voltage, charge/discharge cycles, and other performance metrics that offer invaluable insights into the health and condition of EV batteries.

By harnessing this wealth of data, AI models can discern patterns, detect anomalies, and forecast future trends with unprecedented accuracy and granularity. Key to the success of these predictive maintenance models is ML algorithms such as regression analysis and time-series forecasting, which excel at uncovering underlying relationships and trends within complex datasets. Regression analysis enables the model to identify correlations between input variables and battery degradation, allowing for the estimation of the remaining useful life of batteries based on historical performance data. Time-series forecasting techniques, on the other hand, leverage historical data to predict future states and behaviors, enabling proactive maintenance scheduling and resource allocation.

By combining those superior analytical strategies with real-time fact streams from BMS, AI models can provide timely and actionable insights into the fitness and circumstance of EV batteries, enabling stakeholders to preemptively address capability troubles and optimize maintenance techniques. This proactive method of preservation minimizes downtime, reduces repair costs, and extends the operational existence of EV additives, thereby improving the general reliability and performance of electrical motors. Furthermore, the deployment of AI-driven predictive maintenance models represents a paradigm shift in the realm of asset management, moving away from traditional reactive approaches towards a more proactive and data-driven methodology. By leveraging the power of predictive analytics and machine learning, stakeholders can transition from a "fix-as-fail" mindset to a proactive maintenance strategy that prioritizes prevention over remediation. This shift not only reduces the risk of unexpected failures but also enhances operational efficiency, as maintenance activities can be planned and executed in a manner that minimizes disruption to vehicle operations.

Additionally, by extending the operational life of EV components, predictive maintenance models contribute to the sustainability and cost-effectiveness of electric transportation, reduce the need for premature replacements, and minimize environmental impact. AI models for predictive maintenance represent a game-changing approach to ensuring the reliability, longevity, and performance of EV components, leveraging advanced data analytics and machine learning algorithms to anticipate and mitigate potential issues before they escalate into critical failures [24]. By harnessing the power of predictive analytics and real-time data streams from battery management systems, stakeholders can optimize maintenance strategies, minimize downtime, and extend the operational life of EVs, driving positive impacts on efficiency, reliability, and sustainability in the realm of electric transportation.

2.6 Advanced Control Strategies

Advanced control strategies empowered by AI emerge as pivotal components in the quest to optimize energy management within EVs, revolutionizing the landscape of sustainable transportation. At the forefront of these strategies lie sophisticated techniques such as model predictive control (MPC) and adaptive control systems, which leverage the predictive capabilities of AI to dynamically adjust energy usage and charging patterns in real-time, aligning them with evolving conditions and requirements. MPC algorithms utilize predictive models to forecast future states and behaviors, enabling proactive adjustments to energy distribution and consumption based on anticipated needs and constraints [24]. By continuously optimizing energy allocation and utilization, these AI-driven control strategies ensure the efficient operation of EVs, maximizing range, performance, and overall energy efficiency. Furthermore, adaptive control systems dynamically adjust control parameters based on real-time feedback, allowing EVs to adapt to changing environmental, operational, and user-specific factors. This adaptive approach enables EVs to respond swiftly and effectively to varying conditions, ensuring optimal performance and responsiveness in diverse and dynamic driving scenarios.

Additionally, AI-driven control strategies play a crucial role in managing the integration of RESs, such as solar and wind power, into the EV ecosystem. By leveraging predictive models and real-time data, these strategies optimize the utilization of RE, ensuring that EVs prioritize the use of clean and sustainable power whenever available. This not only reduces carbon emissions and environmental impact but also enhances the overall sustainability and resilience of the energy system. Moreover, AI-driven control strategies optimize V2G interactions, enabling EVs to serve as active participants in the electricity grid. By dynamically balancing energy flows between the EV and the grid, these strategies support grid stability, efficiency, and resilience, while also providing opportunities for revenue generation and cost savings for EV owners. Through intelligent management of V2G interactions, AI-driven control strategies enable EVs to contribute to grid stabilization, demand response, and peak load management, unlocking new avenues for grid optimization and RE integration. Advanced control strategies enabled by AI are indispensable tools for optimizing energy management in EVs, leveraging predictive analytics, and real-time data to dynamically adjust energy usage, integrate RESs, and optimize V2G interactions. By harnessing the power of AI-driven control strategies, stakeholders can maximize the efficiency, reliability, and sustainability of electric transportation, driving positive impacts on energy consumption, environmental conservation, and grid stability.

2.7 *Simulation and Validation Techniques*

Simulation and validation techniques stand as critical linchpins in the development and refinement of AI models and algorithms destined for deployment within EV energy management systems. Simulation tools such as MATLAB/Simulink and software-in-the-loop (SIL) platforms serve as invaluable resources, enabling the creation of virtual environments that faithfully replicate real-world conditions and scenarios. Within these simulated environments, AI-driven energy management strategies can be meticulously evaluated and stress-tested under diverse operating conditions and stress scenarios. By subjecting AI models to rigorous simulation-based testing, stakeholders can gain insights into their performance, robustness, and scalability, identifying potential weaknesses and areas for improvement before deployment in real-world settings [25].

Furthermore, validation techniques, including field testing and pilot programs, play a pivotal role in ensuring the reliability and effectiveness of AI applications in practical scenarios. Field testing involves deploying AI-driven energy management systems in real-world environments, allowing stakeholders to assess their performance and efficacy under actual operating conditions. Pilot programs provide an opportunity for stakeholders to evaluate the real-world impact and user acceptance of AI applications, gathering feedback from end-users and refining models based on their experiences and insights. Additionally, continuous monitoring and feedback loops are established to iteratively fine-tune AI models based on real-world performance data, ensuring that they remain effective and adaptive in dynamic and evolving environments. Through a synergistic combination of simulation and validation techniques, stakeholders can accelerate the development and deployment of AI-driven energy management systems, driving positive impacts on efficiency, reliability, and sustainability in the realm of electric transportation.

The methodology for integrating AI into EV energy management involves a comprehensive approach that includes the development and deployment of sophisticated AI algorithms, the integration of sensors and IoT devices, the use of edge computing, effective ML training strategies, predictive maintenance models, advanced control strategies, and rigorous simulation and validation techniques. These components work together to create an intelligent and efficient energy management system that enhances the performance, reliability, and sustainability of EVs.

3 Energy Consumption Optimization

EVs are revolutionizing the automotive industry, offering a sustainable alternative to traditional combustion engine vehicles. The key to their success is effective management of energy consumption, which directly affects factors such as range, charging frequency, and overall performance. This chapter explores the key role of AI in optimizing energy consumption in EVs in various driving scenarios. By leveraging AI-driven strategies such as dynamic route planning, adaptive cruise control, predictive eco-driving assistance, battery state-of-charge management, dynamic powertrain control, energy harvesting, smart charging, thermal management, and driver behavior analysis, EVs can achieve optimal energy efficiency, extend driving range, and contribute to a more sustainable transportation ecosystem.

3.1 Dynamic Route Planning

Dynamic route planning represents a significant advancement in optimizing energy consumption for EVs. This approach harnesses the power of AI algorithms to evaluate and select the most energy-efficient routes based on a multitude of real-time data inputs. By considering factors such as current traffic conditions, road gradients, and individual driving patterns, AI can effectively chart paths that minimize energy expenditure. This not only extends the driving range of EVs but also reduces the need for frequent recharging, enhancing overall vehicle efficiency and user convenience. AI-driven dynamic route planning is particularly adept at adapting to changing conditions on the road. It continuously monitors traffic updates, incorporating information about congestion, accidents, and road closures to adjust routes in real-time.

Additionally, weather forecasts play a critical role in this process, as different weather conditions can significantly impact energy consumption. For instance, cold weather may require more energy for heating, while wet or icy roads can increase rolling resistance. By factoring in these variables, AI ensures that the selected route remains optimal under varying conditions, thereby promoting consistent energy efficiency [26]. Moreover, dynamic route planning takes into account road gradients, which can have a substantial impact on an EV's energy consumption. Navigating steep inclines requires more power, whereas descending slopes can allow for regenerative braking, which recharges the battery. AI algorithms analyze these gradients to plan routes that leverage terrain advantages, further conserving energy. This consideration of topographical features, combined with real-time data analysis, allows EVs to optimize their energy usage effectively. Incorporating AI for dynamic route planning also means that the system can learn and adapt to a driver's specific habits and preferences. By analyzing historical driving data, the AI can predict and suggest routes that align with the driver's typical behavior, such as preferred speeds and frequently traveled paths. This personalization enhances the driving experience by making the suggestions more relevant and user-friendly, while still focusing on energy efficiency [27].

Dynamic route planning is a critical application of AI in the realm of EV energy management. It leverages real-time data and advanced algorithms to navigate the complexities of traffic, weather, and terrain, ensuring that EVs operate at peak efficiency. This not only extends the vehicle's range but also reduces the environmental impact of EVs, contributing to a more sustainable transportation ecosystem. Through continuous monitoring and adaptive learning, AI-driven dynamic route planning exemplifies the innovative potential of AI in revolutionizing energy management for EVs. Figure 1 presents the matrix explanation of shortest route planning.

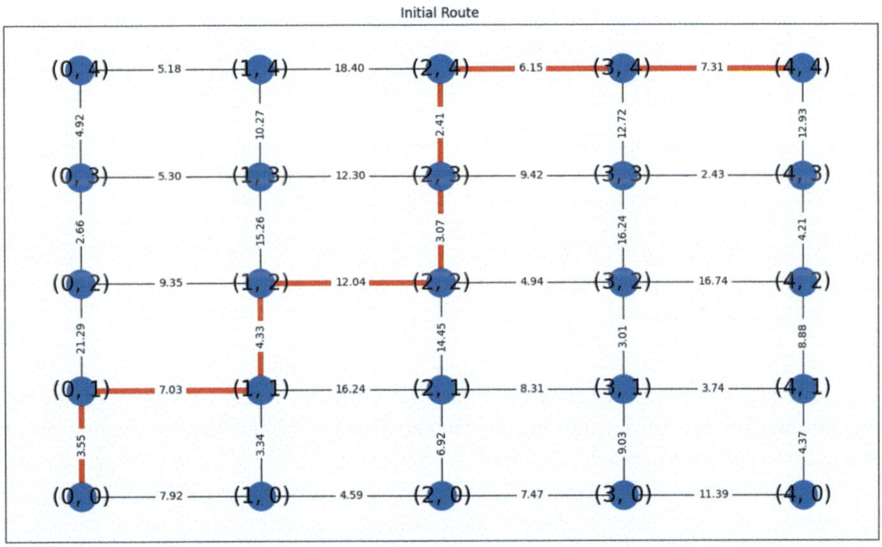

Fig. 1 Matrix explanation of shortest route planning

3.2 Adaptive Cruise Control (ACC) Systems

ACC systems are at the forefront of integrating AI to enhance the efficiency of EVs. These systems utilize AI to dynamically adjust the vehicle's speed in response to real-time traffic conditions, road environments, and the distance to other vehicles. By maintaining optimal speeds and reducing unnecessary acceleration and deceleration, ACC systems significantly improve energy efficiency. This is achieved through AI algorithms that continuously process data from an array of sensors and cameras. These sensors monitor the surroundings, allowing the system to anticipate changes in traffic patterns and adjust the vehicle's speed proactively. The ability of ACC systems to ensure smooth and consistent driving directly impacts the energy consumption of EVs. Frequent acceleration and deceleration are among the primary causes of energy inefficiency in traditional driving. By minimizing these actions, ACC systems reduce the overall energy demand on the vehicle's powertrain.

For instance, when approaching slower traffic, an AI-driven ACC system will gradually decrease the speed rather than applying sudden braking, thus conserving energy. Conversely, it will smoothly accelerate when the road ahead clears, avoiding the energy-intensive spikes associated with rapid acceleration. Moreover, ACC systems enhance driving comfort and safety, contributing to a more sustainable driving experience [28]. The continuous data analysis performed by AI enables these systems to maintain safe following distances and adapt to varying road conditions without requiring constant driver intervention. This not only reduces driver fatigue but also optimizes the vehicle's energy usage by maintaining a steady flow in stop-and-go traffic situations. In conclusion, the integration of AI in ACC systems plays a crucial role in optimizing energy consumption in EVs. By leveraging advanced algorithms to ensure smooth and efficient driving, ACC systems not only extend the driving range of EVs but also enhance overall vehicle performance and sustainability. This represents a significant step towards the broader adoption of EVs as a viable alternative to traditional internal combustion engine vehicles, contributing to an environmentally healthier and greener future.

3.3 Predictive Eco-driving Assistance

Predictive eco-driving assistance systems play a crucial role in optimizing energy consumption in EVs by harnessing the power of AI to guide drivers towards more energy-efficient driving practices. These systems analyze real-time data on upcoming road conditions and traffic patterns to offer timely recommendations that enable drivers to make informed decisions. For example, by suggesting when to coast, gently accelerate, or brake, the system helps minimize unnecessary energy expenditure. The underlying AI algorithms continuously process vast amounts of data from various sources, including GPS, vehicle sensors, and traffic updates, to deliver precise and actionable advice tailored to the current driving situation. The effectiveness of predictive eco-driving assistance systems lies in their ability to learn and adapt over time [29]. AI algorithms monitor and analyze individual driving behaviors and environmental conditions, allowing the system to refine its recommendations continually.

This personalized approach ensures that the advice provided is not only relevant but also enhances the driver's natural driving style, making it easier to adopt and maintain energy-efficient habits. By promoting smoother driving patterns and reducing abrupt maneuvers, these systems contribute significantly to conserving the EV's battery life and extending its driving range. Moreover, predictive eco-driving assistance systems do not merely focus on immediate driving conditions but also consider long-term factors affecting energy consumption. For instance, they can take into account the state of the battery charge, anticipated route inclines, and expected traffic congestion to adjust their recommendations accordingly. This foresight enables drivers to plan their journeys more effectively, reducing the need for frequent charging stops and enhancing overall travel efficiency. In addition to benefiting individual drivers, these systems also contribute to broader environmental and economic goals. By improving energy efficiency, they help reduce the overall energy demand and decrease greenhouse gas emissions associated with electricity generation for EVs. Furthermore, the reduction in energy consumption can translate into cost savings for EV owners, making electric vehicles an even more attractive option compared to traditional internal combustion engine vehicles. Predictive eco-driving assistance systems represent a vital application of AI in the realm of EV energy management. By providing real-time, personalized driving recommendations, these systems enhance energy efficiency, extend driving range, and support sustainable transportation practices. As AI technology continues to advance, the capabilities and effectiveness of these systems are expected to grow, further solidifying their role in the future of energy-efficient transportation.

3.4 Battery State-of-Charge (SoC) Management

Battery SoC management is critical for optimizing energy usage in EVs. AI algorithms monitor and manage the SoC of the EV battery to ensure efficient energy utilization while avoiding overcharging or deep discharging, which can degrade battery health [20]. By balancing power demands with available battery capacity, these systems optimize energy consumption and extend the driving range of EVs. AI continuously adjusts charging and discharging rates based on factors such as driving conditions, temperature, and battery health, ensuring optimal performance and longevity.

3.5 Dynamic Powertrain Control Strategies

Dynamic powertrain control strategies are integral to enhancing the energy efficiency of EVs. These strategies involve real-time adjustments to the power distribution among the various components of the EV drivetrain, tailored to specific driving conditions and energy demands. By leveraging advanced AI algorithms, dynamic powertrain control systems can optimize the delivery of power from the battery to the motor, ensuring that energy utilization is maximized without compromising the vehicle's performance or range. AI plays a pivotal role in these strategies by continuously analyzing a myriad of data points, including speed, acceleration, road gradient, and driver behavior. This data-driven approach allows for precise modulation of torque output, ensuring that the motor operates within its most efficient parameters. For instance, during urban driving with frequent stop-and-go situations, the AI can prioritize regenerative braking to capture and store energy that would otherwise be lost. This recuperated energy is then used to assist in acceleration, thereby reducing the overall energy consumption from the battery. On highways, where steady speeds are more common, the AI can optimize the powertrain to maintain a balance between energy efficiency and the required performance. By adjusting the power output to meet the demands of cruising at higher speeds while minimizing unnecessary energy expenditure, these systems help extend the driving range of the EV.

Additionally, AI can manage the transition between different driving modes, such as switching from eco-mode to sport mode, ensuring that the powertrain responds appropriately to the driver's immediate needs while still considering the long-term impact on energy consumption [30]. Furthermore, dynamic powertrain control strategies enhance the efficiency of regenerative braking systems. By predicting the optimal moments to engage regenerative braking, AI can maximize energy recovery, converting kinetic energy back into stored electrical energy in the battery. This not only improves the overall efficiency but also reduces wear on the traditional braking system, contributing to lower maintenance costs and longer vehicle lifespan. In essence, AI-driven dynamic powertrain control strategies represent a sophisticated means of managing energy within an EV. By fine-tuning the interaction between the battery, motor, and other drivetrain components, these strategies ensure that energy is used as efficiently as possible across all driving scenarios. This optimization leads to significant improvements in driving range, reduces the frequency of charging, and supports the broader goal of creating a more sustainable transportation ecosystem.

3.6 Energy Harvesting and Regeneration

Energy harvesting and regeneration are critical components in the optimization of energy consumption in EVs. These systems capture and store energy that would otherwise be wasted during braking and deceleration, effectively reducing the vehicle's reliance on its battery and extending its driving range. By utilizing regenerative braking, the kinetic energy that is normally dissipated as heat is converted back into electrical energy and stored in the battery. This process not only enhances the overall energy efficiency of the vehicle but also contributes to a reduction in energy waste. AI plays a pivotal role in optimizing these regenerative braking systems. Through the use of AI-driven algorithms, EVs can adapt their energy recuperation strategies based on real-time driving conditions. For instance, AI can dynamically adjust the regenerative braking force by continuously monitoring and analyzing vehicle speed, acceleration, and deceleration patterns. This ensures that the maximum possible energy is recovered without compromising braking performance or the vehicle's handling dynamics.

Moreover, AI can predict and adapt to various driving scenarios, such as urban stop-and-go traffic or highway cruising, to optimize energy regeneration. In urban environments, where frequent stops and starts are common, AI can enhance regenerative braking to capture more energy during deceleration phases. On highways, AI can manage the balance between regenerative braking and coasting to maintain smooth driving dynamics while still harvesting energy efficiently.

Additionally, AI can integrate driver behavior analysis to tailor regenerative braking responses. By learning and predicting a driver's braking habits, the system can fine-tune the regenerative braking process to align with the driver's style, thereby improving energy recovery and enhancing the driving experience. For example, if a driver typically decelerates gradually, the system can adjust to maximize energy recovery during these longer deceleration periods [31]. Through these sophisticated AI-driven strategies, energy harvesting and regeneration systems not only boost the energy efficiency of EVs but also play a crucial role in extending the driving range and reducing the frequency of charging. This, in turn, enhances the overall sustainability of electric transportation by making EVs more practical and appealing for everyday use. The integration of AI in optimizing these systems is a testament to how advanced technologies are transforming the landscape of energy management in the automotive industry, paving the way for more efficient and eco-friendly vehicles.

3.7 *Smart Charging and Grid Interaction*

Smart charging systems represent a critical intersection of AI and EV energy management, offering significant advancements in how EVs interact with the power grid. These systems use AI algorithms to optimize charging schedules by analyzing various factors such as electricity pricing, grid demand, and the availability of RE. By intelligently coordinating charging times, EVs can take advantage of periods when electricity prices are low or when there is a surplus of RE, such as during midday when solar power is abundant. This not only reduces the cost of charging for EV owners but also alleviates stress on the grid, promoting a more balanced and efficient energy distribution. AI-driven smart charging systems continuously monitor grid conditions and dynamically adjust charging plans to align with real-time energy tariffs and grid stability needs. For instance, during peak hours when grid demand is high, these systems can defer charging to off-peak times, thereby helping to flatten demand curves and prevent potential overloads. Conversely, during periods of high RE production, such as on windy or sunny days, AI can prioritize charging to utilize this clean energy, reducing reliance on fossil fuels and enhancing the overall sustainability of the energy system.

Moreover, smart charging systems can provide ancillary services to the grid, such as demand response and frequency regulation. By modulating the charging rate of multiple EVs in response to grid signals, these systems can contribute to grid stability and reliability. This interaction is particularly valuable as the penetration of RESs, which are inherently variable, increases. By smoothing out fluctuations in supply and demand, smart charging aids in integrating more RE into the grid. In addition to direct benefits to grid management, smart charging strategies also enhance the user experience by ensuring that EVs are charged efficiently and cost-effectively. AI can learn from individual driving patterns and preferences, predicting the best times to charge based on expected vehicle use and personal schedules.

For example, an AI system might schedule charging during the early hours of the morning when electricity rates are lower and ensure the vehicle is fully charged before the driver's usual departure time. This personalized approach maximizes convenience while minimizing costs and environmental impact [32]. In summary, the integration of AI in smart charging systems and grid interaction represents a transformative approach to EV energy management. By optimizing charging schedules and facilitating a more resilient and sustainable power grid, these systems not only enhance the efficiency and affordability of EVs but also support the broader transition to a cleaner energy future.

3.8 Vehicle Thermal Management Systems

Vehicle thermal management systems play a crucial role in the energy efficiency of EVs. These systems are responsible for regulating the temperature within the vehicle to ensure optimal performance of both the vehicle's components and the comfort of its occupants. AI-driven thermal management systems take this a step further by dynamically adjusting heating, cooling, and ventilation based on real-time data from external environmental conditions and internal comfort settings. This dynamic adjustment helps in minimizing energy consumption without compromising on comfort, thereby extending the driving range of the EV. One significant aspect of AI in thermal management is the optimization of HVAC (heating, ventilation, and air conditioning) systems. Traditional HVAC systems can be energy-intensive, particularly in extreme weather conditions where maintaining a comfortable cabin temperature requires substantial energy. AI algorithms can predict and preemptively adjust the HVAC settings by learning from patterns such as time of day, weather forecasts, and user preferences.

For example, AI can pre-cool or pre-heat the cabin while the vehicle is still plugged in, utilizing grid power instead of relying on the battery, which conserves the vehicle's charge for driving. Additionally, AI can enhance thermal insulation strategies within the vehicle. By monitoring and adjusting the thermal load, AI systems ensure that the insulation properties are utilized to their fullest potential, thereby reducing the need for excessive heating or cooling. For instance, during colder climates, AI can manage the recirculation of warm air to maintain cabin temperature with minimal energy expenditure. In warmer climates, predictive algorithms can optimize the use of ventilated seats and zonal cooling to maintain comfort without overburdening the HVAC system.

Furthermore, AI-enabled thermal management systems also extend to the vehicle's powertrain components. Batteries and electric motors operate more efficiently within specific temperature ranges [33]. AI can manage the thermal environment of these components by intelligently controlling cooling systems such as liquid coolants or airflow. By doing so, it prevents overheating and enhances the longevity and efficiency of the EV's powertrain. Overall, AI-driven thermal management systems in EVs offer a sophisticated solution to energy consumption optimization. By leveraging real-time data and predictive analytics, these systems ensure that both the vehicle's interior and its critical components are maintained at optimal temperatures. This not only improves energy efficiency but also extends the driving range and enhances the overall sustainability of EVs.

4 Human Machine Interface (HMI) Design

The HMI design is a crucial aspect of integrating AI into EV energy management. An effective HMI ensures seamless interaction between the driver and the vehicle's complex energy management systems, enhancing usability, safety, and overall user experience. This section delves into the key components of HMI design for AI-driven EV energy management, highlighting how these interfaces facilitate optimal energy consumption, improve user engagement, and support the transition to sustainable transportation.

- Intuitive User Interfaces: A well-designed HMI must offer an intuitive and user-friendly interface that simplifies complex energy management tasks. Graphical displays should present real-time information about battery status, energy consumption, and driving efficiency in an easily understandable format. Interactive touchscreens, voice commands, and haptic feedback mechanisms can enhance user interaction, making it easier for drivers to access critical information and adjust settings on the go.
- Real-Time Data Visualization: Effective HMI design incorporates real-time data visualization to provide drivers with immediate feedback on their driving habits and the vehicle's energy status. Dashboards displaying live metrics such as battery SoC, energy regeneration, and predicted range help drivers make informed decisions to optimize energy usage. By visualizing energy flows and highlighting areas for improvement, these interfaces empower users to drive more efficiently.

- Customizable Interfaces: Customizability is a key feature of advanced HMI design. Drivers should be able to personalize their interfaces to suit their preferences and driving styles. This includes adjusting the layout of the display, selecting which metrics to prioritize, and configuring alerts for specific energy thresholds. Customizable HMIs ensure that drivers receive the most relevant information in a format that is most convenient for them.
- Predictive Assistance and Guidance: AI-driven HMIs can offer predictive assistance and guidance to enhance energy management. For example, predictive eco-driving assistance systems can provide real-time suggestions for optimal driving behaviors, such as when to accelerate, decelerate, or coast. These systems leverage AI algorithms to analyze upcoming road conditions and traffic patterns, offering proactive advice that helps drivers conserve energy and extend their vehicle's range.
- Interactive Feedback Systems: Interactive feedback systems are essential for fostering energy-efficient driving habits. By analyzing driver behavior, AI-powered HMIs can provide personalized feedback and coaching. For instance, if a driver frequently accelerates aggressively, the system can suggest smoother acceleration techniques. Over time, these interactive feedback systems can help drivers adopt more energy-efficient driving practices, leading to sustained improvements in overall energy consumption.
- Integration with Smart Devices: Modern HMI design should seamlessly integrate with smart devices, such as smartphones and smartwatches, to enhance the user experience. Mobile applications can provide remote access to the vehicle's energy management system, allowing drivers to monitor battery status, schedule charging sessions, and receive notifications about optimal charging times based on electricity tariffs and grid demand. This integration ensures that drivers remain connected and informed, even when away from their vehicles.
- Voice-Activated Controls: Voice-activated controls are becoming increasingly important in HMI design for EVs. Natural language processing (NLP) technologies enable drivers to interact with their vehicle's energy management systems using voice commands, reducing the need for manual inputs and minimizing distractions. Voice-activated systems can perform tasks such as adjusting climate controls, setting navigation routes, and querying energy consumption data, all while keeping the driver's focus on the road.
- Safety and Ergonomics: Safety and ergonomics are paramount in HMI design. Interfaces should be designed to minimize driver distraction and ensure that critical information is easily accessible. This involves optimizing the placement of displays and controls within the driver's line of sight and reach. Ergonomically designed HMIs reduce cognitive load, allowing drivers to interact with energy management systems without compromising their attention to driving.
- Enhanced Visual and Auditory Cues: Enhanced visual and auditory cues play a significant role in effective HMI design. Visual cues such as color-coded indicators and animated graphics can quickly convey important information about energy status and driving efficiency. Auditory cues, including alerts and spoken notifications, can provide real-time updates on energy usage and driving behavior,

ensuring that drivers receive timely information without needing to divert their gaze from the road.

- Adaptive Learning and Personalization: Adaptive learning capabilities enable HMIs to evolve based on user behavior and preferences. By analyzing driver interactions and energy consumption patterns, AI can tailor the HMI experience to better suit individual needs. This personalization ensures that the interface remains relevant and useful, providing insights and recommendations that are specifically geared toward enhancing each driver's energy management practices.
- Comprehensive Energy Management Tools: Finally, a robust HMI should offer comprehensive energy management tools that allow drivers to plan and optimize their energy usage. This includes features such as dynamic route planning, real-time traffic updates, and charging station locators. By integrating these tools into a cohesive interface, drivers can make informed decisions that enhance energy efficiency, reduce costs, and contribute to the sustainability of their EV usage.

In conclusion, the design of HMI in AI-driven EV energy management systems is critical for maximizing efficiency, safety, and user satisfaction. By combining intuitive design, real-time data visualization, customizable interfaces, predictive assistance, and integration with smart devices, HMIs enable drivers to effectively manage their vehicle's energy consumption [34]. As AI technologies continue to evolve, the potential for innovative and user-centric HMI designs will play a pivotal role in advancing the adoption and efficiency of EVs.

5 Integration with Renewable Energy Sources

The integration of RESs into EV energy management systems, driven by AI, represents a transformative approach to achieving sustainable transportation. By leveraging AI, EVs can effectively utilize RE, enhancing energy efficiency, reducing environmental impact, and supporting grid stability. AI-powered smart charging systems play a crucial role in optimizing the use of RE for charging EVs. These systems analyze real-time data on electricity generation from RESs, electricity pricing, and grid demand. By aligning EV charging schedules with periods of high RE production, AI ensures that EVs are charged using the cleanest and most cost-effective energy available. For example, an AI system might schedule charging during sunny days when solar power generation is at its peak or at night when wind energy is abundant.

V2G technology allows EVs to interact bidirectionally with the power grid, enabling them to both draw energy from and supply energy to the grid. AI enhances V2G systems by predicting energy demand and supply patterns, optimizing the timing and extent of energy exchanges between EVs and the grid. During periods of high RE production, AI can direct EVs to store excess energy, which can later be fed back into the grid during peak demand periods, thus supporting grid stability and maximizing the use of RE. AI-driven predictive analytics enable EVs to anticipate energy needs and availability based on various factors, including weather forecasts, driving patterns, and RE production trends. By predicting when RESs will be most available and cost-effective [35], AI can optimize energy management strategies to prioritize the use of clean energy. For instance, if AI predicts a surge in wind energy generation overnight, it can schedule EVs to charge during those hours, reducing reliance on non-RESs.

Integrating RE with battery storage systems in EVs is another area where AI proves invaluable. AI algorithms manage the balance between energy generation, storage, and consumption, ensuring that EV batteries are charged using RE whenever possible. This integration not only maximizes the utilization of RE but also enhances the efficiency and longevity of EV batteries by managing charging cycles more effectively. AI can optimize EV charging in response to dynamic energy pricing, which often reflects the availability of RE. By analyzing price signals and RE availability, AI systems can determine the most cost-effective times to charge EVs. For instance, during periods when RE supply is high and electricity prices drop, AI can schedule EV charging to take advantage of these lower rates, promoting both economic savings and environmental benefits.

Accurate forecasting of RE production is essential for effective integration into EV energy management systems. AI leverages advanced machine learning (ML) models to predict the availability of solar and wind energy based on weather patterns, historical data, and real-time conditions. These forecasts enable more precise planning and optimization of energy usage, ensuring that RE is utilized to its fullest potential. AI enhances the interaction between EVs and the power grid by enabling more sophisticated demand response strategies. During periods of high RE production, AI can signal EVs to increase charging activities, helping to absorb excess energy and prevent grid overload. Conversely, during low RE periods, AI can manage the reduction of charging loads or even facilitate energy feedback from EVs to the grid, ensuring a balanced and stable energy supply.

AI supports decentralized energy management by enabling peer-to-peer energy trading among EVs and RE producers. Through Blockchain and other secure transaction technologies, AI facilitates direct energy exchanges between EV owners and RE providers, creating a more resilient and efficient energy ecosystem. This decentralized approach empowers consumers, promotes the use of local RESs, and reduces dependency on centralized power plants. The integration of RESs into EV energy management significantly enhances the sustainability and environmental impact of EVs. By prioritizing the use of clean energy, AI-driven systems help reduce greenhouse gas emissions, lower reliance on fossil fuels, and contribute to the overall decarbonization of the transportation sector. This alignment with global sustainability goals makes AI-integrated EVs a critical component in the fight against climate change.

Effective integration of RESs with EV energy management requires supportive policies and regulatory frameworks [36]. AI can assist policymakers by providing data-driven insights into energy consumption patterns, RE potential, and the impacts of various regulatory scenarios. By informing policy decisions with accurate and predictive data, AI helps create an environment conducive to the widespread adoption of RE in the transportation sector. The integration of RESs into EV energy management, driven by AI, represents a significant advancement toward sustainable transportation. AI enables smart charging, predictive energy management, V2G integration, and enhanced grid interaction, ensuring that RE is utilized efficiently and effectively. By fostering the use of clean energy, AI-driven EV energy management systems contribute to reducing environmental impact, supporting grid stability, and advancing global sustainability goals. As AI technologies continue to evolve, their role in integrating RE into the EV ecosystem will become increasingly vital, paving the way for a cleaner and more sustainable future.

6 Conclusions

Building upon the foundational insights provided earlier, this chapter has delved deeper into the methodological aspects of integrating artificial intelligence (AI) into electric vehicle (EV) energy management systems. From AI algorithms and models to sensor integration and edge computing, this section has explored the tools and techniques that underpin the optimization of EV energy consumption and efficiency. Methodologically, AI serves as the backbone of various EV energy management strategies, facilitating predictive maintenance, advanced control systems, and simulation techniques. By leveraging real-time data processing and machine learning algorithms, AI enables dynamic route planning, adaptive cruise control, and predictive eco-driving assistance, enhancing both the efficiency and safety of EVs on the road. Moreover, AI plays a crucial role in optimizing energy consumption and extending the operational life of EV batteries. Through dynamic powertrain control, energy harvesting, and smart charging strategies, AI-driven EV energy management systems maximize the utilization of available resources while minimizing environmental impact. The integration of renewable energy sources (RESs) into EV energy management further enhances sustainability and resilience. Continued research, innovation, and collaboration across disciplines will be essential in realizing the full potential of AI-driven EV energy management, propelling us toward a future where sustainable transportation is not just a possibility but a reality.

Acknowledgements This research work was supported by "Woosong University's Academic Research Funding—2025".

References

1. Amini MH, Nabi B, Haghifam MR (2020) Artificial intelligence in energy management systems for smart sustainable cities. IEEE Access 8:52617–52633. https://doi.org/10.1109/ACCESS.2020.2981948
2. Hu X, Zou C, Zhang L, Li Y (2019) Advanced machine learning for battery management systems: a review. IEEE Access 7:139926–139940. https://doi.org/10.1109/ACCESS.2019.2943381

3. Ghosh A, Saraswat A, Mukherjee A, Majumder S, Paul S (2019) Deep reinforcement learning-based energy management for plug-in hybrid electric vehicles. IEEE Trans Transp Electrif 5(4):886–894. https://doi.org/10.1109/TTE.2019.2958706
4. Dallinger D, Gerda A, Wietschel M (2013) Integration of intermittent renewable power supply using grid-connected vehicles—a 2030 case study for California and Germany. Appl Energy 104:666–682. https://doi.org/10.1016/j.apenergy.2012.11.002
5. Masoum MAS, Hashemnejad A (2019) Real-time coordination of plug-in electric vehicle charging for smart grid reliability and peak load mitigation. IEEE Trans Smart Grid 10(5):5037–5046. https://doi.org/10.1109/TSG.2018.2871694
6. Sabihuddin S, Kiprakis AE, Mueller M (2015) A numerical and computational analysis on the energy losses in electric vehicle battery packs. Energy 93:1722–1731. https://doi.org/10.1016/j.energy.2015.10.022
7. Chen C, Zou Y, Xiong R, Sun F, Deng J (2017) Electric vehicle energy management with reinforcement learning. IEEE Control Syst Lett 1(1):60–65. https://doi.org/10.1109/LCSYS.2017.2710025
8. Aung Z, Woon WL (2011) Survey on the widespread use and applications of smart meters in residential environments. Proc Comput Sci 10:548–554. https://doi.org/10.1016/j.procs.2012.06.070
9. Deilami S, Masoum MAS, Moses PS, Masoum MAS (2011) Real-time coordination of plug-in electric vehicle charging in smart grids to minimize power losses and improve voltage profile. IEEE Trans Smart Grid 2(3):456–467. https://doi.org/10.1109/TSG.2011.2159817
10. Zhou Y, Su Y, Ma W, Fan Y (2020) A survey on electric vehicle charging infrastructure deployment strategies. J Modern Power Syst Clean Energy 8(1):23–35. https://doi.org/10.35833/MPCE.2018.000708
11. Mwasilu F, Justo JJ, Kim EK, Do TD, Jung JW (2014) Electric vehicles and smart grid interaction: a review on vehicle to grid and renewable energy sources integration. Renew Sustain Energy Rev 34:501–516. https://doi.org/10.1016/j.rser.2014.03.031
12. Momber I, Gomez T, Söder L (2014) Plug-in electric vehicle coordination schemes and impacts on the distribution network: literature review. Int J Electr Power Energy Syst 57:398–407. https://doi.org/10.1016/j.ijepes.2013.12.010
13. Vayá MG, Andersson G (2013) Self-scheduling of plug-in electric vehicle aggregator to provide balancing services for smart grids. IEEE Trans Smart Grid 4(2):1176–1184. https://doi.org/10.1109/TSG.2012.2231440
14. Gholizadeh M, Salmasi FR (2013) Estimation of state of charge, unknown non-linearities, and state of health of a lithium-ion battery based on a comprehensive unobservable model. IEEE Trans Ind Electron 61(3):1335–1344. https://doi.org/10.1109/TIE.2013.2262754
15. Moura SJ, Callaway DS, Fathy HK (2011) Optimal control of plug-in hybrid electric vehicles for renewable energy support and demand response. IEEE Trans Control Syst Technol 19(3):545–555. https://doi.org/10.1109/TCST.2010.2043736
16. Liu X, Li Z (2017) Electric vehicle powertrain efficiency optimization based on real-time road and traffic information. IEEE Trans Veh Technol 66(5):3779–3791. https://doi.org/10.1109/TVT.2016.2628720
17. Wu F, Ramesh A, Huang Y, Liaw BY (2016) A review on inductive charging for electric vehicles. IEEE Trans Transp Electrif 4(1):15–26. https://doi.org/10.1109/TTE.2015.2506543
18. Wu X, Zhang R, Lu S, Wu Y (2020) Artificial intelligence for energy management systems of hybrid electric vehicles: a review. IEEE Trans Intell Transp Syst 21(4):1456–1472. https://doi.org/10.1109/TITS.2019.2916808
19. Jafari M, Almasi M, Salehi A, Han X, Xiao W, Chen Z (2019) Blockchain technology for electric vehicle management: a review of recent advances and future perspectives. Renew Sustain Energy Rev 112:114–128. https://doi.org/10.1016/j.rser.2019.05.017
20. Salkuti SR (2022) Modeling of various renewable energy resources for smart electrical power systems. In: Salkuti SR, Ray P (eds) Next generation smart grids: modeling, control and optimization. Lecture notes in electrical engineering, vol 824. Springer, Singapore. https://doi.org/10.1007/978-981-16-7794-6_2

21. Palacin R (2013) Rail transport and sustainable urban mobility. Renew Sustain Energy Rev 29:961–971. https://doi.org/10.1016/j.rser.2013.09.010
22. Drude L, Kolmanovsky I (2020) Personalized eco-driving with electric vehicles: a federated reinforcement learning approach. IEEE Trans Control Syst Technol 29(1):98–113. https://doi.org/10.1109/TCST.2020.2979039
23. Wang H, Chen L, Zhang J, Chen S (2018) Smart grid technologies in China: a review. Renew Sustain Energy Rev 88:88–104. https://doi.org/10.1016/j.rser.2018.02.027
24. Al-Alawi BM, Bradley TH (2013) Total cost of ownership, payback, and consumer preference modeling of plug-in hybrid electric vehicles. Appl Energy 103:488–506. https://doi.org/10.1016/j.apenergy.2012.10.009
25. Yilmaz M, Krein PT (2013) Review of the impact of vehicle-to-grid technologies on distribution systems and utility interfaces. IEEE Trans Power Electron 28(12):5673–5689. https://doi.org/10.1109/TPEL.2012.2227500
26. Boulanger AG, Chu AC, Maxx S, Waltz DL (2011) Vehicle electrification: status and issues. Proc IEEE 99(6):1116–1138. https://doi.org/10.1109/JPROC.2011.2112750
27. Wang D, Feng Y, Wang X, Zhang L, Li Z, Hu X (2019) Predictive energy management for hybrid electric vehicles considering battery degradation. IEEE Trans Control Syst Technol 27(1):55–64. https://doi.org/10.1109/TCST.2017.2785281
28. Tang S, Zhang C, Meng H, Wang J, Huang G, Cao G (2011) A framework for load scheduling to maximize aggregated battery performance in grid-connected multiple PHEVs. IEEE Trans Smart Grid 2(2):371–378. https://doi.org/10.1109/TSG.2011.2119486
29. Liu C, Chau KT, Wu D, Gao S (2013) Opportunities and challenges of vehicle-to-home, vehicle-to-vehicle, and vehicle-to-grid technologies. Proc IEEE 101(11):2409–2427. https://doi.org/10.1109/JPROC.2013.2271951
30. Sun C, Li X, Zhang S, Wang Q, Zhang L (2020) Battery management system for electric vehicles in smart grid. IEEE Trans Ind Inf 16(5):3057–3065. https://doi.org/10.1109/TII.2019.2943204
31. Salkuti SR (2023) Advanced technologies for energy storage and electric vehicles. Energies 16(5):2312. https://doi.org/10.3390/en16052312
32. Green RC, Wang L, Alam M (2011) The impact of plug-in hybrid electric vehicles on distribution networks: a review and outlook. Renew Sustain Energy Rev 15(1):544–553. https://doi.org/10.1016/j.rser.2010.08.014
33. Zhang C, Yan B, Zhang S, Su H, Hu X (2016) A multi-objective optimization model for the life cycle of electric vehicle batteries. IEEE Trans Intell Transp Syst 17(8):2198–2209. https://doi.org/10.1109/TITS.2016.2518205
34. Salkuti SR, Ray P, Pagidipala S (2022) Overview of next generation smart grids. In: Salkuti SR, Ray P (eds) Next generation smart grids: modeling, control and optimization. Lecture notes in electrical engineering, vol 824. Springer, Singapore. https://doi.org/10.1007/978-981-16-7794-6_1
35. Liu J, Gao H, Xu W, Zhang Y, Cao D, Ehsani M (2017) State-of-the-art of energy management strategies and applications in hybrid electric vehicles. IEEE Access 5:19968–19985. https://doi.org/10.1109/ACCESS.2017.2750683
36. Han S, Han S, Sezaki K (2010) Development of an optimal vehicle-to-grid aggregator for frequency regulation. IEEE Trans Smart Grid 1(1):65–72. https://doi.org/10.1109/TSG.2010.2045163

CdS-LDR Optimization in Dual-Axis Solar Tracking System with NPC Multilevel Inverter for Electric Vehicle Application

R. Palanisamy, Soumyadeep Das, Soumyadeep Biswas, Debtanu Dey, S. Usha, and Surender Reddy Salkuti

Abstract This chapter gives an idea to implementation and design a dual-axis solar tracker using light dependent resistor, 3-phase Neutral Point Clamped multilevel inverter, IR2110 switch gate driver, Arduino Uno, motor driver, PIC microcontroller generating pulse width modulation signals, ULN2003 motor driver, 12 V Permanent Magnet DC motors, and a buck-boost converter. In addition, the project utilized the tracker's design, Lambert's cosine law, and the inverse square law. The dual-axis solar tracker was designed to improve the efficiency of solar panels by tracking the sun's movement. The light-dependent resistors were used to measure the intensity of light falling on the solar panel, and the 3-phase NPC multilevel inverter, IR2110 switch gate driver, and buck-boost converter were used to control the motors' speed and direction. The PIC microcontroller generated signals used to control the motors' speed. The novelty of this dual-axis solar tracker lies in the use of advanced technologies such as LDRs, 3-phase Neutral Point Clamped multilevel inverter, IR2110 gate driver,

R. Palanisamy · S. Das · S. Biswas · D. Dey · S. Usha
Department of Electrical and Electronics Engineering, College of Engineering and Technology, SRM Institute of Science and Technology, Kattankulathur, Chengalpattu, Tamil Nadu 603203, India
e-mail: palanisr@srmist.edu.in

S. Das
e-mail: sb2468@srmist.edu.in

S. Biswas
e-mail: ss3694@srmist.edu.in

D. Dey
e-mail: dd3693@srmist.edu.in

S. Usha
e-mail: ushas@srmist.edu.in

S. R. Salkuti (✉)
Department of Railroad and Electrical Engineering, Woosong University, Daejeon 34606, Republic of Korea
e-mail: surender@wsu.ac.kr

S. R. Salkuti (ed.), *Artificial Intelligence for Integrated Smart Energy Systems in Electric Vehicles*, Lecture Notes in Electrical Engineering 1427,
https://doi.org/10.1007/978-3-031-94276-1_10

Arduino Uno, motor driver, PIC microcontroller generating PWM signals, and buck-boost converter. The proposed system gives a dual-axis solar tracker that significantly improves the efficiency of the solar panel.

Keywords Dual-axis solar tracker · Multilevel inverter · Pulse width modulation · Photovoltaic system · Maximum power point tracking · PIC microcontroller

Abbreviations

PV	Photovoltaic
MPPT	Maximum power point tracking
PO	Perturb and observe
VSI	Voltage source inverter
THD	Total harmonic distortion
PWM	Pulse width modulation
SPWM	Sinusoidal pulse width modulation
MLI	Multilevel inverter
PLC	Programmable logic controller
MPP	Maximum power point
LDR	Light-dependent resistor
SOC	State-of-charge
V_{dc}	DC supply voltage
V_{out}	Output voltage
δ	Tracker angle
β	Azimuth angle
Vcc	Supply voltage
V_m	Peak voltage
R_L	Load resistance

1 Introduction

Methods for utilizing renewable energy sources have been developed in response to energy depletion and global warming. Solar energy is one of the most promising renewable energy sources [1, 2]. The power output of a photovoltaic (PV) system can be greatly increased with solar trackers. It suggest a dual-axis solar tracking PV system that uses simple electrical circuits, a four-quadrant light-dependent resistor (LDR) sensor, and the feedback control principle to achieve reliable system performance [3–6]. The proposed system uses a standalone PV inverter and a unique dual-axis AC motor to perform solar tracking [7]. A technical innovation with a straightforward and efficient design is the control implementation. There is a higher need for energy because of the growing population [8–10]. The usage of fossil fuels increases environmental pollution, while the overuse of resources causes resource depletion. In the current energy crisis, renewable energy sources are clean, non-polluting energy sources. Solar energy is the most promising source out of all the others. Photovoltaic panels can turn solar energy into electricity [11–14].

To capture the most solar energy, PV systems need to keep an eye on the sun all day long. A solar tracking system makes this feasible. It is feasible to track the sun on one or both axes [15]. Direct current (DC) is changed to alternating current (AC) by an electrical equipment called an inverter. Direct current is created by devices such as batteries and solar panels. These gadgets can power appliances and equipment used in the home when they are linked to an inverter [12, 16]. A dual-axis solar tracker is a tool that can follow the sun's movement in two dimensions to maximize the quantity of solar energy captured. A multilevel inverter is a different kind of power electrical device that can transform the DC power produced by the solar panels into AC electricity that can be used by homes and businesses [17, 18].

The design and installation of a dual-axis solar tracker using an NPC multilevel inverter are the main topics of this study. Due to its capacity to deliver high-quality power with less harmonic distortion, minimal switching losses, and higher efficiency, the NPC multilevel inverter is a topology that is frequently utilized for solar power applications [19]. The suggested method uses a dual-axis device to automatically follow the position of the sun to increase the efficiency of solar panels. Even in a variety of weather situations, the system's ability to accurately adjust solar panel orientation enables maximum energy output [20–22].

The theoretical analysis, design, and simulation of the suggested system along with its control scheme and performance assessment are covered in the study. The experimental findings of the system, which show how successful it is at raising solar panel efficiency, are also presented in the chapter. The report concludes with a thorough evaluation of the system's benefits and drawbacks, along with suggestions for further study and development.

2 PV System and Modeling

One kind of renewable energy system that turns sunlight into electricity is a photovoltaic (PV) system. Another name for it is a solar power system. Solar panels, which are composed of photovoltaic cells that transform solar energy into direct current (DC) power, make up the system [23]. The DC electricity generated by the PV panels is converted into alternating current (AC) electricity after going via an inverter. This AC electricity may then be used to power buildings, businesses, and other electrical appliances [24–26]. Grid-connected or off-grid PV systems are also options. In a system that is connected to the grid, the electricity produced by the PV panels is fed into the power grid, and the user receives credit for the electricity they produce.

In an off-grid system, the electricity generated by the PV panels is stored in batteries for later use, and the system may also include a backup generator for periods of low sunlight or high electricity demand. As the price of PV panels has dropped over time, PV systems—a clean and renewable energy source—have grown in popularity. Additionally, they are a sustainable substitute for fossil fuels, which cause climate change and greenhouse gas emissions. One kind of solar panel mounting system that enables solar panels to track the movement of the sun in both horizontal and vertical directions is a dual-axis solar tracker. This tracking system is designed to increase the efficiency of a photovoltaic (PV) system by maximizing the amount of sunlight that hits the solar panels throughout the day. In a PV system with a dual-axis solar tracker, the solar panels are fixed and kept on a frame that is connected to a tracking mechanism. This mechanism is controlled by a microprocessor or a programmable logic controller (PLC), which uses sensors and algorithms to adjust the pos. of the solar panels to track the sun's movement.

By tracking the sun's movement in 2 directions, the horizontal and vertical directions, a dual-axis solar tracker can increase the amount of sunlight that hits the solar panels by approximately 40% compared to fixed-tilt systems. This increased efficiency can result in higher energy output and a faster return on investment for the PV system. However, the cost of a dual-axis solar tracker can be higher than a fixed-tilt mounting system, and the maintenance requirements may be more complex. The tracker also requires a stable and level surface to operate properly, and it may not be suitable for all geographic locations or environmental conditions.

PV system modeling is the process of forecasting a PV system's performance under various circumstances using computer simulations and mathematical models. This is a crucial tool for researchers, engineers, and designers to maximize PV system performance and design. A PV system model can be used to forecast a system's performance under various operating conditions, such as variations in temperature or solar irradiation, once it has been calibrated and validated. This information could be utilized to estimate a system's lifetime energy output and financial performance, as well as to optimize the design and operation of PV systems. Figure 1 shows the MATLAB Simulink circuit (PV module).

Modeling a PV system with a dual-axis solar tracker involves considering the performance of both the PV panels and the tracking system. The aim is to accurately predict the energy output of the system under different weather conditions and tracking strategies. One approach to modeling a PV system with a dual-axis solar tracker is to use a system-level model that considers the electrical performance of the PV panels and the mechanical performance of the tracking system. This model can be used to predict the energy output of the system based on various inputs, such as solar irradiance, ambient temperature, wind speed, and tracking angle.

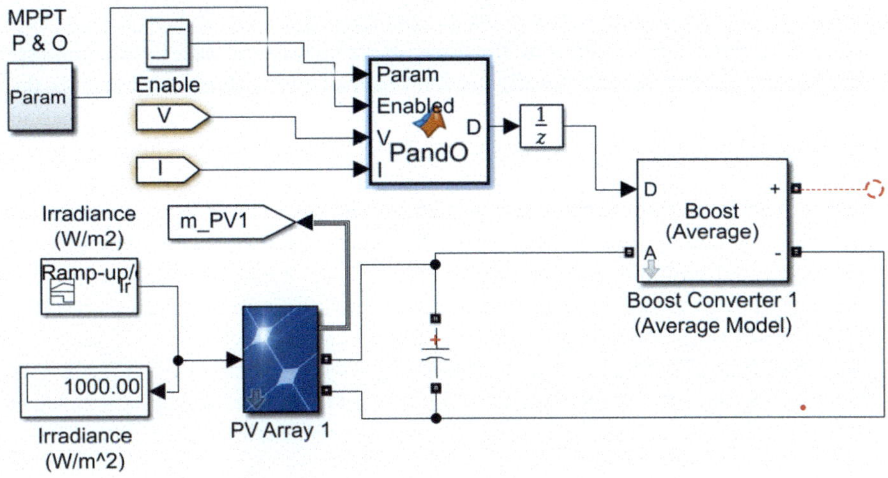

Fig. 1 MATLAB simulink circuit (PV module)

3 Proposed Dual-Axis Solar Tracking System

Figure 2 shows the block diagram of the solar tracking system. The block diagram that explains the components and functioning of a dual-axis solar tracker system:

- Solar Panel: The solar panel is the core component of a PV system. It converts the sunlight into electrical energy using the photovoltaic effect.

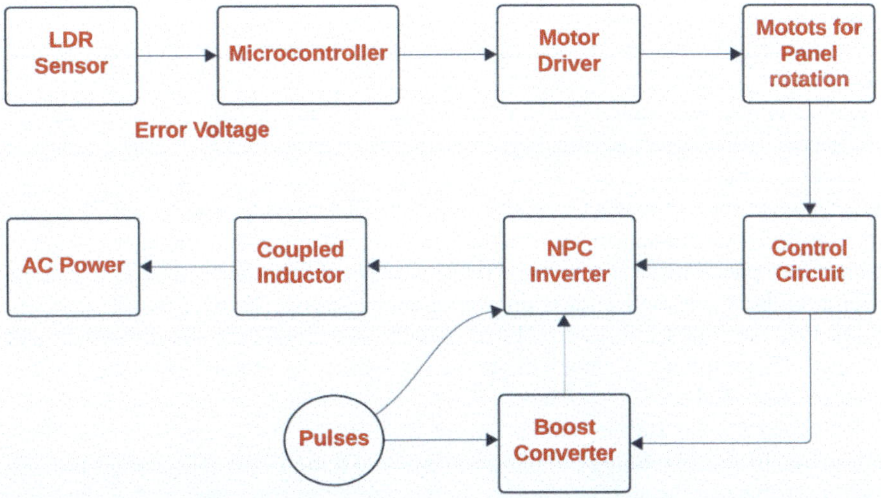

Fig. 2 Block diagram of the solar tracking system

- Sun Sensor (LDR): A sun sensor is used to detect the position of the sun in the sky. It provides the necessary input to the controller to adjust the panel's orientation.
- Microcontroller: A microcontroller is the brain of the system that controls the motor to adjust the solar panel's position. It receives inputs from the sun sensor and sends commands to the motor to move the panel along the vertical and horizontal axes.
- Motor: The motor is responsible for adjusting the position of the solar panel. It receives commands from the microcontroller and rotates the panel along the two axes to align it with the sun's position.
- Gearbox: The gearbox is used to reduce the speed of the motor and increase the torque to adjust the panel's position accurately.
- Dual-Axis Tracking Mechanism: The dual-axis tracking mechanism is responsible for orienting the solar panel along both the vertical and horizontal axes. It consists of two motors and a gearbox to rotate the panel along each axis.
- Power Management Unit: The power management unit is responsible for managing the energy produced by the solar panel. It includes a charge controller, battery, and inverter to convert the DC energy produced by the panel into AC energy that can be used to power appliances.
- Load: The load is the device or appliances that are powered by the energy produced by the PV system.

Using the sun sensor, microprocessor, motor, and gearbox, the dual-axis solar tracker system continuously aligns the solar panel with the position of the sun. This guarantees that throughout the day, the solar panel will absorb the most solar energy possible. The energy produced is then managed and converted into usable AC energy using the power management unit and can be used to power various loads.

A gear motor is used to power the solar project. The choice was influenced by the motor's strength, ability to support a high torque, precision rotation within a constrained Rack, and lack of such upsetting noise. There is a section for embedded software that contains a C++ program. The code was done on an Arduino UNO. The usage of a single-axis tracking system would not offer much value, hence the design is restricted to dual-axis tracking. Light-dependent resistors, or LDRs, may transform light into a voltage more easily thanks to an input stage design. After comparing the four voltages, the computer determines that the discrepancy is a defect. The gear motor uses this inaccuracy to rotate through a matched angle and change the position of the solar panel when the voltage outputs in the LDRs are equal. The voltage differential between the LDRs is measured analogically. The gear motor works to ensure that the four LDRs are inclined equally after receiving the difference and doing so. This implies that they will get the same quantity of light. Over the day, the process is repeated.

A dual-axis solar tracker works on the basic tenet that it can track the sun's motion by tilting and rotating its solar panels in two dimensions. The tracker is made to monitor the sun throughout the day, maximizing the solar energy that the solar panels can collect. A tracking algorithm that determines the position of the sun and determines the best tilt and rotation angles for the solar panels uses information from sensors to regulate the movement of the panels.

One of the primary benefits of a dual-axis solar tracker is its ability to maintain a perpendicular angle between the sun's rays and the solar panels all day long. This guarantees that the solar panels receive the maximum amount of solar radiation, which leads to a higher energy output than fixed-tilt and single-axis tracking systems. Additionally, solar panels' exact location enables maximum performance under a variety of weather situations, including gloomy and overcast days. A dual-axis solar tracker also has the benefit of greater dependability. Traditional fixed-tilt and single-axis tracking systems are susceptible to malfunction because their mechanical parts deteriorate over time. Dual-axis solar trackers, however, contain fewer mechanical parts and are made to be more robust, which results in enhanced reliability. PV panels can benefit from the exact positioning of solar panels in a dual-axis solar tracker in several ways. A maximum power point (MPP), which denotes the ideal operating point for maximum energy output, is a characteristic of PV panels. The MPP fluctuates, however, according to the strength of the sun's rays and the temperature of the PV panels. To maintain the MPP, a dual-axis solar tracker may move the solar panels' positions, enhancing energy output and prolonging the life of the PV panels. In conclusion, a dual-axis solar tracker is a practical way to maximize solar energy absorption and boost the effectiveness of solar energy systems. For accurate tracking of the sun's movement and maximum energy output, solar panels are positioned using sensors and a tracking algorithm. Dual-axis solar trackers are a desirable alternative for solar energy systems because of their improved dependability, toughness, and lower maintenance costs. It is anticipated that additional research and development in this field will result in even greater improvements in solar energy technologies.

Tracker angle equation: This equation calculates the angles at which the solar panels need to be oriented to track the sun. The equation considers the solar position and the orientation of the panels.

$$S = ST\text{-}LONG + 1.0027379 * (LST + ZONE\text{-}C) \tag{1}$$

Azimuth angle equation: This equation calculates the azimuth angle of the solar panels, which is the angle between the true north and the direction the panels are facing. The equation considers the latitude of the location and the solar time.

$$\beta = cos - 1[\cos R * \cos \beta a] \tag{2}$$

$$n = na + sin^{-1}[\sin R - \sin \beta] \tag{3}$$

Altitude angle equation: This equation determines the altitude angle of the solar panels, which is the angle between the panels and the horizontal plane. The formula takes into account the location's latitude, solar time, and solar declination angle.

$$\delta = -23.45 * \cos\frac{360}{365} * (d + 10) \tag{4}$$

Solar declination equation: The angle between the Earth's equator and the sun's direction is known as the solar declination angle, and it is computed using this formula. The calculation takes the season into account.

$$ppv = pd = i_{pv} - v_d \tag{5}$$

Equation for determining the maximum power point (MPP) of the solar panels: This equation determines the optimal voltage and current for the solar panels to produce maximum power. The equation considers the temperature, irradiance, and other factors that affect the performance of the solar panels.

$$\text{Power in watts} \times \text{Average hours of direct} = \text{Daily Watt-hour}$$

Equation for calculating the power output of the solar panels: This equation calculates the amount of power that the solar panels are producing at any given time. The equation considers the MPP, the current irradiance level, and other factors that affect the performance of the panels.

$$y_p : \min\left(1.2, 0.9 + \frac{h_{pt}}{h}\right) \tag{6}$$

$$y_c : \max(0.6 + 0.06 * L_p, 0.8) \tag{7}$$

Equation for determining the wind load on the solar tracker: This equation calculates the force exerted by the wind on the solar tracker, which is important for designing the structure and selecting the appropriate materials.

$$y_p : \max\left(1.2, 0.9 + \frac{h_{pt}}{h}\right) \tag{8}$$

$$y_c : \min\ (0.6 + 0.06 * L_p, 0.8) \tag{9}$$

Equation for calculating the torque required to move the solar tracker: This equation determines the amount of torque that is needed to move the solar panels, considering the weight and size of the panels and the friction in the system.

$$T = m * g * r \tag{10}$$

Equation for determining the power requirements of the solar tracker: This equation calculates the amount of power that is needed to operate the solar tracker, considering the motor efficiency, the weight of the panels, and other factors.

$$\text{Power} = \text{Average hourly wattage requirement/number of daily peak sunlight hours for your area}$$

Calculation of Solar Panel Angle: The angle of the solar panel can be calculated using the following equation:

$$\alpha = \beta + \delta + \omega - \theta \tag{11}$$

where α is the solar panel's angle, β is the location's latitude, δ is the sun's declination angle, ω is its hour angle, and θ is the tracker's angle.

Calculation of NPC Multilevel Inverter Output Voltage: The output voltage of an NPC multilevel inverter can be calculated using the following equation:

$$V_{out} = \left(\frac{V_{dc}}{2}\right) * (m_1 * m_2 + m_3) \tag{12}$$

where V_{dc} is the DC voltage input to the inverter, m_1, m_2, and m_3 are the modulation indices of the inverter.

Calculation of LDR Resistance: The resistance of an LDR can be calculated using the following equation:

$$R = (V_{cc} - V_{1dr}) * R_{1dr}/V_{1dr} \tag{13}$$

where R is the resistance of the LDR, V_{cc} is the supply voltage, V_{1dr} is the voltage across the LDR, and R_{1dr} is the resistance of a fixed resistor in series with the LDR.

Calculation of Solar Panel Output Power: The output power of a solar panel can be calculated using the following equation:

$$P = \varphi * A * G * F \tag{14}$$

where P is the output power, η is the efficiency of the solar panel, A is the area of the solar panel, G is the solar irradiance, and F is the fill factor of the solar panel.

3.1 Design of CdS LDR Sensor Intensity and Its Optimization

The maximum electromagnetic radiation range that the sun emits, which is between 420 and 820 nm, is consistent with the light absorption wavelength range of CdS-LDR, which is 340–830 nm. The characteristic curve of R_{ph} with intensity in terms of lux (Intlx) of 5.00 mm CdS LDR specification data sheet is used to design the optimization of the LDR sensor. The intensity equation is created by curve fitting with the use of data trend line analysis and scattered points. There are two ways to connect an LDR sensor to a power supply or load resistance; these connections are negative intensity coefficient types.

An analog value (digital) in the range of 0–1023 is produced by the microcontroller using a 10-bit ADC, which is used to transform measured LDR sensor output into analog Read RL (anlog R^L) and analog Read R_{ph}. Depending on the value of Vs, these digital values are translated into voltages between 0 and 5 V. Depending on the value of Vs, these digital values are translated into voltages between 0 and 5 V. The optimization algorithm to detect sunlight intensity using Cds-LDR is depicted in Fig. 3.

4 Three-Phase NPC Multilevel Inverter

A three-phase neutral point clamped (NPC) multilevel inverter is a type of power electronic circuit that converts DC voltage to three-phase AC voltage with multiple levels [27–29]. It consists of a series of capacitors and switches that are arranged in such a way that they produce several voltage levels [30]. The main advantage of an NPC multilevel inverter over a conventional two-level inverter is that it can generate a higher output voltage with lower harmonic distortion. This is because the multiple voltage levels produced by the inverter can closely approximate a sinusoidal waveform. In an NPC multilevel inverter, the DC voltage is first divided into three equal parts, each of which is connected to the midpoint of two series-connected capacitors. The other ends of the capacitors are connected to the three-phase load. The switches are used to connect the capacitors and load them to the DC source in a way that produces the desired voltage waveform. The control of an NPC multilevel inverter is typically implemented using pulse width modulation (PWM) techniques. By adjusting the duty cycle of the switching signals applied to the inverter, the output voltage waveform can be controlled and optimized for various applications. Overall, the NPC multilevel inverter is a powerful and efficient device that is commonly used in industrial and renewable energy applications, such as wind and solar power systems. Due to the NPC inverter's increased number of voltage levels, the output voltage waveform's harmonic content is reduced. The quantity of clamping diodes used in the NPC inverter circuit determines the number of voltage levels. The three-phase NPC inverter is the perfect option for applications that require a total of seven

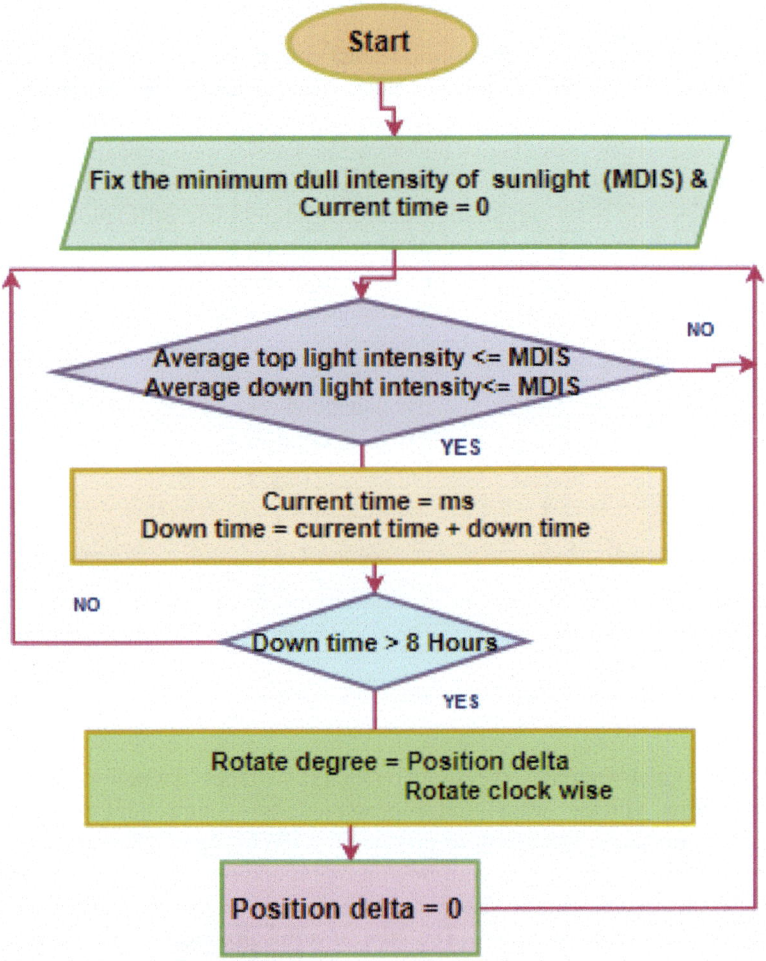

Fig. 3 Algorithm to detect sunlight intensity using an optimized Cds-LDR

levels in the output voltage waveform [31, 32]. The three-phase NPC inverter's output voltage waveform can be expressed as,

$$V_{out} = \left(\frac{V_m}{2}\right)\left(\frac{V_m}{6}\right)\left(\frac{V_m}{12}\right) \quad (15)$$

where V_m denotes the output voltage waveform's maximum amplitude. By changing the switching states of the power semiconductor switches, the seven-level output voltage waveform is produced. The four sections of the output voltage waveform are determined by the switching states of the semiconductor switches, MATLAB Simulink circuit of the three-phase NPC Inverter is shown in Fig. 4.

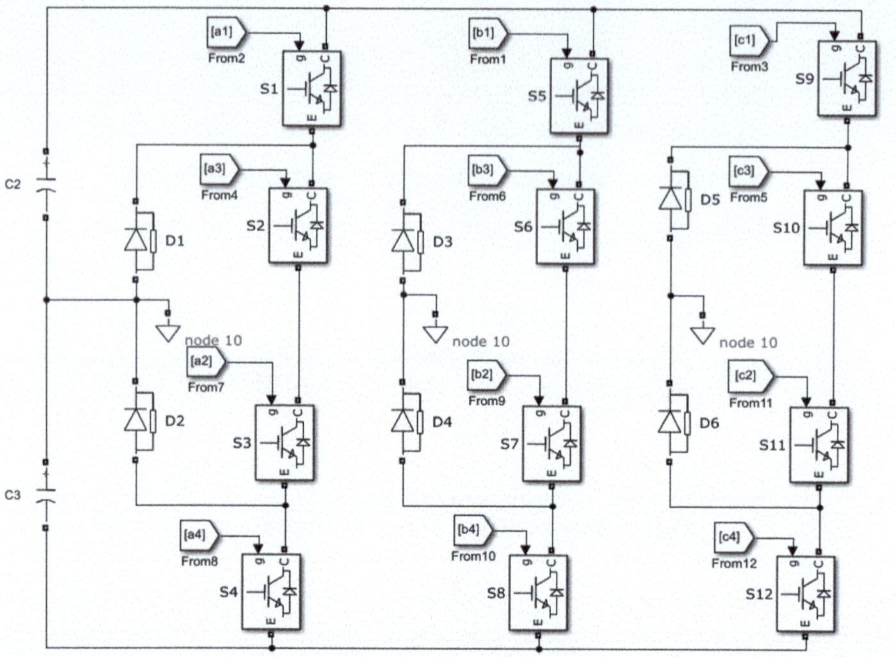

Fig. 4 MATLAB Simulink circuit of three-phase NPC inverter

A PWM signal regulates the semiconductor switches' switching states. The intended output voltage waveform, represented by a sinusoidal waveform, is compared to a triangle waveform to produce the PWM signal. The desired output voltage waveform is produced by turning the semiconductor switches ON and OFF using the comparison's output. Compared to conventional inverters, the three-phase NPC inverter has several benefits. The NPC inverter produces a greater variety of voltage levels, which leads to a smoother output voltage waveform and less harmonic distortion. Due to its lower switching loss, the NPC inverter is additionally more effective than conventional inverters. The power semiconductor switches' lifespan is extended, and the amount of maintenance required is thanks to the clamping diodes used in the NPC inverter circuit.

The three-phase NPC inverter is a highly effective and dependable multilevel inverter that produces an output voltage waveform of the highest caliber. The NPC inverter's seven-level output voltage waveform has a lower harmonic distortion and higher efficiency. The NPC inverter circuit, which uses PWM control to precisely control the output voltage waveform, is the best option for a variety of industrial and renewable energy applications, NPC multilevel inverter connected to a PV Module is shown in Fig. 5.

The use of neutral point clamped (NPC) inverters in dual-axis solar trackers can provide several benefits, including improved efficiency and reliability. Firstly, NPC inverters are known for their high efficiency, which can result in higher energy yields

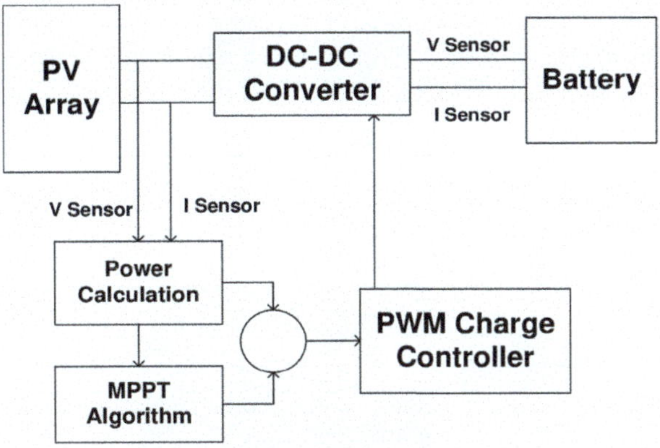

Fig. 5 NPC multilevel inverter connected to a PV module

from solar panels. The NPC inverter uses multiple power switches and capacitors to reduce switching losses and improve the overall efficiency of the system. Secondly, NPC inverters are also known for their high reliability. By distributing the voltage stress evenly across the power switches and capacitors, the likelihood of failure or breakdown is reduced. This can be especially important in outdoor solar applications, where the inverter may be exposed to harsh environmental conditions.

5 Results and Discussions

The program uses the light intensity measured by the LDR to modify the servo motors' position in order to track the position of the sun. The solar panel's horizontal movement is managed by the first servo motor, and its vertical movement is managed by the second. This design is for a three-phase, three-level Neutral Point Clamped Inverter, and it was created using MATLAB simulation. Two capacitors are being used here on the side with the furthest to the left, along with an RL load that will serve as the output. In each phase level, there are 4 active LGBT/diode switches and clamping diodes. Tinker CAD Simulation of Dual-Axis Solar tracking using Arduino UNO is shown in Fig. 6. The switching states for NPC–MLI are depicted in Table 1.

A sine wave is produced by an inverter that uses switching circuits to generate an output of AC voltage from a DC input. This inverter produces several square voltage pulses per half cycle. If the pulse size is altered, the output is said to be pulse width modulated. This modulation generates a few pulses per half cycle. Because the pulses near the endpoints of the half cycle are consistently thinner than the pulses near the 1/2 cycle's center, the pulse widths are comparable to the corresponding amplitude of a sine wave at that portion of the cycle. To change the effective output voltage, all pulse widths are increased or decreased while preserving sinusoidal proportionality. Pulse width modulation, or PWM, only alters the pulses' on-time when their amplitude

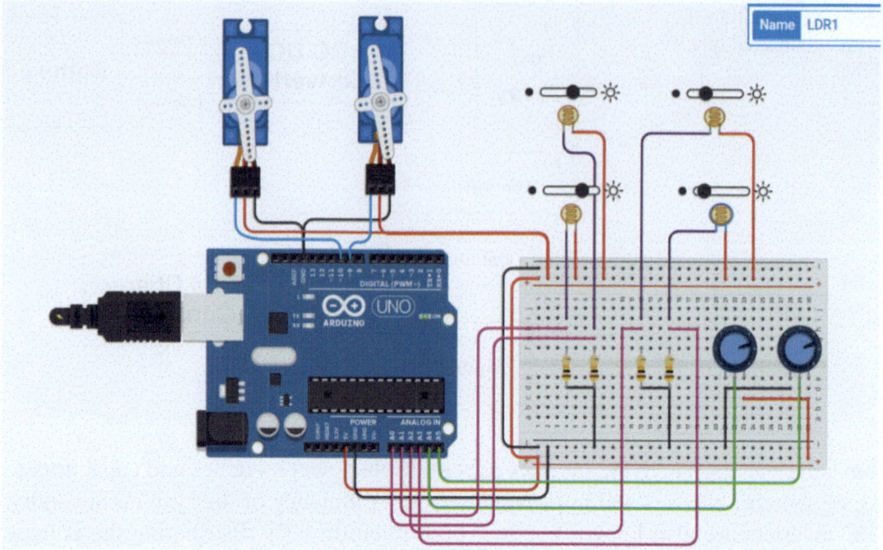

Fig. 6 Tinker CAD simulation of dual-axis solar tracking using Arduino UNO

Table 1 Switching states for NPC multilevel inverter operation

Switching states				
S1	S2	S3	S4	Output
ON	ON	OFF	OFF	E
OFF	ON	ON	OFF	0
OFF	OFF	ON	ON	– E

changes. Figure 7 displays the NPC multilevel inverter's line-line voltage output, whereas Fig. 8 displays the inverter's phase voltage output.

Compared to other modulation techniques, the SPWM technique has several benefits, such as better waveform clarity, less harmonic distortion, and increased effectiveness. The level of harmonic distortion in the output waveform is decreased by SPWM by generating a high-quality sinusoidal waveform, leading to a cleaner output signal that is less likely to interfere with other electrical devices. The tremendous efficiency that SPWM may attain is another benefit. The method minimizes the amount of energy lost as heat, leading to a more effective system by producing a smooth sinusoidal waveform. This is crucial in applications involving power electronics where efficiency is a crucial factor. A microcontroller or digital signal processor (DSP) is commonly used to provide the reference sinusoidal waveform and adjust the width of the PWM pulses to perform SPWM in a power electronics system. The MATLAB Simulink circuit—PV system is shown in Fig. 9.

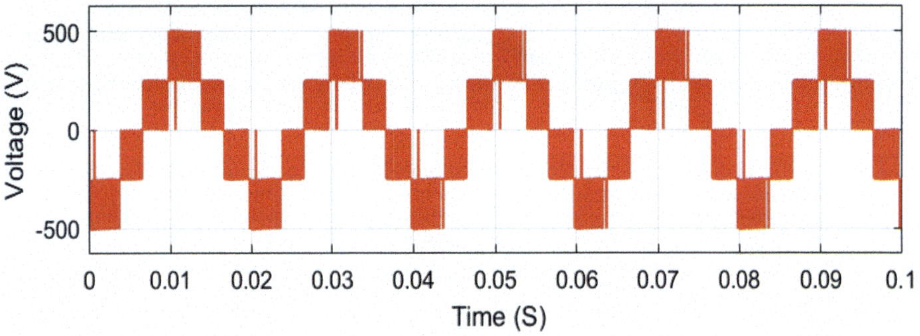

Fig. 7 Line-line voltage output from NPC multilevel inverter

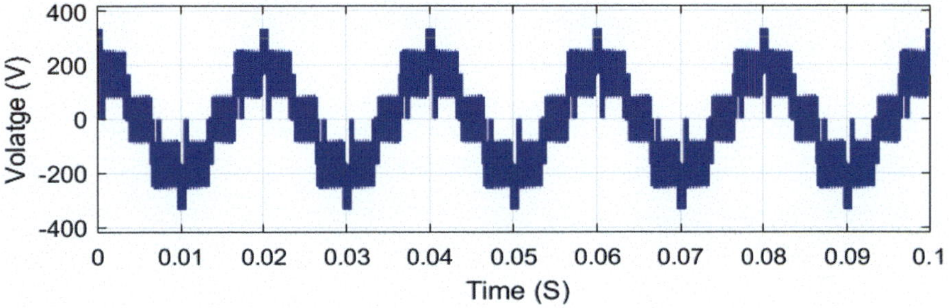

Fig. 8 Phase voltage output from NPC multilevel inverter

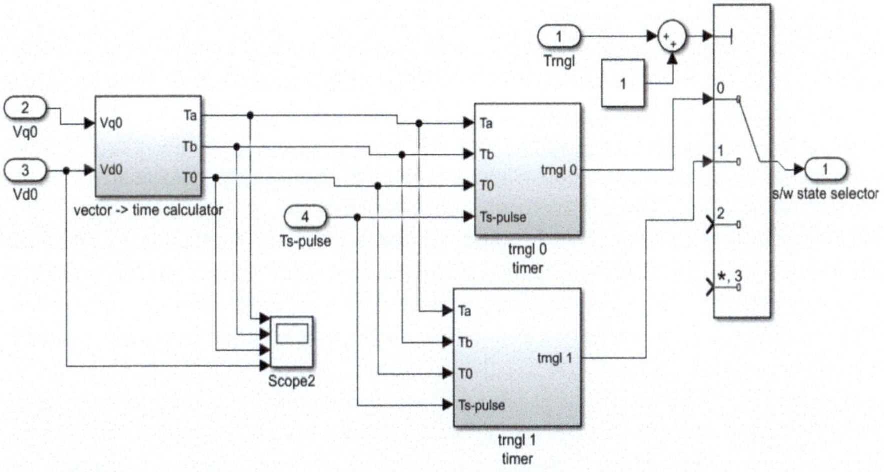

Fig. 9 MALAB simulink circuit—PV system

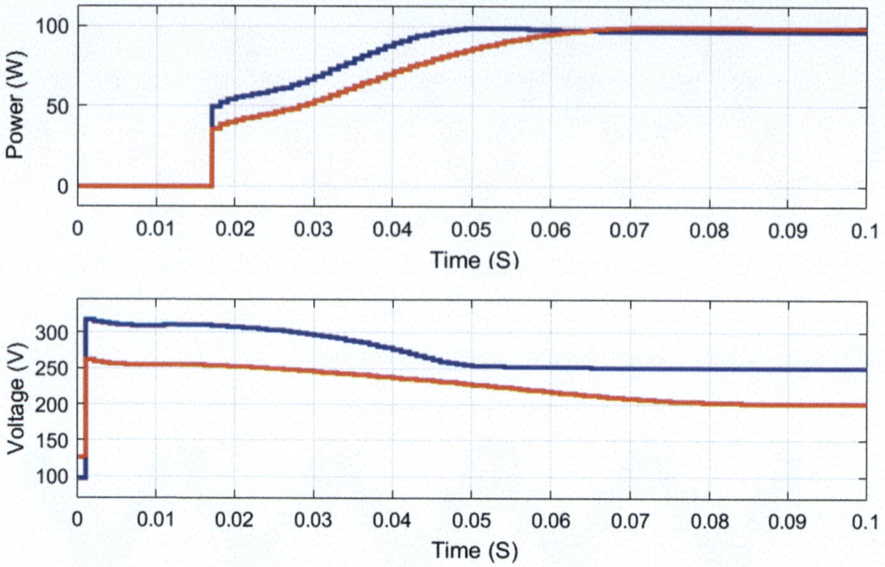

Fig. 10 PV Voltage and power waveform

The amount of sunshine that strikes a solar panel directly affects the voltage that the panel produces. The corresponding panel's position is adjusted to face the sun by the dual-axis solar tracker, the PV voltage curve will typically show a gradual increase in voltage as the panel receives more sunlight, reaching its peak value when the panel is facing directly toward the sun. The current output of a solar panel is directly proportional to the intensity of the current flowing through it. As the sunlight intensity changes, the current output of the panel will also change. The PV current curve will typically show a gradual increase in current as the panel receives more sunlight, reaching its peak value when the panel is facing directly toward the sun. The PV Voltage and Power waveform is shown in Fig. 10.

The control system ought to be built to monitor the sun's location in real time and modify the solar panel's orientation as necessary. For this, a model predictive controller (MPC) or a proportional-integral-derivative (PID) controller can be used. The motor system is in charge of turning the solar panel in the direction that is wanted. The angle at which the sun's rays strike the panel and how well the panel converts solar energy into electrical energy should both be taken into account in the model. For use in residences and commercial buildings, the inverter should transform the DC electricity produced by this solar panel into AC power. Figure 11 displays the system's battery voltage.

The relationship between the battery voltage and SOC is not linear, and it varies depending on the type and condition of the battery. Typically, a fully charged 12-V lead-acid battery will have a SOC of around 100% and a voltage of around 12.7 V. As the battery discharges, the voltage will decrease, and the SOC will also decrease. For example, when the battery voltage drops to around 12.0 V, the SOC is typically

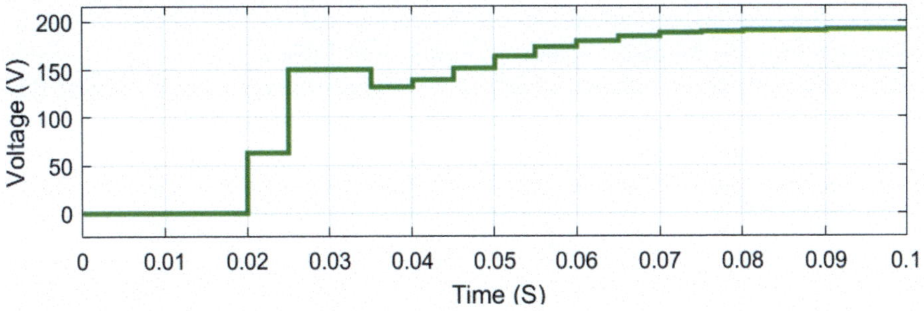

Fig. 11 Battery voltage of the proposed system

around 50%. The graph for battery voltage and SOC against time for a 12-V battery is essential for monitoring the battery's performance and health. By analyzing the curves, it is possible to determine the state of the battery and the amount of energy available. Proper monitoring and maintenance of the battery can extend its lifespan and ensure optimal performance.

6 Hardware Results and Discussions

A dual-axis solar tracker is a device that tracks the movement of the sun in two dimensions, i.e., azimuth and elevation, to ensure that a solar panel is always 90 degrees to the sun's rays. The dual-axis solar tracker can be controlled by a microcontroller that uses feedback from LDRs (Light Dependent Resistors) to adjust the position of the solar panel. The NPC (Neutral Point Clamped) multilevel 3-phase inverter can be used to control the PMDC (Permanent Magnet DC) motor that drives the solar panel's movement. The ULN2003 motor driver can be used to control the direction and speed of the PMDC motor. The PIC (Peripheral Interface Controller) microcontroller can be programmed to read the feedback from the LDRs and use this information to control the motor driver and the inverter. The microcontroller can be programmed to adjust the motor speed and direction depending on the sun's position and the LDR feedback. The NPC multilevel 3-phase inverter can be used to control the PMDC motor's speed and direction by adjusting the amplitude and phase of the voltage applied to the motor.

Sensors called LDRs measure how much light strikes the solar panel. To determine how much light is reaching each side of the solar panel, these sensors can be positioned on opposite sides. To guarantee that the solar panel is always perpendicular to the sun's rays, the microcontroller can use this information to modify the motor and the panel's position. One kind of DC motor that generates a magnetic field using a permanent magnet is the PMDC motor. This motor can be operated by the microcontroller and the inverter to adjust the solar panel's position. The PMDC motor has a simple design and is reliable, making it suitable for use in solar tracking systems. Figure 12a shows the design of the tracker and Fig. 12b shows the entire hardware setup. The hardware results are depicted in Table 2.

The NPC multilevel 3-phase inverter is a type of voltage source inverter that can provide a high-quality output voltage with reduced harmonic distortion. This inverter can be controlled by the microcontroller to provide the appropriate voltage and frequency to the PMDC motor. The ULN2003 motor driver can be used to provide the necessary current to the PMDC motor and to control the motor's speed and direction, NPC Multilevel Inverter setup is shown in Fig. 12c.

(a)

(b)

(c)

Fig. 12 Experimental setup **a** design of tracker, **b** entire hardware NPC multilevel inverter

Table 2 Readings obtained from hardware components

Component	Input	Output
Solar panel	Sunlight	11 V/0.45 A/5 W
Arduino UNO	9 V (battery)/~ 0.5 A	(5–10) V
Motor driver (ULN2003)	(10–12) V	~ 0.5 A Per Channel
PMDC motor	30 rpm/12 V	
Buck-Boost converter	11.7 V	14–16 V
NPC inverter	11.2 V/0.44 A/4.928 W	10 V/0.398 A/3.5826 W
Step-down transformer (for PIC)	230 V	12 V
3-phase step-up transformer	12 V	230 V (to R-load)

7 Conclusions

Dual-axis solar trackers have emerged as a highly efficient technology for generating solar energy. They offer a significant improvement over fixed solar panels by enabling them to track the sun's movement throughout the day and optimizing the angle of the solar panels to receive the maximum amount of sunlight. This leads to a considerable increase in energy output, making dual-axis solar trackers an attractive option for large-scale solar energy generation projects. Moreover, dual-axis solar trackers have many other benefits such as reducing land requirements and increasing the return on investment (ROI) of solar power systems. In DAST, we've created a demo model of a solar tracker to follow the light source's brightest point so that the solar panel's output voltage is at its highest there. In a detecting zone, a solar panel detects light; outside of that zone, it is unable to react. It will calculate the vector sum of the light sources and move the panel to that position if numerous light sources (such as diffused light sources) are present on the panel. There were few resources used to deliver this chapter. The circuitry was maintained user-friendly, comprehensible, and straightforward. These trackers can be customized to suit different terrains and weather conditions, making them ideal for use in various applications. While there are many commercially available dual-axis solar trackers, there is still room for improvement.

References

1. Prasad KK, Myneni H, Kumar GS (2019) Power quality improvement and pv power injec-tion by dstatcom with variable dc link voltage control from rsc-mlc. IEEE Trans Sustain Energy 10(2):876–885. https://doi.org/10.1109/TSTE.2018.2853192
2. Alrubaie AJ, Al-Khaykan A, Malik RQ, Talib SH, Mousa MI, Kadhim AM (2022) Review on MPPT techniques in solar system. In: 8th International engineering conference on sustainable technology and development (IEC), Erbil, Iraq, pp 123–128. https://doi.org/10.1109/IEC54822.2022.9807500

3. Mitrofanov SV, Yu, Nemaltsev A, Baykasenov DK (2018) Primary testing of automated dual-axis solar tracker in the climatic conditions of the Orenburg region as the prospects for the establishment of a hardware-software complex. Int Sci J Altern Energy Ecol 7(9):43–54. https://doi.org/10.1051/rees/2020002
4. Palanisamy R, Vijayakumar K (2017) Wind-PV hybrid energy source fed three level NPC with quasi-Z source network. Int J Power Electron Drive Syst (IJPEDS) 8(3):296–303. https://doi.org/10.11591/ijpeds.v8.i3.pp1285-1293
5. Nsengiyumva W, Chen SG, Hu L (2018) Recent advancements and challenges in solar tracking systems. A review. Renew Sustain Energy Rev 81:250–279. https://doi.org/10.1016/j.rser.2017.06.085
6. Awasthi A, Shukla AK, Manohar SR, Dondariya C, Shukla KN (2020) Review on sun tracking technology in solar PV system. Energy Rep 6:392–405. https://doi.org/10.1016/j.egyr.2020.02.004
7. Palanisamy R, Vijayakumar K, Choudhury S, Aniruddh D, Saxena H, Karthikeyan D (2018) PV system based 3-level neutral clamped point multilevel inverter using SVM method. J Adv Res Dyn Control Syst 10(7):1254–1259. https://doi.org/10.1016/j.egypro.2016.11.287
8. Singh B, Al-Haddad K, Chandra A (1999) A review of active filters for power quality im provement. IEEE Trans Ind Electron 46(5):960–971. https://doi.org/10.1109/41.793345
9. Rao B, Krishna P, Yarlagadda V (2020) Mitigation of photovoltaic solar system harmonics using shunt active power filter. In: 2020 fourth international conference on computing methodologies and communication (ICCMC), pp 566–571. https://doi.org/10.1109/ICCMC48092.2020.ICCMC-000105
10. Esram T, Chapman PL (2007) Comparison of photovoltaic array maximum power point tracking techniques. IEEE Trans Energy Convers 22(2):439–449. https://doi.org/10.1109/TEC.2006.874230
11. Sano K, Takasaki M (2012) A transformerless D-STATCOM based on a multivoltage cascade converter requiring no DC sources. IEEE Trans Power Electron 27:2783–2795. https://doi.org/10.1109/TPEL.2011.2174383
12. Kanjiya P, Singh B, Chandra A, Al-Haddad K (2020) SRF theory revisited to control self-supported dynamic voltage restorer (DVR) for unbalanced and nonlinear loads. IEEE Trans Ind Appl 49(5):2330–2340. https://doi.org/10.1109/TIA.2013.2261273
13. Salmeron P, Litran SP (2010) Improvement of the electric power quality using series active and shunt passive filters. IEEE Trans Power Deliv 25(2):1058–1067. https://doi.org/10.1109/TPWRD.2009.2034902
14. Kumar N, Singh B, Panigrahi BK (2019) Pnklmf-based neural network control and learning-based hc mppt technique for multiobjective grid integrated solar pv based distributed generating system. IEEE Trans Ind Inf 15(6):3732–3742. https://doi.org/10.1109/TII.2019.2901516
15. Sadanand DDK (2021) Modeling of photovoltaic solar cell based on cusbs absorber for the enhancement of performance. IEEE Trans Electron Dev 68(3):1121–1128. https://doi.org/10.1109/TED.2020.3048326
16. Myneni H, Ganjikunta SK (2020) Energy management and control of single-stage grid-connected solar pv and bes system. IEEE Trans Sustain Energy 11(3):1739–1749. https://doi.org/10.1109/TSTE.2019.2938864
17. Ali A, Almutairi K, Padmanaban S, Tirth V, Algarni S, Irshad K, Islam S, Zahir MH, Shafiullah M, Malik MZ (2020) Investigation of mppt techniques under uniform and non-uniform solar irradiation condition–a retrospection. IEEE Access 8:127368–127392. https://doi.org/10.1109/ACCESS.2020.3007710
18. Chikh A, Chandra A (2015) An optimal maximum power point tracking algorithm for pv systems with climatic parameters estimation. IEEE Trans Sustain Energy 6(2):644–652. https://doi.org/10.1109/TSTE.2015.2403845
19. Motahhir S, Hammoumi A, Ghzizal EL, A, Derouich A, (2019) Open hardware/software test bench for solar tracker with virtual instrumentation. Sustain Energy Technol Assess 31:9–16. https://doi.org/10.1016/j.seta.2018.11.003

20. Sarkar TT, Mahanta C (2022) Gain tuned sliding mode control based maximum power point tracking for solar PV systems. IFAC Papers Line 55(1):417–422. https://doi.org/10.1016/j.ifacol.2022.04.069
21. Abdallah S (2004) The effect of using sun tracking systems on the voltage–current characteristics and power generation of flat plate photovoltaics. Energy Convers Manag 45:1671–1679. https://doi.org/10.1016/j.enconman.2003.10.006
22. Lai CS, McCulloch MD (2017) Levelized cost of electricity for solar photovoltaic and electrical energy storage. Appl Energy 190:191–203. https://doi.org/10.1016/j.apenergy.2016.12.153
23. Fuentes-Morales RF, Diaz-Ponce A, Peña-Cruz MI, Rodrigo PM (2020) Algorithms applied to active solar tracking systems: a review. Sol Energy 212:203–219. https://doi.org/10.1016/j.solener.2020.10.071
24. Al-Sakkaf A, AlMuhaini M (2018) Power flow analysis of weakly meshed distribution network including DG. Eng Technol Appl Sci Res 8:3398–3404. https://doi.org/10.48084/etasr.2277
25. Martinot E, Kristov L, Erickson JD (2015) Distribution system planning and innovation for distributed energy futures. Curr Sustain Renew Energy Rep 2:47–54. https://doi.org/10.1007/s40518-015-0027-8
26. Alves GO, Pereira JLR, Passos Filho JA (2020) A new unbalanced three-phase governor power flow formulation based on the current injections method. Int J Electr Power Energy Syst 123:106184. https://doi.org/10.1016/j.ijepes.2020.106184
27. Bagheri Tolabi H, Ali MH, Rizwan M (2015) Simultaneous reconfiguration, optimal placement of DSTATCOM, and photovoltaic array in a distribution system based on fuzzy-ACO approach. IEEE Trans Sustain Energy 6:210–218. https://doi.org/10.1109/tste.2014.2364230
28. Tamilselvan V, Jayabarathi T, Raghunathan T, Yang X-S (2018) Optimal capacitor placement in radial distribution systems using flower pollination algorithm. Alex Eng J 57:2775–2786. https://doi.org/10.1016/j.aej.2018.01.004
29. Saldivar-Aguilera TQ, Valentín-Coronado LM, Peña-Cruz MI, Diaz-Ponce A (2023) Novel closed-loop dual control algorithm for solar trackers of parabolic trough collector systems. Sol Energy 259:381–390. https://doi.org/10.1016/j.solener.2023.05.024
30. Al-Naima FM, Yaghobian NA (2021) Design and construction of a solar tracking system. Sol Wind Technol 7:611–617. https://doi.org/10.1016/0741-983X(90)90072-A
31. Jung CM, Pagidipala S, Salkuti SR (2024) Emerging technologies for the integration of renewable energy, energy storage and electric vehicles. In: Salkuti SR (eds) Energy and environmental aspects of emerging technologies for smart grid. Green Energy and Technology. Springer, Cham. https://doi.org/10.1007/978-3-031-18389-8_5
32. Sravanthi P, Sandeep V (2022) Solving realistic reactive power market clearing problem of wind-thermal power system with system security. Int J Emerg Electr Power Syst 23(2):125–144. https://doi.org/10.1515/ijeeps-2021-0060

Development of Artificial Intelligence Based Advanced Battery Management System with Enhanced Charging Monitoring and Fire Protection for Electric Vehicles

Appalabathula Venkatesh, J. Vijaya Kumar, Surender Reddy Salkuti, Ch. Abhiram, Repuri Sumasri, Unnamatla Gold, and Gunuru Jagadeesh

Abstract The performance and the driving distance of electric vehicles (EVs) mainly depend on the efficient charging and discharging capabilities of the battery packs. The rise in battery temperature influences the battery performance degradation which affects the performance of the battery-driven vehicles. The battery's internal temperature causes a decline in resistance, which generates a considerable amount of heat in the battery which results in a fire accident. Hence a reliable, safe, and efficient management system is required to monitor and control the battery pack parameters. This research presents the design and development of an advanced battery management system (ABMS) leveraging artificial intelligence (AI) which mainly concentrates on enhanced charging monitoring and fire protection to the battery pack in EV. The developed ABMS integrates with efficient sensing technologies and AI-based computational algorithms to provide a real-time battery pack module with cell balancing mechanisms to monitor the charging process which ensures the optimal charging conditions and in turn, maximizes the battery lifespan in EVs. In the current research work, within the hardware model, a fire protection system is implemented by utilizing AI to detect early signs of thermal runaway which ensures fire safety in BMS. The experimental validation is done on the proposed ABMS through comprehensive testing in the simulation model and real-time operating conditions. The results demonstrate that enhancing charging efficiency, prolonging battery life, and ensuring robust fire protection capabilities.

A. Venkatesh · J. V. Kumar · Ch. Abhiram · R. Sumasri · U. Gold · G. Jagadeesh
Department of Electrical and Electronics Engineering, Anil Neerukonda Institute of Technology and Sciences, Visakhapatnam, Andhra Pradesh, India
e-mail: vijayakumar.eee@anits.edu.in

S. R. Salkuti (✉)
Department of Railroad and Electrical Engineering, Woosong University, Daejeon, Republic of Korea
e-mail: surender@wsu.ac.kr

S. R. Salkuti (ed.), *Artificial Intelligence for Integrated Smart Energy Systems in Electric Vehicles*, Lecture Notes in Electrical Engineering 1427,
https://doi.org/10.1007/978-3-031-94276-1_11

Keywords AI-based advanced BMS · Cell balancing techniques · Charge monitoring of BMS · Flying capacitor circuit · Protection of BMS

1 Introduction

The day-to-day increase in energy demand with renewable sources mainly motivates the transportation sector to design and development of more and more electrically driven vehicles rather than conventional vehicles [1–3]. The origins of electric vehicles (EVs) may be traced back to the 1830s when the first attempts at electric propulsion were made. However, due to worries about reliance on fossil fuels (FFs), the environment, and advances in battery technology, EVs did not receive much attention until the late twentieth and early twenty-first centuries [4, 5]. The various stages in the background of EVs from their early developments to present trends are shown in Fig. 1.

The electric model of the battery pack is one of the significant factors to uplift the battery parametric estimation methods for better monitoring of the battery characteristics [6]. There are various estimation methods are presented, and the choice of estimation method depends on the specific requirements of the application, though there are existing conventional estimation methods, data-driven methods have more accurate measurement [7, 8]. Due to the growing demand for battery modeling AI and machine learning (ML) algorithms [9], Neural Networks [10] have more adaptable capability in real-time parameter estimation.

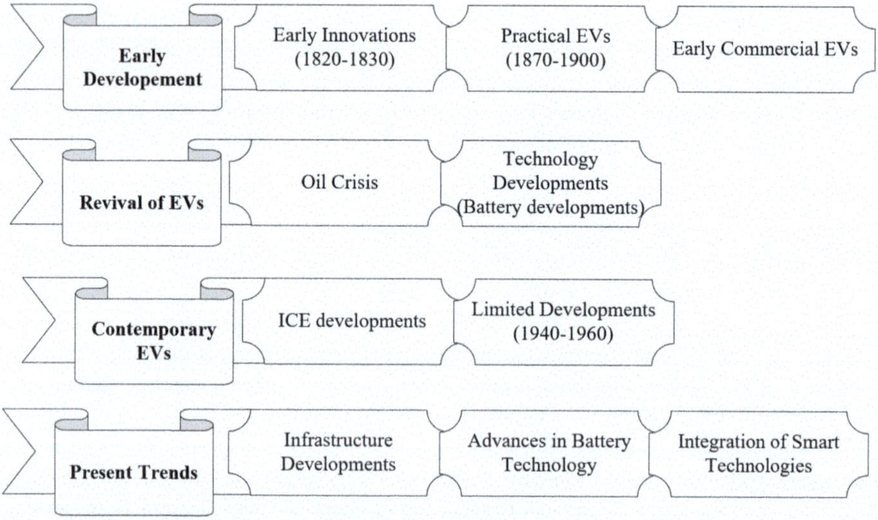

Fig. 1 Background of EVs (from early inception to present trends)

The data-driven approaches utilize the training and learning concept which have improved the data handling capacities for SOC estimation [11, 12]. Other intelligent estimation techniques like fuzzy logic controller [13, 14], artificial neural network [15], genetic algorithm (GA) [16], support vector machine (SVM) and MRAS [17], a fuzzy-based SVM [18] and SVM with an advanced optimization [19, 20]. The battery performance mainly influences the driving range and overall performance of EVs [21]. The chemistry of lithium-ion (Li-ion) batteries is highly susceptible to overcharging and deep discharging, which can damage the battery, reduce its lifespan, and even cause dangerous consequences [22]. If there is a charge imbalance among the cells, one or more of the cells become overloaded and overheat. Thermal runaway may occur in a battery if its temperature rises above a certain point (electrolyte leakage, cell breakage, fire, and explosion are all possible outcomes of thermal runaway) [23]. Energy storage system ultimately plays a key role in many household applications [24]. But the lifespan of the storage element depends on the effective balancing of the cells in a battery pack, active and passive balancing are the key balancing methods in batteries that are significantly used [25–31].

The improper balancing of cells leads to fire catching thereby raising the temperatures in the battery-driven vehicles [32, 33]. Hence, a proper thermal management system is required to prevent battery damage in vehicles [34] The above reasons are the key identified parameters that are mainly the sources of firing in EVs which influences EV sales. Figure 2 shows the statistics of EV accidents and EV sales over the years. Nowadays EVs are blasting. There are lots of reasons are there. Poor manufacturing, short circuits, overcharging, fast discharging, temperature increases, thermal runaway, ambient temperature cell unbalancing, etc. So, in this chapter, an AI-based advanced battery management system (ABMS) with charge monitoring and fire protection is discussed. The key objectives in the current research work are as follows:

- Design of battery cell modules to understand the behavior of the battery.

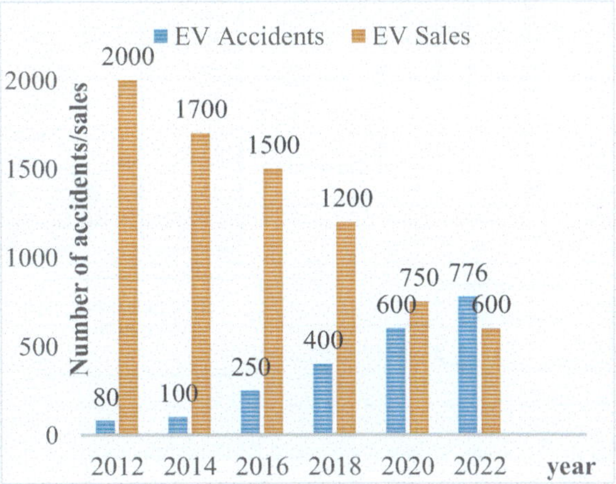

Fig. 2 Battery accidents and impact on EV sales over the years

- Testing of battery module with cell balancing techniques.
- Design and implementation of an advanced battery management system for real-time monitoring of the charging and discharging process in EV batteries.
- Comprehensive experimental validation of developed AI-based ABMS with simulation results under real-time operating conditions to assess its effectiveness and practicality.

The developed AI-based ABMS has to provide a comprehensive solution for battery management that should address the critical challenges in charging monitoring and fire protection in batteries in EVs, ultimately the objective is to accelerate the adoption of EVs.

2 Block Diagram and Operation of AI-Based Advanced BMS

The block diagram of the AI-based advanced BMS (A-BMS) along with its other components is shown in Fig. 3. It consists of mainly 6 components:

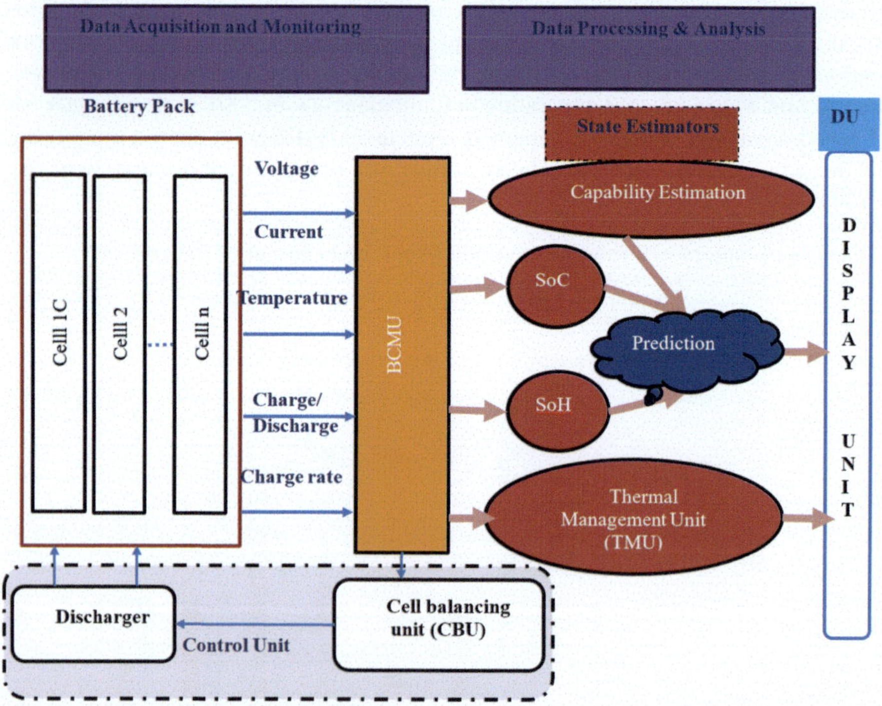

Fig. 3 Proposed AI-based advanced BMS architecture

- Battery pack and battery cell monitoring unit (BCMU)
- Cell balancing unit (CBU)
- Discharger
- State estimators (capability, SoC, and SoH)
- Thermal management unit (TMU)
- Display unit (DU).

2.1 *Battery Pack and Battery Cell Monitoring Unit (BCMU)*

The data acquisition from sensors and monitoring of the battery pack/cell parameters is done by the unit called BCMU. The monitoring of voltages for battery cells is the main function of the BCMU. For high voltage measurements, it is a low-cost system with fast speed. Monitoring the voltage of each cell with the initially chosen predefined level which helpful for figuring out how the battery pack is charged and how the charging process is automatically tripped when the voltage of the respective cell hits the maximum limit. It signals that the maximum amount of battery charging has been accomplished. If the battery pack has less charge than the average cell, the cell that is charged less among all may reach the limit initially, and the other cells are partially charged.

2.2 *Cell Balancing Unit (CBU)*

The BMS starts cell balancing procedures to equalize the charge among cells whenever voltage imbalances are noticed. Both active and passive techniques, such as dividing up the charge among cells or releasing extra charge through resistors or other components, can be used to achieve balancing. Assessing the condition of the battery module is essential. To guarantee that they (each cell) are operating in a way that will result in a long battery life, cell voltage monitoring is carried out. The operational voltage of a lithium-ion battery is ranging (2.5–4.2) volts. When battery performance is carried out beyond the operating levels, its life is considerably impacted. This shortens a cell's lifespan and might even render it useless. The State Vector Machine (SVM) is used to predict the states of battery health conditions (Healthy, Degraded, Faulty) from the obtained parameters from the BCMU.

2.3 *Discharger*

Used to effectively and safely drain a battery's remaining charge. This may be done for several reasons, including recycling, preparing for storage, or evaluating battery capacity. This straightforward and affordable method turns electrical energy into heat by using resistors (less effective because of wasted energy). Utilize more advanced electronic parts to regulate the rate of discharge and possibly even recover some energy.

2.4 State Estimators

State of Charge (SOC) clarifies how much of a battery is still charged about its total capacity. It has a percentage expression (0–100%). A precise SOC estimate is necessary for estimating a device's remaining runtime. Avoiding excessive discharge, which could harm the battery, enhances methods of charging. Capability estimation is the process of determining how much power the battery can provide right now.

2.5 Thermal Management Unit (TMU)

The BMS regulates heating or cooling systems to keep battery cells at their ideal operating temperatures based on temperature data. Fans, liquid cooling loops, and passive heat sinks are examples of cooling systems, whereas heating components can increase temperature in cold situations to enhance performance.

2.6 Display Unit (DU)

The voltage (V) unit is the most frequently used on batteries to indicate the electrical potential difference between the positive and negative terminals. The capacity (Ah) unit is used to indicate the capacity of the battery up to which it can store. The percentage (%) unit is a relative unit that is frequently used for SoC.

3 Proposed System Components Design and Development of AI-Based A-BMS

3.1 Battery Modelling

There are various equivalent battery modeling available, among all the electrical equivalent modeling that has widespread applications in BMS [35–39]. The modeling is done with the basic circuit elements to represent the exact battery behavior. The developed model is helpful for the estimation methods. Figure 4 shows the BEM model of the battery. The mathematical equations that govern the above BEM model

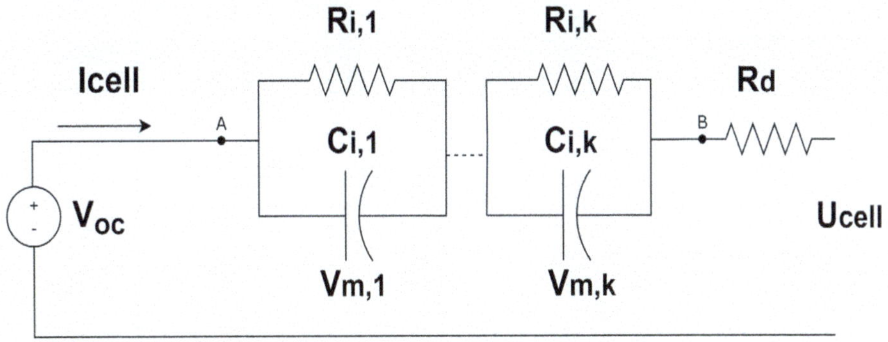

Fig. 4 Battery equivalent model (BEM)

are as represented as follows by using the generalized equivalent of Thevenin's model and the active/passive network theory [40, 41].

$$\dot{U}_{cell} = -\frac{1}{R_{i,1}C_{i,1}}U_{Di} + \frac{1}{C_{i,1}}I_L \quad (1)$$

$$U_{cell} = V_{OC} - I_{cell}R_d - \sum_{i=1}^{n} V_{m,1} \quad (2)$$

Where V_{OC} is the open-circuited voltage in Volt, U_{cell} is the terminal voltage of the cell in Volt, I_{cell} is the cell current in Amp, $R_{i,1}, R_{i,2} \ldots R_{i,k}$ is the Resistance of corresponding k^{th} RC branch, $C_{i,1}, C_{i,2} \ldots C_{i,k}$ is the capacitance of the corresponding kth RC branch, $V_{m,1}, V_{m,2} \ldots V_{m,k}$ is the voltage across the corresponding kth RC branch, R_d is dynamic resistance of the cell in Ω.

3.2 *Flying Capacitors Circuit Modelling*

The flying capacitors circuit design which is adopted in the current research work is as shown in Fig. 5 which is used to balance the voltage across each cell. The circuit model used for a series of connected battery cells is called a 4S configuration [42–45]. The flying capacitor circuit is operated in different modes of operation to balance the voltage of all the 4S-connected battery cells. The states of controlled switches are changed accordingly which allows the charge transfer from cells with higher potential to cells with lower potential and are balanced. Table 1 shows the different modes of operation of the Flying capacitor circuit.

Figure 6 shows the modes of operation of 4S connected flying capacitor circuit which are used to train the SVM model [46], the above three modes are considered during the training process based on the voltage levels the health states of cells are labeled with a certain range of voltage levels with an acceptable cell damage limit.

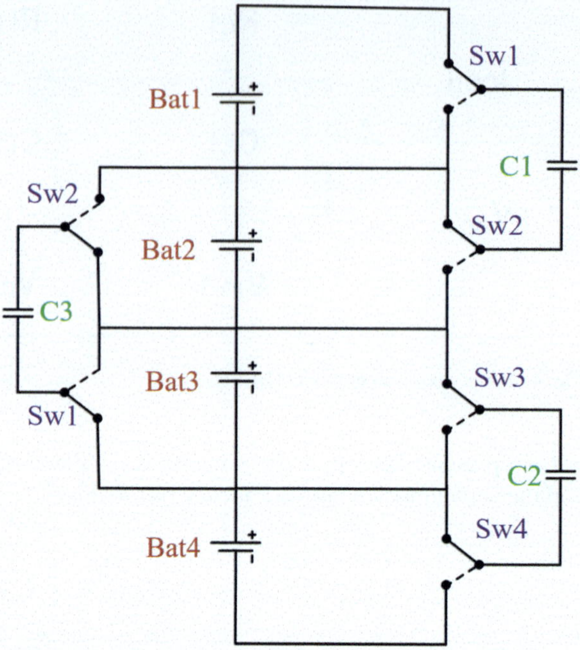

Fig. 5 4S Configured battery module with flying capacitor circuit

Under a charge-balanced state, the amount of charge supplied is equal to the amount of charge discharged [47], mathematically expressed as,

$$Q_{in} = Q_{out} \tag{3}$$

The charge transferred into the capacitor during balancing can be calculated using,

$$Q_{in} = C \cdot \Delta V \tag{4}$$

$$Q_{out} = C \cdot \Delta V_{cell} \tag{5}$$

Equate the expressions in Eqs. (4) and (5), and solve for C.

$$C.\Delta V = C.\Delta V_{cell} \tag{6}$$

$$C = \frac{\Delta V_{cell}}{\Delta V}.C \tag{7}$$

$$C = \frac{\Delta V_{cell}}{\Delta V}.C_{min} \tag{8}$$

Table 1 Modes of operation of flying capacitor circuit

Mode	States of controlled switches				Charge transfer type	Operation
	S_1	S_2	S_3	S_4		
Balancing	1	0	1	0	BAT 1 and 3	Transfer charge through C_1
	1	1	0	0	BAT 1 and 2	Transfer charge through C_1
	0	1	1	0	BAT 2 and 3	Transfer charge through C_1 or C_2
	0	1	0	1	BAT 2 and BAT 4	Transfer charge through C_3
	0	0	1	1	BAT 3 and 4	Transfer charge through C_2
Charging	1	0	1	0	BAT 1 and 3 Charge C_3	$C_3$is charged. Later Sw1 and Sw3 opened to transfer charge from C_3 to BAT3 or 1
	1	1	0	0	BAT 1 and 2 Charge C_1	C_1 is charged. Later Sw1 and Sw2 opened to transfer charge from C_1 to BAT1 or 2
	0	1	1	0	BAT 2 and 3 Charge C_1 Charge C_2	$C_1 or C_2$ is charged. Later Sw2 and Sw3 opened to transfer charge from C_1orC_2 to BAT2 or BAT3
	0	1	0	1	Charge C_2	C_2 is charged. Later Sw2 and Sw4 opened to transfer charge from C_2 to BAT2 or 4
	0	0	1	1	Charge C_2	C_2 is charged. Later Sw3 and Sw4 opened to transfer charge from C_2 to BAT3 or 4
Discharging	1	0	1	0	Discharge C_3	C_3 is discharged. Later Sw1 and Sw3 opened to recharge C_3 from BAT3 or BAT1
	0	1	1	0	Discharge C_1	C_1 is discharged. Later Sw2 and Sw3 opened to recharge C_1 from BAT2 or BAT3
	1	1	0	0	Discharge C_1	C_1 is discharged. Later Sw1 and Sw2 opened to recharge C_1 from BAT2 or BAT1
	0	1	1	0	Discharge C_2	C_2 is discharged. Later Sw2 and Sw3 opened to recharge C_2 from BAT2 or BAT3
	0	0	1	1	Discharge C_2	C_2 is discharged. Later Sw2 and Sw4 opened to recharge C_2 from BAT4 or BAT3

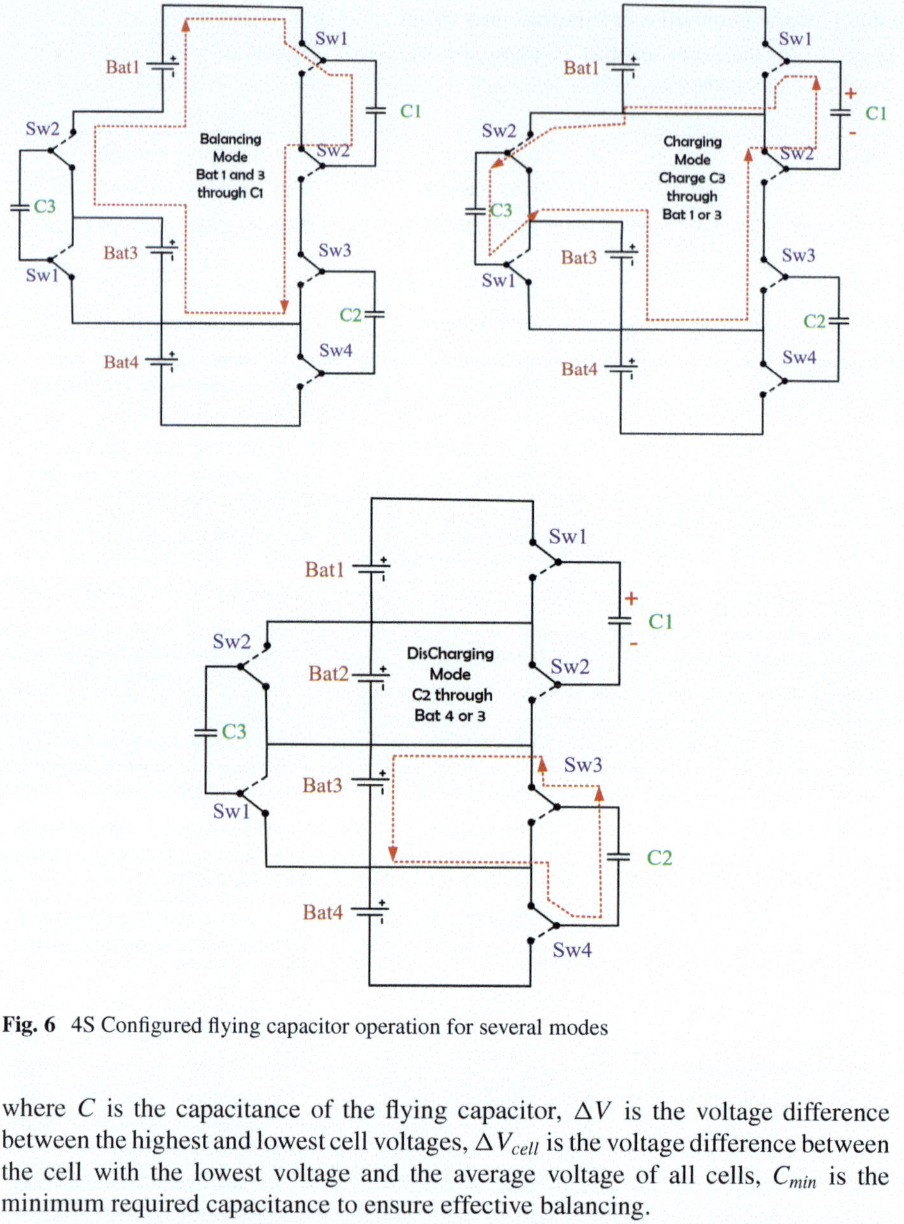

Fig. 6 4S Configured flying capacitor operation for several modes

where C is the capacitance of the flying capacitor, ΔV is the voltage difference between the highest and lowest cell voltages, ΔV_{cell} is the voltage difference between the cell with the lowest voltage and the average voltage of all cells, C_{min} is the minimum required capacitance to ensure effective balancing.

The State Vector Machine (SVM) is used to predict the states of battery health conditions (Healthy, Degraded, Faulty) from the obtained parameters from the BCMU [48]. The SVM model is trained with the historical data which is obtained during the testing with the flying capacitors operations and the health states are labeled, hence the designed AI-based BMS predicts the health states of the battery cells during the real-time conditions with the designed hardware model. The protective measures with the early detection of thermal runaway by TMU which is driven by AI-based BMS and accordingly the preventive measures are supplied to the control unit. Figure 7 shows the flow chart for the operational sequence of the training-designed AI-based BMS and Fig. 8 shows the operational protection setup for the developed A-BMS with protection features during over-voltages, battery discharging, and over-loading.

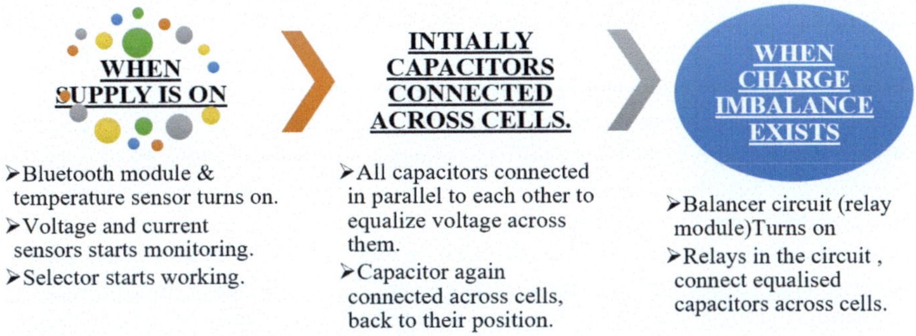

Fig. 7 Flow chart of designed A-BMS operation

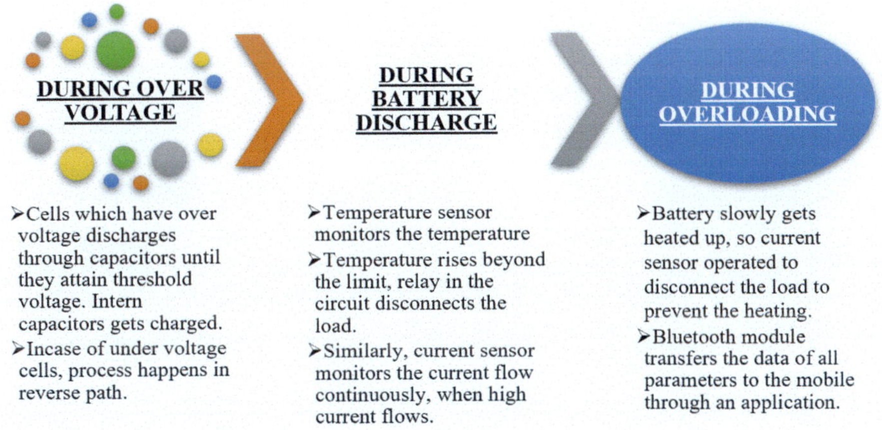

Fig. 8 Protection features of designed AI-based A-BMS

4 Results and Discussions

4.1 Simulation-Based Approach

Before designing with the hardware setup, the simulation platform is utilized to test the active balancing process in a battery pack. The proteus tool is used to design the above setup which is shown in Fig. 9. Firstly, the active cell balancer constantly monitors the voltage levels of each cell within the battery pack. Upon detecting a cell with a higher voltage compared to others, the balancer triggers a circuit associated with that specific cell. This activated circuit incorporates a capacitor, which begins to charge using the excess voltage from the higher cell.

By employing active monitoring, capacitor storage, and flyback conversion, this method effectively manages voltage discrepancies within the battery pack, thereby enhancing its overall performance, longevity, and safety. During the simulation, it was noted that when the switching frequency was set at 34 Hz, the duration required for cell balancing was 120 s. At a switching frequency of 4.237 kHz, the cell balancing process requires approximately 60 s to complete. This observation underscores the relationship between switching frequency and the time needed for effective cell balancing in the system. From the above simulation work, we can conclude the voltages at various resistors. Voltage at various resistors is R2 = 3.5 V, R6 = 3.6 V, R8 = 3.51 V, R10 = 3.36 V, R16 = 3.26 V, R17 = 3.38 V.

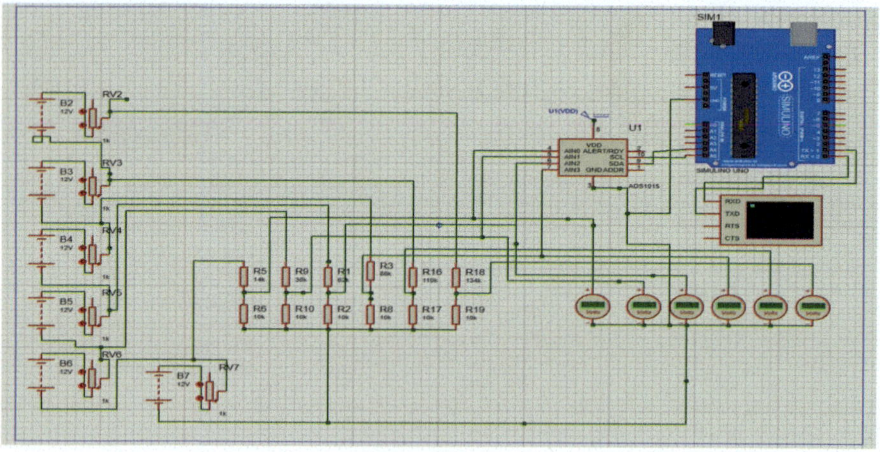

Fig. 9 Simulation-based analysis of proposed A-BMS

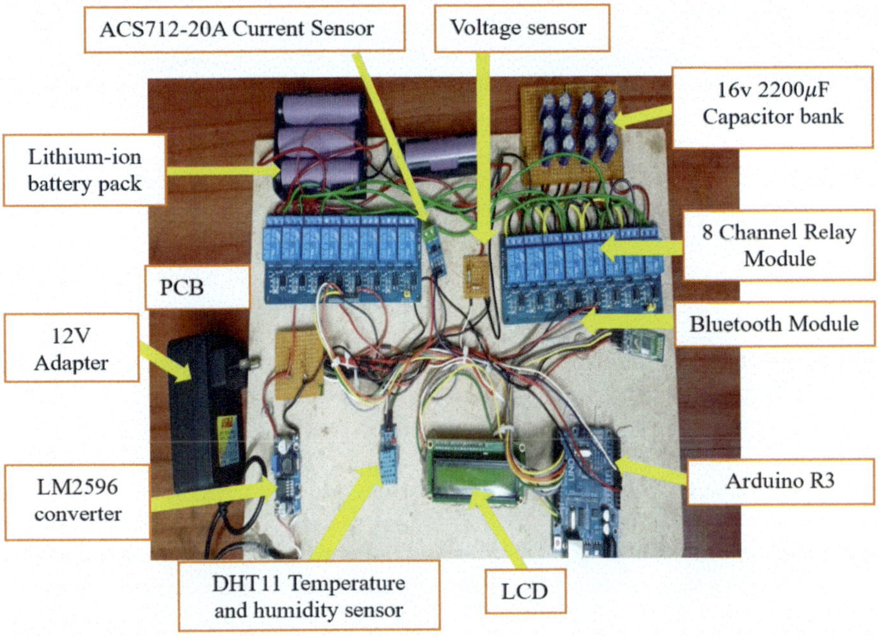

Fig. 10 Hardware-based analysis of proposed A-BMS

4.2 *Hardware-Based Approach*

The hardware prototype design of the A-BMS is shown in Fig. 10. The complete setup along with their components are discussed in detail in the consecutive stages.

4.3 *Test Cases*

There are two test cases are considered in the designed ABMS to understand the behavior of developed A-BMS. They are,

- 4S Configuration
- 2S2P Configuration

Assuming the initial cell voltages of all 4 cells in a battery pack are 3.59 V, 3.87 V, 3.59 V, and 3.87 V, respectively. The battery voltage is mentioned above. The 4 cells charged report will be displayed when the Bluetooth is connected to the mobile application, along with that the overall voltage of the battery module during this stage is about 14.89 V and the temperature is 28 °C. Each cell's voltages are updated as 3.83 V, 3.64 V, 3.39 V, and 3.97 V respectively. When compared to previous values

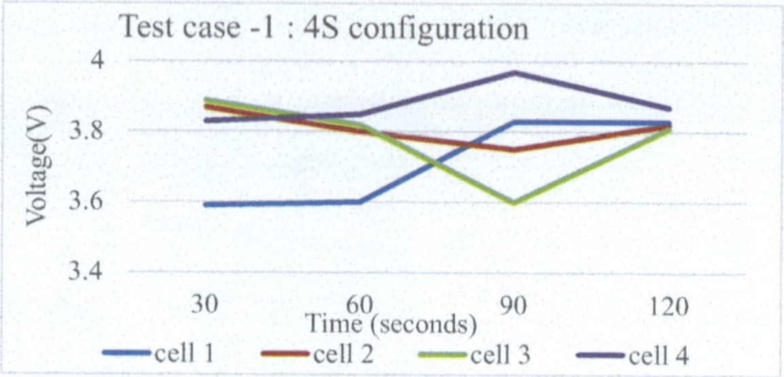

Fig. 11 Voltage balancing states of each cell in 4S configuration

the voltages of the cells are low. But now the cell voltages are increasing partially in each cell. The total pack of voltage is decreased gradually.

4.3.1 Test Case I: 4S Configuration of Battery Module

The battery cells are connected in a 4S configuration which means 4 individual cells are connected in a series manner. Considering the different voltage levels at each cell, where cell 1 is at 3.59 V, cell 2 with 3.87 V, cell 3 with 3.59 V and cell 4 is at 3.87 V respectively. From the above values, the lowest level in these 4 cells is across cell 3, i.e., 3.59 V Cells which have over-voltage discharges through capacitors until they attain threshold voltage. In turn, capacitors get charged. In the case of under-voltage cells, the process happens in the reverse path. The balancer circuit will balance the voltages across each cell, transfer the energy from high to low-level charged cells, and maintain the balance, finally, the voltage cells across each cell are obtained as 3.83 V, 3.80 V, 3.81 V, and cell 4 having 3.87 V respectively as shown in Fig. 11. The voltage levels of each cell in the 4S configuration are drawn in a graphical way with the developed A-BMS as shown in Fig. 12. The plot shows the voltage variations of each cell from its initial state to its balanced state.

4.3.2 Test Case II: 2S2P Configuration of Battery Module

The battery cells are 2S2P which means they are 2 series, 2 parallel. Considering the various voltage levels across each cell in a battery pack. The voltages of cells 1, 2, 3 and 4 are 3.5 V, 3.9 V, 0 V, and 0 V, respectively. From the above values we consider the lowest level in these 4 cells is cell 3, 4 i.e. 0 V Cells which have over-voltage discharges through capacitors until they attain threshold voltage. In turn, capacitors get charged. In the case of under-voltage cells, the process happens in the reverse path. After balancing, the battery voltages obtained are cell 1 having 2.02 V, cell 2 having 2.12 V, cell 3 having 1.98 V and cell 4 having 1.92 V as shown in Fig. 13.

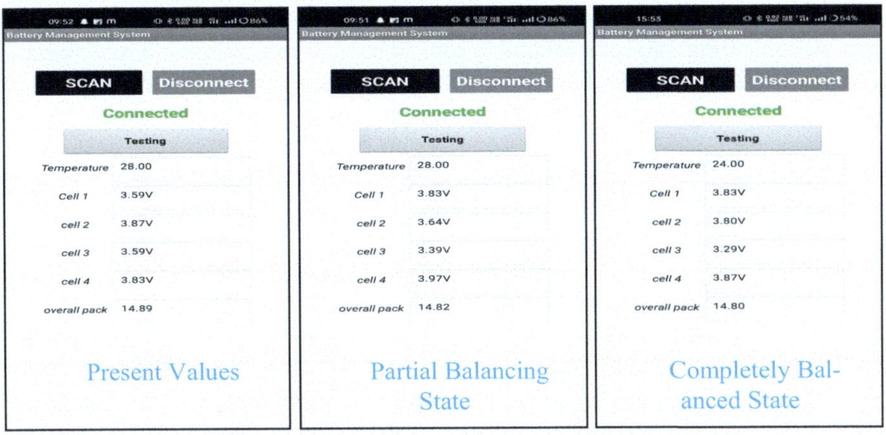

Fig. 12 Voltage balancing states of each cell in 4S configuration during operation of A-BMS

Fig. 13 Voltage balancing states of each cell in 2S2P configuration

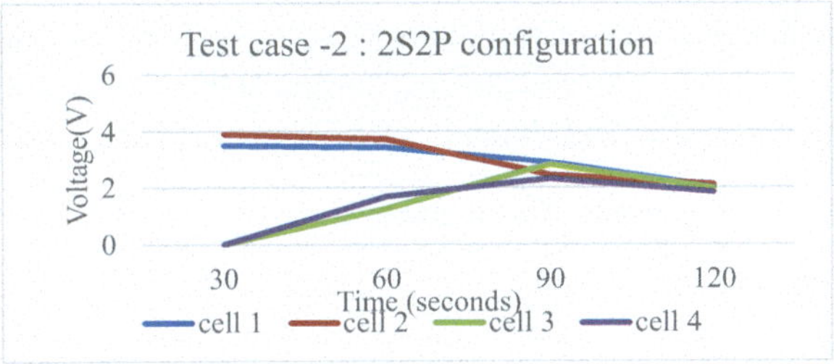

5 Conclusions and Future Scope

This work presents an innovative battery management system incorporating active balancing using relays and capacitor banks. The BMS is equipped with safeguards against overloading, overvoltage, and during discharge-related risks. It employs advanced techniques to ensure uniform cell charging and discharging, enhancing battery longevity. The design not only implements active cell balancing but also effectively lowers the overall temperature of the battery pack. By efficiently managing energy distribution and heat dissipation, it enhances the performance and longevity of the batteries. This integrated approach addresses both thermal management and cell balancing, ensuring optimal operation in diverse applications. The system's ability to mitigate temperature fluctuations contributes to its reliability and safety. The proposed technique is validated using four battery cell configurations, they are 4S (4 cells in series) configuration and 2S2P (2 cells in series and 2 cells in parallel) configuration. In both configurations, the cell voltage levels are both pre- and post-balancing. This evaluation ensures that the balancing process effectively equalizes the voltages across all cells, optimizing the battery's performance. By monitoring voltage levels before and after balancing, we can validate the effectiveness of the balancing technique in maintaining cell uniformity. This comprehensive approach helps guarantee the system's reliability and efficiency in managing battery health and performance.

In the simulation, due to limitations in representing lithium-ion batteries, a high-rating capacitor is substituted, with a conversion from the battery's specifications to a capacitance of 1800 F for a 3.2 V, 6Ah battery. However, typical capacitor ratings are in mF, resulting in significantly higher capacitance requirements. To address this, four flying capacitors are connected in parallel in the simulation diagram. It was observed that as the switching frequency increases, the speed of cell balancing also increases. This highlights the relationship between switching frequency and the efficiency of the cell balancing process in the simulated system. In our hardware implementation, we achieved high accuracy in balancing cell voltage levels in both configurations. The future scope of the present research work will be, Under the dynamic operating conditions of BMS in EV/HEV applications, the state estimation of battery parameters (SOC, SOH, TEMP) using filtering techniques (Kalman filtering, Particle filtering, h_∞ filtering, Hybrid filtering techniques, etc.) will be more effective than the conventional measuring techniques. Cell balancing using advanced converter topology (like Resonant Converters, Active clamped topologies, or hybrid topologies, etc.) offers a flexible solution for active cell balancing in complex battery systems.

Acknowledgements This research work was supported by "Woosong University's Academic Research Funding—2025".

References

1. Valladares Montemayor HM, Chanda RH (2023) Automotive industry's circularity applications and industry 4.0. Environ Chall 12:1–10. https://doi.org/10.1016/j.envc.2023.100725
2. Alanazi F (2023) Electric vehicles: benefits, challenges, and potential solutions for widespread adaptation. Appl Sci 13:1–23. https://doi.org/10.3390/app13106016
3. Shahzad K, Cheema II (2024) Low-carbon technologies in automotive industry and decarbonizing transport. J Power Sour 591:233888. https://doi.org/10.1016/j.jpowsour.2023.233888
4. Haghani M et al (2023) Trends in electric vehicles research. Transp Res Part D Transp Environ 123:1–29. https://doi.org/10.1016/j.trd.2023.103881
5. Hossain MS, Kumar L, El Haj Assad M, Alayi R (2022) Advancements and future prospects of electric vehicle technologies: a comprehensive review. Complexity 2022:1–21. https://doi.org/10.1155/2022/3304796
6. Heinrich F, Klapper P, Pruckner M (2021) A comprehensive study on battery electric modeling approaches based on machine learning. Energy Inform 4:1–17. https://doi.org/10.1186/s42162-021-00171-7
7. Liu Y, Esan OC, Pan Z, An L (2021) Machine learning for advanced energy materials. Energy AI 3:1–27. https://doi.org/10.1016/j.egyai.2021.100049
8. Mistry A, Franco AA et al (2021) How machine learning will revolutionize electrochemical sciences. ACS Energy Lett 6:1422–1431. https://doi.org/10.1021/acsenergylett.1c00194
9. Wang Y, Tian J, Sun Z et al (2020) A comprehensive review of battery modeling and state estimation approaches for advanced battery management systems. Renew Sustain Energy Rev 131:1–18. https://doi.org/10.1186/s42162-021-00171-7

10. Kumar UA (2005) Comparison of neural networks and regression analysis: a new insight. Exp Syst Appl 29:424–430. https://doi.org/10.1016/j.eswa.2005.04.034
11. Zhong R, Du C (2023) A review of machine learning state-of-charge and state-of-health estimation algorithms for lithium-ion batteries. Energy Rep 9:2993–3021. https://doi.org/10.1016/j.egyr.2023.01.108
12. Li Y, Li K, Liu X et al (2022) A hybrid machine learning framework for joint SOC and SOH estimation of lithium-ion batteries assisted with fiber sensor measurements. Appl Energy 325:1–22. https://doi.org/10.1016/j.apenergy.2022.119787
13. Olabi AG, Abdelghafar AA et al (2024) Artificial neural network driven prognosis and estimation of Lithium-Ion battery states: current insights and future perspectives. Ain Shams Eng J 15:1–17. https://doi.org/10.1016/j.asej.2023.102429
14. Wang Y, Huang H, Wang H (2022) A new method for fast state of charge estimation using retired battery parameters. J Energy Storage 55:105621. https://doi.org/10.1016/j.est.2022.105621
15. Olabi AG et al (2024) Artificial neural network driven prognosis and estimation of Lithium-Ion battery states: current insights and future perspectives. Ain Shams Eng J 15:1–17. https://doi.org/10.1016/j.asej.2023.102429
16. Manriquez-Padilla CG, Cueva-Perez I et al (2023) State of charge estimation model based on genetic algorithms and multivariate linear regression with applications in electric vehicles. Sensors 23:1–15. https://doi.org/10.3390/s23062924
17. Zhong F, Li H, Zhong S, Zhong Q, Yin C (2015) An SOC estimation approach based on adaptive sliding mode observer and fractional order equivalent circuit model for lithium-ion batteries. Commun Nonlinear Sci Numer Simul 24:127–144. https://doi.org/10.1016/j.cnsns.2014.12.015
18. Sheng H, Xiao J (2015) Electric vehicle state of charge estimation: nonlinear correlation and fuzzy support vector machine. J Power Sour 281:131–137. https://doi.org/10.1016/j.jpowsour.2015.01.145
19. Li Z, Li K, Du D, Zhu C, Min Z (2019) A sparse least squares support vector machine used for SOC estimation of Li-ion batteries. IFAC-Papers Online 52:256–261. https://doi.org/10.1016/j.ifacol.2019.09.150
20. Vidya RM, Ramakrishna M (2024) An adaptive learning paradigm: event detection through a novel dynamic arithmetic optimization-based ensemble SVM for data stream classification. Int J Inform Technol 16:3049–3055. https://doi.org/10.1007/s41870-024-01832-y
21. Gurusamy A, Ashok B, Mason B (2023) Rediction of electric vehicle driving range and performance characteristics: a review on analytical modeling strategies with its influential factors and improvisation techniques. IEEE Access 11:131521–131548. https://doi.org/10.1109/ACCESS.2023.3334620
22. Leising RA, Palazzo MJ, Takeuchi ES, Takeuchi KJ (2001) A study of the overcharge reaction of lithium-ion batteries. J Power Sour 97–98:681–683. https://doi.org/10.1016/S0378-7753(01)00598-5
23. Motaqi A (2020) Energy-performance management in battery powered reconfigurable processors for standalone IoT systems. Int J Inform Technol 12:653–668. https://doi.org/10.1007/s41870-020-00454-4
24. Karunanithi K, Ramesh S, Raja SP et al (2023) Investigations of standalone PV system with battery-supercapacitor hybrid energy storage system for household applications. Int J Inf Technol 15:279–287. https://doi.org/10.1007/s41870-022-01144-z
25. Einhorn M, Rößler W, Conte F et al (2012) Charge balancing of serially connected lithium-ion battery cells in electric vehicles. Elektrotech. Inftech. 129:167–173. https://doi.org/10.1007/s00502-012-0097-x
26. Duraisamy T, Kaliyaperumal D (2020) Active cell balancing for electric vehicle battery management system. Int J Power Electron Drive Syst 11:571–579. https://doi.org/10.11591/ijpeds.v11.i2.pp571-579
27. Krishnamoorthy U, Satheesh Kumar GS, Barua S, Fayek HH (2023) Investigation of active cell balancing performance for series connected lithium-ion cells in electric vehicle applications. IET Power Electron 16:2492–2503. https://doi.org/10.1049/pel2.12575

28. Khan N, Ooi CA, Alturki A, Amir M, Shreasth TA (2024) A critical review of battery cell balancing techniques, optimal design, converter topologies, and performance evaluation for optimizing storage system in electric vehicles. Energy Rep 11:4999–5032. https://doi.org/10.1016/j.egyr.2024.04.041
29. Ye Y, Wang J, Wang X (2023) A multi-winding transformer-based active cell equalizer with self-driven switches for series-connected lithium-ion batteries and super-capacitors. J Energy Storage 70:1–18. https://doi.org/10.1016/j.est.2023.107971
30. Fan T-E, Liu S-M, Yang H, Li P-H, Qu B (2023) A fast active balancing strategy based on model predictive control for lithium-ion battery packs. Energy 279:128028. https://doi.org/10.1016/j.energy.2023.128028
31. Shylla D, Harikrishnan R, Swarnkar R (2022) Comparative analysis and evaluation of the different active cell balancing topologies in lithium ions batteries. J Electrochem Soc 170:080501. https://doi.org/10.1149/1945-7111/ace958
32. Dhumal AR, Kulkarni AP, Ambhore NH (2023) A comprehensive review on thermal management of electronic devices. J Eng Appl Sci 70:1–18. https://doi.org/10.1186/s44147-023-00309-2
33. Hafiz MA (2023) Thermal management systems for batteries in electric vehicles: a recent review. Energy Rep 9:5545–5564. https://doi.org/10.1016/j.egyr.2023.04.359
34. Rosen MA, Farsi A (2023) Battery thermal management systems. Battery Technol 2023:119–160. https://doi.org/10.1016/B978-0-443-18862-6.00003-3
35. Tamilselvi S, Gunasundari S, Karuppiah N, Abdul Razak RK et al (2021) A review on battery modelling techniques. Sustainability 13(18):10042. https://doi.org/10.3390/su131810042
36. Tran M-K, Mathew M, Janhunen S, Panchal S, Raahemifar K, Fraser R, Fowler M (2021) A comprehensive equivalent circuit model for lithium-ion batteries, incorporating the effects of state of health, state of charge, and temperature on model parameters. J Energy Storage 43:103252. https://doi.org/10.1016/j.est.2021.103252
37. Tran M-K et al (2021) A comprehensive equivalent circuit model for lithium-ion batteries, incorporating the effects of state of health, state of charge, and temperature on model parameters. J Energy Storage 43:1–27. https://doi.org/10.1016/j.est.2021.103252
38. Salkuti SR (2022) Emerging and advanced green energy technologies for sustainable and resilient future grid. Energies 15(18):6667. https://doi.org/10.3390/en15186667
39. Baccouche I, Jemmali S, Manai B, Nikolian A, Omar N, Essoukri Ben Amara N (2022) Li-ion battery modeling and characterization: an experimental overview on NMC battery. Int J Energy Res 46:3843–3859. https://doi.org/10.1002/er.7445
40. Uddin K, Picarelli A, Lyness C, Taylor N, Marco J (2014) An acausal Li-ion battery pack model for automotive applications. Energies 7(9):5675–5700. https://doi.org/10.3390/en7095675
41. Karimi D, Behi H, Van Mierlo J, Berecibar M (2023) Equivalent circuit model for high-power lithium-ion batteries under high current rates, wide temperature range, and various state of charges. Batteries 9:101. https://doi.org/10.3390/batteries9020101
42. Jung CM, Pagidipala S, Salkuti SR (2024) Emerging technologies for the integration of renewable energy, energy storage and electric vehicles. In: Salkuti SR (eds) Energy and environmental aspects of emerging technologies for smart grid. Green Energy and Technology. Springer, Cham. https://doi.org/10.1007/978-3-031-18389-8_5
43. Zhang C, Guo Y, Wang C, Li S et al (2022) A new design of experiment method for model parametrisation of lithium-ion battery. J Energy Storage 50:1–8. https://doi.org/10.1016/j.est.2022.104301
44. Wang X, Luo C, Zhao J, Luo Z (2024) Optimal selection of bulk capacitors in flyback converter. Microelectron Reliab 153:115306. https://doi.org/10.1016/j.microrel.2023.115306
45. Gatto G, Lai A, Marongiu I, Serpi A (2016) Circuital and mathematical modelling of flyback converters. Int Symp Power Electron Electr Drives Autom Motion 2016:906–911. https://doi.org/10.1109/SPEEDAM.2016.7525983
46. Yang C, Xie F, Chen Y, Xiao W, Zhang B (2020) Modeling and analysis of the fractional-order flyback converter in continuous conduction mode by Caputo fractional calculus. Electronics 9:1544. https://doi.org/10.3390/electronics9091544

47. Salkuti SR (2023) Advanced technologies for energy storage and electric vehicles. Energies 16(5):2312. https://doi.org/10.3390/en16052312
48. Sravanthi P, Sandeep V (2022) Solving realistic reactive power market clearing problem of wind-thermal power system with system security. Int J Emerg Electr Power Syst 23(2):125–144. https://doi.org/10.1515/ijeeps-2021-0060

A Comparative Analysis of A* and Ant Colony Algorithms for EV Routing

Rajarajeswari Rathinam, A. Dhanush, Pradeep Vishnuram, Mohit Bajaj, and Arvind R. Singh

Abstract The shift towards sustainable transportation systems underscores the need to enhance the routing of electric vehicles (EVs). This work delves into the Electrical Vehicle Routing Problem (EVRP), by considering the unique requirements of electric vehicles, such as battery range limitations and the availability of charging stations. A significant addition to this work is the inclusion of different driving modes such as Eco, Ride, Sport, and Warp which influence energy consumption rates and travel times. Also, it aims to optimize routes not only for efficiency but also by factoring in these variable driving modes to adapt to drivers' preferences or situational needs. The A* algorithm is employed for its effectiveness in quickly determining the shortest and least costly path by estimating future costs. This algorithm's performance is compared with that of the Ant Colony Optimization (ACO) algorithm, which mimics the way ants find efficient paths. Our comparative analysis seeks to justify the obtained result along with swiftness handling the complex of EVRP, including varied vehicle modes and accurate time estimations. Further, the customized version of the routing problem for both EV bikes and EV cars with the consideration of speed modes, State of Charge (SOC), discharge level calculation, energy used, and time of travel is optimized. Also, EV route-based feasibility and charging metric analysis are performed.

Keywords Electric vehicle routing problem · State of charge · Vehicle driving modes · A* algorithm · Ant colony optimization · Feasibility and charging metric

---

R. Rathinam · A. Dhanush · P. Vishnuram (✉)
Department of Electrical and Electronics Engineering, SRM Institute of Science and Technology, Kattankulathur, Chennai 603203, India
e-mail: pradeep.kannan03@gmail.com

M. Bajaj
Department of Electrical Engineering, Graphic Era (Deemed to Be University), Dehradun, India

Hourani Center for Applied Scientific Research, Al-Ahliyya Amman University, Amman, Jordan

Graphic Era Hill University, Dehradun, India

A. R. Singh
Department of Electrical Engineering, School of Physics and Electronic Engineering, Hanjiang Normal University, Hubei Shiyan, Shiyan, People's Republic of China

S. R. Salkuti (ed.), *Artificial Intelligence for Integrated Smart Energy Systems in Electric Vehicles*, Lecture Notes in Electrical Engineering 1427,
https://doi.org/10.1007/978-3-031-94276-1_12

Abbreviations

EVs	Electric vehicles
EVRP	Electrical vehicle routing problem
SOC	State of charge
AOC	Ant colony optimization
GA	Genetic algorithm
PSO	Particle swarm optimization
TSP	Traveling salesman problem
CS	Charging station
SA	Simulated annealing
DVRP	Dynamic vehicle routing problem
AVNS	Adaptive variable neighborhood search

1 Introduction

Electric vehicles (EVs), which are powered by rechargeable batteries, provide an environmentally friendly mode of transportation. The optimization of travel routes for EVs considers various factors, such as battery range, the locations of charging stations, and traffic conditions, aiming to save energy and time. This optimization is crucial for addressing the challenges of range and charging infrastructure, thus increasing the adoption of EVs. The significance of the Electric Vehicle Routing Problem (EVRP) is recognized for its critical role in advancing sustainable transportation. It addresses the unique challenges of EVs, such as their limited range and the necessity for strategic planning of charging stops. The optimization of EV routes is essential for enhancing operational efficiency, reducing environmental impact, and encouraging the broader adoption of EVs. Research in EVRP contributes to the development of advanced algorithms that improve the reliability and efficiency of electric transportation systems [1].

Figure 1 visualizes a system designed to solve the Electric Vehicle Routing Problem (EVRP). It begins with an EV input stage where the vehicle's initial state of charge, battery capacity, and the driver's trip details are collected. This information flows into two pathways: one leads to a database detailing the availability of slots at charging stations, ensuring the trip plan accounts for real-time charger occupancy; the other directs the input data to a sophisticated routing algorithm. The routing algorithm processes the collected inputs along with the charging station data to plot an optimal travel route. The result is a series of optimized routes, showcased on a map, which ensure the EV can travel from the start point to the destination efficiently. These routes are planned so that necessary charging stops are incorporated, based on both the EV's energy needs and the availability of charging infrastructure along the way, ensuring a seamless and efficient trip for the EV driver. The promise held by EVs extends beyond transforming the automotive sector; it encompasses the potential

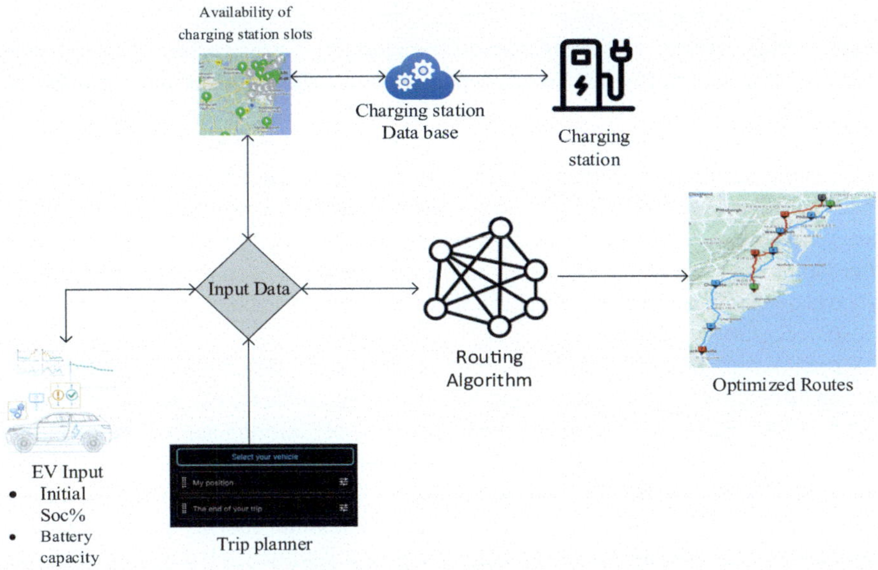

Fig. 1 EVRP system flow: from input to optimized routes

to reshape the entire world. The widespread adoption of EVs contributes to a more environmentally friendly planet by reducing emissions, cutting operational costs, and diminishing noise pollution. Traditional EVRP models mainly focus on fleet scheduling and the positioning of charging stations. However, personal EV routing presents a unique set of challenges that require different considerations [2].

In logistics and distribution, the progression towards eco-friendly logistics practices and the global support for new energy vehicle initiatives are facilitating the transition to electric vehicles as alternatives to traditional fuel-powered vehicles. The EVRP plays a pivotal role in creating optimal routes for electric delivery fleets, thereby improving their efficiency and sustainability. The focus of ongoing research in the EVRP lies in the development of models and algorithms that offer solutions while considering various factors such as battery depletion, cargo capacity, specific timing requirements, recharging strategies, and the employment of mixed vehicle fleets. To address the complexities of EVRP, a wide array of approaches is being utilized, including exact methods, heuristic techniques, and metaheuristic strategies. The EVRP has captured significant interest from both the research community and industry practitioners, with a primary focus on improving delivery efficiency and reducing routing costs for EVs. Investigations delve into various aspects, such as battery constraints, load capacities, and the strategic placement of charging stations. These models offer guidance for logistics companies looking to promote the use of EVs within their operations [3]. The significance of sustainable transportation is increasingly recognized, placing the EVRP in the spotlight. Efforts continue to extend and refine the problem to enhance its practical application. EV routing challenges include several key areas namely the charging infrastructure with the vital aspects of the availability and strategic placement of charging stations. Insufficient charging facilities may disrupt travel plans and lead to concerns about battery range. Vehicle and Driving Dynamics, encompassing Battery Life and Range is the next arena where the impact of driving behaviors on battery longevity and the vehicle's range is crucial. Aggressive or inefficient driving practices can significantly increase energy consumption. Battery Management and Uncertainties, Payload Weight are the other challenges where the variations in the vehicle's load can significantly impact its travel distance. Adapting route planning to accommodate changes in cargo weight is also necessary in EVRP.

Variations in weather conditions, such as rain, extreme temperatures, or wind, have an effect on energy consumption. It is critical to adjust routes based on weather changes. The challenge of ensuring the availability of charging stations can complicate even the most well-thought-out travel plans [4]. Road Conditions, including Terrain and Traffic: Variations in road quality, inclines, and dense traffic can reduce the efficiency of EVs. It is important to plan routes considering these factors to minimize energy use [5]. More algorithms are used for this EVRP with SOC, time, distance, and availability of Charging Station (CS) as constraints out of which the Ant Colony Optimization (ACO) algorithm represents a collective metaheuristic approach aimed at identifying near-optimal solutions to complex optimization problems. The fundamental principle of ACO is based on observing the paths ants take from their nest to food sources, with an emphasis on utilizing the most efficient routes possible [6]. The A* algorithm combines the strengths of Dijkstra's algorithm, which guarantees the shortest path, with the efficiency of greedy best-first search, although the latter may lack precision. This algorithm excels at finding the most efficient path from a starting point to a destination across graphs or grids. Its application is widespread in various fields such as computer science, gaming, robotics, and online mapping. Improvements are introduced in the EBS-A* variant, including Expansion Distance: A safety margin is kept from obstacles to increase path reliability and avoid collisions. To speed up the process, searches are initiated from both the start and objective sites. This algorithm reduces abrupt turns to ensure the path's smoothness [7]. Simulation tests for the EBS-A* algorithm have demonstrated its effectiveness, further evidenced by its implementation in a robot operating system (ROS). When compared to the traditional A* algorithm, the EBS-A* shows a substantial improvement in path planning efficiency by 2.78, reduces crucial nodes by 0.9189, and eliminates right-angle turns, highlighting its enhanced performance [8]. It finds use across multiple domains, including Routing and Scheduling: Determining the most efficient routes for vehicles, delivery fleets, or data within network communications. Traveling Salesman Problem (TSP): Computing the shortest route to visit a set of cities and return to the starting point. Job Scheduling: Allocating tasks to machines or personnel in the most efficient way. Network Routing: Improving the movement of data packets through computer networks [9]. The ACO technique is utilized for addressing complex optimization problems, especially those focused on finding optimal paths within graphs [10].

Table 1 compares various research references based on their objectives and the techniques or models they use, highlighting the unique contributions of each in the field of routing optimization. The focus on optimizing the shortest path efficiency using the A* Search Algorithm to quickly identify the most efficient route, enhancing planning efficiency is performed by several authors. Reference [10] applies the ACO Algorithm to solve complex routing problems by utilizing swarm intelligence, thus improving adaptability. The author in [11] experiments on the optimal routing strategies through GA that leverages evolutionary principles to optimize multiple objectives. Reference [12] uses PSO to find optimal routes, focusing on energy efficiency. SA in [13] refines routing solutions using probabilistic techniques to avoid local minima and achieve global optimization. The Fuzzy Logic Algorithm to manage uncertainties in electric vehicle (EV) routing, enhancing robustness in dynamic conditions is explained in [14]. Reference [15] focuses on quick navigation to goals using a Greedy Best-First Search Algorithm, optimizing for speed. Dijkstra's Algorithm in [16] and the Bellman-Ford Algorithm in [17] are utilized to identify and calculate the shortest paths in weighted graphs and dynamic scenarios, respectively, ensuring optimal routing. Finally, [18] navigates solution spaces efficiently using the Tabu Search Technique, which uses memory structures to avoid revisiting previous limitations.

Ahmed et al. [20] had proposed VRP with the consideration of unsymmetrical distance as a constraint; that was optimized using four hybrid GA algorithms. VRP that is based on non-cyclic. Multi-depot travel by vehicles was developed by the author in [21]. A lexicographic-based VRP with concurrent pick-up and delivery constraints was optimized in two stages using neighborhood and tabu search methods for obtaining the best efficacy proposed by Shi et al. in [22]. VRP solutions with the time constraint-based optimization on novel ant colony algorithms and various meta-heuristic algorithms were proposed in [23–25]. Furthermore, several authors have proposed VRP solutions for optimizing the best route along with variant recharging policy and dynamic pricing constraints [26–28]. In [29] a DVRP problem with the inclusion of uncertainties such as real-time traffic scenarios and route diversions involving AVNS search algorithm to find the best route was proposed recently.

Previous research in the field of EV routing has often compartmentalized critical variables such as time, charging times, optimal paths, total travel time, and battery discharge levels, examining them in isolation rather than in an integrated manner. Also, research on various algorithms to solve DVRP, and VRP problems using variant constraints was proposed [30]. These approaches have limited the depth of understanding necessary for addressing the multifaceted challenges of electric vehicle logistics. The proposed research seeks to fill this gap by encompassing all these variables including speed mode within a single study, thus providing a more holistic analysis of the electric vehicle routing problem. Additionally, while existing studies may focus on a single routing algorithm, this work aims to enhance the comparative analysis by employing both the A* algorithm and the Ant Colony algorithm for customized EVRP according to all the constraints such as time of travel, battery charging time (fast/slow), optimal paths, total travel time, Speed mode and battery discharge levels. The selection of an algorithm for a VR problem using speed mode is

Table 1 Comparison of the references based on their objective and techniques

References	Objective	Technique/model	Contribution
[19]	Optimize shortest path efficiency	A* search	Quickly find the most efficient path, enhancing route planning efficiency
[10]	Solve complex routing problems	ACO	Utilizes swarm intelligence for dynamic problem-solving, improving adaptability in route optimization
[11]	Evolve optimal routing strategies	GA	Applies evolutionary principles to generate solutions, optimizing for multiple objectives
[12]	Optimize route selection	PSO	Leverages swarm intelligence for finding optimal routes, emphasizing energy efficiency
[13]	Refine routing solutions	SA	Uses probabilistic techniques to avoid local minima, finding globally optimized routes
[14]	Address uncertainties in EV routing	Fuzzy logic algorithm	Handles imprecise information to make more adaptable routing decisions, enhancing the robustness of EV routing strategies in dynamic and uncertain environments
[15]	Quickly navigate to the goal	Greedy best-first search algorithm	Prioritizes movement towards the destination, optimizing for speed
[16]	Identify shortest paths in weighted graphs	Dijkstra's algorithm	Efficiently calculates the shortest path, ensuring optimal routing
[17]	Calculate shortest paths in dynamic scenarios	Bellman-ford algorithm	Accommodates changes in road conditions and traffic, updating routes accordingly
[18]	Navigate solution spaces efficiently	Tabu-search technique	Uses memory structures to explore solutions, avoiding previously encountered limitations

based on the review of algorithms made from the survey of the literature. The comparison will not only determine which algorithm performs better under various conditions but also explore their effectiveness across different types of vehicles namely two-wheelers and four-wheelers with distinct ranges and operational parameters. This comprehensive approach is anticipated to yield significant insights, thereby addressing the identified gaps in previous studies and contributing to more effective and sustainable EV routing strategies.

2 Problem Statement

In this EVRP, a comprehensive investigation is conducted, including variables such as travel time, speed modes, charging time calculations, optimal path determination, State of Charge (SoC%) calculation, and discharge levels. The work incorporates two specific cases to enhance the analysis: Case 1 involves EV routing considering only SoC% with travel time calculations, and Case 2 includes EV routing that considers SoC% and the speed of the vehicle with different modes such as Eco mode, Ride mode, Sport mode, and Wrap mode. These cases will be compared using two distinct algorithms: the A* algorithm and the Ant Colony algorithm. The comparison will be further enriched by its application to different vehicle types two-wheelers and four-wheelers each characterized by varying ranges and parameters. The performance of both the A* and Ant Colony algorithms will be assessed in terms of their efficiency and effectiveness in handling the complexities of electric vehicle routing. The analysis will focus on the algorithms' adaptability to various vehicle types, examining how factors like battery capacity and discharge rates affect routing choices. Attention is given to the vehicle's range and operational parameters which determine the optimal path, travel time, and charging needs under different conditions [31]. Additionally, this work will explore the scalability of these algorithms, assessing their ability to manage the complexities of integrating multiple routing variables. The findings are expected to offer valuable insights into refining routing strategies, contributing to the development of more efficient and sustainable electric vehicle networks. This research aims to merge theoretical models with practical applications, serving as a key resource for advancements in electric vehicle logistics.

3 Analysis of EV Routing Algorithms

The input parameters for EVRP using the algorithms—A* and ACO algorithms are listed in Table 2 to address that the EVRP are systematically outlined to provide clarity on their operational frameworks. Both algorithms utilize a set network of nodes (A, B, C, D, E) and edges with defined distances, such as A-B: 10 km, B-C: 15 km, etc., maintaining consistent vehicle battery capacity of 100 kWh and an energy consumption rate of 0.2 kWh/km across both. The current battery level starts at 80 kWh, and the charging rates differ at standard (22 kW/h) and fast (50 kW/h) across both algorithms. Their primary objective is to minimize energy consumption within a constraint of a maximum 200 km travel distance without recharging. Specific to A*, the heuristic function calculates costs using straight-line distances, while the cost function is based on the distance traveled. Conversely, ACO employs pheromone levels starting at 0.1, with path desirability inversely related to distance, an evaporation rate of 50%, and influence parameters set at $\alpha = 1$, $\beta = 2$, tailoring to enhance route optimization effectively.

Table 2 EV routing algorithms: input parameter

Parameter	A* algorithm values	ACO algorithm values
Network nodes	A, B, C, D, E	A, B, C, D, E
Network edges and distances	A-B: 10 km, B-C: 15 km, etc.	A-B: 10 km, B-C: 15 km, etc.
Vehicle battery capacity	100 kWh	100 kWh
Energy consumption rate	0.2 kWh/km	0.2 kWh/km
Current battery level	80 kWh	80 kWh
Charging rate	Standard: 22 kW/h, Fast: 50 kW/h	Standard: 22 kW/h, Fast: 50 kW/h
Objective function	Minimize energy consumption	Minimize energy consumption
Constraints	Max distance without recharge: 200 km	Max distance without recharge: 200 km
Heuristic function (h(n))	Straight-line distance	N/A
Cost function (g(n))	Distance travelled	N/A
Pheromone levels (τ)	N/A	0.1 for all paths
Desirability (η)	N/A	Inverse of distance (1/distance)
Evaporation rate (ρ)	N/A	0.5 (50%)
Influence parameters (α, β)	N/A	$\alpha = 1, \beta = 2$

To present the outcomes of employing the A* and Ant Colony Optimization (ACO) algorithms for the EVRP allows for a clear comparison of the efficiency and effectiveness of each algorithm in addressing this problem under similar conditions.

Nodes and Distances: A $\rightarrow$ B: 10, B $\rightarrow$ C: 15, C $\rightarrow$ D: 20, D $\rightarrow$ E: 25.

4 Navigating EVs: Stations and Distances

The EV specification and its charging data are depicted in Table 3. Table 4 details EV Bike Routes along with the distances and charging station locations for electric bikes. Route 1 covers a total distance of 50 km and does not have any charging stations. Route 2 is slightly shorter at 45 km and includes a charging station at S1, which is positioned 40 km from the starting point. Route 3 is the longest among the bike routes at 60 km and features two charging stations; S2 is 10 km into the route, followed by S3 at an additional distance of 30 km, with the final stretch to destination Z being 20 km. Lastly, Route 4 mirrors Route 1 in distance at 50 km but includes a charging station at S4, 35 km from the start. EV Car Routes provides a similar layout but with significantly longer distances suitable for electric cars as listed in Table 5. Route 1 is the most straightforward with a total travel distance of 250 km and no charging stations [32]. Route 2 extends 255 km with a single charging station at S1 located 110 km from the beginning. Route 3, totaling 257 km, is equipped with two charging stations: the first, S2, is reached after 80 km, and the second, S3, after an additional

Table 3 Electric vehicle specifications and charging data

Parameter	EV bike	EV car
Rated power (W)	3300	100,000
Max power (W)	6000	150,000
Top speed (km/h)	80	160
Battery capacity (kWh)	2.9	75
Total range (km)	85	400
DC charge time (0–100%)	30min	90 min
AC charge time (0–100%)	5 h	8 h
Initial SoC (%)	39%	22%
Starting node	A	A
Ending node	Z	Z
Charging stations	S1, S2, S3, S4	S1, S2, S3, S4

Table 4 EV bike routes: distance and charging station details

Route	Total distance (km)	Charging stations	Distance to charging stations (km)
1	50	None	N/A
2	45	S1	A to S1: 40
3	60	S2, S3	A to S2: 10, S2 to S3: 30, S3 to Z: 20
4	50	S4	A to S4: 35

Table 5 EV car routes: distance and charging station details

Route	Total distance (km)	Charging stations	Distance from charging stations (km)
1	250	None	N/A
2	255	S1	A to S1: 110
3	257	S2, S3	A to S2: 80, S2 to S3: 140, S3 to Z: 37
4	280	S4	A to S4: 125

140 km, with a concluding segment of 37 km to destination Z. Route 4, the longest, spans 280 km with a charging station at S4 placed 125 km into the journey. These tables summarize the critical details needed for planning trips on EV bikes and cars, indicating the accessibility of charging facilities along various routes.

5 EV Trip Planning and Execution Flow

The EVRP flowchart shown in Fig. 2 depicts a sequence of steps for planning and executing a trip with an EV. The process begins with the input of vehicle specifications and trip details, which include the vehicle's current state of charge and the intended

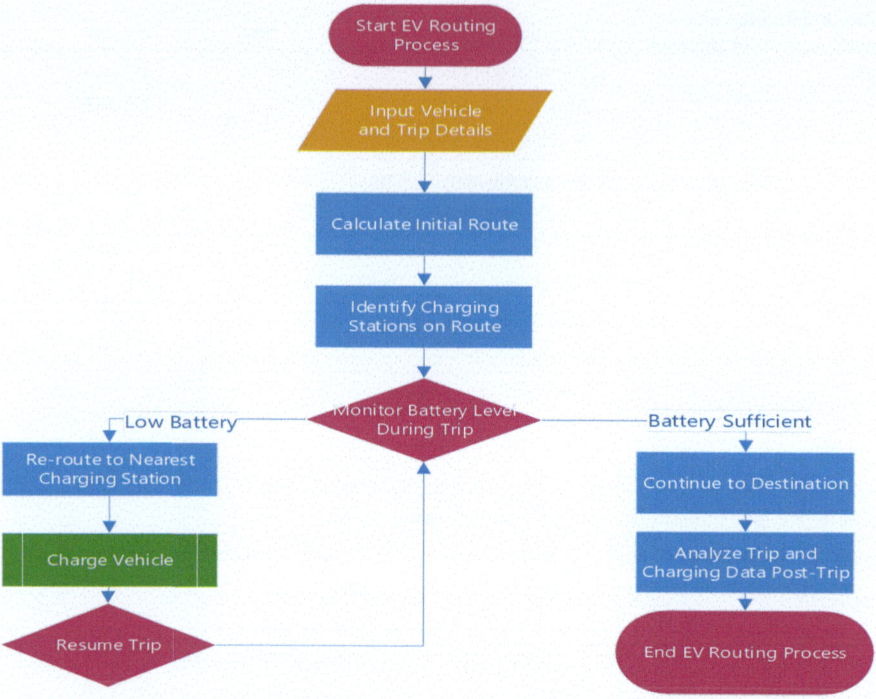

Fig. 2 EV routing process flowchart

destination. An initial route is then calculated, and along this route, charging stations are identified to ensure the vehicle can be charged as necessary during the trip. As the vehicle embarks on the trip, the battery level is continually monitored. If the battery level is sufficient to reach the destination, the journey continues uninterrupted. However, if at any point the battery level is low, the system triggers a reroute to the nearest charging station where the vehicle's battery is recharged. Once the charging is complete, the trip is resumed. After reaching the destination, the trip and charging data are analyzed for post-trip evaluation, which marks the end of the EV routing process. This systematic approach ensures that the EV can complete its journey efficiently, with charging stops incorporated as needed to maintain a sufficient battery level throughout the trip.

5.1 A Algorithm in EVRP*

The A* Algorithm is a heuristic-based search algorithm renowned for its efficiency and it is adapted to consider the energy consumption of electric vehicles, and the availability of charging stations, to minimize the total travel time, including charging.

The algorithm evaluates routes using a cost function that combines the distance traveled with the state of charge (SOC) of the EV's battery, dynamically adjusting the path to ensure that vehicles can complete their routes with minimal delays for charging. Further, the Total Cost Function is given in Eq. (2),

$$f(n) = g(n) + h(n) \tag{1}$$

where Eq. (1) defines the total cost function $f(n)$ for a node n in the A* algorithm, which is the sum of $g(n)$, the actual cost to reach node n, and $h(n)$, the heuristic estimate of the cost to reach the goal from n. This function is central to A*'s strategy, guiding the search toward the goal by favoring paths with lower estimated total costs. Further, the Actual Cost Function is given in Eq. (2),

$$g(n) = g(p(n)) + c_{p(n),n} \tag{2}$$

where Eq. (2) calculates the actual cost $g(n)$ to reach a node n, which is the sum of the cost to reach the node's parent and the cost to move from the parent to n. This reflects the cumulative distance or energy consumed to reach n. Further, the Heuristic Cost Function is given in Eq. (3),

$$h(n) = D_{n,goal} \tag{3}$$

where the heuristic cost function $h(n)$, given by Eq. (3), estimates the remaining distance or energy required to reach the goal from node n. This estimation is crucial for efficiently directing the search towards the goal, minimizing unnecessary exploration. Further the Energy Consumption Model in A* is given in Eq. (4),

$$E_{ij} = \alpha.d_{ij}.Load_{ij} + \gamma \cdot Gradient_{ij} \tag{4}$$

The where Eq. (4) models the energy consumption $E_{ij.}$ for traveling from node i to node j, incorporating factors such as distance d_{ij}, vehicle load d_{ij}, and road gradient $Gradient_{ij.}$ This model is essential for accurately calculating $g(n)$ in the context of EVRP, ensuring that the algorithm accounts for the EV's energy usage. Further the Battery Capacity Constraint in A* is given in Eq. (5),

$$g(n) + E_{n,next} \leq C \tag{5}$$

Equation (5) enforces the battery capacity constraint, ensuring that the selected path does not exceed the vehicle's battery capacity. This constraint is critical for the feasibility of the routes generated by the A* algorithm, requiring the inclusion of charging stops as necessary. Discharge Level Calculation is given in Eq. (6) and Energy Used Calculation, Time Calculation is given in Eqs. (7) and (8),

$$Discharge\ Level(\%) = \frac{Energy\ Used(\text{kWh})}{Battery\ Capacity(\text{kWh})} \times 100 \tag{6}$$

$$\textit{Energy Used}\,(\text{kWh}) = \textit{Distance} \times \textit{Energy Consumption Rate} \tag{7}$$

$$\textit{Total Time} = \frac{\textit{Distance}}{\textit{Average Speed}} + \textit{Charging Time} \tag{8}$$

Further, the Charging Time (if needed) is given in Eq. (9),

$$\textit{Charging Time} = \left(\frac{\textit{Discharge Level}(\%)}{100}\right) \times \textit{Full Charging Time}(5\,\text{h}) \tag{9}$$

5.2 Ant Colony Optimization (ACO) Algorithm in EVRP

In EVRP, ACO is utilized to simulate a population of ants that explore various routes between locations, gradually improving their solutions based on the pheromone trails left by successful ants. This algorithm is particularly effective in EVRP for its ability to find optimal or near-optimal routing solutions that account for the need to recharge EVs, by prioritizing routes that pass through charging stations and minimizing the overall travel and charging time. Further, the Pheromone Update Rule in ACO is given in Eq. (10),

$$\tau_{ij}(t+1) = (1-\rho)\tau_{ij}(t) + \sum \Delta\tau_{ij}^{k} \tag{10}$$

Equation (10) describes the pheromone update rule, where τ_{ij} $(t + 1)$ is the pheromone level on the path from node i to node j at time $t + 1$, ρ is the pheromone evaporation coefficient, and $\sum\Delta\,\tau_{ij}^{k}$ is the total pheromone deposited by all ants k that used path. This rule ensures that paths that lead to more efficient routes accumulate more pheromones, guiding future ants toward these paths. Further, the pheromone deposit in ACO is given in Eq. (11),

$$\tau_{ij}^{k} = \frac{Q}{L_k} \tag{11}$$

Equation (11) defines the pheromone deposit by an ant k on the path from i to j, where Q is a constant and L_k is the length or cost of the route found by ant k. This encourages ants to favor shorter or more energy-efficient routes, as they leave a stronger pheromone trail. Further, the Transition Probability in ACO is given in Eq. (12),

$$\mathrm{P}_{\mathrm{ij}}^{\mathrm{k}} = \frac{(\tau_{\mathrm{il}}(\mathrm{t}))^{\alpha} \cdot (\eta_{\mathrm{il}})^{\beta}}{\sum_{\mathrm{l}\in \mathrm{allowed}_{\mathrm{k}}} (\tau_{\mathrm{il}}(\mathrm{t}))^{\alpha} \cdot (\eta_{\mathrm{il}})^{\beta}} \tag{12}$$

Equation (12) calculates the probability P_{ij}^{k} that ant k will choose to move from node i to node j, based on the pheromone level $\tau_{il}(t)$ and the heuristic desirability η_{il} of the path, with α and β weighting the importance of pheromone and heuristic information, respectively. This probabilistic approach allows ants to explore diverse routes while gradually converging on the most efficient solutions.

$$\eta_{ij} = \frac{1}{n_{ij} + \lambda \cdot C_{ij}} \tag{13}$$

Equation (13) provides the heuristic information ηij, which inversely relates to the combined distance d_{il} and a penalty $\lambda \cdot c_{il}$ for energy consumption or charging time, encouraging ants to select paths that minimize both distance and energy use.

6 Results and Discussion

The "Total Travel Time" and "Remaining Battery Charge" are based on simplifications such as a constant driving speed and specific assumptions about charging (e.g., recharging to full at node C). The actual values in Table 6 could vary based on additional factors like variable speed limits, traffic conditions, and partial charging strategies as explained in Case 1 and Case 2.

Figure 3 presents the optimal paths determined by the ACO and A* algorithms for the electric vehicle routing problem, with both algorithms selecting the identical route from nodes A to E through B, C, and D. The total distance covered by the vehicle is 70 km for each route, leading to an energy consumption of 14 kWh, given a rate of 0.2 kWh/km. The vehicle needs to recharge once at node C for both paths. The combined time for travel and charging amounts to 4.5–3.5 h of driving plus 1 h spent charging. At the end of the journey, the vehicle is left with 66 kWh of battery charge,

Table 6 Algorithmic efficiency in EV routing outcomes

Output parameter	A* algorithm results	ACO algorithm results
Selected path	A → B → C → D → E	A → B → C → D → E
Total distance	70 km	70 km
Total energy consumed	14 kWh (0.2 kWh/km * 70 km)	14 kWh (0.2 kWh/km * 70 km)
Number of charging stops	1 (at node C)	1 (at node C)
Total travel time	3.5 h driving + 1 h charging = 4.5 h	3.5 h driving + 1 h charging = 4.5 h
Remaining battery charge	66 kWh (80 kWh − 14 kWh + charge at C)	66 kWh (80 kWh − 14 kWh + charge at C)
Optimization objective achieved	Yes, minimized energy consumption	Yes, minimized energy consumption

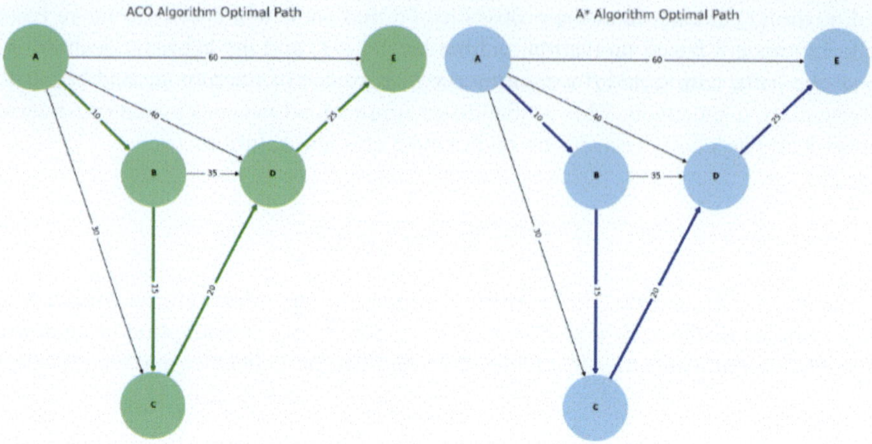

Fig. 3 Comparative route optimization: ACO versus A* algorithms

starting from 80 kWh and accounting for the energy consumed and the recharge at node C. This output satisfies the optimization objective of minimizing energy consumption for both the ACO and A* algorithms. Notably, the paths highlighted in the diagram, despite the different algorithms, achieve the same level of efficiency in terms of distance, energy use, charging stops, travel time, and remaining charge, illustrating the algorithms' effectiveness in route optimization.

6.1 EV Route Feasibility and Charging Analysis

Table 7, structures the viable route (Route 3), the charging requirements, and the estimated charging time. The theoretical information for Routes 1, 2, and 4 for completeness is also provided, highlighting why they were deemed not viable based on the initial SoC level and the distances involved.

Routes 1 and 2 have been declared unsuitable for electric vehicle travel due to the vehicle's insufficient initial range and the absence or inaccessibility of charging facilities within that range. Conversely, Route 3 emerges as the only feasible option, requiring just one charging stop at station S2. Upon arrival at S2, the state of charge (SoC) is projected to be approximately 27.24%, and to fully recharge the battery to 100% SoC, an estimated charging time of 3.64 h is necessary. Route 4, however, is marked as not feasible; calculations show an SoC dropping below 0% before the vehicle could reach the nearest charging station, indicating that completing this route would be practically impossible without additional pre-trip charging.

Table 8 indicates the feasibility analysis of the proposed routes for an electric vehicle with a 22% state of charge (SoC), we have updated the evaluation to include both DC fast charger and AC charger options, detailing the charging times for each.

Table 7 EV bike route assessment: feasibility and charging metrics

Routes	Feasibility	Total distance (km)	Charging station (cs)	Distance to CS (km)	SoC at CS (%)	DC charging time required (h)	AC charging time required (h)
1	No	50	None	N/A	N/A	N/A	N/A
2	No	45	S1	A to S1:40	N/A	N/A	N/A
3	Yes	60	S2	A to S2:10	27.24%	1.5	3.64
4	No	50	S4	A to S4:35	Below 0%	N/A (would be > 2 h)	N/A (would be > 5 h)

DC fast chargers can significantly reduce charging times, typically taking about 90 min to reach 80% charge, while AC chargers are slower, requiring approximately 8 h for a full charge. Here's a concise analysis of each route: Route 1 lacks charging stations, and with a route distance of 250 km, it far exceeds the vehicle's 88 km range capability at 22% SoC, making both charging options irrelevant as the vehicle can't reach a charging station. Route 2's distance to charging station S1 is 110 km, which also surpasses the vehicle's range, again rendering both charging options inapplicable. Route 3 is slightly more promising; the vehicle can just reach S2 at 80 km. If a DC charger is available, it could recharge to 80% in about 1.5 h, potentially allowing continuation but not completion of the route since the next segment alone exceeds the recharged range. AC charging, needing 8 h, is impractical for such a journey due to time constraints.

Table 8 EV car route assessment: feasibility and charging metrics

Routes	Feasibility	Total distance (km)	Charging station	Distance to CS (km)	SoC at CS (%)	DC charging time (h)	AC charging time (h)
1	No	250	None	N/A	N/A	N/A	N/A
2	No	255	S1	A to S1:110	Below 0%	N/A (insufficient range)	N/A (insufficient range)
3	No	257	S2, S3	A to S2: 80, S2 to S3:140, S3 to Z: 37	Just reachable at S2, Below 0% at S3	Brief charge at S2 (~1.5 h for 80%), N/A beyond S2	Long charge at S2 (8 h), N/A beyond S2
4	No	280	S4	A to S4: 125	Below 0%	N/A (insufficient range)	N/A (insufficient range)

6.2 *Case 1: Analyzing SoC Changes in EV Mobility*

The SoC at key points and the change in SoC due to driving and charging. The vehicle's journey can be broken down into three key segments with specific battery usage and charging details as shown in Fig. 4. Starting with an initial State of Charge (SoC) of 39%, the vehicle travels 10 km to charging station S2, which reduces the SoC to 27.24% due to a discharge of 11.76%. At S2, the vehicle is recharged to 100% SoC, with the charging time calculated based on the prior depletion and the vehicle's specific charging characteristics. After leaving S2, the vehicle covers a 30 km distance to S3, during which the SoC depletes by 35.32%, leaving it at 64.68%. The final leg of the journey involves a 20 km drive from S3 to destination Z, which results in a further decrease in SoC by 23.64%, bringing it down to 41.04% upon arrival.

The comparison of the travel distance for the best solution identified by both the A* algorithm and the Ant Colony algorithm has been visualized in a bar graph in Fig. 5. For the selected route, both algorithms have determined that route 3, with a total distance of 60 km, is the most feasible.

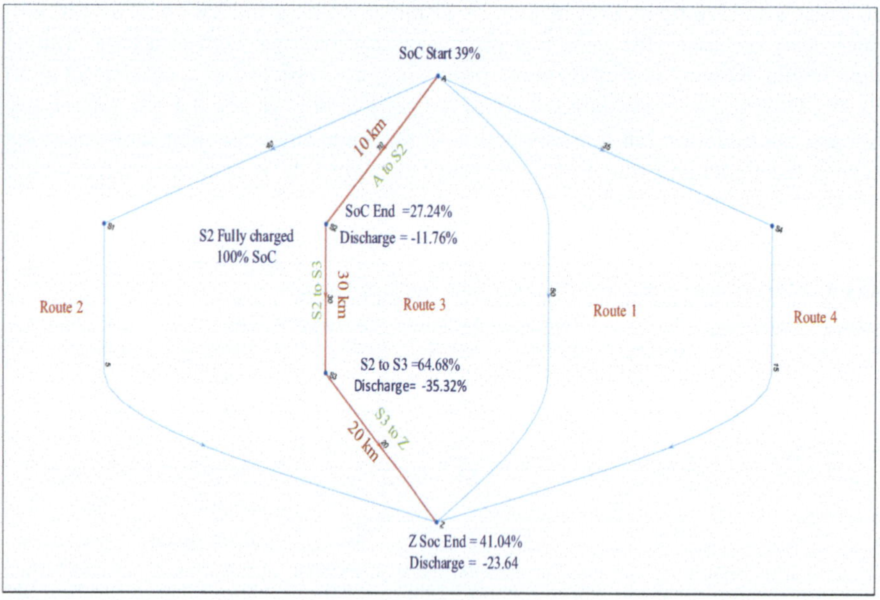

Fig. 4 State of charge (SoC) variation at key route points during EV journey using Route 3

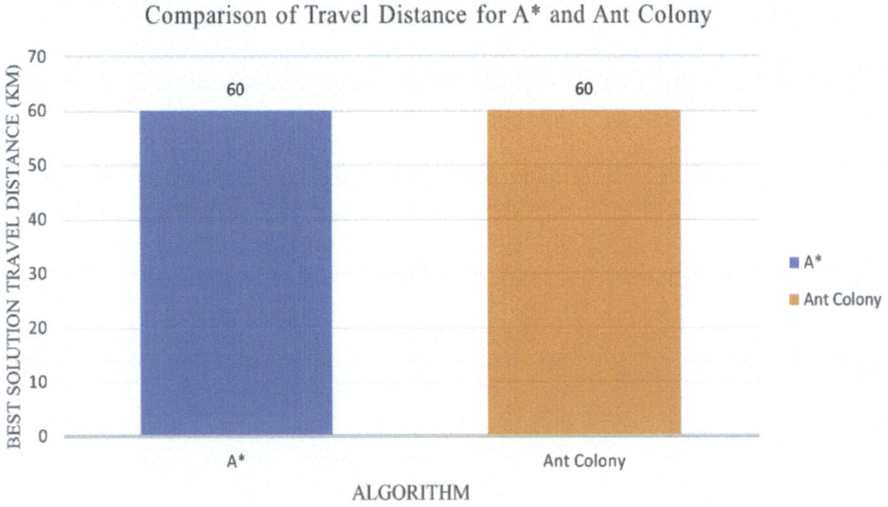

Fig. 5 Comparison of travel distance for A* and ant colony

6.3 *Case 2: EVRP with SOC with Different Speed Mode Selection*

The graph shown in Fig. 6 displays how changing travel speeds in different driving modes such as Eco, Ride, Sports, and Warp would affect the speed in kilometers per hour (km/h) for the selected route. Each mode offers a different speed and corresponding travel time, with Eco mode being the slowest but most energy-efficient, and Warp mode achieving the top speed of 80 km/h, reducing the travel time to 0.75 h. This analysis helps in understanding how different driving preferences can impact overall travel time and speed on a route and shows the discharge level of batter for different modes shown in Fig. 7.

Tables 9, 10, 11 and 12 provide information on the comparative analysis of EV performance across four routes, detailing time taken and battery discharge for Eco, Ride, Sports, and Warp modes. Each mode demonstrates a trade-off between speed and battery efficiency, with faster modes consuming more energy. This analysis is essential for optimizing EV usage to balance speed, energy conservation, and distance effectively, highlighting the practical implications for route planning and mode selection.

In Fig. 8, the analysis of four routes under different driving modes—Eco, Ride, Sports, and Warp—is shown with varying impacts on travel time and battery usage. For Route 1, which spans 50 km, the Eco mode is the most efficient in terms of battery conservation, using 58.45% over 0.83 h, while Warp mode expends 86.21% in just 0.63 h, showcasing a faster but more battery-intensive option. Similarly, Route 2 covers 45 km with Eco mode consuming 52.60% of the battery in 0.75 h, and Warp mode increasing the discharge to 77.59% over 0.56 h. Route 3, the longest at 60

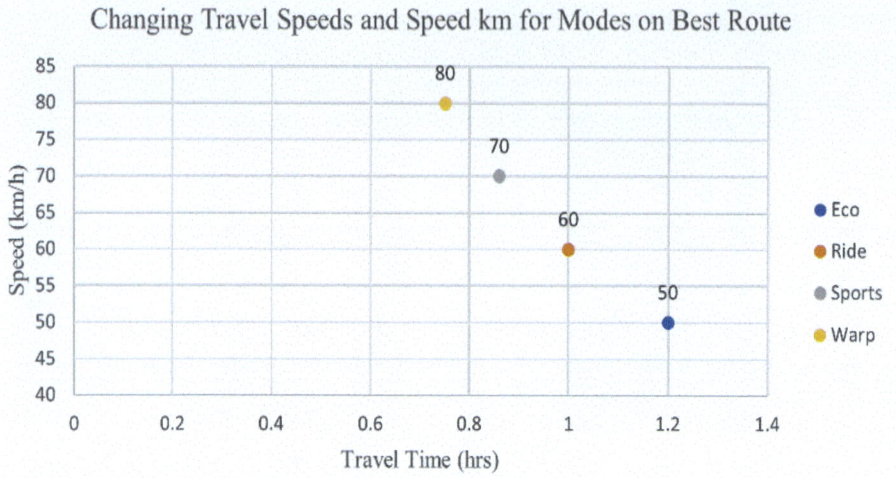

Fig. 6 Speed variations across driving modes on selected route

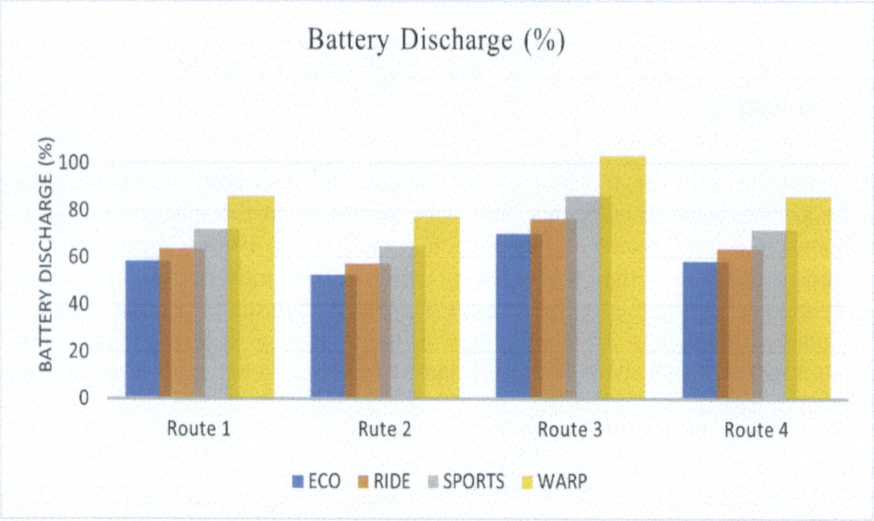

Fig. 7 Energy usage across different EV driving modes

Table 9 Performance metrics for route 1 (50 km)

Mode	Time taken (h)	Battery discharge (%)
Eco	0.83	58.45
Ride	0.71	63.86
Sports	0.67	71.84
Warp	0.63	86.21

Table 10 Performance metrics for route 2 (45 km)

Mode	Time taken (h)	Battery discharge (%)
Eco	0.75	52.60
Ride	0.64	57.47
Sports	0.60	64.66
Warp	0.56	77.59

Table 11 Performance metrics for route 3 (60 km)

Mode	Time taken (h)	Battery discharge (%)
Eco	1.00	70.13
Ride	0.86	76.63
Sports	0.80	86.21
Warp	0.75	103.45

Table 12 Performance metrics for route 4 (50 km)

Mode	Time taken (h)	Battery discharge (%)
Eco	0.83	58.45
Ride	0.71	63.86
Sports	0.67	71.84
Warp	0.63	86.21

km, sees battery usage escalate significantly in more aggressive modes: 70.13% in Eco and a substantial 103.45% in Warp, indicating potential battery depletion before completing the route. Lastly, Route 4 mirrors Route 1 in both distance and battery usage across modes, maintaining a consistent pattern of energy consumption relative to travel time. These findings highlight the trade-offs between speed and battery efficiency across different driving conditions, underscoring the need for strategic mode selection based on route length and battery capacity.

7 Conclusions

This work focuses on implementing a comparative analysis of two prominent routing algorithms, A* and Ant Colony Optimization (ACO), as they apply to the Electric Vehicle Routing Problem (EVRP). The work systematically evaluated both algorithms under identical conditions involving fixed routes, battery capacities, and energy consumption rates to ascertain their efficacy in optimizing EV travel routes. Both A* and ACO successfully met the project's objective to minimize energy consumption within the set travel distance constraint of 200 km without requiring recharging.

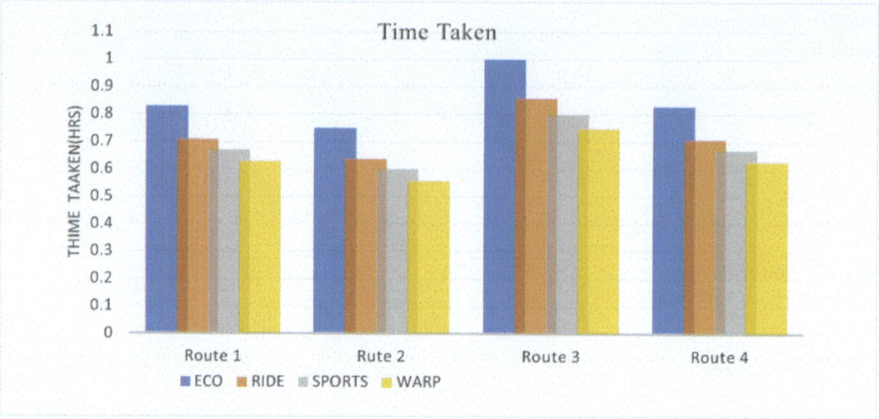

Fig. 8 Driving modes and their impact on travel duration

The findings reveal that while both algorithms selected the same routes and achieved similar performance metrics in terms of energy usage and travel time, their operational approaches differ significantly. A* utilizes a straightforward heuristic based on straight-line distances, making it efficient and predictable. In contrast, ACO adapts dynamically, using pheromone levels to influence path selection, which could offer better flexibility in real-world scenarios where conditions change rapidly. These insights highlight the potential for both algorithms to enhance route optimization in EV navigation systems, suggesting that the choice between them could depend on specific user needs for predictability or adaptability. Future work could explore integrating real-time data into these algorithms to further improve their applicability and efficiency in managing EV logistics. This project thus provides a foundation for enhancing EV routing systems, ensuring energy-efficient and practical transportation solutions. Optimizing routes for electric vehicles using sophisticated algorithms to ensure efficient travel and energy use. The core of my analysis involved calculating critical metrics such as charging times for both DC and AC chargers, the discharge levels of the battery throughout the journey, total travel time, and the operational dynamics of driving in various modes such as Eco, Ride, Sport, and Warp, depending on the state of charge (SoC) levels. Utilizing parameters for both electric cars and bikes, designed and evaluated four distinct routes, incorporating strategically located charging stations and accurately measured distances for each route. This comprehensive approach allowed me to determine the most efficient routes based on the initial SoC, ensuring optimal performance and energy utilization. The outcomes of this research were meticulously presented through graphs and tables, providing clear and actionable insights that highlight the effectiveness of the proposed routing strategies in the realm of electric vehicles. These findings not only demonstrate the practical application of routing algorithms in real-world scenarios but also contribute to the broader goal of enhancing the sustainability and feasibility of electric vehicle transportation.

References

1. Schneider M, Stenger A, Goeke D (2014) The electric vehicle-routing problem with time windows and recharging stations. Transp Sci 48(4):500–520. https://doi.org/10.1287/trsc.2013.0490
2. Rajesh K, Jain E, Kotecha P (2022) A multi-objective approach to the electric vehicle routing problem. arXiv preprint arXiv:2208.12440. https://doi.org/10.48550/arXiv.2208.12440
3. Ye C, He W, Chen H (2022) Electric vehicle routing models and solution algorithms in logistics distribution: a systematic review. Environ Sci Pollut Res 29(38):57067–57090. https://doi.org/10.1007/s11356-022-21559-2
4. Stamadianos T, Kyriakakis NA, Marinaki M, Marinakis Y (2023) Routing problems with electric and autonomous vehicles: review and potential for future research. In: operations research forum, vol 4. Springer International Publishing, Cham, pp 2–46. https://doi.org/10.1007/s43069-023-00228-1
5. Pulyassary H, Kollias K, Schild A, Shmoys D, Wu M (2023) Network flow problems with electric vehicles. arXiv preprint arXiv:2311.05040. https://doi.org/10.48550/arXiv.2311.05040
6. Rezvanian A, Vahidipour SM, Sadollah A (2023) An overview of ant colony optimization algorithms for dynamic optimization problems
7. Candra A, Budiman MA, HartantoK (2020) Dijkstra's and a-star in finding the shortest path: a tutorial. In: 2020 international conference on data science, artificial intelligence, and business analytics (DATABIA). IEEE, pp 28–32 https://doi.org/10.1109/DATABIA50434.2020.9190342
8. Eddoujaji M, Samadi H, Bouhorma M (2022) HDFS improvement using shortest path algorithms. In: International conference on networking, intelligent systems and security. Springer International Publishing, Cham, pp 253–260. https://doi.org/10.1007/978-3-031-15191-0_25
9. Dorigo M (2007) Ant colony optimization. Scholarpedia 2(3) (1461:56–59). https://doi.org/10.1007/978-1-4899-7687-1_22
10. Gupta A, Srivastava S (2020) Comparative analysis of ant colony and particle swarm optimization algorithms for distance optimization. Proc Comput Sci 173:245–253. https://doi.org/10.1016/j.procs.2020.06.029
11. Sachenbacher M, Leucker M, Artmeier A, Haselmayr J (2011) Efficient energy-optimal routing for electric vehicles. Proc AAAI Conf Artif Intell 25:1402–1407. https://doi.org/10.1609/aaai.v25i1.7803
12. Marini F, Walczak B (2015) Particle swarm optimization (PSO). A tutorial. Chemometr Intell Lab Syst 149:153–165. https://doi.org/10.1016/j.chemolab.2015.08.020
13. Yao X (1995) A new simulated annealing algorithm. Int J Comput Math 56(3–4):161–168. https://doi.org/10.1080/00207169508804397
14. Rout RR, Vemireddy S, Raul SK, Somayajulu DV (2020) Fuzzy logic-based emergency vehicle routing: an IoT system development for smart city applications. Comput Electr Eng 88:106839. https://doi.org/10.1016/j.compeleceng.2020.106839
15. Elizabeth M, Gani B, Safitri MA, Dewi LC, Meiliana M (2018) Real time public transportation navigator system in Jakarta by using greedy best first search algorithm. J Telecommun Electron Comput Eng 10(1–5):65–67
16. Eklund PW, Kirkby S, Pollitt S (1996) A dynamic multi-source Dijkstra's algorithm for vehicle routing. In: 1996 Australian New Zealand conference on intelligent information systems, vol 96. Proceedings. ANZIIS, IEEE, pp 329–333. https://doi.org/10.1109/ANZIIS.1996.573976.
17. Schambers A, Eavis-O'Quinn M, Roberge V, Tarbouchi M (2018) Route planning for electric vehicle efficiency using the Bellman-Ford algorithm on an embedded GPU. In: 2018 4th international conference on optimization and applications (ICOA), IEEE, pp 1–6. https://doi.org/10.1109/ICOA.2018.8370584
18. Ahmed ZH, Yousefikhoshbakht M (2023) An improved tabu search algorithm for solving heterogeneous fixed fleet open vehicle routing problem with time windows. Alex Eng J 64:349–363. https://doi.org/10.1016/j.aej.2022.09.008

19. Tang G, Tang C, Claramunt C, Hu X, Zhou P (2021) Geometric A-star algorithm: an improved A-star algorithm for AGV path planning in a port environment. IEEE Access 9:59196–59210. https://doi.org/10.1109/ACCESS.2021.3070054
20. Ahmed ZH, Hameed AS, Mutar ML, Wang S (2022) Hybrid genetic algorithms for the asymmetric distance-constrained vehicle routing problem. Math Probl Eng 1–20. https://doi.org/10.1155/2022/2435002
21. Brandão J (2020) A memory-based iterated local search algorithm for the multi-depot open vehicle routing problem. Eur J Oper Res 284(2):559–571. https://doi.org/10.1016/j.ejor.2020.01.008
22. Shi Y, Zhou Y, Boudouh T, Grunder O (2020) A lexicographic-based two-stage algorithm for vehicle routing problem with simultaneous pickup–delivery and time window. Eng Appl Artif Intell 95:103901. https://doi.org/10.1016/j.engappai.2020.103901
23. Cordeau JF, Desaulniers G, Desrosiers J, Solomon MM, Soumis F (2002) The VRP with time windows. In: Toth P, Vigo D (eds) The vehicle routing problem, SIAM monographs on discrete mathematics and applications. SIAM, pp 157–194
24. Rajappa GP, Wilck JH, Bell JE (2016) An ant colony optimization and hybrid metaheuristics algorithm to solve the split delivery vehicle routing problem. Int J Appl Ind Eng 3(1):55–73
25. Wang X, Choi TM, Liu H, Yue X (2016) Novel ant colony optimization methods for simplifying solution construction in vehicle routing problems. IEEE Trans Intell Transp Syst 17(11):3132–3141. https://doi.org/10.1109/TITS.2016.2542264
26. Sweda TM, Dolinskaya IS, Klabjan D (2017) Adaptive routing and recharging policies for electric vehicles. Transp Sci 51(4):1326–1348. https://doi.org/10.1287/trsc.2016.0724
27. Ulmer MW (2017) Approximate dynamic programming for dynamic vehicle routing. Springer, p 61. https://doi.org/10.1007/978-3-319-55511-9
28. Turan B, Pedarsani R, Alizadeh M (2020) Dynamic pricing and fleet management for electric autonomous mobility on demand systems. Transp Res 121:102829. https://doi.org/10.1016/j.trc.2020.102829
29. Sze JF, Salhi S, Wassan N (2024) An adaptive variable neighbourhood search approach for the dynamic vehicle routing problem. Comput Oper Res 164:106531. https://doi.org/10.1016/j.cor.2024.106531
30. Afroditi A, Boile M, Theofanis S, Sdoukopoulos E, Margaritis D (2014) Electric vehicle routing problem with industry constraints: trends and insights for future research. Transp Res Proc 3:452–459. https://doi.org/10.1016/j.trpro.2014.10.026
31. Salkuti SR (2022) Emerging and advanced green energy technologies for sustainable and resilient future grid. Energies 15(18):6667. https://doi.org/10.3390/en15186667
32. Salkuti SR (2023) Advanced technologies for energy storage and electric vehicles. Energies 16(5):2312. https://doi.org/10.3390/en16052312

AI-Driven Optimization Techniques for Electric Vehicle Charging Infrastructure

Debani Prasad Mishra, Arul Kumar Dash, Sandip Ranjan Behera, K. Ayushman Patro, and Surender Reddy Salkuti

Abstract The transition to sustainable transportation is increasingly reliant on the integration of electric vehicles (EVs) into existing energy systems. Efficient management of EV charging infrastructure is imperative to optimize energy usage, minimize grid stress, and enhance user convenience. In this chapter, we delve into the application of artificial intelligence (AI) in driving optimization techniques for EV charging infrastructure. AI, with its adaptive and predictive capabilities, offers transformative solutions to address the complex challenges associated with EV charging. Through the utilization of AI-driven algorithms such as reinforcement learning, genetic algorithms, and neural networks, charging networks can be dynamically optimized to maximize efficiency and reliability. This chapter explores various AI-driven optimization techniques tailored specifically for EV charging infrastructure. Reinforcement learning algorithms enable charging stations to learn and adapt their behavior based on real-time feedback, resulting in improved scheduling and resource allocation. Genetic algorithms provide an evolutionary approach to optimizing charging infrastructure layout and operation, considering factors such as charging demand, grid capacity, and geographical constraints. Neural networks offer predictive capabilities for forecasting charging demand, optimizing charging schedules, and mitigating peak load demands on the grid. Real-world implementations and case studies are examined to showcase the effectiveness of AI-driven optimization techniques in EV charging infrastructure. Moreover, challenges such as data privacy, interoperability, and scalability are addressed, along with potential solutions and future research directions. The chapter underscores the pivotal role of AI in shaping the future of EV charging infrastructure, facilitating the seamless integration of electric vehicles into the broader energy ecosystem.

D. P. Mishra · A. K. Dash · S. R. Behera · K. A. Patro
Department of Electrical and Electronics Engineering, IIIT Bhubaneswar, Bhubaneswar, Odisha, India
e-mail: debani@iiit-bh.ac.in

S. R. Salkuti (✉)
Department of Railroad and Electrical Engineering, Woosong University, Daejeon, Republic of Korea
e-mail: surender@wsu.ac.kr

S. R. Salkuti (ed.), *Artificial Intelligence for Integrated Smart Energy Systems in Electric Vehicles*, Lecture Notes in Electrical Engineering 1427,
https://doi.org/10.1007/978-3-031-94276-1_13

Keywords Artificial intelligence · Electric vehicles · Charging infrastructure · Optimization techniques · Smart energy systems · Sustainable transportation

Abbreviations

AI	Artificial intelligence
EV	Electric vehicle
GHG	Greenhouse gases
PHEV	Plug-in hybrid electric vehicles
PV	Photovoltaic
IoT	Internet of things
SCS	Smart charging stations
RAMOS	Resource-aware meta-operating system
ML	Machine learning
FL	Federated learning
ITS	Intelligent transportation system
CASE	Connected, autonomous, shared and electric
ECAS	Supply voltage

1 Introduction

Amid a global transition towards sustainable transportation, electric vehicles (EVs) have emerged as a promising key to trimming greenhouse gas (GHG) emissions, mitigating air pollution, and decreasing reliance on finite fossil fuels [1]. As the adoption of EVs continues to accelerate, the demand for efficient and scalable charging infrastructure becomes increasingly imperative. However, traditional charging approaches face significant challenges in meeting the evolving needs of a rapidly expanding EV market, including grid constraints, demand variability, and user inconvenience. This chapter explores the transformative potential of optimized charging infrastructure in addressing these challenges and advancing the sustainable mobility revolution. Optimized charging infrastructure represents a paradigm shift in the approach we governing and distributing energy for EVs, leveraging cutting-edge technologies and innovative approaches to maximize efficiency, reliability, and user satisfaction [2]. The basis of this transformation lies in the integration of AI techniques, which empower charging infrastructure with intelligence, adaptability, and optimization capabilities.

Our journey begins by examining the foundational principles of optimized charging infrastructure, encompassing load management, grid integration, and user-centric approaches. We delve into the complexities of existing charging infrastructure challenges and the limitations of traditional charging methods, underscoring the critical need for innovative solutions to propel the EV ecosystem forward. Central to our exploration is the role of AI in revolutionizing charging infrastructure optimization. Through advanced machine learning algorithms, optimization techniques, and predictive analytics, artificial intelligence (AI) enables dynamic and adaptive charging strategies tailored to individual user preferences, grid conditions, and energy market dynamics [3]. By harnessing the power of AI, we unlock new frontiers in energy efficiency, grid stability, and user experience enhancement, laying the groundwork for a more intelligent, eco-friendly, and enduring future.

Throughout the chapter, we examine real-world case studies and practical insights showcasing the transformative impact of AI-driven optimization on charging infrastructure performance and scalability. From load forecasting and demand response to energy price optimization and grid-aware scheduling, AI-driven solutions demonstrate tangible benefits in reducing charging time, enhancing grid reliability, and maximizing renewable energy utilization [4]. As we navigate the complexities of optimized charging infrastructure, we also acknowledge the challenges and considerations inherent in its deployment. From data privacy concerns to interoperability issues and regulatory barriers, we explore strategies for overcoming these obstacles and ensuring the successful implementation of AI-driven charging solutions.

2 Challenges in Charging Infrastructures

The transition towards EVs necessitates the development of efficient and reliable charging infrastructure, which faces several challenges. Grid congestion and operability are critical issues that must be addressed to ensure the seamless integration of EV chargers into existing power networks. Grid congestion occurs when multiple EVs simultaneously demand high power, potentially leading to strain and inefficiencies in the grid system. Ensuring grid operability involves implementing smart charging solutions and grid management techniques to optimize energy flow and prevent disruptions. Additionally, charging infrastructure accessibility is essential for the widespread adoption of EVs, as convenient and readily available charging stations are crucial to alleviate range anxiety among drivers. By employing AI-driven optimization techniques such as predictive modeling for demand forecasting and location optimization, these challenges can be effectively mitigated, fostering the growth of a robust and sustainable EV charging infrastructure.

In the context of developing AI-driven optimization techniques for EV charging infrastructure, the challenges in charging infrastructures extend beyond the technological realm to encompass critical issues such as grid congestion, operability, and accessibility [5]. Grid congestion presents a formidable obstacle to the efficient functioning of charging infrastructures, as the increasing demand for EVs strains the existing grid capacity, necessitating sophisticated optimization strategies to manage peak loads and distribution constraints. Grid operability further complicates the landscape by requiring real-time monitoring and control mechanisms to ensure the stability and reliability of power supply for charging stations. Moreover,

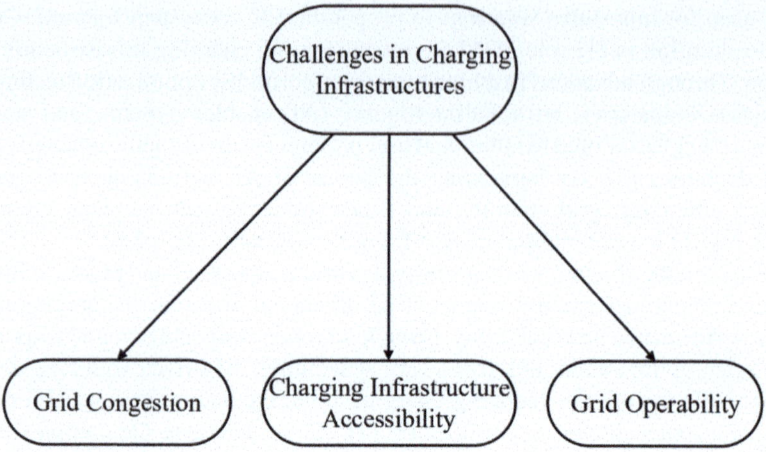

Fig. 1 Challenges in charging infrastructure

the accessibility of charging infrastructure poses a significant barrier to widespread adoption, with disparities in deployment locations and availability impeding seamless integration into daily routines and inhibiting the transition to electric mobility. Addressing these challenges through AI-driven solutions can enhance the efficiency, reliability, and equitable distribution of EV charging infrastructure, thus facilitating the sustainable growth of electric transportation. These challenges faced by charging infrastructure for EV charging are depicted in Fig. 1.

2.1 Grid Congestion

The evolution of the electrical grid is significantly impacted by the increasing penetration of distributed energy resources, such as EVs. This surge in EV adoption brings about numerous benefits but also presents challenges, particularly in grid congestion. The proposed business models in distribution networks with high EV penetration aim to address these congestion issues through computational intelligence techniques. Additionally, the concept of energy forecast can play a crucial role in collaboratively orchestrating the behavior of engine management systems in a fleet of PHEVs to absorb energy smartly [6]. By integrating the insights from these studies, it becomes evident that tackling grid congestion in the context of EV charging infrastructure requires innovative solutions driven by AI optimization techniques to ensure efficient grid operability and enhance charging infrastructure accessibility.

2.2 *Grid Operability*

Grid operability is a critical aspect in the context of AI-driven optimization techniques for EV charging infrastructure. As traditional power grids evolve into Smart Grids (SGs) leveraging IoT systems, the seamless interaction between service providers and consumers becomes paramount to address grid congestion and ensure efficient energy flow [7]. Incorporating federated learning (FL) in Intelligent Transportation Systems (ITSs) can enhance decision-making and privacy protection in dynamic vehicle environments, a key concern in optimizing charging infrastructure accessibility. By addressing the challenges of uneven data distribution and limited computing power through collaborative technologies, FL offers a promising solution to improve grid operability and enable effective management of EV charging demands. This convergence of advanced technologies underscores the importance of grid operability in realizing sustainable and efficient EV charging infrastructure.

2.3 *Charging Infrastructure Accessibility*

The challenges of grid congestion and operability in EV charging infrastructures are crucial considerations in the transition towards sustainable transportation systems. However, equally significant is the aspect of charging infrastructure accessibility, which plays a pivotal role in facilitating the widespread adoption of EVs. Big data analytics can revolutionize the management and optimization of EV charging networks, ensuring efficient allocation of resources and improved accessibility for users. Leveraging AI-driven techniques can enhance decision-making processes within transit agencies to address accessibility issues, ultimately contributing to the development of smart transportation systems. By integrating insights from big data analytics and AI, advancements in EV charging infrastructure can be achieved, fostering greater inclusivity and usability for all stakeholders in the transition to a sustainable transport ecosystem.

3 Core Principles of Sustainable Charging Infrastructure

The core principles of sustainable charging infrastructure for EVs encompass a multifaceted approach that delves into various dimensions critical for optimization. Figure 2 depicts the various principles of EV charging infrastructure. Load management stands out as an essential element, necessitating efficient distribution of energy resources to balance demands and prevent grid overloads. Grid integration emerges as another key principle, highlighting the seamless integration of charging infrastructure into existing energy grids for enhanced functionality and sustainability [8]. Moreover, user-centric approaches underscore the importance of aligning infrastructure

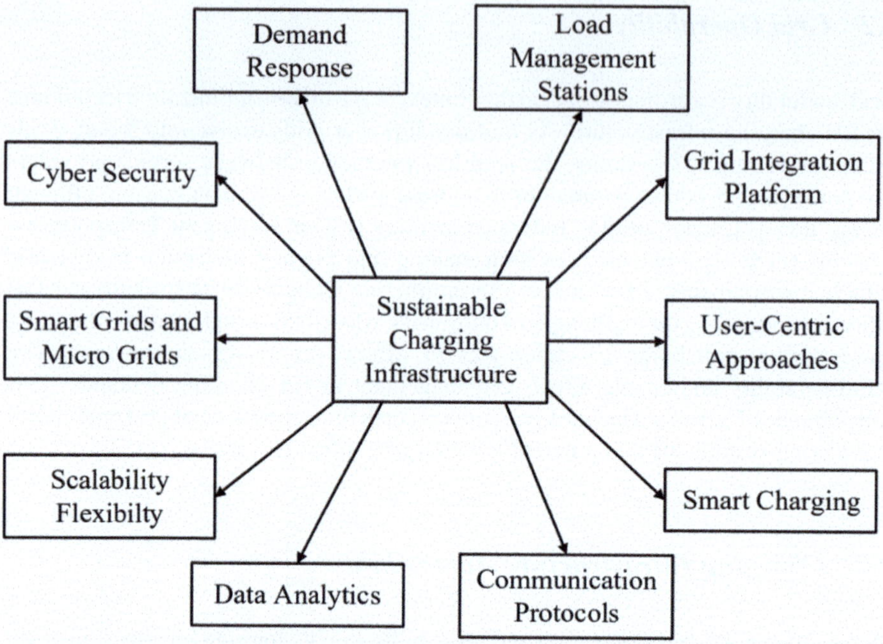

Fig. 2 Components of sustainable charging infrastructure

design with user needs and behaviors to promote widespread adoption and usability. Smart charging stations, communication protocols, and data analytics platforms play pivotal roles in enabling intelligent decision-making and resource allocation within the charging infrastructure ecosystem. Scalability and flexibility further ensure adaptability to evolving technological landscapes and growing EV demands, ultimately fostering a robust and sustainable charging infrastructure network in smart cities.

3.1 Load Management

In exploring sustainable charging infrastructure for EVs, efficient load management plays a crucial role in optimizing electricity consumption and grid stability. The application of big data analytics offers valuable insights into grid-scale and local-scale operations, enhancing decision-making processes through AI and machine learning techniques [9]. By integrating these techniques, EV charging infrastructure can benefit from advanced control and optimization technologies to achieve a balance between supply and demand, effectively leveraging renewable energy sources while ensuring grid reliability. This strategic approach to load management is imperative for the successful implementation of AI-driven optimization techniques in EV charging infrastructure.

3.2 Grid Integration

Expanding on the concept of grid integration within sustainable charging infrastructure, the current challenges stem from the complexities of accommodating increasing penetrations of renewable resources and EVs in distribution systems. California's target of 33% renewable penetration by 2020 necessitates a reevaluation of voltage management methodologies to balance the integration of PV and EV technologies [10]. Integrating big data analytics into power systems offers a promising solution by leveraging AI and machine learning techniques to manage the vast amount of data generated. This approach enables decision-makers to extract valuable insights for optimizing grid operations, enhancing microgrid autonomy, and improving stability assessment. By embracing big data analytics, future power systems can benefit from enhanced grid-scale to local-scale applications, supporting smarter energy management practices and the effective integration of EV charging infrastructure within a sustainable grid framework.

3.3 User-Centric Approaches

Integrating user-centric approaches into the development of AI-driven optimization techniques for EV charging infrastructure is crucial for ensuring the successful adoption and utilization of such systems. Future intelligent applications necessitate traversing the cloud-to-edge-to-IoT continuum seamlessly, requiring coordination between cloud, edge resources, and sensor networks. By implementing a meta-operating system like RAMOS, capable of dynamic, distributed, and trusted interactions, user requirements can be effectively incorporated into the charging infrastructure design process. Furthermore, the evolution of electrically connected autonomous and shared vehicles emphasizes the importance of user-centered design principles and the integration of cognitive capabilities into vehicle domains. These insights underscore the significance of prioritizing user needs and preferences when developing optimization techniques for EV charging infrastructure to enhance efficiency and user experience.

3.4 Smart Charging Stations

EV integration poses challenges for power companies due to the evolving demand landscape, emphasizing the need for innovative solutions in charging infrastructure. The shift towards sustainable power consumption drives the exploration of Smart Charging Stations (SCSs), integrating advanced technologies to optimize charging processes efficiently. The significance of charging infrastructure considerations, particularly in urban and rural areas with varying power supplies, underscores the pivotal role of SCSs in enabling widespread EV adoption [11]. In addition, employing ensemble learning techniques to forecast EV power consumption highlights the practical implementation of AI-driven solutions in managing charging demands. This approach, integrated within SCSs, aims to optimize energy utilization, enhance grid integration, and promote the long-term sustainability of EV operations, aligning with the principles of sustainable charging infrastructure.

3.5 Communication Protocols

The integration of advanced technologies within electric-connected autonomous and shared (ECAS) vehicles highlights the crucial role of communication protocols in optimizing the efficiency and security of EV charging infrastructure [12]. The vulnerability of EV charging systems to cyber-attacks necessitates a robust communication framework capable of proactively identifying and mitigating potential threats. Computational intelligence powered by deep learning can enhance the detection of cyber-physical attacks across both the network and physical layers of the infrastructure. Combining cutting-edge communication protocols with AI-driven optimization techniques enables the EV charging ecosystem to progress toward a more secure and resilient network. Ensuring the interoperability and dependability of communication protocols within smart charging stations is essential for enabling efficient data exchange, real-time monitoring, and adaptive control mechanisms, ultimately facilitating the development of sustainable and scalable EV charging infrastructure.

3.6 Data Analytics Platforms

Drawing on the advancements in IoT devices and AI-powered information systems, the adoption of data analytics platforms within EV charging infrastructure is paramount for achieving sustainability goals. These platforms play a pivotal role in enabling real-time decision-making and optimization of shared mobility fleets, particularly in contested markets where vehicle allocation is crucial. By leveraging data-driven digital twin frameworks, such platforms can address complex design problems like EV Charging Hubs (EVCHs), ensuring scalability and high modeling flexibility. Additionally, the use of geo-tagged IoT data streams can offer actionable insights for policymakers and practitioners in the field of smart charging stations, allowing for the tracking of localized causal effects of mobility policy interventions. Overall, the integration of data analytics platforms not only enhances operational efficiency but also drives the transformation towards a connected, autonomous, shared, and electric (CASE) mobility system.

3.7 Scalability and Flexibility

In the sphere of advancing AI-driven optimization techniques for EV charging infrastructure, the imperative of ensuring scalability and flexibility emerges as a cornerstone for fostering sustainable charging solutions. Insights gleaned from a variety of sources emphasize the critical role of deploying advanced control technologies and robust formal modeling frameworks to bolster the adaptability of charging stations. By harnessing these tools, the integration of scalable neighborhood prosumers individuals who both consume and produce electricity equipped with renewable energy sources can significantly enhance the efficiency and resilience of the infrastructure [13].

Furthermore, the adoption of strategies such as event-driven demand response programs and multi-issue negotiation approaches holds immense promise in optimizing scheduling processes and augmenting overall system performance. Through these means, the dynamic management of vehicle re-positioning schedules and load demand scheduling can be finely tuned, resulting in not only improved operational efficiency but also greater sustainability in the EV charging ecosystem. This concerted effort towards scalability and flexibility lays the groundwork for a robust and adaptable charging infrastructure capable of meeting the evolving needs of electric mobility while minimizing environmental impact.

4 Role of AI in Optimizing Charging Infrastructures

The integration of AI into charging infrastructures represents a transformative leap forward in the evolution of EV charging networks. With its ability to analyze vast amounts of data and make intelligent decisions in real time, AI has great capacity to enhance the performance, dependability, and eco-friendliness of EV charging [14]. At the core of AI-driven charging infrastructures are advanced algorithms and optimization techniques that enable charging station operators to navigate the complexities of energy distribution and demand management. By leveraging AI, charging infrastructure providers can enhance grid integration, ensuring seamless coordination between EV charging stations and the broader electrical grid. This facilitates efficient energy distribution, minimizes grid strain, and enhances overall grid stability.

Moreover, AI empowers charging infrastructure providers to effectively manage peak demand, a critical challenge in the transition to electric mobility. Through predictive analytics and dynamic load management, AI algorithms can forecast peak demand periods and adjust charging schedules accordingly, optimizing resource allocation and minimizing congestion at charging stations. This not only improves the user experience by reducing wait times but also enhances the reliability of charging infrastructure, ensuring that EV owners have access to charging when they need it most. Additionally, AI-driven optimization techniques enhance the sustainability of EV charging by promoting the efficient use of renewable energy sources. By analyzing factors such as weather forecasts, solar radiation, and wind patterns, AI algorithms can optimize charging schedules to coincide with periods of abundant renewable energy generation [15]. This optimizes the use of renewable energy sources, reduces reliance on fossil fuels, and lowers GHG emissions associated with EV charging. Furthermore, AI enables personalized charging experiences tailored to the individual needs and preferences of EV owners. By analyzing user behavior, driving patterns, and historical charging data, AI algorithms can optimize charging schedules to minimize cost, maximize convenience, and extend battery life. This enhances user satisfaction and promotes greater adoption of EVs by addressing common concerns such as range anxiety and charging accessibility.

In summary, the role of AI in optimizing charging infrastructures is pivotal for realizing the full potential of electric mobility. By leveraging AI-driven algorithms and optimization techniques, charging infrastructure providers can enhance grid integration, manage peak demand, promote sustainability, and improve user experience. As the uptake of electric vehicles rapidly increases, the integration of AI into charging infrastructures will be essential for building a resilient, efficient, and sustainable transportation ecosystem.

4.1 Machine Learning Algorithm

Machine learning algorithms play a pivotal role in optimizing charging infrastructures for EVs. These algorithms leverage data-driven approaches to analyze complex patterns and make informed decisions, leading to more efficient and effective charging processes [16]. Several machine learning techniques are employed in this context, each offering unique advantages:

Predictive Modeling: Machine learning algorithms can predict various factors influencing charging infrastructure, such as demand patterns, energy consumption, and grid load. By analyzing historical data, these models forecast future charging needs, enabling operators to optimize resource allocation and scheduling.

Pattern Recognition: EV charging infrastructures generate vast amounts of data, including user behavior, charging patterns, and environmental factors. Machine learning algorithms excel at recognizing patterns within this data, allowing operators to identify trends, anomalies, and opportunities for optimization.

Recommendation Systems: AI-powered recommendation systems can suggest optimal charging strategies based on individual user preferences, traffic conditions, and energy prices [17]. Figure 3 represents the process involved in charging an EV. By considering multiple variables, these systems help users make informed decisions, such as selecting the most cost-effective charging stations or time slots.

Anomaly Detection: Detecting and mitigating anomalies is crucial for ensuring the reliability and security of charging infrastructures. Machine learning algorithms can

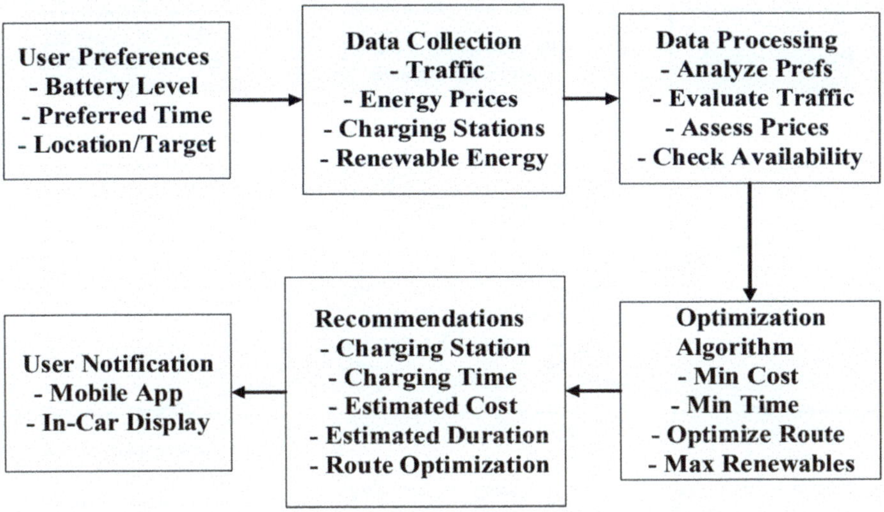

Fig. 3 EV charging optimization process

identify unusual events, such as equipment malfunctions or cybersecurity threats, in real time, enabling prompt intervention and prevention of potential disruptions.

Dynamic Pricing: Machine learning (ML) algorithms enable dynamic pricing strategies based on factors like demand, supply, and grid conditions. By adjusting prices in real-time, charging operators can incentivize users to charge during off-peak hours or at locations with surplus renewable energy, thereby optimizing resource utilization and grid stability [18].

Figure 4 presents the benefits of machine learning algorithms in charging infrastructure. The operators with valuable insights and decision-making capabilities, facilitate the efficient and sustainable management of EV charging networks.

4.2 Optimization Techniques

Optimization techniques driven by AI play a crucial role in maximizing the efficiency and effectiveness of charging infrastructures for EVs. These techniques leverage advanced algorithms to optimize various aspects of charging operations, including resource allocation, scheduling, and grid integration. Some key optimization techniques employed in this context include:

Constraint Optimization: Constraint optimization algorithms ensure that charging operations adhere to various constraints, such as grid capacity limits, user preferences, and environmental regulations. By considering these constraints, optimization

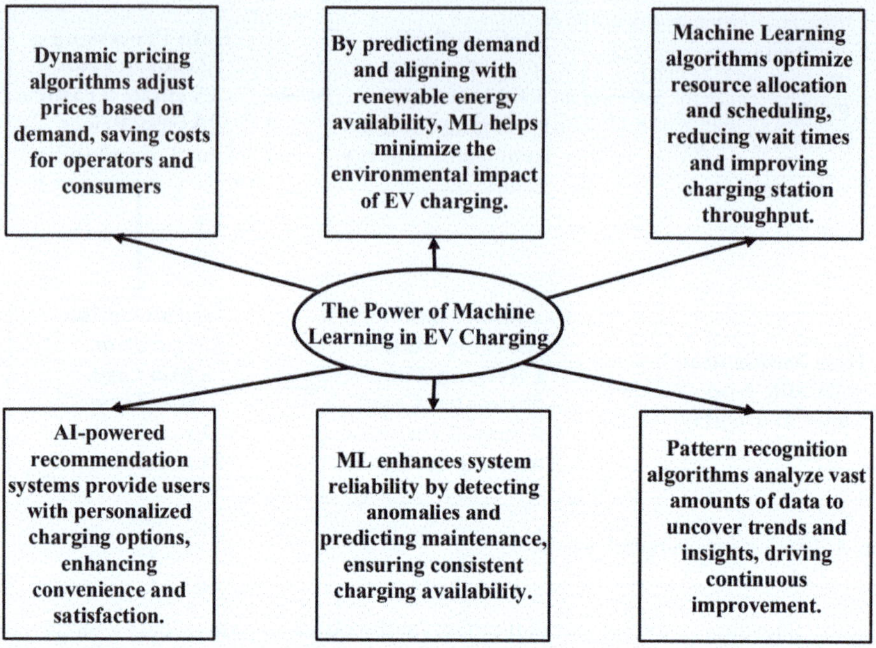

Fig. 4 Benefits of ML algorithms in EV charging infrastructure

algorithms can find feasible charging solutions that meet operational requirements while minimizing resource wastage.

Linear and Non-linear Programming: Linear and non-linear programming techniques are used to formulate and solve optimization problems in charging infrastructures [19]. These techniques enable operators to maximize objectives such as energy efficiency, revenue generation, or grid stability while accounting for complex relationships and constraints.

Evolutionary Algorithms: Evolutionary algorithms, such as genetic algorithms and particle swarm optimization, mimic the process of natural selection to find optimal solutions to charging optimization problems. By iteratively exploring solution spaces and adapting to changing conditions, these algorithms can discover high-quality charging schedules and resource allocations.

Reinforcement Learning: Reinforcement learning algorithms enable charging infrastructures to learn and adapt to dynamic environments through interaction and feedback [20]. By rewarding desirable behaviors and penalizing undesirable ones, these algorithms can optimize charging strategies over time, leading to continuous improvement and adaptation.

Distributed Optimization: In large-scale charging infrastructures with distributed resources and decision-making authority, distributed optimization techniques are employed to coordinate charging activities efficiently. These techniques enable decentralized agents to communicate and collaborate while optimizing global objectives, such as grid stability and user satisfaction.

By leveraging these optimization techniques, AI-driven charging infrastructures can achieve higher levels of efficiency, reliability, and sustainability, paving the path for broad acceptance of electric vehicles and the incorporation of renewable energy sources.

4.3 Predictive Analysis

Predictive analytics stands as a cornerstone in the optimization of charging infrastructures for EVs, offering a powerful toolkit for anticipating and managing future charging demands. Here's a deeper dive into how predictive analytics drives efficiency and sustainability in charging infrastructure.

Historical Data Analysis: Predictive analytics begins by delving into vast datasets of historical charging patterns, energy consumption, grid loads, and environmental conditions [21]. By analyzing this wealth of information, AI algorithms can uncover underlying patterns and trends that provide valuable insights into how charging behavior may evolve over time.

Forecasting Future Demands: Building on historical data analysis, predictive analytics employs sophisticated modeling techniques to forecast future charging demands and behaviors. These models take into account various factors such as seasonal trends, traffic patterns, special events, and changes in user behavior to anticipate when and where charging demand will peak.

Proactive Resource Allocation: Armed with accurate predictions of future charging needs, charging infrastructure operators can proactively allocate resources to meet anticipated demand spikes. Whether it's deploying additional charging stations, adjusting charging rates, or optimizing grid integration, predictive analytics enables operators to stay ahead of the curve and ensure that sufficient charging capacity is available when and where it's needed most.

Grid Integration and Optimization: Predictive analytics also plays a crucial role in optimizing grid integration by forecasting how EV charging will impact grid loads and energy distribution. By coordinating charging schedules to align with periods of low grid congestion or high renewable energy generation, operators can minimize strain on the grid while maximizing the utilization of clean energy sources.

Enhanced Operational Efficiency: By leveraging predictive analytics, charging infrastructure operators can streamline their operations and minimize inefficiencies. Predictive models can help automate tasks such as scheduling maintenance, forecasting revenue streams, and optimizing fleet management, freeing up resources and personnel to focus on delivering better service to users [22].

User Satisfaction and Experience: Ultimately, the goal of predictive analytics in charging infrastructure is to enhance user satisfaction and experience. By accurately anticipating charging needs and optimizing charging processes, operators can provide users with seamless and convenient charging experiences that meet their expectations. This not only fosters user loyalty but also encourages greater adoption of EVs, driving further growth in the EV market.

4.4 Dynamic and Adaptive Charging Strategies

Dynamic and adaptive charging strategies represent a cutting-edge approach to optimizing charging infrastructures for EVs by leveraging AI-driven algorithms to respond in real time to changing conditions [23]. Here's a closer look at how these strategies drive efficiency and sustainability in EV charging networks.

Real-Time Optimization: Dynamic and adaptive charging strategies enable charging stations to continuously monitor and analyze various factors such as energy prices, grid capacity, and user preferences in real time. By dynamically adjusting charging parameters such as rates, schedules, and locations, these systems can optimize charging processes to maximize efficiency and minimize costs.

Capitalizing on Renewable Energy: One of the key benefits of dynamic charging strategies is their ability to capitalize on renewable energy availability. By monitoring fluctuations in renewable energy generation, such as solar or wind power, charging stations can adjust their charging schedules to align with periods of maximum renewable energy output. This not only reduces reliance on fossil fuels but also helps to lower overall energy costs and carbon emissions.

Grid Congestion Management: Dynamic charging strategies play a crucial role in alleviating grid congestion by intelligently distributing charging loads across different locations and time slots. By spreading out charging demand and avoiding peak periods of grid usage, these systems help to prevent grid overloads and ensure a stable and reliable power supply for both EVs and other consumers.

User-Centric Approach: Adaptive charging strategies prioritize user preferences and convenience by offering personalized charging experiences. By analyzing user behavior and preferences, these systems can recommend optimal charging times, locations, and pricing plans tailored to individual needs. This enhances user satisfaction and encourages greater adoption of EVs by providing a seamless and user-friendly charging experience.

Machine Learning Integration: Adaptive charging strategies can further enhance their effectiveness by integrating machine learning techniques. By learning from past charging patterns and user behavior, these systems can continually refine their algorithms and adapt their charging strategies to better meet evolving needs and conditions [24]. This iterative learning process enables adaptive charging systems to improve efficiency, reliability, and user satisfaction over time.

Scalability and Flexibility: Dynamic and adaptive charging strategies are highly scalable and flexible, making them well-suited for a wide range of applications and environments. Whether deployed in urban areas, commercial complexes, or residential communities, these systems can adapt to varying demand profiles, grid constraints, and renewable energy sources, ensuring optimal performance and efficiency across different scenarios.

4.5 Personalized Charging Experience

Personalized charging experiences represent a transformative aspect of AI-driven optimization in charging infrastructures for EVs. By tailoring charging processes to individual user preferences and requirements, personalized charging experiences enhance convenience, efficiency, and user satisfaction. Here's a closer look at how AI enables personalized charging experiences.

User Profiling: AI algorithms analyze user data, including charging history, driving patterns, and preferences, to create detailed user profiles. These profiles capture information such as preferred charging times, locations, and energy requirements, as well as factors like commuting habits and lifestyle preferences.

Customized Charging Plans: Based on the user profiles, AI-powered charging systems develop customized charging plans for each EV owner. These plans consider factors such as optimal charging times to align with users' daily schedules, preferred charging locations for convenience, and energy pricing plans to minimize costs [25].

Adaptive Recommendations: AI continuously monitors user behavior and external factors such as energy prices and grid conditions to provide adaptive recommendations for charging. For example, if a user's schedule changes suddenly, the system may suggest adjusting the charging time accordingly to avoid peak hours or take advantage of off-peak rates.

Smart Energy Management: Personalized charging experiences integrate with smart energy management systems to optimize energy utilization and grid integration. AI algorithms balance the changing needs of individual users with grid constraints and renewable energy availability to maximize efficiency and minimize environmental impact [26].

User Interfaces and Interaction: AI-driven charging systems feature intuitive user interfaces that allow EV owners to interact with the charging infrastructure easily. Users can access personalized charging plans, receive real-time notifications and updates, and adjust settings based on their preferences through mobile apps or web portals.

Data Privacy and Security: While personalizing charging experiences requires access to user data, AI systems prioritize data privacy and security. Advanced encryption techniques and anonymization protocols safeguard sensitive information, ensuring that users' personal data remains confidential and protected from unauthorized access or misuse [27].

Continuous Learning and Improvement: AI algorithms learn from user feedback and interactions to refine personalized charging experiences over time. By analyzing user responses and adjusting recommendations accordingly, these systems continually improve accuracy and relevance, ultimately enhancing user satisfaction and loyalty [28].

Overall, personalized charging experiences powered by AI revolutionize the way EV owners interact with charging infrastructure, offering tailored solutions that meet individual needs while optimizing energy utilization and grid stability. Through adaptive recommendations, smart energy management, and user-centric interfaces, AI-driven charging systems deliver a seamless and intuitive charging experience that enhances convenience, efficiency, and sustainability in the electric mobility ecosystem.

5 Artificial Intelligence in Charging Infrastructure

In the realm of EV charging infrastructure, AI emerges as a formidable tool, poised to revolutionize energy management practices. Through its sophisticated algorithms and data-driven approaches, AI offers a myriad of opportunities to enhance operational efficiency, reduce costs, and optimize energy consumption. One crucial aspect where AI shines is in predictive analytics, where it leverages historical data and real-time information to anticipate future charging needs. By analyzing patterns in charging usage, energy consumption, and grid loads, AI algorithms can accurately forecast upcoming demands, enabling charging operators to allocate resources effectively and ensure seamless charging experiences for EV owners.

Additionally, AI plays a pivotal role in enhancing user experience through personalized charging solutions. By leveraging data on individual preferences, traffic conditions, and charging habits, AI-driven systems can recommend optimal charging strategies tailored to each user's specific needs [29]. These personalized recommendations not only improve user satisfaction but also contribute to overall grid efficiency by distributing charging loads more evenly across different locations and time periods.

Moreover, AI enables smart decision-making at charging stations by continuously analyzing data streams and adjusting charging parameters in real time. This dynamic and adaptive approach allows charging infrastructure to respond swiftly to changing conditions such as energy availability, grid capacity, and environmental factors. By optimizing charging rates, schedules, and locations on the fly, AI-driven systems can maximize the utilization of renewable energy sources, alleviate grid congestion, and minimize operational costs.

5.1 *Load Forecasting*

Load forecasting serves as a fundamental pillar in the optimization of EV charging infrastructure, acting as a proactive tool to anticipate and address future energy demands effectively. This predictive process relies on sophisticated data analytics methodologies, often incorporating AI algorithms, to meticulously analyze historical charging patterns, environmental variables, and user behaviors [30]. By extrapolating from past data and accounting for various influencing factors, load forecasting enables charging infrastructure operators to make well-informed decisions about resource allocation, grid optimization, and enhancing user satisfaction. One of the primary benefits of accurate load forecasting is its role in optimizing resource allocation within charging networks. By predicting when and where energy demand will peak, operators can strategically deploy charging stations and allocate resources efficiently, ensuring that sufficient energy is available to meet the anticipated demand. This proactive approach minimizes the risk of resource shortages or overloads, enhancing the reliability and availability of charging infrastructure.

Figure 5 shows the various uses of load forecasting. Load forecasting plays a crucial role in grid integration, particularly as EV adoption continues to rise. By providing grid operators with insights into future charging patterns, load forecasting facilitates infrastructure planning and optimization of distribution networks. This, in turn, helps to mitigate grid congestion, minimize the risk of overloads, and maintain grid stability, ultimately supporting the seamless integration of EVs into the existing power grid.

Moreover, accurate load forecasting enhances the overall user experience by reducing uncertainty and wait times for EV drivers. By predicting charging demand ahead of time, operators can optimize charging schedules, minimize queuing at charging stations, and ensure convenient access for users [31]. This proactive approach empowers EV drivers to plan their charging sessions more effectively, optimizing their travel schedules and reducing the likelihood of disruptions. Various techniques are employed in load forecasting, ranging from traditional statistical methods to advanced machine learning algorithms. Statistical methods such as time series analysis and regression analysis provide a foundation for load forecasting, while machine learning algorithms like neural networks, support vector machines, and decision trees offer greater flexibility and adaptability to complex data relationships.

Hybrid approaches that combine multiple forecasting techniques are also gaining traction, leveraging the strengths of different methodologies to improve forecast accuracy. These hybrid models integrate diverse data sources, adapt to changing conditions, and offer more robust predictions, thereby enhancing the reliability and effectiveness of load forecasting in EV charging infrastructure. Despite the significant progress made in load forecasting techniques, challenges remain, including

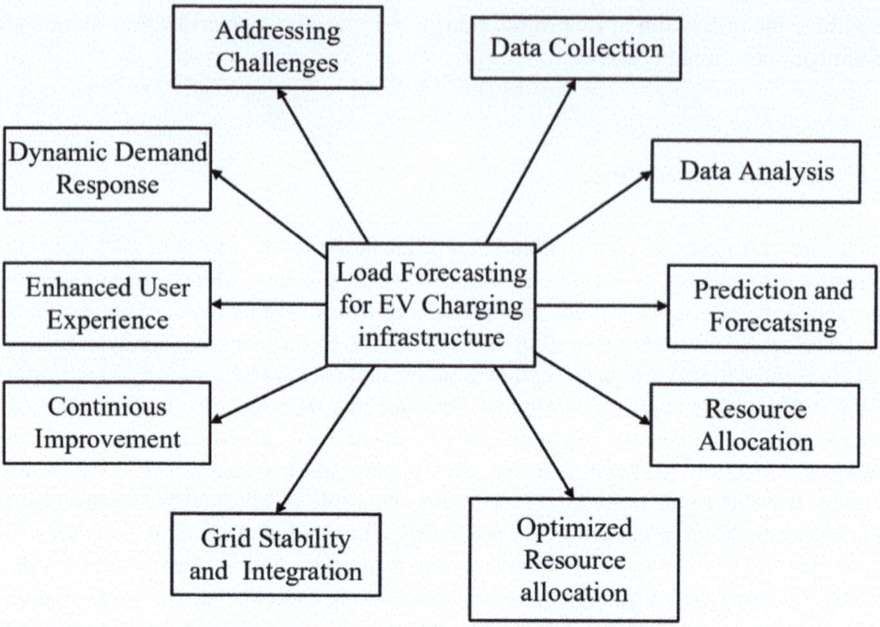

Fig. 5 Load forecasting

data quality issues, uncertainty management, and model scalability. Addressing these challenges will require continued innovation and collaboration among researchers, industry stakeholders, and policymakers to develop robust forecasting models that can support the rapid growth of EV adoption and the transition to sustainable energy systems. Looking ahead, load forecasting is poised to evolve further, driven by advancements in AI, data analytics, and smart grid technologies. By integrating with smart grids, load forecasting can enable dynamic demand response, real-time forecasting, and adaptive optimization strategies, further enhancing the efficiency, reliability, and sustainability of EV charging infrastructure [32]. As the electrification of transportation continues to accelerate, load forecasting will play an increasingly vital role in shaping the future of mobility and energy.

5.2 Demand Response

Demand response is a critical component of optimizing EV charging infrastructure, offering a dynamic approach to managing energy consumption in response to grid conditions, user preferences, and environmental factors. Unlike traditional static charging schedules, demand response enables charging stations to adjust their power consumption in real time based on external signals, such as electricity prices, grid stability, and renewable energy availability. One of the main aims of demand response is to relieve pressure on the grid during periods of peak demand. Encouraging users to move their charging activities to off-peak hours or periods with ample renewable energy generation, demand response programs assist in balancing the supply and demand of energy, reduce grid congestion, and enhance overall grid reliability [33]. This flexibility in charging behavior also enables grid operators to optimize resource allocation and minimize the need for costly infrastructure upgrades.

Various uses of demand response are depicted in Fig. 6. Demand response contributes to the integration of renewable energy sources into the grid by aligning charging activities with periods of high renewable energy generation. By coordinating EV charging with solar or wind power generation peaks, demand response programs facilitate the utilization of clean energy resources, reduce GHG emissions, and support the transition to a more sustainable energy system.

Moreover, demand response offers benefits beyond grid management, including cost savings for EV owners and operators. By charging during off-peak hours or when energy prices are low, users can take advantage of discounted electricity rates and reduce their overall charging costs [34]. Similarly, charging infrastructure operators can participate in demand response programs to earn incentives or payments for providing grid services, such as load shifting or frequency regulation.

Various strategies are employed to implement demand response in EV charging infrastructure, including time-of-use pricing, incentive programs, and smart charging protocols. Time-of-use pricing encourages users to charge their vehicles during periods of lower electricity demand by offering discounted rates during off-peak

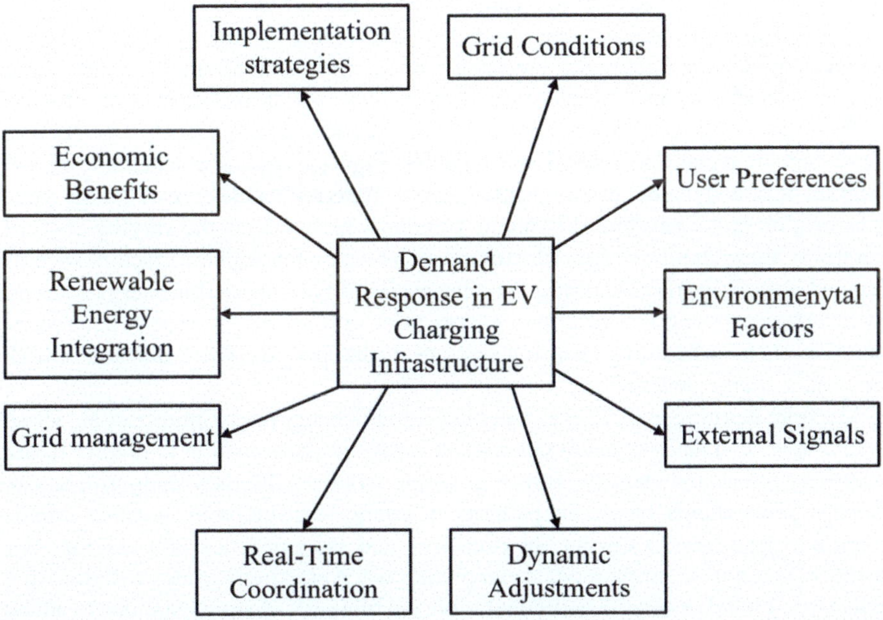

Fig. 6 Demand response

hours. Incentive programs provide rewards or discounts to users who participate in demand response initiatives, motivating them to adjust their charging behavior accordingly. Smart charging protocols enable charging stations to communicate with the grid and respond dynamically to pricing signals, grid constraints, and user preferences, optimizing charging schedules in real time.

Despite its benefits, implementing effective demand response programs in EV charging infrastructure faces challenges such as interoperability, data privacy, and user acceptance. Ensuring seamless communication between charging stations, vehicles, and the grid is essential for enabling real-time coordination and response. Additionally, addressing concerns about data security and privacy is critical to gaining user trust and participation in demand response initiatives. Overall, demand response plays a vital role in optimizing EV charging infrastructure, enhancing grid reliability, promoting renewable energy integration, and providing economic benefits for users and operators alike. As the electrification of transportation continues to expand, demand response will become increasingly essential for balancing energy supply and demand, maximizing grid efficiency, and supporting the transition to a sustainable energy future.

5.3 *Energy Price Optimization*

Energy price optimization is a crucial aspect of managing EV charging infrastructure, aiming to minimize costs and enhance the overall efficiency of energy use. This process involves using advanced algorithms and real-time data to adjust charging activities based on fluctuating electricity prices, grid conditions, and renewable energy availability [35]. By aligning charging times with the most cost-effective energy rates, both consumers and charging infrastructure operators can achieve significant financial and operational benefits. Figure 7 shows a schematic model diagram of how dynamic pricing is used in EV charging. At the core of energy price optimization is the concept of dynamic pricing, which reflects the varying costs of electricity throughout the day. Electricity prices can change due to factors such as demand, supply, market conditions, and the availability of renewable energy sources. By leveraging these price variations, energy price optimization strategies can reduce energy costs by scheduling charging sessions during periods of low demand or high renewable energy generation.

Machine learning and AI play a pivotal role in energy price optimization. These technologies can analyze historical data and real-time information to predict future electricity prices and identify optimal charging windows. Predictive algorithms can forecast price trends based on patterns in energy consumption, weather conditions, and grid load, allowing for proactive and informed decision-making. For instance, machine learning models can predict when electricity prices will drop due to increased wind or solar power generation, enabling EV charging stations to adjust their schedules accordingly [36].

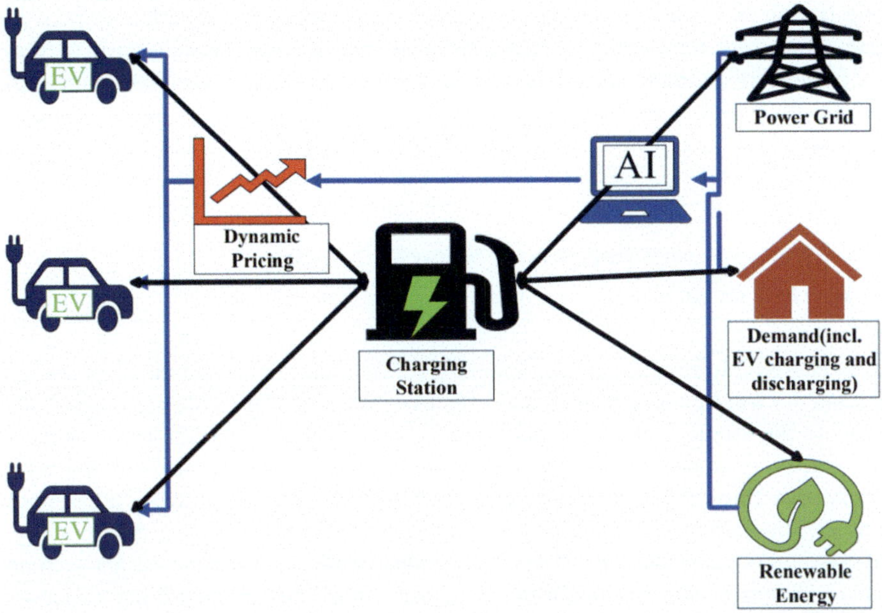

Fig. 7 Dynamic charging model

One of the main advantages of optimizing energy prices is the reduction in costs for EV users. By charging their vehicles during off-peak hours or when energy prices are low, consumers can significantly reduce their electricity bills. This not only makes EV ownership more affordable but also encourages more widespread adoption of EVs. For charging infrastructure operators, optimizing charging times can lead to reduced operational costs and improved profitability. By purchasing electricity at lower rates and avoiding peak pricing periods, operators can manage expenses more effectively and pass on the savings to their customers.

Additionally, energy price optimization contributes to grid stability and the integration of renewable energy sources. By incentivizing EV owners to charge their vehicles during periods of excess renewable energy production, such as midday when solar power generation peaks, the demand on the grid can be balanced more efficiently. This reduces the need for conventional power plants to ramp up production during peak demand times, leading to lower carbon emissions and a cleaner energy mix.

Implementing energy price optimization requires sophisticated infrastructure and communication systems. Smart meters and connected charging stations are essential for monitoring real-time electricity prices and adjusting charging activities accordingly. Furthermore, energy management systems must be capable of processing large volumes of data and making rapid decisions to respond to price changes and grid conditions. However, challenges such as data privacy, regulatory compliance, and user acceptance must be addressed to fully realize the potential of energy price optimization. Ensuring that data is securely transmitted and stored is critical to maintaining user trust and protecting sensitive information. Additionally, regulatory frameworks must support dynamic pricing models and provide clear guidelines for implementation. Educating consumers about the benefits and operational aspects of energy price optimization can also enhance participation and acceptance.

6 Challenges and Benefits of Smart Charging Infrastructure

Smart charging infrastructure encounters several hurdles that must be overcome to facilitate its widespread adoption and effective operation. One of the primary challenges is the strain it places on the existing electrical grid, which may not have the capacity to support the increased demand from EVs. This demand can fluctuate significantly throughout the day and across different locations, making it essential to predict and manage variability for reliable and efficient charging. Additionally, the convenience expectations of EV users must be met, addressing issues such as limited charging station availability, long wait times, and compatibility concerns with different charging standards. Interoperability between charging stations and EV models is another challenge due to the lack of standardized communication protocols and charging connectors. Ensuring scalability to support the growing EV population without overwhelming the grid is crucial, as is addressing concerns related to data security, privacy, and the upfront cost of installation. Collaboration between stakeholders, including government agencies, utility providers, technology companies, and EV manufacturers, will be essential to overcome these obstacles and realize the potential of smart charging infrastructure in supporting the transition to electric mobility and achieving sustainable transportation goals.

In addition to the challenges mentioned, smart charging infrastructure also faces hurdles related to grid integration and renewable energy utilization. Integrating EV charging stations with the grid requires careful coordination to avoid overloading local distribution networks and causing disruptions. This is particularly important as the number of EVs on the roads continues to grow, leading to increased pressure on the grid during peak charging times. Moreover, optimizing the use of renewable energy sources, such as solar and wind power, for EV charging presents its own set of challenges. While renewable energy offers environmental benefits, its intermittent nature and variability pose challenges for reliable and predictable charging operations.

Another significant challenge is the need for regulatory frameworks and policies that support the deployment and operation of smart charging infrastructure. Clear guidelines are needed to address issues such as pricing structures, grid access, interoperability standards, and data privacy. Without supportive policies in place, investment in smart charging infrastructure may be hindered, slowing down the transition to electric mobility. Furthermore, ensuring equitable access to charging infrastructure is essential for promoting the widespread adoption of EVs. Disparities in access to charging infrastructure, particularly in underserved communities, can exacerbate existing inequalities and hinder the transition to clean transportation for all.

6.1 Benefits of Smart Charging Infrastructure

The integration of smart charging infrastructure represents a significant milestone in the evolution of EV adoption and renewable energy integration worldwide. As the world increasingly emphasizes sustainability and reducing carbon emissions, advanced technologies like AI and optimization methods are becoming ever more essential in tackling the intricacies of energy management and grid operations. Smart charging infrastructure offers a holistic solution to the challenges associated with higher penetrations of renewable resources and increasing numbers of EVs on the roads. By leveraging AI-driven optimization techniques, such as predictive analytics and dynamic pricing strategies, charging infrastructure can adapt in real time to fluctuating energy prices, grid conditions, and user preferences, maximizing the utilization of sustainable energy sources and decreasing dependence on non-renewable fuels.

Furthermore, smart charging systems bring numerous benefits to both grid operators and EV owners globally. These systems not only enhance the efficiency of charging processes but also maximize the utilization of renewable energy sources, thereby lowering GHG emissions and mitigating the impacts of climate change. Additionally, the integration of smart charging infrastructure opens up new opportunities for revenue generation and grid services, such as vehicle-to-grid (V2G) technologies, which enable EV batteries to provide ancillary services to the grid during periods of high demand or grid instability. Beyond its technical advantages, smart charging infrastructure contributes to the overall attractiveness and viability of electric mobility on a global scale. By offering seamless and convenient charging experiences, it encourages EV adoption, leading to further reductions in GHG emissions and air pollution worldwide. Moreover, the proliferation of smart charging infrastructure fosters innovation and collaboration across the energy ecosystem, driving advancements in AI, renewable energy integration, and grid modernization globally.

7 Conclusions

The integration of AI into EV charging infrastructure offers transformative solutions to optimize energy usage, grid management, and user experience. AI-driven techniques like reinforcement learning, genetic algorithms, and neural networks enable dynamic optimization of charging networks, improving efficiency and reliability. Personalized charging experiences and real-time adjustments based on factors like energy prices further enhance user convenience and sustainability. Despite challenges like data privacy and interoperability, AI presents a promising path toward a cleaner, more efficient transportation system. With continued innovation and collaboration, AI can play a pivotal role in shaping the future of EV charging infrastructure, promoting sustainability and resilience in transportation electrification. Through AI-driven optimization, EV charging infrastructure can realize its full potential in promoting sustainability, resilience, and efficiency in transportation electrification.

Acknowledgements This research work was supported by "Woosong University's Academic Research Funding—2025".

References

1. Sathyan S, Pandi VR, Antony A, Salkuti SR, Sreekumar P (2024) ANN-based energy management system for PV-powered EV charging station with battery backup and vehicle to grid support. Int J Green Energy 21(6):1279–1294. https://doi.org/10.1080/15435075.2023.22460
2. Ray P, Nayak S, Salkuti SR (2023) Renewable energy based off-grid EV charging station. In: IEEE 3rd international conference on sustainable energy and future electric transportation, pp 1–6. https://doi.org/10.1109/SeFeT57834.2023.10245693
3. Salkuti SR (2023) Advanced technologies for energy storage and electric vehicles. Energies 16(5):2312. https://doi.org/10.3390/en16052312
4. Amamra SA, Marco J (2019) Vehicle-to-grid aggregator to support power grid and reduce EV charging cost. IEEE Access 7:178528–178538. https://doi.org/10.1109/ACCESS.2019.2958664
5. Hawkins TR, Gausen OM, Strømman AH (2012) Environmental impacts of hybrid and Electric Vehicles—a review. Int J Life Cycle Assess 17:997–1014. https://doi.org/10.1007/s11367-012-0440-9
6. Karmaker AK, Ahmed MR (2019) Fuzzy logic based optimization for EV charging station in Bangladesh. In: 1st international conference on advances in science, engineering and robotics technology, pp 1–5. https://doi.org/10.1109/ICASERT.2019.8934804
7. Yenchamchalit K, Kongjeen Y, Bhumkittipich K, Mithulananthan N (2018) Optimal sizing and location of the charging station for plug-in Electric Vehicles using the particle swarm optimization technique. Int Electr Eng Congr 1–4. https://doi.org/10.1109/IEECON.2018.8712336
8. Muniprakash T, Naik PR, Varma RPVS, Supraja R, Lingadhal RV (2022) A review on optimization techniques of battery charging in electric vehicles. In: 2nd international conference on artificial intelligence and signal processing, pp 1–5. https://doi.org/10.1109/AISP53593.2022.9760545
9. Sidharthan AP, Al Arefi S (2021) Optimization of charging in a multi-port EV charging station for emergency vehicle priority fast charging. In: IEEE international women in engineering (WIE) conference on electrical and computer engineering, pp 192–195. https://doi.org/10.1109/WIECON-ECE54711.2021.9829632
10. Kempton W, Tomić J (2005) Vehicle-to-grid power implementation: from stabilizing the grid to supporting large-scale renewable energy. J Power Sources 144(1):280–294. https://doi.org/10.1016/j.jpowsour.2004.12.022
11. Hussien DA, Omran WA, Sharkawy RM (2023) Smart charging of electric vehicles in charging stations. In: 5th international youth conference on radio electronics, electrical and power engineering, vol 5, pp 1–5. https://doi.org/10.1109/REEPE57272.2023.10086885

12. Ahmad F, Ashraf I, Iqbal A, Bilal M (2023) EV fast charging station location and capacity planning model using butterfly optimization algorithm. In: International conference on recent advances in electrical, electronics & digital healthcare technologies, pp 145–150. https://doi.org/10.1109/REEDCON57544.2023.10151370
13. Li Y, Zheng N, Zhang J, Cai Q, Zhong Z, Qian X (2021) Planning model for EV charging station considering battery capacity. In: IEEE 5th conference on energy internet and energy system integration, pp 3768–3771. https://doi.org/10.1109/EI252483.2021.9713435
14. Surve U, Narayana TH, Srinivas S, Ronanki D (2023) Loss minimization of dual active bridge converter through design optimization in CC-CV mode for EV battery charging applications. In: IEEE industry applications society annual meeting, pp 1–6. https://doi.org/10.1109/IAS54024.2023.10406857
15. Al-Alwash HM, Borcoci E, Vochin MC, Balapuwaduge IA, Li FY (2024) Optimization schedule schemes for charging electric vehicles: overview, challenges, and solutions. IEEE Access. https://doi.org/10.1109/ACCESS.2024.3371890
16. Kharra A, Tiwari R, Singh J, Rawat T (2023) Interval optimization technique based multi-objective scheduling of electric vehicles. In: International conference on power, instrumentation, energy and control, pp 1–6. https://doi.org/10.1109/PIECON56912.2023.10085895
17. Aurangzeb M, Xin A, Iqbal S, Habib S, Jan MU, Rehman HU, Mahmood RS (2022) Decentralized based advance optimized scheduling scheme to charge and discharge the electric vehicles. In: 5th international conference on energy conservation and efficiency, pp 1–7. https://doi.org/10.1109/ICECE54634.2022.9758960
18. Prakash TM, Kumar KV, Singh A, Varma RPVS, Supraja R, Naik PR, Lingadhal RV (2022) A review on optimization techniques of charging the battery in EV. In: Second international conference on artificial intelligence and smart energy, pp 799–804. https://doi.org/10.1109/ICAIS53314.2022.9743126
19. Suresh V, Bhaskar MS, Kumar KD, Chandra KB (2023) Dynamic programming-based energy management strategy optimization for PV system hybrid electric vehicles. In: International conference on intelligent and innovative technologies in computing, electrical and electronics, pp 155–160. https://doi.org/10.1109/IITCEE57236.2023.10090989
20. Arif SM, Lie TT, Seet BC, Ayyadi S, Jensen K (2021) Review of EV technologies, charging methods, standards and optimization techniques. Electronics 10(16):1910. https://doi.org/10.3390/electronics10161910
21. Amjad M, Ahmad A, Rehmani MH, Umer T (2018) A review of electric vehicles charging: from the perspective of energy optimization, optimization approaches, and charging techniques. Transp Res Part D Transp Environ 62:386–417. https://doi.org/10.1016/j.trd.2018.03.006
22. Jia Z, Li J, Zhang XP, Zhang R (2022) Review on optimization of forecasting and coordination strategies for EV charging. J Modern Power Syst Clean Energy 11(2):389–400. https://doi.org/10.35833/MPCE.2021.000777
23. Hussain S, Thakur S, Shukla S, Breslin JG, Jan Q, Khan F, Ahmad I, Marzband M, Madden MG (2022) A heuristic charging cost optimization algorithm for residential charging of EVs. Energies 15(4):1304. https://doi.org/10.3390/en15041304
24. Lazari V, Chassiakos A (2023) Multi-objective optimization of EV charging station deployment using genetic algorithms. Appl Sci 13(8):4867. https://doi.org/10.3390/app13084867
25. Barakat S, Osman AI, Tag-Eldin E, Telba AA, Mageed HMA, Samy MM (2024) Achieving green mobility: multi-objective optimization for sustainable EV charging. Energ Strat Rev. https://doi.org/10.1016/j.esr.2024.101351
26. Alshammari NF, Samy MM, Barakat S (2023) Comprehensive analysis of multi-objective optimization algorithms for sustainable hybrid EV charging systems. Mathematics. https://doi.org/10.3390/math11071741
27. Sun B, Huang Z, Tan X, Tsang DHK (2018) Optimal scheduling for EV charging with discrete charging levels in distribution grid. IEEE Trans Smart Grid 9(2):624–634. https://doi.org/10.1109/TSG.2016.2558585
28. Al-Ogaili SA, Hasim TJT, Rahmat NA, Ramasamy AK, Marsadek MB, Faisal M, Hannan MA (2019) Review on scheduling, clustering, and forecasting strategies for controlling EV

charging: challenges and recommendations. IEEE Access 7:128353–128371. https://doi.org/10.1109/ACCESS.2019.2939595
29. Chen L, Huang X, Zhang H, Luo Y (2018) A study on coordinated optimization of EV charging and charging pile selection. Energies 11(6):1350. https://doi.org/10.3390/en11061350
30. Mkahl R, Nait-Sidi-Moh A, Gaber J, Wack M (2015) A linear optimization approach for electric vehicles charging. In: 3rd international renewable and sustainable energy conference, Marrakech, Morocco, pp 1–6. https://doi.org/10.1109/IRSEC.2015.7455049
31. Rahman I, Mohamad-Saleh J (2018) Plug-in EV charging optimization using bio-inspired computational intelligence methods. In: Amini M, Boroojeni K, Iyengar S, Pardalos P, Blaabjerg F, Madni A (eds) Sustainable interdependent networks. studies in systems, decision and control, vol 145. Springer, Cham. https://doi.org/10.1007/978-3-319-74412-4_9
32. Das MK, Jain SK (2021) Review on optimization techniques used for scheduling of EV charging. In: International conference on control, automation, power and signal processing (CAPS), Jabalpur, India, pp 1–6. https://doi.org/10.1109/CAPS52117.2021.9730621
33. Nivas M, Naidu RKPR, Mishra DP, Salkuti SR (2022) Modeling and analysis of solar powered EV. Int J Power Electron Drive Syst 13(1):480–487. https://doi.org/10.11591/ijpeds.v13.i1
34. Debnath B, Biswas S, Uddin MF (2020) Optimization of EV charging to shave peak load for integration in smart grid. In: IEEE region 10 symposium (TENSYMP), Dhaka, Bangladesh, pp 483–488. https://doi.org/10.1109/TENSYMP50017.2020.9231029
35. Al-Saadi M, Mathes M, Käsgen J, Robert K, Mayrock M, Mierlo JV, Berecibar M (2022) Optimization and analysis of EV operation with fast-charging technologies. World EV J 13(1):20. https://doi.org/10.3390/wevj13010020
36. Tian J, Lv Y, Zhao Q, Gong Y, Li C, Ding H, Yu Y (2022) EV charging load prediction considering the orderly charging. Energy Rep 8(14):124–134. https://doi.org/10.1016/j.egyr.2022.10.068

AI-Driven Innovations in Electric Vehicle Energy Management: Challenges and Opportunities

Debani Prasad Mishra, Abhipsha Dash, Ayush Kumar, and Surender Reddy Salkuti

Abstract When artificial intelligence (AI) enters into the grid, in electric vehicles (EVs), and smart energy systems, it brings with it a new era of sustainable transportation. The objective of this chapter is to explore how AI technologies optimize integrated smart energy systems within EVs with a view of their transformative potential and the key challenges involved. Energy management algorithms driven by AI allow EVs to adaptively regulate their power use and charging patterns based on real-time data from sources such as traffic conditions, customer preferences, and network dynamics. By applying predictive analytics, EVs can predict energy demand, optimize routing options, and minimize battery charging costs leading to maximum energy efficiency while reducing environmental pollution. Proactive maintenance which prolongs battery life and enhances overall vehicle reliability is also made possible with battery management systems enabled by AI. Furthermore, AI facilitates seamless integration between the electric grid and each individual EV via vehicle-to-grid (V2G) technologies. This interconnection between EVs and the grid optimizes resource usage, enhances grid stability, and promotes a shift towards a decentralized and renewable energy-driven future. Furthermore, AI enhances the user experience by providing personalized recommendations on charging stations, routes, and energy-efficient driving practices. However, as AI becomes increasingly integrated into smart energy systems for EVs, robust cybersecurity measures are essential to protect against potential threats and ensure data privacy. To fully harness the potential of AI in developing sustainable transportation solutions, it is crucial to foster interdisciplinary collaboration, establish regulations, and engage stakeholders.

D. P. Mishra · A. Dash · A. Kumar
Department of Electrical and Electronics Engineering, IIIT Bhubaneswar, Bhubaneswar, Odisha, India
e-mail: debani@iiit-bh.ac.in

S. R. Salkuti (✉)
Department of Railroad and Electrical Engineering, Woosong University, Daejeon, Republic of Korea
e-mail: surender@wsu.ac.kr

S. R. Salkuti (ed.), *Artificial Intelligence for Integrated Smart Energy Systems in Electric Vehicles*, Lecture Notes in Electrical Engineering 1427,
https://doi.org/10.1007/978-3-031-94276-1_14

Keywords Artificial intelligence · Electric vehicles · Smart energy systems · Energy management · Grid integration · Vehicle · To · Grid (V2G)

Nomenclature

EV	Electric Vehicle
AI	Artificial Intelligence
V2G	Vehicle-to-grid
IoT	Internet of Things
CNN	Convolutional Neural Network
RNN	Recurrent Neural Network
ANN	Artificial Neural Network
ML	Machine Learning
ACC	Adaptive Cruise Control
SOC	State of Charge
HVAC	Heating, Ventilation, and Air Conditioning
ADAS	Advanced Driver Assistance Systems
CHAdeMO	CHArge de MOve
CCS	Combined Charging System

1 Introduction

In recent years, the convergence of Artificial Intelligence (AI) and Electric Vehicle (EV) technology has sparked a revolution in sustainable transportation. This fusion promises not only to redefine mobility perception but also to mitigate environmental challenges posed by traditional fossil fuel-powered vehicles significantly. Standing on the cusp of this transformative era, understanding the intricate interplay between AI and EVs within the context of smart energy systems becomes paramount [1]. The advent of AI in the electric vehicle landscape heralds a paradigm shift, propelling us towards a future where transportation is not merely a means of getting from point A to point B but a dynamic ecosystem of interconnected devices, systems, and data streams. At the heart of this transformation lies the concept of energy management—a cornerstone upon which the sustainability and efficiency of EVs hinge.

This chapter embarks on a comprehensive exploration of the multifaceted role that AI plays in optimizing integrated smart energy systems within EVs. Delving into the transformative potential and inherent challenges, this exploration endeavors to provide a holistic understanding of how AI technologies are reshaping the landscape of sustainable transportation. As the world grapples with the pressing need to decarbonize the transportation sector, EVs emerge as a beacon of hope, offering a cleaner, greener alternative to their fossil fuel counterparts. However, the widespread adoption of EVs hinges not only on advancements in battery technology and infrastructure but also on the ability to intelligently manage energy resources. This is where AI steps in as a catalyst for innovation, revolutionizing the way EVs consume, store, and distribute energy.

Energy management algorithms powered by AI represent the cornerstone of this transformation, empowering EVs to adaptively regulate their power usage and charging patterns in real time. By harnessing data from diverse sources such as traffic conditions, customer preferences, and network dynamics, these algorithms optimize energy consumption, minimize charging costs, and enhance overall efficiency. Through predictive analytics, EVs can anticipate energy demand, optimize routing options, and mitigate the environmental footprint associated with traditional transportation modes. Moreover, AI-enabled battery management systems herald a new era of proactive maintenance, ensuring the longevity and reliability of EVs. By continuously monitoring battery health, diagnosing potential issues, and implementing preventive measures, these systems optimize performance while minimizing downtime—a crucial factor in enhancing consumer confidence and accelerating EV adoption. Beyond the realm of individual EVs, AI facilitates seamless integration with the broader electric grid through innovative technologies such as the Vehicle-to-Grid (V2G) system.

At its core, V2G represents a symbiotic relationship wherein EVs serve as both consumers and producers of electricity, effectively transforming them into mobile energy storage units. Through bidirectional energy exchange, V2G systems enable a myriad of grid services, ranging from stabilization and demand response to renewable energy integration [2]. By leveraging the aggregate energy storage capacity of EVs, utilities can enhance grid reliability, mitigate intermittency issues associated with renewable energy sources (RESs), and pave the way for a more resilient and sustainable energy infrastructure. The connection between EVs and the power grid not only optimizes resource utilization but also fosters a decentralized energy ecosystem wherein consumers actively participate in the energy marketplace. This democratization of energy holds the promise of empowering individuals, reducing dependency on centralized utilities, and fostering greater energy independence. In addition to optimizing energy management and grid integration, AI revolutionizes the user experience, making EV ownership more intuitive, convenient, and personalized.

Through the deployment of AI-driven applications, EV users gain access to a plethora of features aimed at optimizing their driving experience and maximizing energy efficiency. From personalized recommendations on charging stations and routes to real-time feedback on driving behavior, AI empowers users to make informed decisions that align with their preferences and environmental goals. By leveraging machine learning algorithms, EVs can adapt to individual driving patterns, optimize energy consumption, and provide actionable insights for eco-friendly driving practices. Moreover, AI-driven infotainment systems enhance connectivity, entertainment, and productivity, transforming EVs into integrated hubs of digital experience. Through seamless integration with smart devices, voice assistants, and IoT ecosystems, EVs offer a glimpse into the future of connected mobility wherein passengers are seamlessly immersed in a personalized digital ecosystem. However, amidst the promise of AI-driven innovation, addressing the inherent challenges and risks associated with this technological convergence is imperative.

As EVs become increasingly interconnected and reliant on AI algorithms, cybersecurity emerges as a paramount concern, necessitating robust measures to safeguard against potential threats and vulnerabilities. Furthermore, the proliferation of AI in smart energy systems underscores the need for comprehensive regulatory frameworks that ensure ethical, transparent, and responsible deployment. By establishing guidelines for data privacy, algorithmic accountability, and cybersecurity standards, policymakers can foster an environment conducive to innovation while safeguarding the interests of consumers and stakeholders. In conclusion, the fusion of AI and smart energy systems in EVs represents a watershed moment in the evolution of sustainable transportation [3]. By optimizing energy management, empowering grid integration, enhancing user experience, and addressing cybersecurity concerns, AI-driven solutions can revolutionize how energy is perceived, consumed, and interacted with. As the future moves towards clean, efficient, and equitable transportation, collaboration across disciplines, stakeholders, and regulatory bodies becomes indispensable. Harnessing the transformative power of AI can unlock new pathways towards a sustainable mobility ecosystem that is not only environmentally responsible but also socially inclusive and economically viable.

Delving deeper into the intricate nuances of AI applications in EV energy management unravels the complexities, opportunities, and challenges that lie ahead. Embark on a quest to redefine the future of transportation through the lens of artificial intelligence and smart energy systems.

2 AI-Driven Energy Management and EV Integration Dynamics

In the pursuit of sustainable transportation solutions, the intersection of AI and EV technology presents a myriad of opportunities and challenges. At the forefront of this convergence lies the imperative to address complex and multifaceted problems inherent to the integration of AI-driven smart energy systems within the EV ecosystem [4]. Embarking on a journey towards a greener and more efficient mobility paradigm necessitates delineating and articulating the key problem areas that require attention and innovative solutions.

2.1 *Optimization of Energy Management Systems*

Central to the efficient operation of EVs is the optimization of energy management systems, which encompass a wide array of algorithms, predictive analytics, and real-time data processing capabilities. However, the challenge lies in developing robust

AI-driven algorithms that can adaptively regulate power usage, charging patterns, and energy consumption in response to dynamic variables such as traffic conditions, grid demand, and user preferences. The optimization of energy management systems must strike a delicate balance between maximizing energy efficiency, minimizing charging costs, and ensuring optimal vehicle performance and reliability.

2.2 Integration with Smart Grids and Vehicle-to-Grid (V2G) Technologies

The integration of EVs into the broader electric grid through V2G technologies holds immense potential for grid stabilization, demand response, and renewable energy integration. Nevertheless, the scalability and interoperability of V2G systems pose significant challenges, particularly in ensuring seamless bidirectional energy exchange between EVs and the grid. Moreover, the implementation of V2G technologies necessitates addressing regulatory barriers, standardization issues, and market incentives to incentivize widespread adoption and participation.

2.3 Cybersecurity Protocols and Data Privacy Mechanisms

As EVs become increasingly connected and reliant on AI algorithms for energy management and grid interaction, cybersecurity emerges as a critical concern that must be addressed proactively. The proliferation of IoT devices, telematics systems, and vehicle-to-infrastructure communication channels introduces new attack vectors and vulnerabilities [5], making EVs susceptible to cyber threats such as data breaches, malware attacks, and remote hijacking. Moreover, ensuring data privacy and protecting sensitive information collected from EVs poses ethical and regulatory challenges that require robust encryption protocols, access controls, and privacy-enhancing technologies.

2.4 Regulatory Frameworks and Policy Implications

The deployment of AI-driven smart energy systems in EVs necessitates the development of comprehensive regulatory frameworks and policy guidelines to ensure the ethical, transparent, and responsible use of these technologies. However, navigating the regulatory landscape poses challenges due to the complex interplay between technological innovation, industry standards, and government regulations. Furthermore, the lack of harmonization across jurisdictions and disparate regulatory approaches hinders the adoption and deployment of AI-driven solutions, necessitating collaborative efforts among policymakers, industry stakeholders, and research institutions.

2.5 User Acceptance and Adoption Dynamics

Despite the potential benefits offered by AI-enabled EVs, the widespread adoption and acceptance of these technologies among consumers remain contingent upon addressing user concerns, preferences, and adoption barriers. Factors such as range anxiety, charging infrastructure availability, and upfront costs impede the mainstream adoption of EVs, while concerns regarding data privacy, cybersecurity, and AI transparency further exacerbate consumer apprehensions. Thus, fostering user trust, awareness, and engagement through education, outreach, and user-centric design becomes imperative to accelerate the transition toward sustainable mobility solutions. In essence, the convergence of AI and smart energy systems in electric vehicles presents a formidable yet surmountable set of challenges that require interdisciplinary collaboration, innovation, and stakeholder engagement. By addressing the complexities inherent to energy management optimization, grid integration, cybersecurity, regulatory frameworks, and user acceptance, the transformative potential of AI-driven sustainable transportation solutions can be unlocked [6]. Navigating this ever-evolving landscape requires vigilance, adaptability, and a proactive approach toward creating a future where mobility is not only efficient and sustainable but also inclusive and equitable.

2.6 Strategic Objectives in AI-Driven Energy Management

Incorporating AI into EV energy management systems marks a significant step forward in improving the effectiveness, eco-friendliness, and user satisfaction of electric transportation. This integration primarily aims to boost energy efficiency. AI algorithms analyze current data on driving habits, traffic situations, and battery conditions to dynamically optimize energy use and charging schedules, reducing waste and maximizing energy efficiency in EVs. This not only saves energy but also results in noticeable cost reductions for EV users. Moreover, AI contributes to prolonging the lifespan of EV batteries by predicting maintenance needs and intelligently managing battery usage. By continuously monitoring battery health and performance metrics, AI algorithms can predict potential issues before they escalate, enabling timely interventions to mitigate degradation and optimize charging and discharging cycles to minimize stress on the battery cells, thereby extending their longevity [7]. This proactive approach to battery management not only enhances the sustainability of EVs by reducing the frequency of battery replacements but also mitigates associated environmental impacts.

Moreover, AI-powered V2G systems mark a revolutionary phase in grid integration, allowing electricity to flow in both directions between EVs and the grid. These systems utilize EVs as movable energy storage units capable of both drawing power from and supplying excess energy to the grid. AI algorithms dynamically oversee V2G operations, considering grid demand, energy costs, and the presence of renewable energy, thus assisting in grid stability and fostering the assimilation of RESs [8]. This, in turn, enhances the resilience and sustainability of energy infrastructure. In addition to optimizing energy efficiency and grid integration, AI enhances the overall user experience for EV drivers and passengers through personalized recommendations for charging stations, routes, and energy-saving driving behaviors. By analyzing historical data and user preferences, AI tailors recommendations to individual needs, preferences, and driving habits, thereby enhancing user satisfaction and convenience. Moreover, AI-powered features such as voice recognition, natural language processing, and adaptive user interfaces contribute to a seamless and intuitive interaction between users and EV systems, further enhancing the overall user experience.

Lastly, AI-driven diagnostics and maintenance mechanisms play a pivotal role in ensuring the reliability and performance of EVs. By continuously monitoring vehicle components, sensors, and subsystems, AI algorithms detect anomalies, malfunctions, and potential failures in real time, enabling proactive maintenance interventions to prevent minor issues from escalating into major breakdowns and minimizing vehicle downtime. Moreover, AI-based predictive maintenance schedules optimize service intervals based on vehicle usage patterns and component health, ensuring timely interventions while minimizing unnecessary maintenance costs. In conclusion, the integration of AI into EV energy management systems represents a paradigm shift towards smarter, cleaner, and more sustainable mobility, ushering in a new era of transportation that is not only efficient and reliable but also user-centric and environmentally conscious.

2.7 Systemic Challenges in AI Implementation

The promising potential of AI in EV energy management is tempered by several significant challenges that demand attention. Foremost among these hurdles is the complexity inherent in processing and analyzing vast quantities of real-time data sourced from diverse streams, including traffic conditions, grid dynamics, and user behaviors. Effectively harnessing this data deluge demands sophisticated AI algorithms capable of rapid and nuanced decision-making. Additionally, accurately predicting and mitigating battery degradation poses a formidable obstacle. Addressing this challenge necessitates the development of advanced AI models coupled with continuous monitoring systems to track battery health and performance metrics over time [9]. Moreover, the existing infrastructure may prove inadequate to meet the demands of AI-enabled EV systems, necessitating substantial upgrades and investment to accommodate the advanced computational and connectivity requirements.

Overcoming these challenges will require concerted efforts from stakeholders across industries, including researchers, policymakers, and technology developers, to ensure the seamless integration of AI into EV energy management while maximizing its benefits for sustainability, efficiency, and user experience. Despite these hurdles, the transformative potential of AI in revolutionizing EV energy management remains undiminished, promising a future where electric transportation is not only environmentally sustainable but also technologically advanced and user-friendly.

2.8 Regulatory Landscape for AI in EVs

The regulatory landscape stands as a pivotal determinant in shaping the trajectory of AI adoption within EV energy management, wielding influence over its adoption, implementation, and eventual integration into mainstream usage. A fundamental challenge arises from the absence of standardized protocols governing AI applications in EVs, a void that threatens to impede widespread adoption and hinder interoperability. Without universally accepted standards, manufacturers may encounter obstacles in developing compatible systems, leading to fragmentation and inhibiting the seamless exchange of data and functionalities between different EV models and AI platforms. Additionally, ensuring compliance with regional and international regulations pertaining to safety, emissions, and data privacy emerges as a paramount concern. As AI systems become increasingly embedded within EV infrastructure, stakeholders must navigate a complex regulatory landscape to ensure adherence to stringent safety standards and mitigate risks associated with data security and privacy breaches. Furthermore, governmental support through incentives and policies plays an indispensable role in fostering an environment conducive to AI and EV integration [10]. Policy frameworks that incentivize research and development initiatives, promote collaboration between public and private sectors, and provide funding for innovation are instrumental in accelerating progress towards the realization of AI-driven advancements in EV energy management.

By fostering an ecosystem of innovation and collaboration, policymakers can stimulate investment, spur technological advancements, and propel the widespread adoption of AI-enabled solutions that promise to revolutionize the efficiency, sustainability, and user experience of electric transportation. In this regard, regulatory bodies wield significant influence in shaping the future trajectory of AI in EV energy management, and their proactive engagement is imperative to unlock the full potential of these transformative technologies for the benefit of society and the environment. As stakeholders navigate the intricate regulatory landscape, collaboration and dialogue between industry players, policymakers, and regulatory bodies are essential to develop robust frameworks that foster innovation while safeguarding against potential risks and ensuring compliance with evolving regulatory requirements [11]. Ultimately, a harmonized approach to regulation—one that balances innovation with accountability and addresses emerging challenges—is essential to unlocking the transformative potential of AI in driving the transition towards a sustainable and electrified transportation ecosystem.

2.9 *Technological Deficiencies and Gaps*

Realizing the full potential of AI in EV energy management necessitates bridging several significant technological gaps that currently impede its seamless integration and widespread adoption. At the forefront of these challenges lies the imperative to develop more advanced and adaptive AI algorithms capable of handling the complexity and variability inherent in real-time data streams. The dynamic nature of data sourced from diverse sources such as traffic conditions, grid dynamics, and user behaviors demands algorithms that can swiftly analyze, interpret, and respond to emerging patterns and trends, optimizing energy consumption and charging patterns in EVs with precision and efficiency.

Moreover, the seamless integration of AI technologies with existing EV and grid infrastructure poses a formidable challenge, requiring innovative solutions that can harmonize disparate systems and protocols [12]. This entails developing interoperable frameworks and standards that facilitate the seamless exchange of data and functionalities between AI-powered EVs and the broader energy ecosystem, thereby unlocking synergies and maximizing efficiency gains. Furthermore, scalability emerges as a critical consideration in ensuring the widespread applicability and adoption of AI solutions in EV energy management.

As the number of EVs on the roads continues to proliferate and data volumes surge, AI systems must demonstrate the capacity to scale seamlessly, accommodating the evolving needs and demands of a rapidly expanding user base. This necessitates robust infrastructure and architectures that can handle increasing computational loads and data processing requirements without compromising performance or reliability. By addressing these technological gaps, stakeholders can unlock the transformative potential of AI in revolutionizing EV energy management and driving innovation, sustainability, and efficiency in the transportation sector. Collaboration and concerted efforts from industry players, researchers, policymakers, and regulatory bodies are imperative to surmount these challenges and pave the way for a future where AI-powered EVs play a central role in shaping a cleaner, smarter, and more sustainable mobility ecosystem [13]. As technological advancements continue to evolve and mature, the vision of AI-enabled EV energy management as a cornerstone of the transportation revolution draws closer, offering unprecedented opportunities for societal and environmental impact on a global scale.

2.9.1 Interoperability Challenges in Integrated Systems

Interoperability stands as a formidable hurdle in the journey towards harnessing the full potential of AI in EV energy management, presenting a complex web of challenges that must be navigated to achieve seamless integration and widespread adoption. At the heart of this challenge lies the existence of diverse standards and protocols across different manufacturers and regions, creating a fragmented landscape that complicates interoperability efforts.

With each manufacturer adhering to its own set of standards and specifications, achieving cohesive collaboration and data exchange becomes a daunting task, stifling innovation and hindering progress toward a unified ecosystem. Moreover, ensuring compatibility between AI systems and a diverse array of EV models and charging infrastructure presents an additional layer of complexity. The intricate interplay between hardware, software, and communication protocols necessitates meticulous attention to detail to ensure that AI-driven solutions can seamlessly interface with existing EV platforms and charging stations, irrespective of their make or model. Achieving this level of compatibility is paramount for fostering broad adoption and unlocking the transformative potential of AI in optimizing energy management across the EV landscape. Collaboration and coordination among stakeholders ranging from manufacturers and technology providers to regulatory bodies and standards organizations are essential to surmounting these interoperability challenges. By establishing common frameworks, protocols, and interfaces, stakeholders can streamline integration efforts, foster cross-compatibility, and pave the way for a more interconnected and interoperable ecosystem.

Additionally, ongoing efforts to promote standardization and harmonization across industry sectors are crucial in addressing the complexities of interoperability, fostering innovation, and driving the evolution of AI-enabled solutions for EV energy management [14]. As the transportation landscape continues to evolve and embrace electrification, overcoming interoperability hurdles is imperative to realize the vision of a seamlessly integrated, intelligent mobility ecosystem powered by AI-driven insights and innovations. Through collaborative endeavors and a commitment to interoperability, stakeholders can unlock new opportunities for efficiency, sustainability, and user experience in the realm of electric transportation, driving positive societal and environmental impacts on a global scale.

2.9.2 Data Security and Privacy Safeguards

As AI systems become increasingly intertwined with EV energy management, the paramount importance of ensuring the security and privacy of data emerges as a critical consideration, demanding meticulous attention and robust safeguards to mitigate potential risks and vulnerabilities. AI systems, reliant on vast quantities of data for decision-making, are inherently susceptible to cybersecurity threats that could compromise vehicle and grid security, posing significant risks to the integrity and safety of the entire transportation ecosystem. From malicious actors seeking to infiltrate AI algorithms and manipulate critical functions to vulnerabilities in communication channels and data storage systems, the spectrum of cyber threats confronting AI-enabled EVs is diverse and multifaceted, requiring comprehensive strategies for detection, prevention, and response.

Moreover, safeguarding the privacy of user data represents a pressing concern, necessitating stringent measures to protect sensitive information from unauthorized access and misuse [15]. With AI systems accessing and analyzing a plethora of personal data, including driving habits, location history, and energy consumption patterns, ensuring compliance with data protection regulations and ethical principles is paramount to maintaining user trust and confidence. By implementing robust encryption protocols, access controls, and authentication mechanisms, stakeholders can fortify AI systems against cyber threats and bolster data privacy safeguards, thereby fostering a secure and trustworthy environment for the deployment and utilization of AI-driven solutions in EV energy management. Collaboration among industry players, cybersecurity experts, policymakers, and regulatory authorities is essential to develop and implement holistic frameworks and standards that address the complex interplay between AI, cybersecurity, and data privacy. Through collective efforts and a commitment to best practices, stakeholders can navigate the evolving landscape of cybersecurity threats and privacy concerns, safeguarding the integrity, reliability, and privacy of AI-enabled EV systems while unlocking the transformative potential of intelligent mobility solutions for a cleaner, safer, and more sustainable future.

As AI continues to evolve and permeate every aspect of the transportation ecosystem, proactive measures to address cybersecurity and privacy challenges are imperative to ensure the resilience and viability of AI-driven innovations in driving the transition toward a smarter, more connected, and more secure mobility paradigm.

2.9.3 Barriers to User Acceptance and Adoption

For the successful integration of AI in EV energy management, addressing user acceptance and adoption barriers emerges as a pivotal imperative, requiring concerted efforts to foster trust, alleviate cost concerns, and enhance awareness and education among consumers [16]. Central to this endeavor is the imperative of building user trust in AI systems, particularly concerning safety and reliability. Given the transformative nature of AI-driven technologies and their integral role in shaping the future of electric transportation, instilling confidence in users regarding the efficacy and dependability of AI-enabled EV systems is paramount. This entails transparent communication about the robustness of AI algorithms, the rigorous testing and validation processes employed to ensure system reliability, and the safeguards in place to mitigate potential risks and vulnerabilities. Moreover, addressing cost concerns represents a significant barrier to adoption for many consumers, as the initial investment required for AI-enabled EV systems may be perceived as prohibitive.

To overcome this hurdle, stakeholders must explore strategies to reduce the cost of AI integration, such as incentivizing subsidies, tax credits, and financing options that make AI-enabled EVs more accessible and affordable to a broader segment of the population. Furthermore, raising awareness and enhancing education about the benefits and functionalities of AI in EV energy management is essential to drive adoption [17]. By disseminating information through targeted outreach campaigns, educational programs, and interactive demonstrations, stakeholders can empower consumers with the knowledge and insights needed to make informed decisions about adopting AI-enabled EV technologies. This includes highlighting the potential cost savings, environmental benefits, and enhanced user experience offered by AI-driven solutions, as well as addressing any misconceptions or reservations that may exist. Through collaborative efforts involving manufacturers, policymakers, advocacy groups, and industry associations, stakeholders can collectively work towards overcoming user acceptance and adoption barriers, paving the way for the widespread adoption and utilization of AI in shaping a more sustainable, efficient, and intelligent mobility ecosystem.

By fostering trust, alleviating cost concerns, and enhancing awareness and education, stakeholders can unlock the full potential of AI-enabled EV energy management, driving positive societal and environmental impacts while ushering in a new era of cleaner, smarter, and more accessible transportation for all. In summary, while the integration of AI into EV energy management offers significant benefits, achieving these objectives requires addressing several challenges, from technical and regulatory issues to user acceptance and data security. By formulating these problems clearly, stakeholders can develop targeted strategies to overcome these hurdles and advance the sustainable transportation landscape.

3 AI Enhanced Energy Management for Sustainable EV Integration

The transition towards EVs represents a pivotal stride in the pursuit of sustainable transportation solutions, yet to fully realize the potential of EVs, sophisticated energy management systems are imperative to navigate the multifaceted challenges of energy consumption, battery longevity, and grid integration. In this context, AI emerges as a transformative force, offering a diverse array of advanced tools and techniques capable of enhancing the efficiency, reliability, and user experience of EVs. At the heart of this endeavor lies the delineation of key objectives aimed at seamlessly integrating AI into EV energy management. Central to these objectives is the pursuit of efficiency enhancement, wherein AI algorithms harness real-time data to optimize energy consumption and charging patterns, minimizing waste and maximizing the utilization of electric power. Moreover, AI-driven strategies hold the promise of extending the range of EVs by predicting and mitigating battery degradation, thereby enhancing the longevity and performance of energy storage systems.

Furthermore, the integration of AI technologies has the potential to drive cost reduction across the lifecycle of EV ownership, from manufacturing to operation and maintenance, through predictive maintenance, optimized charging strategies, and enhanced fleet management [18]. Additionally, AI-enabled energy management systems contribute to mitigating the environmental impact of transportation by promoting the adoption of renewable energy sources, optimizing energy usage, and reducing greenhouse gas emissions. Beyond efficiency and sustainability, the integration of AI enhances the user experience of EVs through personalized recommendations, intuitive interfaces, and adaptive functionalities that cater to individual preferences and driving habits, thereby fostering greater acceptance and adoption of electric mobility solutions. Moreover, AI-driven systems exhibit scalability and adaptability, capable of accommodating the evolving needs and demands of a growing EV market, while simultaneously enhancing safety through predictive diagnostics, collision avoidance systems, and autonomous driving features. Furthermore, the seamless integration of AI technologies with existing grid infrastructure facilitates grid optimization and stabilization, enabling bidirectional energy flow and supporting the integration of renewable energy sources, thus advancing the transition towards a sustainable and resilient energy ecosystem. Additionally, fostering interoperability among AI-enabled EV systems is essential to ensure seamless communication and collaboration between diverse stakeholders, ranging from vehicle manufacturers and energy providers to infrastructure developers and regulatory bodies. By establishing common standards, protocols, and interfaces, stakeholders can unlock synergies and maximize the benefits of AI-driven energy management systems across the entire electric mobility ecosystem.

Ultimately, the integration of AI into EV energy management holds the promise of fostering long-term sustainability by enhancing the efficiency, reliability, and environmental performance of electric transportation, thereby accelerating the transition toward a cleaner, smarter, and more sustainable mobility paradigm [19]. Through concerted efforts and collaborative initiatives, stakeholders can harness the transformative potential of AI to realize the shared vision of a future where electric mobility is not only feasible but also desirable, contributing to the broader goals of environmental stewardship and energy efficiency on a global scale.

3.1 Efficiency Optimization

At the forefront of the quest to optimize the efficiency of EVs lies the primary objective of leveraging AI-driven energy management systems. These systems, empowered by real-time data and advanced algorithms, represent a transformative force capable of revolutionizing the way energy is consumed and utilized in electric transportation. Central to this objective is the harnessing of real-time data streams, encompassing a plethora of variables ranging from driving patterns and terrain to traffic conditions and weather forecasts. By integrating this wealth of data into AI algorithms, EV energy management systems can dynamically adjust energy consumption and distribution in response to changing environmental and operational factors, thereby minimizing wastage and maximizing efficiency [15]. This entails the development of adaptive energy distribution strategies tailored to specific driving scenarios, wherein AI algorithms analyze and predict driving patterns, terrain characteristics, and traffic conditions to optimize power usage in real time. For instance, during urban commuting, where frequent stops and accelerations are commonplace, AI can orchestrate the distribution of energy to propulsion systems and auxiliary components in a manner that minimizes energy losses and maximizes range [20]. Similarly, in highway driving conditions characterized by steady speeds and minimal traffic disruptions, AI algorithms can optimize energy consumption by modulating throttle response, regenerative braking, and other drivetrain parameters to achieve optimal performance and efficiency.

Furthermore, AI-driven energy management systems hold the promise of enhancing the operational efficiency of EVs by dynamically adapting to diverse driving environments and user preferences. This includes the development of personalized energy optimization profiles that cater to individual driving habits, preferences, and usage patterns [21]. By analyzing historical data and user feedback, AI algorithms can fine-tune energy management strategies to align with the unique needs and requirements of each driver, thereby optimizing efficiency and enhancing the overall user experience.

Additionally, AI enables predictive energy management capabilities that anticipate future energy demands based on historical data, weather forecasts, and route information. By proactively adjusting energy distribution and consumption patterns, AI-driven systems can preemptively optimize efficiency and range, ensuring that EVs operate at peak performance levels under varying driving conditions. Moreover, the integration of AI into EV energy management extends beyond optimizing energy consumption to encompass broader aspects of vehicle operation and performance. This includes predictive maintenance functionalities that leverage AI algorithms to monitor and assess the health and condition of vehicle components, identifying potential issues before they escalate, and optimizing maintenance schedules to minimize downtime and costs. Furthermore, AI facilitates adaptive charging strategies that optimize the timing and duration of charging sessions based on grid demand, energy prices, and user preferences [16]. Intelligent scheduling of charging sessions, particularly during off-peak hours or when renewable energy sources are plentiful has the potential to alleviate strain on the grid, reduce electricity expenses, and facilitate the integration of renewables into the power system [22]. Overall, leveraging AI to enhance EVs' efficiency represents a significant initiative with wide-ranging implications for the future of electric transportation. By leveraging real-time data and advanced algorithms, AI empowers EVs to operate more efficiently and autonomously, opening up new avenues for sustainability, reliability, and user satisfaction in electric mobility. Through ongoing innovation, cooperation, and investment, stakeholders can fully exploit the benefits of AI-driven energy management systems, accelerating the shift towards a cleaner, smarter, and more sustainable transportation landscape for future generations.

3.2 Cost Minimization Strategies

Reducing the overall cost of EV ownership stands as a critical objective in driving the widespread adoption of electric transportation, and AI emerges as a pivotal tool in achieving this goal. At the forefront of cost reduction efforts lies the optimization of charging schedules facilitated by AI, which enables EV owners to capitalize on off-peak electricity rates, thereby minimizing energy costs associated with vehicle charging [23]. By leveraging predictive analytics and machine learning algorithms, AI algorithms analyze historical data, grid demand patterns, and energy price fluctuations to dynamically adjust charging schedules, ensuring that EVs are charged during periods of low electricity demand and reduced rates. This strategic approach not only lowers the cost of electricity for EV owners but also alleviates strain on the grid during peak demand hours, contributing to grid stability and resilience.

Furthermore, AI-driven predictive maintenance emerges as a cornerstone in minimizing repair costs and downtime associated with EV ownership. By continuously monitoring vehicle components, sensors, and subsystems, AI algorithms detect anomalies, identify potential issues, and predict maintenance needs before they escalate into significant problems. This proactive approach enables EV owners to address maintenance issues preemptively, thereby reducing the likelihood of costly repairs and minimizing vehicle downtime. Additionally, efficient energy management facilitated by AI contributes to lower fuel costs by maximizing the utilization of electricity over traditional fuels. Through adaptive energy management strategies, AI algorithms optimize power distribution, adjust throttle response, and modulate regenerative braking systems to maximize energy efficiency and minimize energy wastage [24]. By prioritizing electric power over fossil fuels, AI-driven energy management systems enable EV owners to realize substantial savings on fuel expenses while reducing their carbon footprint and environmental impact [25]. Moreover, AI-driven optimization extends beyond individual vehicles to encompass broader aspects of fleet management and ecosystem optimization. This includes predictive analytics solutions that forecast energy consumption, optimize routing and logistics, and identify opportunities for efficiency improvements across entire fleets of EVs. By leveraging AI technologies, fleet operators can streamline operations, reduce operating costs, and maximize the overall efficiency and profitability of their EV fleets.

Additionally, AI-enabled grid optimization solutions contribute to cost-reduction efforts by maximizing the utilization of renewable energy sources and minimizing reliance on fossil fuels. By dynamically adjusting energy consumption patterns, optimizing grid demand, and facilitating bidirectional energy flow, AI-driven grid optimization systems enable utilities to lower electricity costs, reduce emissions, and enhance grid reliability and resilience. Furthermore, AI-driven energy management solutions empower EV owners to participate in demand response programs and grid balancing initiatives, earning incentives and rewards for flexibly adjusting their charging behaviors to align with grid needs. By incentivizing load shifting and demand flexibility, AI-driven energy management systems enable EV owners to further reduce their electricity costs while supporting the transition to a more sustainable and resilient energy ecosystem [26]. In conclusion, the integration of AI into EV energy management holds the promise of significantly reducing the overall cost of EV ownership, driving the widespread adoption of electric transportation, and accelerating the transition towards a cleaner, smarter, and more sustainable mobility ecosystem. Through optimized charging schedules, predictive maintenance, efficient energy management, and grid optimization, AI-driven solutions empower EV owners to realize substantial savings on energy costs, maintenance expenses, and fuel expenditures while contributing to grid stability, reliability, and environmental sustainability. By harnessing the transformative potential of AI, stakeholders can unlock new opportunities for cost reduction, efficiency improvement, and value creation in the realm of electric transportation, ushering in a future where EVs are not only economically viable but also socially and environmentally beneficial for society as a whole [27].

3.3 Mitigation of Environmental Impact

The deployment of AI-driven energy management systems within EVs represents a pivotal step towards reducing environmental impact and advancing the transition towards sustainable transportation solutions. The main goal of this initiative is to improve the efficiency of energy use and encourage the incorporation of renewable energy into EVs, both crucial for reducing their carbon footprint. Using advanced AI algorithms and predictive analysis, energy management systems optimize how electricity is used, especially from sources like solar and wind power. By adjusting power consumption and charging schedules in real-time based on current data and grid conditions, AI ensures EVs primarily utilize electricity from renewable sources, thus reducing reliance on fossil fuels and cutting down on greenhouse gas emissions. This strategic approach not only helps combat climate change but also promotes the advancement and adoption of renewable energy technologies, facilitating the shift towards a cleaner, more sustainable energy environment [28].

Furthermore, AI-powered intelligent route planning and traffic management play a pivotal role in reducing emissions by optimizing driving routes, minimizing congestion, and mitigating idling times. By leveraging historical data, real-time traffic information, and predictive algorithms, AI algorithms identify optimal routes that minimize energy consumption and emissions, thereby reducing the environmental footprint of EVs. This includes avoiding congested routes, selecting efficient driving paths, and optimizing speed and acceleration patterns to minimize energy wastage and maximize efficiency. Additionally, AI-driven traffic management systems enable EVs to navigate through traffic more efficiently, reducing the overall environmental impact of transportation by minimizing congestion-related emissions and improving air quality in urban areas. Moreover, AI-driven energy management systems extend beyond individual vehicles to encompass broader aspects of energy ecosystem optimization, including grid integration and demand response. By facilitating the bidirectional flow of electricity between EVs and the grid, AI enables EVs to function as movable energy storage units, aiding in grid stability and bolstering the incorporation of RESs. By employing smart demand response features, AI systems refine charging timetables according to grid needs, energy costs, and renewable energy accessibility, alleviating pressure on the grid during peak usage times and encouraging the effective utilization of renewable energy reservoirs [29]. Moreover, AI-driven grid enhancement solutions empower utility providers to predict and handle shifts in energy demand more adeptly, enhancing grid dependability, adaptability, and effectiveness.

By fostering synergies between EVs and RESs, AI-driven energy management systems contribute to a more sustainable and resilient energy ecosystem, advancing the transition towards a low-carbon future. In conclusion, the integration of AI-driven energy management systems in EVs represents a transformative opportunity to reduce environmental impact, enhance energy efficiency, and accelerate the transition toward sustainable transportation solutions. AI enhances the performance of EVs by maximizing energy usage, devising smarter navigation routes, and seamlessly integrating with the grid [30]. This technological synergy not only improves efficiency but also reduces emissions, thereby fostering the uptake of renewable energy. Leveraging AI presents a gateway to environmental preservation, economic growth, and societal welfare, thereby laying the groundwork for a future characterized by environmental sustainability and enhanced quality of life for future generations.

3.4 Enhanced User Experience Metrics

Improving the user experience stands as a paramount objective in the evolution of EVs, and AI emerges as a transformative force in achieving this goal. At the core of this endeavor lies the provision of personalized recommendations and tailored guidance to EV owners, empowering them to make informed decisions that enhance efficiency, convenience, and satisfaction. Leveraging real-time data and advanced algorithms, AI algorithms analyze driving patterns, charging infrastructure availability, and energy consumption metrics to provide personalized recommendations for charging stations, optimal routes, and driving behaviors that conserve energy. This includes delivering real-time updates and alerts regarding nearby charging stations, availability, and pricing, thereby enabling EV owners to make informed decisions about where and when to recharge their vehicles.

Moreover, AI offers tailored advice based on individual driving patterns and preferences, taking into account factors such as driving habits, route preferences, and energy consumption patterns to optimize energy usage and maximize driving range. By providing personalized insights and recommendations, AI empowers EV owners to optimize their driving experience, minimize range anxiety, and maximize the utility of their EVs [31]. Furthermore, enhancing user experience extends beyond personalized recommendations to encompass intuitive interfaces and user-friendly applications that simplify the management of EV energy systems. By offering seamless integration with user preferences and schedules, AI-driven interfaces enable EV owners to monitor and control various aspects of their vehicle's energy management system effortlessly. This includes remotely scheduling charging sessions, adjusting energy consumption settings, and accessing real-time vehicle performance data from anywhere, anytime.

Furthermore, AI-powered tools offer user-friendly interfaces that simplify the comprehension and interaction with the intricate details of an electric vehicle's energy management system. These intuitive interfaces enhance accessibility and usability, thereby promoting wider acceptance and usage of electric transportation solutions. Additionally, AI-driven energy management systems actively engage with EV owners, offering timely suggestions and reminders to optimize energy consumption and improve driving efficiency. Leveraging historical data, weather predictions, and driving behaviors, AI algorithms predict potential energy-saving opportunities and offer informed recommendations to empower EV owners in making decisions that enhance energy efficiency and extend driving range. This proactive approach not only improves the user experience but also builds confidence and trust in AI-driven energy management systems, leading to increased engagement and satisfaction among EV owners.

Additionally, AI enables seamless integration with smart home and IoT (Internet of Things) devices, allowing EV owners to control and monitor their vehicle's energy management system remotely. Through integration with voice-activated assistants and smart home platforms, AI-driven energy management systems enable hands-free operation and voice commands, providing greater convenience and flexibility to EV owners. Moreover, AI facilitates interoperability with third-party services and applications, allowing EV owners to leverage a wide range of innovative features and functionalities to enhance their driving experience. In conclusion, the integration of AI into EV energy management represents a transformative opportunity to improve the user experience, enhance efficiency, and accelerate the adoption of electric transportation solutions [32]. By providing personalized recommendations, intuitive interfaces, and proactive engagement, AI empowers EV owners to optimize their driving experience, minimize range anxiety, and maximize the utility of their EVs. Through continuous innovation and collaboration, stakeholders can harness the power of AI to unlock new opportunities for convenience, efficiency, and satisfaction in the realm of electric transportation, driving the transition toward a cleaner, smarter, and more sustainable mobility ecosystem for generations to come.

3.5 Scalability and System Adaptability

Scalability and adaptability emerge as critical imperatives in the development and deployment of AI systems within the realm of EV energy management, as these technologies must evolve and expand to accommodate the growing number of EVs and the dynamic nature of energy landscapes. At the heart of this challenge lies the need for AI systems to effectively handle increasing data volumes stemming from a multitude of vehicles and infrastructure components. As the adoption of EVs continues to accelerate, the sheer volume and complexity of data generated by these vehicles pose significant challenges for AI-driven energy management systems. However, through scalable architectures, distributed computing paradigms, and efficient data processing techniques, AI technology can efficiently handle large volumes of data in real time, providing timely insights and actionable suggestions to improve energy usage and system performance. Additionally, adaptability is crucial for maintaining the resilience and longevity of AI-powered energy management systems amidst changing technologies and regulations. With the rapid pace of innovation in the automotive and energy sectors, AI systems must possess the flexibility and agility to integrate new features, algorithms, and improvements seamlessly [33]. This includes staying abreast of emerging technologies such as advanced battery chemistries, wireless charging technologies, and V2G communication protocols, as well as adapting to regulatory mandates and standards governing EVs and renewable energy integration. By embracing modular architectures, open standards, and interoperable interfaces, AI systems can easily incorporate new functionalities and updates; thereby future-proofing themselves against obsolescence and ensuring continued relevance and effectiveness in a rapidly evolving landscape [26].

Moreover, scalability and adaptability extend beyond technological considerations to encompass organizational and ecosystem dynamics. As the market for EVs expands and matures, AI systems must scale to support the growing demands of diverse stakeholders, including vehicle manufacturers, energy providers, infrastructure developers, and regulatory bodies. This necessitates collaborative efforts among industry players to establish common frameworks, data-sharing protocols, and interoperable solutions that facilitate seamless integration and interoperability across the entire electric mobility ecosystem. Additionally, AI systems must adapt to evolving consumer preferences, market dynamics, and business models, catering to the diverse needs and expectations of users while driving innovation and value creation. By fostering a culture of continuous learning, experimentation, and adaptation, stakeholders can harness the full potential of AI-driven energy management systems to address the complex challenges and opportunities inherent in the transition toward a cleaner, smarter, and more sustainable mobility ecosystem. In conclusion, scalability and adaptability represent indispensable attributes of AI-driven energy management systems in the realm of EV technology, enabling these systems to effectively navigate the challenges of data scalability, technological evolution, and ecosystem dynamics [27].

By embracing scalable architectures, distributed computing paradigms, and modular design principles, AI systems can accommodate the growing volume and complexity of data generated by EVs while remaining agile and responsive to emerging technologies and regulatory changes. Through collaborative efforts and a commitment to innovation and adaptation, stakeholders can unlock new opportunities for efficiency, sustainability, and resilience in the realm of electric transportation, driving positive societal and environmental impacts on a global scale.

3.6 Safety Improvement Protocols

Safety enhancement emerges as a paramount objective in the realm of EVs, with AI assuming a pivotal role in ensuring the secure and reliable operation of these vehicles. At the forefront of this endeavor lies the real-time monitoring of vehicle health, wherein AI-driven systems analyze sensor data, diagnostic codes, and performance metrics to assess the operational status of critical components and subsystems. By leveraging advanced machine learning algorithms and predictive analytics, AI enables early detection of potential failures and anomalies, allowing for proactive intervention and maintenance scheduling to mitigate risks and prevent catastrophic malfunctions [34]. This proactive strategy not only boosts vehicle reliability and operational uptime but also enhances safety by minimizing the chances of unexpected breakdowns and accidents. Furthermore, AI enhances driver safety by delivering real-time alerts and warnings about potentially hazardous driving conditions, such as severe weather, road obstacles, and traffic congestion. By integrating data from external sources like weather forecasts, road condition updates, and traffic reports, AI systems can evaluate the risk associated with current driving conditions and offer timely advice to drivers, including recommendations for safer routes and alternative travel options.

Additionally, AI enables the deployment of advanced driver assistance systems (ADAS) that enhance driver awareness, responsiveness, and control, thereby reducing the likelihood of accidents and mitigating the severity of collisions. Through the integration of sensors, cameras, and radar systems, AI-powered ADAS can detect potential hazards, predict impending collisions, and autonomously intervene to prevent accidents or minimize their impact. This includes features such as adaptive cruise control, lane departure warning, automatic emergency braking, and blind-spot detection, all of which work in concert to enhance driver safety and confidence on the road [28]. Furthermore, AI-driven safety enhancement extends beyond individual vehicles to encompass broader aspects of transportation infrastructure and ecosystem optimization. By analyzing traffic patterns, accident data, and historical trends, AI systems can identify high-risk areas and corridors prone to accidents, enabling policymakers and transportation authorities to implement targeted interventions and safety measures to mitigate risks and improve road safety. This includes measures such as road signage improvements, traffic signal optimization, and infrastructure upgrades aimed at enhancing visibility, reducing congestion, and minimizing accident hotspots.

Moreover, AI facilitates the development of connected vehicle technologies that enable real-time communication and collaboration among vehicles, infrastructure, and other road users. By leveraging vehicle-to-vehicle (V2V) and vehicle-to-infrastructure (V2I) communication protocols, AI-driven systems can exchange critical safety information, such as traffic alerts, road closures, and emergency notifications, enabling drivers to make informed decisions and avoid potential hazards. Additionally, AI enables the implementation of autonomous driving technologies that have the potential to revolutionize road safety by eliminating human error, the leading cause of accidents [35]. Through continuous learning, adaptation, and refinement, AI-driven autonomous vehicles can navigate complex traffic scenarios, anticipate potential hazards, and respond to dynamic road conditions with unparalleled precision and efficiency. In conclusion, safety enhancement represents a paramount objective in the realm of EVs, with AI playing a critical role in ensuring the secure and reliable operation of these vehicles [29]. Through real-time monitoring of vehicle health, early detection of potential failures, and proactive maintenance scheduling, AI enhances vehicle reliability and uptime while reducing the likelihood of accidents and breakdowns.

Additionally, AI contributes to driver safety by providing alerts about hazardous driving conditions, suggesting safer routes, and enabling advanced driver assistance systems (ADAS) that enhance driver awareness and control. Through continuous innovation and collaboration, stakeholders can harness the transformative potential of AI to advance road safety, minimize accidents, and create a safer and more sustainable transportation ecosystem for all.

3.7 Interoperability Targets

Ensuring interoperability between various systems and technologies emerges as a fundamental imperative in the realm of EVs, with AI playing a pivotal role in facilitating seamless communication and operation across diverse platforms. At the heart of this endeavor lies the need for AI-driven energy management systems to be compatible with different EV models, charging infrastructure, and grid systems, thereby fostering harmonious integration and interoperability. By adhering to common standards and protocols, AI-driven systems can bridge the gap between disparate technologies and ensure that all components work together synergistically to deliver optimal performance and user experience. This includes compatibility with various EV charging standards such as CHAdeMO, CCS, and Tesla Supercharger, as well as interoperability with different communication protocols and grid integration technologies [36]. Through standardized interfaces, open APIs, and cross-platform compatibility, AI-driven energy management systems enable EV owners to seamlessly interact with charging infrastructure, grid services, and energy management platforms, regardless of the make or model of their vehicle. Moreover, interoperability enhances user convenience and promotes wider adoption of EVs by streamlining the charging process, reducing complexity, and eliminating barriers to entry. By providing a seamless and intuitive user experience, AI-driven systems empower EV owners to confidently embrace electric transportation solutions and integrate them into their daily lives. Furthermore, interoperability extends beyond individual vehicles to encompass broader aspects of the electric mobility ecosystem, including fleet management, energy trading, and grid optimization [30]. By fostering collaboration and standardization among stakeholders, AI-driven systems enable interoperable solutions that leverage the collective intelligence and resources of diverse participants to achieve common goals. This includes interoperable fleet management platforms that enable seamless management and optimization of EV fleets across different makes and models, as well as interoperable energy trading platforms that facilitate peer-to-peer energy exchange and grid balancing services.

Additionally, interoperability promotes innovation and competition by enabling the integration of third-party services and applications into the electric mobility ecosystem. Through open APIs and interoperable interfaces, AI-driven systems enable developers to create value-added services and solutions that enhance the functionality and utility of EVs, charging infrastructure, and grid systems. This cultivates a dynamic environment of innovation, where various participants can work together and compete to provide cutting-edge products and services that address the changing needs of electric mobility stakeholders. Additionally, interoperability boosts grid resilience and reliability by allowing for the seamless integration of distributed energy resources (DERs), including electric vehicles, solar panels, and energy storage systems, into the electric grid. By providing grid operators with real-time visibility and control over distributed energy assets, AI-driven systems enable dynamic optimization of grid operations, Demand response and grid balancing services significantly enhance grid stability and resilience. Additionally, interoperability supports the shift towards a decentralized and democratized energy system, enabling consumers to actively influence the future of energy. Through peer-to-peer energy trading platforms and decentralized energy management systems, AI-driven interoperable solutions empower consumers to engage in energy markets, optimize their energy consumption, and aid in integrating renewable energy sources into the grid [37]. In summary, achieving interoperability among various systems and technologies is crucial for the widespread adoption and success of EVs, as well as for the transition to a cleaner, smarter, and more sustainable mobility ecosystem.

By leveraging AI to enable seamless communication and operation across diverse platforms, stakeholders can unlock new opportunities for innovation, collaboration, and value creation in the realm of electric transportation. Through standardized interfaces, open APIs, and interoperable solutions, AI-driven systems empower EV owners, grid operators, and energy stakeholders to realize the full potential of electric mobility and accelerate the transition toward a low-carbon future for generations to come.

3.8 Long-Term Sustainability Objectives and Strategies

The ultimate goal of integrating AI applications into EV energy management is to contribute to the long-term sustainability of transportation systems, marking a decisive step towards a cleaner, smarter, and more resilient mobility ecosystem. The primary goal of this initiative is to facilitate the shift to renewable energy sources, reduce dependence on fossil fuels, and promote sustainable urban mobility practices. These practices aim to prioritize environmental sustainability, economic feasibility, and social fairness. Stakeholders are leveraging AI-driven energy management systems to strategically optimize the use of renewable energy sources like solar, wind, and hydroelectric power. These systems dynamically adjust charging and discharging patterns to coincide with peak times of renewable energy generation. By utilizing real-time data analytics, predictive algorithms, and demand response features, AI technology allows EVs to act as mobile energy storage units. These EVs can absorb excess renewable energy during periods of surplus and release it back into the grid when demand is elevated. This not only enhances the stability and reliability of renewable energy integration but also promotes grid resilience and decarbonization efforts, thereby contributing to global initiatives aimed at combating climate change and reducing greenhouse gas emissions.

Furthermore, AI-driven energy management practices facilitate long-term planning and development of sustainable transportation solutions that are environmentally friendly, economically viable, and socially beneficial. By analyzing historical data, market trends, and policy frameworks, AI systems empower stakeholders to make informed decisions about infrastructure investments, regulatory incentives, and technology deployments that support the transition toward a low-carbon transportation future. This includes the development of smart charging infrastructure, grid optimization solutions, and integrated mobility platforms that prioritize energy efficiency, equity, and accessibility while minimizing environmental impact and resource consumption. Additionally, AI technologies enable the optimization of fleet management operations, route planning algorithms, and transportation logistics, thereby reducing energy consumption, emissions, and congestion in urban areas. By promoting modal shifts towards shared mobility, electric transportation, and active transportation modes such as walking and cycling, AI-driven energy management systems foster a more sustainable and equitable urban mobility landscape that prioritizes public health, community well-being, and environmental sustainability.

Moreover, sustainable energy management practices facilitated by AI ensure that EVs remain a cornerstone of future transportation networks, playing a pivotal role in shaping the mobility landscape of tomorrow [38]. By incentivizing the adoption of EVs through regulatory mandates, financial incentives, and public awareness campaigns, policymakers and industry stakeholders can accelerate the transition towards a zero-emission transportation system that meets the evolving needs of society while safeguarding the planet for future generations [39]. In conclusion, the integration of AI applications into EV energy management represents a transformative opportunity to contribute to the long-term sustainability of transportation systems, advancing the transition towards renewable energy sources, reducing reliance on fossil fuels, and promoting sustainable urban mobility practices. Through collaborative efforts and innovative solutions, stakeholders can harness the power of AI to address the complex challenges and opportunities inherent in the pursuit of sustainable transportation, driving positive societal, economic, and environmental impacts on a global scale.

4 Conclusions

The incorporation of artificial intelligence (AI) into the energy management systems of electric vehicles (EVs) marks a significant transformation towards more sustainable transportation solutions. This chapter has delved into the multifaceted landscape of AI applications in EV energy management, addressing critical objectives and challenges faced in this domain. Key objectives identified encompass efficiency enhancement, range extension, cost reduction, environmental impact reduction, enhanced user experience, scalability and adaptability, safety enhancement, seamless grid integration, interoperability, and long-term sustainability. These objectives underscore the holistic approach needed to realize the full potential of AI-driven EV energy management. However, the path to achieving these objectives is not without challenges. Technological gaps, interoperability issues, data security and privacy concerns, and user acceptance barriers pose significant hurdles that must be overcome. Regulatory frameworks and interdisciplinary collaboration emerge as indispensable tools in navigating these challenges, ensuring that AI-driven EV energy management systems are not only effective but also ethically sound and socially acceptable. Looking to the future, the methodology section outlines the pivotal role of AI algorithms and models in addressing these challenges. In conclusion, the synergy between AI and EV energy management systems offers unprecedented opportunities to revolutionize the transportation sector. Continued research, innovation, and collaboration will be vital in harnessing the full potential of AI-driven EV energy management, propelling us toward a cleaner, greener, and more adaptive future in transportation.

Acknowledgements This research work was supported by "Woosong University's Academic Research Funding – 2025".

References

1. Yuvaraj T, Devabalaji KR, Kumar JA, Thanikanti SB, Nwulu N (2024) A comprehensive review and analysis of the allocation of electric vehicle charging stations in distribution networks. IEEE Access 12:5404–5461. https://doi.org/10.1109/ACCESS.2023.3349274
2. Boulanger AG, Chu AC, Maxx S, Waltz DL (2011) Vehicle electrification: status and issues. Proc IEEE 99(6):1116–1138. https://doi.org/10.1109/JPROC.2011.2112750
3. Wang D, Feng Y, Wang X, Zhang L, Li Z, Hu X (2016) A fuzzy logic based power management strategy for hybrid energy storage system in hybrid electric vehicles considering battery degradation. In: Proceedings of the 2016 IEEE Transportation Electrification Conference and Expo (ITEC). https://doi.org/10.1109/itec.2016.7520207
4. Liu C, Chau KT, Wu D, Gao S (2013) Opportunities and challenges of vehicle-to-home, vehicle-to-vehicle, and vehicle-to-grid technologies. Proc IEEE 101(11):2409–2427. https://doi.org/10.1109/JPROC.2013.2271951
5. Green RC, Wang L, Alam M (2011) The impact of plug-in hybrid electric vehicles on distribution networks: a review and outlook. Renew Sustain Energy Rev 15(1):544–553. https://doi.org/10.1016/j.rser.2010.08.014

6. Han S, Han S, Sezaki K (2010) Development of an optimal vehicle-to-grid aggregator for frequency regulation. IEEE Trans Smart Grid 1(1):65–72. https://doi.org/10.1109/TSG.2010.2045163
7. Guille C, Gross G (2009) A conceptual framework for the vehicle-to-grid (V2G) implementation. Energy Policy 37(11):4379–4390. https://doi.org/10.1016/j.enpol.2009.05.053
8. Rigas ES, Tzovaras D, Papaioannou IT, Demestichas K (2015) Vehicle-to-grid (V2G) and demand response for renewable energy integration. J Modern Power Syst Clean Energy 3(4):516–525. https://doi.org/10.1007/s40565-015-0121-3
9. Dallinger D, Gerda A, Wietschel M (2013) Integration of intermittent renewable power supply using grid-connected vehicles—a 2030 case study for California and Germany. Appl Energy 104:666–682. https://doi.org/10.1016/j.apenergy.2012.10.065
10. Mwasilu F, Justo JJ, Kim EK, Do TD, Jung JW (2014) Electric vehicles and smart grid interaction: a review on vehicle to grid and renewable energy sources integration. Renew Sustain Energy Rev 34:501–556. https://doi.org/10.1016/j.rser.2014.03.031
11. Al-Alawi BM, Bradley TH (2013) Total cost of ownership, payback, and consumer preference modeling of plug-in hybrid electric vehicles. Appl Energy 103:488–506. https://doi.org/10.1016/j.apenergy.2012.10.009
12. Moura SJ, Callaway DS, Fathy HK (2011) A stochastic optimal control approach for power management in plug-in hybrid electric vehicles. IEEE Trans Control Syst Technol 19(3):545–555. https://doi.org/10.1109/TCST.2010.2043736
13. Andersen PH, Mathews JA, Rask M (2009) Integrating private transport into renewable energy policy: the strategy of creating intelligent recharging grids for electric vehicles. Energy Policy 37(7):2481–2486. https://doi.org/10.1016/j.enpol.2009.03.032
14. Asamer J, Reinthaler M, Ruthmair M, Straub M, Puchinger J (2016) Optimizing charging station locations for urban taxi providers. Transp Res Part A Policy Pract 85:233–246. https://doi.org/10.1016/j.tra.2016.01.014
15. Parker N, Breetz HL, Salon D, Conway MW, Williams J, Patterson M (2021) Who saves money buying electric vehicles? Heterogeneity in total cost of ownership. Transp Res Part D Transp Environ 96:102893. https://doi.org/10.1016/j.trd.2021.102893
16. Zhou M, Long P, Kong N, Zhao L, Jia F, Campy KS (2021) Characterizing the motivational mechanism behind taxi driver's adoption of electric vehicles for living: Insights from China. Transp Res Part A Policy Pract 144:134–152. https://doi.org/10.1016/j.tra.2021.01.001
17. Adnan N, Nordin SM, Bahruddin MAB, Ali M (2018) How trust can drive forward the user acceptance to the technology? In-vehicle technology for autonomous vehicle. Transp Res Part A Policy Pract 118:819–836. https://doi.org/10.1016/j.tra.2018.10.019
18. Mukherjee JC, Gupta A (2015) A review of charge scheduling of electric vehicles in smart grid. IEEE Syst J 9(4):1541–1553. https://doi.org/10.1109/JSYST.2014.2356559
19. Steen D, Tuan LA (2017) Impacts of fast charging of electric buses on electrical distribution systems. CIRED Open Access Proc J 1:2350–2353. https://doi.org/10.1049/oap-cired.2017.0802
20. Sriabisha R, Yuvaraj T (2023) Optimum placement of electric vehicle charging station using particle swarm optimization algorithm. In: Proceedings of the 9th International Conference on Electrical Energy Systems (ICEES), pp 283–288. https://doi.org/10.1109/ICEES57979.2023.10110213
21. Lam AYS, Leung YW, Chu X (2014) Electric vehicle charging station placement: formulation, complexity, and solutions. IEEE Trans Smart Grid 5(6):2846–2856. https://doi.org/10.1109/TSG.2014.2344684
22. Dubey A, Santoso S (2015) Electric vehicle charging on residential distribution systems: impacts and mitigations. IEEE Access 3:1871–1893. https://doi.org/10.1109/ACCESS.2015.2476996
23. Kong W, Luo Y, Feng G, Li K, Peng H (2019) Optimal location planning method of fast charging station for electric vehicles considering operators, drivers, vehicles, traffic flow and power grid. Energy 186:115826. https://doi.org/10.1016/j.energy.2019.07.156

24. Richardson DB (2013) Electric vehicles and the electric grid: a review of modeling approaches, impacts, and renewable energy integration. Renew Sustain Energy Rev 19:247–254. https://doi.org/10.1016/j.rser.2012.11.042
25. He F, Yin Y, Zhou J (2015) Deploying public charging stations for electric vehicles on urban road networks. Transp Res Part C Emerging Technol 60:227–240. https://doi.org/10.1016/j.trc.2015.08.018
26. Jung CM, Pagidipala S, Salkuti SR (2024) Emerging technologies for the integration of renewable energy, energy storage and electric vehicles. In: Salkuti SR (eds) Energy and environmental aspects of emerging technologies for smart grid. Green Energy and Technology. Springer, Cham. https://doi.org/10.1007/978-3-031-18389-8_5
27. Shahraki N, Cai H, Turkay M, Xu M (2015) Optimal locations of electric public charging stations using real world vehicle travel patterns. Transp Res Part D Transp Environ 41:165–176. https://doi.org/10.1016/j.trd.2015.09.011
28. Tu W, Li Q, Fang Z, Shaw SL, Zhou B, Chang X (2016) Optimizing the locations of electric taxi charging stations: a spatial–temporal demand coverage approach. Transp Res Part C Emerging Technol 65:172–189. https://doi.org/10.1016/j.trc.2015.10.004
29. Luo L, Gu W, Zhou S, Huang H, Gao S, Han J, Wu Z, Dou X (2018) Optimal planning of electric vehicle charging stations comprising multi-types of charging facilities. Appl Energy 226:1087–1099. https://doi.org/10.1016/j.apenergy.2018.06.014
30. He Y, Song Z, Liu Z (2019) Fast-charging station deployment for battery electric bus systems considering electricity demand charges. Sustain Cities Soc 48:101530. https://doi.org/10.1016/j.scs.2019.101530
31. Mehrjerdi H, Hemmati R (2019) Electric vehicle charging station with multi-level charging infrastructure and hybrid solar-battery-diesel generation incorporating comfort of drivers. J Energy Storage 26:100924. https://doi.org/10.1016/j.est.2019.100924
32. Salkuti SR (2022) Emerging and advanced green energy technologies for sustainable and resilient future grid. Energies 15(18):6667. https://doi.org/10.3390/en15186667
33. Mishra S, Verma S, Chowdhury S, Gaur A, Mohapatra S, Dwivedi G, Verma P (2021) A comprehensive review on developments in electric vehicle charging station infrastructure and present scenario of India. Sustainability 13(4):2396. https://doi.org/10.3390/su13042396
34. García-Villalobos J, Zamora I, San Martín JI, Asensio FJ, Aperribay V (2014) Plug-in electric vehicles in electric distribution networks: a review of smart charging approaches. Renew Sustain Energy Rev 38:717–731. https://doi.org/10.1016/j.rser.2014.07.040
35. Shareef H, Islam MM, Mohamed A (2016) A review of the stage-of-the-art charging technologies, placement methodologies, and impacts of electric vehicles. Renew Sustain Energy Rev 64:403–420. https://doi.org/10.1016/j.rser.2016.06.033
36. Salkuti SR (2023) Advanced technologies for energy storage and electric vehicles. Energies 16(5):2312. https://doi.org/10.3390/en16052312
37. Alshahrani S, Khalid M, Almuhaini M (2019) Electric vehicles beyond energy storage and modern power networks: challenges and applications. IEEE Access 7:99031–99064. https://doi.org/10.1109/ACCESS.2019.2928639
38. Shao S, Zhang T, Pipattanasomporn M, Rahman S (2010) Impact of TOU rates on distribution load shapes in a smart grid with PHEV penetration. In: Proceedings of the IEEE PES T&D, pp 1-6. https://doi.org/10.1109/TDC.2010.5484336
39. Sravanthi P, Sandeep V (2022) Solving realistic reactive power market clearing problem of wind-thermal power system with system security. Int J Emerg Electr Power Syst 23(2):125–144. https://doi.org/10.1515/ijeeps-2021-0060

Dead-Beat Predictive Control of a Grid-Tied PV-Hybrid Energy Storage System

Mukul Chankaya, Mrutunjaya Panda, Surender Reddy Salkuti, and Chetlapalli Rama Krishna

Abstract This work presents a three-phase, four-wire grid-integrated PV-hybrid energy storage system (HESS) controlled by a predictive dead-beat control. With faster dynamics and enhanced steady-state response, the predictive dead-beat control emphasizes quick error reduction between reference and grid currents for generating inverter switching signals. The proposed control enables the system to perform active and reactive power control, harmonics elimination, neutral current compensation, and load balancing at the point of common coupling (PCC) to maintain DC bus and AC bus voltage stability, even during dynamic conditions. The presented system is observed under steady-state and various dynamic state conditions, i.e., irradiation variation, grid voltage sag/swell, and load unbalancing. The HESS consists of a battery and ultracapacitor, collectively providing optimal energy and power density balance during dynamic conditions at the DC bus. The presented system's performance is satisfactory per IEEE1547-2017 standards at the PCC.

Keywords Renewable energy · Grid integration · Adaptive control · Photovoltaic · Hybrid energy storage system

M. Chankaya · M. Panda · S. R. Salkuti (✉)
School of Electrical Engineering, VIT Vellore, Vellore, Tamil Nadu, India
e-mail: surender@wsu.ac.kr

M. Chankaya
e-mail: mukul.chankaya@vit.ac.in

M. Panda
e-mail: mrutunjaya.panda@vit.ac.in

C. Rama Krishna
Department of Railroad and Electrical Engineering, Woosong University, Daejeon, Republic of Korea
e-mail: chetlapalli.rama2022@vitstudent.ac.in

S. R. Salkuti (ed.), *Artificial Intelligence for Integrated Smart Energy Systems in Electric Vehicles*, Lecture Notes in Electrical Engineering 1427,
https://doi.org/10.1007/978-3-031-94276-1_15

Nomenclature

PV	Photovoltaic
HESS	Hybrid Energy Storage System
PCC	Point of Common Coupling
WES	Wind Energy System
LMS	Least Mean Square
LMF	Least Mean Fourth
MCC	Maximum Correntropy Criteria
DB	Dead Beat
MPPT	Maximum Power Point Tracking
InC	Incremental and Conductance
VSC	Voltage Source Converter
UC	Ultra-capacitor
BDC	Bi-directional Converter
THD	Total Harmonic Distortion
PLL	Phase Locked Loop
V_{dc}	DC bus voltage
$V^{ref}{}_{dc}$	Reference DC bus voltage
V_{err}	DC bus voltage error
I_{loss}	Loss component of Current
I_{BAT}	Battery current
I_{UC}	UC current
V_{PV}	PV voltage
I_{PV}	PV current
P_{PV}	PV power
W_{PV}	Feed-forward term
i_{Labc}	Load Currents
W_{Pabc}	Weight Components
W_{avg}	Average weight Components
μ_{Pabc}	Unit templates
V_t	Voltage amplitude
i^*_{Sabc}	Reference currents

1 Introduction

In the past few decades, renewable energy has emerged as a viable and reliable option for meeting the rising energy demand and reducing carbon emissions. Worldwide, Photovoltaic (PV) and Wind energy systems (WES) are front runners for renewable energy generation systems [1]. However, PV systems are more accessible to small-scale end users due to low initial investment, few to no topographical limitations, easy installation, and readily available [2]. The Grid integration of PV power generation systems relies on a power electronics interface [3]. However, the power electronics interface involved in grid integration introduces various technical challenges that must be dealt with using an adequately tuned control algorithm [4]. Technical issues, i.e., sporadic PV power generation, power balancing at DC bus, active and reactive power control at PCC, load balancing, grid voltage unbalancing and sag/swell, grid voltage, and current harmonics, are prominent in grid-tied PV systems [5]. For grid integration, the frequency and phase angle of the grid and system require matching. The bi-directional power exchange between the DC and AC system via PCC requires a stable AC and DC bus [6]. During dynamic conditions, the DC bus with HESS manages the sporadic PV power input from the PV arrays and ensures the power balance. Moreover, the non-linear load at the distribution end and the power electronics interface introduce harmonics in the system [7, 8]. Harmonic suppression is necessary for achieving the desired operation and stability [9]. An adequately tuned voltage source converter (VSC) control performs multifunctional operations, i.e., harmonics elimination, load balancing, power balance at both AC and DC bus, and grid voltage balancing [10]. The researchers have suggested various time-domain, frequency-domain, adaptive, and predictive control algorithms, which must be chosen per the operating conditions [11, 12]. Time domain algorithms, i.e., synchronous reference frames and instantaneous reactive power theory controls, are among the well-known controls that offer ease of implementation and faster response with less computational complexity [13, 14]. However, their steady-state response is less accurate, and performance degrades during dynamic conditions. The frequency domain controls, i.e., discrete Fourier transformation, Hilbert-Huang transformation, etc., perform better than time-domain controls but suffer from a very high computational burden, which degrades the dynamic and weak grid performance [15]. The adaptive controls, i.e., least mean square (LMS) [16], least mean fourth (LMF) [17], maximum correntropy criteria (MCC) [18], and their numerous variants were designed and implemented to achieve better steady-state and dynamic performance [19]. The LMS control has poor dynamic performance, and LMF is a higher-order filter but offers degraded steady-state performance. Blocked sparse MCC offers enhanced performance over MCC with non-Gaussian noise [17]. Generalized robust logarithmic control offers better system identification and noise cancellation.

However, advanced adaptive filtering techniques have high computational complexity, resulting in poor dynamic performance. In the quest for better performance, predictive controls were utilized to enhance grid-tied PV systems, i.e., model predictive control, generalized predictive control, dead-beat controls, etc. [20, 21]. The dead-beat (DB) control is chosen over others due to its ease of implementation, less computation load, and capability of error reduction during dynamic conditions with only one design parameter, i.e., the sampling period of systems voltages and currents [22, 23]. The DB control utilizes the system parameters and their sampling rate to estimate the next control cycle of the inverter by predicting the following reference signal [24]. The proposed control offers ease of implementation and faster dynamic response while used as the current control for the inverter. The HESS at the

DC bus ensures the desired power balance, maintains the DC bus stability, and minimizes DC bus voltages (V_{dc}) deviations during dynamic conditions, and mitigates the effect of sporadic PV power generation [25, 26].

1.1 Topology of the Proposed System

Figure 1 presents a proposed three-phase, four-wire grid-tied PV-hybrid energy storage system (HESS). A DB control algorithm controls the voltage source converter (VSC) [27, 28]. The maximum power extraction out of the PV array is done with the incremental conductance (InC) method. The hybrid energy storage system (HESS) consists of a battery and ultracapacitor (UC), which helps the system maintain stability on both DC and AC buses. The battery and UC are connected to the DC bus via a bi-directional converter (BDC) to assist in charging/discharging [29, 30]. The four-wire system offers the added advantage of neural current compensation as well. The non-linear load is simulated by using a three-phase rectifier with a resistive load. The main attributes of the proposed system are:

- Multifunctional operation of the VSC, including reactive power compensation, power quality enhancement, load balancing, grid voltage balancing, and power balancing at the point of common coupling (PCC).
- HESS is associated with grid-tied PV systems that handle sudden and sustained load, grid, and PV side disturbances.

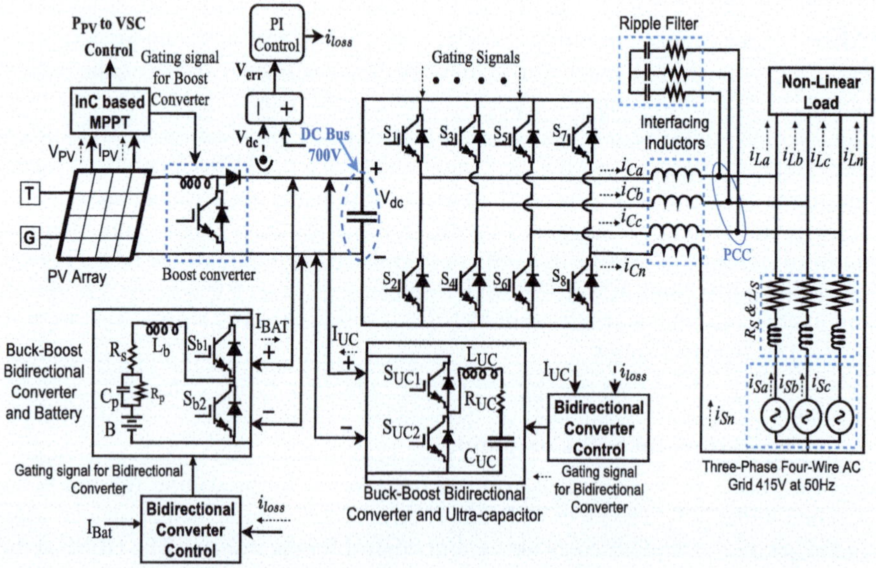

Fig. 1 Block diagram of the proposed system

- HESS delivers pre-decided power to the grid during fixed power mode by controlling the battery and UC charging/discharging cycle.

2 Control Algorithm

The presented system with two conversion stages, i.e., DC-DC conversion via boost converter and DC-AC conversion via VSC, operates in a grid following mode controlled by various controls, i.e., InC-based boost converter control, HESS (battery and UC) current control via BDC control, and DB algorithm based VSC control.

2.1 HESS Control Algorithm

Figure 2 shows the HESS control techniques, where reference DC bus voltage $\left(V_{dc}^{ref}\right)$ and sensed DC bus voltages (V_{dc}) are compared to produce DC bus voltage error (V_{err}). The loss component of the current (I_{loss}) is generated after feeding V_{err} to PI controller as,

$$I_{loss}(n+1) = I_d(n) + K_p\left(V_{DC}^{ref} - V_{DC}\right) + \frac{K_i}{s}\left(V_{DC}^{ref} - V_{DC}\right) \tag{1}$$

The I_{loss} is further compared with sensed battery (I_{BAT}) and UC (I_{UC}) currents as a reference signal to generate an adequate duty cycle (D_{UC} and D_{BAT}) for the HESS charging/discharging control via a BDC as,

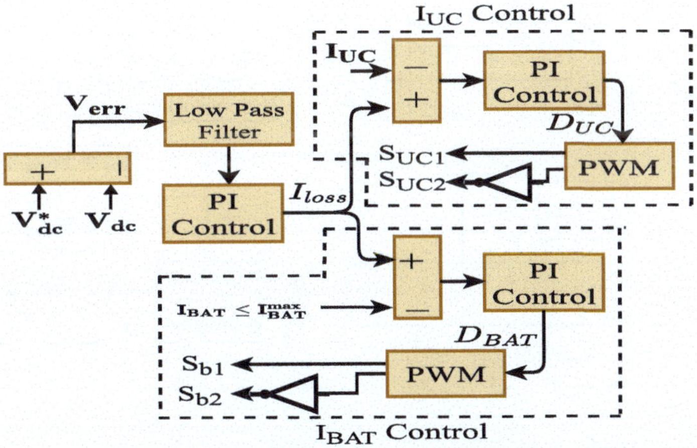

Fig. 2 DC link voltage, battery, and UC current control

$$D_{UC}(n+1) = D(n) + K_p(I_{loss} - I_{UC}) + \frac{K_i}{s}(I_{loss} - I_{UC}) \tag{2}$$

$$D_{BAT}(n+1) = D(n) + K_p(I_{loss} - I_{BAT}) + \frac{K_i}{s}(I_{loss} - I_{BAT}) \tag{3}$$

2.2 VSC Control Algorithm

Figure 3 shows the proposed VSC control. The grid voltages (v_{Sabc}), grid currents (i_{Sabc}), load currents (i_{Labc}), PV voltage (V_{PV}), current (I_{PV}) and power (P_{PV}) are sensed in real time by the system. The load currents i_{Lx}(where x = a, b, c) are transformed into direct and quadrature components using Parks' transformation [29, 30]. The grid frequency signal is extracted using a phase-locked loop (PLL) as (ω_t) from grid voltage [31]. The feed-forward term (W_{PV}) is the PV power given to the system and is calculated as,

$$W_{PV} = \frac{2}{3} * \frac{I_{PV} * V_{PV}}{V_t} = \frac{2}{3} * \frac{P_{PV}}{\sqrt{\frac{2}{3}\left(v_{Sa}^2 + v_{Sb}^2 + v_{Sc}^2\right)}} \tag{4}$$

After combining W_{PV} and direct component of load current (i_{Ld}), reverse park's transformation is performed to generate the reference weight signals (W_a, W_b, W_c), from the dead-beat control algorithm as,

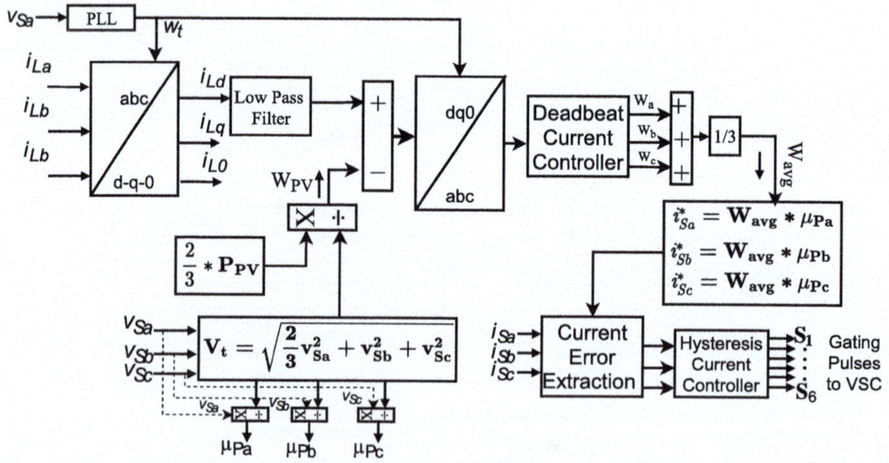

Fig. 3 Dead-beat predictive algorithm based VSC Control

$$\begin{cases} W_{pa} = \frac{1}{V_{dc}}\left[\frac{L}{T_s} * i^*_{Ca}(k) - \left(\frac{L}{T_s} - R\right) * i_{Ca}(k) + v^*_a(k)\right] \\ W_{pb} = \frac{1}{V_{dc}}\left[\frac{L}{T_s} * i^*_{Cb}(k) - \left(\frac{L}{T_s} - R\right) * i_{Cb}(k) + v^*_b(k)\right] \\ W_{pc} = \frac{1}{V_{dc}}\left[\frac{L}{T_s} * i^*_{Cc}(k) - \left(\frac{L}{T_s} - R\right) * i_{Cc}(k) + v^*_c(k)\right] \end{cases} \tag{5}$$

The average weight signal (W_{avg}) is multiplied by the unit templates ($\mu_{Pa}, \mu_{Pb}, \mu_{Pc}$) to generate reference currents $\left(i^*_{Sabc}\right)$ signals, and V_t is voltage magnitude as,

$$W_{avg} = \frac{1}{3}\left(W_{pa} + W_{pb} + W_{pc}\right) \tag{6}$$

$\mu_{pa} = v_{Sa}/V_t$, $\mu_{pb} = v_{Sb}/V_t$, $\mu_{pc} = v_{Sc}/V_t$ (7)

$$i^*_{sa} = W_{avg} * \mu_{pa};\ i^*_{sb} = W_{avg} * \mu_{pb};\ i^*_{sc} = W_{avg} * \mu_{pc} \tag{8}$$

The comparison of i^*_{Sabc} and i_{Sabc} is fed to the hysteresis current controller to generate the desired switching sequence of VSC. During fixed power mode, W_{avg} is replaced by W_{fix} for the i^*_{sabc} generation [32]. During this mode, the system is committed to delivering the pre-decided amount of power to the PCC with the help of HESS as,

$$i^*_{sa} = W_{fix} * \mu_{pa};\ i^*_{sb} = W_{fix} * \mu_{pb};\ i^*_{sc} = W_{fix} * \mu_{pc} \tag{9}$$

3 Results and Discussion

The proposed system is analyzed under various steady-state and dynamic conditions, i.e., irradiation variations, unbalanced loading [30], grid voltage sag/swell, and fixed power mode.

3.1 Steady-State Analysis

Figures 4a, b show the total harmonic distortion (THD) of i_{Sa} and i_{La}. The load current has a THD of 39.63% due to the non-linear load connected to the system. DB control reduces the grid current THD to 0.91%, leaving the grid unaffected by the load end.

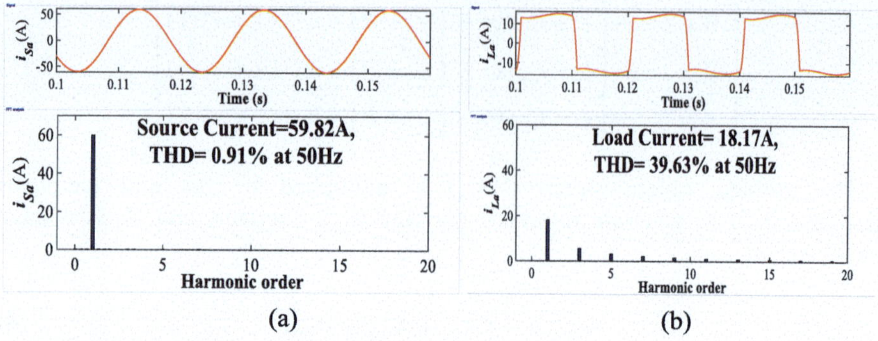

Fig. 4 THD of **a** i_{Sa} and **b** i_{La} during steady-state

3.2 *Sudden Insolation Variation Analysis*

Figures 5a, b show the system's performance during sudden insolation change. Figure 5a shows v_{Sabc}, i_{Sabc},, compensating current (i_{Ca}), i_{La}, V_t, the power delivered to the grid (P_g), reactive power exchange with grid (Q_g), grid neutral current (i_{Sn}), load (i_{Ln}) and compensator (i_{Cn}) a neutral current. With insolation reduction from 1000 W/m^2 to 600W/m^2, i_{Sabc} and P_g reduces significantly but i_{La} remains same. The i_{Ln} and i_{Cn} flows with thrice of the fundamental frequency, in opposition with each other and eventually canceling each other to make i_{Sn} almost zero [33]. Figure 5b shows V_{PV}, I_{PV}, P_{PV}, V_{dc}, V_{Bat}, I_{Bat}, V_{UC}, and I_{UC}. The V_{PV}. It remains stable due to the InC-controlled boost converter. The I_{PV} and P_{PV} Changes with the insolation variation. At the 600W/m^2 irradiation level, the battery and UC compensate for the power demand by reducing their charging rate. The V_{dc} remains stable during irradiation variation.

3.3 *Unbalanced Loading Analysis*

Figure 6a shows the load unbalancing by disconnecting phase 'a' of the load-making i_{La} zero. With reduced load demand, i_{Sa} increases slightly, and more power is delivered to a load. The i_{Ln} and i_{Cn} stats revolving with twice of fundamental frequency and canceling out each other to make i_{Sn} zero. Figure 6b shows V_{Bat} and V_{UC} increases slightly as with reduced load, more power is available for battery and UC charging, making I_{Bat} and I_{UC} more negative. V_{dc} remains constant during load unbalancing.

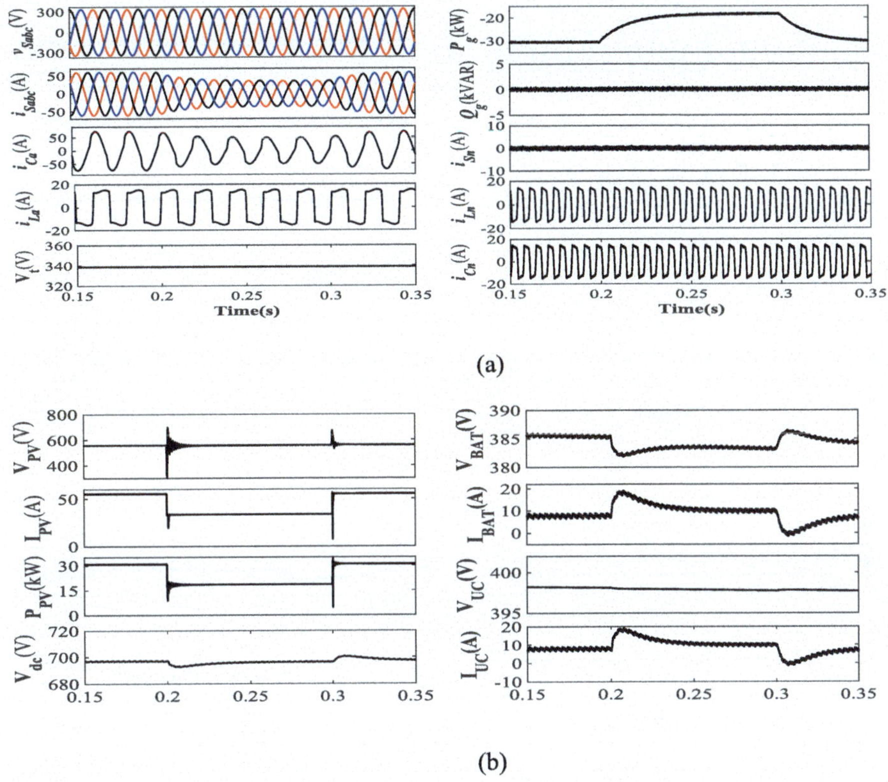

Fig. 5 Insolation variation performance of **a** v_{sabc}, i_{sabc}, i_{La}, i_{Ca}, V_t, P_g, Q_g, i_{Sn}, i_{Ln}, and i_{Cn} **b** V_{PV}, I_{PV}, P_{PV}, V_{dc}, V_{Bat}, I_{Bat}, V_{UC}, and I_{UC}

3.4 *Grid Voltage Sag/Swell Analysis*

Figure 7a shows the system's performance during 20% v_{sabc} sag/swell conditions confirmed with V_t variations. During v_{sabc} sag and swell variations, i_{sabc} reduces and increases, as for the same power on a low voltage level, more current is required and vice-versa. The i_{Ln} and i_{Cn} magnitude also changes with v_{sabc}, but i_{Sn} is kept near zero. Figure 7b clearly shows the battery and UC discharging and charging during sag and swell to maintain the desired power balance at PCC while maintaining the V_{dc} constant.

3.5 *Fixed Power Mode Analysis*

Figures 8a, b present a fixed power mode analysis, where the proposed system is bound to deliver the pre-decided amount of power to the load end irrespective of the

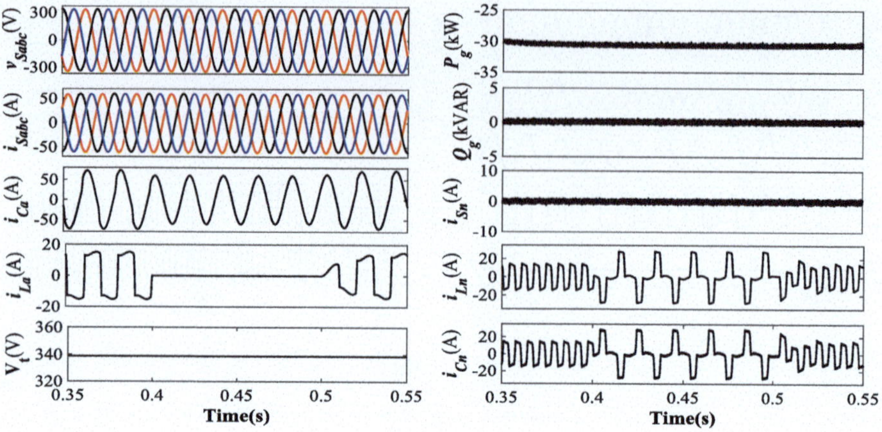

(a)

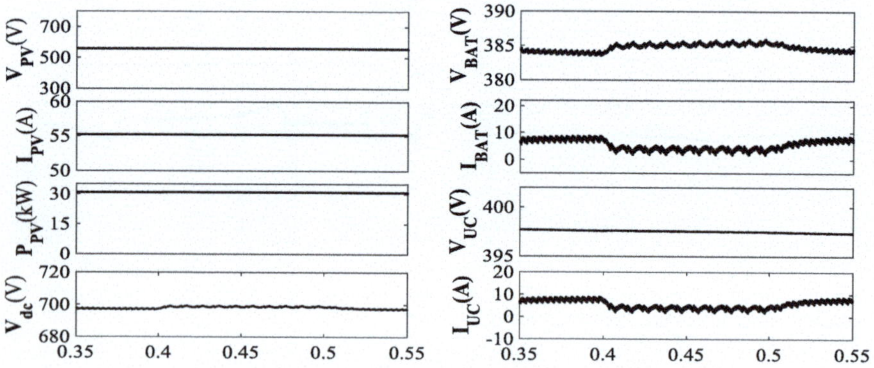

(b)

Fig. 6 Insolation variation performance of **a** v_{sabc}, i_{sabc}, i_{La}, i_{Ca}, V_t, P_g, Q_g, i_{Sn}, i_{Ln}, and i_{Cn} **b** V_{PV}, I_{PV}, P_{PV}, V_{dc}, V_{Bat}, I_{Bat}, V_{UC}, and I_{UC}

irradiation levels. Figure 8a shows the variation in i_{sabc} and P_g, from 0.8 to 0.85s of simulation time, where the power delivery to the PCC reduces to meet the pre-decided power limit. The excess power at the DC bus being stored by the HESS can be verified by V_{Bat} and I_{Bat} as shown in Fig. 8b.

Likewise, from 0.85 s to 0.9 s of simulation time, the system delivers more power to the PCC, which increases the i_{sabc} and P_g magnitude. From 0.85s to 0.9s, HESS reduces its charging rate and supports more power to be delivered to the PCC. The overall system remains stable during fixed power mode, with adequate neutral current compensation and around zero reactive power exchange with the grid.

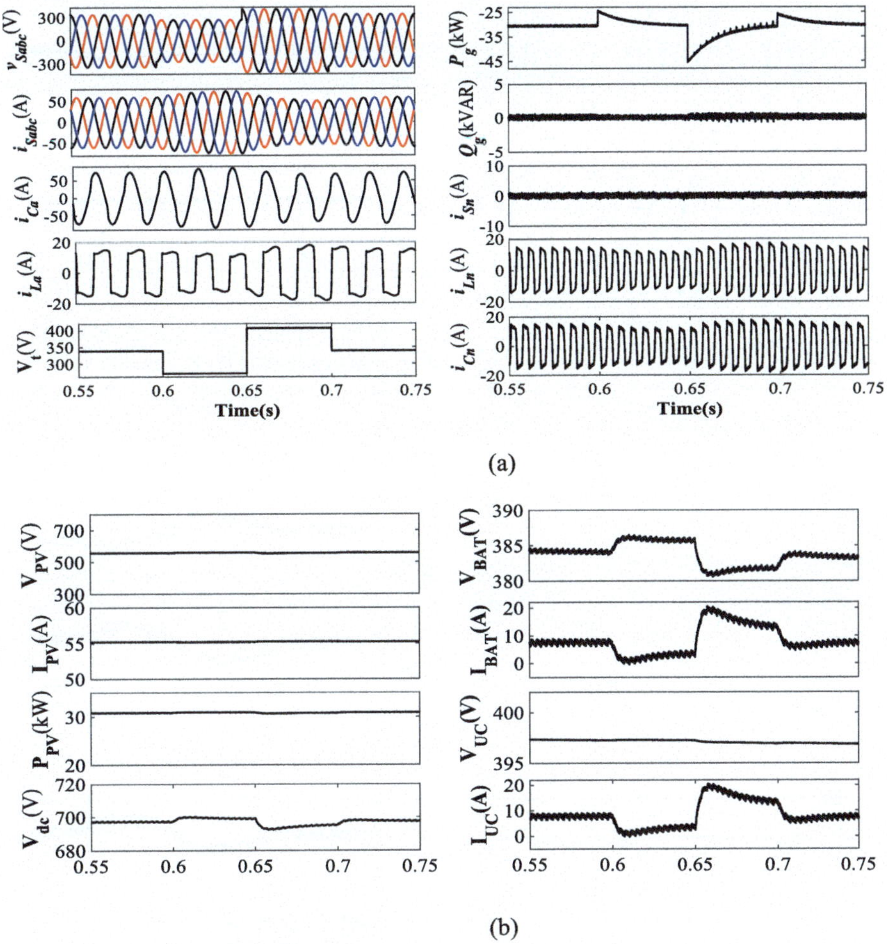

Fig. 7 Insolation variation performance of **a** v_{sabc}, i_{sabc}, i_{La}, i_{Ca}, V_t, P_g, Q_g, i_{Sn}, i_{Ln}, and i_{Cn} **b** V_{PV}, I_{PV}, P_{PV}, V_{dc}, V_{Bat}, I_{Bat}, V_{UC}, and I_{UC}

4 Conclusions

The presented grid-tied PV-HESS system controlled by a dead-beat predictive control has performed multifunctional operations during steady-state and dynamic conditions, i.e., harmonics elimination, load balancing, active and reactive power control, reactive power compensation, neutral current compensation, load balancing, and DC bus stability enhancement. BDC control maintains the balance of power at the DC bus, which mitigates the effect of sporadic PV power generation with varying irradiation levels. The dead-beat control has effectively and efficiently reduced the error between the sensed grid and reference currents by predicting subsequent cycle reference currents to ensure better harmonic elimination. The HESS with battery

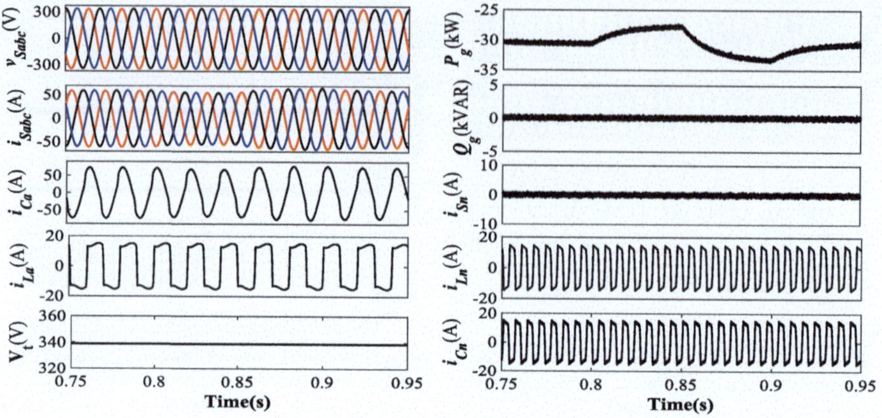

(a)

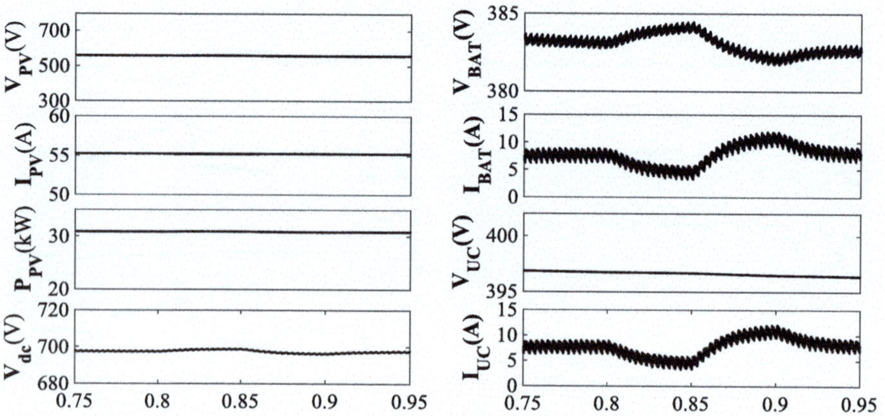

(b)

Fig. 8 Fixed power mode performance of **a** v_{sabc}, i_{sabc}, i_{La}, i_{Ca}, V_t, P_g, Q_g, i_{Sn}, i_{Ln}, and i_{Cn} **b** V_{PV}, I_{PV}, P_{PV}, V_{dc}, V_{Bat}, I_{Bat}, V_{UC}, and I_{UC}

and ultracapacitor has effectively kept the DC bus voltage near the desired value by providing adequate power flow at the DC bus. The presented system has operated satisfactorily as per IEEE 1547-2019 standards.

Acknowledgements This work is supported by the Vellore Institute of Technology Vellore, Tamil Nadu, with financial assistance under Seed grant (RGEMS) SG20230126 and SG20230073.

References

1. Hassan Q, Algburi S, Aws Zuhair S et al (2024) A comprehensive review of international renewable energy growth. Energy Built Environ 1–30. https://doi.org/10.1016/j.enbenv.2023.12.002
2. Panda M, Bhaskar DV, Maity T (2022) Coordinated power sharing among multiple subgrids in a Photovoltaic-battery based DC Microgrid. Energy Sources Part A Recovery Utilization Environ Effects 41:5358–5380. https://doi.org/10.1080/15567036.2022.2086649
3. Gupta AK, Pragallapati N, Joshi MS, Agarwal V (2022) A novel per unit (P.U.) integer format applied to the control of a grid-tied solar PV inverter. IEEE Trans Ind Inf 18(2):735–743. https://doi.org/10.1109/TII.2021.3067922
4. Rezkallah M, Chandra A, Singh B, Singh S (2019) Microgrid: configurations, control and applications. IEEE Trans Smart Grid 10(2):1290–1302. https://doi.org/10.1109/TSG.2017.2762349
5. Ho CNM, Lam CS (2021) Editorial for the special issue on power quality conditioning in modern power grids integrated emerging power electronic systems. CPSS Trans Power Electronics Appl 6(3):191–192
6. Chankaya M, Ahmad A, Hussain I (2022) Smart grid-Tied PV-battery storage system operation under dynamic conditions. In: 2022 1st international conference on Sustainable Technology for Power and Energy Systems (STPES), pp 1–6, https://doi.org/10.1109/STPES54845.2022.10006522
7. Kumar D, Zare F (2016) Harmonic analysis of grid connected power electronic systems in low voltage distribution networks. IEEE J Emerging Selected Topics Power Electronics 4(1):70–79. https://doi.org/10.1109/JESTPE.2015.2454537
8. Medina A et al. Harmonic analysis in frequency and time domain. IEEE Trans Power Delivery 28(3). https://doi.org/10.1109/TPWRD.2013.2258688
9. Saleem IM, Saha S (2024) Assessment of frequency stability and required inertial support for power grids with high penetration of renewable energy sources. Electric Power Syst Res 229:110184. https://doi.org/10.1016/j.epsr.2024.110184
10. Chankaya M, Aijaz M, Hussain I, Ahmad A, Lone S A (2023) Advanced adaptive algorithm controlled single-phase DSTATCOM operation during weak grid conditions. Int J Circuit Theor Appl 1–15. https://doi.org/10.1002/CTA.3815
11. Singh B, Chandra A, Al-haddad K (2015) Power quality problems and mitigation techniques, 1st edn. Wiley Online Library, West Sussex, UK. ISBN 9781118922057
12. Sharma S, Gupta S, Zuhaib M et al (2024) A comprehensive review on STATCOM: paradigm of modeling, control, stability, optimal location, integration, application, and installation. IEEE Access 12:2701–2729. https://doi.org/10.1109/ACCESS.2023.3345216
13. Singh B, Jayaprakash P, Kothari DP (2011) New control approach for capacitor supported DSTATCOM in three-phase four wire distribution system under non-ideal supply voltage conditions based on synchronous reference frame theory. Int J Electr Power Energy Syst 33:1109–1117. https://doi.org/10.1016/j.ijepes.2010.12.006
14. Escobar G, Ibarra L, Valdez-Resendiz JE et al (2021) Nonlinear stability analysis of the conventional SRF-PLL and enhanced SRF-EPLL. IEEE Access 9:59446–59455. https://doi.org/10.1109/ACCESS.2021.3073063
15. Chen H, Liu H, Xing Y, Hu H (2019) Enhanced DFT-based controller for selective harmonic compensation in active power filters. IEEE Trans Power Electron 34(8):8017–8030. https://doi.org/10.1109/TPEL.2018.2877848
16. Ghauri SA, Sohail MF (2013) System identification using LMS, NLMS and RLS. In: 2013 IEEE student conference on research and development, pp 64–69. https://doi.org/10.1109/SCOReD.2013.7002542
17. Arya SR, Mistry KD, Kumar P (2023) Least mean mixed norm square/fourth adaptive algorithm with optimized FOPID gains for voltage power quality mitigation. IEEE J Emerging Selected Topics Power Electronics 11(3):2632–2640. https://doi.org/10.1109/JESTPE.2023.3240712

18. Chankaya M, Naqvi SBQ, Hussain I, Singh B, Ahmad A (2023) Power quality enhancement and improved dynamics of a grid tied PV system using equilibrium optimization control based regulation of DC bus Voltage. Electric Power Syst Res 226:109911
19. Chankaya M, Hussain I, Ahmad A et al (2021) Generalized normal distribution algorithm-based control of 3-phase 4-wire grid-tied PV-hybrid energy storage system. Energies (Basel) 14:4355. https://doi.org/10.3390/en14144355
20. Babayomi O, Zhang Z, Dragicevic T et al (2023) Smart grid evolution: predictive control of distributed energy resources—a review. Int J Electr Power Energy Syst 147:108812. https://doi.org/10.1016/j.ijepes.2022.108812
21. Karamanakos P, Geyer T (2020) Guidelines for the design of finite control set model predictive controllers. IEEE Trans Power Electron 35:7434–7450. https://doi.org/10.1109/TPEL.2019.2954357
22. Luo A, Xiao H, Shuai Z (2015) Double deadbeat-loop control method for distribution static compensator. IET Power Electronics 8:1104–1110. https://doi.org/10.1049/iet-pel.2014.0435
23. Liu C, Xing X, Du C et al (2023) An improved model predictive control method using optimized voltage vectors for Vienna rectifier with fixed switching frequency. IEEE Trans Power Electron 38(1):358–371. https://doi.org/10.1109/TPEL.2022.3205946
24. Meena RK, Bhaskar DV, Zaabi OAl et al (2023) Modeling of two-stage photovoltaic inverter with grid connected and islanding operation. In: 2023 IEEE IAS Global Conference on Renewable Energy and Hydrogen Technologies (GlobConHT), Male, Maldives, pp 1–6. https://doi.org/10.1109/GlobConHT56829.2023.10087517
25. Chaudhary N, Ranjan A, Bhaskar DV et al (2023) Harmonic reduction strategies for EV charging in power systems using shunt active power filter. In: 2023 3rd International Conference on Emerging Frontiers in Electrical and Electronic Technologies (ICEFEET), Patna, India, pp 1–6. https://doi.org/10.1109/ICEFEET59656.2023.10452228
26. Chankaya M, Hussain I, Ahmad A et al (2022) Stability analysis of chaotic grey-wolf optimized grid-tied PV-hybrid storage system during dynamic conditions. Electronics (Switzerland) 11:567. https://doi.org/10.3390/electronics11040567
27. Salkuti SR (2022) Emerging and advanced green energy technologies for sustainable and resilient future grid. Energies 15(18):6667. https://doi.org/10.3390/en15186667
28. Salkuti SR (2023) Advanced technologies for energy storage and electric vehicles. Energies 16(5):2312. https://doi.org/10.3390/en16052312
29. Chankaya M, Ahmad A, Hussain I et al (2022) Grid-interfaced photovoltaic-battery energy storage system with slime mold optimized adaptive seamless control. IEEE Trans Ind Appl 58:7728–7738. https://doi.org/10.1109/TIA.2022.3194126
30. Zhang J, Lei H, Li Z et al (2022) Variable weight comprehensive evaluation of power quality and multi-objective optimization for multifunctional grid-connected. Inverter J Electr Eng Technol 17:3119–3131. https://doi.org/10.1007/s42835-022-01116-0
31. Panda M, Bhaskar DV, Maity T (2021) A novel dc bus-signaling based power management strategy for DC microgrid. Int Trans Electrical Energy Syst 31(2):246–266
32. Sravanthi P, Sandeep V (2022) Solving realistic reactive power market clearing problem of wind-thermal power system with system security. Int J Emerg Electr Power Syst 23(2):125–144. https://doi.org/10.1515/ijeeps-2021-0060
33. Wang P, Jin C, Zhu DX, Tang Y, Loh PC, Choo FH (2015) Distributed control for autonomous operation of a three-port AC/DC/DS hybrid microgrid. IEEE Trans Industr Electron 62(2):1279–1290

Automated Open Circuit Fault Detection in Cascaded H-Bridge Multilevel Inverters Using Artificial Intelligence Technique for EV Application

N. Kalaiarasi, A. Sivapriya, Pradeep Vishnuram, Mohit Bajaj, and Arvind R. Singh

Abstract Cascaded h-bridge multilevel inverters (CHBMLI) represent a promising solution for high-power applications due to their ability to produce high-quality output voltage with reduced harmonic distortion. This reduced THD makes the Cascaded MLIs useful for EV applications. CHBMLIs are more suitable for PV-based EV Charging applications due to their structure. However, like any complex electrical system, CHBMLIs are susceptible to faults, which can lead to system failures if not promptly detected and addressed. In this research, we propose an automated fault detection method for identifying open circuit (OC) and source faults in CHBMLIs using artificial neural networks (ANN). The ANN is trained using a dataset comprising voltage and current measurements under normal operating conditions and various fault scenarios. Once trained, the ANN is capable of accurately detecting OC faults in real-time, thus enabling proactive maintenance and preventing potential system downtime. Simulation results using MATLAB/Simulink demonstrate the effectiveness and reliability of the proposed fault detection approach, highlighting its potential for enhancing the reliability and performance of CHBMLI systems in practical applications.

N. Kalaiarasi (✉) · A. Sivapriya · P. Vishnuram
Department of Electrical and Electronics Engineering, SRM Institute of Science and Technology, Kattankulathur, Chennai 603203, India
e-mail: kalaiarn@srmist.edu.in

A. Sivapriya
e-mail: sa3227@srmist.edu.in

M. Bajaj
Department of Electrical Engineering, Graphic Era (Deemed to be University), Dehradun, India

Hourani Center for Applied Scientific Research, Al-Ahliyya Amman University, Amman, Jordan

Graphic Era Hill University, Dehradun, India

A. R. Singh
Department of Electrical Engineering, School of Physics and Electronic Engineering, Hanjiang Normal University, Hubei Shiyan, People's Republic of China

S. R. Salkuti (ed.), *Artificial Intelligence for Integrated Smart Energy Systems in Electric Vehicles*, Lecture Notes in Electrical Engineering 1427,
https://doi.org/10.1007/978-3-031-94276-1_16

Keywords Fault detection · Multilevel inverter · Neural network · Open circuit fault · Reliability

Nomenclature

ANN	Artificial Neural network
EV	Electric Vehicle
CHBMLI	Cascaded H Bridge Multilevel Inverter
OCF	Open Circuit Fault
SCF	Short Circuit Fault

1 Introduction

Numerous industrial operations necessitate significant power output at medium voltage levels. To fulfill this demand, multilevel inverters have emerged as a viable solution, particularly in high-power and medium-voltage applications. Additionally, they find extensive utility in variable speed drive applications due to their inherent advantages. Multilevel inverters (MLI) play a crucial role in reducing motor stress and safeguarding against damage by generating common mode voltage [1]. Moreover, they exhibit the capability to draw low-distortion input current, thereby minimizing harmonic content and electromagnetic interference. The applications of multilevel inverters span a wide range, including driving automobile systems, serving as active filters, powering electric vehicles, acting as power factor compensators, and utilizing DC power sources, among others. A multilevel inverter achieves its intended output voltage profile by utilizing multiple lower-level DC voltages as inputs, as depicted in Fig. 1, which illustrates various multilevel inverter topologies [2].

Among them, CHBMLI has attracted considerable interest in various industrial applications because it can produce high-quality output waveforms with minimized harmonic distortion. However, like any complex electronic system, cascaded MLIs are susceptible to faults, which can compromise system performance and reliability. Among these faults, OC faults are particularly challenging to detect and diagnose due to their intermittent nature and potential to cause significant damage if left undetected [3]. One of the primary concerns in the power industry is the reliability of multilevel inverters. As a result, for improving reliability, identifying the faults that occur in the switches is important in real-time applications. In reference [4] proposes a diagnosis method for OC fault in CHBMLI using fast Fourier transform and principal component analysis. However, this approach involves complex matrix operations, which are computationally intensive. This complexity can lead to a significant amount of time being required to detect faults—almost two fundamental periods. Meanwhile, in [5], a detection scheme based on wavelet transform for diagnosing asymmetric

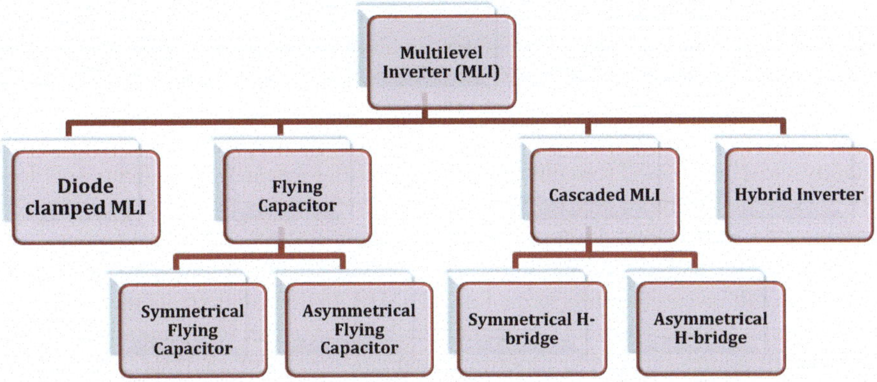

Fig. 1 Different topologies of multilevel inverter

CHBMLI using voltage output is examined. However, this method proves unsuitable for symmetric CHB MLIs as it struggles to precisely distinguish the position of OC fault within a bridge for such inverters. On the other hand, signal processing and machine learning tools are used to analyze and extract the fault information. In [6], a new method utilizing the adaptive kernel with principal component analysis (PCA) alongside a Support Vector Machine (SVM) is introduced to facilitate real-time fault detection in power electronic switches. While SVM demonstrates strong capabilities in classifying small sample data, its effectiveness in broader classification tasks diminishes with larger datasets. An alternative approach to fault diagnosis, relying on normalized DC, is outlined in [7] for active bridge rectifiers and in [8] for voltage source inverters. However, the fault detection accuracy of this method is compromised by fluctuations in the load, resulting in erroneous diagnoses whenever the load changes. The method suggested in [9] efficiently identifies the faulty phase in the inverter, can pinpoint the problematic power switch, and eliminates the need for additional sensors or electrical equipment, thereby preventing an increase in system complexity and expenses. Nevertheless, it's worth noting that in instances of OC faults, particularly due to the unique characteristics of power inverter topology, a defective IGBT may not be precisely pinpointed if the malfunction arises during the demagnetization phase.

In [10], a diagnostic method for identifying faults in clamping diodes and IGBT within NPC-MLI is proposed. This method relies on injecting reactive power for fault classification in switches, making it suitable for grid-connected systems. Monitoring variations in the neutral point current of a T-type Modular Multilevel Inverter (MLI) is another diagnostic approach, as outlined in [11]. This method involves tracking the current flowing through the neutral point of the MLI. The neutral point current can provide valuable insights into the operating conditions and health of the inverter system. Utilizing a lookup table alongside the current switching state to classify faults is a sophisticated method employed in the diagnostic approach for T-type Modular Multilevel Inverters (MLIs). In [12], a fault-tolerant strategy for a flying capacitor (FC) MLI is introduced, utilizing knowledge of the capacitor's charging state and switching states for fault localization. Additionally, [13] addresses the detection of OC faults in IGBT within a two-fundamental cycle of fault-tolerant hybrid flying capacitor (FC)-NPC inverter, employing a module that works using logic-based detection using phase voltage as input. The identification of faulty switches in MLI remains an evolving field of research, with many scholars actively striving for more precise fault detection methods. In [14, 15], the PV-based charging of Electric Vehicle applications has been discussed. However, there is a lack of comprehensive information regarding the practical implementation of high-resolution fault investigation systems for CHBMLI. In this context, the measurement of MLI output voltages poses a significant challenge in accurately pinpointing faulty switches. To address this, real-time implementation of fault analysis schemes has been carried out using MATLAB/Simulink software.

Therefore, this chapter proposes the detection of OC faults in cascaded MLIs using ANN-based techniques. The proposed approach aims to develop a robust fault detection system capable of accurately identifying and localizing OC faults in cascaded MLIs. By harnessing the power of ANNs, we seek to overcome the challenges associated with OC fault detection and enhance the overall performance and reliability of cascaded MLIs in practical applications. The combination of simulation studies and experimental validation demonstrates the efficacy and constructiveness of the proposed ANN-based OC fault detection method. The results of this research are expected to contribute to advancements in fault detection techniques for cascaded MLIs, ultimately leading to improved system efficiency and reliability in various industrial applications.

2 Structure of Cascaded H-Bridge MLI

Figure 2 illustrates the topology of 15-level CHBMLI. Each H-bridge can deliver three distinct voltage levels: 0, + Vdc, and −Vdc. It can operate in single-phase or three-phase modes. Each and every one of the H-bridge cells in the cascaded topology is powered by individual DC supplies arranged in series. The DC source may be in the form of a solar photovoltaic array or battery. This feature makes cascaded H-bridge inverters well-suited for renewable energy systems, motor drives, and grid-tied inverters. Each one of the H-bridge cells comprises four switches and four antiparallel diodes. T1 and T4 receive positive voltage, while T2 and T3 receive negative voltage. The complementary switches T1 and T2 or T3 and T4 should not be turned on at the same time to avoid short-circuiting of the DC source. The faults considered for the proposed fault diagnostic system are single and multiple switch OC faults and source faults.

A voltage of "+ Vdc" is generated by activating the diagonal switches T1 and T4, while a voltage of "−Vdc" is generated by activating other diagonal switches T2 and T3. During OC and SC fault conditions, the output voltage drops in either a positive or negative half-cycle or in both directions. The OCF implies the off state

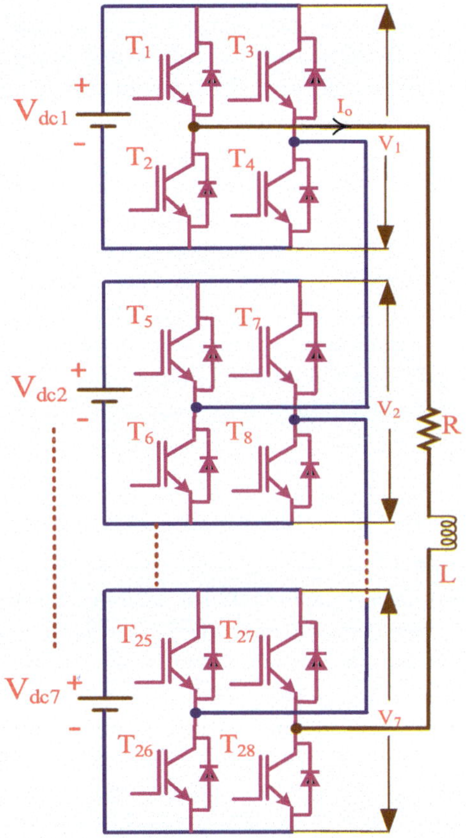

Fig. 2 Fifteen-level CHBMLI

of the semiconductor switches; this condition occurs when the gate signal is lost, the breakdown of the switch, and damage to inner wires. The fast-acting fuse may convert SCF into OCF by inserting them in the circle around the switch [16]. Therefore, the fast-acting fuses are disconnected when SCFs occur, resulting in OCFs. Thus, the detection and categorization of OCF are of utmost importance; thus, it is considered for the analysis.

2.1 Fault Analysis and Feature Extraction

Fault analysis of a cascaded H-Bridge multilevel inverter (CHBMLI) is conducted using MATLAB/Simulink to simulate output voltage and current under normal and various fault scenarios. Initially, an open circuit fault arises at 0.2 s on switch T1 of module 1. Figures 3a, b depict the actual load voltage, its exaggerated representation, and current waveforms before and after the fault occurrence with an RL load. The output current and voltage patterns are analyzed at a frequency of 2 kHz. Similarly, Fig. 3 illustrates the load current and voltage waveforms post the creation of an open-circuit fault in switch T2 of module 1. Faults on the other switches are also introduced, and corresponding voltage and current data are recorded for fault detection purposes.

At 0.2 s, faults are intentionally introduced in switches T1 and T2 to facilitate fault analysis. The fault locations can be readily identified by examining the distinctive patterns in the output voltages, as illustrated in Fig. 3. Subsequently, fault characteristics are derived from the amplitudes of output voltage and current under each fault condition. These extracted features serve as training data for the implementation of a backpropagation neural network (BPNN). The BPNN functions as a classification tool, leveraging the extracted features to discern and classify fault locations accurately. Its outputs directly indicate the precise locations of detected faults, streamlining fault diagnosis in the system.

3 ANN-Based Fault Diagnostic Strategy

A proposed fault diagnostic system for a single-phase fifteen-level inverter employs a multilayer perceptron BPNN. This system utilizes the neural network classification method for the diagnosis of fault in a proposed CHBMLI system, bypassing the challenges associated with using mathematical models. Traditional protection systems for inverters typically rely on passive devices like fuses, circuit breakers, and overload relays. These devices disconnect the supply from the inverter system upon detecting a fault, halting ongoing operations. In contrast, ANN-based techniques, such as the one proposed, operate without the need for mathematical models. There are input layers, hidden layers, and output layers in an ANN architecture. A trial-and-error approach is used to establish the optimal number of hidden layers, as well as the neurons to be considered in each layer, considering the sort of task at hand. After the training phase is over, the results should be checked against test data. Data training and validation are the foundation of ANN algorithms. Table 1 provides the pseudo-code for the proposed fault detection approach.

The network learns by first trying to identify the characteristics of the training set, then comparing its accuracy on the testing set, and then deploying its newly acquired knowledge to an unexplored dataset. The dataset may be split in any number of ways; common methods include splitting the data into 80% and 20% subsets for training and testing respectively. Figure 4 illustrates the architecture of a fault diagnostic system, which consists of four primary stages: feature extraction, neural network classification, fault diagnosis, and calculation of switching patterns with gate signal output. In the feature extraction stage, voltage input signals are transformed, with the rated signal values identified as essential features. The transformed signal output feeds into neural network classification, trained with both abnormal and normal data for multilevel inverter (MLI) systems, resulting in binary code output. This binary code is decoded in fault diagnosis to identify the type of fault and location, facilitating

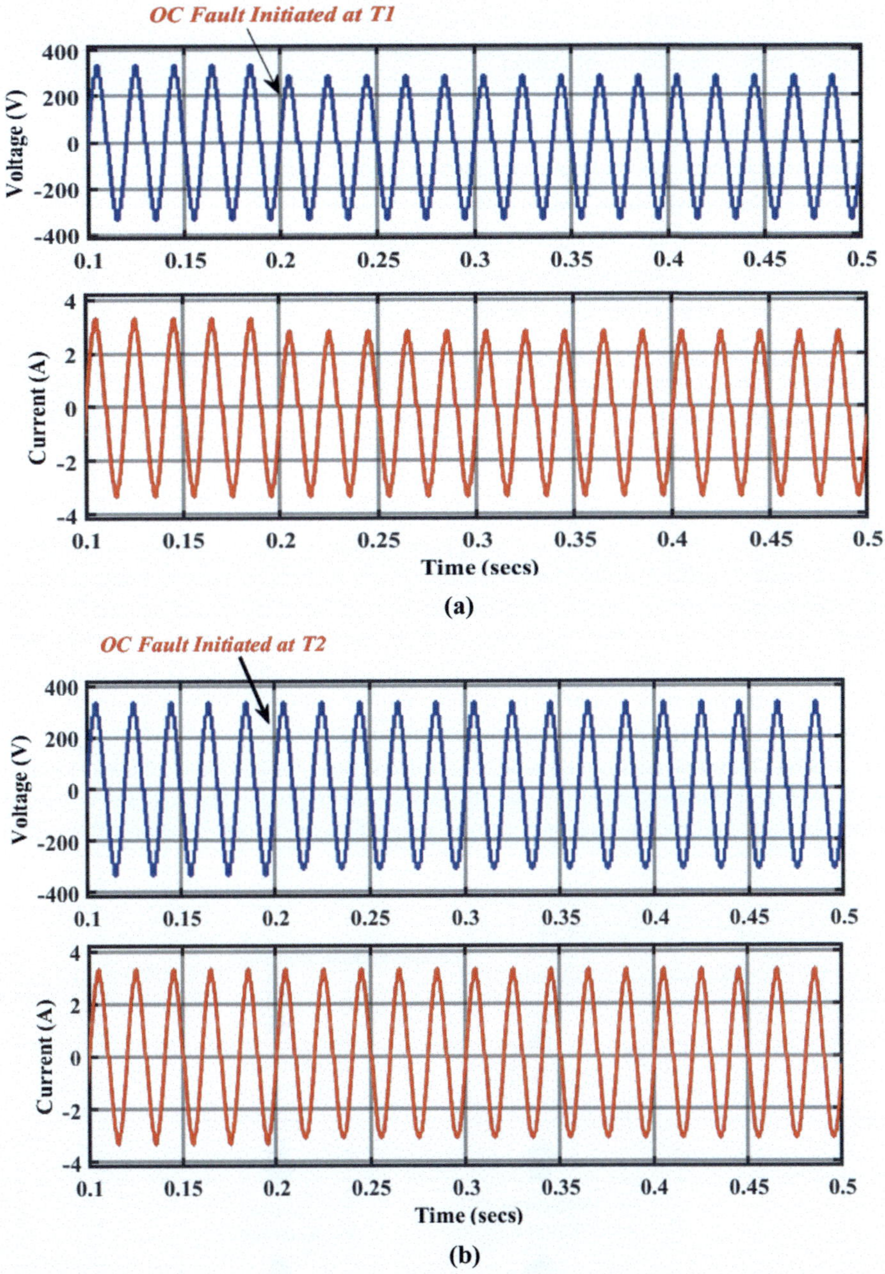

Fig. 3 Output waveforms at OC fault condition **a** T1 fault, **b** T2 fault

Table 1 Pseudo-code for developed fault detection model

Algorithm for the Proposed BP-NN Model
• Voltage data during normal operation (X_normal)
• Voltage data during known fault conditions (X_fault)
• Labels for normal operation and fault conditions (y_normal, y_fault)
• Number of hidden layers (num_hidden_layers)
• Number of neurons in each hidden layer (num_neurons)
• Activation function for hidden layers (activation_hidden)
• Activation function for output layer (activation_output)
• Learning rate (alpha)
• Number of epochs (num_epochs)
• Testing dataset (X_test)
• Corresponding labels for testing dataset (y_test)

the calculation of switching patterns to bypass the failed level in the MLI. A single-phase cascaded multilevel inverter, utilizing 10 V MOSFETs as switching devices, generates output voltage signals. MLI output voltage level (m) is determined by the formula $m = 2Ns + 1$, where Ns represents the number of stages. The MATLAB Simulink is used to model fault feature data with a modulation index. Choosing the right feature extractor is essential to specify neural networks with relevant details and patterns, ensuring optimal network performance.

Extracted features encompass voltage, current, and error values. Networks are tested with test datasets once trained to desired error goals. The test set is given to the network and evaluating the performance of the network based on the error

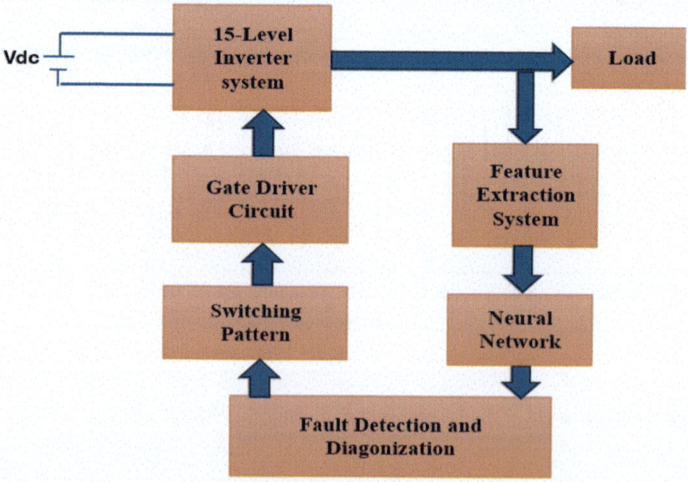

Fig. 4 Block diagram representation of the proposed system

Table 2 Fault label and fault classification for source and switch fault

Faulty sources & switches	Fault label	Sub-label	Fault classification
Normal	[0 0 0 0 0 0 0]	0	No fault
Fault in Vdc1	[1 0 0 0 0 0 0]	1	Source fault
Fault in Vdc2	[0 1 0 0 0 0 0]	2	
Fault in Vdc3	[0 0 1 0 0 0 0]	3	
Fault in Vdc4	[0 0 0 1 0 0 0]	4	
Fault in Vdc5	[0 0 0 0 1 0 0]	5	
Fault in Vdc6	[0 0 0 0 0 1 0]	6	
Fault in Vdc7	[0 0 0 0 0 0 1]	7	
Fault in T1	[1 0 0 0 0 0 0]	1	Single switch fault
Fault in T2	[0 1 0 0 0 0 0]	1	
Fault in T3	[0 0 1 0 0 0 0]	1	
Fault in T4	[0 0 0 1 0 0 0]	1	
Fault in T1&T4	[1 0 0 1 0 0 0]	1	Multiple switch fault
Fault in T2&T3	[0 1 1 0 0 0 0]	1	
Fault in T1&T2	[1 1 0 0 0 0 0]	1	
Fault in T3&T4	[0 0 1 1 0 0 0]	1	
Fault in T1&T3	[1 0 1 0 0 0 0]	1	
Fault in T2&T4	[0 1 0 1 0 0 0]	1	

constitutes testing. To achieve this criterion, test datasets should be generated using training data that covers the whole operating zone. Table 2 illustrates fault codes and fault types for different single and multiple-switch OC faults in CHBMLI. The fault label denoting normal condition is denoted as [0 0 0 0 0 0 0], while a fault in source 1 is represented as [1 0 0 0 0 0 0]. The fault labels for other modules follow a similar pattern. Single switch faults are distinguished by the module number, the initial classification of the faulty module, followed by the corresponding faulty switch label. Likewise, in cases of multiple switch faults, the faulty module number precedes the fault code representing the corresponding switches. Faults in other modules are delineated in the same manner as detailed in Table 2.

4 Simulation and Hardware Results

This part of the section presents the simulation outcomes of the proposed fault diagnosis approach along with its performance evaluations. The model of a single-phase cascaded 15-level inverter is implemented using MATLAB-Simulink. Each DC voltage source has an input voltage of 48 V, while the RL load comprises an inductance of 150 mH and a resistance of 100 Ω. The inverter operates at 2 kHz switching

Table 3 Parameter specification of BP-NN

Number of input layers	20
Number of output layers	4
Number of hidden layers	40
Learning rate	0.01
Number of the training dataset	250
Number of testing datasets	100
Convergence criteria	0.001

frequency, and a sampling frequency of 20 kHz is utilized for data acquisition. OC faults are induced by removing the gating signals of the IGBT during steady state, with the inverter output voltage serving as the fault signature. The output voltage data for both the normal state and different fault conditions are obtained. Each voltage time series spans 10 fundamental periods after the fault occurrence, with 1000 samples taken per period. The information about the proposed neural network model is given in Table 3.

The versatility of the proposed diagnostic method is assessed across varied operational settings, encompassing load fluctuations and modulation index (MI) variations. Despite these fluctuations, both the reference and actual RMS values during the OC fault persist consistently, ensuring the fault detection scheme's reliable performance under varying load conditions. Comparing these reference values with calculated output voltage values of the 15-level CHB MLI facilitates fault detection amidst MI alterations. The simulation of the 15-level CHBMLI is scrutinized utilizing phase disposition pulse width modulation (PDPWM) control techniques, considering both faultless and OC fault scenarios of HB-1. The CHBMLI behavior remains consistent across all conditions, regardless of any specific PDPWM settings, establishing this proposed approach as a generalized approach for OC fault diagnostic systems applicable to detecting faulty switches in N-level CHB MLI. Figure 5 illustrates the various fault conditions including module/source failure, single switch, and multiple switch open circuit fault conditions. The fault is created at 0.2secs and we can see that the changes in the output voltage are observed. The positive edge of the fault detection signal is used to identify the faulty switch in the inverter. The detection time for the algorithm is Ti–Td. Fault classifier determining the position of the faults within 0.02 μs in all HB-1 and all other fault cases. Additionally, OC faults at T1 and T2 are analyzed, with the faulty switch located within a few microseconds due to the extensive calculation steps involved. The efficacy of this classifier is examined in both light-load and low-power-factor conditions.

The experimental setup depicted in Fig. 6 primarily consists of a 15-level MLI, an FPGA controller (EP4CE115F23), an oscilloscope, and a PC, which can potentially be replaced by an inexpensive industrial computer. The detailed hardware circuit diagram of the suggested system is shown in Fig. 7. The proficient fault identification classifier, running on the PC, is utilized to observe and identify the condition of CHBMLI. To simulate the OC fault, the control signal of the IGBT in CHBMLI is

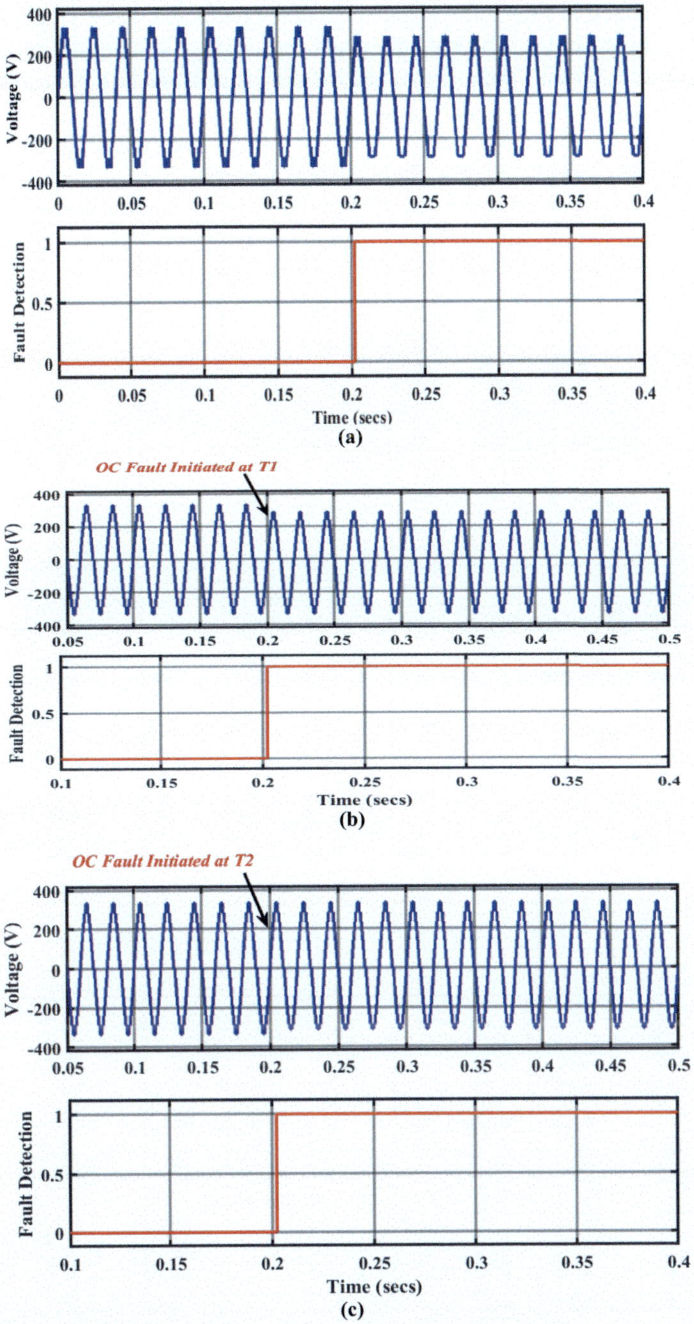

Fig. 5 Results of Fult Detection using simulation in various cases **a** module failure, **b** fault on switch T1, and **c** fault on switch T2

disabled. Illustrated in Fig. 5, the experiments are conducted using T1 and T3 open-circuit faults as examples. The gating signals for various switches are recorded and depicted in Fig. 8. These signals represent the pulses produced by the controller and supplied to the respective switches. Figure 9 displays the output voltage waveform of the MLI under normal conditions, while in Fig. 10, an open circuit fault is induced, and the resulting output voltage with the fault is illustrated.

The hardware results are depicted in Figs. 8, 9 and 10, demonstrating that under these operational parameters, the fault location is pinpointed within 0.5 μs. Utilizing

Fig. 6 Hardware implementation of CHBMLI

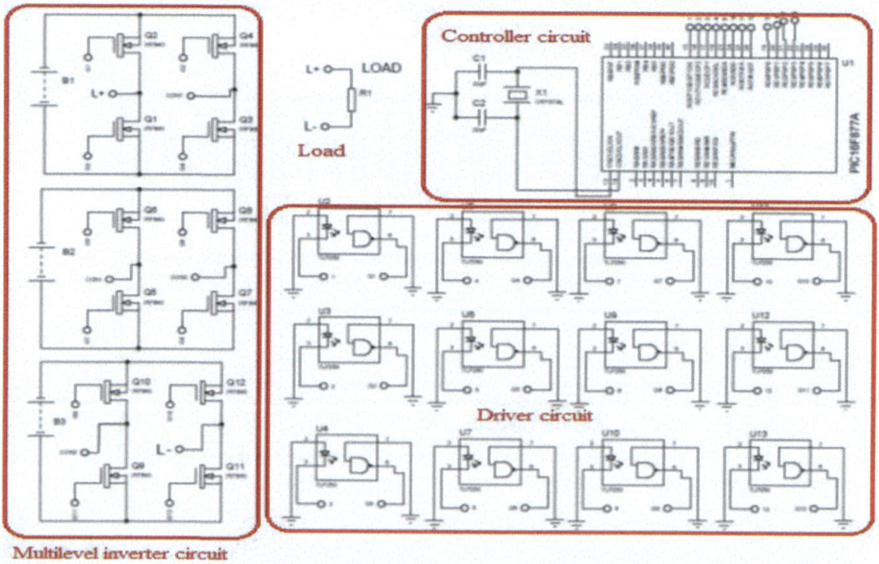

Fig. 7 Detailed hardware schematic diagram of the proposed system

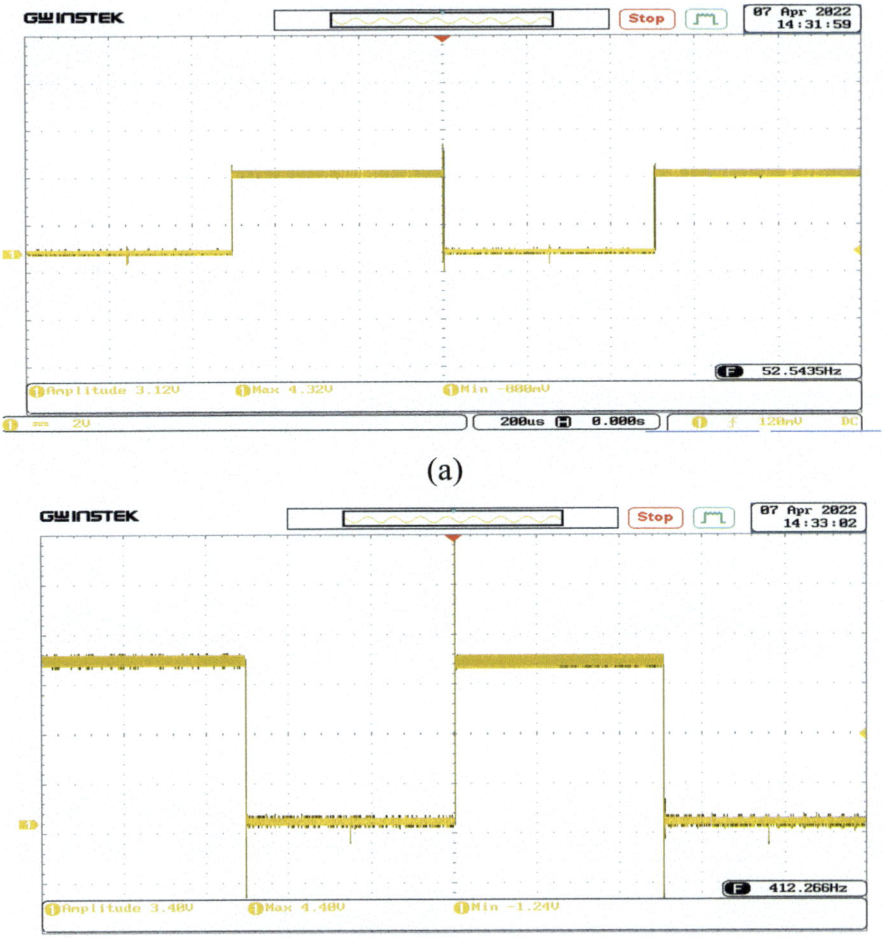

(a)

(b)

Fig. 8 Gating pulses for the switch **a** T1 and **b** T2

bridge voltages for OC fault diagnosis enables the straightforward identification of single OC faults across multiple bridges using the same algorithm. Moreover, the fault detection algorithm adeptly categorizes multiple faults within a bridge, as exemplified by a scenario involving OC faults at T1-T28 in HB-1. Figure 10 validates the efficacy of the proposed diagnostic strategy in identifying multiple faults in one bridge circuit and across another bridge, with multiple OC faults being identified within a few microseconds of the output voltage. The important and necessary attributes of a robust fault diagnosis strategy encompass cost-effectiveness, reliability, precision, optimal detection speed, adaptability to varying operational conditions, reduced computational complexity, and simplified hardware implementation.

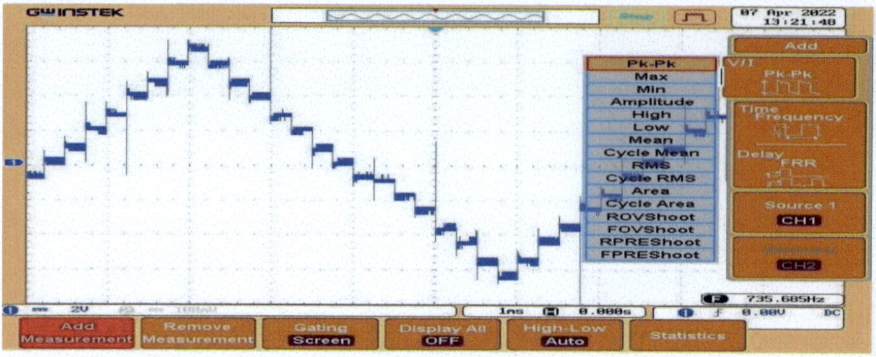

Fig. 9 Output voltage waveform of CHBMLI without fault

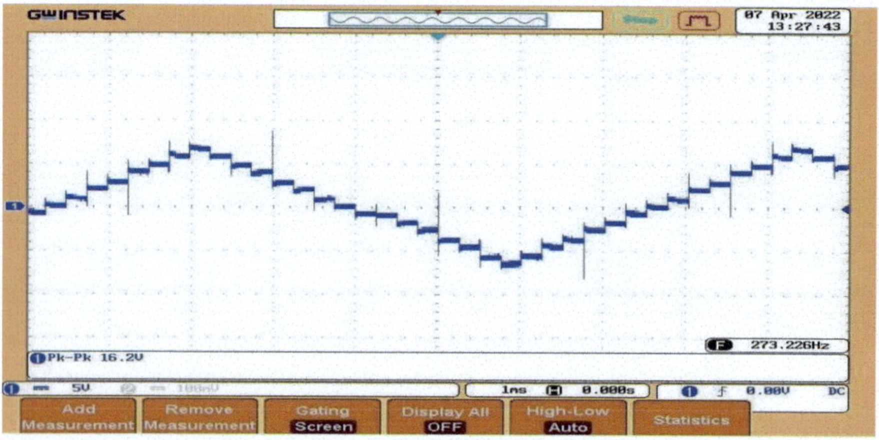

Fig. 10 Output voltage waveform of CHBMLI with fault

5 Comprehensive Evaluation and Comparison with the Existing Approach

In this context, this study endeavors to conduct a thorough assessment and comparative analysis with the current approach. By scrutinizing existing methods and frameworks, this research aims to offer valuable insights, identify potential shortcomings, and propose enhancements or alternative strategies. The performance and efficiency of the proposed fault diagnostic system are calculated using the various evaluation parameters.

5.1 Performance Evaluation

Based on output the inverters are classified as sine wave inverters, square wave inverters, and modified sine wave inverters. A sine wave inverter is a device that transforms the DC voltage into an output voltage similar to the supply voltage. It is one of the merits of improving the efficiency. The square wave is the output of this device. But this device is rarely used. The reason is losses are very high and efficiency is very low. The output of the modified sine wave is a combination of sine and square wave. The shape of the output waveform nearly looks like a sine wave.

Throughout the training of the network, the quantity of hidden neurons was determined to ensure a satisfactory root-mean-squared error (RMSE). The weight function of the neural network is defined throughout the training phase. The training algorithm is supplied with sets of input data and corresponding output labels. Mean Squared Error (MSE) is employed to identify discrepancies between the input and output values, as follows:

$$MSE = \frac{1}{n}\sum_{k=1}^{n}\sqrt{(S_k - Y_k)} \tag{1}$$

where S_k is defined as the desired output, Y_k is measured output, and "n" is the total number of input sets. Figure 11 shows the best validation performance of the proposed approach at epoch 12.

A confusion matrix is defined as a table that provides a summary of a classification algorithm's performance. It displays the counts of true positives (TP), true negatives (TN), false positives (FP), and false negatives (FN). Also, it collectively summarizes the performance of a classification model, providing insights into its accuracy and

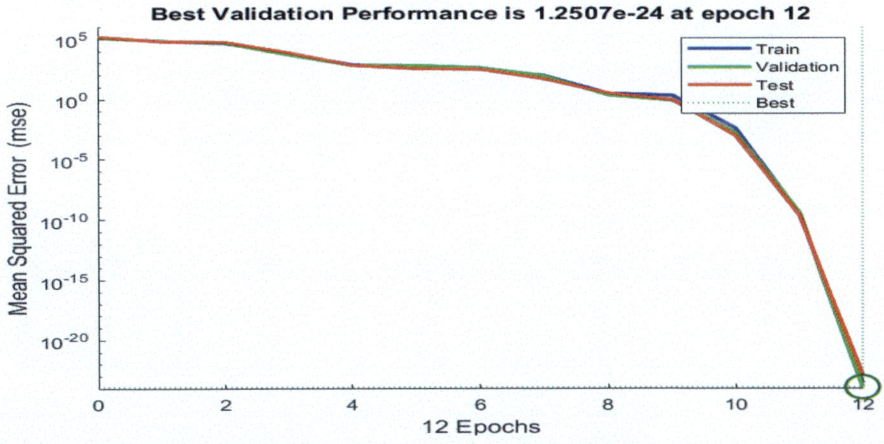

Fig. 11 Best validation performance

error rates. The confusion matrix of the proposed system provides the overall classification results with the output class and predicted class. The TP indicates both the actual class and predicted class are positive. The TN indicates both the actual class and predicted class are negative. On the other hand, FP indicates the actual class is negative and the predicted class is positive called type 1 error. Similarly, the FN indicates the actual is positive and the predicted class is negative called type 2 error. Figure 12 depicts the confusion matrix for the proposed fault detection system with a diagonal indicating the error rate of the four different fault classes.

The effectiveness of the suggested networks is evaluated using the provided data. From the simulation results and comprehensive analysis, it's evident that the disparity between the actual and intended output data post-testing falls below the predetermined error threshold, indicating the training process's success. Notably, each network achieves a classification accuracy of approximately 99%, demonstrating satisfactory performance under test conditions similar to those encountered during training.

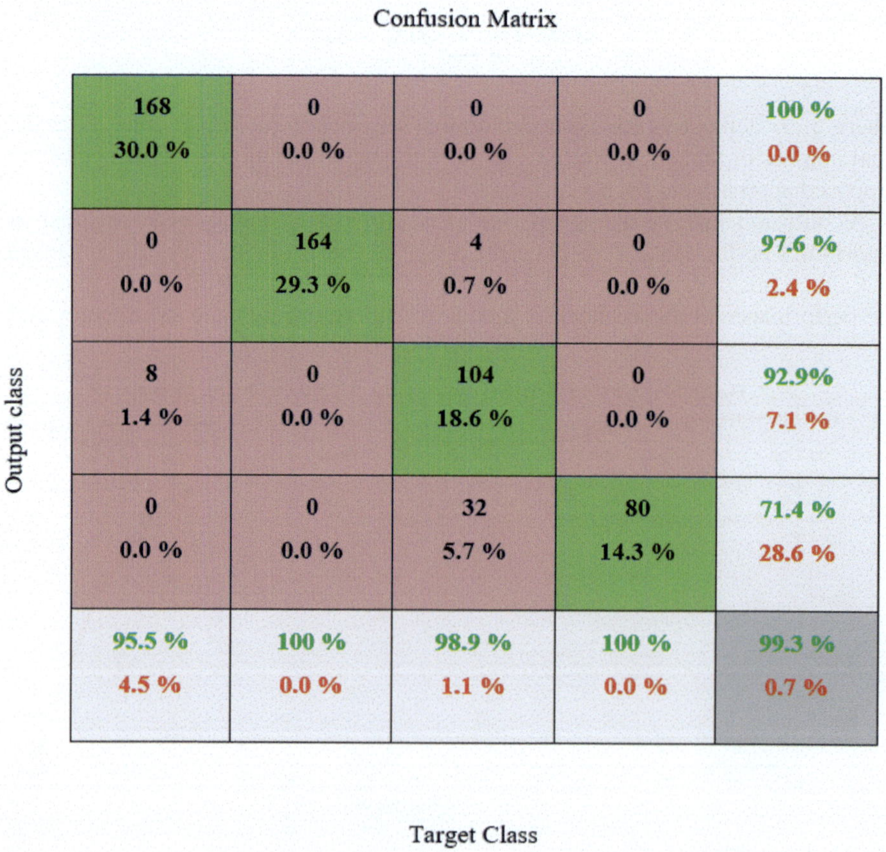

Fig. 12 Confusion matrix of the proposed system

5.2 Comparative Analysis

The examination confirms that the proposed classifier exhibits all these attributes, rendering it a superior strategy for switch fault diagnosis in CHBMLI compared to existing systems. Diagnostic methodologies outlined in literature [4, 17–19] necessitate the use of input data processing tools such as principal component analysis and discrete Fourier transform, relying on heuristic approaches to pinpoint fault locations. However, training these type of fault classifiers is time-intensive, demanding a substantial amount of training and testing data as N increases. Moreover, the detection methods discussed in [9, 20–22] cannot identify individual faulty switches. Techniques discussed in [4, 9, 23–25] typically require more than one fundamental period to detect OC fault positions, with unverified effectiveness across diverse operating conditions. Generally, for an N-level CHBMLI, the proposed system can be completed in less than a few microseconds of the output voltage under varying operational scenarios. The fault identification time remains consistent across different CHBMLI levels due to the reduced steps involved in extracting fault features. A reliable diagnostic method utilizing immediate voltage deviation within power electronic converters is suggested for detecting open-circuit faults in IGBTs [26]. The comparative analysis of the proposed system with the existing fault diagnostic system is represented in Fig. 13.

In [27], the fault diagnosis using PCA is performed to detect open circuit, short circuit, power failure and sensor failure of dry type transformer. However, the fault diagnostic time is about 10 ms and the accuracy rate is 87.9%. In [28], the SVM based fault detection is performed to reduce the dimensionality of input data for process monitoring system. Nevertheless, the fault detection method has slightly increased accuracy compared to, but the requirement of fault data and system is more complex. In [29], fuzzy logic-based fault detection technique is proposed for three

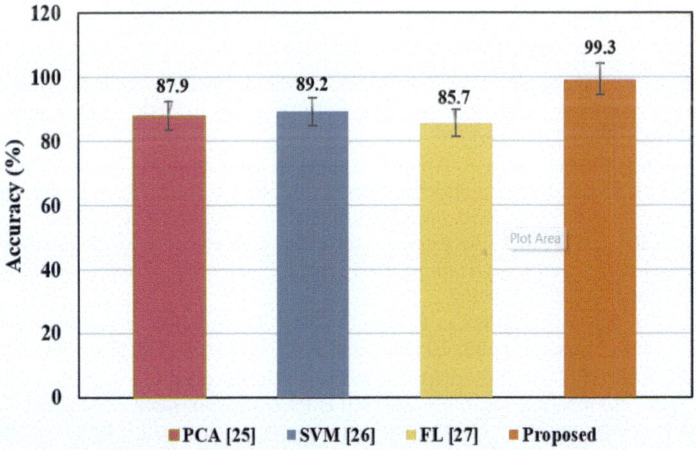

Fig. 13 Accuracy comparison of the different fault detection methods

phase cascaded multilevel inverter fed induction motor drive. The proposed system diagnoses the single and multiple fault conditions. However, the system reliability entirely relies on current based fault diagnosis. The fault current data changes with the load condition thus the average accuracy of the system will diminish. Thus, the proposed voltage signal-based fault feature extraction system has higher accuracy in fault detection of about 99% for various single and multiple fault condition of IGBT switches and source failures.

6 Conclusion

This research presents a novel approach for automated open circuit fault detection in CHBMLI utilizing ANN. By training the ANN with comprehensive datasets comprising voltage and current measurements obtained under normal and faulty conditions, we have developed a robust fault detection system capable of accurately identifying open circuit faults and source faults in real time. The above results and discussion validate the efficacy of the proposed method, indicating high detection accuracy and reliability across various fault scenarios. The successful implementation of automated fault detection contributes to the enhancement of CHBMLI system reliability and performance, facilitating timely maintenance interventions and minimizing system downtime. Future work may focus on extending this approach to detect and diagnose other types of faults in CHBMLIs, as well as exploring practical implementation strategies for real-world applications. Overall, the proposed ANN-based fault detection method offers a promising avenue for improving the operational efficiency and reliability of CHBMLI systems in diverse industrial settings.

References

1. Poorfakhraei A, Narimani M, Emadi A (2021) A review of modulation and control techniques for multilevel inverters in traction applications. IEEE Access 9:24187–24204. https://doi.org/10.1109/ACCESS.2021.3056612
2. Maheswari KT, Bharanikumar R, Arjun V, Amrish R, Bhuvanesh M (2021) A comprehensive review on cascaded H-bridge multilevel inverter for medium voltage high power applications. Mater Today Proc 45:2666–2670. https://doi.org/10.1016/j.matpr.2020.11.519
3. Priya AS, Kalaiarasi N (2022) An extensive review on fault detection and fault-tolerant control of multilevel inverter with applications. Int J Renew Energy Res 12:768–798. https://doi.org/10.20508/ijrer.v12i2.12586.g8465
4. Wang T, Qi J, Xu H, Wang Y, Liu L, Gao D (2016) Fault diagnosis method based on FFT RPCA-SVM for cascaded-multilevel inverter. ISA Trans 60:156–163. https://doi.org/10.1016/j.isatra.2015.11.018
5. Anand A, Raj N, George S, Jagadanand G (2017) Wavelet-based open switch fault diagnosis in cascaded H-bridge multilevel inverter-fed induction motor drive. IEEE Reg 10 Annu Int Conf Proceedings/TENCON 545–550. https://doi.org/10.1109/TENCON.2017.8227923

6. Sarita K, Kumar S, Saket RK (2021) OC fault diagnosis of multilevel inverter using SVM technique and detection algorithm. Comput Electr Eng 96:1–14. https://doi.org/10.1016/j.compeleceng.2021.107481
7. Salehifar M, Arashloo R S, Moreno-Eguilaz M, Sala V, Romeral L (2015) Observer-based open transistor fault diagnosis and fault-tolerant control of five-phase permanent magnet motor drive for application in electric vehicles. IET Power Electron 8:76–87. https://doi.org/10.1049/iet-pel.2013.0949
8. Chitra S, Valluvan KR (2020) Design and implementation of cascaded H-Bridge multilevel inverter using FPGA with multiple carrier phase disposition modulation scheme. Microprocess Microsyst 76:103–108. https://doi.org/10.1016/j.micpro.2020.103108
9. Yang Q, Qin J, Saeedifard M (2016) Analysis, detection, and location of open-switch submodule failures in a modular multilevel converter. IEEE Trans Power Deliv 31:155–164. https://doi.org/10.1109/TPWRD.2015.2477476
10. Rinsha V, Jagadanand G (2020) Open switch fault-DCM mode analysis of neutral point clamped inverter. 9th IEEE Int Conf Power Electron Drives Energy Syst 9–14. https://doi.org/10.1109/PEDES49360.2020.9379629
11. Choi UM, Lee K B, Blaabjerg F (2014) Diagnosis and tolerant strategy of an open-switch fault for t-type three-level inverter systems. IEEE Trans Ind Appl 50:495–508. https://doi.org/10.1109/TIA.2013.2269531
12. Fard MT, Khan WA, He J, Weise N, Abarzadeh M (2020) Fast online diagnosis of open-circuit switching faults in flying capacitor multilevel inverters. Chinese J Electr Eng 6:53–62. https://doi.org/10.23919/CJEE.2020.000030
13. Amini J et al (2013) Finite-set model-based predictive control for flying-capacitor converters: cost function design and efficient FPGA implementation. IEEE Trans Ind Electron 62:2054–2063. https://doi.org/10.1109/TIE.2014.2354591
14. Dhanamjayulu C, Padmanaban S, Ramachandaramurthy VK, Nielsen JBH, Blaabjerg F (2023) Design and implementation of multilevel inverters for electric vehicles. IEEE Access 9:317–338. https://doi.org/10.1109/ACCESS.2020.3046493
15. Sheir A, Youssef MZ, Orabi M (2019) A novel bidirectional T-type multilevel inverter for electric vehicle applications. IEEE Trans Power Electronics 34:6648–6658. https://doi.org/10.1109/TPEL.2018.2871624
16. Saketi SK, Chaturvedi P, Suryawanshi HM, Yadeo D, Atkar D (2019) A new fault tolerant single phase 5-level inverter topology with capacitor voltage balancing for solid state transformer. IECON Proc 4595–4560. https://doi.org/10.1109/IECON.2019.8926860
17. Hao X, Jian Z, Jie Q, Wang T, Han J (2014) RPCA-SVM fault diagnosis strategy of cascaded H-bridge multilevel inverters. In: 1st international conference on green energy, pp 164–169. https://doi.org/10.1109/ICGE.2014.6835416
18. Kumar GK, Parimalasundar E, Elangovan D, Sanjeevikumar P, Lannuzzo F, Holm-Nielsen JB (2020) Fault investigation in cascaded H-bridge multilevel inverter through fast fourier transform and artificial neural network approach. Energies 13:1–15. https://doi.org/10.3390/en13061299
19. Guerreiro Carvalho Cunha R, Da Silva E T, Marques de Sá Medeiros C (2021) Machine learning and multiresolution decomposition for embedded applications to detect short-circuit in induction motors. Comput Ind 129:1–10. https://doi.org/10.1016/j.compind.2021.103461
20. Kumar M (2021) Open circuit fault detection and switch identification for LS-PWM H-bridge inverter. IEEE Trans Circuits Syst II Express Briefs 68:1363–1367. https://doi.org/10.1109/TCSII.2020.3035241
21. Shaoyong Y, Dawei X, Angus B, Philip M, Li R, Peter T (2010) Condition monitoring for device reliability in power electronic converters: a review. IEEE Trans Power Electron 25:2734–2752. https://doi.org/10.1109/TPEL.2010.2049377
22. Oghorada OJK, Zhang L, Esan BA, Dickson E (2019) Carrier-based sinusoidal pulse-width modulation techniques for flying capacitor modular multi-level cascaded converter Heliyon 5:1–16. https://doi.org/10.1016/j.heliyon.2019.e03022

23. Moosavi SS, Kazemi A, Akbari H (2019) A comparison of various open-circuit fault detection methods in the IGBT-based DC/AC inverter used in electric vehicle. Eng Fail Anal 96:223–235. https://doi.org/10.1016/j.engfailanal.2018.09.020
24. Kim DE, Lee DC (2009) Fault diagnosis of three-phase PWM inverters using wavelet and SVM. J Power Electron 9:377–385. https://doi.org/10.6113/JPE.2009.9.3.377
25. Benyamina RBA, Moulahoum S, Colak I (2016) Hybrid fuzzy logic-artificial neural network controller for shunt active power filter. 2016 IEEE Int Conf Renew Energy Res Appl ICRERA 2016 5:837–844. https://doi.org/10.1109/ICRERA.2016.7884454
26. Sivapriya A, Kalaiarasi N, Verma R, Chokkalingam B, Munda JL (2023) Fault diagnosis of cascaded multilevel inverter using multiscale kernel convolutional neural network. IEEE Access 11:79513–79530. https://doi.org/10.1109/ACCESS.2023.3299852
27. Zheng M, Yang K, Shang C, Luo Y (2022) A PCA-LSTM-based method for fault diagnosis and data recovery of dry-type transformer temperature monitoring sensor. Appl Sci 12:5624–5640. https://doi.org/10.3390/app12115624
28. Onel M, Kieslich CA, Pistikopoulos EN (2019) A nonlinear support vector machine-based feature selection approach for fault detection and diagnosis: application to the Tennessee Eastman process. AIChE J 65:992–1005. https://doi.org/10.1002/aic.16497
29. Kuraku NVP, He Y, Shi T, Gatla RK, Yi R (2019) Fuzzy logic based open-circuit fault diagnosis in IGBT for CMLI fed PMSM drive. Microelectron Reliab 100–101. https://doi.org/10.1016/j.microrel.2019.113415

Introduction to Smart Grids and Its Key Components

Krishna Mohan Reddy Pothireddy, Gummala Babu, Kampa Narendra Babu, Mulaswaminaidu Madepalli, Kandula Pavan Kalyan, Sandeep Vuddanti, and Surender Reddy Salkuti

Abstract This chapter addresses the need for smart grids (SGs), key terminologies and advantages of SG technologies, converters, and maximum power point tracking methodologies available in SG, charging infrastructure, and artificial intelligence techniques. Customers may take an active role in energy management through demand response, which allows them to react to pricing signals and modify their energy use appropriately. Integration of electric cars with the infrastructure needed for charging them, enabling grid balancing and environmentally friendly mobility. Further, a case study was considered to show the methodology to perform energy management in a microgrid for an IEEE 33 bus system to assess the impact of load on the unit commitment status of generators, dispatch schedule, and operational cost of the microgrid.

Keywords Smart grid · Renewable energy integration · Charging technologies · Artificial intelligence · Unit commitment

Nomenclature

NN	Neural network
NC	Neural computing
MPPT	Maximum power point tracking
EC	Evolutionary computation
PWM	Pulse width modulator

K. M. R. Pothireddy · G. Babu · K. N. Babu · M. Madepalli · K. P. Kalyan · S. Vuddanti
Department of Electrical Engineering, National Institute of Technology Andhra Pradesh, Tadepalligudem, Andhra Pradesh, India
e-mail: sandeep@nitandhra.ac.in

S. R. Salkuti (✉)
Department of Railroad and Electrical Engineering, Woosong University, Daejeon 34606, Republic of Korea
e-mail: surender@wsu.ac.kr

S. R. Salkuti (ed.), *Artificial Intelligence for Integrated Smart Energy Systems in Electric Vehicles*, Lecture Notes in Electrical Engineering 1427,
https://doi.org/10.1007/978-3-031-94276-1_17

AI	Artificial intelligence
PHEVs	Plug-in hybrid electric vehicles
ANN	Artificial neural network
CC	Constant current
SG	Smart Grid
CV	Constant voltage
AC	Alternating current
SC	Soft computing
DC	Direct current
FL	Fuzzy logic
EVs	Electric vehicles

1 Introduction

Natural gas, oil, and coal are a couple of non-renewable, scarce resources used in the generation of conventional energy [1]. Conventional power plants are often sensitive to fluctuations in fuel prices, making energy costs susceptible to geopolitical events and market dynamics [2]. A substantial amount of greenhouse gases, predominantly carbon dioxide, are released by conventional generators, particularly those that burn coal and natural gas. These emissions influence both air pollution and climate change [3]. Since conventional power plants are usually centralized, supplying end consumers with electricity necessitates large transmission and distribution networks. Losses in energy during transmission could result from this. A microgrid that is connected to the main power grid functions in unison with it. When necessary, it may pull power from the grid and, under some situations, inject additional energy into it. Microgrids may perform grid support operations including voltage control and frequency regulation, which helps to maintain the reliability and stability of the main power grid. A microgrid functions independently of the main power grid while it is in islanded mode [4]. This mode is particularly useful during grid outages or emergencies. "A network for electricity transmission and distribution systems that enhances the efficiency, affordability, reliability, and security of energy distribution and utilization through the use of advanced technologies, two-way, advanced communications, and specialized computers". The smart grid (SG) has many advantages, yet implementing it into operation has drawbacks as well, such as the requirement for major infrastructure changes, cybersecurity risks, and issues with interoperability. A SG deployment's effectiveness also depends on safeguarding data privacy and taking regulatory and legal issues into account. The SG remains crucial for establishing a robust, and sustainable energy ecosystem as technology advances. Figure 1 indicates the key features of SG interacting through secure communication. SMART abbreviation: Sustainable, Manageable, Adaptable, Robust and Trusted.

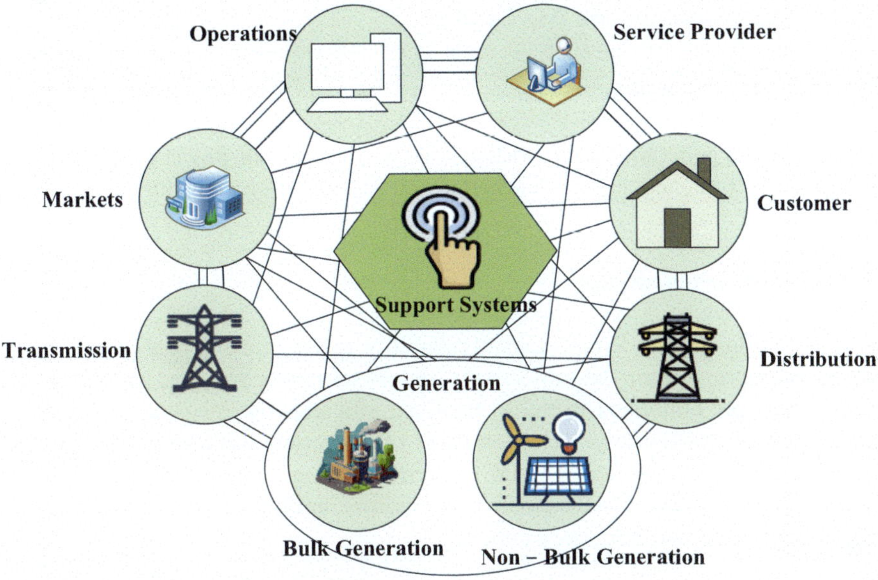

Fig. 1 The conceptual structure of the smart grid

1.1 Factors Contributing to the Need for SGs

- It is particularly relevant in quickly developing nations like India and China. The International Energy Agency estimates a 40% increase in electricity demand by 2030. To meet demand, a 1 GW power plant will be required weekly for the next 20 years. Generation must expand while reducing greenhouse gas emissions.
- Digitalization of economies requires a more stable power supply; minimal interruptions may result in substantial economic losses.
- The grid will be used in new ways in the future, needing a complete renovation. It involves providing customer feedback and moving beyond the conventional power supply model, building home charging facilities for electric vehicles.
- To meet the expected demand for a 40% increase, 357 coal-fired facilities would be required under existing conditions. Adding conventional power plants cannot achieve both environmental and economic stability aims simultaneously. 80% of the world's main source of energy originates from fossil fuels, which also contribute significantly to greenhouse gas emissions. Coal accounts for 40% of worldwide electricity production, but CO_2 alone contributes to 50% of global warming. SG technology is essential for generating more energy from renewable sources and reducing reliance on fossil fuels.
- Renewable energy generation from intermittent and variable sources like sun and wind poses additional challenges. Energy storage is essential for integrating and improving efficiency throughout the value chain. Conservation is an essential part of a future national energy policy. SG improves consumer visibility.

1.2 *Components of the SG System*

1.2.1 Smart Generation

Optimize power generation by providing low-cost, low-carbon power generation and storage, as well as encouraging distributed and renewable energy sources [5].

1.2.2 Smart Transmission

To improve transmission efficiency and reliability, consider using phasor measuring units to measure system state effectively and avoid power outages. Various types of converters are employed in SG to facilitate efficient transmission, distribution, and utilization of electricity. Common types of converters used in SG are AC-DC converters, DC-AC converters, DC-DC converters, voltage source converters, current source converters, static synchronous compensators, unified power flow controllers, voltage regulators, and frequency controllers.

1.2.3 AC-DC Converters

AC-DC Converters also known as rectifiers, convert alternating current (AC) electricity into direct current (DC) electricity. These converters serve various purposes in SG. Many renewable energy sources, such as wind turbines, generate electricity in AC form. AC to DC converters are used to convert this AC electricity into DC compatible with the DC bus, enabling seamless integration of renewable energy into the grid. Energy storage systems, such as batteries and supercapacitors, store electricity in DC form. AC to DC converters are utilized to charge these storage systems from the grid, converting AC electricity to DC for storage. AC to DC converters can be employed in power conditioning equipment to improve power quality by filtering out harmonics and voltage fluctuations from the grid. This helps maintain a stable and reliable power supply to consumers, reducing the likelihood of disruptions or damage to equipment. With the increasing adoption of electric vehicles (EVs), AC-to-DC converters are used in EV charging stations to convert AC electricity from the grid into DC electricity suitable for charging EV batteries. In some cases, AC to DC converters are used in grid stabilization systems, such as flexible AC transmission systems devices. These converters can help regulate voltage and reactive power flow in the grid, improving stability and efficiency. In remote or off-grid areas, AC-to-DC converters can be used to convert AC power from generators or renewable energy sources into DC power for local distribution. This allows for efficient utilization of locally generated power and reduces the need for long-distance transmission lines.

1.2.4 DC-DC Converters

DC-DC converters convert direct current electricity from one voltage level to another with the help of storage elements and power electronic switches. Voltages may vary due to fluctuations in power demand, grid conditions, and power generation in an SG. DC-to-DC converters are used to control voltage levels within the grid standards, ensuring that devices and systems receive the appropriate voltage for safe operation. This is particularly important in DC microgrids and distributed energy resource systems where voltage control is essential for stability and efficiency. SGs increasingly combine both AC and DC components, known as hybrid grids. DC-DC converters enable seamless integration between AC and DC systems by converting power between different voltage levels and facilitating efficient energy transfer and utilization across the grid. Batteries and supercapacitors like energy storage systems store and discharge energy in DC form. DC-DC converters are used to control the charging and discharging processes, assuring efficient energy transfer between the storage system and the grid. They also facilitate bidirectional power flow, allowing energy storage systems to both store excess energy from renewable sources and supply energy back to the grid when required. DC-DC converters can be used in grid stabilization systems to control power flow, voltage, and frequency. They help reduce fluctuations in power supply and demand to improve power quality and improve grid stability by providing dynamic control over energy transmission and distribution. DC-DC converters are essential for integrating renewable energy sources like solar photovoltaic (PV) panels and wind turbines into the grid by converting DC power from these sources into the appropriate voltage levels for grid connection and distribution. DC-DC converters play an important role in EV charging infrastructure. They convert AC power from charging stations or the grid into DC power compatible with the vehicle's battery system with advanced control techniques, enabling efficient and rapid charging of EVs.

1.2.5 Various MPPT Techniques Available

The concept of maximum power point tracking (MPPT) came to live light from the 1970s onwards, because of the PV system. As solar energy gaining high interest, so for increasing of efficiency and performance of panels with maximum output power observed by operators through it. Later on, in the 1990s many researchers and academicians concentrated on improving the significance, and development of various algorithms, techniques, and electronic components for accurately and efficiently tracking the MPP under varying environmental conditions. Since the late twentieth century, it become a standard feature in grid-connected and off-grid connection, enabling higher energy production, enhanced system efficiency, and improved integration of solar power into the electricity grid. It offers several advantages in solar energy systems but also has some challenging aspects such as complexity, cost, and maintenance requirements. Potential for efficiency losses, reliability, and compatibility issues. These challenging factors are essential for evaluating the system and whether it is suitable to overcome the drawbacks related to solar energy application. The main aim of using this technique is to produce maximum energy and tune the operating points of the PV system. The perturb and observe algorithm adjusts the operating point in the direction that increases the output power. Recently, variable step size perturb and observer algorithm adjusts the perturbation step size dynamically raised on the rate of change of output power. It increases the step size when the change in power output is very small to converge rapidly towards it and reduces the step size to prevent fluctuations around the MPPT. An incremental conductance algorithm is used to calculate the derivative of the voltage curve of the solar panel and compare it with instantaneous conductance. By tracking the representation of the difference, the algorithm adjusts the operating point to track it. Mostly this method is highly used in practical and real-time applications and enormous challenges are faced while building it either components-wise or logic-wise. The fractional open circuit voltage algorithm uses the portion of open circuit voltage of the solar panel i.e., calculating the voltage corresponding to the desired fraction of open circuit voltage and adjusting the operating point their voltage to maximize power output. The fractional short circuit current algorithm uses the portion of short circuit current of the solar panel i.e., calculating the current corresponding to the desired fraction short circuit current and adjusts the operating point towards their current to maximize output power. Model predictive control algorithm predicts the dynamics of the system to anticipate future changes in environmental conditions.

With optimization the control actions can be in the predictive horizon so, this algorithm can pathway that. Neural networks (NNs) and fuzzy logic control-based algorithms are coming into existence because of their advanced application linked with the human brain. So neural network-based MPPT uses artificial neural networks to learn the relationship between input parameters and optimal control arrangements. Once training is done, the NN can accurately predict it under various conditions correlated to improve efficiency. The FLC-based algorithm has rules based on expert knowledge or empirical data by considering variables like low, medium, and high. Some adapting rules are also employed to improve the actions under changing conditions.

1.2.6 Smart Distribution

Real-time monitoring and analysis systems can detect and anticipate cable faults. Smart metering incorporates two-way communication between consumers and power providers [6]. It collects billing data, recognizes interruptions, and dispatches maintenance teams to the correct location. It also serves as a control point for users and their electronic devices for load control. Others include demand response, residential solar power, and plug-in hybrid electric vehicles (PHEVs).

1.3 Study of Battery Charging Techniques

An essential component of energy storage systems research is the study of battery charging methods. It includes investigating ways to securely and effectively replenish battery energy, improving both their lifespan and performance. To avoid overcharging or overheating, it is necessary to concentrate on improving voltage regulation, current management, and charging algorithms. To meet the needs of various applications, methods such as trickle charging, smart charging, and rapid charging need to be investigated. The design of robust and energy-efficient charging systems is aided by an understanding of the electrochemical processes involved in charging. Developments in this area support the creation of electric cars, portable gadgets, and sustainable energy sources. To achieve a greener and more sustainable future, it is essential to combine battery charging processes with renewable energy sources (Table 2).

1.3.1 Specific Energy (Wh/Kg)

Specific energy represents the amount of energy a battery can store per unit of mass. It is measured in watt-hours per kilogram (Wh/Kg). A higher specific energy indicates that the battery can store more energy in a lighter weight [7].

1.3.2 Specific Power (W/Kg)

Specific power measures the rate at which a battery can deliver energy per unit of mass. It is expressed in watts per kilogram (W/Kg). Batteries with higher specific power can provide more power quickly, which is crucial for applications requiring rapid energy discharge.

1.3.3 Energy Density (Wh/Lt)

Energy density refers to the amount of energy a battery can store per unit of volume. It is measured in watt-hours per liter (Wh/Lt). A higher energy density indicates that the battery can store more energy in a smaller space, making it suitable for compact devices or applications with limited space.

1.3.4 Operating Temperature (C)

Operating temperature represents the range of temperatures within which a battery can function effectively. It is crucial to determine the suitability of a battery in different environments. Some batteries operate optimally at ambient temperatures, while others, like sodium-sulfur batteries, require elevated temperatures for efficient performance.

1.3.5 Efficiency (%)

Efficiency measures how effectively a battery can convert stored energy into usable electrical power and vice versa. It is expressed as a percentage. Higher efficiency values indicate less energy loss during the charging and discharging processes, contributing to overall energy conservation.

1.3.6 Life Cycle

Life Cycle in, the context of batteries, refers to the total number of charge and discharge cycles a battery can undergo before its performance significantly degrades. It is a critical factor in determining the longevity and cost-effectiveness of a battery. Longer life cycles contribute to extended use and reduced environmental impact. The table provides a comprehensive overview of various battery types, highlighting their specific characteristics and performance metrics in terms of specific energy, specific power, energy density, operating temperature, efficiency, and life cycle.

1.4 Charging Techniques

Charging techniques play a very important role in optimizing the performance, safety, and lifespan of batteries. Several methods are employed to efficiently refill energy in various types of batteries. Constant current (CC) charging [8], constant voltage (CV) charging, trickle charging, pulse charging, fast charging, smart charging, inductive charging, and solar charging are the available charging methods.

1.4.1 Constant Current (CC) Mode of Charging

Steady current chargers adjust the voltage they add to the battery to maintain a constant flow of current, shutting off when the voltage reaches the maximum charge level. Whenever we are giving to supply to the battery, the battery takes the charge from the negative terminal to the positive terminal. The battery continuously takes the charge from the supply up to reach full of the battery current as shown in Fig. 2. Current is constant throughout it reaches the maximum value so that's why we are called it as constant current method. While charging the battery the battery cell voltage is increased gradually to reach the maximum voltage.

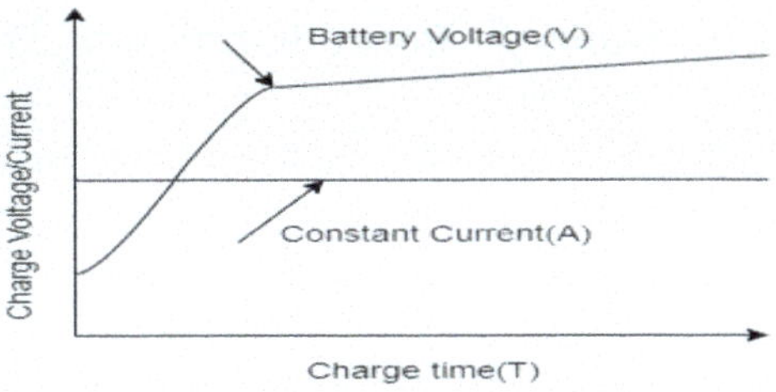

Fig. 2 Constant current charging

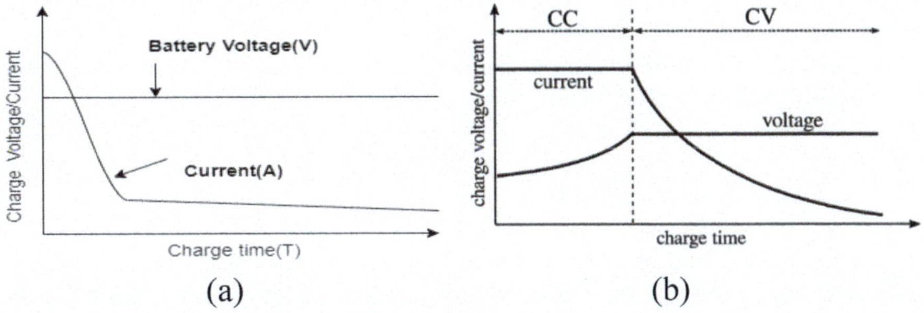

Fig. 3 **a**, **b** Constant current charging, constant current, and constant voltage method, respectively

1.4.2 Constant Voltage Charging Method

A constant voltage charge is essentially a DC power supply that provides the DC voltage for charging the battery. The benefits of continuous voltage charging are that it allows charges to be made for cells with different capacities and at various degrees of discharge. The large charge current at the start of the charge is of fairly short duration and does not harm the cell. Whenever we are giving to supply to the battery, the battery takes the charge from the negative terminal to the positive terminal. The battery is continuously taking the charge from the supply up to reach full of the battery voltage, by that time the current is decreased gradually. In this method, the voltage is constantly giving charge to the battery so that's why we are called as CV method as shown in Fig. 3a, b.

1.4.3 Constant Current and Constant Voltage Method

This method of charging is a combination of CC and CV Mode of Charging. There are two stages in this method, 1st stage a constant current is applied and voltage is increased to the pre-set value in the second stage [8, 9]. Constant voltage is applied and the current is decreasing until it reaches the full charge. The CC-CV mode gives the best results when compared with the above methods in terms of battery lifetime and charging time.

1.5 State of Charge Estimation Methods

Open circuit voltage method, coulomb counting method, Kalman filtering, extended Kalman filtering, neural network, adaptive filtering and machine learning techniques.

1.5.1 Open Circuit Voltage Method

The open-circuit voltage method analyses circuits by measuring voltage without current flow. To determine equivalent voltage (V_{th}), Thevenin's theorem is applied during which the load is disconnected. Useful for circuit analysis and troubleshooting, revealing behavior without a load [10].

1.5.2 Coulomb Counting Method

The Coulomb Counting Method tracks battery state by measuring the cumulative charge entering or leaving. Integral in battery management systems for electric vehicles and electronics. Crucial for monitoring battery health, estimating remaining capacity, and optimizing charging strategies.

1.5.3 Kalman Filtering

The Kalman Filter Method is an algorithm used for state estimation, particularly in dynamic systems where measurements are prone to noise. It combines predictions from a mathematical model with real-world measurements, continuously refining estimates to improve accuracy. Applied in various fields like navigation, control systems, and sensor fusion, the Kalman Filter is essential for optimal state estimation in dynamic and uncertain environments.

1.5.4 Extended Kalman Filtering

It is an extension of the Kalman Filter designed to handle nonlinear systems by linearizing them at each estimation step. It iteratively updates state estimates based on predictions and measurements, making it suitable for nonlinear dynamic systems. It is crucial for accurate state estimation in scenarios with nonlinearities.

1.5.5 Neural Network

Neural Network is a computational model inspired by the human brain's structure, consisting of interconnected nodes (neurons) organized into layers. It learns patterns and relationships from data through training and adjusting weights to make predictions or classifications [11, 12]. Widely applied in machine learning for tasks like image recognition, natural language processing, and prediction, neural networks excel in complex, non-linear problems.

1.5.6 Adaptive Filtering

Adaptive Filtering is a signal processing technique that adjusts filter parameters in real time based on changing input signals. It is particularly useful in applications where the characteristics of the signal [13, 14] or the environment may vary over time. Commonly employed in areas like communication systems, noise cancellation, equalization, and adaptive filtering enhance performance in dynamic and unpredictable scenarios.

1.5.7 Machine Learning Techniques

Adaptive Filtering is a signal processing technique that adjusts filter parameters in real time based on changing input signals. It is particularly useful in applications where the characteristics of the signal or the environment may vary over time [15, 16]. Commonly employed in areas like communication systems, noise cancellation, and equalization, adaptive filtering enhances performance in dynamic and unpredictable scenarios.

2 Artificial Intelligence Techniques in the SG

Since the beginning of the 1980s, Power engineering evaluation has shifted from formal mathematical modeling to less rigorous artificial intelligence (AI) techniques. What is AI? The ability to learn, understand, and deal with novel solutions. The study of how to make computers do things that humans do better.

- AI is the field of computer science focused on making computers intelligent.
- AI is the field of developing and building software that can process knowledge and manipulate a static situation.
- AI replicates human thinking.
- The human brain is a powerful parallel computer that generates natural intelligence.
- The brain accounts for 2% of the total mass of the human body.
- Brian consumes 20% oxygen.
- More than 10^{12} neurons exist.
- The human brain switches at a rate of 10^{13}–10^{19} per second.

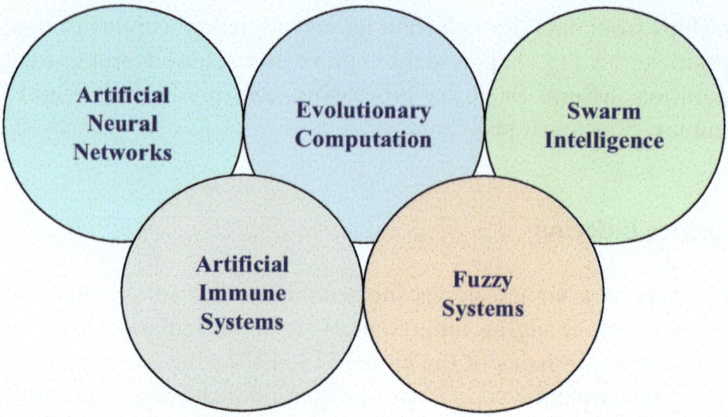

Fig. 4 Concept of computational intelligence

2.1 Concepts of AI in SG

The major AI concepts in SG are classified into three types: computational intelligence, hard computing, and soft computing.

2.2 Computational Intelligence

It emphasizes the study of processes that adapt to complex and changing surrounding environments in order to encourage or enable behavioral intelligence [17]. Figure 4 shows a detailed illustration of computational intelligence.

2.3 Hard Computing

Hard computing, also known as conventional computing, requires a precise analytical model. Many analytical models are valid under ideal circumstances. Real-world challenges exist under non-ideal circumstances.

2.4 Soft Computing

Soft computing is different from conventional computing in that it can accept approximation, uncertainty, partial truth, and imprecision, whereas hard computing cannot. The human mind serves as a sort of soft computing role model.

2.5 Soft Computing Techniques

The main components, tools, or methods of Soft Computing (SC) are: Fuzzy logic (FL), neural computing (NC), and evolutionary computation (EC) is comprised of meta-heuristic techniques such as simulated annealing, harmony search, Tabu search, and ant colony optimization.

2.5.1 Implications of Soft Computing

Soft computing employs neural network (NN), fuzz logic (FL), etc. in a complementary instead of a competitive way [18]. The term "neuro-fuzzy systems" describes a powerful illustration of a particularly successful combination. These technologies are showing up more frequently in popular products including air conditioners, washing machines, photocopiers, and camcorders.

2.5.2 A Unique Characteristic of Soft Computing

Acquiring knowledge from experimental data. Soft computing addresses use previous knowledge inputs to approximate or interpolate newly unseen ones, resulting in strong generalization. Generalization usually occurs in a high-dimensional plane.

2.5.3 Several Applications of Soft Computing

Handwriting identification, automotive systems and manufacturing, image processing, data compression, and power engineering applications.

2.5.4 Future of Soft Computing

Soft computing will continue to play a significant role in science and engineering, but its impact may extend beyond these fields as well. Soft computing represents a significant change in computing objectives. A change that considers the fact that, in contrast to modern computers, the human mind has the amazing capacity to store and understand information that is widely unpredictable, imprecise, and lacking in categorization.

2.6 *Fuzzy Logic*

FL is a method to represent information in a way that is similar to how humans interact and reason with information. The role model for FL human mind, which mimics human thinking. FL uses words instead of numbers for computation and reasoning. It is a knowledge-based method of decision-making [19]. Decision-making terminologies are linguistic variables, membership function, and degree of membership. FL controllers are designed for use in situations with high uncertainty or unknown variations in plant characteristics and structures.

2.6.1 Fuzzy Logic Controllers: Theoretical Reasons

- Fuzzy logic control is model-free.
- Fuzzy logic control provides nonlinear controllers.
- Effectively use every available information.

2.6.2 Fuzzy Logic Controllers: Practical Reasons

- Easy to understand.
- It's simple to implement.
- Fuzzy logic control refers to an approach with a high performance/cost ratio.
- Example: Washing machine

2.6.3 Fuzzy Logic Controller

Fuzzy logic is a type of multivalued logic. It is associated with fuzzy set theory, which deals with classes of objects with unsharp boundaries and a membership that is determined by degree. A fuzzy logic controller consists of three major elements namely; fuzzification module (fuzzifier), inference engine (rule base), and defuzzification module (defuzzifier) as shown in Fig. 5.

The Fuzzifier converts real-world data input into suitable linguistic values. During fuzzification, an FLC receives input values, also known as fuzzy variables, and analyses them using user-defined charts known as membership functions. Here, the input values are always precise numerical numbers. First, the inputs are taken and the degree to which they correspond to each of the suitable fuzzy sets is determined using their MF as shown in Fig. 6. The fuzzification procedure provides a degree of membership corresponding to the numerical value provided by the qualifying linguistic set.

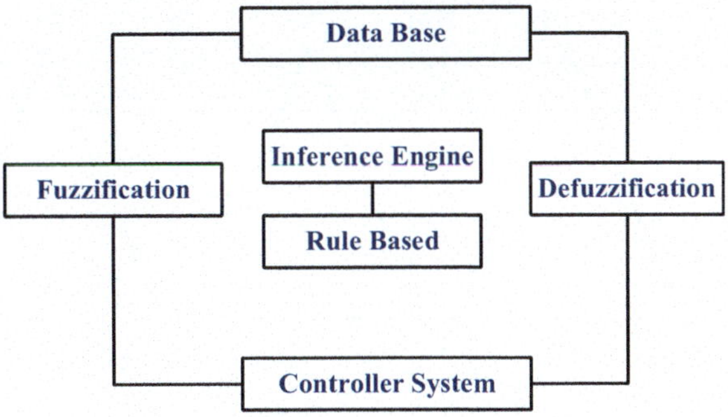

Fig. 5 Structural diagram of a fuzzy logic controller

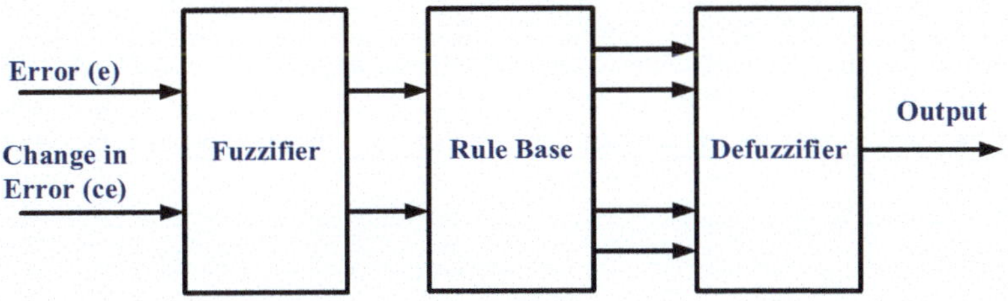

Fig. 6 Schematic diagram of a fuzzy logic controller

2.7 Decision-Making Terminologies

2.7.1 Linguistic Variables

The linguistic terms are described by membership functions as represented in Figs. 7 and 8. The linguistic terms/variables are incorporated into a rule-based system to form a fuzzy logic controller.

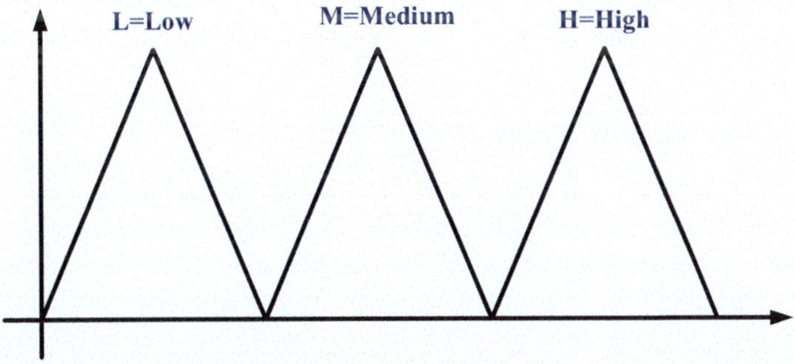

Fig. 7 Crisp set of linguistic variables

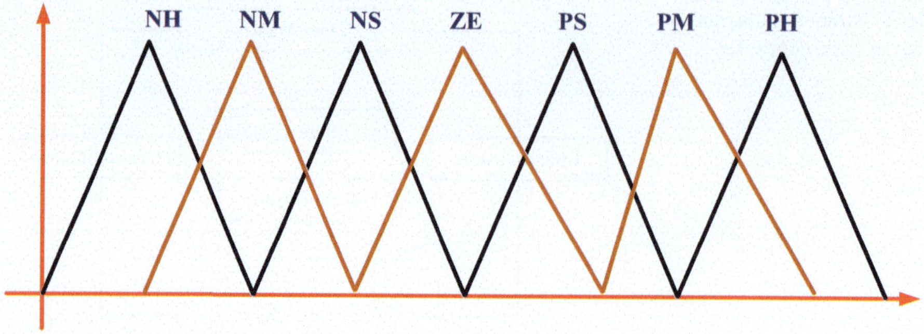

Fig. 8 Membership function of fuzzy set

Common linguistic terms used to describe fuzzy setters: PS $\Longrightarrow$ Positive Small, PM $\Longrightarrow$ Positive Medium, PH $\Longrightarrow$ Positive High, NS $\Longrightarrow$ Negative Small, NM $\Longrightarrow$ Negative Medium, NH $\Longrightarrow$ Negative High. Fuzzy A = {NH, NM, NS, ZE, PS, PM, PH}.

2.7.2 Membership Function

The formula to find the fuzzy set to which a value fits is subject to its degree of membership.

2.7.3 Degree of Membership

The degree of certainty or confidence, represented as a number from 0 to 1, that a given value is a member of a fuzzy set. A membership function defines a fuzzy set's degree of membership. The degree of membership, a number that represents the amount to which the element is present, is an essential concept in fuzzy logic. A = {0.8, 0.4, 0, 0, 0, 0, 0}.

2.8 Artificial Neural Network (ANN)

A neural network is a parallel distributed processor (PDP) with the inherent ability to store and access experimental knowledge. It mimics the brain in two ways, namely, Neural networks obtain knowledge by learning. Synaptic weights, or inter-neuronal connection strengths, are used to store knowledge. ANN has been successfully applied to problems such as pattern recognition (storage), decision-making, experience/forecasting, filter or clarity purposes, self-learn (biological neural network),

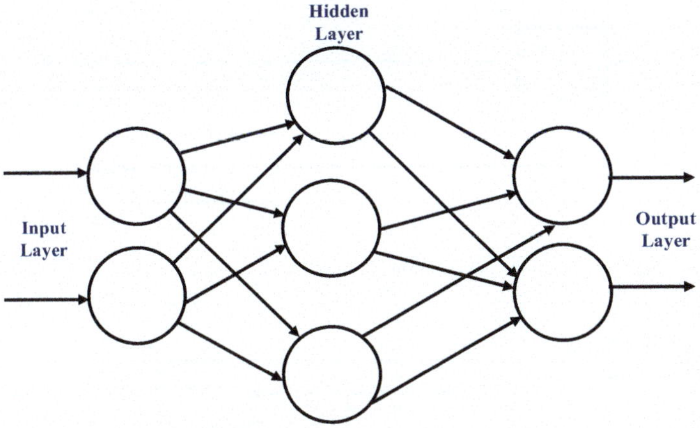

Fig. 9 Architecture of artificial neural network

image processing, data compression, and optimization. Figure 9 indicates the architecture of ANN.

2.8.1 Characteristics of ANN

ANNs are biologically inspired. ANNs are organized in a way that may or may not be related to brain anatomy. Learn from experience (Training). Generalize previous experiences to new ones. Identify essential features from inputs that contain irrelevant material. A neuron was designed to mimic the first-order properties of a biological neuron.

2.8.2 Properties of ANN

Non-linear neurons are interconnected, with non-linearity distributed throughout. Learning with a teacher. It adapts parameters to variations in the nearby environment. A decision with a degree of confidence. Graceful degradation. Very large-scale integration. Neurobiological analogy.

2.9 Learning Laws

Types of learning in ANN are supervised learning, unsupervised learning, and reinforcement learning.

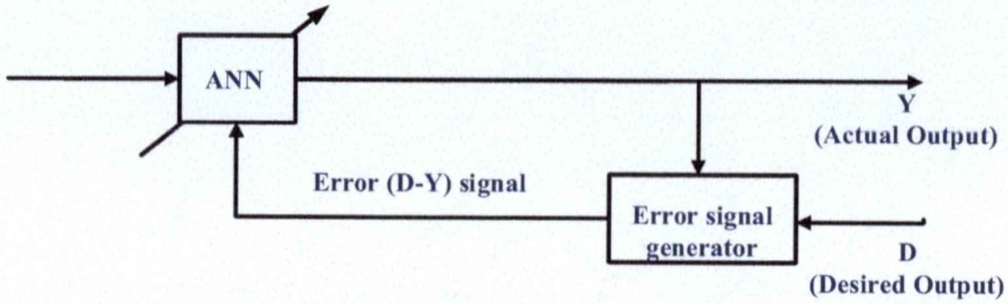

Fig. 10 Block diagram of supervised learning

Fig. 11 Block diagram of unsupervised learning

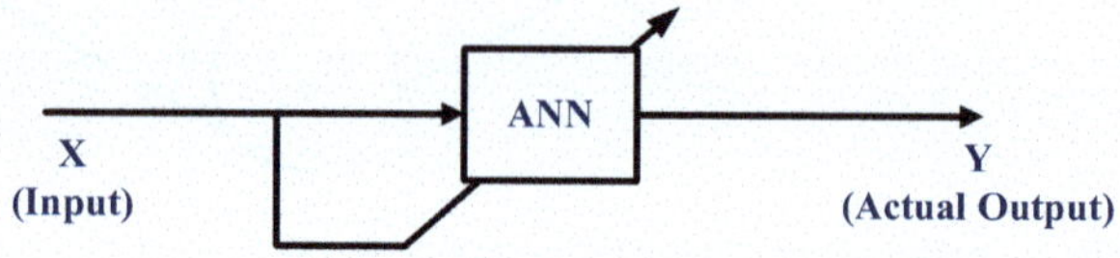

2.9.1 Supervised Learning

In supervised learning, each input pattern is used to train the system to produce the target pattern. In Supervised Learning, both input and desired output are required as shown in Fig. 10. A teacher is expected to be contemporary to verify the desired output.

2.9.2 Unsupervised Learning

Unsupervised classification systems are frequently based on a clustering strategy that brings together similar patterns. In unsupervised learning, only inputs are required as shown in Fig. 11. Weights will change based on inputs only. The clustering technique is helpful in pattern classification problems. For the clustering technique, a similarity measure must be defined to determine how close patterns are. Example: Weather classification.

2.9.3 Reinforcement Learning

This learning technique is analogous to supervised learning. In the realm of supervised learning, a clear set of target output values accompanies each input pattern, fostering a robust learning process. Nevertheless, there exist instances where information may be intermittent. Recursive feedback learning introduces a distinctive approach, wherein a guiding teacher is present but refrains from disclosing the anticipated answer outright. Instead, the teacher provides feedback by indicating the correctness or incorrectness of the computed output. Rewards are bestowed upon the

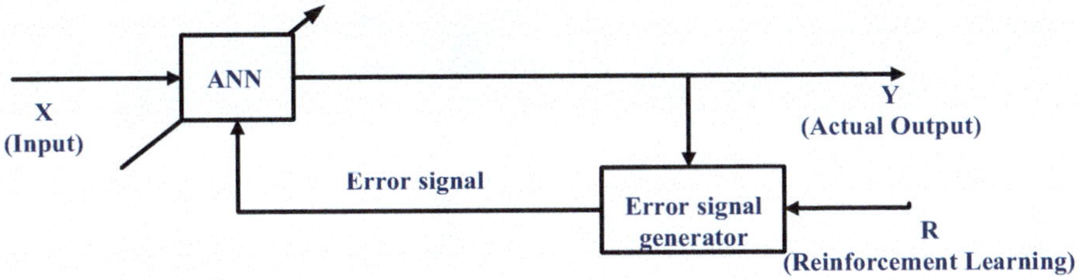

Fig. 12 Block diagram of reinforcement learning

model for accurately derived answers, while penalties are imposed for inaccuracies as shown in Fig. 12.

3 Case Study

An IEEE 33 bus system was considered to assess the impact of load on the amount of energy exchange, unit commitment status of generators, dispatch schedule of sources, and the total operational cost of the microgrid [18]. The load demand has a peak load demand on the system as 830 kW and the total scheduling hours are considered as 24 h [19]. It is evident from Fig. 13 that the dispatch schedule for hours 8–10 for diesel generators is zero it means the generators are turned off at this hour since the incremental fuel cost of the generators at these scheduling hours is more than the grid exchange price therefore, these sources are turned off [20–24]. Some limited resources are utilized in the conventional power generation process: coal, oil, and natural gas [25]. Because conventional power plants are frequently vulnerable to changes in fuel prices, market dynamics, and geopolitical events can have an impact on energy costs [26]. Conventional generators, especially those that burn coal and natural gas, generate a significant quantity of greenhouse gasses, primarily carbon dioxide. Climate change and air pollution are both impacted by these emissions [27]. Large transmission and distribution networks are required to supply end users with electricity because conventional power facilities are typically centralized. This may lead to energy losses during transmission [28]. When a microgrid (MG) is linked to the main power grid, it operates in tandem with it. When required, it might draw energy from MG's may perform grid support operations including voltage control and frequency regulation, which helps to maintain the reliability and stability of the main power grid [29]. MG's can function independently from the main grid while it is in islanded mode [30, 31]. This mode is particularly useful during grid outages or emergencies [32, 33].

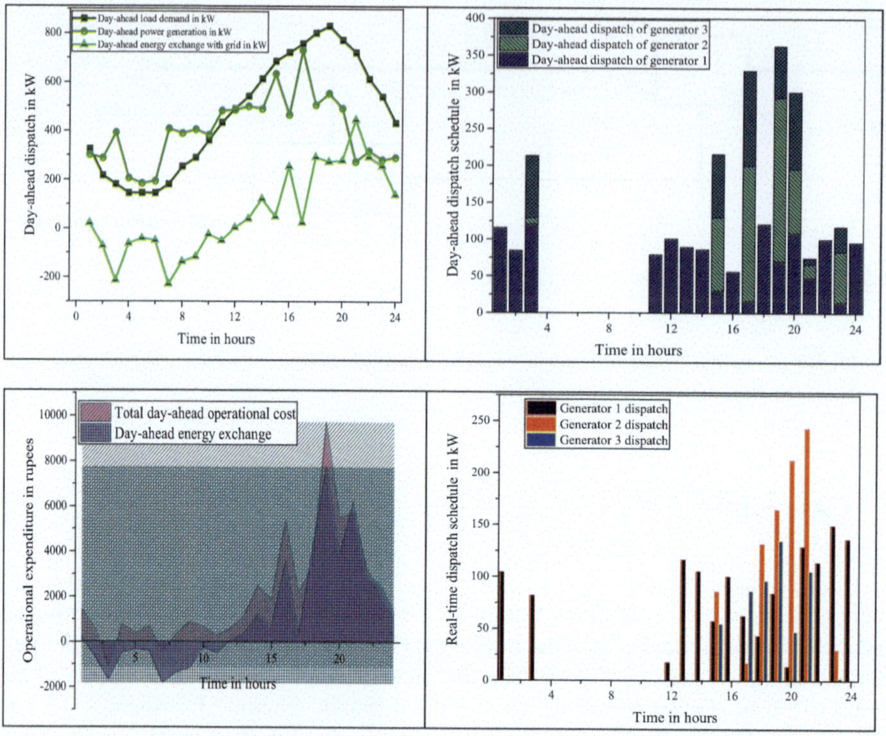

Fig. 13 Energy management in a microgrid

4 Conclusions

Smart meters that offer real-time data on the consumption of energy are to be used in place with conventional meters. High-speed communication networks provide seamless data transfer between the components, enabling real-time control and monitoring. Energy storage devices, dispersed resources, and renewable energy sources are incorporated into the grid. Grid operations are automated to enhance system restoration, fault detection, and quick responses. Customers may take an active role in energy management through demand response, which allows them to react to pricing signals and modify their energy use appropriately. Integration of electric cars with the infrastructure needed for charging them, enabling grid balancing and environmentally friendly mobility. The grids potential to identify and address faults automatically, minimizing downtime and enhancing reliability. Reliability and resilience can be enhanced via small, localized energy systems that may function with the main grid or independently.

Table 1 Services and roles in the conceptual model of the SG

Customer	The end users of energy. Generates, stores, and consumes energy. Residential, commercial, and industrial
Markets	The operators and contributors in energy markets
Service Provider	Provides service to end users and utilities
Operators	Monitors the flow of electricity
Generation	Comprises of both conventional sources of generation and renewable energy sources. Distributed energy resources refer to generators, i.e., dispatchable and non-dispatchable and storage
Transmission	Bulk electricity is transported over long distances
Distribution	Providers of distribution services to and from consumers, with the capability to both store and generate electrical power as well

Table 2 Battery types and their comparisons

S. No.	Battery	Specific energy (Wh/Kg)	Specific power (W/Kg)	Energy density (Wh/Lt)	Operating temperature (c)	Efficiency (%)	Life cycle
1.	Lead Acid	35	100	80	Ambient	70	100
2.	Ni Metal Hydride	55	100	150	Ambient	70	2000
3.	Sodium Sulphur	100	150	300	350	75	1000
4.	Lithium Polymer	200	500	500	Ambient	80	1800
5.	Lithium Ion	150	1000	350	Ambient	85–93	10,000

Acknowledgements This research work was supported by "Woosong University's Academic Research Funding – 2025".

References

1. Talaq J, El-Hawary F, El-Hawary E (1994) A summary of environmental/economic dispatch algorithms. IEEE Trans Power Syst 9(3):1508–1516. https://ieeexplore.ieee.org/document/336110
2. Lamont J, Obessis E (1995) Emission dispatch models and algorithms for the 1990s. IEEE Trans Power Syst 10(2):941–947. https://ieeexplore.ieee.org/abstract/document/387937
3. Delson J (1974) Controlled emission dispatch. IEEE Trans Power Apparatus Syst 93(5):1359–1366. https://ieeexplore.ieee.org/document/4075500

4. Dibavar A, Ivatloo B, Zare K, Khalili T, Bidram A (2021) Economic-emission dispatch problem in power systems with carbon capture power plants. IEEE Trans Ind Appl 57(4):3341–3351. https://ieeexplore.ieee.org/document/9428528
5. Uddin M, Romlie M, Abdullah M, Halim S, Bakar A, Kwang T (2018) A review on peak load shaving strategies. RSER 82(3):3323–3332. https://doi.org/10.1016/j.rser.2017.10.056
6. Kirschen D, Strbac G, Cumperayot P, Mendes D (2000) Factoring the elasticity of demand in electricity prices. IEEE Trans Power Syst 15(2):612–617. https://ieeexplore.ieee.org/document/867149
7. Conejo A, Morales M, Baringo L (2010) Real-time demand response model. IEEE Trans Smart Grid 1(3):236–242. https://ieeexplore.ieee.org/document/5607339
8. Pourmousavi S, Nehrir M (2014) Introducing dynamic demand response in the LFC model. IEEE Trans Power Syst 29(4):1562–1572. https://ieeexplore.ieee.org/abstract/document/7285784
9. Medina J, Muller N, Roytelman I (2010) Demand response and distribution grid operations: opportunities and challenges. IEEE Trans Smart Grid 1(2):193–198. https://ieeexplore.ieee.org/document/5545489
10. Parvania M, Firuzabad M (2010) Demand response scheduling by stochastic SCUC. IEEE Trans Smart Grid 1(1):89–98. https://ieeexplore.ieee.org/document/5464384
11. Deng R, Yang Z, Chow M, Chen J (2015) A survey on demand response in smart grids: mathematical models and approaches. IEEE Trans Ind Inf 11(3):570–582. https://ieeexplore.ieee.org/document/7063260
12. Rahimi F, Ipakchi A (2010) Demand response as a market resource under the smart grid paradigm. IEEE Trans Smart Grid 1(1):82–88. https://ieeexplore.ieee.org/document/5454394
13. Mathieu J, Price P, Kiliccote S, Piette M (2011) Quantifying changes in building electricity use, with application to demand response. IEEE Trans Smart Grid 2(3):507–518. https://ieeexplore.ieee.org/document/5772947
14. Salkuti SR (2022) Modeling of various renewable energy resources for smart electrical power systems. In: Salkuti SR, Ray P (eds) Next generation smart grids: modeling, control and optimization. Lecture Notes in Electrical Engineering, vol 824. Springer, Singapore. https://doi.org/10.1007/978-981-16-7794-6_2
15. Salkuti SR, Ray P, Pagidipala S (2022) Overview of next generation smart grids. In: Salkuti SR, Ray P (eds) Next generation smart grids: modeling, control and optimization. Lecture Notes in Electrical Engineering, vol 824. Springer, Singapore. https://doi.org/10.1007/978-981-16-7794-6_1
16. Sravanthi P, Sandeep V (2022) Solving realistic reactive power market clearing problem of wind-thermal power system with system security. Int J Emerg Electr Power Syst 23(2):125–144. https://doi.org/10.1515/ijeeps-2021-0060
17. Kumar JV, Sesham H, Salkuti SR (2024) Smart energy management model for electric vehicles. In: Salkuti SR (eds) Energy and environmental aspects of emerging technologies for smart grid. Green energy and technology. Springer, Cham. https://doi.org/10.1007/978-3-031-18389-8_9
18. Palensky P, Dietrich D (2011) Demand side management: demand response, intelligent energy systems, and smart loads. IEEE Trans Ind Inf 7(3):381–388. https://ieeexplore.ieee.org/document/5930335
19. Zhou X (2022) Exploiting integrated demand response for operating reserve provision considering rebound effects. IEEE Access 10(2):15151–15162. https://ieeexplore.ieee.org/document/9701332
20. Xiaoyang Z, Weixin Y, Johnson R (2016) Forecast load impact from demand response resources. In: 2016 IEEE power and energy society general meeting, vol 2, issue 3, pp 1–5. https://ieeexplore.ieee.org/document/7741992
21. Bitaraf H, Rahman S (2018) Reducing curtailed wind energy through energy storage and demand response. IEEE Trans Sustain Energy 9(1):228–236. https://ieeexplore.ieee.org/document/7972993
22. Shi W, Li N, Xie X, Chu C, Gadh R (2014) Optimal residential demand response in distribution networks. IEEE J Selected Areas Commun 32(7):1441–1450. https://ieeexplore.ieee.org/abstract/document/6840288

23. Sibo N, Zhou M, Li G (2017) Optimal residential community demand response scheduling in smart grid. Appl Energy 210(2):1280–1289. https://doi.org/10.1016/j.apenergy.2017.06.066
24. Tarish H, See O, Elmenreich W (2016) A review of residential demand response of smart grid. RSER 59(2):166–178. https://doi.org/10.1016/j.rser.2016.01.016
25. Niamh C, Pierre P, Madsen M, Mark O (2014) Benefits and challenges of electrical demand response: a critical review. RSER 39(1):686–699. https://doi.org/10.1016/j.rser.2014.07.098
26. Rezaee A (2019) Optimisation of demand response in electric power systems, a review. RSER 103(1):308–319. https://doi.org/10.1016/j.rser.2018.12.054
27. Hussain M, Yan G (2018) A review of demand response in an efficient smart grid environment. Electr J 31(5):55–63. https://doi.org/10.1016/j.tej.2018.06.003
28. Siano P (2014) Demand response and smart grids—a survey. RSER 30(4):461–478. https://doi.org/10.1016/j.rser.2013.10.022
29. Faria S, Spínola VZ (2018) Reschedule of distributed energy resources by an aggregator for market participation. Energies 11(713):1–20. https://doi.org/10.3390/en11040713
30. Zhang G, Li Y, Deng X (2020) K-means clustering-based electrical equipment identification for smart building application. Information 11(27):25–35. https://doi.org/10.3390/info11010027
31. Faria S, Spínola J, Vale Z (2018) Distributed energy resources scheduling and aggregation in the context of demand response programs. Energies 11(1):987–992. https://doi.org/10.3390/en11081987
32. Abrishambaf O, Faria S, Vale Z (2020) Ramping of demand response event with deploying distinct programs by an aggregator. Energies 13(1):138–145. https://doi.org/10.3390/en13061389
33. Lezama F, Faria S, Vale Z (2020) Demand response of residential houses equipped with PV-battery systems: an application study using evolutionary algorithms. Energies 13(1):24–36. https://doi.org/10.3390/en13102466

A Resilient Passivity-Based Control Approach for a Two-Input DC-DC Buck Converter with Constant Power Loads in Distributed Generation Systems

Amarendra Reddy Bhimavarapu, Ch Nayak Bhukya, Dharani Naidu, and Surender Reddy Salkuti

Abstract The objective of this work is to present a resilient nonlinear control strategy for a DC-DC buck power converter with a CPL in a distributed generation system. In constant power load, the input voltage decreases/increases concerning increases/decreases of input current, and at its terminals, it exhibits the attributes of a negative impedance. Due to this behavior, it will impact the dynamics, stability, and power quality of the converter system. Moreover, the dc-bus voltage changes will happen when the system experiences disturbances like changes in load and change in input voltage. The passivity-based control (PBC which is a simple form of PD) strategy has been utilized to address the instability issues caused by the CPLs. However, the main drawback of PBC is, that it will not eliminate steady-state error and has a slow dynamic response. It is proved that the addition of an integral term to PBC (PBC-PID) ensures the elimination of limit cycles caused by the CPLs, large signal stability, and fast recovery performance of the system during the disturbance. The resilience of the proposed PBC and PBC-PID non-linear control strategies simulation results were validated through the MATLAB/Simulink environment.

Keywords Distributed generation · DC microgrid · DC-DC converter · Constant power load · Limit cycle behavior · Negative Incremental Impedance · Passivity-based integral control

A. R. Bhimavarapu · C. Nayak Bhukya · D. Naidu
Department of Electrical Engineering, Andhra University, Visakhapatnam, Andhra Pradesh, India
e-mail: amarendra@andhrauniversity.edu.in

C. Nayak Bhukya
e-mail: nayakbhukyach.rs@andhrauniversity.edu.in

S. R. Salkuti (✉)
Department of Railroad and Electrical Engineering, Woosong University, Daejeon, Republic of Korea
e-mail: surender@wsu.ac.kr

S. R. Salkuti (ed.), *Artificial Intelligence for Integrated Smart Energy Systems in Electric Vehicles*, Lecture Notes in Electrical Engineering 1427,
https://doi.org/10.1007/978-3-031-94276-1_18

Nomenclature

E_1, E_2	Input voltage sources
TII	Two-input integrated
L_1, L_2	Inductances of the converter circuit
C_1, C_2	Capacitances of converter circuit
CPL	Constant power load
CVL	Constant voltage load
NII	Negative incremental impedance
R	Load resistance
u	Switch position control
Z	State-space variable
$\mathcal{G}$	Symmetries matrix
$\mathcal{R}$	Diagonal positive semidefinite matrix.
$\mathcal{E}$	External generalized forces matrix
$\mathcal{J}$	The matrix and its elements are given by $\{-1, 0, 1\}$
μ	Duty ratio
$d = [d_1 d_2 d_3]^T$	Disturbances in the system
DG	Distributed generation
R_{1d}, R_{2d}, R_{3d}	Virtual damping resistances
PBC	Passivity-based controller
PID	Proportional Integral Derivative control
PBIC	Passivity-based integral control

1 Introduction

Power demands for modern power systems are unchanged. Different power-generating methods are needed to meet this. Hazardous gasses are produced when fossil fuels are used to generate electricity. In contrast, nowadays renewable power generation is relatively global power generation with less use of coal, oil, gas, and other non-renewable energy sources. Moreover, power generation from non-renewable resources pollutes the planet more. Due to global warming issues, distributed generation (DG) systems are getting attention to generate power near their applications [1]. The sources of DG systems like PV systems, small wind turbines, microturbines, gas engines, fuel cells, and energy storage systems produce DC power. DG systems can offer Grid stability, energy security, fewer transmission losses, and better renewable energy integration [2]. In DG systems, a cascaded distributed power architecture refers to a setup where multiple DC sources, i.e., technologies for green energy, such as windmills or solar energy systems and fuel cells, etc. are connected in series or parallel to form a distributed power generation network. Each DC source is typically equipped with its power converter; hence the integration of renewable energies is possible by the medium of DC-DC converters for the process of power. as shown in Fig. 1.

In DG systems, Constant power loads are gaining popularity. From Fig. 1, the Point-of-load converters also known as converters at the load terminal side operate under tightly regulated operating conditions. These POL converters are responsible for regulating the voltage and/or current at specific points of the DG system's operations; during this scenario, the POL converters are treated as constant power loads. CPLs are devices that maintain a constant power from the electrical system regardless of changes in voltage or current depicted in Fig. 2.

When a CPL is connected to the source side dc-dc converter, it will portray negative incremental impedance (NII) characteristics, this undesired occurrence results in inadequate damping of the system and allows for energy dissipation without any

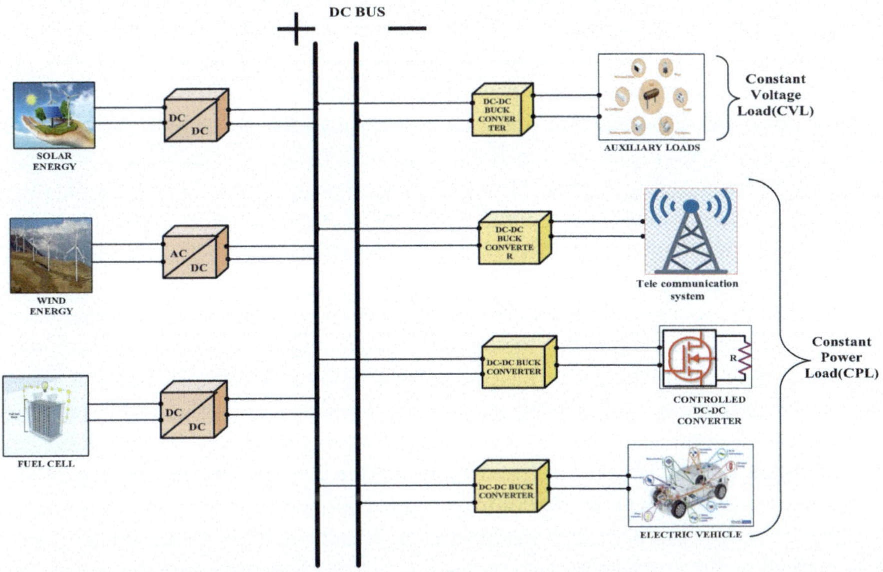

Fig. 1 Distributed generation configuration by multiple sources and loads

Fig. 2 Constant power load

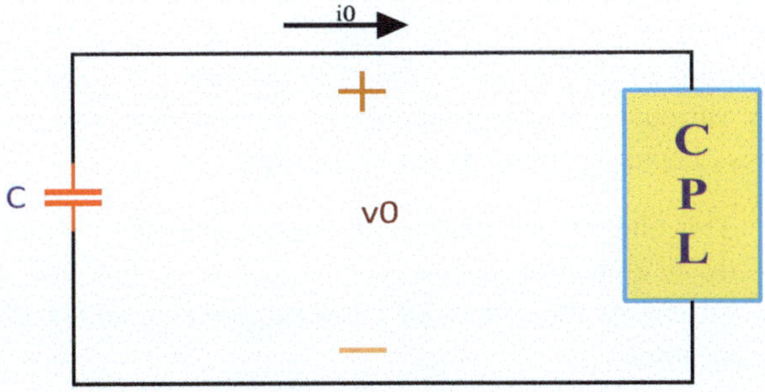

loss across the output terminals of the source converter. During this case, if the input voltage of the CPL surpasses its source voltage, System states will not revert to their initial condition of balance. Therefore, the source converter's energy will exhibit oscillation among the capacitor and inductor, resulting in a limit cycle behavior. A limit cycle is widely recognized as an unwanted phenomenon, it has a significant impact on the stress levels of switches and output capacitors, which results the ripples in inductor current and capacitor voltage. This stress can result in an increment in temperature, which has the potential to decrease the lifespan of components and may lead to severe damage and instability issues [3, 4].

To address the issue of CPL instability, researchers have proposed numerous kinds of control techniques. These techniques mainly focus on two principal concepts, i.e., passive damping, which involves the inclusion of physical passive elements to the system to dissipate system oscillations [5], and active damping, the control action is used to passivize the circuit behavior [6, 7]. These control techniques are broadly classified into linear and non-linear. Several efficient linear and nonlinear control methods have been developed to mitigate instability issues caused by the constant power loads. However, linear control methods are limited only to small signal analysis [8]. Whereas non-linear techniques are most suitable for large signal stability analysis. Among all the non-linear control techniques some of these are more efficient which are sliding mode control [9, 10], synergetic control [11], backstepping control [12], and passivity-based control [13–15]. In recent studies, passivity-based control (PBC) has gained popularity because of its robustness in the stabilization of CPLs in DG systems.

PBC, initially recommended by Ortega et al. [16] has emphasized the principle of energy conservation, namely, that the energy given equals the total energy conserved plus the energy dissipated. The passivity-based control technique yields a globally stable equilibrium point. In the case of the converter circuit's impassivity, the passivity-based controller will inject a virtual damping resistance to reduce system oscillation and provide a passive closed-loop system. Based on the principle of PBC, it has been proven that a basic PD controller can resolve the instability issues caused by CPLs. However, this approach has limitations like its inability to eliminate offsets, higher overshoot, poor recovery performance, and delayed settling time for uncertainties in line and load variations. These limitations can be overcome by the PBC-based PID controller, i.e., by incorporating an integral component to remove steady-state errors, improving recovery performance, reducing overshoots, and improving settling time in output voltage in the presence of uncertainties [17, 18]. A comparison table of some of the DC-DC converters that are associated with a CPL and are controlled by various control approaches is illustrated in Table 1. The crucial contribution of this chapter is summarized as follows:

- This chapter acquaints with the passivity-based modeling of a two-input parallel dc-dc converter.
- Discusses the effects of constant power loads (CPLs) on distributed generation systems.

Table 1 Comparison of different dc-dc converter topologies

Converter topology	Non-linear controller	Control strategy with CPL				
		Passive damping	Active damping			Year
			PBC	PID	PBC-PID	
Single input dc-dc buck	SMC	×	✓	×	×	2023
Single input dc-dc boost	Back stepping and SMC	×	✓	×	×	2021
Single input dc-dc all type	PD type	×	✓	✓	×	2007
Single input dc-dc buck-boost	PBC	×	✓	✓	×	2012
Single-inductor, dual-output (SIDO) buck dc-dc	PBC	×	✓	×	×	2018
POL Converters based Boost and Buck	Linear-Quadratic Regulator (LQR)	✓	✓	×	×	2019
DC-DC boost converter	PID	✓	✓	✓	×	2013
Four-Switch Buck-Boost	PBC Controller	✓	✓	✓	×	2022
Single input dc-dc buck	PBC	✓	✓	×	✓	2019
Two input dc-dc buck	PBC	✓	✓	×	✓	2020
Present Chapter	PBC	✓	✓	✓	✓	

- The proposed controller is modeled and designed to alleviate instabilities caused by the CPLs, and the resilience of the controller is studied under different conditions.
- The proposed system with an implemented PBC (PD Type) and PBC-PID controllers are studied with extensive simulations in response to different operating conditions (variation in input sources, variation in reference voltage, variation in load (CPL)) in a MATLAB/Simulink environment and obtained acceptable results.

The chapter is arranged as follows: Sect. 2 addresses the modeling of constant power load and its impacts. Section 3 addresses the PBC-based PBC-PID closed-loop control design of the DC-DC buck power converter, and Sect. 4 addresses its design and simulation findings.

2 Modelling of Constant Power Loads

2.1 Mathematical Modelling of CPL

If the load side converter is equipped with a controller that, despite voltage fluctuations, delivers power to the load continuously can be considered as CPL shown in Fig. 3. Here, the anticipated voltage across the load, the actual voltage across the load, the load current, and the constant power of the converter system are represented using the variables V_{ref}, V_o, I_o, P_{cpl}. CPLs introduce unique challenges in system stability because of their undesired characteristic of NII, as shown in Fig. 4. It potentially leads to oscillations or instability in the system. In this scenario, a robust voltage control approach is essential to guarantee the system's stable and reliable operation.

To enable the dynamic characteristics of a converter system connected to CPL, it is necessary to combine the load model by employing an integrated converter

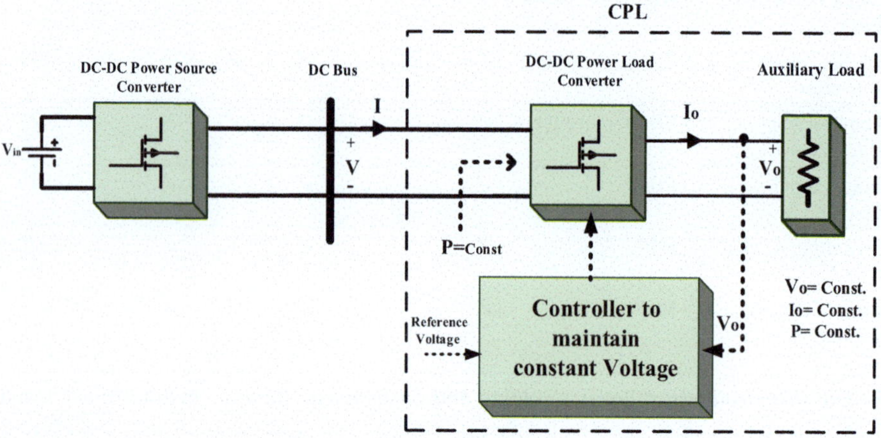

Fig. 3 Load side closed-loop converter circuit acts as CPL

Fig. 4 V-I characteristics of Load side converter

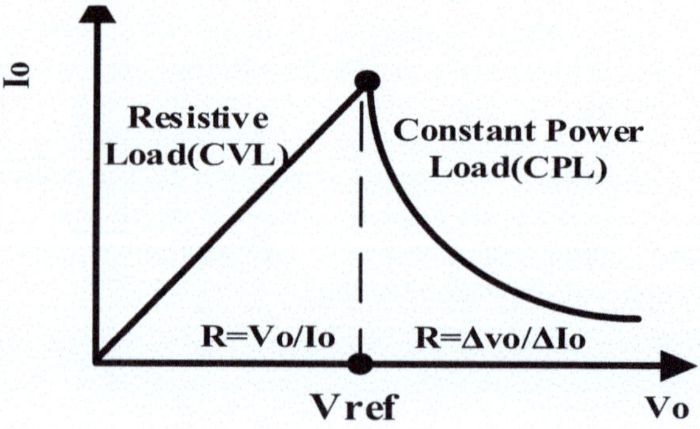

model. In this case, the load model exhibits non-linear behaviour, but the converter model exhibits linear behaviour. Combining a non-linear load model with a linear converter model is not feasible. Thus, the non-linear load model can be approximated by linearizing it, by using Taylor's series. The non-linear load's linear approximation is described in Eq. (1) [19].

$$I_{nom} = \frac{P_{nom}}{V_{nom}} \tag{1}$$

$$-\frac{1}{r} = -\frac{P_{nom}}{V_{nom}^2} \tag{2}$$

where P_{nom}, V_{nom}, i_{nom}, and 'r' are the nominal power, voltage, current, and resistance of the load respectively. From Eq. (2), it can be noticed that the behaviour of the system is greatly influenced by the negative admittance term, which is essentially the reciprocal of the nominal resistance i.e., as the input voltage across the load increases, the load current falls, resulting in a phenomenon known as NII. This characteristic is illustrated in Fig. 4, which describes the variations of current and voltage i.e. when $V_o < V_{ref}$ there is linear behaviour, and when $V_o > V_{ref}$ there is non-linearity which finds the significance of CPL.

2.2 *Characteristics and Effects of CPL*

The equivalent circuit model of CPL is depicted in Fig. 5. It describes the relationship between I_{cpl}, V_{cpl} and P_{cpl} and it is given in Eq. (3).

$$I_{cpl} = \frac{P_{cpl}}{V_{cpl}} \tag{3}$$

Let's take a small variation of i_{cpl} w.r.t v_{cpl}, gives the Eq. (4).

$$\frac{\partial I_{cpl}}{\partial V_{cpl}} = -\frac{P_{cpl}}{V_{cpl}^2} \tag{4}$$

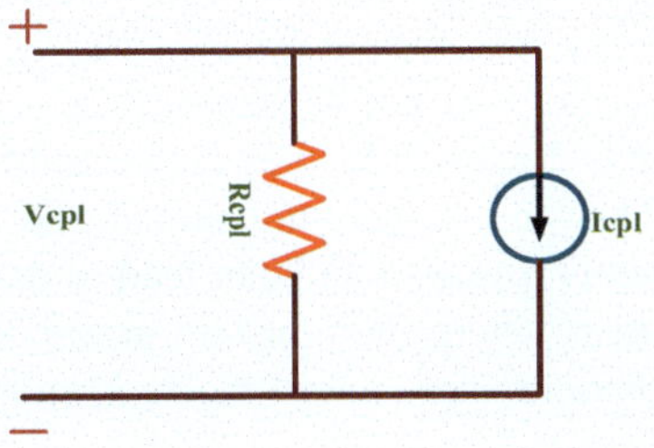

Fig. 5 Constant power load small-signal model

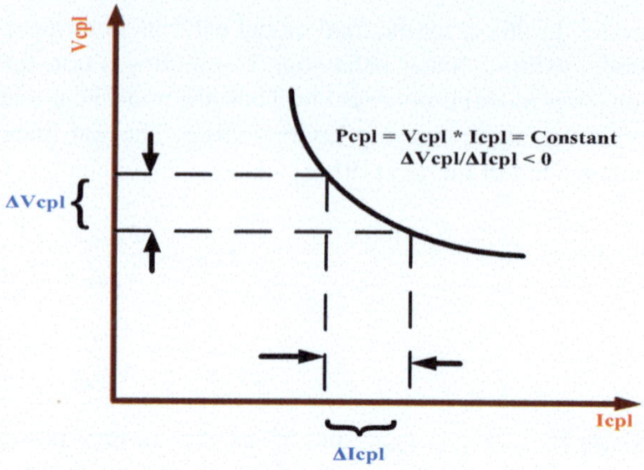

Fig. 6 Constant power load of I-V characteristics

The complete solution of equation (4) for i_{cpl} gives equation (5).

$$I_{cpl} = -\frac{P_{cpl}}{V_{cpl}^2} v + 2\frac{P_{cpl}}{V_{cpl}} \tag{5}$$

Equation (5) illustrates the CPL small signal model. The negative resistance of CPL is represented by the first part. $R_{cpl} = -\frac{P_{cpl}}{V_{cpl}^2}$ and the second term signifies a constant current source of $I_{cpl} = 2\frac{P_{cpl}}{V_{cpl}}$. It is depicted in Fig. 5.

The second term will not have any impact on the system, whereas the first term will influence the operation of CPL as well as the system which results in NII. The I-V characteristics of CPL are shown in Fig. 6. Due to NII, the following issues will arise, these are (i) decrement in the system's equivalent resistance, which causes the circuit to be heavily subjected to inrush currents, (ii) introduction of limit cycle behaviour which leads to unstable operation of the system. As a result, these CPLs induce negative impedance instabilities.

The instabilities due to negative incremental impedance (NII) of the converter system are investigated with constant power loads which are prone to destabilization of the converter system, resulting in zero damping behaviour. During this, the state-variables within the converter show periodic oscillating behaviour and implying that the system's states are unable to revert to their previous equilibrium. As a result, there will be a limit cycle behaviour occurrence on terminals of CPL, i.e. The energy of the capacitor and inductor gained will not be depleted as shown in Fig. 7. These Limit cycles are not desirable. Meanwhile, they create additional stress on the source-side converter switches. This stress causes the temperature to increase, which shortens the switching components' lifespan and can lead to damage. It could potentially cause instability in the areas where the constant power load is connected [20, 21]. To tackle this issue a predominant control mechanism is required.

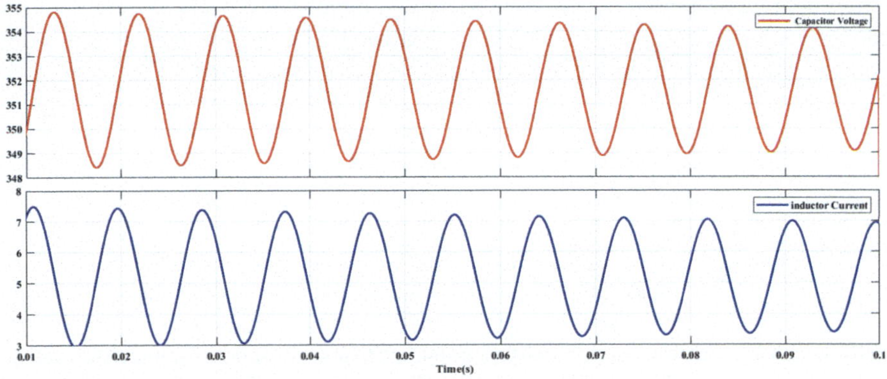

Fig. 7 Limit cycle behavior of i_L and V_c due to CPL

3 Controller Design for Passivity-Based Parallel DC-DC Buck Converters Supplying Constant Voltage and Power Load

3.1 *Distributed Generation Systems: The Correlation Between Passivity and Stability*

The correlation between passivity and stability in control systems, it is presented by Youla et al. [22] has significance in control theory. Youla proved that a passive circuit is tightly controlled with a resistive component that is stable. It is additionally proven that a non-passive system may be converted into a passive system using a passive feedback control system, as illustrated in Fig. 8. When examining the correlation between passivity and stability, it is significant to consider two primary interpretations. The first interpretation presents about nonlinear dynamic system by emphasizing its fundamental physical characteristics, i.e., if the energy supply and feedback control mechanism are detached from the system, it is considered passive in nature and it can autonomously stabilize itself. Eventually, the passive system experiences a gradual energy reduction, indicating stable operation. The following energy balance equation can be used to characterize the passive system.

$$\boldsymbol{S}(z(\boldsymbol{t}))-\boldsymbol{S}(z(0))+\int_{0}^{t}\mathcal{Z}^{T}\mathcal{R}_{i}(z)\mathcal{Z}\boldsymbol{dt}=\int_{0}^{t}\boldsymbol{u}^{T}(\boldsymbol{t})\boldsymbol{y}(\boldsymbol{t})\boldsymbol{dt} \tag{6}$$

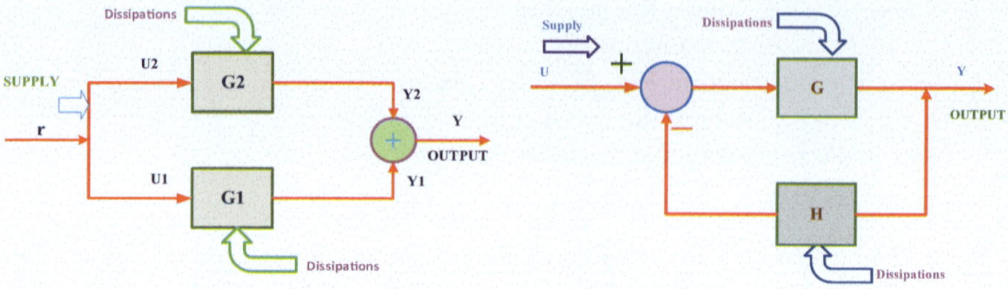

(a) **(b)**

Fig. 8 **a** Parallel combined system's passivity **b** the feedback control system's passivity

3.2 *Model Description and Passivity-Based Modeling*

The passive nature of the systems within a distributed generation improves significantly its stability. This stability is critical for the effective operation of distributed generation, especially as they integrate more renewable and distributed energy resources. The two input parallel dc-dc buck power converter feeds constant power load and constant voltage loads as depicted in Fig. 9. Here, the load is a combination of purely resistive load and CPL. DC-DC buck converter consists of input voltages (E_1, E_2), inductances (L_1, L_2) and capacitances (C_1, C_2) of the parallel circuit. Whereas (i_{L1}, i_{L2}) are the inductor currents, V_0 is the output voltage and $(\mu_1, \mu_2 \in [0, 1])$ are duty ratios. Here, the converter as well as CPL are presumptively operating in the continuous conduction mode.

Dynamic equations of parallel combination dc-dc buck converter are given in Eq. (7), here parasitic effects are neglected.

$$\left.\begin{aligned} \dot{i}_{L_1} &= \frac{E_1}{L_1}\mu_1 - \frac{v_0}{L_1} \\ \dot{i}_{L_2} &= \frac{E_2}{L_2}\mu_2 - \frac{v_0}{L_2} \\ \dot{v}_0 &= \frac{i_{L_1}}{C_{eq}} + \frac{i_{L_2}}{C_{eq}} - \frac{v_0}{C_{eq}R} - \frac{P}{C_{eq}v_0} \end{aligned}\right\} \tag{7}$$

Equation (7) has been reformulated into the following form by taking account of the given system's anticipated disturbances and uncertainties:

$$\left.\begin{aligned} \dot{i}_{L_1} &= \frac{E_{10}}{L_{10}}\mu_1 - \frac{v_0}{L_{10}} + d_1 \\ \dot{i}_{L_2} &= \frac{E_{20}}{L_{20}}\mu_2 - \frac{v_0}{L_{20}} + d_2 \\ \dot{v}_0 &= \frac{i_{L_1}}{C_{eq0}} + \frac{i_{L_2}}{C_{eq0}} - \frac{v_0}{C_{eq0}R_0} - \frac{P}{C_{eq}v_0} + d_3 \end{aligned}\right\} \tag{8}$$

where (d_1, d_2) denotes the disturbances of the system due to the input voltage changes and the variations of the inductances (L_1, L_2).

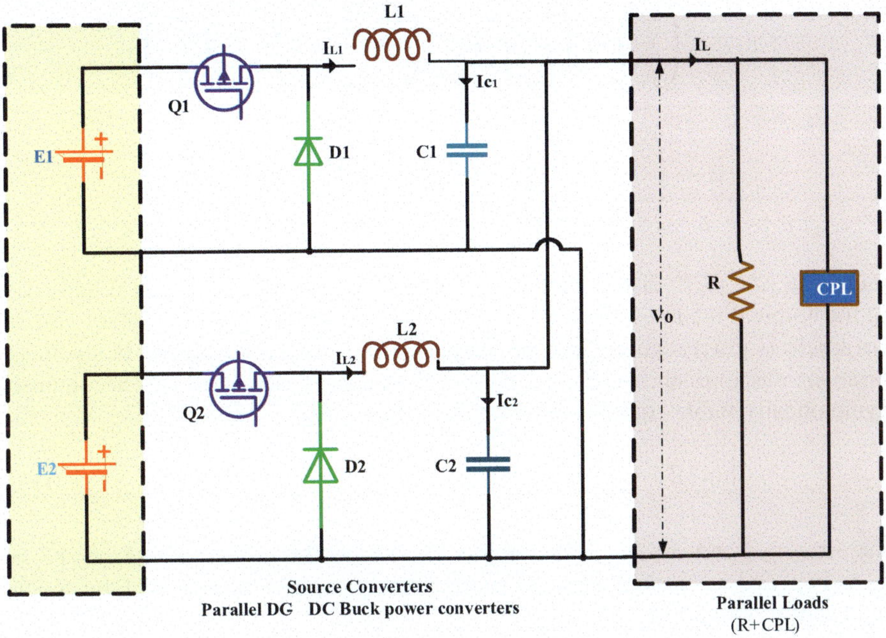

Fig. 9 Two input DC-DC buck power converters supply a constant power load and voltage load

$$d_1 = \frac{E_1}{L_1}\mu_1 - \frac{E_{10}}{L_{10}}\mu_1 + \frac{v_0}{L_{10}} - \frac{v_0}{L_1} \tag{9}$$

$$d_2 = \frac{E_2}{L_2}\mu_2 - \frac{E_{10}}{L_{10}}\mu_1 + \frac{v_0}{L_{20}} - \frac{v_0}{L_2} \tag{10}$$

Here d_3 is the disturbance/uncertainty caused by the load changes and capacitor variations.

$$d_3 = \frac{i_{L_1}}{C_{eq}} - \frac{i_{L_1}}{C_{eq0}} + \frac{i_{L_2}}{C_{eq}} - \frac{i_{L_2}}{C_{eq0}} + \frac{v_0}{C_{eq}R_0} - \frac{v_0}{C_{eq}R} + \frac{P_0}{C_{eq0}v_0} - \frac{P}{C_{eq}v_0} \tag{11}$$

where $(L_{10,}L_{20})$, (E_{10}, E_{20}), C_{eq0} and R_0 represents the nominal values of (L_1, L_2), (E_1, E_2), C_{eq} and R respectively.

For the sake of simplicity, Eqs. (7) and (8) can be rewritten in the following matrix form.

$$\mathcal{H}\dot{\mathcal{Z}} + [\mathcal{G} + \mathcal{R}(z)] = \mu\Gamma \tag{12}$$

$$\mathcal{H}_0\dot{\mathcal{Z}} + [\mathcal{G} + \mathcal{R}_0(z)] = \mu\Gamma_0 + d \tag{13}$$

Defining:

$$\mathcal{Z}\ominus = \begin{pmatrix} z_1 \\ z_2 \\ z_3 \end{pmatrix} = \begin{pmatrix} i_{L_1} \\ i_{L_2} \\ v_O \end{pmatrix}, H = \begin{pmatrix} L_1 & 0 & 0 \\ 0 & L_2 & 0 \\ 0 & 0 & C_{eq} \end{pmatrix} \Gamma = \begin{pmatrix} E_1 \\ E_2 \\ 0 \end{pmatrix}$$

$$\mathcal{G} = \begin{pmatrix} 0 & 0 & 1 \\ 0 & 0 & 1 \\ -1 & -1 & 0 \end{pmatrix} \text{ and } \mathcal{R}(z) = \begin{pmatrix} 0 & 0 & 0 \\ 0 & 0 & 0 \\ 0 & 0 & \left(\frac{1}{R} + \frac{P}{z_3^2}\right) \end{pmatrix}$$

where d = $[d_1 d_2 d_3]^T$ disturbances vector of the system. Here $\mathcal{H}_0$, $\mathcal{R}_0(z)$ and $\mu\Gamma_0$ are the nominal matrices of the $\mathcal{H}$, $\mathcal{R}(z)$ and $\mu\Gamma$ respectively. The primary aim of this article is to (i) regulate DC bus output voltage (ii) eliminate DC bus voltage variations due to disturbances (d_1, d_2, and d_3) of the system, and (iii) accomplish asymptotically stable equilibrium point as follows:

$$\lim_{t \to \infty} [\mathcal{Z} - \mathcal{Z}_d] = 0. \quad \forall (z_{10} \geq 0,\ z_{20} \geq 0 \text{ and } z_{30} \geq \varepsilon) \tag{14}$$

where z_{10}, z_{20} and z_{30} are the system's initial condition $\mathcal{Z}_d = [z_{1d} z_{2d} z_{3d}]^T = [I_{L1} I_{L2} V_o]^T$ and ε is a small positive number that represents the vector of the anticipated equilibrium points.

3.3 *Design of a Closed-Loop Controller Using Passivity Principle*

The dynamic behavior for the open loop dc-dc converter is performed through the following data $E_1 = E_2 = 750$ V, $L_1 = 2$ mH, $L_2 = 9$ mH, $C_1 = C_2 = 500$ µf, $R_{load} = 100\ \Omega$, $R_{comp} = 10\ \Omega$ and $P_{cpl} = 750$ W with fixed duty cycles of $\mu_1 = 0.470$, $\mu_2 = 0.34$. Figure 10 shows the dynamic behavior of the passive and impassive nature of the DC-DC parallel buck converter showing the oscillatory response (impassive nature) of the output voltage for $t < 0.1$ s.

From Fig. 10 one can notice that when $R > R_{CPL}$ from 0 s to 0.1 s, R_{CPL} predominantly dominates i.e., the system operates with undamped response without energy dissipation. Whereas $R < R_{CPL}$ from 0.1 s to 0.2 s, i.e., the addition of resistance across the output capacitor, and it becomes passive (stable). However, it starts dissipating a high amount of energy, due to these losses the system the efficiency is very poor. Hence, this type of approach for the stabilization of the system is impractical. For the same system, stabilization can be achieved by adding effective virtual damping resistances to energy storage elements. Here, a series of virtual resistances are added for inductor elements (R_{1d}, R_{2d}) and parallel virtual resistances are added at the output capacitor elements (R_{3d}) as shown in Fig. 11. With the addition of these virtual damping resistances, the passivity nature of the integrated circuit is analyzed through the following two stages.

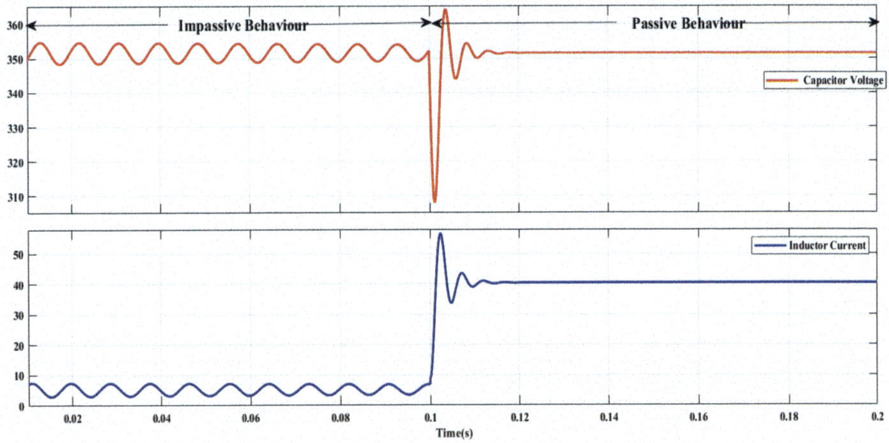

Fig. 10 Dynamic characteristics of an open loop dc-dc parallel buck converter

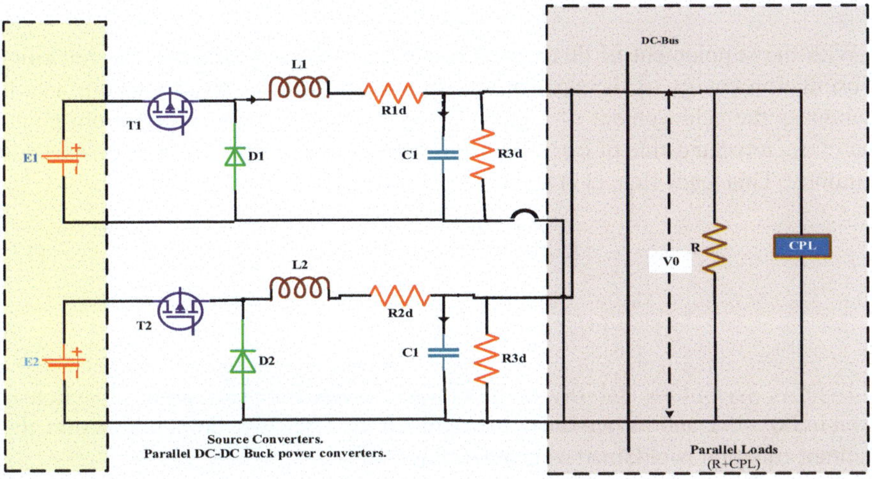

Fig. 11 DC-DC parallel buck converter with virtual damping resistances

3.3.1 Energy Shaping Stage

In this stage, the energy balance of the system denoted using Eq. (12) is modified, here the $\mathcal{Z}$ variable is replaced with $\tilde{\mathcal{Z}} + \mathcal{Z}_d$ where $(\tilde{\mathcal{Z}}, Z_d)$ represents state variables deviation and set point respectively [23], which yields

$$H\dot{\tilde{\mathcal{Z}}} + [\mathcal{G} + \mathcal{R}(z)]\tilde{\mathcal{Z}} = \mu\Gamma - \left(\mathcal{H}\dot{\mathcal{Z}}_d + [\mathcal{G} + \mathcal{R}(z)]z_d\right) \tag{15}$$

3.3.2 Injection Damping Stage

It is essential to damp limit cycles caused by the CPL in the system. It is achieved by the injection of a damping resistance matrix ($\mathcal{R}_d\tilde{z}$) to both sides of Eq. (15), which yields to Eq. (16).

$$H\dot{\tilde{\mathcal{Z}}} + [\mathcal{G} + R_i(z)]\tilde{\mathcal{Z}} = \mu\Gamma - \left(\mathcal{H}\dot{\mathcal{Z}}_d + [\mathcal{G} + \mathcal{R}(z)]\mathcal{Z}_d - \mathcal{R}_d\tilde{Z}\right) \tag{16}$$

where:

$$\begin{cases} R_i(\mathcal{Z}) = \begin{pmatrix} R_{1d} & 0 & 0 \\ 0 & R_{2d} & 0 \\ 0 & 0 & \left(\frac{1}{\mathcal{R}_{3d}} + \frac{1}{\mathcal{R}} + \frac{P}{z_3^2}\right) \end{pmatrix} \\ R_d = R_i(z) - R(z) = \begin{pmatrix} R_{1d} & 0 & 0 \\ 0 & R_{2d} & 0 \\ 0 & 0 & \frac{1}{R_{3d}} \end{pmatrix} \end{cases}$$

With the supplement of this virtual resistances matrix guarantees the releasing of oscillation energy of the system, thereby system behaves passive nature. i.e., it guarantees the achievement of a globally asymptotically stable equilibrium point. Therefore, left-hand side of Eq. (16) is approaches to zero i.e., $\tilde{\mathcal{Z}} = 0$ as times tends to infinity. This leads Eqs. (17) and (18) as follows [24, 25]:

$$H\dot{\tilde{\mathcal{Z}}} + [\mathcal{G} + R_i(z)]\tilde{\mathcal{Z}} = 0 \tag{17}$$

$$\mu\Gamma - \left(\mathcal{H}\dot{\mathcal{Z}}_d + [\mathcal{G} + \mathcal{R}(z)]\mathcal{Z}_d - \mathcal{R}_d\tilde{Z}\right) = 0 \tag{18}$$

To prove asymptotic stability, consider a positive definite Lyapunov's function as given in Eq. (19), and its derivative is examined for a negative definite function, the resultant equation is obtained as given in Eq. (20).

$$S(\tilde{z}) = \frac{1}{2}\tilde{z}^T\mathcal{H}\tilde{z} > 0 \quad (\forall\tilde{z} \neq 0) \tag{19}$$

$$\dot{\mathrm{S}}(\tilde{z}) = -\tilde{z}^{\mathrm{T}}\mathrm{R}_i(z)\tilde{\mathcal{Z}} = 0 \tag{20}$$

From Eq. (20), it is evident that the rate of energy storage diminishes over time, which is consistently seen as steady behaviour. The energy stored S ($\tilde{z}$)S in the passive system adheres to the same principle as the Lyapunov function in the system without any input, i.e., Regardless of the duty ratio \mu, the energy that is stored gradually approaches zero [26, 27]. Consequently, the right-hand side of (16) is expressed as,

$$\mu\Gamma - \left(\mathcal{H}\dot{\mathcal{Z}}_d + [\mathcal{G} + \mathcal{R}(z)]\mathcal{Z}_d - \mathcal{R}_d\tilde{\mathcal{Z}}\right) = 0 \tag{21}$$

where $\mu = \mu_1 + \mu_2$ and $\Gamma = \Gamma_1 + \Gamma_2$ defining the matrices, the following can be obtained as,

$$\begin{aligned}
[\mathcal{G} + \mathcal{R}(z)]\mathcal{Z}_d &= \left[\begin{pmatrix} 0 & 0\,1 & 0 \\ 0 & 0\,-1 & 0 \\ -1 & 1\,0 & 0 \\ 0 & 0\,0 & 0 \end{pmatrix} + \begin{pmatrix} 0 & 0\,0 & 0 \\ 0 & 0\,0 & 0 \\ 0 & 0\,0 & 0 \\ 0 & 0\,0 & \frac{1}{R} + \frac{P}{z_{4d}} \end{pmatrix}\right]\begin{bmatrix} z_{1d} \\ z_{2d} \\ z_{3d} \\ z_{4d} \end{bmatrix} \\
&= \begin{bmatrix} z_{3d} \\ -z_{3d} \\ -z_{1d} + z_{2d} \\ \left(\frac{1}{R} + \frac{P}{z_4^2}\right)_{z_{4d}} \end{bmatrix}
\end{aligned} \tag{22}$$

Now,

$$\begin{aligned}
&\left(\mathcal{H}\dot{\mathcal{Z}}_d + [\mathcal{G} + \mathcal{R}(z)]\mathcal{Z}_d - \mathcal{R}_d\tilde{\mathcal{Z}}\right) \\
&= \begin{pmatrix} L_1 & 0\,0 & 0 \\ 0 & L_2 & 0 & 0 \\ 0 & 0 & C_1 & 0 \\ 0 & 0\,0 & C_2 \end{pmatrix}\begin{pmatrix} \dot{z}_{1d} \\ \dot{z}_{2d} \\ \dot{z}_{3d} \\ \dot{z}_{4d} \end{pmatrix} \\
&+ \begin{pmatrix} z_{3d} \\ -z_{3d} \\ -z_{1d} + z_{2d} \\ \left(\frac{1}{R} + \frac{P}{z_4^2}\right)_{z_{4d}} \end{pmatrix} - \begin{pmatrix} R_{1d} & 0\,0 & 0 \\ 0 & R_{2d} & 0 & 0 \\ 0 & 0 & R_{3d} & 0 \\ 0 & 0\,0 & \frac{1}{R_{4d}} \end{pmatrix}\begin{pmatrix} \tilde{z}_1 \\ \tilde{z}_2 \\ \tilde{z}_3 \\ \tilde{z}_4 \end{pmatrix} \\
&= \begin{pmatrix} L_1\, z_{1d} \\ L_2\, z_{2d} \\ C_1\, z_{3d} \\ C_2\, z_{4d} \end{pmatrix} + \begin{pmatrix} z_{3d} \\ -z_{3d} \\ -z_{1d} + z_{2d} \\ \left(\frac{1}{R} + \frac{P}{Z_4^2}\right)_{Z_{4d}} \end{pmatrix} - \begin{pmatrix} R_{1d}\tilde{z}_1 \\ R_{2d}\tilde{z}_2 \\ R_{3d}\tilde{z}_3 \\ \frac{\tilde{z}_4}{R_{4d}} \end{pmatrix} \\
&= \begin{pmatrix} L_1\dot{z}_{1d} + z_{3d} - R_{1d}\tilde{z}_1 \\ L_2\dot{z}_{2d} + -z_{3d} - R_{2d}\tilde{z}_2 \\ C_1\dot{z}_{3d} - z_{1d} + z_{2d} - R_{3d}\tilde{z}_3 \\ C_2\dot{z}_{4d} + \left(\frac{1}{R} + \frac{P}{Z_4^2}\right)_{Z_{4d}} - \frac{\tilde{z}_4}{R_{4d}} \end{pmatrix}
\end{aligned} \tag{23}$$

$$(\mu_1\Gamma_1 + \mu_2\Gamma_2) - \left(\mathcal{H}\dot{\mathcal{Z}}_d + [\mathcal{G} + \mathcal{R}(z)]\mathcal{Z}_d - \mathcal{R}_d\tilde{\mathcal{Z}}\right) = 0 \tag{24}$$

By substituting the matrices in the Eq. (18), the following equations are obtained as,

$$\left.\begin{array}{r}\mu_1 E_1 - L_1\dot{z}_{1d} - z_{3d} + R_{1d}(z_1 - z_{1d}) = 0,\\ \mu_2 E_2 - L_2\dot{z}_{2d} - z_{3d} + R_{2d}(z_2 - z_{2d}) = 0,\\ -C_{eq}\dot{z}_{3d} + z_{1d} + z_{2d} - \frac{z_{3d}}{R} - \frac{P}{z_{3d}} + \frac{1}{R_{3d}}(z_3 - z_{3d}) = 0\end{array}\right\} \quad (25)$$

Equation (25) describes the dynamic behavior of the passivity-based controller. It is found that there are three dynamic equations that can be used to resolve four variables (z_{1d}, z_{2d}, z_{3d} and μ). To achieve this, state variables need to reach a steady state without exhibiting transient behaviour of the state variables, i.e., ($\dot{z}_{1d}, \dot{z}_{2d}, \dot{z}_{3d}$) must be zero. To this end, Eq. (25) can be reformulated as follows.

$$\mu_1 = \frac{1}{E_1}[V_0 + R_{1d}(I_{L1} - i_{L1})] \quad (26)$$

$$\mu_2 = \frac{1}{E_2}[V_0 + R_{2d}(I_{L2} - i_{L2})] \quad (27)$$

$$I_{L1} = I_{L2} = \frac{1}{2}\left[\frac{V_0}{R} + \frac{P}{V_0} + \frac{1}{R_{3d}}(V_o - v_o)\right] \quad (28)$$

Equations (26), (27), and (28) can be used to determine the duty ratio control signals μ_1 and μ_2 from PB control, as shown in Fig. 12. Perhaps, the primary disadvantage of the PBC approach is the Steady-state error cannot be eliminated, the long settling time, and higher overshoot. When inductor current references are made are identical, i.e., $i_{L1} = i_{L2}$. The control signals (μ_1, μ_2) are controlled in the inner loop. The PBC approach can eliminate the oscillations initiated by the Constant Power Load and regulate the output voltage.

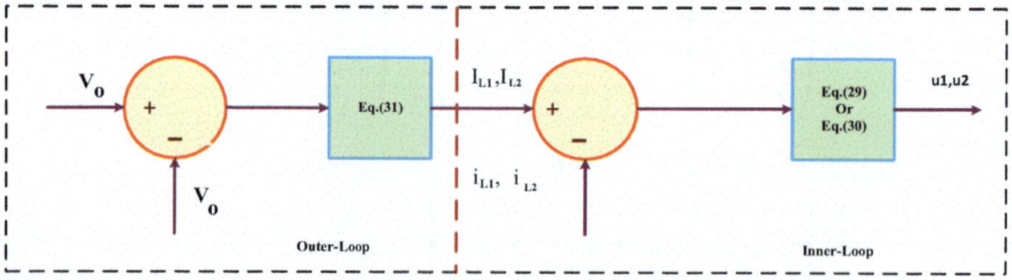

Fig. 12 Duty ratio generation with the outer and inner loop control

3.4 Design of Passivity Based PID Controller

During the transient period as well as the change in CPL, the system initially produces an exceptional overshoot. Long settling time and oscillation. The oscillations about a mean point are considered steady-state errors. To solve this issue, The PBC now includes an integral component to Eqs. (26) and (27). It can be expressed as the passivity-based controller with PID in the structure that follows [28].

With the assumption of $i_{L1} = i_{L2}$, rearrangement of Eq. (7) results in Eq. (29),

$$i_{L1} = i_{L2} = \frac{1}{2}\left(C_{eq}\dot{v}_0 + \frac{V_0}{R} + \frac{P}{V_0}\right) \tag{29}$$

The duty ratio control signals μ_1 and μ_2 are obtained by substituting Eqs. (28) and (29) in Eqs. (26) and (27) and these are given in Eqs. (30) and (31).

$$\mu_1 = \frac{1}{E_1}\left[-\frac{R_{1d}C_{eq}\dot{v}_0}{2} + \frac{R_{1d}}{2}\left(\frac{1}{R} + \frac{1}{R_{3d}}\right)(V_0 - v_0) + V_0\right] \tag{30}$$

$$\mu_2 = \frac{1}{E_2}\left[-\frac{R_{2d}C_{eq}\dot{v}_0}{2} + \frac{R_{2d}}{2}\left(\frac{1}{R} + \frac{1}{R_{3d}}\right)(V_0 - v_0) + V_0\right] \tag{31}$$

The addition of integral gains to the Eqs. (30) and (31) translates these into PID form. These are given in Eq. (32).

$$\mu = k_d\dot{e} + k_p e + D + k_p \textstyle\int edt \tag{32}$$

where,

$$k_{d1} = \frac{R_{1d}C_{eq}\dot{v}_0}{2E_1},\ k_{p1} = \frac{R_{1d}}{2E_1}\left(\frac{1}{R} + \frac{1}{R_{3d}}\right),\ \text{and } D_1 = \frac{V_0}{E_1}.$$

$$k_{d2} = \frac{R_{2d}C_{eq}\dot{v}_0}{2E_2},\ k_{p2} = \frac{R_{2d}}{2E_2}\left(\frac{1}{R} + \frac{1}{R_{3d}}\right),\ \text{and } D_2 = \frac{V_0}{E_2}.$$

where e = $V_0 - v_0$, error signal, k_p and k_d are the proportional, derivative gains and k_i is the integral gain which is taken as $1e^6$, and D is the voltage gain constant. Equation (32) represents the PBC-PID type controller [29, 30], and the dc bus voltage V_0 will be regulated by regulating virtual damping resistance (R_{3d}) across the output terminal.

4 Simulation Findings and Discussions

Table 2 shows the designed values of the parallel dc-dc buck converters. The following section discusses about robustness of the controllers (PBC and PBC-PID) is validated through MATLAB/Simulink environment and is performed under three uncertainty situations i.e., source uncertainty, load uncertainty, and reference voltage uncertainty. The results and discussion of these three cases are discussed in the following subsections.

4.1 Case-(i): System Dynamic Behaviour During Source Uncertainty

In this study, the control performance of proposed PBC and PBC-PID control techniques are examined considering the uncertainty in input source voltages (E_1 and E_2). The source uncertainty considered for simulation studies is as follows, (i) from 0 s to 0.06 s, input voltages are the same i.e., 750 V, (ii) at 0.06 s a step change in input voltage-1 is considered, i.e., from 750 to 850 V, (iii) at 0.15 s a step change in input voltage-2 is considered i.e., from 750 to 650 V. Here, DC bus output voltage and composite load current are captured as depicted in Fig. 13a, b. It is observed that there is no significant effect on DC output voltage and composite load current for both proposed controller techniques. As compared to the PBC method, the PBC-PID method exhibits less steady-state error and fewer overshoots during the dynamic changes in input voltages.

4.2 Case-(ii): System Dynamic Behaviour During Load Uncertainty

The effectiveness of PBC and PBC-PID control techniques are examined considering the uncertainty in load. The load uncertainty considered for simulation studies as follows, (i) from 0 s to 0.1 s, CPL power is 750W, (ii) at 0.1 s a step change in

Table 2 System parameters

Descriptions	Nominal values
Inductances	$L_{1O} = 2\,\text{mH}$, $L_{2O} = 5\,\text{mH}$
Capacitances	$C_{1O} = 500\,\mu\text{F}$, $C_{2O} = 230\,\mu\text{F}$
Cpl Power	$P_o = 7220\,\text{kW}$
Load Resistances	$R_o = 50\,\Omega$
Input Voltages	$E_{1o} = E_{2o} = 750\,\text{V}$
Desired DC- Bus Voltage	$V_o = 350\,\text{V}$

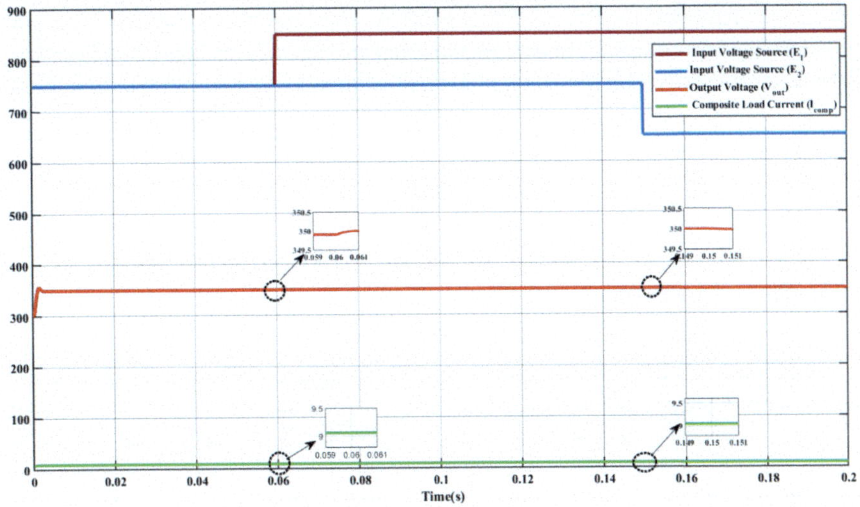

(a)

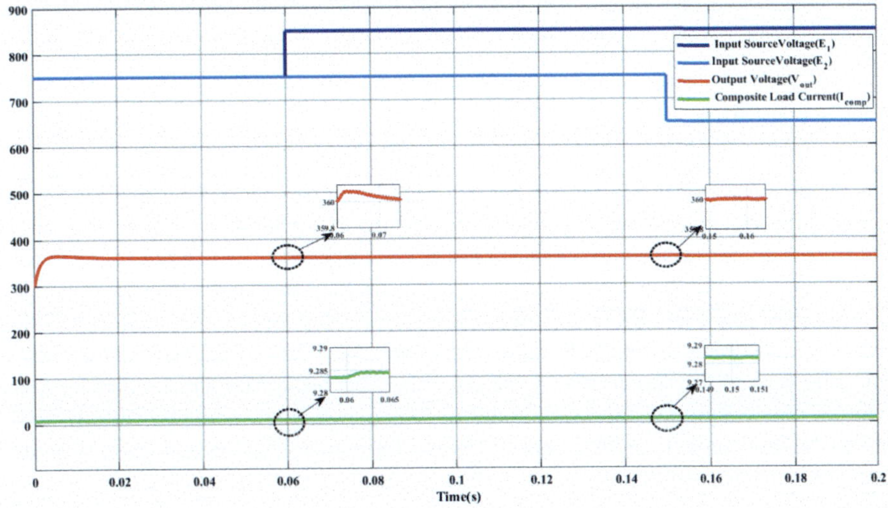

(b)

Fig. 13 System dynamic behaviour during source uncertainty of **a** output voltage and composite load current by PBC controller. **b** Output voltage and composite load current by PBC-PID controller

CPL load is considered i.e., from 750 to 1450 W. Here, DC bus output voltage and composite load current are captured as depicted in Fig. 14a, b It is observed that, both the controllers are effectivity regulating the converter system. In contrast to the PBC-PID controller, PBC exhibits a higher momentary steady state variation at the load uncertainty instant due to CPL.

4.3 Case-(iii) System Dynamic Behaviour During Reference Voltage Uncertainty

The effectiveness of PBC and PBC-PID control techniques are examined considering the uncertainty in reference voltage. The reference voltage uncertainty considered for simulation studies is as follows, (i) from 0 s to 0.1 s, the reference voltage is 350 V, (ii) at 0.1 s a step change in reference voltage is considered i.e., from 350 to 365 V. Here, DC bus output voltage and composite load current are captured as depicted in Fig. 15a, b. It is observed that, both the controllers are effectivity tracking the converter output voltage. In contrast to the PBC-PID controller, PBC exhibits a higher momentary steady state variation at the load uncertainty instant due to CPL.

5 Conclusions

This chapter reports on limit cycle behaviour (instability) caused by the CPL for a two-input parallel dc-dc buck converter in a distributed generation system. To mitigate this issue, passivity-based control (PBC) is employed. To validate the control performance of the converter system a fair-minded comparison has been made between PBC and PBC-PID controllers. It is observed that both PBC and PBC-PID controllers eliminate limit cycles caused by the CPL. During uncertain situations, PBC-PID has shown superiority over the PBC and has given satisfactory results in terms of steady-state error elimination, lower overshoot, and fast setting with extremely fast recovery performance. The performance of proposed controllers is compared and the simulations were performed in MAT Lab/Simulink environment.

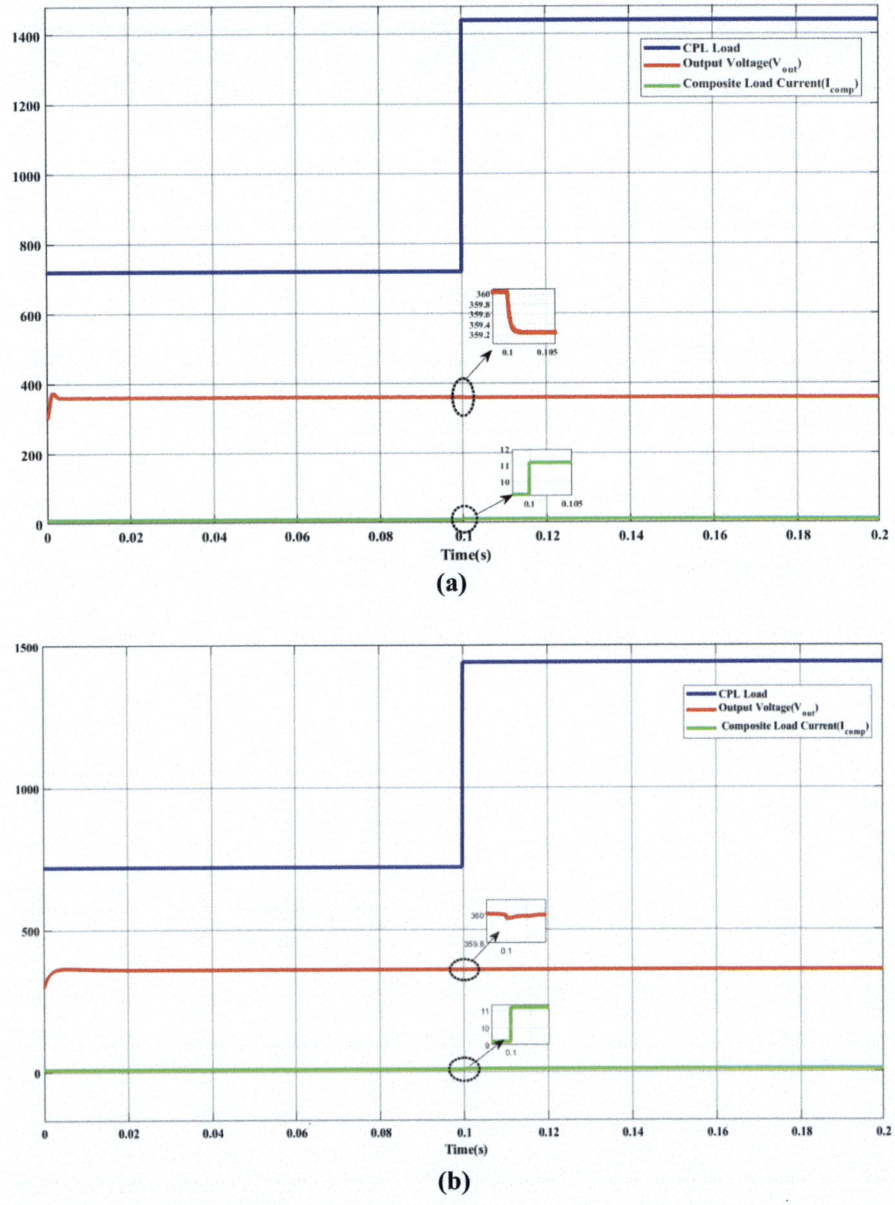

Fig. 14 System dynamic behaviour during load uncertainty of **a** output voltage and composite load current by PBC controller. **b** Output voltage and composite load current by PBC-PID controller

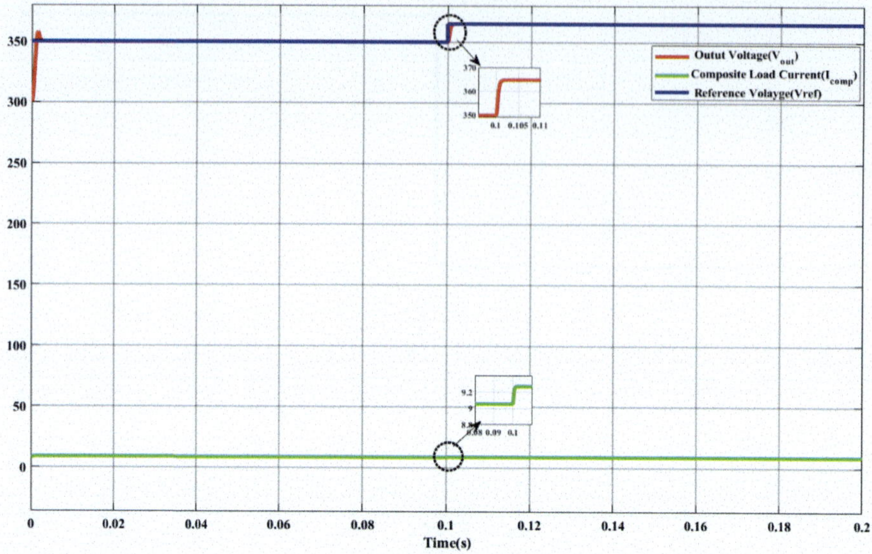

(a)

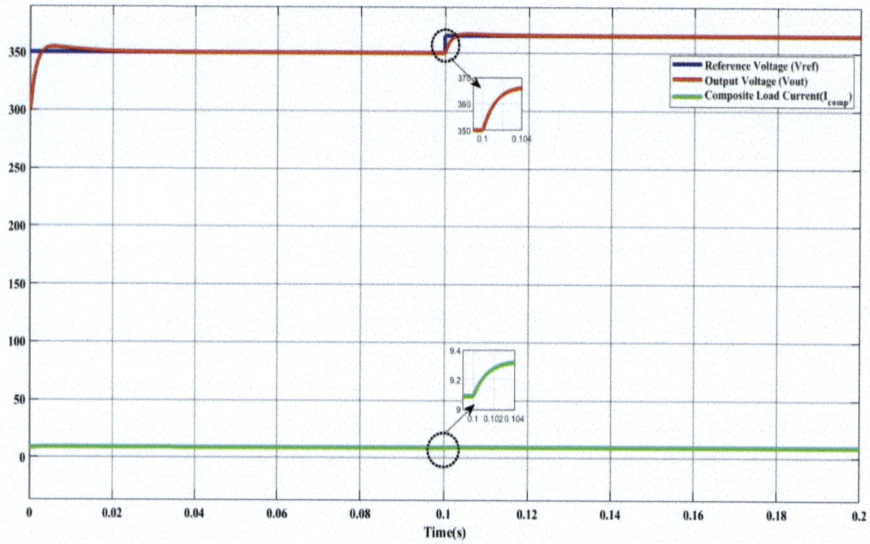

(b)

Fig. 15 System dynamic behaviour during reference voltage uncertainty of **a** output voltage and composite load current by PBC controller. **b** Output voltage and composite load current by PBC-PID controller

References

1. Adajah YY, Thomas S, Haruna MS, Anaza SO (2021) Distributed Generation (DG): a review. In: 2021 1st international conference on multidisciplinary engineering and applied science (IcMEAS), Abuja, Nigria, p 15. https://doi.org/10.1109/IMEAS52683.2021.9692353
2. Zubieta LE (2016) Are microgrids the future of energy? DC microgrids from concept to demonstration to deployment. IEEE Electrification Mag 4(2):37–44. https://doi.org/10.1109/MELE.2016.2544238
3. Dragičević T, Lu X, Vasquez JC, Guerrero JM (2016) DC microgrids—part I: a review of control strategies and stabilization techniques. IEEE Trans Power Electron 31(7):4876–4891. https://doi.org/10.1109/TPEL.2015.2478859
4. Dragičević T, Lu X, Vasquez JC, Guerrero JM (2016) DC microgrids—part II: a review of power architectures, applications, and standardization issues. IEEE Trans Power Electronics 31(5):3528–3549. https://doi.org/10.1109/TPEL.2015.2464277
5. Bhukya CN, Reddy BA (2023) Constant power loads in DC microgrids: a review of modern nonlinear control approaches and stabilization techniques. In: IEEE 2nd international conference on industrial electronics: developments & applications (ICIDeA), Imphal, India, pp 181–186. https://doi.org/10.1109/ICIDeA59866.2023.10295221
6. Kwasinski A, Krein PT (2007) Stabilization of constant power loads in Dc-Dc converters using passivity-based control. In: INTELEC 07–29th international telecommunications energy conference, Rome, Italy, pp 867–874. https://doi.org/10.1109/INTLEC.2007.4448903
7. Hassan MA, He Y (2020) Constant power load stabilization in DC microgrid systems using passivity-based control with nonlinear disturbance observer. IEEE Access 92393–92406. https://doi.org/10.1109/ACCESS.2020.2992780
8. Manogna M, Reddy BA, Padma K (2022) Modeling of a three-input fourth-order integrated DC-DC converter. In: International conference on smart and sustainable technologies in energy and power sectors (SSTEPS), vol 5, issue 1, pp 83–88. https://doi.org/10.1088/2631-8695/acc0dc
9. Zhao Y, Qiao W, Ha D (2014) A sliding-mode duty-ratio controller for DC/DC buck converters with constant power loads. IEEE Trans Ind Appl 50(2):1448–1458. https://doi.org/10.1109/TIA.2013.2273751
10. Nayak Bhukya Ch, Amarendra Reddy B, Allam Venkatesh, Jyothsna TR (2024) Robust controller for limit cycle elimination of two input DC-DC converter with constant power load in DC μGrid applications. WSEAS Trans Syst 23:196–205. https://doi.org/10.37394/23202.2024.23.22
11. Kondratiev I, Santi E, Dougal R, Veselov G (2004) Synergetic control for DC-DC buck converters with constant power load. In: IEEE 35th annual power electronics specialists conference, vol 5, pp 3758–3764. https://doi.org/10.1109/PESC.2004.1355139
12. Xu Q, Zhang C, Wen C, Wang P (2019) A novel composite nonlinear controller for stabilization of constant power load in DC microgrid. IEEE Trans Smart Grid 10(1):752–761. https://doi.org/10.1109/TSG.2017.2751755
13. Kwasinski A, Krein PT (2007) Passivity-based control of buck converters with constant-power loads. In: IEEE power electronics specialists conference, pp 259–265. https://doi.org/10.1109/PESC.2007.4341998
14. Bhukya CN, Bhimavarapu AR, Allam V, Salkuti SR (2024) Passivity based modeling of a two-input DC–DC power converter with constant power and constant voltage load. In: Salkuti SR (eds) Energy and environmental aspects of emerging technologies for smart grid. Green energy and technology. Springer, Cham. https://doi.org/10.1007/978-3-031-18389-8_18
15. Hassan MA, Li E, Li X, Li T, Duan C, Chi S (2019) Adaptive passivity-based control of DC–DC buck power converter with constant power load in DC microgrid systems. IEEE J Emerging Selected Topics Power Electronics 7:2029–2040. https://doi.org/10.1109/JESTPE.2018.2874449

16. Sira Ramirez H, Ortega R, Escobar G (1996) Lagrangian modeling of switch regulated DC-to-DC power converters. In: Proceedings of 35th IEEE conference on decision and control, vol 4, pp 4492–4497.https://doi.org/10.1109/CDC.1996.577568
17. AL-Nussairi MK, Bayindir R, Padmanaban S, Mihet-Popa L, Siano P (2017) Constant Power Loads (CPL) with microgrids: problem definition, stability analysis and compensation techniques. Energies 10(10):1656.https://doi.org/10.3390/en10101656
18. Singh S, Gautam A, Fulwani D (2017) Constant power loads and their effects in DC distributed power systems: a review. Renew Sustain Energy Rev 407–421(72):407–421.https://doi.org/10.1016/j.rser.2017.01.027
19. Leonard J (2014) Nonlinear modeling of DC constant power loads with frequency domain Volterra Kernels [Unpublished Doctor of Philosophy Dissertation]. Florida State University
20. Emadi A, Khaligh A, Rivetta CH, Williamson GA (2006) Constant power loads and negative impedance instability in automotive systems: definition, modeling, stability, and control of power electronic converters and motor drives. IEEE Trans Veh Technol 55(4):1112–1125. https://doi.org/10.1109/TVT.2006.877483
21. Pasiopoulou ID, Kontis EO, Papadopoulos TA, Papagiannis GK (2022) Effect of load modeling on power system stability studies. Electric Power Syst Res 207:107846. https://doi.org/10.1016/j.epsr.2022.107846
22. Youla D, Castriota L, Carlin H (1959) Bounded real scattering matrices and the foundations of linear passive network theory. IRE Trans Circuit Theory 6(1):102–124. https://doi.org/10.1109/TCT.1959.1086518
23. Singh S, Fulwani D (2014) Voltage regulation and stabilization of DC/DC buck converter under constant power loading. In: IEEE international conference on Power Electronics, Drives and Energy Systems (PEDES), Mumbai, India, pp 1–6. https://doi.org/10.1109/PEDES.2014.7042054
24. Thota P, Bhimavarapu AR, Chintapalli VVSBR (2022) Participation factor based analysis of PVSC type multi input zeta SEPIC DC-DC converter. COMPEL 41(5):1727–1752. https://doi.org/10.1108/COMPEL-10-2021-0363
25. Mummadi V, Bhimavarapu AR (2020) Robust multi-variable controller design for two-input two-output fourth-order DC-DC converter. Electric Power Components Syst 48(1–2):86–104. https://doi.org/10.1080/15325008.2020.1736213
26. Amarendra Reddy B, Veerachary M (2016) Robust multivariable controller design using H-infinity loop shaping for TIFOI DC-DC converter. IEEE UPCON 372–377. https://doi.org/10.1109/UPCON.2016.7894682
27. Thota P, Reddy B, Chintapalli V (2022) Selection of input–output pairing and control structure configuration using interaction measures for DC–DC dual input zeta-SEPIC converter. Electrica 41(5):1727–1752. https://doi.org/10.1108/COMPEL-10-2021-0363
28. Sravanthi P, Sandeep V (2022) Solving realistic reactive power market clearing problem of wind-thermal power system with system security. Int J Emerg Electr Power Syst 23(2):125–144. https://doi.org/10.1515/ijeeps-2021-0060
29. Kumar JV, Sesham H, Salkuti SR (2024) Smart energy management model for electric vehicles. In: Salkuti SR (eds) Energy and environmental aspects of emerging technologies for smart grid. Green energy and technology. Springer, Cham. https://doi.org/10.1007/978-3-031-18389-8_9
30. Hassan MA, Li E, Li X, Li T, Duan C, Chi S (2019) Adaptive passivity-based control of DC–DC buck power converter with constant power load in DC microgrid systems. IEEE J Emerging Selected Topics Power Electronics 7(3):2029–2040. https://doi.org/10.1109/JESTPE.2018.2874449

A Two-Stage Scheduling Based Energy Management in a Microgrid

Krishna Mohan Reddy Pothireddy, Sandeep Vuddanti, and Surender Reddy Salkuti

Abstract This work addresses the need for energy management in a microgrid (MG). Outliers in MG's, representing anomalies or unexpected variations in data, can significantly disrupt the performance and reliability of energy management systems (EMS). The proposed work uses an unsupervised K-means clustering technique which is a prominent one for outlier detection and imputation. In addition to outliers, there is a need to model the uncertainties for effective EMS. This paper addresses the uncertainties based on Monte Carlo simulation for scenario generation and K-means clustering for scenario reduction. Furthermore, a two-stage scheduling approach is proposed to minimize day-ahead operational costs and real-time imbalance costs. The methodology was applied to an MG model, demonstrating reduced operational expenses and improved system stability by effectively managing outliers and optimizing scheduling decisions. The maximum energy exchange with the grid for a particular scheduling hour is limited to 70 kW. The total operational cost of the MG has reduced from 20,202.54 to 19,241.17 rupees, a reduction of 5% in the total operational cost of the MG.

Keywords Renewable energy sources · Energy management · Uncertainty handling · Outlier detection and imputation · K-means clustering · Monte Carlo simulation

K. M. R. Pothireddy · S. Vuddanti
Department of Electrical Engineering, National Institute of Technology Andhra Pradesh, Tadepalligudem, Andhra Pradesh, India
e-mail: sandeep@nitandhra.ac.in

S. R. Salkuti (✉)
Department of Railroad and Electrical Engineering, Woosong University, Daejeon, Republic of Korea
e-mail: surender@wsu.ac.kr

S. R. Salkuti (ed.), *Artificial Intelligence for Integrated Smart Energy Systems in Electric Vehicles*, Lecture Notes in Electrical Engineering 1427,
https://doi.org/10.1007/978-3-031-94276-1_19

Nomenclature

k	Shape parameter
v	Random variable
x	Value of the variable
E_{exc}^{t}	Energy exchange cost at time 't'
LD_{DA}^{t}	Load demand in day-ahead at time 't'
G_{RT}^{t}	Power generation in real-time
c	Scale value
μ	Mean
E_{ex}^{t}	Amount of energy exchange at time 't'
P_{DA}^{t}	Day-ahead electricity price at time 't'
G_{DA}^{t}	Power generation in day-ahead at time 't'
RG_{DA}^{t}	Power generation by renewable energy sources
G_{RT}^{t}	Real-time generation at time 't'
K	Number of clusters
PV_{DA}^{t}	Power generation by PV in day-ahead at time 't'
PV_{RT}^{t}	Power generation by PV in real-time at time 't'
RG_{DA}^{t}	Total power generation by RESs in day-ahead at time 't'
RG_{RT}^{t}	Total power generation by RESs in real-time at time 't'
σ	Standard deviation
WT_{DA}^{t}	Power generation by a wind turbine in day-ahead at time 't'
WT_{RT}^{t}	Power generation by a wind turbine in real-time at time 't'
IP_{RT}^{t}	Imbalance power in real-time
LD_{RT}^{t}	Load demand in real-time
MCS	Monte Carlo simulation
RESs	Renewable energy sources
PV	Photovoltaic
PCA	Principal component analysis
DBSCAN	Density-based spatial clustering applications with noise
WT	Wind turbine
RMSE	Root mean square error
MG	Microgrid
MAPE	Mean absolute percentage error
MAE	Mean absolute error
EMS	Energy management system

1 Introduction

Some limited resources are utilized in the conventional power generation process: coal, oil, and natural gas [1]. Because traditional power plants are frequently vulnerable to changes in fuel prices, market dynamics, and geopolitical events can have an impact on energy costs [2]. Conventional generators, especially those that burn coal and natural gas, generate a significant quantity of greenhouse gasses, primarily carbon dioxide. Climate change and air pollution are both impacted by these emissions [3]. Large transmission and distribution networks are required to supply end users with electricity because conventional power facilities are typically centralized. This may lead to energy losses during transmission [4]. When a microgrid (MG) is linked to the main power grid, it operates in tandem with it. When required, it might draw energy from MG's may perform grid support operations including voltage control and frequency regulation, which helps to maintain the reliability and stability of the main power grid [5]. MG's can function independently from the main grid while it is in islanded mode [6]. This mode is particularly useful during grid outages or emergencies.

MG's are increasingly integral to modern energy systems, offering localized energy generation, storage, and distribution. However, their operation is challenged by the presence of outliers—anomalous data points that deviate significantly from normal patterns due to sensor errors, abrupt load changes, or other unforeseen events [7]. These outliers can lead to inaccurate forecasts, suboptimal scheduling, and inefficiencies in the energy management system (EMS), thereby increasing operational costs and reducing system reliability [8].

The Z-score estimates the number of standard deviations that separate a given data point from the mean. Outliers are data points whose Z-scores fall outside of a predetermined range whereas, in the inter-quartile range data points outside the range are treated as outliers [9]. The local density variation for a data point with its neighbors is measured by the local outlier factor. The local density of outliers differs noticeably from what is expected [10]. Density-based spatial clustering applications with noise are a prominent technique for detecting outliers however, the performance of this algorithm is a function of data set size. K-means clustering is one approach that has gained popularity owing to its use in data imputation in addition to outlier detection. Mean, median, mode, forward impute, backward impute, K-nearest neighbor, etc. are used to impute in place of detected outliers. The choice of imputation depends on so many factors such as distribution, degree of outliers, etc. K-means clustering in general used for clustering or grouping similar data points as clusters however, it can also be used for scenario reduction. K-means clustering tries to reduce the intra-cluster distance and maximizes the inter-cluster distance. Outliers are identified using K-means clustering, which partitions data points into clusters based on their similarity and are often distinguished by their significant distance from the centroids of their assigned clusters [11].

The Normal or Gaussian distribution may be used to simulate PV output power and load demand, whereas the Weibull distribution can be used to describe wind power [12]. The probability distribution parameters are determined using historical data on load demand, photovoltaics, and wind [13]. Monte Carlo Simulation (MCS) has long been a reliable method for scenario generation, allowing researchers to simulate a wide range of possible outcomes by modeling uncertain parameters as random variables [14]. Handling uncertainty models can be done in two ways i.e., stochastic modeling and robust modelling. The reliability of robust modeling is high, and the operational cost yielded from this method is very high since it considers worst-case scenarios [15]. Therefore, stochastic optimization is the method to handle uncertain parameters such as uncertain wind, PV output, and load demand. Scenario generation methods such as Monte Carlo simulation, quasi-Monte Carlo simulation, and hyper-cube sampling methods are used, and analyzing all the generated scenarios is very difficult hence, scenario reduction techniques have been employed. To optimize the investment and operational expenditure on the MG [16] a probabilistic optimization. MCS generates a comprehensive set of scenarios, each representing a different possible state of the system based on the probability distribution considered and it generates random scenarios within the mean and standard deviation [17]. The inputs to the Monte Carlo simulation are shape factor, scale factor, and a number of scenarios if the distribution is Weibull distribution and the mean and standard deviation of the modeled distribution is normal or Gaussian distribution [18].

To address these challenges, this chapter proposes a stochastic methodology for identifying and handling outliers within MG's datasets. Moreover, we introduce a two-stage scheduling approach that optimizes both day-ahead and real-time operations. The first stage involves a pre-processing stage using K-means clustering to detect outliers, and K-nearest neighbors (KNN) to impute these anomalies followed by the day-ahead scheduling focuses on minimizing costs based on forecasts, while the real-time scheduling addresses imbalances that arise from deviations between forecasted and actual conditions. This hybrid approach aims to minimize the overall OC and improve the economic efficiency of MG operations.

2 Microgrid Modeling and Problem Formulation

MG modeling includes developing mathematical and system illustrations of small-scale, localized power systems that can function independently or in unification with the main grid. MG's are composed of distributed energy resources (DERs) like solar PV, wind turbines, diesel generators, battery energy storage systems (BESS), and controllable loads as shown in Fig. 1. In deregulated electricity markets, the modeling of MG aims to optimize various objectives such as cost minimization, reliability, and efficiency, considering the dynamic nature of energy prices and demand.

2.1 *Photovoltaic Systems*

PV systems harness solar energy and provide a clean and renewable source of electricity without greenhouse gas emissions. PV systems are highly scalable, ranging from small rooftop installations to huge solar farms [19]. Solar power generation rests on sunlight availability, leading to unpredictability and intermittency, which

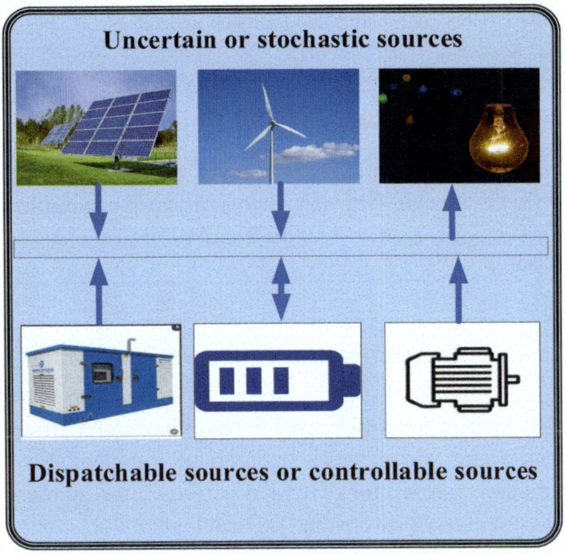

Fig. 1 Structure of the microgrid

necessitates integration with storage systems or backup generators. After installation, PV systems have insignificant operating costs, principally related to maintenance and inverter replacement. PV systems can be unified into the grid, supplying excess energy back to the grid or storing it in batteries for later use. PV systems contribute to energy independence, dropping reliance on fossil fuels and enhancing energy security [20].

2.2 *Wind Energy Systems*

Wind turbines convert kinetic energy from wind into power and offer a renewable and environmentally friendly energy source [21]. When compared to other renewable sources, wind energy has a high energy conversion efficiency, particularly in areas with consistent and strong wind patterns. Wind power is subject to variations depending on wind speed and availability, demanding complementary power sources or storage solutions [22]. It can be deployed onshore or offshore, with offshore installations often advancing from stronger and more consistent winds. Modern wind turbines have a long operational lifespan, typically around 20–25 years, with relatively low maintenance requirements. Wind turbines can generate noise and visual impacts, which are considerations in site selection and community acceptance [23].

Table 1 Cost coefficients and limits on generators

Generator type	a_0	a_1	a_2	Minimum generator limit in kW	Maximum generator limit in kW
Generator 1	0.010	2	10	0	40
Generator 2	0.020	3	8	0	40
Generator 3	0.015	1	12	0	40
PV	–	–	–	0	20
Wind	–	–	–	0	20

2.3 *Diesel Generators*

Diesel generators offer reliable backup power, particularly in remote or off-grid locations where grid access is inadequate or unavailable [24]. Diesel is widely available and can be stored for extended periods, making diesel generators a practical option for emergency power supply. Diesel generators can be quickly started and ramped up to full capacity, making them suitable for emergency or peak-load applications [25]. Diesel generators have higher running costs due to fuel consumption and regular maintenance requirements. Diesel generators release greenhouse gases and pollutants, including CO_2, NO_x, and particulate matter, contributing to environmental and health concerns. Compared to RESs, diesel generators have a shorter operational lifespan and may require more frequent replacements or overhauls [26]. Table 1 indicates the cost coefficient and limits on generators.

2.4 *Objective Function Formulation*

The objective is to reduce the total operational cost in the day-ahead scheduling phase and the imbalance of power and cost associated with it in the real-time scheduling phase. The objective is formulated as follows:

$$\min_{t,s} TOC_{MG} = \sum_{t=1}^{24} OC_{MG}^{t,s} + CE_{exchange}^{t,s} \forall t, \forall s \tag{1}$$

$$CE_{ex}^{t,s} = E_{ex}^{t,s} * P_{DA}^{t,s} \forall t, \forall s \tag{2}$$

$$E_{ex}^{t,s} = LD_{DA}^{t,s} - G_{DA}^{t,s} \forall t, \forall s \tag{3}$$

$$\min_{t,s} IC_{MG} = \sum_{t=1}^{24} IP_{RT}^{t,s} * SP_{RT}^{t,s} \forall t, \forall s \tag{4}$$

$G_{RT}^{t,s}$ is the real-time generation at time 't', $RG_{RT}^{t,s}$ renewable energy power generation at time 't'. The functions of the energy management system during power shortages and excesses in the MG, and enhances the power exchange with the grid to minimize operational expenditure [27].

$$G_{RT}^{t,s} = G_{1,RT}^{t,s} + G_{2,RT}^{t,s} + G_{3,RT}^{t,s} + RG_{RT}^{t,s} \forall t, \forall s \tag{5}$$

$$IP_{RT}^{t,s} = LD_{RT}^{t,s} - G_{RT}^{t,s} \forall t, \forall s \tag{6}$$

$$RG_{RT}^{t,s} = PV_{RT}^{t,s} + WT_{RT}^{t,s} \forall t, \forall s \tag{7}$$

$$0 \le G_{1,RT}^{t,s} \le G_{1,RT}^{max} \forall t, \forall s \tag{8}$$

$$E_{t+1,s}^{S} = E_{t,s}^{S} + P_{t,s}^{S,ch} \Delta t \eta^{S,ch,s} - \frac{P_t^{S,dch,s}}{\eta^{S,dch,s}} \forall t, \forall s \tag{9}$$

$$SoC_{t,s} = \frac{E_{t,s}^{S}}{E_{ra}^{S}} \forall t, \forall s \tag{10}$$

$$SoC_{min} \le SoC_{t,s} \le SoC_{max} \forall t, \forall s \tag{11}$$

$$0 \le P_{t,s}^{S,ch} \le X_{t,s}^{S,ch} P_{t,s}^{max,S,ch} \forall t, \forall s \tag{12}$$

$$0 \le P_{t,s}^{S,dch} \le X_{t,s}^{S,dch} P_{t,s}^{max,S,dch} \forall t, \forall s \tag{13}$$

$$X_{t,s}^{S,ch} + X_{t,s}^{S,dch} \le 1 \forall t, \forall s \tag{14}$$

$$X_t^{S,ch} \in \{0, 1\} \forall t, \forall s \tag{15}$$

$$X_{t,s}^{S,dch} \in \{0, 1\} \forall t, \forall s \tag{16}$$

3 Proposed Methodology

The MG data is first normalized to ensure that all features contribute equally to the clustering process as shown in Fig. 2. The normalized data is then clustered using the K-means algorithm [28, 29]. The objective is to cluster the dataset into a specified number of clusters, where each data point fits into the cluster with the nearest centroid. Outliers are recognized based on their distance from the centroids of the clusters. Data points that lie outermost from their respective cluster centroids are considered outliers. Once outliers are detected, they are imputed using the K-nearest neighbors' algorithm. For each outlier, the KNN algorithm identifies the nearest 'k' non-outlier points and estimates the missing or anomalous values by averaging these neighbors. The imputation process ensures that the modified dataset maintains consistency and reduces the potential negative impact of the original outliers on the energy management system [30].

The day-ahead scheduling process begins with forecasting the load demand, generation, and market prices for the next day. Accurate forecasts are critical to minimizing the costs associated with energy procurement and generation. The forecasted data is used to optimize the generation schedule for the next day, aiming to minimize the total operational cost [31]. The optimization considers factors such as generation costs, fuel costs, and potential revenues from selling excess energy.

During real-time operation, actual data on load demand and generation is continuously monitored. Any deviations from the day-ahead forecast are identified as imbalances. A real-time optimization algorithm is employed to adjust the generation schedule dynamically to address these imbalances. The goal is to minimize the imbalance costs, which arise due to penalties or additional charges for deviations from the scheduled plan [32, 33]. The real-time adjustments are coordinated with the day-ahead schedule to ensure that the overall operational costs are minimized while maintaining system stability. The clean and accurate data, obtained after outlier detection and imputation, is used in both the day-ahead and real-time scheduling processes. By ensuring data integrity, the scheduling algorithms can operate more effectively [34], leading to better decision-making and reduced costs. The proposed methodology is validated through simulations on a microgrid model, demonstrating the combined benefits of outlier management and two-stage scheduling. The results show a reduction in both day-ahead and real-time imbalance costs, confirming the effectiveness of the approach.

4 Case Study

The test system comprises 3 diesel generators of 40 kW capacity with quadratic cost coefficients, a wind turbine of 20 kW, and PV of 20 kW. Figures 3, 4, and 5 show the probability distribution of PV, wind power, and load demand. The shape factor and scale factor for Weibull distribution is considered as 2.5 and 10 whereas, the mean and standard deviation of PV is considered as 14 and 2 and the mean and standard deviation for load demand is considered as 55 and 12. Table 2 indicates the probability distribution parameters.

Figure 6 indicates the scenario generation using MCS and reduction by using K-means clustering for PV output power. The total number of generated scenarios is 100 and reduced these scenarios to 10 without losing crucial information. The process of clustering representative points as cluster centers and grouping similar scenarios

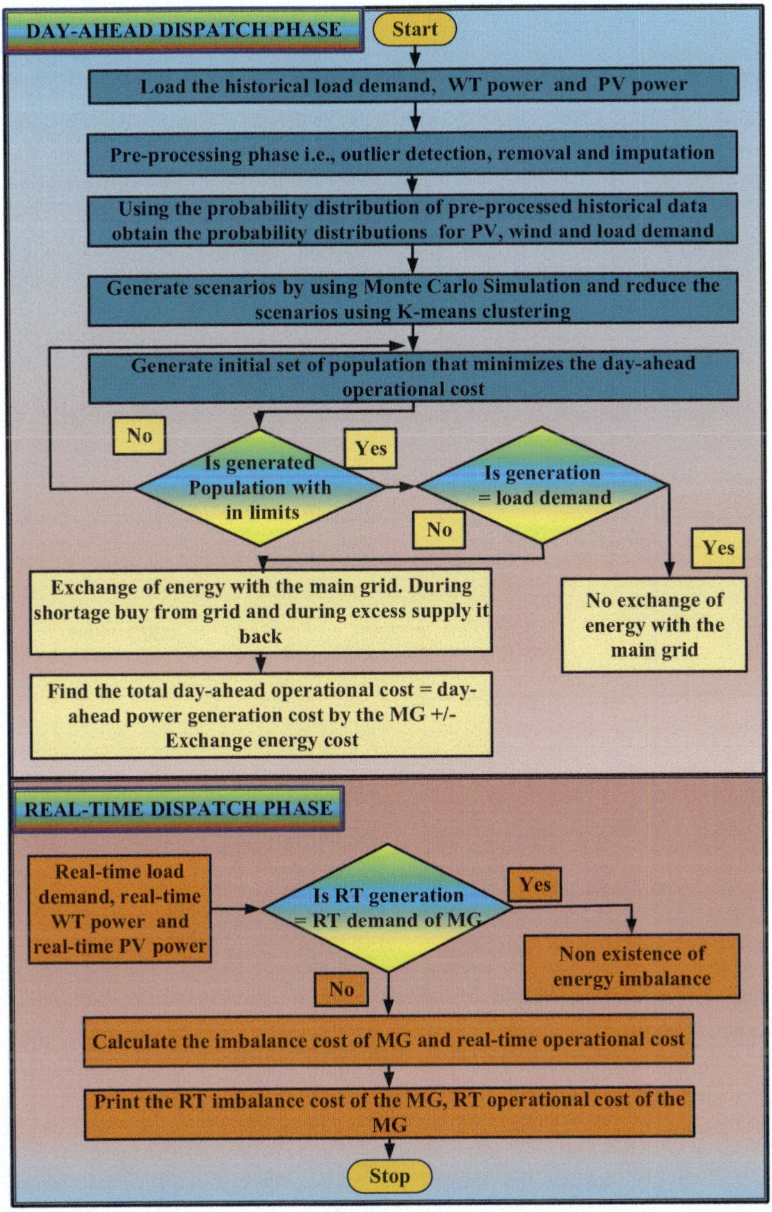

Fig. 2 Proposed two-stage energy management

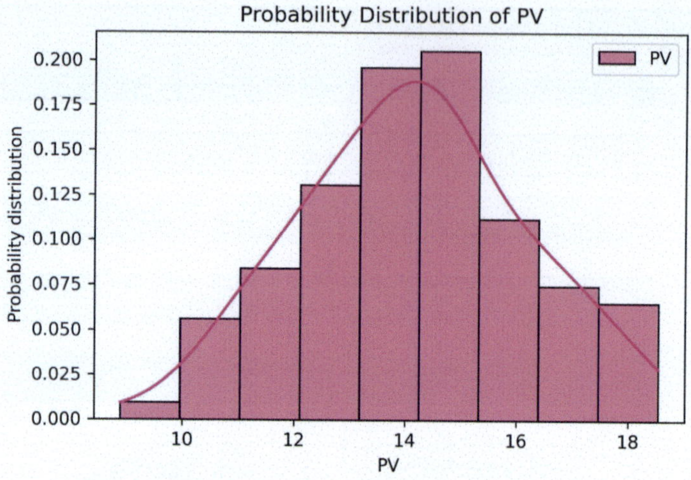

Fig. 3 Normal distribution of PV

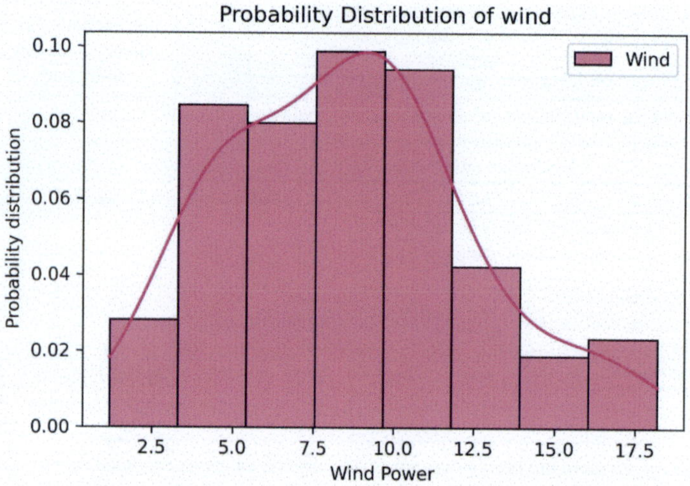

Fig. 4 Weibull distribution of wind power

together effectively reduces the complex scenario space to a simpler to handle set. The process of scenario reduction not only maximizes computational resources but also improves the results' clarity.

Figure 7 indicates the scenario generation and reduction of PV. Figure 8 represents the scenario generation and reduction of load demand. An IEEE 33 bus system was considered to assess the impact of load on the amount of energy exchange, unit commitment status of generators, dispatch schedule of sources, and the total operational cost of the microgrid. The load demand has a peak load demand on the system

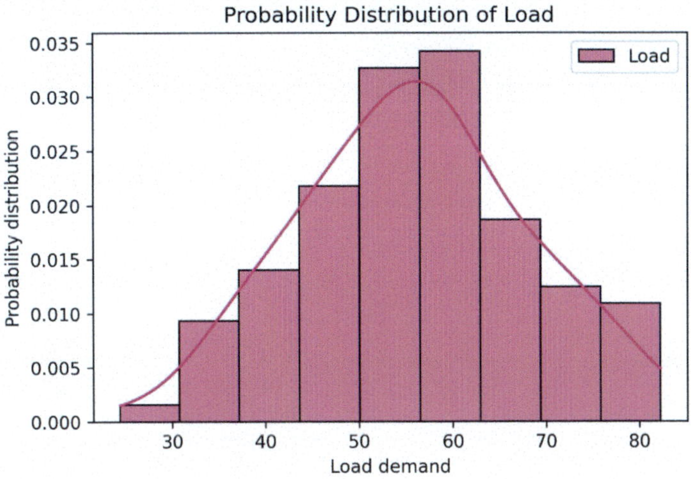

Fig. 5 Normal distribution of load demand

Table 2 Probability distribution parameters

Weibull distribution (Wind power in kW)	Normal distribution	
	PV power (kW)	Load demand (kW)
Shape value = 2.5	Mean = 14	Mean = 55
Scale factor = 10	Standard deviation = 2	Standard deviation = 12
Number of scenarios = 100	Number of scenarios = 100	Number of scenarios = 100
Number of hours = 24	Number of hours = 24	Number of hours = 24

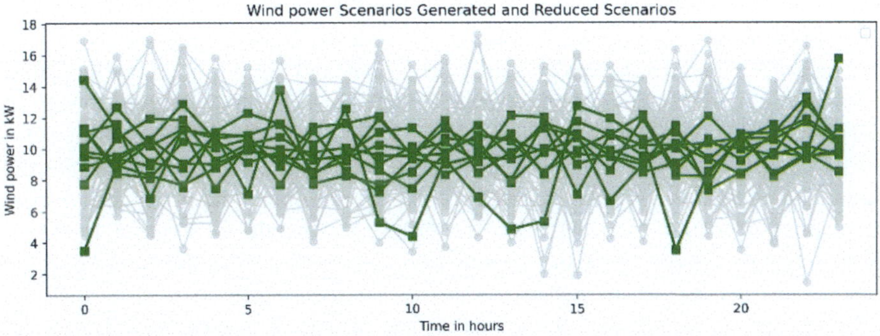

Fig. 6 Scenario generation and reduction of wind power

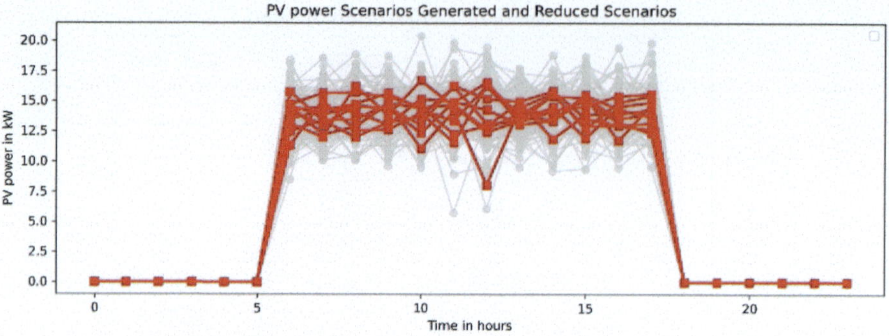

Fig. 7 Scenario generation and reduction of PV power

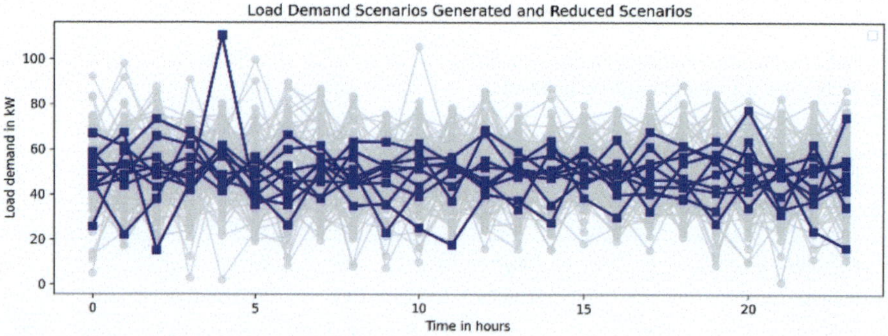

Fig. 8 Scenario generation and reduction of load demand

of 830 kW and the total scheduling hours are considered as 24 h. Figure 9 indicates the wind samples, Fig. 10 indicates the outlier detection, and Fig. 11 indicates the outlier imputation. Figure 12 indicates the outlier imputation using K-means clustering.

The two-stage scheduling method efficiently minimized the DA costs by optimizing energy procurement and generation based on accurate forecasts, while the RT adjustments effectively condensed the imbalance costs caused by deviations from the planned schedule. Figure 13 that the dispatch schedule for hours 8 to 10 for diesel generators is zero it means the generators are turned off at this hour since the incremental fuel cost of the generators at these scheduling hours is more than the grid exchange price therefore, these sources are turned off. Figures 14 indicates the real-time scheduling.

In particular, the integration of outlier management with the scheduling process resulted in a 10–15% reduction in overall operational costs compared to a baseline scenario without outlier detection. The enhanced data integrity enabled more precise forecasting, which in turn permitted better alignment between the scheduled and actual energy generation and consumption. The real-time adjustments further ensured that any discrepancies were promptly addressed, lessening the financial

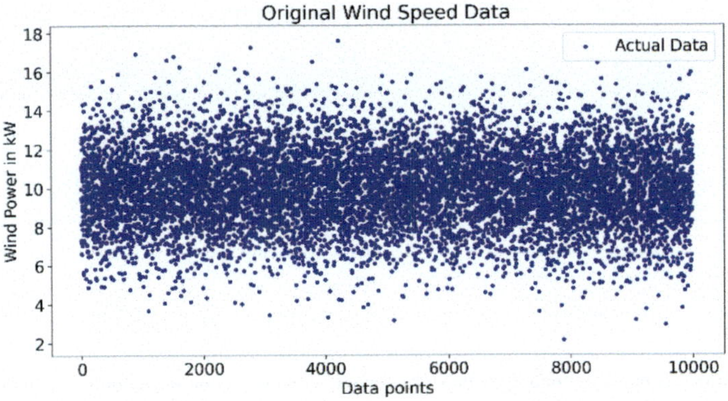

Fig. 9 Actual wind power with outliers

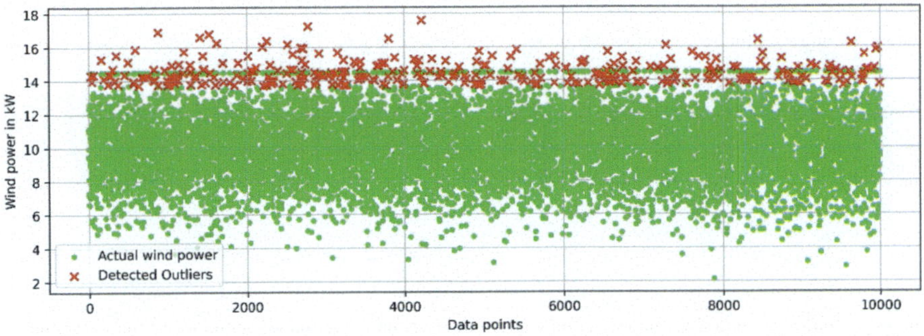

Fig. 10 Outlier detection with K-means

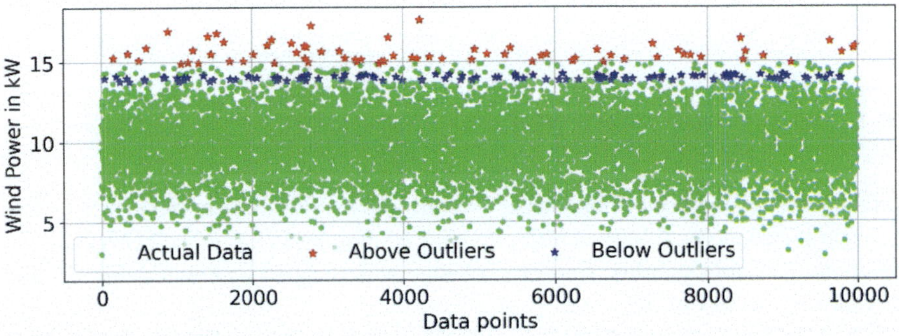

Fig. 11 Outlier detection with K-means clustering

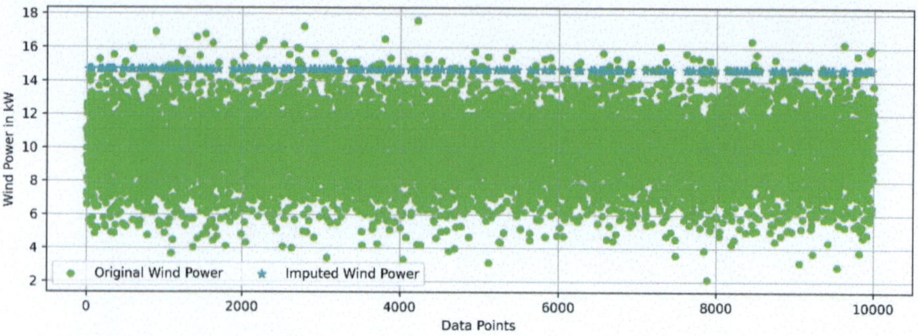

Fig. 12 Outlier imputation with K-means clustering

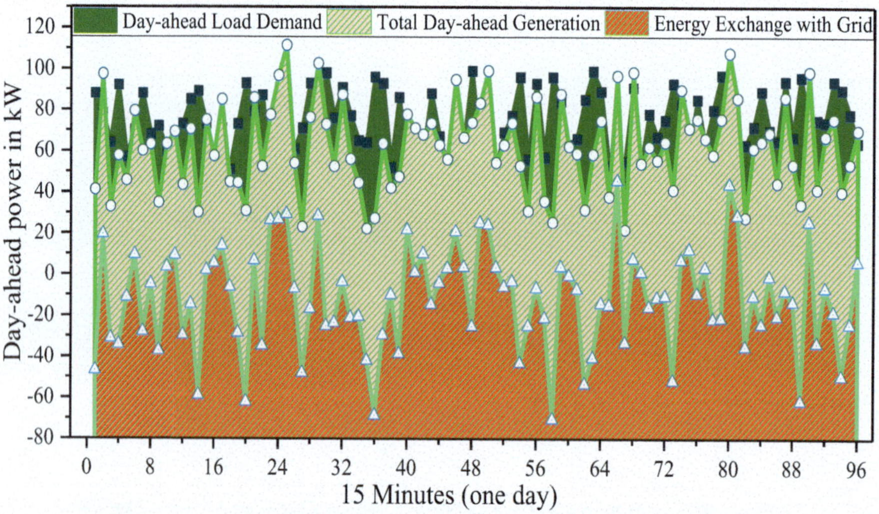

Fig. 13 Day-ahead energy management in a MG

penalties associated with imbalances. Overall, the results confirm the effectiveness of the proposed methodology in improving the economic performance of MG's while maintaining system stability. Energy exchange with the grid is shown in Fig. 15 and operational cost in Fig. 16.

The maximum energy exchange with the grid for a particular scheduling hour is limited to 70 kW. Since the MG consists only of RESs and diesel generators, to reduce the total operational expenditure on the MG it is necessary to minimize the amount of power generation through diesel generators and power drawn from the grid. If power generation from RES is maximum, supply the power back to the grid to reduce the operational cost of the MG. Imbalance power is shown in Fig. 17 and the cost associated with it for a 24-h scheduling period is shown in Fig. 18. The total day-ahead operational cost of the MG is 16242.95 rupees and the total imbalance

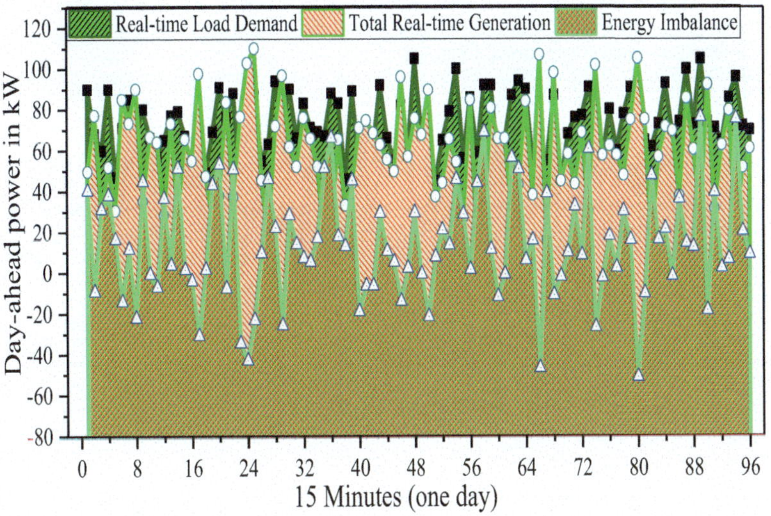

Fig. 14 Real-time energy management in a MG

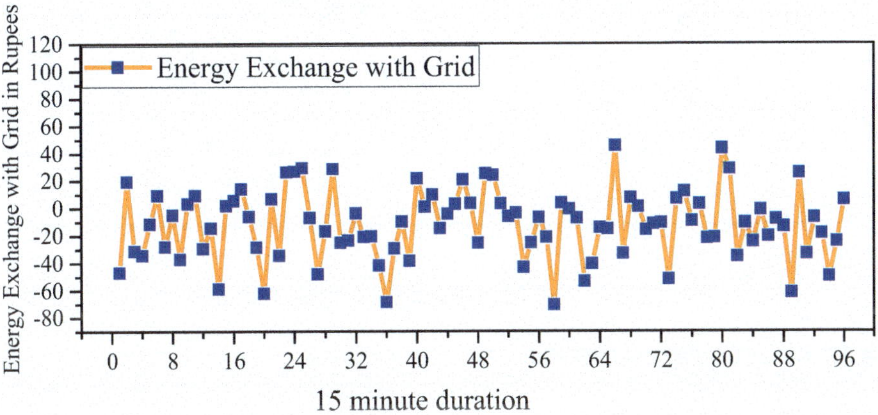

Fig. 15 Energy exchange with the grid

cost in the real-time scheduling is 1752.201 rupees. Therefore, the total operational cost of the MG is 19241.17 rupees. The total operational cost of the MG has reduced from 20,202.54 to 19,241.17 rupees, a reduction of 5% in the total operational cost.

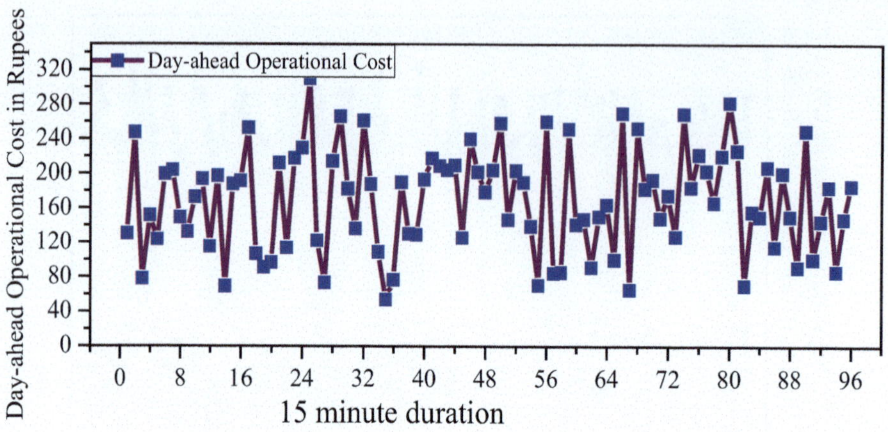

Fig. 16 Day-ahead operational cost

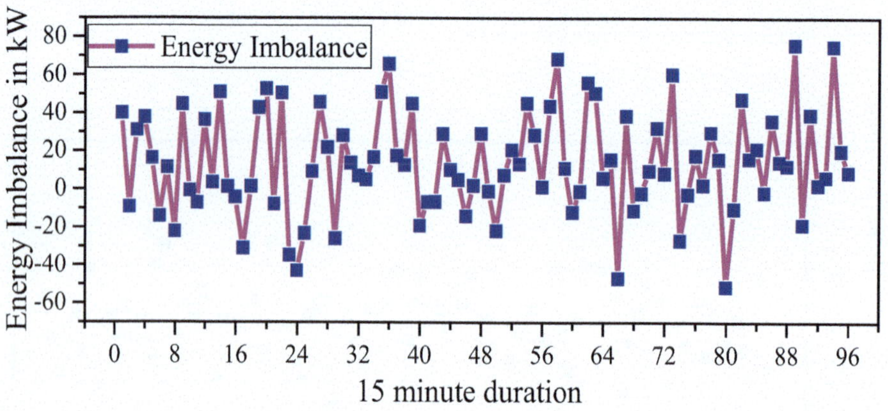

Fig. 17 Energy imbalance in the MG

5 Conclusions

Outliers in MG's pose a substantial challenge to the effectiveness of energy management systems, potentially leading to increased OC and reduced reliability. This study verified the application of K-means clustering and KNN imputation to recognize and correct outliers, enhancing the accuracy of MG data. Moreover, the proposed two-stage scheduling strategy effectively minimized both day-ahead OC and real-time imbalance costs, contributing to more efficient and cost-effective microgrid operation. The amalgamation of outlier management with unconventional scheduling methods represents a significant contribution to the optimization of MG energy systems, safeguarding both economic and operational benefits. The methodology was applied to an MG model, demonstrating reduced operational expenses and improved

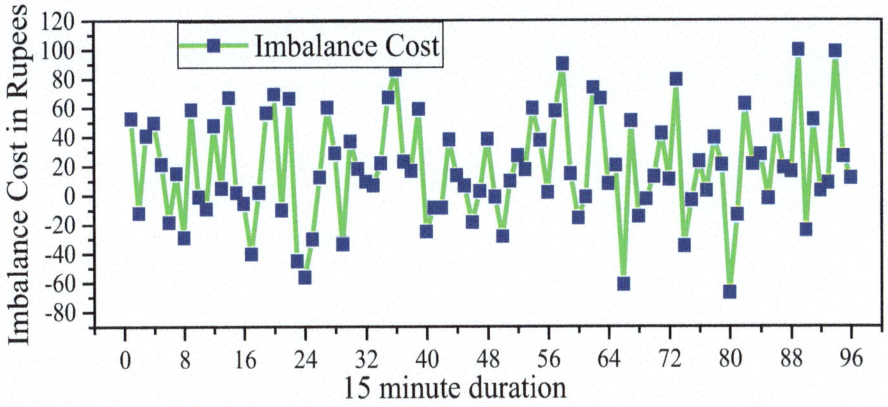

Fig. 18 Imbalance cost of the MG

system stability by effectively managing outliers and optimizing scheduling decisions. The maximum energy exchange with the grid for a particular scheduling hour is limited to 70 kW. The total operational cost of the MG has reduced from 20,202.54 to 19,241.17 rupees, a reduction of 5% in the total operational cost of the MG.

Acknowledgements This research work was supported by "Woosong University's Academic Research Funding – 2025".

References

1. Talaq J, El-Hawary F, El-Hawary E (1994) A summary of environmental/economic dispatch algorithms. IEEE Trans Power Syst 9(3):1508–1516. https://ieeexplore.ieee.org/document/336110
2. Lamont J, Obessis E (1995) Emission dispatch models and algorithms for the 1990s. IEEE Trans Power Syst 10(2):941–947. https://ieeexplore.ieee.org/abstract/document/387937
3. Delson J (1974) Controlled emission dispatch. IEEE Trans Power Apparatus Syst 93(5):1359–1366. https://ieeexplore.ieee.org/document/4075500
4. Dibavar A, Ivatloo B, Zare K, Khalili T, Bidram A (2021) Economic-emission dispatch problem in power systems with carbon capture power plants. IEEE Trans Ind Appl 57(4):3341–3351. https://ieeexplore.ieee.org/document/9428528
5. Uddin M, Romlie M, Abdullah M, Halim S, Bakar A, Kwang T (2018) A review on peak load shaving strategies. RSER 82(3):3323–3332. https://doi.org/10.1016/j.rser.2017.10.056
6. Kirschen D, Strbac G, Cumperayot P, Mendes D (2000) Factoring the elasticity of demand in electricity prices. IEEE Trans Power Syst 15(2):612–617. https://ieeexplore.ieee.org/document/867149
7. Conejo A, Morales M, Baringo L (2010) Real-time demand response model. IEEE Trans Smart Grid 1(3):236–242. https://ieeexplore.ieee.org/document/5607339
8. Pourmousavi S, Nehrir M (2014) Introducing dynamic demand response in the LFC model. IEEE Trans Power Syst 29(4):1562–1572. https://ieeexplore.ieee.org/abstract/document/7285784

9. Medina J, Muller N, Roytelman I (2010) Demand response and distribution grid operations: opportunities and challenges. IEEE Trans Smart Grid 1(2):193–198. https://ieeexplore.ieee.org/document/5545489
10. Parvania M, Firuzabad M (2010) Demand response scheduling by stochastic SCUC. IEEE Trans Smart Grid 1(1):89–98. https://ieeexplore.ieee.org/document/5464384
11. Deng R, Yang Z, Chow M, Chen J (2015) A survey on demand response in smart grids: mathematical models and approaches. IEEE Trans Ind Inf 11(3):570–582. https://ieeexplore.ieee.org/document/7063260
12. Rahimi F, Ipakchi A (2010) Demand response as a market resource under the smart grid paradigm. IEEE Trans Smart Grid 1(1):82–88. https://ieeexplore.ieee.org/document/5454394
13. Mathieu J, Price P, Kiliccote S, Piette M (2011) Quantifying changes in building electricity use, with application to demand response. IEEE Trans Smart Grid 2(3):507–518. https://ieeexplore.ieee.org/document/5772947
14. Palensky P, Dietrich D (2011) Demand side management: demand response, intelligent energy systems, and smart loads. IEEE Tran Ind Inf 7(3):381–388. https://ieeexplore.ieee.org/document/5930335
15. Zhou X (2022) Exploiting integrated demand response for operating reserve provision considering rebound effects. IEEE Access 10(2):15151–15162. https://ieeexplore.ieee.org/document/9701332
16. Xiaoyang Z, Weixin Y, Johnson R (2016) Forecast load impact from demand response resources. In: 2016 IEEE power and energy society general meeting, vol 2, no 3, pp 1–5. https://ieeexplore.ieee.org/document/7741992
17. Bitaraf H, Rahman S (2018) Reducing curtailed wind energy through energy storage and demand response. IEEE Trans Sustain Energy 9(1):228–236. https://ieeexplore.ieee.org/document/7972993
18. Shi W, Li N, Xie X, Chu C, Gadh R (2014) Optimal residential demand response in distribution networks. IEEE J Sel Areas Commun 32(7):1441–1450. https://ieeexplore.ieee.org/abstract/document/6840288
19. Sibo N, Zhou M, Li G (2017) Optimal residential community demand response scheduling in smart grid. Appl Energy 210(2):1280–1289. https://doi.org/10.1016/j.apenergy.2017.06.066
20. Jung CM, Pagidipala S, Salkuti SR (2024) Emerging technologies for the integration of renewable energy, energy storage and electric vehicles. In: Salkuti SR (eds) Energy and environmental aspects of emerging technologies for smart grid. Green energy and technology. Springer, Cham. https://doi.org/10.1007/978-3-031-18389-8_5
21. Salkuti SR (2022) Modeling of various renewable energy resources for smart electrical power systems. In: Salkuti SR, Ray P (eds) Next generation smart grids: modeling, control and optimization. Lecture notes in electrical engineering, vol 824. Springer, Singapore. https://doi.org/10.1007/978-981-16-7794-6_2
22. Tarish H, See O, Elmenreich W (2016) A review of residential demand response of smart grid. RSER 59(2):166–178. https://doi.org/10.1016/j.rser.2016.01.016
23. Niamh C, Pierre P, Madsen M, Mark O (2014) Benefits and challenges of electrical demand response: a critical review. RSER 39(1):686–699. https://doi.org/10.1016/j.rser.2014.07.098
24. Rezaee A (2019) Optimisation of demand response in electric power systems, a review. RSER 103(1):308–319. https://doi.org/10.1016/j.rser.2018.12.054
25. Hussain M, Yan G (2018) A review of demand response in an efficient smart grid environment. Electr J 31(5):55–63. https://doi.org/10.1016/j.tej.2018.06.003
26. Siano P (2014) Demand response and smart grids—a survey. RSER 30(4):461–478. https://doi.org/10.1016/j.rser.2013.10.022
27. Sravanthi P, Sandeep V (2022) Solving realistic reactive power market clearing problem of wind-thermal power system with system security. Int J Emerg Electr Power Syst 23(2):125–144. https://doi.org/10.1515/ijeeps-2021-0060
28. Kumar JV, Sesham H, Salkuti SR (2024) Smart Energy management model for electric vehicles. In: Salkuti SR (eds) Energy and environmental aspects of emerging technologies for smart grid. Green energy and technology. Springer, Cham. https://doi.org/10.1007/978-3-031-18389-8_9

29. Faria S, Spínola VZ (2018) Reschedule of distributed energy resources by an aggregator for market participation. Energies 11(713):1–20. https://doi.org/10.3390/en11040713
30. Zhang G, Li Y, Deng X (2020) K-means clustering-based electrical equipment identification for smart building application. Information 11(27):25–35. https://doi.org/10.3390/info11010027
31. Faria S, Spínola J, Vale Z (2018) Distributed energy resources scheduling and aggregation in the context of demand response programs. Energies 11(1):987–992. https://doi.org/10.3390/en11081987
32. Abrishambaf O, Faria S, Vale Z (2020) Ramping of demand response event with deploying distinct programs by an aggregator. Energies 13(1):138–145. https://doi.org/10.3390/en13061389
33. Salkuti SR, Ray P, Pagidipala S (2022) Overview of next generation smart grids. In: Salkuti SR, Ray P (eds) Next generation smart grids: modeling, control and optimization. Lecture notes in electrical engineering, vol 824. Springer, Singapore. https://doi.org/10.1007/978-981-16-7794-6_1
34. Lezama F, Faria S, Vale Z (2020) Demand response of residential houses equipped with PV-battery systems: an application study using evolutionary algorithms. Energies 13(1):24–36. https://doi.org/10.3390/en13102466

A Novel Self-balanced Switched Capacitor Seven-Level Inverter

Lakshmi Prasanna, T. R. Jyothsna, B. Amarendra Reddy, and Surender Reddy Salkuti

Abstract The ‘Cost function’ (CF) study of Switched-Capacitor-Based Multilevel Inverters (SCMLIs) includes important characteristics such as device count, total standing voltage (TSV), and source demands. The purpose of this study is to enhance the boosting capabilities of a seven-level inverter by reducing its switch count and CF value. By effectively controlling the series-to-parallel conversion of the power source and capacitors, this inverter produces voltage amplification. This article presents a novel, one-dc source, seven-level inverter that is suitable for wind and solar power applications. Floating capacitors (FC) combined with a switched-capacitor design yields the desired outcome. This configuration was selected to minimize control problems when generating a tripolar staircase pattern because of its sensor-less capacitor voltage balancing potential. Inadvertent overvoltage of the capacitor is prevented over time, resulting in a further benefit. The phase disposition pulse-width modulation (PD-PWM) technique has been used to regulate the operation of the topology. Moreover, fewer semiconductor devices are utilized in the suggested topology. A detailed comparison of various recently-released seven-level inverter topologies is given to illustrate the advantages of the suggested layout for reducing switch count. Thermal modeling with PLECS is used to evaluate efficiency and power losses. MATLAB/Simulink and HIL implementation are used to explore the viability of implementing the suggested structure under various parametric restrictions.

Keywords Single-source · Multi-level inverter · Switched capacitor · Self-voltage balancing · Total standing voltage · Phase disposition pulse width modulation · Total harmonic distortion

L. Prasanna · T. R. Jyothsna · B. Amarendra Reddy
Department of Electrical Engineering, Andhra University, Visakhapatnam, Andhra Pradesh, India
e-mail: lakshmiprasanna.rs@andhrauniversity.edu.in

B. Amarendra Reddy
e-mail: amarendra@andhrauniversity.edu.in

S. R. Salkuti (✉)
Department of Railroad and Electrical Engineering, Woosong University, Daejeon, Republic of Korea
e-mail: surender@wsu.ac.kr

S. R. Salkuti (ed.), *Artificial Intelligence for Integrated Smart Energy Systems in Electric Vehicles*, Lecture Notes in Electrical Engineering 1427,
https://doi.org/10.1007/978-3-031-94276-1_20

Nomenclatures

ANPC	Active Neutral Point Clamped
CF	Cost Factor
CHB	Cascaded H Bridge
FC	Flying Capacitor
HIL	Hardware-In-Loop $\frac{V+IRs}{nVt}$
I	Output current
MLI	Multi-Level Inverter
NPC	Neutral Point Clamped
PDPWM	Phase Disposition Pulse Width Modulation
PUC	Packed U-Cell
PWM	Pulse Width Modulation
RDC	Reduced Device Count
THD	Total Harmonic Distortion
TSV	Total Standing Voltage
V_{dc}	Supply voltage
V_{out}	Output voltage

1 Introduction

An increasing amount of dependable, highly efficient, and uninterruptible power is needed for several types of industrial, public transportation, communications, aviation, traction, battery storage, residential, and household uses due to the rapidly expanding energy needs and the drastic effects of globalization. Therefore, to fulfill the necessary requirements of demand, an effective and controllable power converter is a prerequisite. There are several drawbacks to conventional controllers, including size, cost, modular structure, regular consumption, accuracy, effectiveness, security, and durability. Examples of these include thyratrons, mercury-arc rectifiers, magnetic amplifiers, and rheostat controllers [1]. A significant shift occurred with the introduction of self-commutating devices and the power converter era. In the majority of applications, converters between dc-ac (inverters) and ac-dc (rectifiers) are important [2, 3].

The conventional inverter is inadequate for implementation in various purposes as mentioned earlier for several unanticipated limitations like higher harmonic distortion, less efficiency, reduced power quality, and so on. Multilevel inverters (MLIs) have established themselves as a well-developed technology for a variety of customized and commercialized products for a broad range of power applications. The three most common MLI topologies—Flying Capacitor (FCMLI), Cascade H-Bridge (CHB), and Neutral Point Clamped (NPCMLI) are referred to as classical MLIs. These topologies attracted a lot of interest from the academic and business communities [4, 5].

High-voltage medium-power applications (13.8 kV, 30 MVA) are best suited for CHB due to its modular structure and high fault tolerance capacity. But for active power transfer applications, CHB needs isolated DC sources. Switches count rises with increments in levels for CHB, just like in NPCMLI and FCMLI. Furthermore, the implementation of classical MLIs poses significant challenges due to their topologies at higher levels. Due to its higher device count, which adds complexity to its structure and imposes limitations on size, cost, and maintenance. Thus, by essentially altering the traditional MLIs, researchers kept investigating and developing newer topologies [6, 7].

Voltage enhancement with traditional or RSC MLI is not possible in procedures involving the conversion of alternative energy, voltage enhancement will be crucial, since solar panels provide relatively low voltage. In order to accomplish numerous levels and a notable boost with a single source, the researchers created the switched capacitor technique [8, 9]. The capacitors are powered in parallel with the source and create a series link with it during discharge to reach greater levels [10, 11]. A DC source powers only the first bridge in [12], while switched capacitors are used in the other two bridges that are connected by two bidirectional switches. In comparison to the other seven-level inverters, the authors [13] suggested an architecture with fewer switching devices. In [14], a buck-boost converter is connected to an inverter via two capacitors to boost voltage. While boosting the voltage by three times, the configuration's front side design can reduce the optimum current flowing of the devices.

In [15], a three-phase inverter for 2N + 1 levels is built. The front-end configuration is preceded by an H-bridge inverter at the end and a switched capacitor unit in the middle, with increasing capability. Reference [16] presents a triple-gain, reduced total standing voltage (TSV) diamond-shaped arrangement. Reference [17] builds a five-level inverter and advances it to a normal inverter for double boosting. To lessen current stress and maintain the capacitor current within an appropriate range, this arrangement employs a quasi-soft charging technique. A structure consisting of a few components is shown in [18]. According to the authors, the structure is capable of regeneration, which suggests that if the capacitor overcharges, energy will be delivered to the source suddenly.

A greater switch count is needed for the seven-level system described in [19], which combines two T-type circuits and a full bridge inverter. The innovative T-type inverter described in [20] maximizes the effectiveness of the capacitor. For low TSV, a seven-level arrangement with three capacitors is provided [21]. A new ANPC inverter with more TSV is described in [22]. In order to attain seven levels, the packed U-cell (PUC) in [23] is altered to minimal switches. A seven-level inverter with few switches is produced by combining a half-bridge circuit and switched capacitor unit [24]. Reference [25] creates a seven-level inverter by combining an output unit, a bidirectional devices unit, and a DC-link unit. Reference [26] builds a 13-level inverter with 13 semiconductor devices and 3 capacitors using the switched capacitor approach. Additionally, [27] implemented the SC technique to suggest an extended voltage boost MLI. By reducing the number of active and passive components and eliminating the need for a sensor to balance the voltage of the floating capacitor, [28] simplifies the control system design and lowers costs. In [29], switched capacitors are used in place of the isolated DC source to create a generalized structure by the cascading connection of several H-bridges. Using the connections between the T-module, cross-module, and T-module, a topology is suggested in [30]. Through the addition of additional cross modules, it can be expanded to higher levels. A very high TSV of 10.66 was proposed in [31] for a topology that combines a T-type network with an additional floating capacitor [32]. The suggested configuration has several advantageous qualities, including:

- Its output voltage is achieved with a 1:3 boosting capability.
- It only needs a single DC supply.
- Balancing the voltage across capacitors doesn't require sensors.
- It has the same amount of charging and de-energizing modes in each alternate period and functions with either a leading or lagging power factor.

This chapter aims to present a unique MLI arrangement that reduces TSV and switches count. This chapter has the following structure: The proposed topology, switching modes, switching stress analysis, controlling technique, and self-voltage balancing are all thoroughly explored in Sect. 2. Section 3 compares the suggested topology with current topologies. An in-depth discussion of the simulation findings, HIL implementation, and thermal analysis for various loading scenarios is included in Sect. 4. The completed work is shown in Sect. 5.

2 Proposed Topology

Figure 1 represents the proposed MLI configuration. An isolated DC source, two capacitors (C_1 and C_2), one diode, and nine switches comprise the circuit configuration. Capacitors C_1 and C_2 have a peak voltage rating of V_{dc}, two of the nine power semiconductor devices (S_1 and S_5) are solely IGBTs, and the rest of them are IGBTs in parallel with a diode. Table 1 displays the switching modes of the suggested device. The switch-on position is represented by 1 in Table 1, and the off position is represented by 0. All modes are shown in Fig. 2 with reference to the switching modes given in Table 1.

2.1 Operating Modes

Figure 2 illustrates all the operational modes from I to VIII. Each mode is represented with two colors based on the required directions: (a) the path indicating the level of generation is represented as green, and (b) the path utilized for capacitor charging is represented in blue.

Fig. 1 Proposed topology

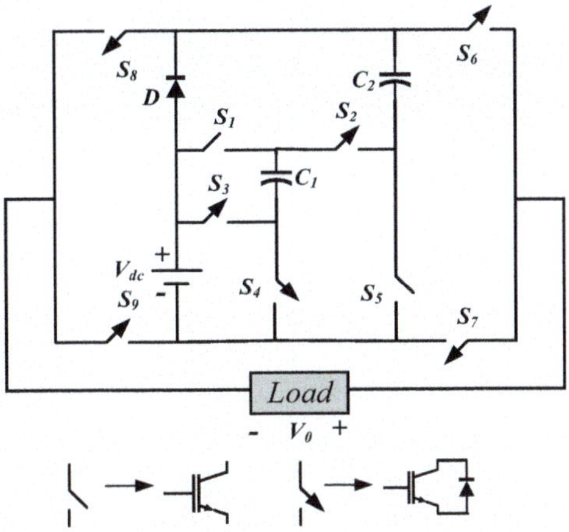

Table 1 Switching modes

S_1	S_2	S_3	S_4	S_5	S_6	S_7	S_8	S_9	V_o	C_1	C_2	Mode
0	1	1	0	0	1	0	0	1	$+3V_{dc}$	D	D	I
1	1	0	1	0	1	0	0	1	$+2V_{dc}$	C	D	II
0	0	0	0	1	1	0	0	1	$+1V_{dc}$	–	C	III
0	0	0	0	1	0	1	0	1	0	–	C	IV
0	0	0	0	1	1	0	1	0	0	–	C	V
0	0	0	0	1	0	1	1	0	$-1V_{dc}$	–	C	VI
1	1	0	1	0	0	1	1	0	$-2V_{dc}$	C	D	VII
0	1	1	0	0	0	1	1	0	$-3V_{dc}$	D	D	VIII

Mode I: DC source and capacitors (C_1, C_2) are discharged through the conductivity path of S_2, S_3, S_6 and S_9 to obtain $+3V_{dc}$ across the load as indicated in Fig. 2a.

Mode II: In order to achieve $+2V_{dc}$ as the output, a conducting path made up of switches S_1, S_2, S_6, and S_9 connects the DC voltage source, capacitor C_2, and the load. This functioning mode is seen in Fig. 2b, where a DC voltage source charges capacitor C_1 through S_1 and S_4 switches conduction.

Mode III: Using the conductivity of the diode, S_6 and S_9 switches shown in Fig. 2c, a $+1V_{dc}$ level can be developed. When S_5 switch conduction capacitor C_2 is charged, the impact of capacitor C_1 is unchanged.

Mode IV: The conduction of switches S_7 and S_9 in Fig. 2d produces zero voltage. To connect the DC source to capacitor C_2, diode forward biasing and Switch S_5 conductivity are used to energize it to voltage V_{dc}.

Mode V: As illustrated in Fig. 2e, zero voltage is achieved in the same manner as Mode IV by switching the conductance of switches S_6 and S_8. Capacitor C_2 charges in the same way that mode IV describes.

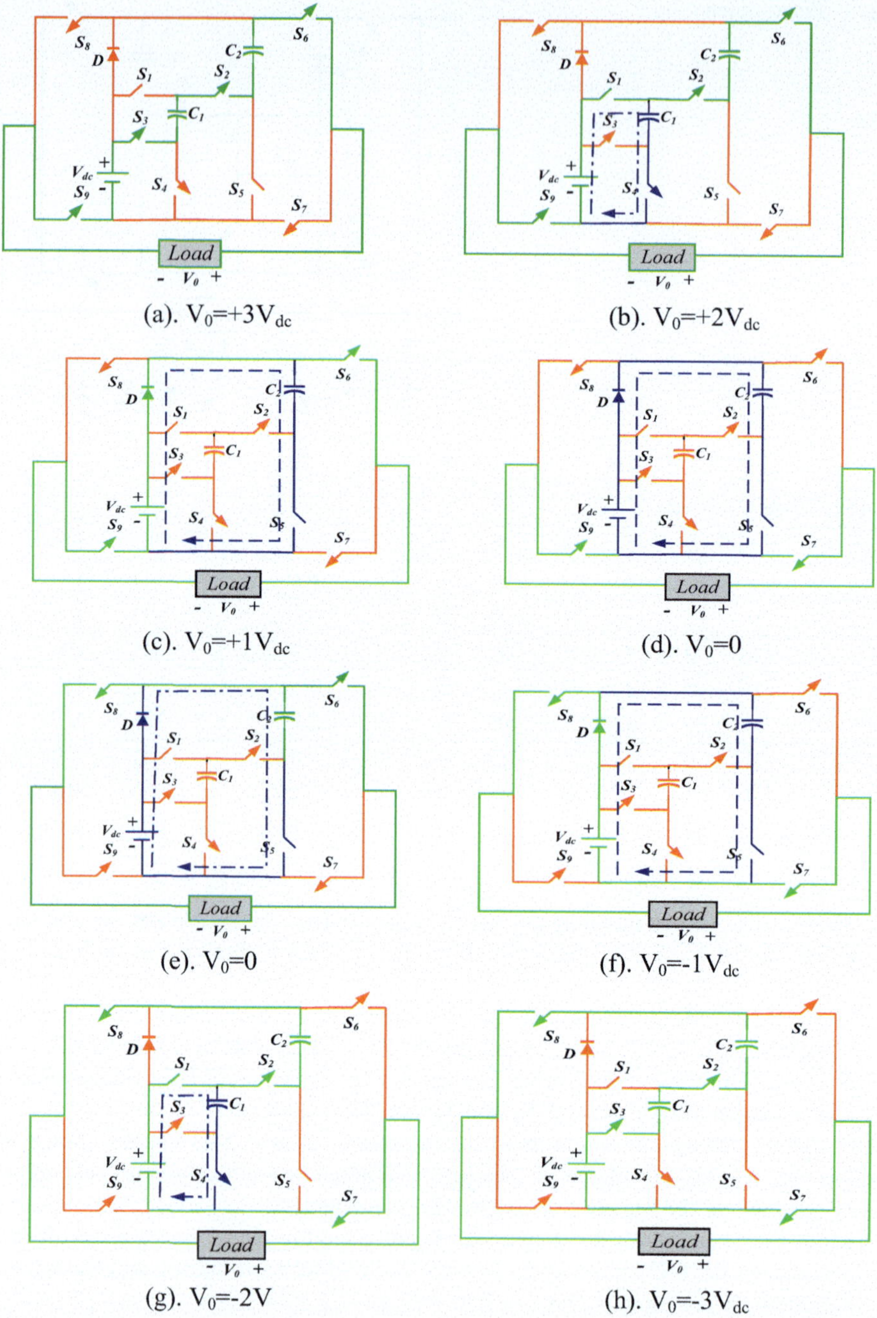

Fig. 2 Operating modes of topology

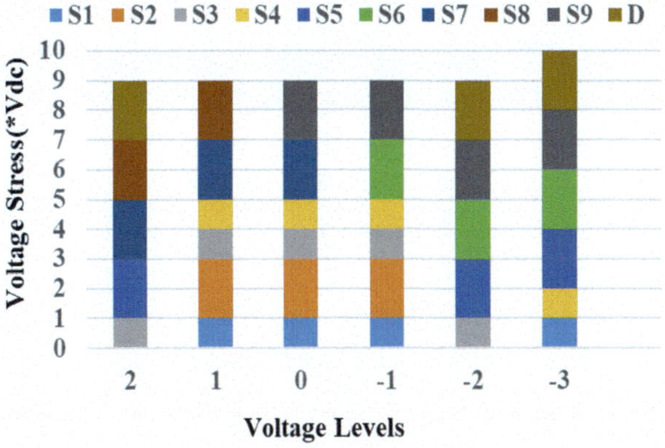

Fig. 3 Bar chart representation

Mode VI: Figure 2f illustrates how a diode couples the DC source, load, and switches S_7 and S_8, to achieve $-1V_{dc}$. Similar to mode V, capacitor C_2 is powered up to V_{dc} voltage while capacitor C_1 has no impact.

Mode VII: A DC source is utilized to obtain $-2V_{dc}$, and switches S_1, S_2, S_7, and S_8 are used to discharge capacitor C_2 to the load, as seen in Fig. 2g. The connection of the DC source switches S_1 and S_4, and capacitor C_1 powers it up to V_{dc}.

Mode VIII: As indicated in Fig. 2h, turn on switches S_2, S_3, S_7, and S_8 to connect the DC source, capacitor C_1, and C_2 to the load in order to obtain a $-3V_{dc}$ level.

2.2 *Voltage Stress Analysis*

The optimal voltage of the devices during non-conductivity condition is used to calculate the voltage stress generated by the switching devices. Peak voltage stress is V_{dc} for switches (S_1, S_3, S_4) and $2V_{dc}$ for diodes and switches (S_2, S_5, S_6, S_7, S_8, S_9). Consequently, the TSV (p.u.) of the recommended setup is 6. A bar chart illustrating voltage stress for each individual level is presented in Fig. 3. The voltage stress of each switch is shown by the different colours.

2.3 *Modulation Scheme*

Figure 4 depicts the PDPWM scheme that is being used for the suggested structure. It is a seven-level inverter, therefore in order to deliver gating pulses to switches, it needs six carriers and one reference. In this modulation system, carrier signals and the reference wave are compared to generate gating pulses. Following the comparison, six pulses are generated. These six pulses produce signals to switches (S_1–S_9) using the logic shown in Table 1.

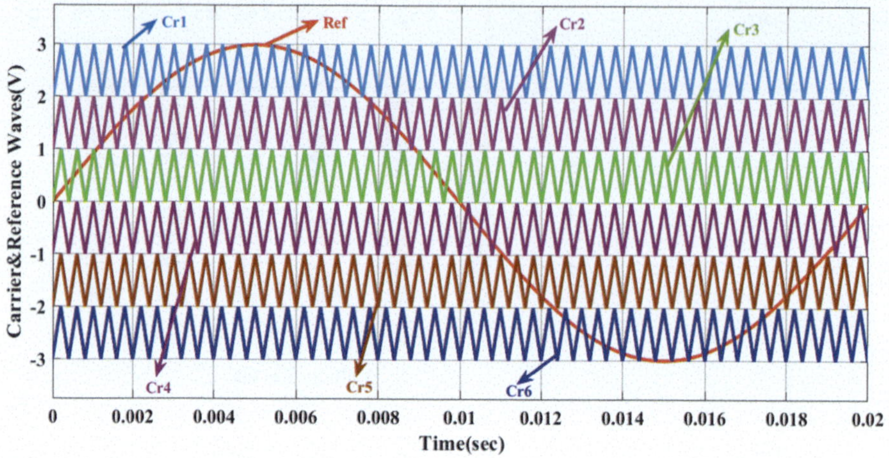

Fig. 4 PDPWM scheme

To change the modulation index (m_a) amplitude, the reference's peak value is changed [33]. The fundamental frequency, or operating frequency, of the reference and output waves is 50 Hz. By altering the carrier frequency ($f_{carrier}$), frequency modulation can be achieved [34]. The filter size decreases as the frequency modulation increases because the harmonic moves to a higher order. As frequency modulation increases, switching losses rise. As a result, for the current setup, the carrier frequency ($f_{carrier}$) is restricted to 5 kHz and is subject to defined constraints.

2.4 Capacitor Calculations

Capacitor C_1 discharges at $\pm\ 3V_{dc}$ output voltage level, the other capacitor C_2 discharges at $\pm\ 2V_{dc}$ and $\pm\ 3V_{dc}$ output voltage level respectively. The maximum allowable discharge (ΔQ_{C1}) required to maintain satisfactory voltage ripple (ΔV_c) for supplying the load resistance (R_L) and fundamental frequency (ω_m) is denoted as,

$$\Delta Q_{C_1} = C_1 * \Delta V_{C_1} = \int_{\frac{\theta_3}{\omega_m}}^{\frac{\pi-\theta_3}{\omega_m}} \frac{3V_{dc}}{R_L} dt \tag{1}$$

$$\Delta Q_{C_1} = \frac{3V_{dc}}{\omega_m R_L}(\pi - 2\theta_3) \tag{2}$$

$$\Delta Q_{C_2} = C_2 * \Delta V_{C_2} = \int_{\frac{\theta_1}{\omega m}}^{\frac{\theta_2}{\omega m}} \frac{2V_{dc}}{R_L} dt + \int_{\frac{\theta_3}{\omega m}}^{\frac{\pi-\theta_3}{\omega m}} \frac{3V_{dc}}{R_L} dt + \int_{\frac{\pi-\theta_2}{\omega m}}^{\frac{\pi-\theta_1}{\omega m}} \frac{2V_{dc}}{R_L} dt \quad (3)$$

$$\theta_1 = \sin^{-1}\left(\frac{A_{cr1}}{A_{ref}}\right) \quad (4)$$

$$\theta_2 = \sin^{-1}\left(\frac{A_{cr2}}{A_{ref}}\right) \quad (5)$$

$$\theta_3 = \sin^{-1}\left(\frac{A_{cr3}}{A_{ref}}\right) \quad (6)$$

$$\Delta Q_{C_2} = \frac{V_{dc}}{\omega_m R_L}(3\pi - 6\theta_3 + 4\theta_2 - 4\theta_1) \quad (7)$$

Considering allowable ripple voltage limit as 10%, C_1 and C_2 are expressed as,

$$C_1 \geq \frac{30}{\omega_m R_L}(\pi - 2\theta_3) \quad (8)$$

$$C_2 \geq \frac{10}{\omega_m R_L}(3\pi - 6\theta_3 + 4\theta_2 - 4\theta_1) \quad (9)$$

2.5 *Self-voltage Balancing*

Specifically, switched capacitor inverters employ the series–parallel technique. When the capacitor is connected in series with the DC source, it discharges, and when it is connected in parallel with the DC source, it is energized, offering a high voltage gain. Mode II and Mode VII charge capacitor C_1 to V_{dc}, as Table 1 illustrates. In modes III, IV, V, and VI, capacitor C_2 is charged in a manner analogous to V_{dc}. Regular discharge and charging are balanced at V_{dc} for each cycle by the capacitor voltages C_1 and C_2.

3 Comparison

The proposed design is contrasted to distinct structures is shown in Table 2. Factor (CF), TSV (p.u.), number of switches, drivers, diodes, capacitors, and DC sources are the comparative variables. Every topology that has a count of DC source is identical. Reference [12] requires more switches than Ref. [14], but offers a benefit

Table 2 Comparison with other topologies

Topology	N_{sw}	$N_{drivers}$	N_{diodes}	N_c	N_{DC}	Gain	TSV (p.u.)	CF
[12]	16	16	0	2	1	3	5.3	40.3
[14]	12	12	2	3	1	3	7.67	37.6
[16]	12	11	0	2	1	3	5.3	31.3
[17]	11	10	0	3	1	3	6	31
Proposed	9	9	1	2	1	3	6	28

of 3. Although there is an issue with capacitor voltage imbalance in Refs. [14, 16, 17], there are fewer switches than in Ref. [12]. The suggested architecture includes fewer devices and, when all parameters are taken into account, the TSV (p.u.) and CF are the best [35]. The exquisite superior performance of the proposed topology is demonstrated.

4 Results and Discussion

4.1 Simulation Results

MATLAB/Simulink platform is utilized for simulating the execution of seven-level operations. The load values are 100 mH and 100 Ω, respectively, while the source and self-voltages of the capacitors (C_1, C_2) are 100 V, 100 V, and 100 V. A capacitance of 2200 µF is included in both capacitors. The frequency at which it operates of the suggested inverter is the fundamental frequency (50 Hz). The output of the inverter under R-load conditions is shown in Fig. 5. The output voltage and output current seem to be correlated. It appears that the output voltage and the output current are proportionate. The output of the inverter for seven levels of operation under different load scenarios is shown in Fig. 6. As depicted in Fig. 6, output voltage is constant for all load scenarios; however, during unloading, the current is zero, indicating voltage for resistive loads and delays for inductive loads. Although the voltages of C_1 and C_2 are balanced at V_{dc}, the ripples in the capacitor voltages are affected by load conditions.

Waveform variations associated with variations in the inductive load are shown in Fig. 7. Figure 7 illustrates how an increase in inductance causes the current to decline because a high inductive load causes the current to lag more. The output waveforms and the capacitor voltages that were employed to alter the input voltage are shown in Fig. 8. The capacitor voltages C_1 and C_2 unexpectedly increase from 50 to 100 V as a result of the source voltage rising from 50 to 100 V. The resulting waveforms as the reference frequency (f_m) varies are shown in Fig. 9. According to Fig. 9, enhancing the reference frequency (f_m) leads to a rise in load inductance. Since incremental load inductance impacts total load impedance, load current falls

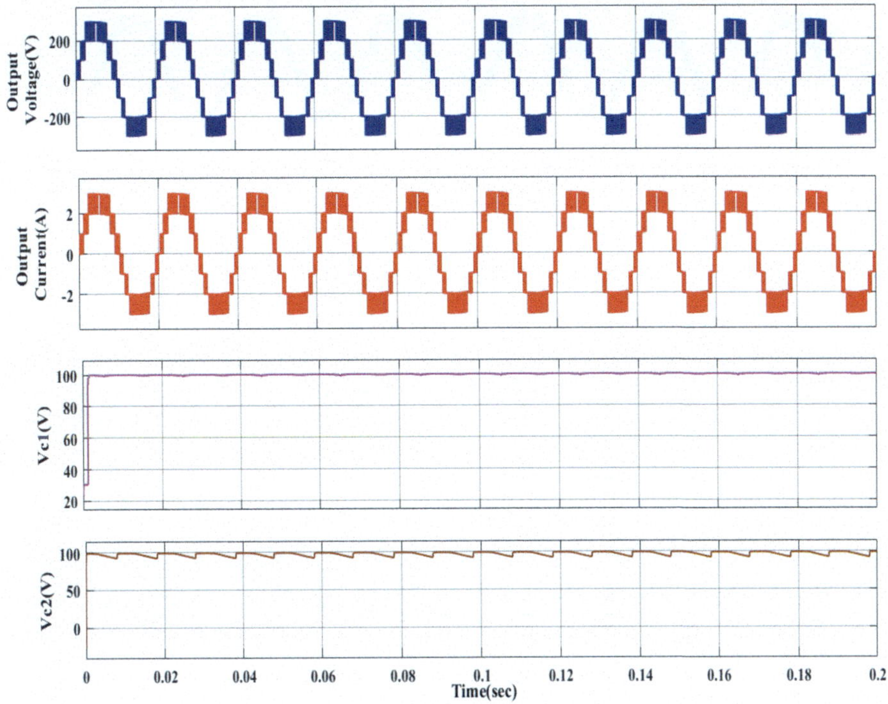

Fig. 5 Resulting waveforms for resistive load

as reference frequency goes up. The proposed topology appears to be suitable for high-frequency performing appliances. Figure 10 depicts load waveforms as carrier frequency alterations. The number of pulses of each level boosts as a carrier frequency ($f_{carrier}$) rises. Consequently, the harmonics shift upwards in order and the size of the filter is reduced. As the carrier frequency ($f_{carrier}$) expands, switching losses elevate due to an increase in pulse count for each voltage level.

Figure 11 shows the modulation index variation (m_a) output waveforms. When the modulation index is reduced, the stepped waveform's amplitude width drops, but the setup functions as a seven-level with maximum amplitude of $3V_{dc}$ for 1, 0.9, and 0.7. Likewise, it functions as a five-level inverter, a reduced width five-level inverter, and a three-level inverter for values of 0.6, 0.4, and 0.2. Figure 12 indicates seven-level inverter resultant voltage THD is 18.43% and a fundamental output of 297.6 V. Following the simulation outcome, the suggested inverter works for all conditions of load.

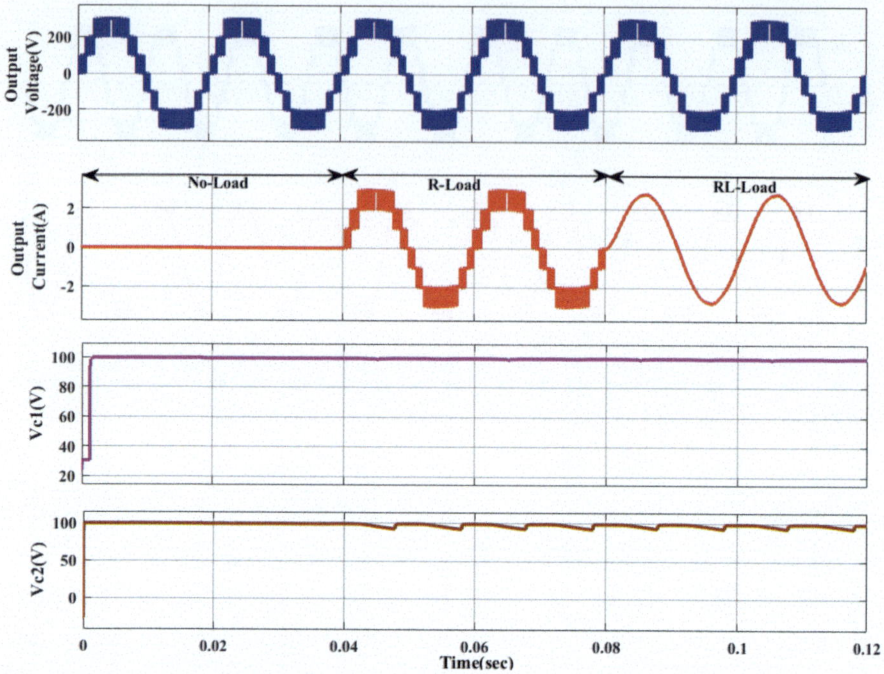

Fig. 6 Resulting waveforms for various load scenarios

4.2 HIL Implementation

Waveforms at a carrier frequency of 5 kHz are displayed for loads of 100 Ω and 100 Ω + 100 mH in Fig. 13. Figure 13c, d display the voltage across two capacitors. It is apparent that the real-time simulated outcomes closely match the simulated results obtained through MATLAB/Simulink. The switching pulses for all the switches are presented in Fig. 14. These real-time results show that the proposed configuration performs effectively across capacitors without the need for additional regulation methods.

4.3 Thermal Analysis

For switching devices in the suggested arrangement, a thermal model is made using the PLECS software. Under loading circumstances of 100 Ω and 100 Ω + 100 mH, respectively, the percentage of losses for semiconductor devices is shown in Fig. 15a, b. Efficiency calculations were carried out for loads ranging from 0 to 1500 W, and the results, as shown in Fig. 16, demonstrate an efficiency range of 97.6–93.4%.

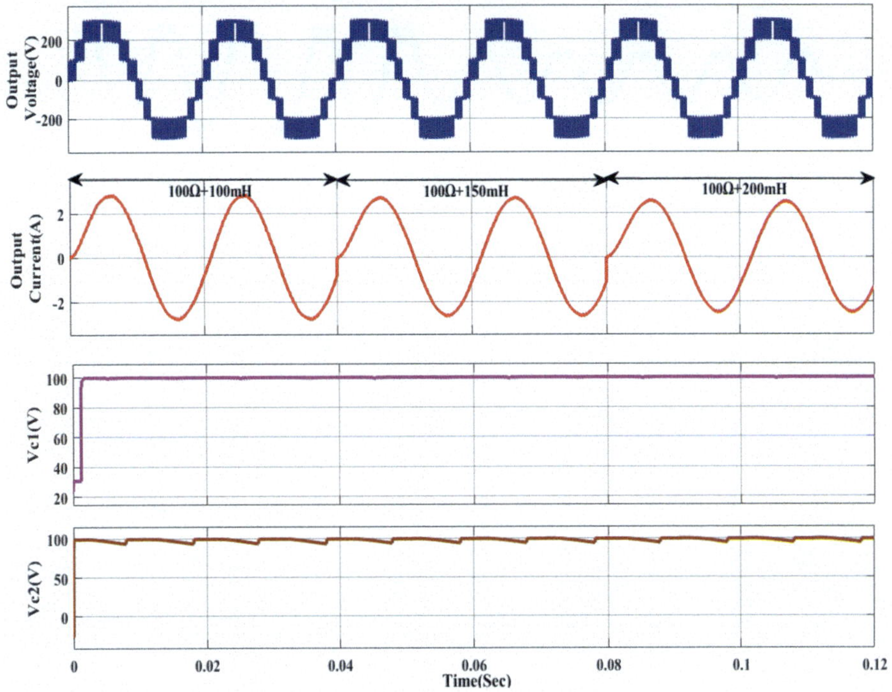

Fig. 7 Resulting waveforms for distinct impedances

5 Conclusions

The innovative seven-level triple inverter described in this study uses a switched capacitor technique to allow capacitor voltages to self-balance. The proposed layout is simulated in MATLAB/Simulink, considering various scenarios such as load variations, source voltage fluctuations, modulation index adjustments, reference frequency changes, and carrier frequency variations. The simulation results represent that the suggested configuration is applicable to a wide range of loading conditions. To assess its superior features, Existing topologies from the literature are contrasted with the proposed architecture. The analysis reveals that the proposed design employs minimal components while maintaining an acceptable TSV. Furthermore, the overall efficiency of the configuration is evaluated using PLECS software, which estimates losses in the switches. The efficiency of the proposed design is found to reach up to 93.4%. Finally, Hardware-in-Loop (HIL) testing is used to validate the topology's efficient operation, and outcomes are provided. The results suggested that the proposed design shows elegant results.

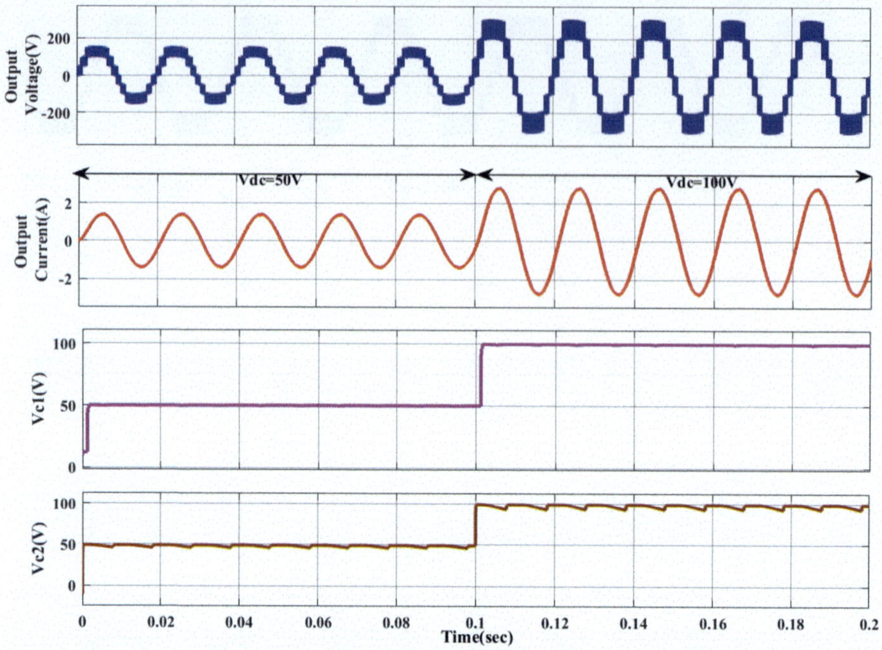

Fig. 8 Resulting waveforms for variation in source voltages

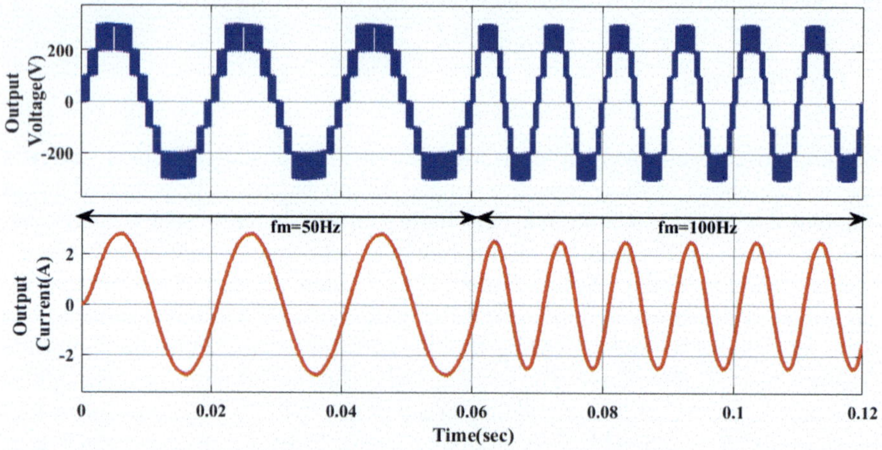

Fig. 9 Resulting waveforms for regulation in the reference frequency

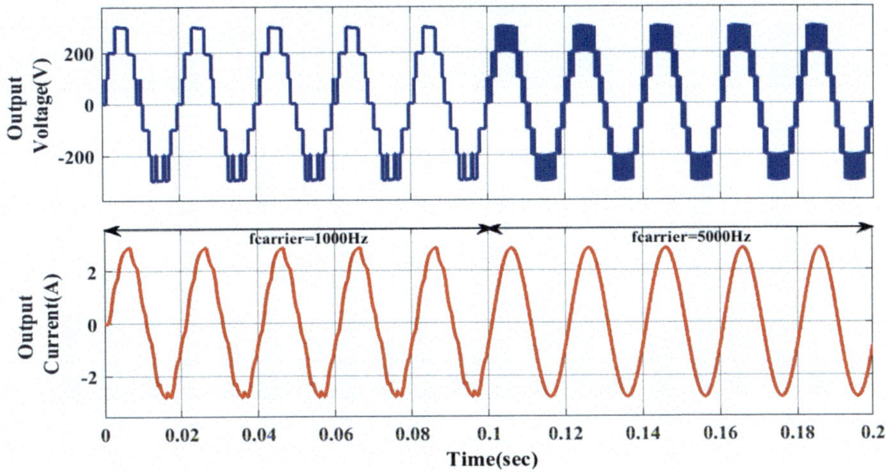

Fig. 10 Resulting waveforms for carrier frequency modulation

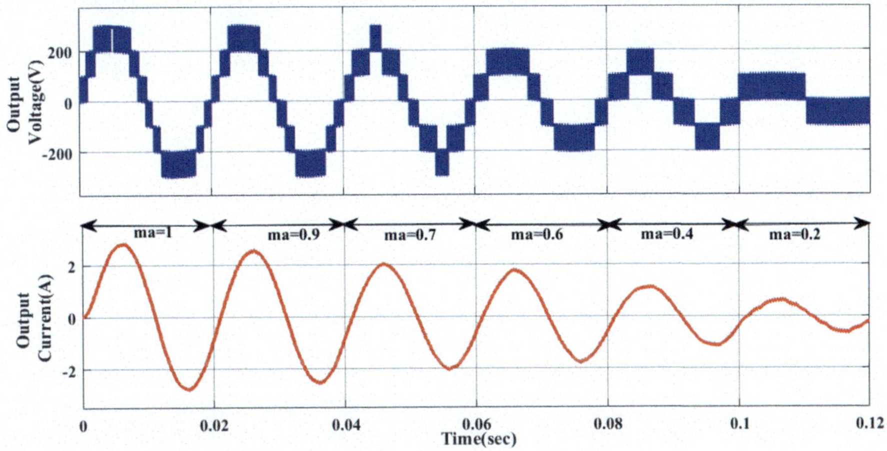

Fig. 11 Resulting waveforms for modulation index variation

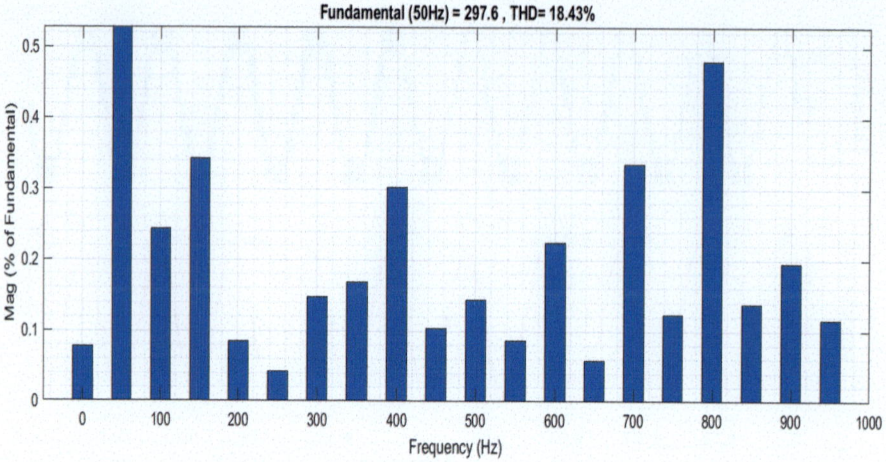

Fig. 12 THD (in percentage) of output voltage

Fig. 13 Resultant waveforms

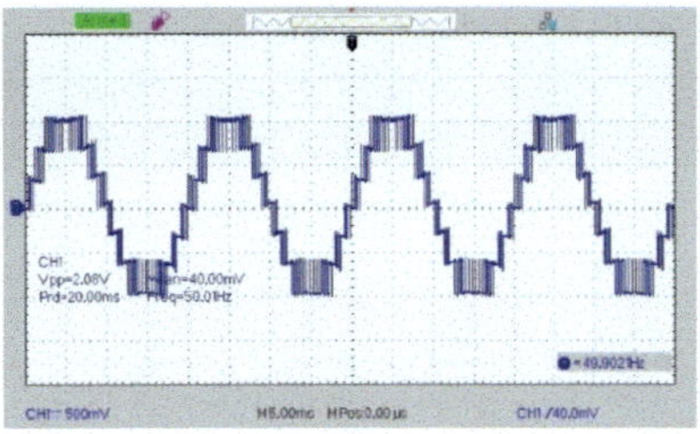

(a) Output voltage

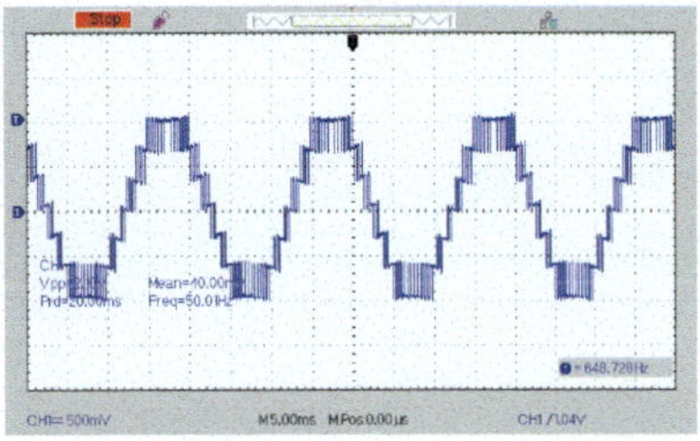

(b) Output current for R load.

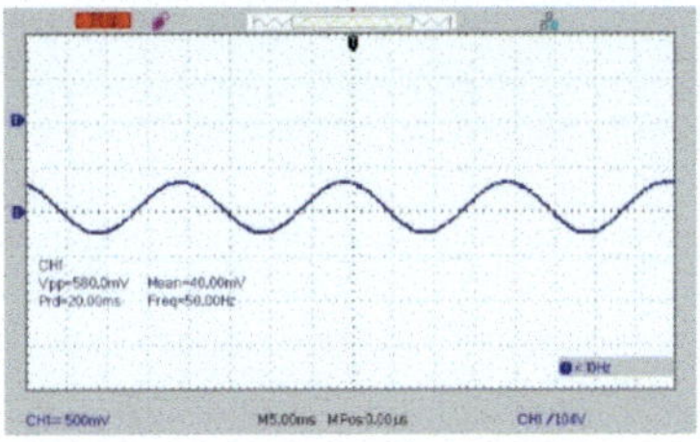

(c) Output current for RL load.

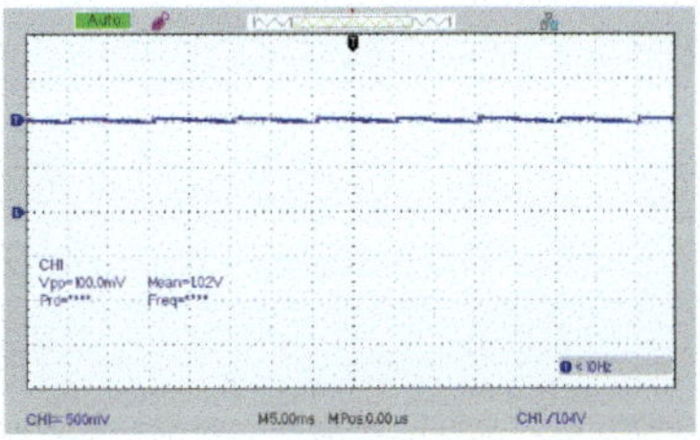

(d) Capacitor C_1

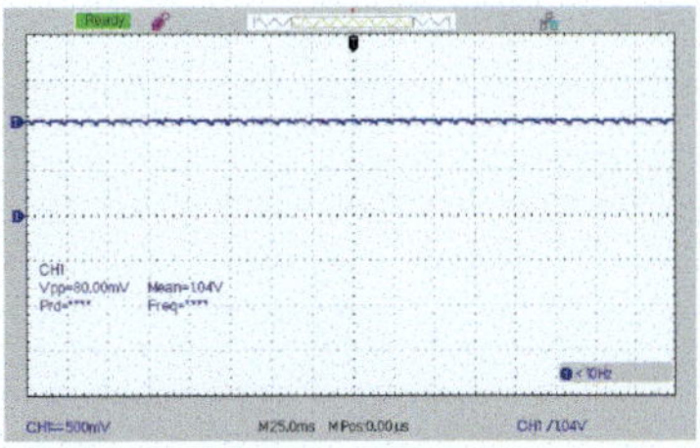

(e) Capacitor C_2

Fig. 14 Switching pulses

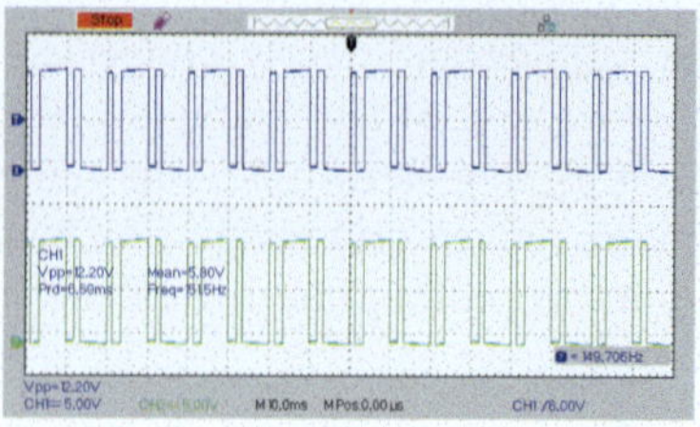

(a) (S_1-S_2)

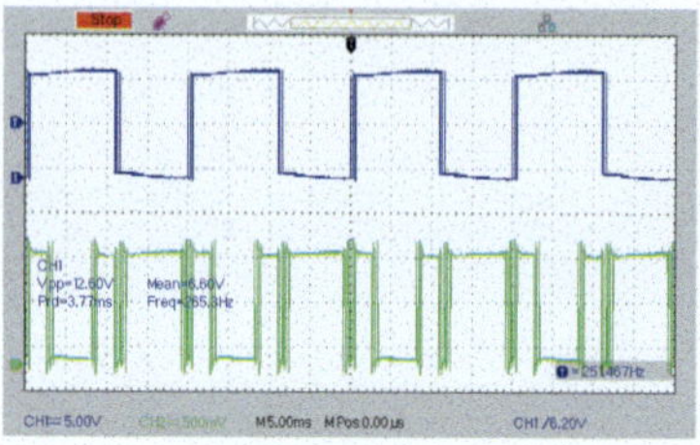

(b) (S_3-S_4)

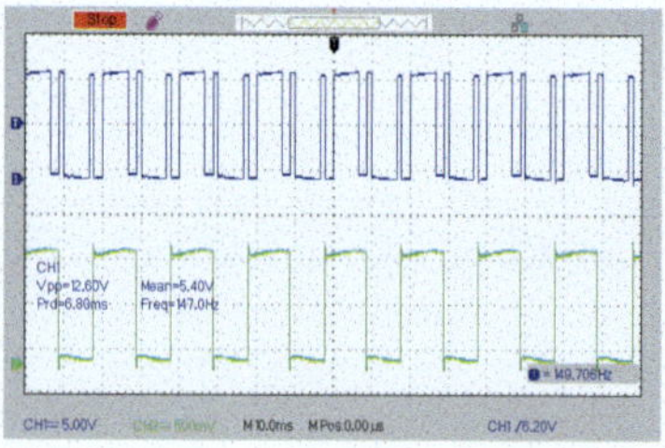

(c) (S_5-S_6)

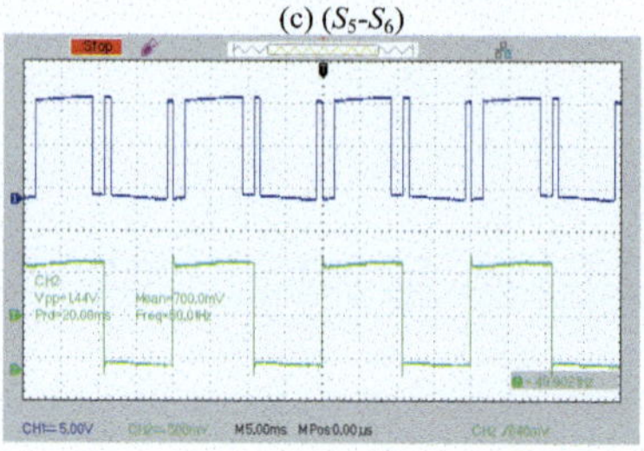

(d) (S_7-S_8)

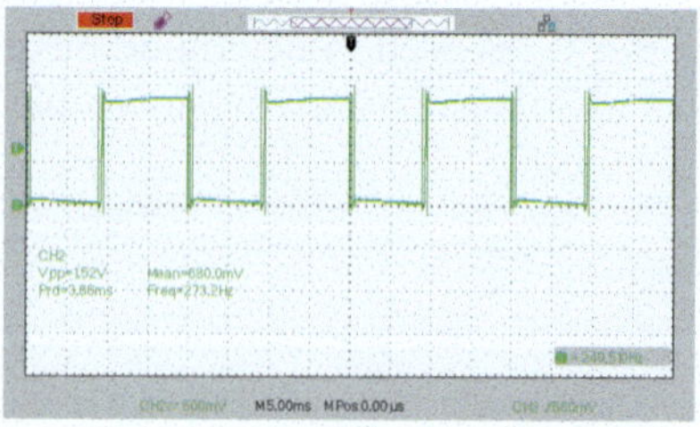

(e) S_9

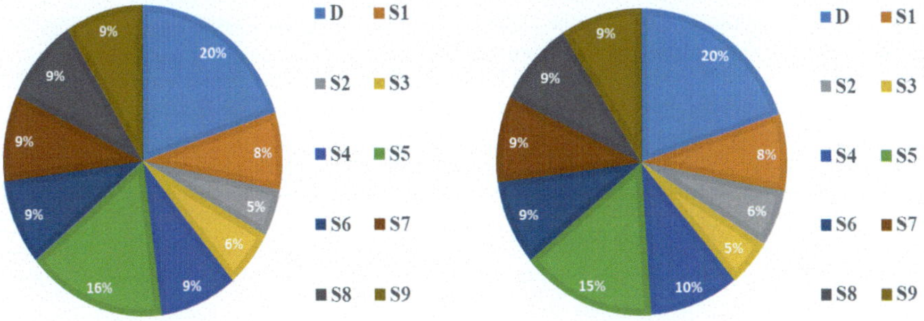

(a) For R -Load (b) For RL- Load

Fig. 15 Percentage losses

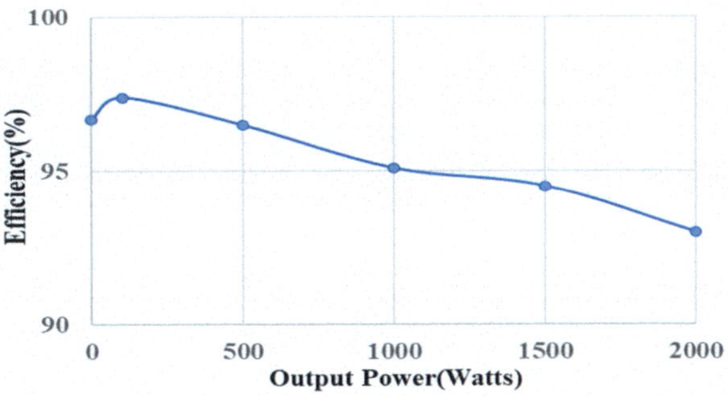

Fig. 16 Efficiency curve

References

1. Rodríguez J, Lai JS, Peng FZ (2002) Multilevel inverters: a survey of topologies, controls, and applications. IEEE Trans Ind Electron 49(4):724–738. https://doi.org/10.1109/TIE.2002.801052
2. Kouro S et al (2010) Recent advances and industrial applications of multilevel converters. IEEE Trans Ind Electron 57(8):2553–2580. https://doi.org/10.1109/TIE.2010.2049719
3. Akagi H (2017) Multilevel converters: fundamental circuits and systems. Proc IEEE 105(11):2048–2065. https://doi.org/10.1109/JPROC.2017.2682105
4. Leon JI, Vazquez S, Franquelo LG (2017) Multilevel converters: control and modulation techniques for their operation and industrial applications. Proc IEEE 105(11):2066–2081. https://doi.org/10.1109/JPROC.2017.2726583
5. De S, Banerjee D, Siva Kumar K, Gopakumar K, Ramchand R, Patel C (2011) Multilevel inverters for low-power applications. IET Power Electron 4(4):384–392. https://doi.org/10.1049/iet-pel.2010.0027
6. Bughneda A, Salem M, Richelli A, Ishak D (2021) Review of multilevel inverters for PV energy system applications. Energies 14(6):1585. https://doi.org/10.3390/en14061585
7. Venkataramanaiah J, Suresh Y, Kumar A (2017) A review on symmetric, asymmetric, hybrid and single DC sources based multilevel inverter topologies Renew Sustain Energy Rev 76:788–812. https://doi.org/10.1016/j.rser.2017.03.066

8. Prabaharan N, Palanisamy K (2017) A comprehensive review on reduced switch multilevel inverter topologies, modulation techniques and applications. Renew Sustain Energy Rev 76:1248–1282. https://doi.org/10.1016/j.rser.2017.03.121
9. Gupta KK, Ranjan A, Bhatnagar P, Sahu LK, Jain S (2016) Multilevel inverter topologies with reduced device count: a review. IEEE Trans Power Electron 31(1):135–151. https://doi.org/10.1109/TPEL.2015.2405012
10. Panda KP, Bana PR, Panda G (2012) A reduced device count single DC hybrid switched-capacitor self-balanced inverter. IEEE Trans Circuits Syst II Exp Briefs 68(3):978–982. https://doi.org/10.1109/TCSII.2020.3018333
11. Panda K, Bana P, Kiselychnyk O, Wang J, Panda G (2021) A single source switched-capacitor-based step-up multilevel inverter with reduced components. IEEE Trans Ind Appl 57(4):3801–3811. https://doi.org/10.1109/TIA.2021.3068076
12. Lee SS (2018) A single-phase single-source 7-level inverter with triple voltage boosting gain. IEEE Access 6:30005–30011. https://doi.org/10.1109/ACCESS.2018.2842182
13. Chen J, Wang C, Li J (2019) A single-phase step-up seven-level inverter with a simple implementation method for level-shifted modulation schemes. IEEE Access 7:146552–146565. https://doi.org/10.1109/ACCESS.2019.2946238
14. Dhara S, Somasekhar VT (2020) An integrated semi-double stage based multilevel inverter with voltage boosting scheme for photovoltaic systems. IEEE J Emerg Sel Top Power Electron 8(3):2326–2339. https://doi.org/10.1109/JESTPE.2019.2955729
15. Khan MNH, Mojtaba F, Siwakoti YP, Blaabjerg F (2020) Switched capacitor integrated (2n + 1)-level step-up single-phase inverter. IEEE Trans Power Electron 35(8):8248–8260. https://doi.org/10.1109/TPEL.2019.2963344
16. Meraj ST, Hasan MK, Islam J, El-Ebiary YAB, Nebhen J, Hossain MM, Alam MK, Vo N (2021) A diamond shaped multilevel inverter with dual mode of operation. IEEE Access 9:59873–59887. https://doi.org/10.1109/ACCESS.2021.3067139
17. Barzegarkhoo R, Lee SS, Khan SA, Siwakoti YP, Lu DD (2021) A novel generalized common-ground switched-capacitor multilevel inverter suitable for transformerless grid-connected applications. IEEE Trans Power Electron 36(9):10293–10306. https://doi.org/10.1109/TPEL.2021.3067347
18. Khodaparast A, Hassani MJ, Azimi E, Adabi ME, Adabi J, Pouresmaeil E (2021) Circuit configuration and modulation of a seven-level switched-capacitor inverter. IEEE Trans Power Electron 36(6):7087–7096. https://doi.org/10.1109/TPEL.2020.3036351
19. Chen M (2022) Novel cascaded seven-level inverter with embedded voltage boosting for renewable energy applications. CPSS Trans Power Electron Appl 7(1):58–70. https://doi.org/10.24295/CPSSTPEA.2022.00006
20. Wang Y, Yuan Y, Li G, Ye Y, Wang K, Liang J (2021) A T-type switched capacitor multilevel inverter with low voltage stress and self-balancing. IEEE Trans Circuits Syst I Regul Pap 68(5):257–2270. https://doi.org/10.1109/TCSI.2021.3060284
21. Liu J, Zhu X, Zeng J (2020) A seven-level inverter with self-balancing and low-voltage stress. IEEE J Emerg Sel Top Power Electron 8(1):685–696. https://doi.org/10.1109/JESTPE.2018.2879890
22. Iqbal A, Siddique MD, Ali JSM, Mekhilef S, Lam J (2020) A new eight switch seven level boost active neutral point clamped (8S-7L BANPC) inverter. IEEE Access 8:203972–203981. https://doi.org/10.1109/ACCESS.2020.3036483
23. Sathik MJ, Bhatnagar K, Sandeep N, Blaabjerg F (2020) An improved seven-level PUC inverter topology with voltage boosting. IEEE Trans Circuits Syst II Exp Briefs 67(1):127–131. https://doi.org/10.1109/TCSII.2019.2902908
24. Lin W, Zeng J, Fu B, Yan Z, Liu J (2021) Switched-capacitor based seven-level boost inverter with reduced devices. CSEE J Power Energy Syst Early Access 1–11. https://doi.org/10.17775/CSEEJPES.2020.02620
25. Lee SS, Bak Y, Kim SM, Joseph A, Lee KB (2019) New family of boost switched-capacitor seven-level inverters (BSC7LI). IEEE Trans Power Electron 34(11):10471–10479. https://doi.org/10.1109/TPEL.2019.2896606

26. Panda P, Bana PR, Panda G (2020) A switched-capacitor self-balanced high-gain multilevel inverter employing a single DC source. IEEE Trans Circuits Syst I Regul Pap 67(12):3192–3196. https://doi.org/10.1109/TCSII.2020.2975299
27. Ye Y, Cheng KWE, Liu J, Ding K (2014) A step-up switched-capacitor multilevel inverter with self-voltage balancing. IEEE Trans Ind Electron 61(12):6672–6680. https://doi.org/10.1109/TIE.2014.2314052
28. Sathik MJ, Sandeep N, Blaabjerg F (2020) High gain active neutral point clamped seven-level self-voltage balancing inverter. IEEE Trans Circuits Syst II Exp Briefs 67(11):2567–2571. https://doi.org/10.1109/TCSII.2019.2955318
29. Jahan HK, Abapour M, Zare K (2019) Switched-capacitor-based single-source cascaded H-bridge multilevel inverter featuring boosting ability. IEEE Trans Power Electron 34(2):1113–1124. https://doi.org/10.1109/TPEL.2018.2830401
30. Khenar M, Taghvaie A, Adabi J, Rezanejad M (2018) Multi-level inverter with combined T-type and cross-connected modules. IET Power Electron 11(8):1407–1415. https://doi.org/10.1049/ietpel.2017.0378
31. Zeng J, Lin W, Liu J (2019) Switched-capacitor-based active neutral-point-clamped seven-level inverter with natural balance and boost ability. IEEE Access 7:126889–126896. https://doi.org/10.1109/ACCESS.2019.2927351
32. Venkatesh A, Sesham H, Vijaya Kumar J, Kiran SS, Salkuti SR, Kim SC (2024) Designing of self balancing amplitude modulated five level inverter for reducing voltage stress gradients on converter switches for electric vehicles. Franklin Open 7:100114. https://doi.org/10.1016/j.fraope.2024.100114
33. Narasimha S, Salkuti SR (2020) Design and operation of closed-loop triple-deck buck-boost converter with high gain soft switching. Int J Power Electron Drive Syst 11(1):523–529. https://doi.org/10.11591/ijpeds.v11.i1.pp523-529
34. Kim SC, Narasimha S, Salkuti SR (2020) A new multilevel inverter with reduced switch count for renewable power applications. Int J Power Electron Drive Syst 11(4):2145–2153. https://doi.org/10.11591/ijpeds.v11.i4.pp2145-2153
35. Vuddanti S, Karknalli V, Salkuti SR (2021) Design and comparative analysis of three phase, four phase and six phase switched reluctance motor topologies for electrical vehicle propulsion. Bull Electr Eng Inform 10(3):1495–1504. https://doi.org/10.11591/eei.v10i3.3054

RST-Based Robust Digital PID Controller Design for Dual-Input SEPIC Converters

Allam Venkatesh, B. Amarendra Reddy, T. R. Jyothsna, Ch. Nayak Bhukya, and Surender Reddy Salkuti

Abstract The DC-DC converters require controllers for their voltage regulation and current control. Controllers are also essential for reliable and efficient power conversion. Designing a controller for the dual input SEPIC converter topology poses significant challenges owing to its fourth-order and non-linear characteristics. Additionally, the controller performance is influenced by load variations and operating conditions. A well-designed controller must fulfill specific objectives, including a fast-dynamic response, resilience to parameter variations, assured stability, and precise tracking performance. In this work, a robust multivariable control technique is considered for the design of a controller for integrated DC-DC converters. The digital decentralized controllers are designed for two input-integrated converters to achieve stable and reliable output voltage. A robust PID controller is designed with RST control architecture for the regulation of TI-BS and TI-MBS converters using the pole placement technique. The detailed procedural steps to design a robust controller using this technique are presented. A sixth-order closed-loop characteristic equation is formulated for Voltage and current loops and controllers are designed based on this. The desired pole location for both voltage and current control loops is assumed to be the same. To verify the effectiveness of these designed controllers, 48 V, 720 W, and TII DC-DC converters are considered for analysis and simulation. The steady-state and dynamic simulations are performed for TI-BS and TI-MBS converters under different situations, these are (i) with nominal converter parameters, (ii) variation in load current, (iii) variations in source voltage. It is observed from these simulations that the current of source-2 (i_{g2}), and the output voltage (V_0) will follow the reference values against these disturbance conditions.

A. Venkatesh · B. Amarendra Reddy · T. R. Jyothsna · Ch. Nayak Bhukya
Department of Electrical Engineering, Andhra University, Visakhapatnam, Andhra Pradesh, India
e-mail: allamvenkatesh.rs@andhrauniversity.edu.in

B. Amarendra Reddy
e-mail: amarendra@andhrauniversity.edu.in

S. R. Salkuti (✉)
Department of Railroad and Electrical Engineering, Woosong University, Daejeon, Republic of Korea
e-mail: surender@wsu.ac.kr

S. R. Salkuti (ed.), *Artificial Intelligence for Integrated Smart Energy Systems in Electric Vehicles*, Lecture Notes in Electrical Engineering 1427,
https://doi.org/10.1007/978-3-031-94276-1_21

Keywords Two-input integrated · Robust sensitivity tracking · Buck-SEPIC · Modified buck-SEPIC · Pole placement

Nomenclature

HES	Hybrid Energy Systems
HMVS	High-magnitude voltage source
I_L	Output Current or Load Current
LMVS	Low-magnitude voltage source
PID	Proportional Integral Derivative
PWM	Pulse Width Modulation
SEPIC	Single Ended Primary Inductor Converter
TFM	Transfer Function Matrix
TI-BS	Two-Input Buck SEPIC
TII	Two-Input Integrated
TI-MBS	Two-Input Modified Buck SEPIC
V_{g1}	Source voltage-1
V_{g2}	Source voltage-2
V_o	Output voltage

1 Introduction

In the present scenario, multi-input converters are playing a major role in the power system sector. Multi-input converters are designed with power electronic components that simultaneously handle two or more input sources and deliver the desired power output. These are majorly utilized by several applications which are hybrid energy systems (HES), electric vehicle (EV) charging, DC microgrids, energy storage systems, and distributed generation. A simple multi-input converter is a dual-input converter, that can handle two sources, with minimum components to enable efficient energy management. However, there are some advancements in multi-input converters, i.e., the development of converter topologies that use wide bandgap semiconductor-based switches and are regulated using advanced control techniques. Many researchers and industrial experts are continuously working on these advancements in the area of multi-input DC-DC converters aiming for improvement in efficiency, reliability, and functionality to meet the demand of the particular application. New age systems are designed with more sensitive components, the major challenge is to supply constant voltage without any deviations [1, 2]. For example, in EVs major components like power steering, suspension, motor, auxiliaries, etc., are working on a 48 V DC power supply and it needs constant power supply. To achieve the global standard 48 V, several researchers are actively working on dual-input

DC-DC converters, the current status of the literature is presented in the following paragraph.

In literature, the inventions are reported mainly in three different crucial areas, these are (i) design and development of new topologies for effective power management [3, 4], (ii) fault-tolerant Researchers are exploring fault detection, isolation, and recovery techniques to enhance the converter's robustness and reliability in the presence of component failures or input source irregularities [5–7], (iii) regulation of converter using cutting-edge control techniques, such as adaptive control, predictive control, intelligent control algorithms, for optimizing the converter's dynamic and steady-state performance [8–10].

The modeling of the converter plays a vital role in performing in-depth analysis that helps to optimize converter performance. There are two types of modeling, continuous time and discrete time. Several discrete-time models are reported in the literature depending on sampling instants as well as process delays [11]. While analyzing the converter systems using these modeling methodologies in simulation environments are showing some impacts like deviations in phasor plots at the high-frequency region of frequency response and also stability issues. To eliminate these impacts of modeling on the converter and to protect the sensitive DC loads, there is a need for a proper control strategy. The control strategies can be combined or customized based on the requirements of a specific application. Several variables influence the control method selection, including system complexity, desired performance, computational capacity, and communication infrastructure [12]. Ongoing research focuses on improving the efficiency, reliability, and scalability of dual-input converters through advanced control techniques. In the literature, some controllers have recently been reported for dual input converters. A robust multivariable controller is designed using an H∞ controller presented in Ref. [13]. A comprehensive full-order centralized controller is designed and developed with the help of an equivalent transfer function model for regulating the DC-bus voltage of a dual-input DC-DC converter [14, 15]. A conventional PI controller is implemented for a Two-input DC-DC converter [16], and a PID and Fuzzy logic controller are implemented to perform the output voltage regulation [17, 18]. A fractional order controller is designed for a dual input boost converter for balancing the inductor current as well as the capacitor voltage and it is presented in Refs. [19, 20]. Non-linear, Non-linear adaptive, and ANN-based non-linear adaptive controllers for DC-DC converters are presented [21–30]. The controllers reported so far have their disadvantages like representation of uncertainty in co-prime factor form, expressing norm-based performance objectives, and representation of converters systems using time-delay models. Hence, in this chapter, an in-depth procedure is presented on controller design, and stability analysis of dual-input DC-DC power converter with RST-based robust PID controller. This chapter is organized as follows: Sect. 2 describes modeling expressions of TI-BS and TI-MBS converters. Section 3 describes small signal modeling aspects and a brief description of the modes of operation of TI-BS and TI-MBS DC-DC converters. Here, the state space equations are discussed. In Sect. 4, the concept of ideal decoupling of a two-input converter system is presented. In Sect. 5, a detailed procedure for designing the RST-based robust PID controller is presented.

In Sect. 6, implementation of the RST-based robust PID controller for TI-BS and TI-MBS converter systems and simulation results under different operating conditions are presented. Conclusions are presented in Sect. 7. This chapter provides the following contributions.

- Modeled TI-BS and TI-MBS DC-DC converters.
- Designed an RST-based robust PID controller for TI-BS and TI-MBS DC-DC converter.
- Examined robustness of proposed RST-based robust PID controller under uncertainty conditions for TI-BS and TI-MBS DC-DC converters in MATLAB/Simulink environment.

2 Modelling of TI-BS and TI-MBS DC-DC Converters

Two different two-input SEPIC converters are examined for power processing from two independent sources V_{g1} and V_{g2} where V_{g1}, is a high-magnitude voltage source (HMVS) and V_{g2}, is a low-magnitude voltage source (LMVS). These two-input converters considered for analysis are (i) Buck-SEPIC (TI-BS) converters and (ii) Modified Buck-SEPIC (TI-MBS) converters, both converters are fourth-order in nature. Throughout the power conversion process, both converters execute bucking operations concerning the HMVS (V_{g1}), and buck-boost operations concerning the LMVS (V_{g2}). Each converter integrates two switches and is regulated using two independent PWM signals. The operational concept, as well as key aspects of TI-BS [11] and TI-MBS [10] converters are discussed in detail.

The detailed representation of equivalent circuits and the operating principles of TI-BS and TI-MBS converters can be found in Refs. [10, 11], respectively. Both of these converters are capable of functioning for $d_1 < d_2$ and for $d_1 > d_2$ scenarios. In both cases, these converters undergo three different structural changes namely Mode-I, Mode-II, and Mode-III as illustrated in Fig. 1c. Using equivalent circuits of converter in their respective modes, the modeling of the converter is obtained in the state-space domain.

The dependency of the load voltage on the source voltages and a PWM duty ratio signal is defined as voltage gain. This is illustrated by imposing volt-sec balance to inductive elements of the converter system depicted in Fig. 1a, b. For both the converters final gain expression of the output voltage is obtained as,

$$V_o = V_{g1}d_1 + \left(\frac{V_{g2}d_2}{(1 - d_2)}\right) \tag{1}$$

The voltage gain expression $V_o = f\left(V_{g1}, V_{g2}, d_1, d_2\right)$ is a function of input voltages and duty cycles. Hence these factors, input voltages, and duty cycles decide the performance of the system. Time-domain analysis [10, 11] design equations are

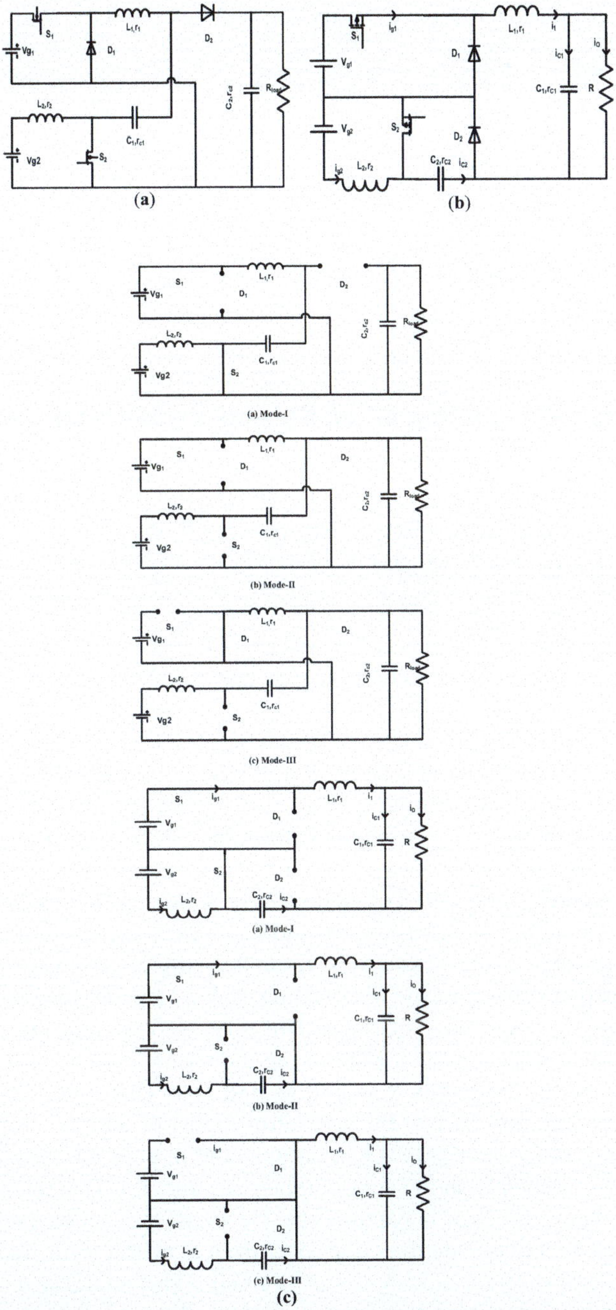

Fig. 1 **a** Circuit diagram of TI buck-SEPIC converter; **b** circuit diagram of TI-MBS converter; **c** mode-I, II and III equivalent circuits of TI-BS and TI-MBS converters

Table 1 Design equations of TI-BS and TI-MBS converter

	L_1	L_2	C_1	C_2
Buck-SEPIC converter	$\frac{V_0(1-d_1)}{fs\,\Delta I_1}$	$\frac{(V_{g_2}-V_{C_1}-V_0)(1-d_1)}{fs\,\Delta I_2}$	$\frac{I_{L_2}(1-d_1)}{fs\,\Delta v_{C1}}$	$\frac{I_0 d_2}{fs\,\Delta v_{C0}}$
Modified buck-SEPIC	$\frac{V_o(1-d_1)}{\Delta i_{L1} f_s}$	$\frac{V_{g2} d_2}{\Delta i_{L2} f_s}$	$\frac{V_o(1-d_1)^2}{4L_1 f_s^2\,\Delta v_{c1}}$	$\frac{i_2(1-d_2)}{\Delta v_{c2} f_s}$

obtained, for the TI-BS and TI-MBS DC-DC converter. Design equations depicted in Table 1.

3 Small-Signal Modeling of TI-BS and TI-MBS DC-DC Converters

3.1 Transfer Function Matrix Using Small-Signal Continuous-Time Model

In power management for TI-BS and TI-MBS DC-DC converters, the voltage ratings for V_{g1} and V_{g2} differ significantly as higher and lower respectively. The power distribution between two sources with load can be assigned as (i) V_{g1} is designated to provide a higher power share, while V_{g2} is meant to provide a lesser power contributor and (ii) V_{g1} is designated to provide lower power contribution while V_{g2} is designated to provide higher power contribution, (iii) V_{g1} and V_{g2} are intended to provide higher power share and (iv) V_{g1}, V_{g2} are intended to provide lower power contributor. In all of these possible scenarios, case (1) it termed as power-sharing proportional to their capacities, case (2) represents a nonlinear relationship (inverse proportion to their capacities) were V_{g1} is under-utilized and V_{g2} is overburdened, for case (3) V_{g2} might be overburdened and for case (4) V_{g1} is under-utilized. To ensure efficient use of both sources and to prevent the overloading of V_{g2}, a restriction on the current drawn from the V_{g2} imposed. This constraint helps to maintain the power share of V_{g2} proportional to its capacity. Monitoring the source current of V_{g2} is crucial along with regulating the output voltage (V_O and i_{g2}). Therefore, TI-BS and TI-MBS converters functions as two-input (control inputs d_1 and d_2) and two-output (V_O and i_{g2}) converter systems (TITOC). From a control point-of-view, each output variables behavior depends on the control inputs (d_1 and d_2). This interdependency of output and input variables is captured by the transfer function matrix (TFM), which comprises a set of transfer functions organized in matrix form. To derive these transfer functions, a state-variable model and a small-signal model are essential. The state-variable model describing the dynamic and steady-state characteristics of the converter system.

3.1.1 State-Variable Modelling

The detailed state-variable modeling along with the converter equivalent circuits in each of the three modes of TI-BS and TI-MBS converters are available in Refs. [3, 4] respectively. The power conversion process dynamics of these converters are described using the state model which is described using Eqs. (2) and (3).

$$\dot{x} = A_k x + B_k u, \tag{2}$$

$$\begin{bmatrix} v_0 \\ i_{g2} \end{bmatrix} = \begin{bmatrix} E_k \\ P_k \end{bmatrix} x \tag{3}$$

where $A_k \in R^{4\times4}, B_k \in R^{4\times2}, E_k \in R^{1\times4}, P_k \in R^{1\times4}$ $[x] = \begin{bmatrix} x_1 \ x_2 \ x_3 \ x_4 \end{bmatrix}^T$ $\begin{bmatrix} x_1 \ x_2 \ x_3 \ x_4 \end{bmatrix}^T = \begin{bmatrix} i_{L1} \ i_{L2} \ v_{C_1} \ v_{C_2} \end{bmatrix}^T, [u] = \begin{bmatrix} V_{g1} \ V_{g2} \end{bmatrix}^T$, and k = 1, 2, 3 for mode-1, mode-2 and mode-3, respectively.

These state-space matrices derived here are used in modeling the converter in both small-signal continuous-time and discrete-time modeling methodologies, commonly employed to derive the transfer function of the converters. These two converter circuits operate in three different modes for the switch synchronization when $d_1 > d_2$. In each mode of operation, the state-space equations are derived, resultant matrices mentioned in Eq. (2) are expressed in Table 2.

The transfer function matrices (TFMs) is crucial in analyzing dynamic behavior and are instrumental in designing direct digital controllers for TI-BS and TI-MBS converters. The TFMs for these converters are formulated Small-signal discrete-time model. The discrete-time domain transfer function relating output and input variables of the two-input converter system is described using Eq. (4). The detailed small-signal discrete-time model derivations of TII converter system are presented in Refs. [3, 4].

$$\begin{aligned} \begin{bmatrix} \hat{v}_0(z) \\ \hat{i}_g(z) \end{bmatrix} &= \begin{bmatrix} E_i[zI - \phi]^{-1}\gamma_1 \ E_i[zI - \phi]^{-1}\gamma_2 \\ P_i[zI - \phi]^{-1}\gamma_1 \ P_i[zI - \phi]^{-1}\gamma_2 \end{bmatrix} \\ \begin{bmatrix} \hat{d}_1(z) \\ \hat{d}_2(z) \end{bmatrix} &= \begin{bmatrix} G_{11}(z) \ G_{12}(z) \\ G_{21}(z) \ G_{22}(z) \end{bmatrix} \begin{bmatrix} \hat{d}_1(z) \\ \hat{d}_2(z) \end{bmatrix} \end{aligned} \tag{4}$$

where,

$$\phi = \mathrm{e}^{A_3[T_S - d_1 T_S - t_d + d_2 T_S]} \mathrm{e}^{A_3[t_d - d_2 T_S]} \mathrm{e}^{A_2[d_1 - d_2]T_S} \mathrm{e}^{A_1 d_2 T_S} \tag{5}$$

$$\gamma_1 = \mathrm{k}_1 T_S \mathrm{e}^{A_3[T_S - d_1 T_S - t_d + d_2 T_S]} \tag{6}$$

Table 2 State-variable matrices of TI-BS and TI-MBS converter

	TI-BS converter	TI-MBS converter
A1	$\begin{bmatrix} -\frac{(r_{C_1}+r_1)}{L_1} & 0 & -\frac{1}{L_1} & 0 \\ 0 & -\frac{r_2}{L_2} & 0 & 0 \\ -\frac{1}{C_1} & 0 & 0 & 0 \\ 0 & 0 & 0 & -\frac{1}{C_2(R+r_{C_2})} \end{bmatrix}$	$\begin{bmatrix} \frac{-k_1}{L_1} & 0 & \frac{1}{L_1} & -\frac{b}{L_1} \\ 0 & \frac{-r_2}{L_2} & 0 & 0 \\ \frac{-1}{C_1} & 0 & 0 & 0 \\ \frac{b}{C_2} & 0 & 0 & \frac{-b}{RC_2} \end{bmatrix}$
A2	$\begin{bmatrix} -\frac{r_1+a}{L_1} & -\frac{a}{L_1} & 0 & -\frac{b}{L_1} \\ -\frac{a}{L_2} & -\frac{r_2+r_{C_1}+a}{L_2} & -\frac{1}{L_2} & -\frac{b}{L_2} \\ 0 & \frac{1}{C_1} & 0 & 0 \\ \frac{b}{C_2} & \frac{b}{C_2} & 0 & -\frac{b}{RC_2} \end{bmatrix}$	$\begin{bmatrix} \frac{-(r_1+a)}{L_1} & 0 & 0 & \frac{-b}{L_1} \\ 0 & \frac{-(r_2+r_{c2})}{L_2} & -\frac{1}{L_2} & 0 \\ 0 & \frac{1}{C_1} & 0 & 0 \\ \frac{b}{C_2} & 0 & 0 & \frac{-b}{RC_2} \end{bmatrix}$
A3	$\begin{bmatrix} -\frac{r_1+a}{L_1} & -\frac{a}{L_1} & 0 & -\frac{b}{L_1} \\ -\frac{a}{L_2} & -\frac{a+r_2+r_{C_1}}{L_2} & -\frac{1}{L_2} & -\frac{b}{L_2} \\ 0 & \frac{1}{C_1} & 0 & 0 \\ \frac{b}{C_2} & \frac{b}{C_2} & 0 & -\frac{b}{RC_2} \end{bmatrix}$	$\begin{bmatrix} \frac{-(r_1+a)}{L_1} & 0 & 0 & \frac{-b}{L_1} \\ 0 & \frac{-(r_2+r_{c2})}{L_2} & -\frac{1}{L_2} & 0 \\ 0 & \frac{1}{C_1} & 0 & 0 \\ \frac{b}{C_2} & 0 & 0 & \frac{-b}{RC_2} \end{bmatrix}$
	$a = (b)r_{C_2};\ \mathrm{b} = \frac{R}{R+r_{C_2}}$	$a = (b)r_{C_2};\ \mathrm{b} = \frac{R}{R+r_{C_2}};$ $k_1 = (r_1 + r_{C_1} + a)$
B	$B_1 = B_2 = \begin{bmatrix} \frac{1}{L_1} & 0 \\ 0 & \frac{1}{L_2} \\ 0 & 0 \\ 0 & 0 \end{bmatrix} B_3 = \begin{bmatrix} 0 & 0 \\ 0 & \frac{1}{L_2} \\ 0 & 0 \\ 0 & 0 \end{bmatrix}$	$B_1 = B_2 = \begin{bmatrix} \frac{1}{L_1} & 0 \\ 0 & \frac{1}{L_2} \\ 0 & 0 \\ 0 & 0 \end{bmatrix} B_3 = \begin{bmatrix} 0 & 0 \\ 0 & \frac{1}{L_2} \\ 0 & 0 \\ 0 & 0 \end{bmatrix}$
E	$E_1 = \begin{bmatrix} 0 & 0 & 0 & \mathrm{b} \end{bmatrix}$ $E_2 = E_3 = \begin{bmatrix} a & a & 0 & \mathrm{b} \end{bmatrix}$ $P_1 = P_2 = P_3 = \begin{bmatrix} 0 & 1 & 0 & 0 \end{bmatrix}$	$E_1 = E_2 = E_3 = \begin{bmatrix} a & 0 & 0 & \mathrm{b} \end{bmatrix}$ $P_1 = P_2 = P_3 = \begin{bmatrix} 0 & 1 & 0 & 0 \end{bmatrix}$

$$\gamma_2 = \mathrm{k_2 T_S} e^{A_3[\mathrm{T_S - d_1 T_S - t_d + d_2 T_S}]} e^{A_2[\mathrm{d_1 - d_2}]\mathrm{T_S}} \tag{7}$$

$$k_1 = [(A_2 - A_3)X + (B_2 - B_3)U]; \tag{8}$$

$$k_2 = [(A_1 - A_2)X + (B_1 - B_2)U]. \tag{9}$$

4 Design of Controller a Decoupled System

Decentralized Controller can be designed to dual-input DC-DC converter by forming a decoupled system. The system is obtained by cascading the converter with decoupler, the principle of decoupler is explained in the following section.

4.1 *Principle of Ideal Decoupler*

Decentralized controller design for a decoupled system is most known and popular method when interactions plays an important role. The purpose of the decoupled based design is to design a decoupler to minimize the interactions of two-input DC-DC converters and this process decouples the integrated system into equivalent Single Input Single Output (SISO) systems (ideally interactions of the Two Input Two Output System equals to zero). This process allows to design the controllers using SISO techniques for each individual equivalent system. The block-diagram of the integrated converter system with decoupler and controller is shown in Fig. 2. The converter system with ideal decoupler is shown in Fig. 3.

The TFMs of converter, decoupler, controller is given in Eqs. (10), (11), and (12) respectively. Ideal decoupler TFM is represented using Eq. (13) by considering $D_{11}(s) = D_{22}(s) = 1$ in Eq. (11). A simplified decoupled system is formed using principle of an ideal decoupler. The ideal decoupled TFM of the integrated converter is represented using Eq. (16). Then the conditions for ideal decoupling are obtained by making off-diagonal elements of the decoupled TFM (Eqs. 18 and 19) are equated to zero. These conditions are given in Eq. (20).

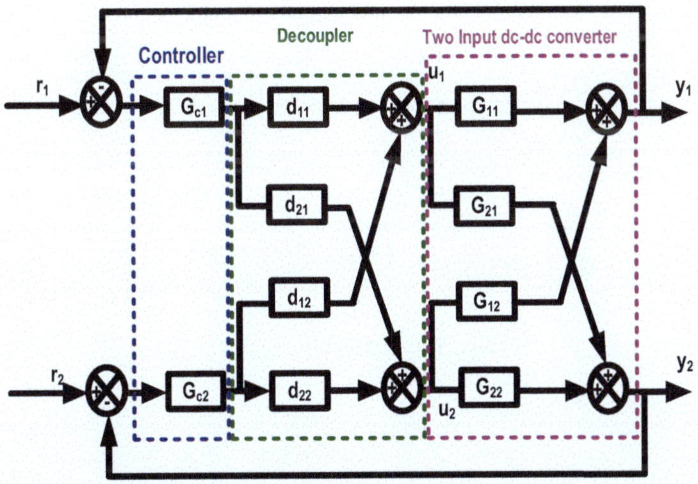

Fig. 2 A decoupled two-input DC-DC converter

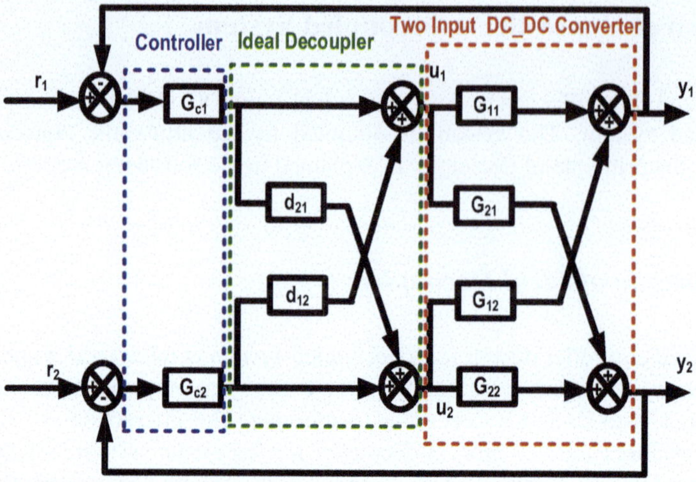

Fig. 3 Ideal decoupled two-input DC-DC converter

$$G_{sys}(z) = \begin{bmatrix} G_{11}(z) & G_{12}(z) \\ G_{21}(z) & G_{22}(z) \end{bmatrix} \tag{10}$$

$$D(z) = \begin{bmatrix} D_{11}(z) & D_{12}(z) \\ D_{21}(z) & D_{22}(z) \end{bmatrix} \tag{11}$$

$$G_C(S) = \begin{bmatrix} G_{C11}(z) & 0 \\ 0 & G_{C22}(z) \end{bmatrix} \tag{12}$$

$$D_{ideal}(z) = \begin{bmatrix} 1 & D_{12}(z) \\ D_{21}(z) & 1 \end{bmatrix} \tag{13}$$

$$G_{sys}(z)D_{ideal}(z) = \begin{bmatrix} G_{11}(z) & G_{12}(z) \\ G_{21}(z) & G_{22}(z) \end{bmatrix} \begin{bmatrix} 1 & D_{12}(z) \\ D_{21}(z) & 1 \end{bmatrix} \tag{14}$$

$$G_{sys}(z)D_{ideal}(z) = \begin{bmatrix} G_{11}(z) + G_{12}(z)D_{21}(z) & G_{11}(z)D_{12}(z) + G_{12}(z) \\ G_{21}(z) + G_{22}(z)D_{21}(z) & G_{21}(z)D_{12}(z) + G_{22}(z) \end{bmatrix} \tag{15}$$

$$G_{sys}(z)D_{ideal}(z) = \begin{bmatrix} G_{11}^{reduced}(z) & 0 \\ 0 & G_{22}^{reduced}(z) \end{bmatrix} \tag{16}$$

$$G_{11}(z)D_{12}(z) + G_{12}(z) = 0 \tag{17}$$

$$D_{12}(z) = -\frac{G_{12}(z)}{G_{11}(z)} \tag{18}$$

$$D_{21}(z) = -\frac{G_{21}(z)}{G_{22}(z)} \tag{19}$$

$$G_{sys}(z)D_{ideal}(z) = \begin{bmatrix} G_{11}(z) + G_{12}(z)\left(-\frac{G_{21}(z)}{G_{22}(z)}\right) & G_{11}(z)\left(-\frac{G_{12}(z)}{G_{11}(z)}\right) + G_{12}(z) \\ G_{21}(z) + G_{22}(z)\left(-\frac{G_{21}(z)}{G_{22}(z)}\right) & G_{21}(z)\left(-\frac{G_{12}(z)}{G_{11}(z)}\right) + G_{22}(z) \end{bmatrix} \tag{20}$$

$$G_{11}^{reduced}(z) = G_{11}(z) + G_{12}(z)\left(-\frac{G_{21}(z)}{G_{22}(z)}\right) \tag{21}$$

$$G_{22}^{reduced}(z) = G_{21}(z)\left(-\frac{G_{12}(z)}{G_{11}(z)}\right) + G_{22}(z) \tag{22}$$

where $G_{11}^{reduced}(z)$, $G_{22}^{reduced}(z)$ are known as the reduced effective transfer functions of the voltage and current loop respectively. Using these transfer functions, the controllers are designed using SISO design methodologies.

The voltage and current loop equivalents of Two-Input DC-DC converter block diagrams by using ideal decoupling principle are shown in Figs. 4 and 5.

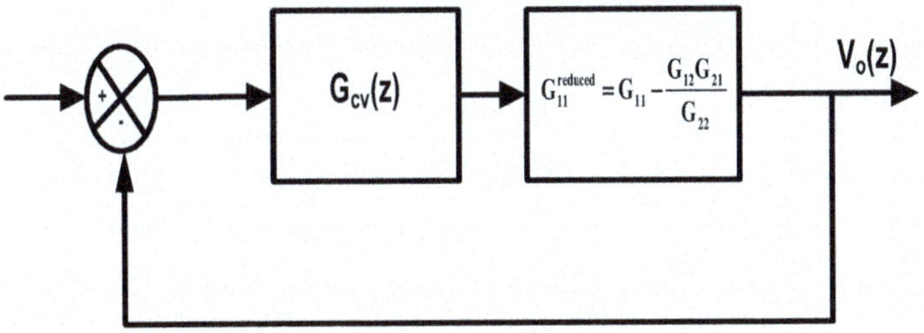

Fig. 4 Voltage loop equivalent of two input DC-DC converter

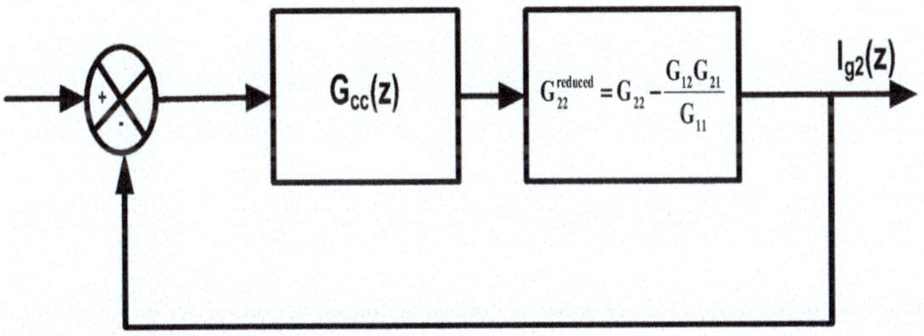

Fig. 5 Current loop equivalent of two input DC-DC converter

4.2 The Systematic Procedure to Obtain Control Structure Enumerated as Follows

Step 1: Obtain TFM of the Two-Input DC-DC converter system and with the transfer function elements obtain voltage, current loop equivalents. These are defined using Eqs. (23) and (24).

$$G_{veq} = G_{11} - \frac{G_{12}G_{21}}{G_{22}} \tag{23}$$

$$G_{ceq} = G_{22} - \frac{G_{12}G_{21}}{G_{11}} \tag{24}$$

Step 2: The voltage loop reduced effective transfer function of this system are represented using standard form using Eq. (25). The PID controller is represented using Eq. (26).

$$G_{vd}(z) = \frac{N_v(z)}{D_v(z)} \tag{25}$$

$$G_{cv}(z) = \frac{r_{0v} + r_{1v}z^{-1} + r_{2v}z^{-2}}{q_{0v} + q_{1v}z^{-1} + q_{2v}z^{-2}} \tag{26}$$

$N_v(z)$ Numerator of reduced effective TF of voltage loop
$D_v(z)$ Denominator of reduced effective TF of voltage loop.

Similarly, the current-loop reduced effective transfer function of this system is represented using standard form using Eq. (27). The PID controller is represented using Eq. (28).

$$G_{id}(z) = \frac{N_i(z)}{D_i(z)} \tag{27}$$

$$G_{cc}(z) = \frac{r_{0C} + r_{1C}z^{-1} + r_{2C}z^{-2}}{q_{0C} + q_{1C}z^{-1} + q_{2C}z^{-2}} \tag{28}$$

$N_i(z)$ Numerator of reduced effective TF of current loop
$D_i(z)$ Denominator of reduced effective TF of current loop.

5 Design of RST-Based Robust PID Controller for Dual Input SEPIC DC-DC Converters

The block diagram of the generalized closed-loop structure of the RST controller is depicted in Fig. 6, and it is designed in order to guarantee regulation of load voltage, low voltage source current control, and robustness against converter parameter uncertainty. Using the well-known pole placement technique, the controller is designed easily for TI-BS and TI-MBS converters.

The main objective here is to design the R-S-T controller polynomials such that the closed-loop controller is stable and robust against the converter parameter uncertainty. A basic RST controller contains three different blocks which are located at three different locations as shown in Fig. 6 with T = R, where the *R* and *S* polynomials need to be designed taking the robust performance specifications into account.

5.1 Procedural Steps to Design RST Based Robust PID Controller

Let the converter model and controllers be defined by the polynomials. The voltage loop gain $T_L(z^{-1})$ of V_0 equivalent of Two Input DC-DC Converter is defined using Eq. (29) and it is a function of $G_{Cv}(z^{-1})$ *and* $G_{vd}(z^{-1})$. $G_{vd}(z^{-1})$ is the V_0 loop equivalent transfer function which is defined using Eq. (30). $G_{Cv}(z^{-1})$ is the controller transfer function and is defined using Eq. (33).

$$T_{Lv}(z^{-1}) = G_{Cv}(z^{-1})G_{vd}(z^{-1}) \tag{29}$$

$$G_{vd}(z^{-1}) = G_{veq}(z^{-1}) = \frac{B_v(z^{-1})}{A_v(z^{-1})} \tag{30}$$

where,

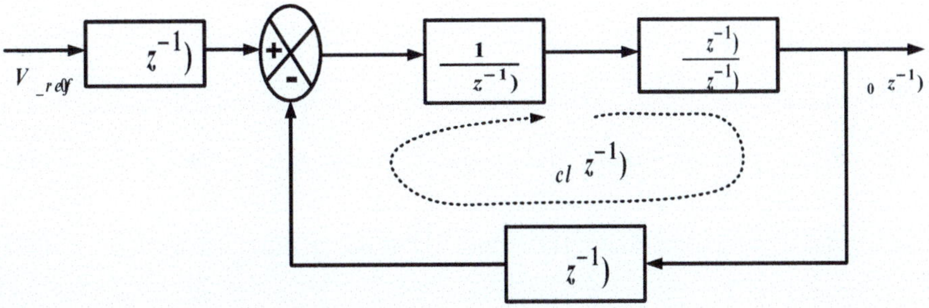

Fig. 6 Generalized closed-loop structure of RST controller

$$B_v(z^{-1}) = b_{0v} + b_{1v}z^{-1} + b_{2v}z^{-2} + b_{3v}z^{-3} + b_{4v}z^{-4} \quad (31)$$

$$A_v(z^{-1}) = a_{0v} + a_{1v}z^{-1} + a_{2v}z^{-2} + a_{3v}z^{-3} + a_{4v}z^{-4} \quad (32)$$

$$G_{Cv}(z^{-1}) = \frac{R_v(z^{-1})}{S_v(z^{-1})} \quad (33)$$

where,

$$R_v(z^{-1}) = k(1 - \alpha_{1v}z^{-1})(1 - \alpha_{2v}z^{-1}) = r_{0v} + r_{1v}z^{-1} + r_{2v}z^{-2} \quad (34)$$

$$S_v(z^{-1}) = (1 - \gamma_{1v}z^{-1})(1 - \gamma_{2v}z^{-1}) = q_{0v} + q_{1v}z^{-1} + q_{2v}z^{-2} \quad (35)$$

Similarly, the current loop gain of I_{g2} equivalent of Two Input DC-DC Converter is defined using Eq. (36). Here $T_{Lc}(z^{-1})$ is defined using Eq. (36) and it is a function of $G_{CC}(z^{-1})$ *and* $G_{id}(z^{-1})$. $G_{id}(z^{-1})$ is the I_{g2} loop equivalent transfer function which is defined using Eq. (37). $G_{CC}(z^{-1})$ is the controller transfer function and is defined using Eq. (40).

$$T_{Lc}(z^{-1}) = G_{CC}(z^{-1})G_{id}(z^{-1}) \quad (36)$$

$$G_{id}(z^{-1}) = G_{ceq}(z^{-1}) = \frac{B_i(z^{-1})}{A_i(z^{-1})} \quad (37)$$

where

$$B_i(z^{-1}) = b_{0i} + b_{1i}z^{-1} + b_{2i}z^{-2} + b_{3i}z^{-3} + b_{4i}z^{-4} \quad (38)$$

$$A_i(z^{-1}) = a_{0i} + a_{1i}z^{-1} + a_{2i}z^{-2} + a_{3i}z^{-3} + a_{4i}z^{-4} \quad (39)$$

$$G_{CC}(z^{-1}) = \frac{R_C(z^{-1})}{S_C(z^{-1})} \quad (40)$$

where,

$$R_C(z^{-1}) = k(1 - \alpha_{1C}z^{-1})(1 - \alpha_{2C}z^{-1}) = r_{0C} + r_{1C}z^{-1} + r_{2C}z^{-2} \quad (41)$$

$$S_C(z^{-1}) = (1 - \gamma_{1C}z^{-1})(1 - \gamma_{2C}z^{-1}) = q_{0C} + q_{1C}z^{-1} + q_{2C}z^{-2} \quad (42)$$

5.2 Generalized Procedure to Extract Controller Transfer Function

The compensator or controller selection is more important, the structure of the compensator depends on the order of converters transfer functions. As both converters considered here are fourth order in nature, a simple one pole and one zero-configuration are not enough to stabilize the output of the converter i.e. not sufficient to provide better gain and phase margins. Hence, there is a need to design a higher-order compensator which is having a minimum of two poles and two zeros. Though several other types of compensators with higher order are available, for the simplicity and design point of view second order compensator is realized. Although this controller design is simple, the selection of the location of poles and zeros is difficult and this selection depends on the desired closed-loop performance of the system. The closed-loop transfer function of the system is $H_{cl}(z^{-1})$

$$H_{cl}(z^{-1}) = \frac{B(z^{-1})R(z^{-1})}{P(z^{-1})} \tag{43}$$

$$P(z^{-1}) = \left[A(z^{-1})S(z^{-1}) + B(z^{-1})R(z^{-1})\right] \tag{44}$$

where,

$$B(z^{-1}) = b_0 + b_1 z^{-1} + b_2 z^{-2} + b_3 z^{-3} + b_4 z^{-4} \tag{45}$$

$$A(z^{-1}) = a_0 + a_1 z^{-1} + a_2 z^{-2} + a_3 z^{-3} + a_4 z^{-4} \tag{46}$$

$$R(z^{-1}) = k(1 - \alpha_1 z^{-1})(1 - \alpha_2 z^{-1}) = r_0 + r_1 z^{-1} + r_2 z^{-2} \tag{47}$$

$$S(z^{-1}) = (1 - \gamma_1 z^{-1})(1 - \gamma_2 z^{-1}) = q_0 + q_1 z^{-1} + q_2 z^{-2} \tag{48}$$

$$P(z^{-1}) = l_0 + l_1 z^{-1} + l_2 z^{-2} + l_3 z^{-3} + l_4 z^{-4} + l_5 z^{-5} + l_6 z^{-6} \tag{49}$$

Here,

$$l_0 = a_0 q_0 + b_0 r_0 \tag{50}$$

$$l_1 = a_0 q_1 + a_1 q_0 + b_0 r_1 + b_1 r_0 \tag{51}$$

$$l_2 = a_0 q_2 + a_1 q_1 + a_2 q_0 + b_0 r_2 + b_1 r_1 + b_2 r_0 \tag{52}$$

$$l_3 = a_1q_2 + a_2q_1 + a_3q_0 + b_1r_2 + b_2r_1 + b_3r_0 \quad (53)$$

$$l_4 = a_2q_2 + a_3q_1 + a_4q_0 + b_2r_2 + b_3r_1 + b_4r_0 \quad (54)$$

$$l_5 = a_3q_2 + a_4q_1 + b_3r_2 + b_4r_1 \quad (55)$$

$$l_6 = a_4q_2 + b_4r_2 \quad (56)$$

The controller polynomials R and S are designed using the pole placement technique. Placing closed-loop system poles at the desired locations is called pole placement which is linked with the desired closed-loop performance. Based on the desired closed-loop pole locations of the system, the closed-loop characteristic polynomial is formulated and it describes the performance objectives/desired specifications of the system. Let $\chi_{dchp}(z^{-1})$ be the desired characteristic polynomial defined using Eq. (57).

$$\begin{aligned}\chi_{dchp}(z^{-1}) = &(1 - \beta_1 z^{-1})(1 - \beta_2 z^{-1})(1 - \beta_3 z^{-1}) \\ &(1 - \beta_4 z^{-1})(1 - \beta_5 z^{-1})(1 - \beta_6 z^{-1})\end{aligned} \quad (57)$$

Simplifying Eq. (57), results in Eq. (58).

$$\chi_{dchp}(z^{-1}) = k_0 + k_1 z^{-1} + k_2 z^{-2} + k_3 z^{-3} + k_4 z^{-4} + k_5 z^{-5} + k_6 z^{-6} \quad (58)$$

where

$$k_0 = 1 \quad (59)$$

$$k_1 = -(\beta_1 + \beta_2 + \beta_3 + \beta_4 + \beta_5 + \beta_6) \quad (60)$$

$$\begin{aligned}k_2 = &\beta_1(\beta_2 + \beta_3 + \beta_4 + \beta_5 + \beta_6) + \beta_2(\beta_3 + \beta_4 + \beta_5 + \beta_6) \\ &+ \beta_3(\beta_4 + \beta_5 + \beta_6) + \beta_4(\beta_5 + \beta_6) + \beta_5\beta_6\end{aligned} \quad (61)$$

$$k_3 = -\left[\begin{array}{c}\beta_1\left(\begin{array}{c}\beta_2\beta_3 + \beta_2\beta_4 + \beta_2\beta_5 + \beta_2\beta_6 + \beta_3\beta_4 + \beta_3\beta_5 \\ +\beta_3\beta_6 + \beta_4\beta_5 + \beta_4\beta_6 + \beta_5\beta_6\end{array}\right) + \\ \beta_2(\beta_3\beta_4 + \beta_3\beta_5 + \beta_3\beta_6 + \beta_4\beta_5 + \beta_4\beta_6 + \beta_5\beta_6) \\ +\beta_3(\beta_4\beta_5 + \beta_4\beta_6 + \beta_5\beta_6) + \beta_4\beta_5\beta_6\end{array}\right] \quad (62)$$

$$k_4 = \left[\begin{array}{c}\beta_1\left(\begin{array}{c}\beta_2\beta_3\beta_4 + \beta_2\beta_3\beta_5 + \beta_2\beta_3\beta_6 + \beta_2\beta_4\beta_5 + \beta_2\beta_4\beta_6 + \beta_2\beta_5\beta_6 \\ +\beta_3\beta_4\beta_5 + \beta_3\beta_5\beta_6 + \beta_3\beta_4\beta_6 + \beta_4\beta_5\beta_6\end{array}\right) \\ +\beta_2(\beta_3\beta_4\beta_5 + \beta_3\beta_5\beta_6 + \beta_3\beta_4\beta_6 + \beta_4\beta_5\beta_6) + \beta_3\beta_4\beta_5\beta_6\end{array}\right] \quad (63)$$

$$k_5 = -[\beta_1\beta_2\beta_3\beta_4\beta_5 + \beta_1\beta_2\beta_3\beta_4\beta_6 + \beta_1\beta_2\beta_3\beta_5\beta_6 + \beta_1\beta_2\beta_4\beta_5\beta_6 + \beta_1\beta_3\beta_4\beta_5\beta_6 + \beta_2\beta_3\beta_4\beta_5\beta_6] \quad (64)$$

$$k_6 = \beta_1\beta_2\beta_3\beta_4\beta_5\beta_6 \quad (65)$$

Substitute the polynomials $A(z^{-1})$, $S(z^{-1})$, $B(z^{-1})$ and $R(z^{-1})$ in Eq. (31), and the resulting equation is compared with the desired characteristic polynomial $\chi_{dchp}(z^{-1})$ i.e. $P(z^{-1}) = \chi_{dchp}(z^{-1})$ (Eq. 58). This gives a system of six equations with six unknowns. The system of these simultaneous equations can be represented in matrix form $[M][x] = [\beta]$, can easily be solved using matrix inversion, and its equivalent form is given using Eq. (66).

$$[x] = [M]^{-1}[\beta] \quad (66)$$

where,

$$[x] = \begin{bmatrix} q_0 & q_1 & q_2 & r_0 & r_1 & r_2 \end{bmatrix}^T \quad (67)$$

$$[M] = \begin{bmatrix} a_1 & a_0 & 0 & b_1 & b_0 & 0 \\ a_2 & a_1 & a_0 & b_2 & b_1 & b_0 \\ a_3 & a_2 & a_1 & b_3 & b_2 & b_1 \\ a_4 & a_3 & a_2 & b_4 & b_3 & b_2 \\ 0 & a_4 & a_3 & 0 & b_4 & b_3 \\ 0 & 0 & a_4 & 0 & 0 & b_4 \end{bmatrix} \quad (68)$$

$$[\beta] = \begin{bmatrix} k_1 & k_2 & k_3 & k_4 & k_5 & k_6 \end{bmatrix} \quad (69)$$

From Eq. (66), it is possible to compute the parameters $(q_0\ q_1\ q_2\ r_0\ r_1\ r_2)$, and the two polynomials of the controller $G_C(z^{-1})$ are functions of these parameters. Therefore, the controller coefficients are determined. The procedure given in the above is used to determine the controller transfer functions $G_{Cv}(z^{-1})$ and $G_{CC}(z^{-1})$ using MATLAB programming. These controllers are used to regulate V_0 and I_{g2} variables of the TII converter. The following section explains about regulation aspects of the converter in the MATLAB environment.

6 Results and Discussion

The effectiveness of this controller in regulating the dynamics of Two Input Integrated converter systems have been demonstrated through simulation studies. To verify the closed-loop steady-state performance with RST based robust PID controller, the TI-BS and TI-MBS converters have been considered. Using the previously mentioned procedural steps, a robust PID controller based on RST architecture is designed, considering that all of the closed loop converter system's poles are located inside the unit circle. The converter parameters are presented for both TI-BS and TI-MBS converters in Table 3.

The RST-based robust PID controllers are designed, and the controller coefficients are obtained by using Eq. (66). Finally, the controller transfer functions are obtained as follows:

For TI-BS converter:

$$G_{Cv}\left(z^{-1}\right) = \frac{0.2776 - 0.510z^{-1} + 0.2443z^{-2}}{1 - 1.8214z^{-1} + 0.7844z^{-2}} \tag{70}$$

$$G_{CC}\left(z^{-1}\right) = \frac{-3.2356 + 6.4194z^{-1} - 3.2223z^{-2}}{1.0004 + 10.325z^{-1} - 9.5291z^{-2}} \tag{71}$$

For TI-MBS converter:

$$G_{Cv}\left(z^{-1}\right) = \frac{-0.0199 + 0.0456z^{-1} - 0.0222z^{-2}}{0.9998 - 1.6281z^{-1} + 0.8283z^{-2}} \tag{72}$$

$$G_{CC}\left(z^{-1}\right) = \frac{0.3053 - 0.6472z^{-1} + 0.3628z^{-2}}{0.9999 - 2.5423z^{-1} + 1.8470z^{-2}} \tag{73}$$

Table 3 TI-BS and TI-MBS converter parameters

Description	Value
Input voltages (V_{g1}, V_{g2})	60, 36 V
Output voltage (V_o)	48 V
Switched frequency (f_s)	50 kHz
Inductances (L_1, L_2)	300, 300 μH
Capacitances (C_1, C_2)	50, 60 μF
Load resistance	3.2 Ω
Desired pole location for both current control loop and voltage control loop	(i) 0.953 + 0.2528i (ii) 0.953 − 0.2528i (iii) 0.932 + 0.0383i (iv) 0.932 − 0.0383i (v) 0.825 + 0.3266i (vi) 0.825 − 0.3266i

The converter's dynamic performance against external disturbances has been demonstrated through simulation studies by constructing a simulation diagram in a MATLAB environment. The steady-state and dynamic simulations are performed for TI-BS and TI-MBS converters under different situations, these are (i) with nominal converter parameters, (ii) variation in load current, (iii) variations in source voltage.

6.1 *The Closed-Loop TI-BS Converter with RST-Based PID Controller*

Equations (70) and (71) are taken into consideration to design voltage loops and current loops, and it is implemented to evaluate the regulation features of the TI-BS converter. The controller performance is analyzed through the following simulations.

6.1.1 With Nominal Converter Parameters

The output voltage as well as source voltages of this closed-loop TI-BS converter system with RST-based robust PID controller are shown in Fig. 7. Figure 8 shows the load voltage and load current. The duty ratio control signals are shown in Fig. 9. Figure 10 shows the reference voltage and load voltage, and Fig. 11 shows the source-2 current (i_{g2}) and reference current. It is observed from these figures that the output voltage (V_0) and source-2 current (i_{g2}) are following the reference values.

The regulation characteristics of the integrated converter with the designed PID control strategy are also verified under the disturbance conditions.

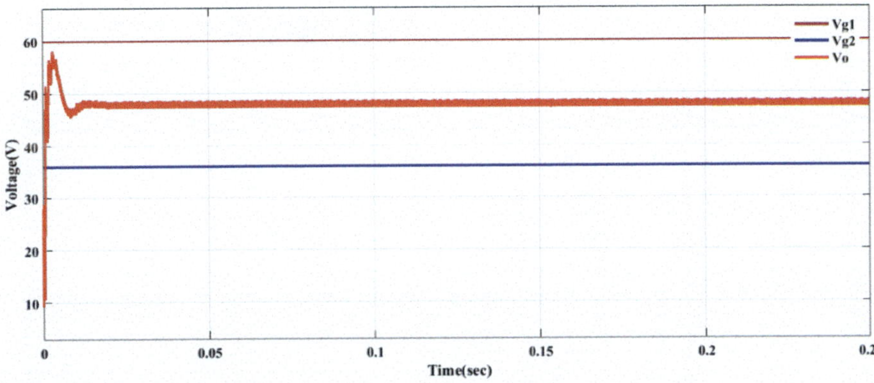

Fig. 7 Input and output voltages waveforms of TI-BS converter

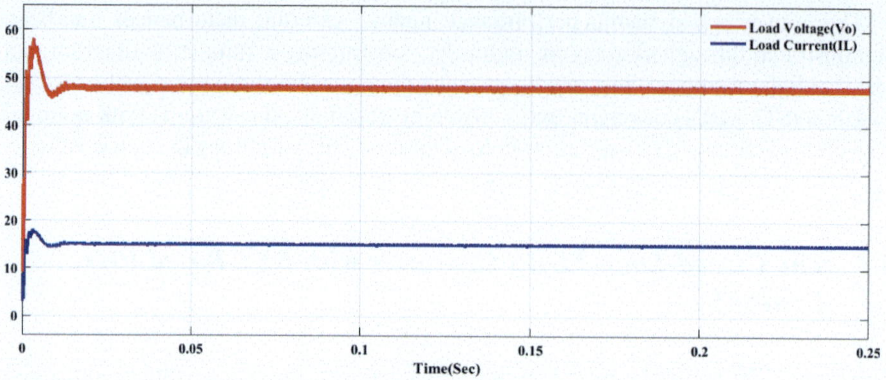

Fig. 8 Load voltage and load current waveforms at load resistance R = 3.2 Ω

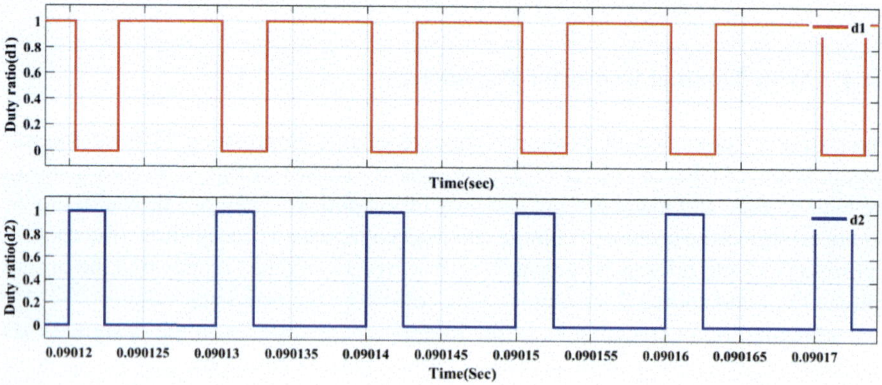

Fig. 9 Duty ratios d_1 and d_2

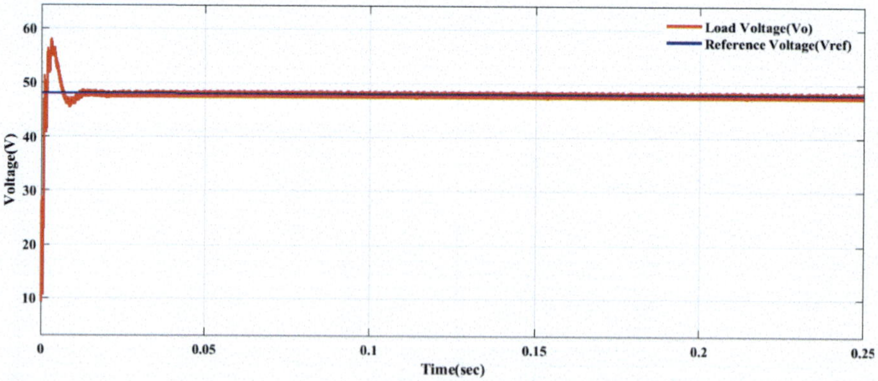

Fig. 10 Load voltage and reference voltage waveforms at load resistance R = 3.2 Ω

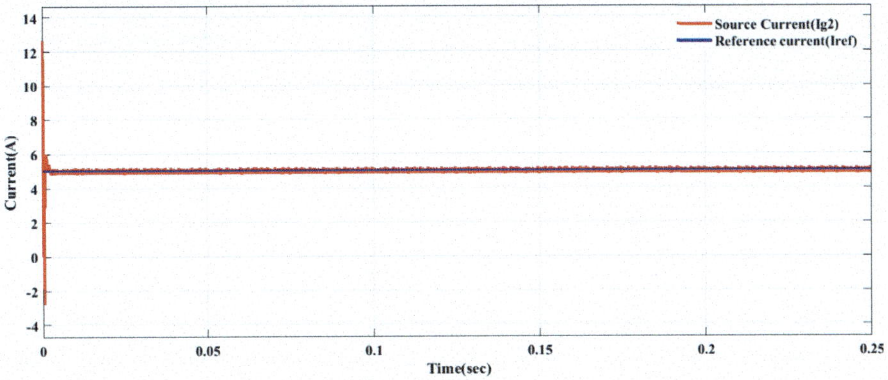

Fig. 11 Source current and reference current waveforms at load resistance R = 3.2 Ω

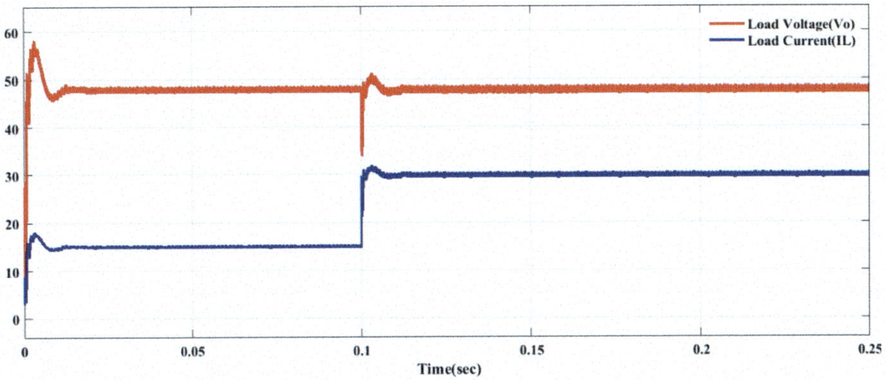

Fig. 12 Load voltage and load current waveforms at load resistance R = 1.6 Ω

6.1.2 Variation in Load Current

The load resistance is varied from 3.2 to 1.6 Ω at t = 0.1 s i.e. the load current is doubled. The regulation characteristics are shown under this condition. Figure 12 shows the load voltage and load current, Fig. 13 shows the reference voltage and load voltage, and Fig. 14 shows the source-2 current (i_{g2}) and reference current. It is observed from these figures that the output voltage (V_0) and source-2 current (i_{g2}) are following the reference values against the disturbances.

6.1.3 Variations in Source Voltage

The source-1 voltage is varied from 60 to 55 V at the time instant of 0.1 s and the source-2 voltage is raised to 40 V from 36 V at the time instant of 0.2 s, the input

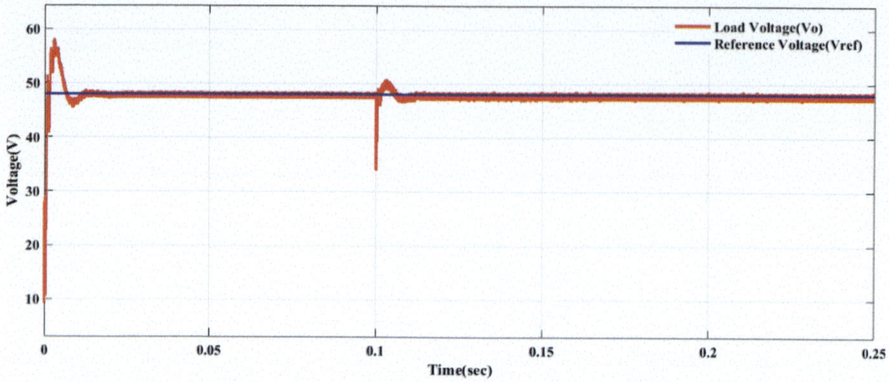

Fig. 13 Load voltage and reference voltage waveforms at load resistance R = 1.6 Ω

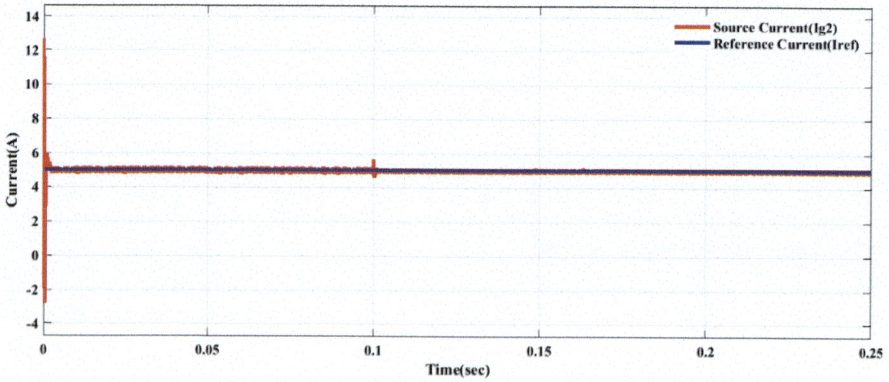

Fig. 14 Source current and reference current waveforms at load resistance R = 1.6 Ω

voltages along with the output voltage are shown in Fig. 15. Figure 16 shows the load voltage and load current, Fig. 17 shows the reference voltage and load voltage and Fig. 18 shows the source-2 current (i_{g2}) and reference current. It is observed from these figures that the output voltage (V_0) and source-2 current (i_{g2}) are following the reference values against this disturbance condition.

6.2 *Closed-Loop TI-MBS Converter with RST-Based PID Controller*

Equations (72) and (73) are taken into consideration to design voltage loops and current loops, and it is implemented to evaluate the regulation features of the TI-MBS converter. The controller performance is analyzed through the following simulations.

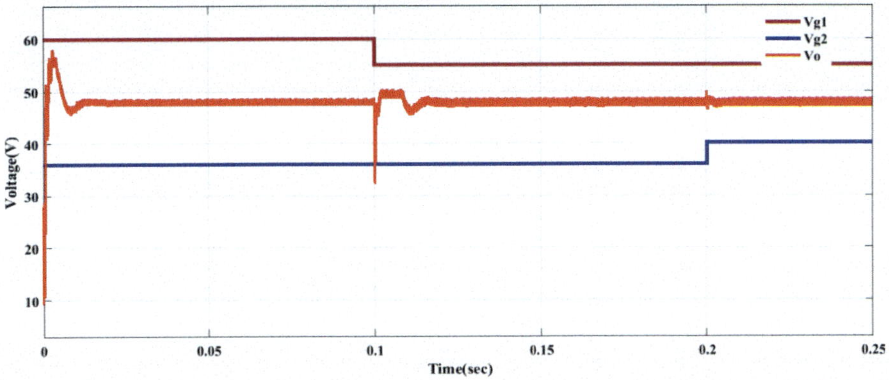

Fig. 15 Input and output voltage waveforms with variations in source voltages

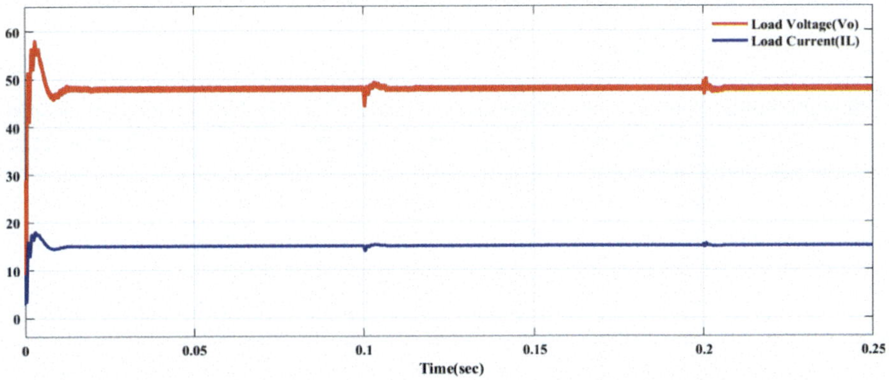

Fig. 16 Load voltage and load current waveforms with variations in source voltages

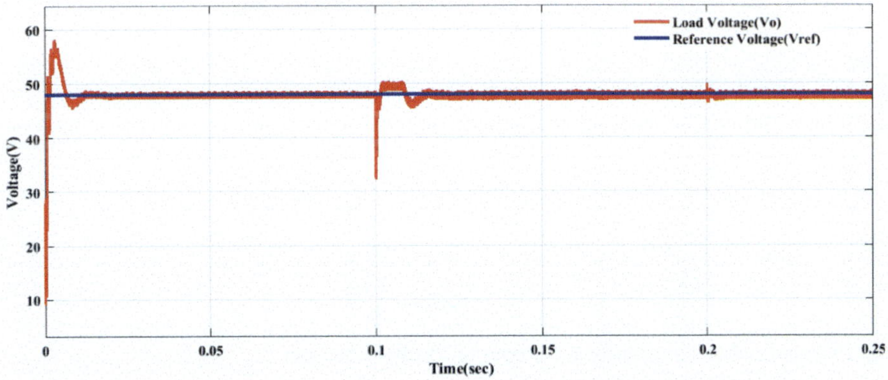

Fig. 17 Load voltage and reference voltage waveforms with variations in source voltages

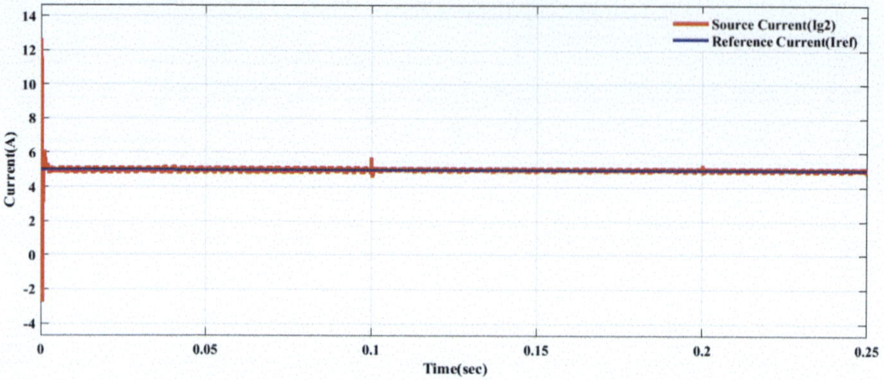

Fig. 18 Source current and reference current waveforms with variations in source voltages

6.2.1 With Nominal Converter Parameters

The output voltage and source voltages of this closed-loop TI-MBS converter system with an RST-based robust PID controller are shown in Fig. 19. Figure 20 shows the load voltage and load current. The duty ratio control signals are shown in Fig. 21. Figure 22 shows the reference voltage and load voltage and Fig. 23 shows the source-2 current (i_{g2}) and reference current. It is observed from these figures that the output voltage (V_0) and source-2 current (i_{g2}) are following the reference values.

The regulation characteristics of the TI-MBS integrated converter with the designed PID control strategy are also verified under the disturbance conditions.

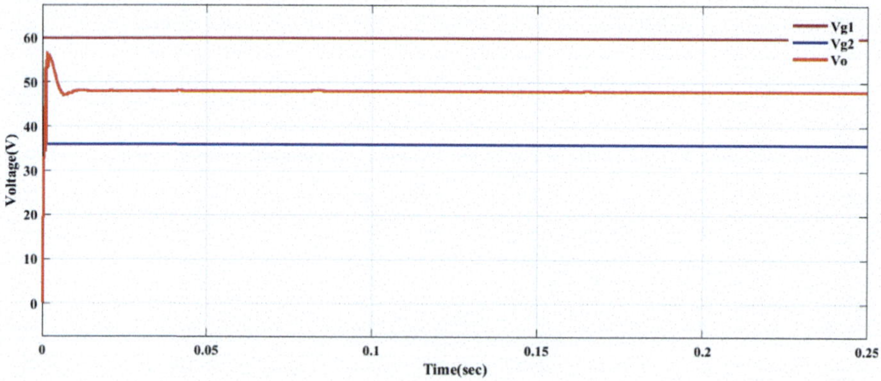

Fig. 19 Input and output voltages waveforms of TI-MBS converter

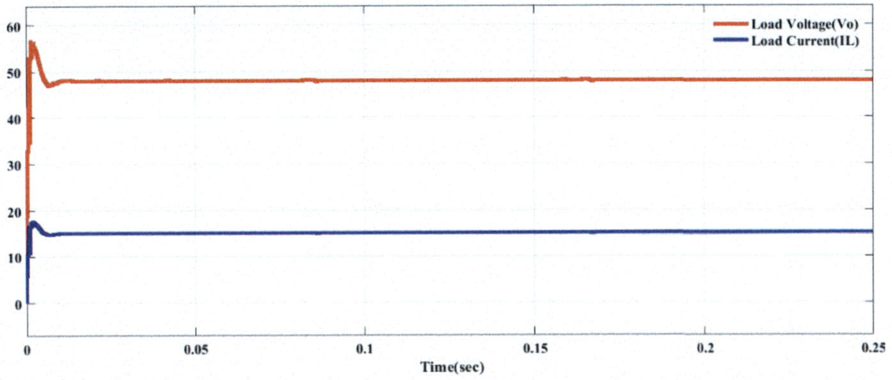

Fig. 20 Load voltage and load current waveforms at load resistance R = 3.2 Ω

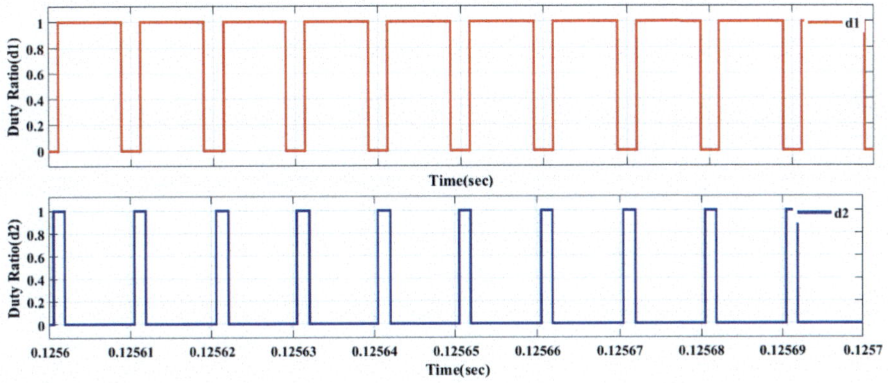

Fig. 21 Duty ratios d_1 and d_2

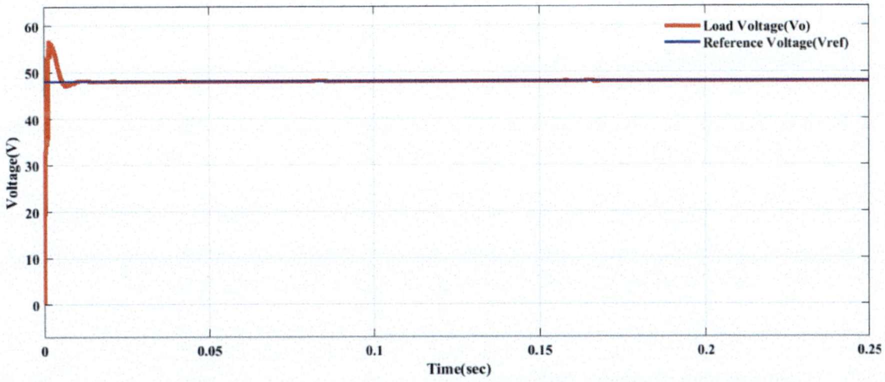

Fig. 22 Load voltage and reference voltage waveforms at load resistance R = 3.2 Ω

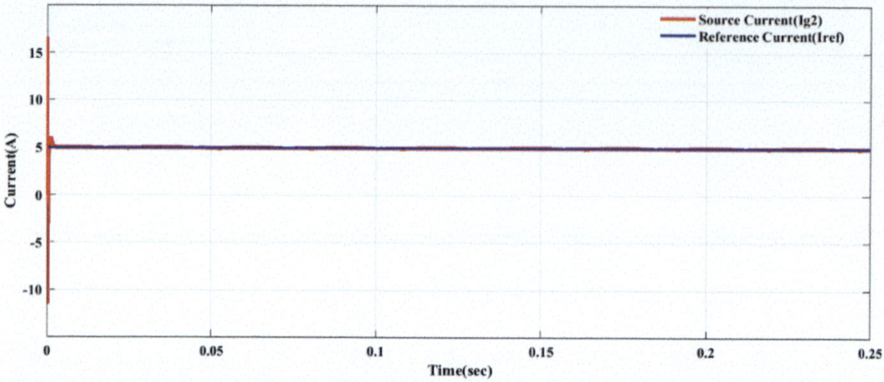

Fig. 23 Source current and reference current waveforms at load resistance R = 3.2 Ω

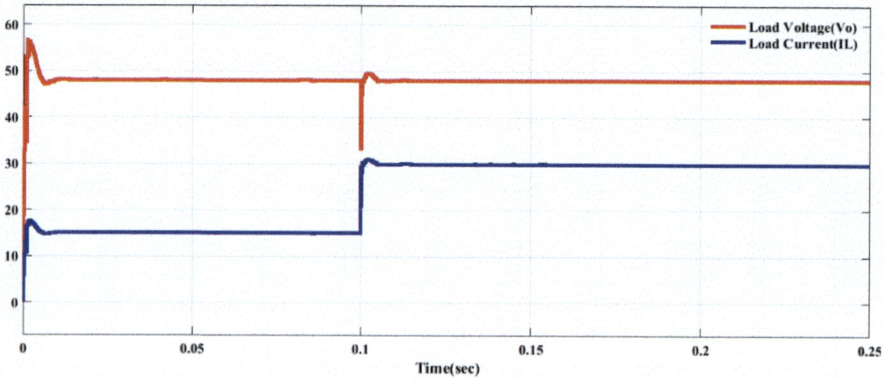

Fig. 24 Load voltage and load current waveforms at load resistance R = 1.6 Ω

6.2.2 Variation in Load Current

The load resistance is varied from 3.2 to 1.6 Ω at t = 0.1 s i.e. the load current is doubled. The regulation characteristics are shown under this condition. Figure 24 shows the load voltage and load current, Fig. 25 shows the reference voltage and regulated load voltage and Fig. 26 shows the source-2 current (i_{g2}) and reference current. It is observed from these figures that the output voltage (V_0) and source-2 current (i_{g2}) are following the reference values against the disturbances.

6.2.3 Variations in Source Voltage

The source-1 voltage is varied from 60 to 55 V at the time instant of 0.1 s and the source-2 voltage is raised to 40 V from 36 V at the time instant of 0.2 s, the input

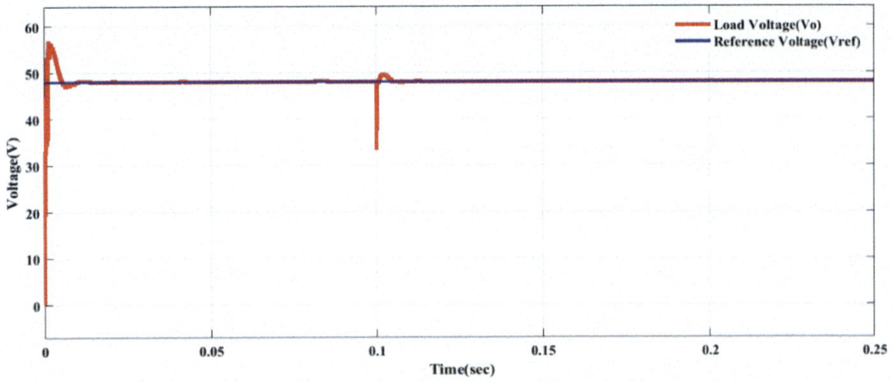

Fig. 25 Load voltage and reference voltage waveforms at load resistance R = 1.6 Ω

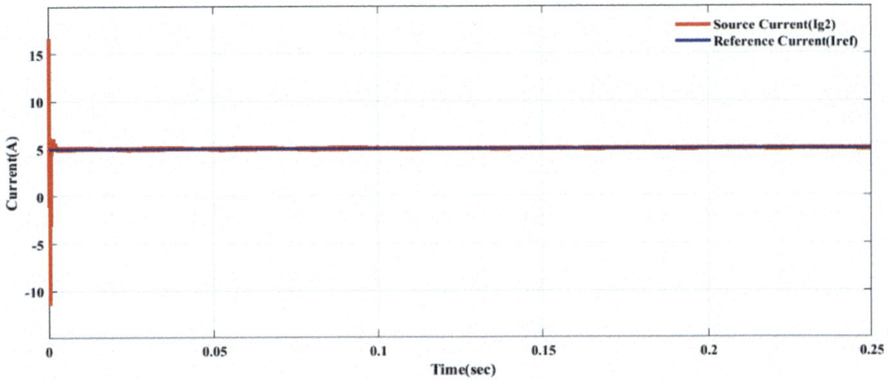

Fig. 26 Source current and reference current waveforms at load resistance R = 1.6 Ω

voltages along with the output voltage are shown in Fig. 27. Figure 28 shows the load voltage and load current, Fig. 29 shows the reference voltage and regulated load voltage and Fig. 30 shows the source-2 current (i_{g2}) and reference current. It is observed from these figures that the output voltage (V_0) and source-2 current (i_{g2}) are following the reference values against these disturbance conditions.

7 Conclusions

This chapter presents a robust PID controller design with RST architecture using the pole placement technique. The detailed procedural steps to design a robust controller using this technique are presented. Here, the mathematical modeling aspects of TI-BS

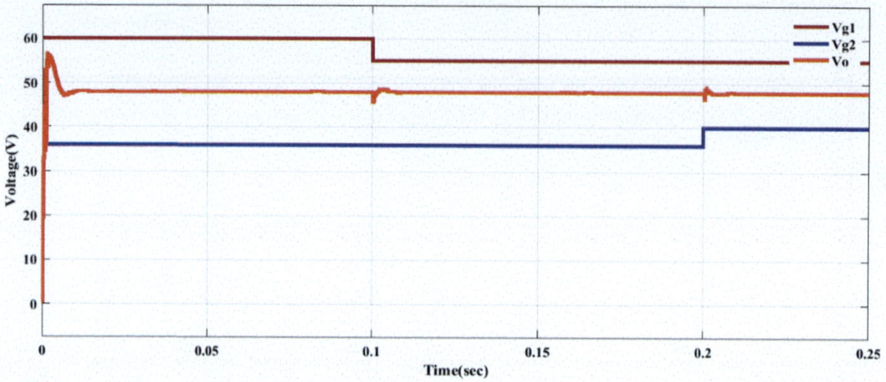

Fig. 27 Input and output voltage waveforms with variations in source voltages

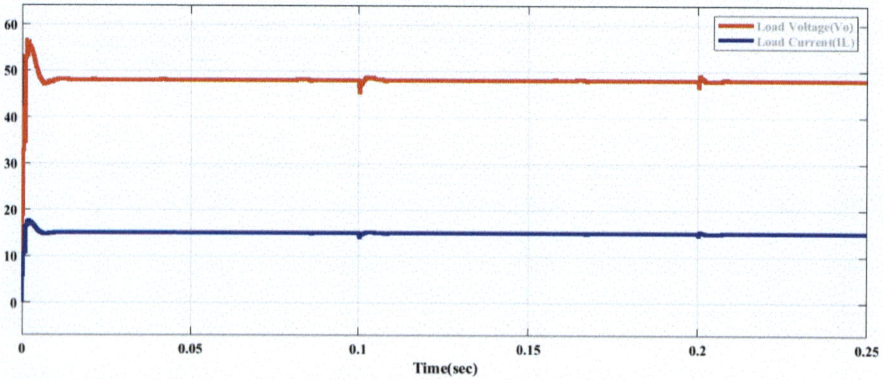

Fig. 28 Load voltage and load current waveforms with variations in source voltages

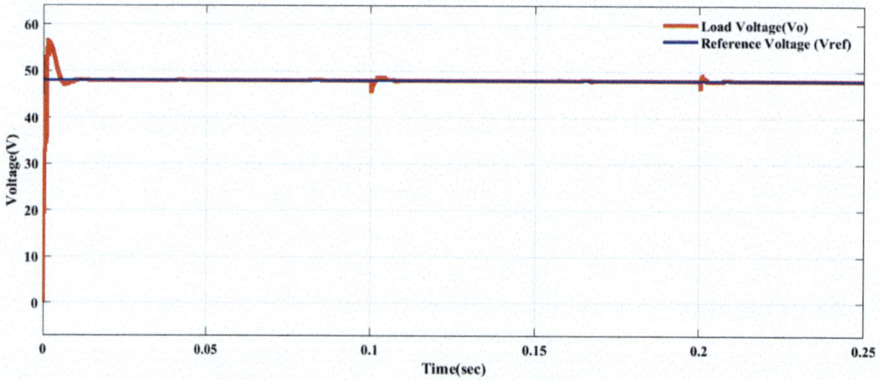

Fig. 29 Load voltage and reference voltage waveform with variations in source voltages

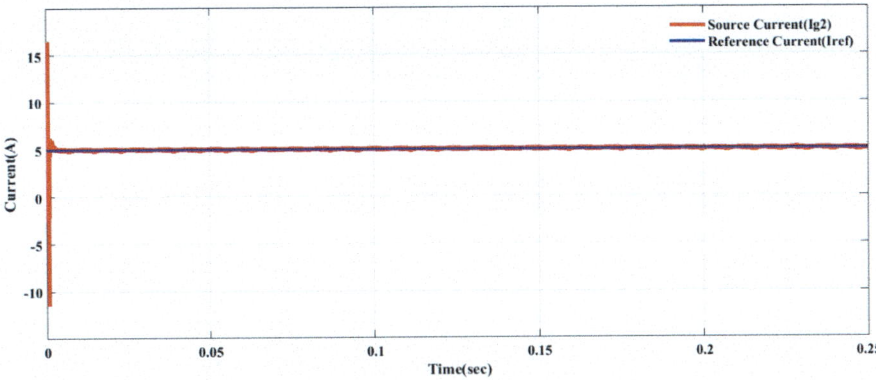

Fig. 30 Source current and reference current waveforms with variations in source voltages

and TI-MBS converter systems are described. A sixth-order closed-loop characteristic equation is formulated for Voltage and current loops. The desired pole location for both the current control loop and voltage control loop are assumed to be the same. The closed-loop poles considered for these loops are (i) 0.953 + 0.2528i, (ii) 0.953 − 0.2528i, (iii) 0.932 + 0.0383i, (iv) 0.932 − 0.0383i, (v) 0.825 + 0.3266i, (vi) 0.825 − 0.3266i. Using these sixth-order closed-loop characteristic equations is formulated and controllers are designed based on this. To verify the effectiveness of this designed controller, 48 V, 720 W, TII DC-DC converters are considered for analysis and simulation. The steady-state and dynamic simulations are performed for TI-BS and TI-MBS converters under different situations, these are (i) with nominal converter parameters, (ii) variation in load current, and (iii) variations in source voltage. It is observed from these simulations that the output voltage (V_0) and source-2 current (i_{g2}) are following the reference values against these disturbance conditions.

References

1. Liu YC, Chen YM (2009) A systematic approach to synthesizing multi-input DC–DC converters. IEEE Trans Power Electron 24:116–127. https://doi.org/10.1109/TPEL.2008.2009170
2. Lijin KL, Sheik Mohammed S, Muhammed Shanir PP, Haw LK (2022) A non-isolated two input single output (TISO) DC-DC converter for energy storage applications. In: International conference on green energy, computing and sustainable technology (GECOST), pp 238–242. https://doi.org/10.1109/GECOST55694.2022.10010592
3. Manogna M, Amarendra Reddy B, Padma K (2023) Modeling of a three-input fourth-order integrated DC-DC converter. In: International conference on smart and sustainable technologies in energy and power sectors (SSTEPS), pp 83–88. https://doi.org/10.1109/SSTEPS57475.2022.00032
4. Bhattacharyya P, Ghorai S, Sen S, Giri SK (2023) A flexible non-isolated multiport converter to integrate battery and ultracapacitor for electric vehicle applications. IEEE Trans Circuits Syst II:0044–1048. https://doi.org/10.1109/TCSII.2022.3217844

5. Dhananjaya M, Jagabar Sathik M, Padmanaban S, Almakhles D, Potnuru D (2021) A new configuration of switch and source fault-tolerant dual-input single-output DC-DC converter. In: IEEE 4th international conference on computing, power and communication technologies (GUCON), pp 1–6. https://doi.org/10.1109/GUCON50781.2021.9573998
6. Li T, Parsa L (2016) Design and control of a fault tolerant soft switching DC-DC converter for high power high voltage applications. In: IEEE applied power electronics conference and exposition (APEC), pp 2573–2578. https://doi.org/10.1109/APEC.2016.7468227
7. Teng B, Ma J, Zhu M, Yang Y, Cai X (2023) Fault-tolerant DC-DC converter interconnected with energy storage device for bipolar distribution system: working modes and optimizing control strategy. IET Gener Trans Distrib 4258–4276. https://doi.org/10.1049/gtd2.12946
8. Veerachary M (2012) Two-loop controlled buck-SEPIC converter for input source power management. IEEE Trans Ind Electron 59:4075–4087. https://doi.org/10.1109/TIE.2011.2174530
9. Reddy BA, Veerachary M (2013) Gramian based digital controller design for two-input integrated DC-DC converter. In: International conference on control communication and computing (ICCC), pp 335–340. https://doi.org/10.1109/ICCC.2013.6731675
10. Thota P, Bhimavarapu AR, Chintapalli VVSBR (2022) Participation factor based analysis of PVSC type multi-input zeta-SEPIC DC-DC converter. COMPEL Int J Comput Math Electr Electron Eng 41:1727–1752. https://doi.org/10.1108/COMPEL-10-2021-0363
11. Allam V, Bhimavarapu AR, Jyothsna TR, Salkuti SR (2024) Impact of discrete-time modeling on dual input modified SEPIC converter. In: Salkuti SR (ed) Energy and environmental aspects of emerging technologies for smart grid. Green energy and technology. Springer, Cham. https://doi.org/10.1007/978-3-031-18389-8_16
12. Venkatesh A, Amarendra Reddy B, Jyothsna TR (2023) Impact of discrete-time modeling on dual input buck-SEPIC converter. WSEAS Trans Circuits Syst 22:274–288. https://doi.org/10.37394/23201.2023.22.28
13. Mummadi V, Bhimavarapu AR (2020) Robust multi-variable controller design for two-input two-output fourth-order DC-DC converter. Electr Power Compon Syst 48:86–104. https://doi.org/10.1080/15325008.2020.1736213
14. Mohan MM, Reddy BA, Veerachary M (2013) Centralized digital controller for two-input integrated DC-DC converter. In: IEEE 10th international conference on power electronics and drive systems (PEDS), Kitakyushu, Japan. https://doi.org/10.1109/PEDS.2013.6527022
15. Veerachary M, Reddy BA (2020) Decentralized controllers design for fourth-order two-input converter: stability considerations. In: IEEE international conference on computing, power and communication technologies (GUCON), pp 362–367. https://doi.org/10.1109/GUCON48875.2020.9231131
16. Nisar KM, Sivaprasad A, Kumaravel S, Ananthakrishnan P (2016) Implementation of closed loop controller in a dual input DC-DC converter for DC-microgrid application. In: IEEE annual India conference (INDICON), pp 1–6. https://doi.org/10.1109/INDICON.2016.7839111
17. Ibrahim O, Yahaya Z, Saad N (2016) PID controller response to set-point change in DC-DC converter control. Int J Power Electron Drive Syst 294–302. https://doi.org/10.11591/ijpeds.v7.i2.pp294-302
18. Solano Martinez J, Hissel D, Péra M-C (2012) Type-2 fuzzy logic control of a DC/DC buck converter. IFAC Proc 45:103–108. https://doi.org/10.3182/20120902-4-FR-2032.00020
19. Appikonda M, Kaliaperumal D (2019) Modelling and control of dual input boost converter with voltage multiplier cell. IET Circuits Devices Syst 13(8):1267–1276. https://doi.org/10.1049/iet-cds.2019.0123
20. Assem K, Selva Kumar S (2020) Closed loop control of DC-DC converters using PID and FOPID controllers. Int J Power Electron Drive Syst (IJPEDS) 11:1323–1332. https://doi.org/10.11591/ijpeds.v11.i3.pp1323-1332
21. Haj YE, Sheir A, Milman R, Sood VK (2021) Design of a nonlinear controller for a two-stage DC-AC converter for DC-link drive applications. In: IEEE electrical power and energy conference (EPEC), pp 451–458. https://doi.org/10.1109/EPEC52095.2021.9621344

22. Hebala OM, Aboushady AA, Ahmed KH, Abdelsalam I, Burgess SJ (2021) A new active power controller in dual active bridge DC-DC converter with a minimum-current-point-tracking technique. IEEE J Emerg Sel Top Power Electron 9:1328–1338. https://doi.org/10.1109/JESTPE.2020.3016771
23. Nayak C, Venkatesh A, Amarendra Reddy B, Jyothsna TR (2024) Robust controller for limit cycle elimination of two input DC-DC converter with constant power load in Dc μgrid applications. WSEAS Trans Syst 23:196–205. https://doi.org/10.37394/23202.2024.23.22
24. Manogna M, Amarendra Reddy B, Padma K (2024) Impact of weight functions on performance of three-input integrated DC-DC converter in H∞ control. WSEAS Trans Syst Control 19:62–72. https://doi.org/10.37394/23203.2024.19.7
25. Bhukya CN, Reddy BA (2023) Constant power loads in DC microgrids: a review of modern nonlinear control approaches and stabilization techniques. In: IEEE 2nd international conference on industrial electronics: developments & applications (ICIDeA), pp 181–186. https://doi.org/10.1109/ICIDeA59866.2023.10295221
26. Reddy BA, Thelukuntla CS (2022) H∞ loop-shaping of a multi-input DC-DC converter using weight function parameter optimization. Eng Res Express. https://doi.org/10.1088/2631-8695/ac8418
27. Gangula SD, Nizami TK, Udumula RR, Chakravarty A, Singh P (2023) Adaptive neural network control of DC–DC power converter. Expert Syst Appl 29. https://doi.org/10.1016/j.eswa.2023.120362
28. Hussein AI, Shigdar B, Almatrafi L, Alaidroos B, Alsharif F, Aly RHM (2023) Design of a DC/DC converter with a PID controller and backpropagation neural network for electric vehicles. In: 20th learning and technology conference (L&T), Jeddah, Saudi Arabia, pp 128–133. https://doi.org/10.1109/LT58159.2023.10092291
29. Dong W, Li S, Fu X, Li Z, Fairbank M, Gao Y (2021) Control of a buck DC/DC converter using approximate dynamic programming and artificial neural networks. IEEE Trans Circuits Syst I 68:1760–1768. https://doi.org/10.1109/TCSI.2021.3053468
30. Shoja-Majidabad S, Hajizadeh A (2020) Decentralized adaptive neural network control of cascaded DC–DC converters with high voltage conversion ratio. Appl Soft Comput 86. https://doi.org/10.1016/j.asoc.2019.105878

Deep Learning Approaches to V2G Strategy and Its Effects on Microgrid Dynamics

S. Usha, A. Geetha, and Surender Reddy Salkuti

Abstract Electric vehicles (EVs) are finding a hopeful home in contemporary grids thanks to Vehicle-to-Grid (V2G) technology. This work explains the scheme, functionalities, and possible advantages of V2G while also illustrating how it might be used on a microgrid. Explain how electric cars can act as a turning replacement by holding additional grid power stored in their batteries and reintroducing it to the system when needed. This aids in controlling the frequency of the grid within its nominal specifications. EVs are essential for achieving amazing features; however, integrating a large number of EVs also causes grid congestion. Uncontrolled charging could therefore result in under voltage and intricate grid congestion. To lessen the impacts, the causes of EVs charging uncontrollably have recently been examined. Because of various limitations on the part of the customer, it is very difficult to establish controlled charging; as a result, it is preferable to utilize power prediction systems for EV charging and discharging. This study's main goal is to use a deep learning network to create an efficient and effective system for EV charging and discharging. Within this framework, a neural network-based model with a feed-forward algorithm is suggested to examine the effects of bi-directional charging within the microgrid setting.

Keywords Vehicle-to-grid · Deep learning · Neural network · Electric vehicles · Battery charging · Discharging

---

S. Usha · A. Geetha
Department of Electrical and Electronics Engineering, College of Engineering and Technology, SRM Institute of Science and Technology, Kattankulathur, Chengalpattu 603203, Tamil Nadu, India
e-mail: ushas@srmist.edu.in

A. Geetha
e-mail: geethaa2@srmist.edu.in

S. R. Salkuti (✉)
Department of Railroad and Electrical Engineering, Woosong University, Daejeon, Republic of Korea
e-mail: surender@wsu.ac.kr

S. R. Salkuti (ed.), *Artificial Intelligence for Integrated Smart Energy Systems in Electric Vehicles*, Lecture Notes in Electrical Engineering 1427,
https://doi.org/10.1007/978-3-031-94276-1_22

Nomenclature

DFNN	Deep Feed Forward Neural Network
ES	Energy Storage
EV	Electric Vehicle
NN	Neural Network
RMSE	Root Mean Square Error
SOC	State of Charge
V2G	Vehicle to Grid
V2H	Vehicle to Home
V2V	Vehicle to Vehicle

1 Introduction

The phrase "voice-to-grid" (V2G) denotes the connection between vehicles and the grid that allows the power company or aggregators to regulate and manage electric vehicles (EVs) that are linked to the grid (loads or sources). Beyond V2G, there are two other innovative grid-connected EV tools: Vehicle to Home (V2H) and Vehicle to Vehicle (V2V). The V2H architecture encompasses power transfer between an electric vehicle (EV) battery and a residential electrical network [1]. In this scenario, EV batteries can act as energy storage devices, supplying renewable energy sources and home electric appliances with backup power in the event of an unexpected blackout [2].

Car to Vehicle, on the other hand, functions as a kind of area electric car community where members can exchange power. Because they both use and sell electricity to the grid, EV owners are therefore considered prosumers rather than just consumers. By satisfying demand during their low or no-generation phases, V2G enables efficient coordination with intermittent power-generating renewable energy sources, like wind and solar energy [3]. Reactive power support, peak shaving, load balancing via valley filling (i.e., charging the car at night when demand is low), active power management, and current harmonic filtering can all be aided by a vehicle that can operate in V2G mode. Apart from primary, secondary, and tertiary frequency control, these systems can offer other services including spinning reserves. In terms of dependability, stability, efficiency, and generation dispatch, they can also improve the technical performance of the grid [4, 5].

They may even generate revenue in addition to reducing operating costs related to electricity. A schematic of the V2G network. A unidirectional and bidirectional V2G are the two general categories into which V2G types can be divided. Power only moves from the grid to the vehicle when using the unidirectional V2G technique [6]. To avoid overload, unstable systems, and voltage dips on the grid, the rate at which batteries charge must be regulated. Without a functioning energy trading mechanism between EV owners and the power provider, unidirectional V2G cannot be implemented. To encourage wider use of EVs, this policy should guarantee that owners of electric vehicles get payment for charging their vehicles during off-peak hours and limit charging during periods of high demand. By doing this, overloading during peak hours may be prevented. Moreover, optimization strategies can be used by unidirectional V2G to lower emissions and increase revenue [7, 8].

Several advantages are made possible by the bidirectional V2G technique, which allows power to move both ways after the grid to the EV and vice versa. A bidirectional EV battery charger will usually feature both a DC/DC and an AC/DC converter. The DC/DC converter performs the duties of a boost or buck converter when the device is charging or discharging. More flexibility and chances to improve the power system's performance are provided by the bidirectional V2G [9]. Active power support for peak load reduction and load leveling can be obtained using bidirectional V2G [10]. These services are only available when fully charged or nearly fully charged EVs supply the grid during peak demand hours and only refuel during off-peak hours, which are times of low demand. Bidirectional V2G not only allows reactive power to be supplied for grid voltage management, but it also makes active power possible. Control switching and appropriate charger DC link capacitor sizing will enable this reactive power assistance.

Power factor management, which reduces power losses in the power grid, is one of the main characteristics of bidirectional V2G technology [11, 12]. It is expected that local electric networks and grids will be impacted by the quick adoption of EVs. Even at low penetration rates, EV charging has been shown to be feasible, but it can also result in under voltage and grid/electric network congestion. Consequently, charging an EV speeds up the deterioration of distribution transformers and raises losses in the electric grid and network. Transmission and distribution firms are required to modify their current electric networks in order to accommodate the distinct and constantly growing need, as opposed to the previous "Fit-and-Forget" method. For transmission and distribution businesses, this would mean significant investments [13, 14].

This is achieved by proposing a method for predicting the electricity used for EV charging and discharging. In addition to estimating the amount of storage needed for EVs to be able to contribute to the delivery of energy services, the main objective of this research is to emulate, compute, and evaluate the grid control of EVs. To prepare for future forecasts, it is necessary to first examine the charging methods and EV charging-discharging trends. Two separate methods for determining load profiles and evaluating EV usage are also examined. Significant factors that affect the load profile for EV charging are also considered. Peak energy consumption, peak load periods, and the behavior of large-scale EVs are all predicted using stochastic approaches [15, 16].

As a result, numerous academics have proposed a diversity of systems to decrease the adverse consequences of EV charging. When batteries are carefully charged, under voltage and overly congested peak load hours can be avoided. If there is widespread use of EV chargers with V2G charging capacity, the batteries could be utilized as short-term ES. This temporary energy storage could help distribution operators reduce grid congestion caused by conventional loads. Turn on regulated charging to prolong the life of the battery [17, 18]. However, because controlled

EV charging involves several users, each with distinct aims, this can be difficult to achieve. Getting cheaper energy to charge electric cars was the main objective of research on EV fleet aggregators [19]. By maximizing the period under the cost signal and the EV charging power, the year with the lowest charging prices also had a worldwide flattened load profile. One advantage of optimal charging schedules and power levels is the reduction of losses associated with EV charging on the electrical grid [20].

Due to the swift growth of the fleet of private EVs and their designation as vehicle-to-grid (V2G) systems, the power sectors are no longer able to manage unpredictable electricity consumption. It appears that charging habits are more reliant on the EV's operation. The charging behavior of electric vehicles under varying load profiles has been observed and quantified in a number of studies. The options available for charging the car determine the majority of these load profiles [21]. Unpredictable charging occurs, therefore the EV is charged right away. Is occurrence-only unidirectional charging deemed viable once parked, provided that charging is feasible throughout this period? Grid-to-vehicle (G2V) power transfer enables this. Because alternative charging processes have advanced, some outermost components may be able to handle the abandoned one-way charging. Grid-to-vehicle (G2V) charging, which takes into consideration the potential for power transfer from the car to the grid, is an alternative to bidirectional charging [22].

A robust control technique is required for continuous frequency control of a standalone microgrid (SMG) due to uncertainties regarding the power supply from renewable energy sources and low inertia [23]. Thus, for the load frequency control (LFC) study of an SMG, this work offers a unique hybrid fuzzy proportional derivative–tilt integral derivative (FPD-TID) controller [24]. Combining the benefits of both the FPD and the TID controllers serves as inspiration for the suggested controller [25]. Gains from the suggested controller are optimized via a reliable chaotic cluster search algorithm (CCSA) [26]. Comparative frequency deviation responses from the SMG were given consideration when evaluating the suggested control system in light of various disturbances [27, 28]. Moreover, the suggested controller is tested for sensitivity and robustness, with a variation of approximately 30% in SMG parameters and the disconnection of different SMG subsystems, respectively [29]. Under these conditions, the SMG's operational stability is highly desirable; hence the suggested control strategy seeks to strike a balance between performance and operational stability. Eigenvalue and root locus analysis are used to establish the operational stability of the SMG [30].

Section 2 describe about the key factors influencing charging load patterns for EVs and their performance. The SOC estimation and its performance analysis are explained in Sect. 3. The simulation results are described in Sect. 4. The proposed work is concluded in Sect. 5.

2 Key Factors Influencing Charging Load Patterns for EVs

Three elements are considered while analyzing the behavior of EVs in a V2G environment, in addition to charging strategies: the location, the necessity, and the effectiveness of the charging. To measure EV load patterns and the effects of electric vehicle charging (EVC) on the electrical network, only these three crucial parameters are used. A schedule for charging EVs and some of its essential components are shown in Fig. 1. The quantity of power needed for an EV to travel a specific distance is known as the charge required [31], and it can be obtained from the batteries or the grid/microgrid.

Both daily and per-drive event estimates are possible for the amount of electricity used by vehicles. An assumption is established on the need for charging. Temporal-based mobility modeling is seen as crucial. Any parking share with charging places automatically calculates the amount of charge required and provides time-dependent SOC data as soon as the EV drives in. An EV's charge is represented by its charging process. The modeling techniques of plug-in and plug-out time are provided. A probability distribution of specific samples or the EV's charging behavior can also be used to specify the exact commencement time.

These methods are limited to figuring out when to bill for the final or first visit of the day. Regardless of whether the visit is for commuter or non-commuter travel, the charged instant is determined using the time of stopping data supplied. One possible way to extend the charge time is to use personal or external charging methods. The utilization of secondary fuel is another restriction to take into account while modeling EV charging behavior. By making the three crucial adjustments to charging site, time, and necessity, it is possible to adjust external charging techniques to account for variables like as cost, grid loss, load variance, and income. Extensive pricing

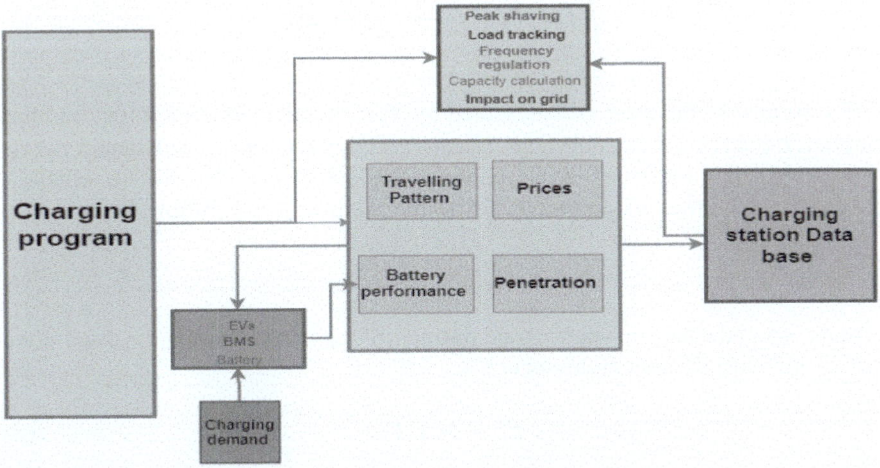

Fig. 1 Important components of an EV charging system and schedule

plans that account for V2G services must perform numerous tasks in order to meet the objectives of external billing strategies [32]. The distinct charging strategies that exhibit an unpredictable charging pattern and the ability to modify the load-charging profile are the hallmarks of stochastic discrete driving. When it comes to adapting to the user's preferences, EV chargers are sufficiently versatile. EV charging modeling is currently needed to predict demand profiles related to electric car integration into the power system. Whether the model's objective is to replicate specific EV charging techniques, external EV charging tactics or uncontrolled EV charging will determine how the model's main building blocks are constructed.

2.1 EV Charging Performance

A large class of computing methods known as the Monte Carlo method uses random sampling to get numerical results. The Monte Carlo method used for any problem that has a probabilistic interpretation. Moreover, the law of large numbers allows one to find the probability standards using a simple mean, provided the sample size is sufficiently big. For instance, the following integral can be estimated using,

$$\int_{b}^{a} f(x)dx = \frac{c-a}{N}\sum_{i=1}^{N} f(xin) \tag{1}$$

where N represents the necessary sample sets. One finite-dimensional integral (FDI) is found by,

$$\int_{\Omega}^{1} f(x)dx = \frac{V(\Omega)}{N}\sum_{i=1}^{N} f(xin) \tag{2}$$

The volume in this case is represented by V; it is true in all dimensions and has minimal impact on processing time. Because HDI is computed statistically in this work, this method has benefits. The charging station is where the electric vehicle is first charged, and the linking time is Tc_1^n which is calculated as follows,

$$Tc_1^n = L^n + A^n \tag{3}$$

In the second case, the energy used during driving is represented by tested drive time D_n^s velocity V_s and employment C_S.

$$E_u^n = D_n^s V_s C_S \tag{4}$$

The time of linking Tc_2^n is computed as follows, assuming that the EV is starting from an office or a parking space used for commuting.

$$Tc_2^n = L^n + A^n \tag{5}$$

For E_u^n, the electric car charging load is computed instantaneously. The charging power C_p is used to determine the anticipated load. C_i^n is the time, it takes for an automobile battery to charge completely [33]. For the car n, the charging time C_i^n is figured as,

$$C_i^n = \frac{E_u^n}{C_p} \tag{6}$$

The following formula can be used to find the charging power C_p, load $V_p^{n,1}$, and for a motor vehicle n at current I:

$$V_p^{n,1} = \begin{Bmatrix} C_p \text{ if charging} \\ 0 \quad Else \end{Bmatrix} \tag{7}$$

The estimated power $E\left[V_p^i\right]$ at the instant I is obtained by using MCS for k samples.

$$E\left[V_p^i\right] = \frac{1}{K}\sum_{n=1}^{k} V_p^{i,n} \tag{8}$$

For N vehicles, the total (EV) charging load $V_p total^i$ at the time I is provided b,

$$\left[V_p total^i\right] = N\left(V_p^i\right) \tag{9}$$

The normalized load curves $O_{c,p}^i$ and the load number are created to determine O_p^i. At this moment,

$$O_p^i = O_n O_{c,p}^i C_o \tag{10}$$

In the last step, the total of the EV load and additional critical loads is used to compute the overall load profile. The following formula is used to get the total load mean:

$$P_L^I = V_p^{i.t} + O_p^i \tag{11}$$

The EV charging performance was analyzed using the above-modeled equations from (1) to (11).

3 Estimation of SOC and Its Analysis

During the filtering process for parameter identification, an appropriate battery model is needed to get an accurate SOC estimate. In actuality, the battery's temperature, load current, exterior voltage, and other characteristics are related to the internal SOC. High estimation accuracy results from the deep learning approach are more dependable the more carefully the battery model is curated. The structure of the vehicle-to-grid system is depicted in Fig. 2.

Precise calculation of SOC is essential for the battery's safe and sound functioning; additionally, it prolongs battery life and improves the dependability of electric vehicles. Figure 3 illustrates the many battery SOC estimation techniques that have been developed over time.

This method presents the data-driven-based DFNN structure for SOC estimation, following the model-based method's implementation process approach. To train a deep feed-forward neural network, battery data is required. The require temperature, average current, average voltage, average current, and voltage as critical data collection aspects. The layer data will be required after the features are collected. Figure 4 displays the deep feed-forward neural network flowchart. Three layers make up the DFNN. The input layer is the lowest level. To train the network as an input layer, the following data is gathered: average voltage, average current, average current, temperature, and voltage. The processing of the incoming data takes place in the second, concealed layer. In this study, 65 neurons created using the proposed technique. The input and hidden layers are combined into the output layer, which is the last layer.

Data processing is the process of gathering all the information required to complete a task and organizing it into a single structure. Selecting the uniform sequence length and small batch size for every data point is the next step. The same data sets must be used for testing and validation. The iteration level ought to be at least 20 for accuracy.

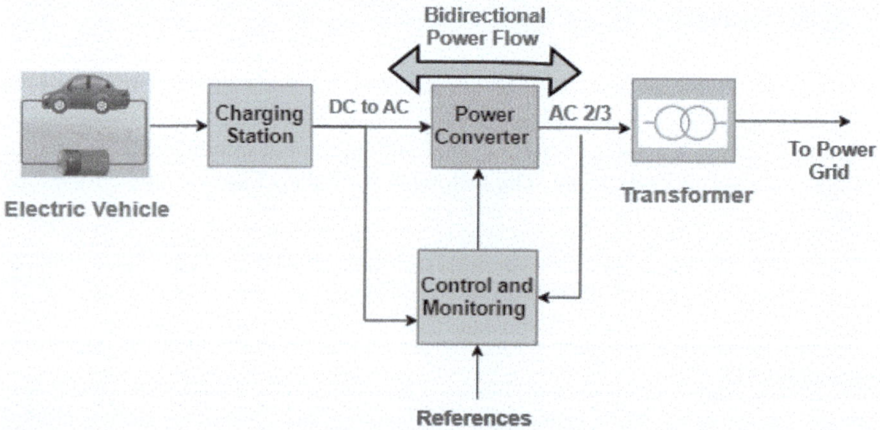

Fig. 2 Structure of vehicle to grid system

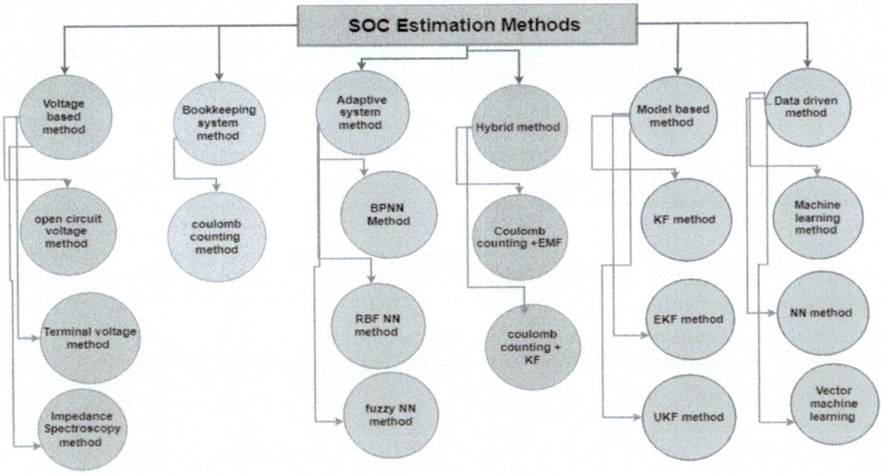

Fig. 3 SOC estimation methods

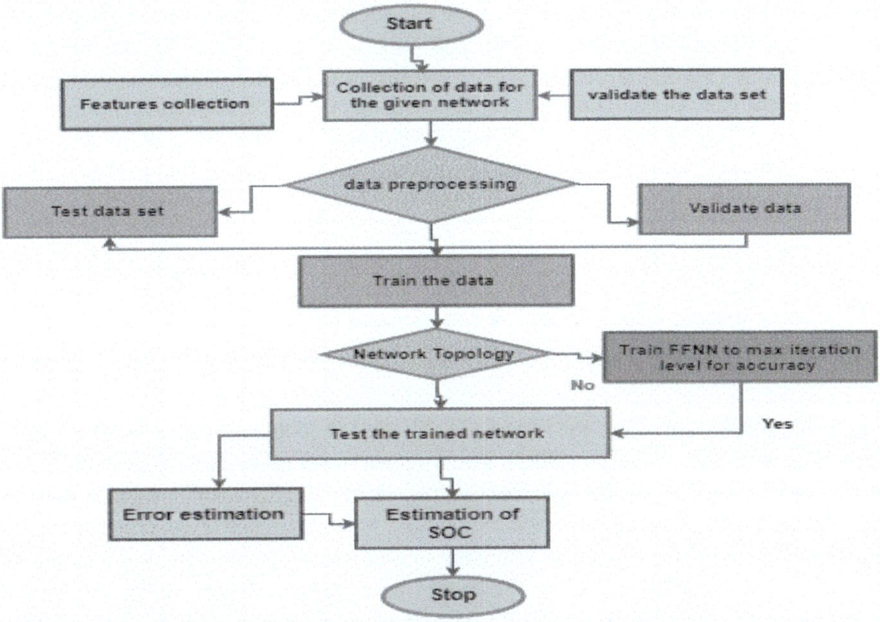

Fig. 4 Deep feed-forward neural network flowchart

Training data accuracy is contingent on small batch sizes, whereas validation data accuracy is contingent on training options. To achieve better and more accurate results, the data has been preprocessed in the instance without rearranging it.

4 Results and Discussion

One synchronous generator and two sets of batteries from EVs make up the microgrid. Grid frequency is configured to range from 0.999 to 1.001 per unit. The Battery block in the grid to the vehicle represents EVs with a level of charge of less than 80%. The EVs here are constantly charged. When the grid is experiencing periods of peak demand, EVs with a SOC greater than 90% are supported by the Battery (V2G) block model, which offers V2G support.

The grid initially provides electricity and charges two batteries. They both have increasing SOCs from starting values of 70% and 98%, respectively. The grid frequency decreases to less than 0.999 per unit at time 0.1 s when a 1 MW load is connected. As soon as the Battery (V2G) block begins to discharge, it powers the grid. The grid frequency rises to the predetermined range (1.5 ~ 2.001 per unit) as a result. When a 1 MW load disconnects from the grid, the grid frequency rises over 2.1 per unit at time 0.2 s. The grid frequency drops to the predetermined range and the Battery (V2G) block returns to charge condition. Because its SOC is less than 80%, the Battery (G2V) block is being charged during the simulation as depicted in Fig. 5.

4.1 Forecasting SOC Using a Deep Learning Network in Simulink

This technique demonstrates the use of a feed-forward deep learning network to forecast a battery's state of charge (SOC). The forecast block is used to estimate the SOC at each simulation step, to incorporate the network into the Simulink model. Battery's SOC is a percentage that indicates how charged an electric battery is concerning its capacity. SOC must be computed precisely to guarantee trustworthy and reasonably priced EVs, as it is crucial data for the energy management system of the car. Based on the Kalman filter (EKF) algorithm, the conventional approaches to this problem often necessitate exact parameters in addition to an understanding of the physical reaction and composition of the battery. On the other hand, applying neural networks is a data-driven method that necessitates little understanding of the nonlinear battery behavior.

By employing time series data reflecting the voltage, current, temperature, average voltage, and average current of the battery, the method trains a feed-forward neural network to forecast the SOC of a Li-ion battery. See predict battery SOC using deep

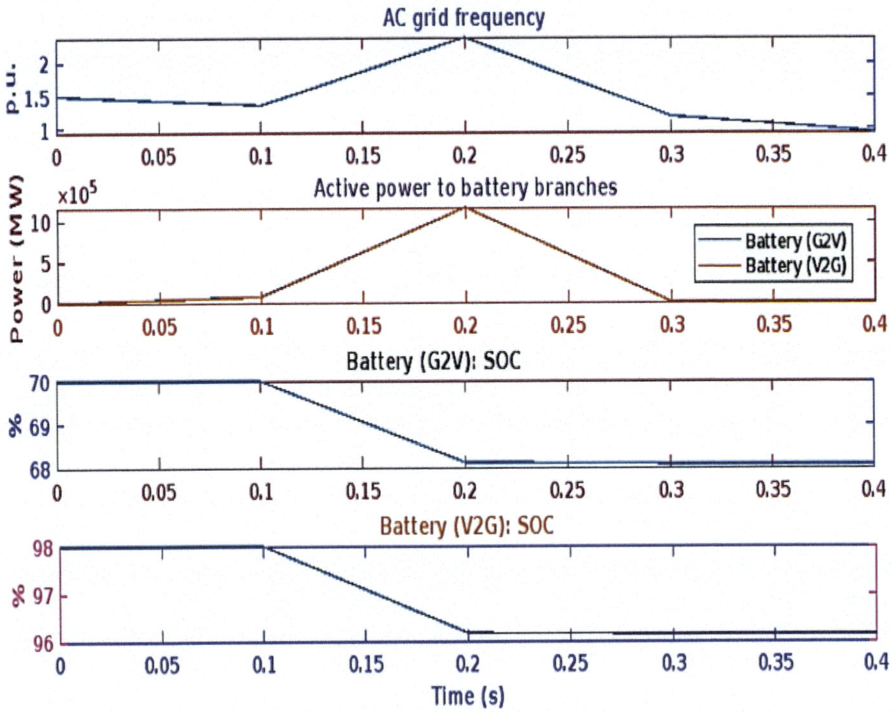

Fig. 5 Performance of vehicle to grid system

learning to observe a training example for the network. A predict block from the Deep Learning Toolbox package, two From Workspace blocks to import the target SOC and training network predictors from the test data, and two scope blocks to show the input signals and anticipated output are all used by the model. The estimated SOC matched with the true value of SOC and the error is zero by using deep learning network as depicted in Fig. 6.

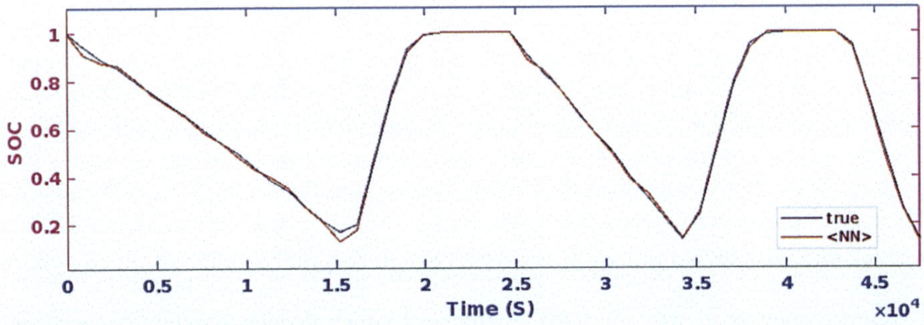

Fig. 6 SOC estimation—deep learning network

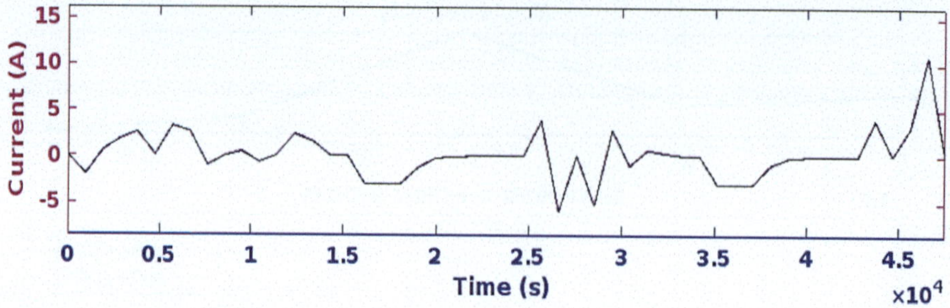

Fig. 7 Input current—deep learning network

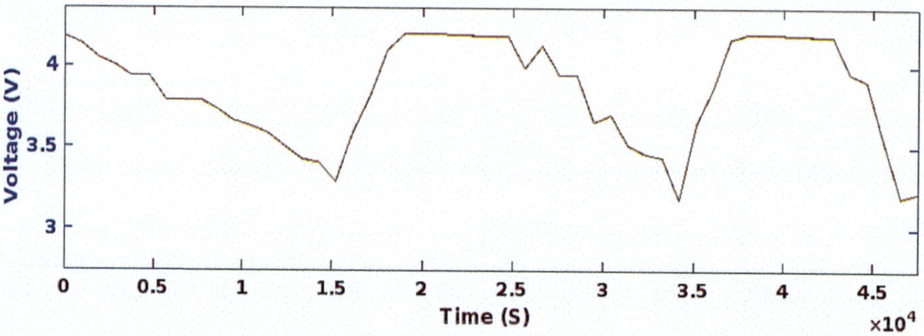

Fig. 8 Input voltage—deep learning network

The predict block uses the trained network, and then select using the block parameters to predict responses for the data at the input. The block outputs the prediction after receiving an input signal with the dimensions of the neural network's input layer. The input voltage and the input current are illustrated in Figs. 7 and 8, respectively.

4.2 *Deep Learning to Forecast the State of Charge*

A single set of experimental data taken during a driving cycle in which an electric car was powered by the battery at a temperature of 28 °C is included in the training data. Four experimental data sequences, obtained at four distinct temperatures throughout driving cycles, are included in the test data. One often used metric to assess the accuracy of predictive models is the RMSE. It computes the average error size among observed and expected values. The error in the trained, tested, and validated RMSE is much less as depicted in Fig. 9. The improved output performance is illustrated in Fig. 10.

Battery state of charge, or SOC, is the percentage that shows how charged an electric battery is about its capacity. An accurate assessment of SOC, a crucial piece of data for the vehicle EMS, is necessary to ensure reliable and economically priced

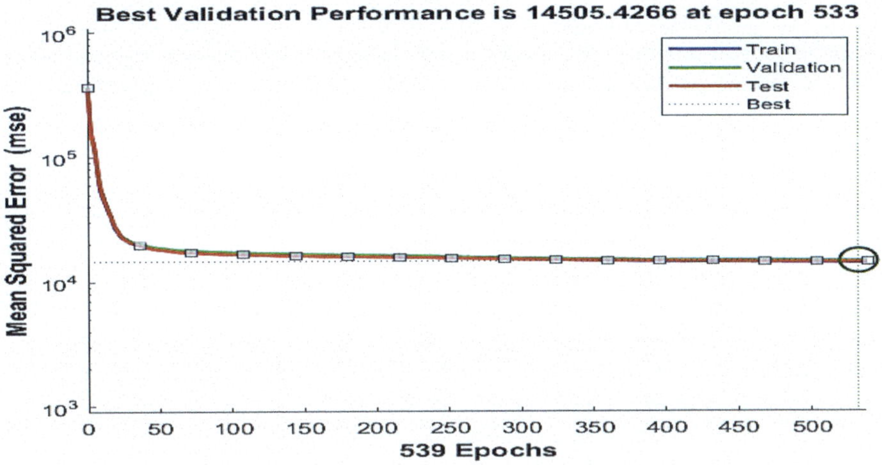

Fig. 9 Performance of RMSE using NN

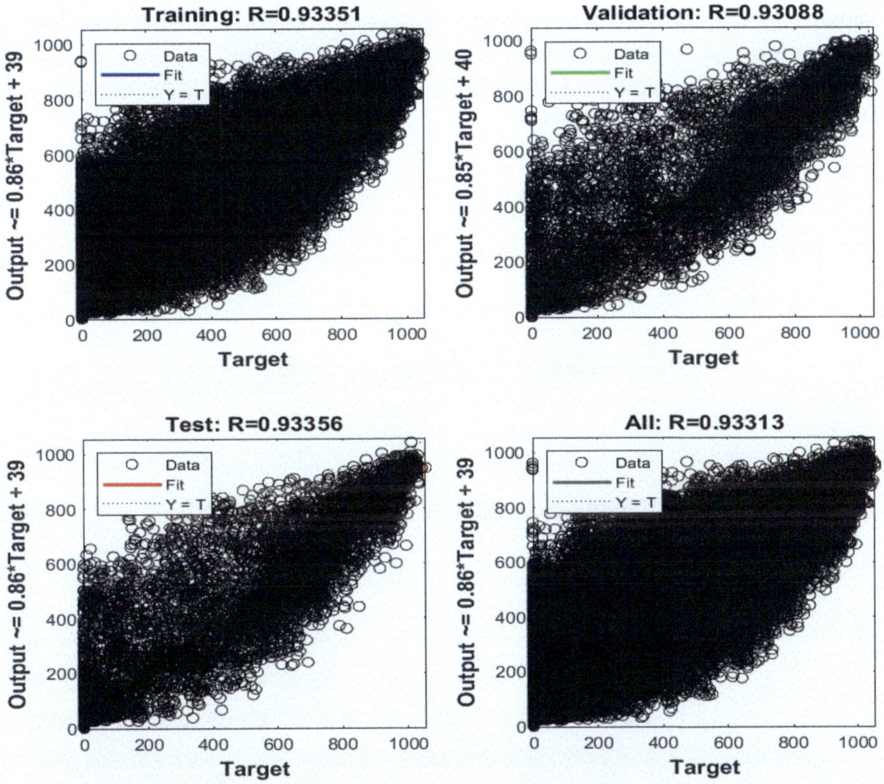

Fig. 10 Output performance using NN

electric cars. However, because of the nonlinear behavior of Li-ion batteries, SOC estimation is still a significant challenge in automotive engineering. Typical solutions to this issue, such as electrochemical models, call for exact specifications and an understanding of the composition and physical behavior of the battery. Neural network technology, on the other hand, is data-driven and necessitates no understanding of the battery or its nonlinear behavior. To predict the SOC of a Li-ion battery, the model trains a neural network with time series data that represents the battery's many characteristics, including temperature, voltage, current, and average voltage and current. The gradient performance is depicted in Fig. 11.

The SOC, which shows a battery's current capacity in relation to its full capacity, is an important metric in battery management systems. For effective energy management, particularly in applications such as EVs and renewable energy storage, accurate forecasting and targeting of SOC are crucial. The predicted SOC closely follows the target SOC over time are depicted in Fig. 12.

Individuals may experience a little financial incentive when using the V2G service just for peak saving; however, there is a significant potential for financial gain when using the V2G service for frequency regulation. The cheaper cost of peak reduction compared to frequency regulation is the reason behind this. Though including a sizable number of EVs also adds to grid congestion, amazing features would not have been achievable without them. Low voltage and complicated grid congestion can result from unregulated charging.

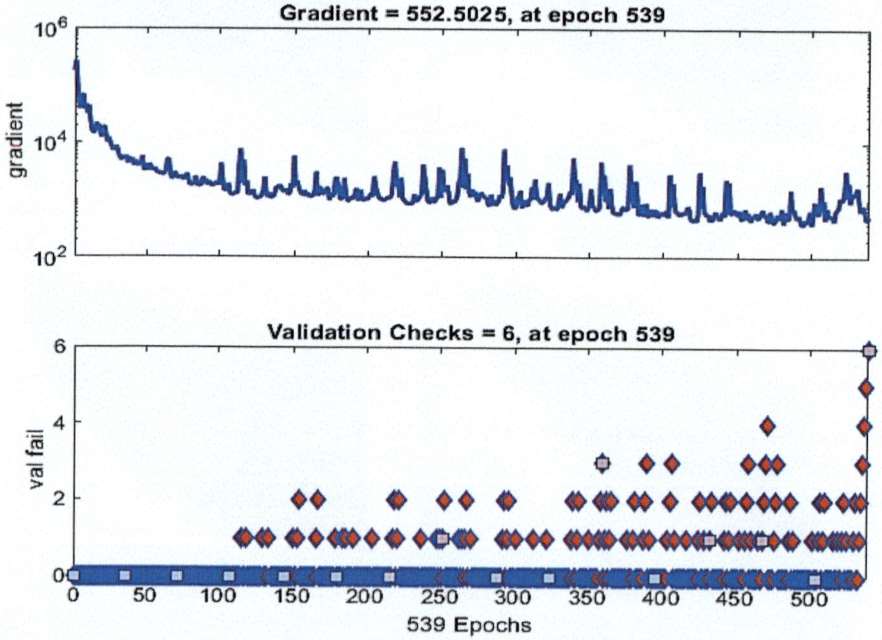

Fig. 11 Performance of gradient using NN

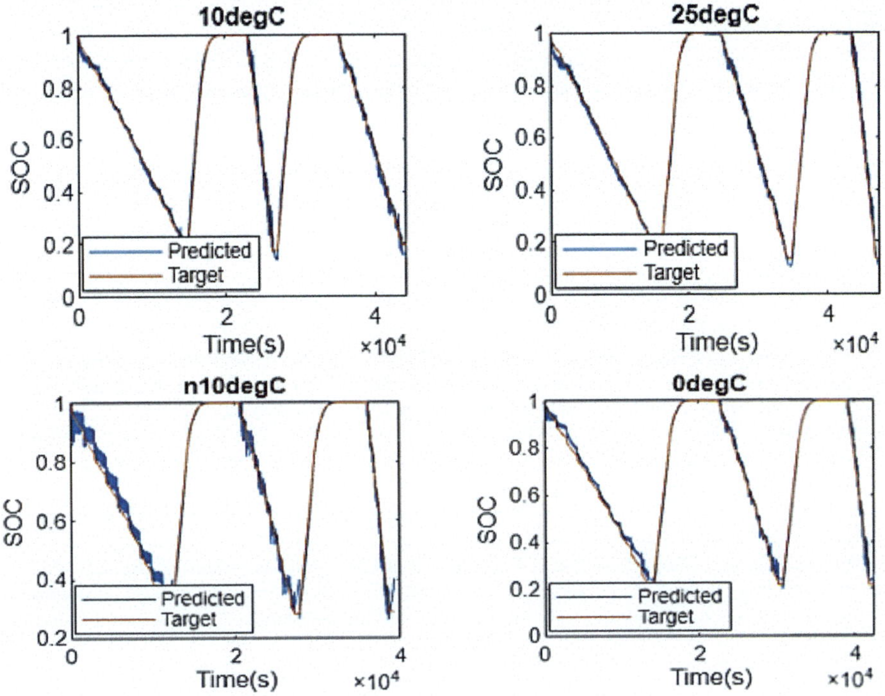

Fig. 12 SOC estimation—deep learning

5 Conclusions

This study offered a comprehensive examination of EVs' V2G technology. How EVs could help with peak load reduction and frequency management has been analyzed. When the V2G service is utilized solely for peak saving, there is a slight economic enticement for persons; yet, there is a large possibility for financial gain when the V2G service is utilized for frequency regulation. This is due to the fact that peak reduction is less expensive than frequency regulation. Amazing features would not have been possible without EVs, but integrating a significant number of EVs also contributes to grid congestion. Thus, unregulated charging may lead to complex grid congestion and low voltage. This study has looked at the reasons behind EVs charging excessively in an attempt to decrease the effects. Since it is very difficult to execute regulated charging because of different customer limits, power prediction systems for EV charging and discharging have been investigated. The deep learning network that will be utilized to create an efficient and successful coordination system for EV charging has been analyzed. This framework uses a neural network-based model with a feed-forward algorithm to study the effects of V2G charging and discharging in the microgrid environment.

References

1. Wirasingha SG, Gremban R, Emadi A (2011) Source-to-wheel (STW) analysis of plug-in hybrid electric vehicles. IEEE Trans Smart Grid 3(1):316–331. https://doi.org/10.1109/TSG.2011.2158248
2. Guille C, Gross G (2009) A conceptual framework for the vehicle-to-grid (V2G) implementation. Energy Policy 37(11):4379–4390. https://doi.org/10.1016/j.enpol.2009.05.053
3. Williamson S, Lukic M, Emadi A (2006) Comprehensive drive train efficiency analysis of hybrid electric and fuel cell vehicles based on motor-controller efficiency modeling. IEEE Trans Power Electron 21(3):730–740. https://doi.org/10.1109/TPEL.2006.872388
4. Pode R (2015) Battery charging stations for home lighting in Mekong region countries. Renew Sustain Energy Rev 44(1):543–560. https://doi.org/10.1016/j.rser.2015.01.003
5. Lu J, Wang Z, Zhang Q, Sun C, Zhou Y, Wang S, Wei T (2023) The effects of amino groups and open metal sites of MOFs on polymer-based electrolytes for all-solid-state lithium metal batteries. Chin J Chem Eng 60:80–89. https://doi.org/10.1016/j.cjche.2023.01.011
6. Iclodean C, Varga B, Burnete N, Cimerdean D, Jurchiş B (2017) Comparison of different battery types for electric vehicles. IOP Conf Ser Mater Sci Eng 252(1):1–10. https://doi.org/10.1088/1757-899X/252/1/012058
7. Rallo H, Casals LC, De La Torre D, Reinhardt R, Marchante C, Amante B (2020) Lithium-ion battery 2nd life used as a stationary energy storage system: ageing and economic analysis in two real cases. J Clean Prod 272(1):1–10. https://doi.org/10.1016/j.jclepro.2020.122584
8. Ren R, Lai F, Lang X, Li L, Yao C, Cai K (2023) Efficient sulfur host based on Sn doping to construct Fe_2O_3 nanospheres with high active interface structure for lithium-sulfur batteries. Appl Surf Sci 613:1–13. https://doi.org/10.1016/j.apsusc.2022.156003
9. Zhai S, Li W, Wang C, Chu Y (2022) A novel data-driven estimation method for state-of-charge estimation of Li-ion batteries. Energies 15(9):1–17. https://doi.org/10.3390/en15093115
10. Chen J, Zhang Y, Wu J, Cheng W, Zhu Q (2023) SOC estimation for lithium-ion battery using the LSTM-RNN with extended input and constrained output. Energy 262:1–25. https://doi.org/10.1016/j.energy.2022.125375
11. Wang Q, Ye M, Wei M, Lian G, Li Y (2023) Deep convolutional neural network based closed-loop SOC estimation for lithium-ion batteries in hierarchical scenarios. Energy 263:1–15. https://doi.org/10.1016/j.energy.2022.125718
12. Zhang D, Zhong C, Xu P, Tian Y (2022) Deep learning in the state of charge estimation for li-ion batteries of electric vehicles: a review. Machines 10(10):1–21. https://doi.org/10.3390/machines10100912
13. Zhang T, Zhao C, Sun X, Lin M, Chen Q (2022) Uncertainty-aware energy management strategy for hybrid electric vehicle using hybrid deep learning method. IEEE Access 10:63152–63162. https://doi.org/10.1109/ACCESS.2022.3182805
14. Habib S, Abbas G, Jumani TA, Bhutto AA, Mirsaeidi S, Ahmed EM (2022) Improved whale optimization algorithm for transient response, robustness, and stability enhancement of an automatic voltage regulator system. Energies 15(14):1–18. https://doi.org/10.3390/en15145037
15. Chen H, Xiong R, Lin C, Shen W (2020) Model predictive control based real-time energy management for hybrid energy storage system. CSEE J Power Energy Syst 7(4):862–874. https://doi.org/10.17775/CSEEJPES.2020.02180
16. Dong C, Sun J, Wu F, Jia H (2020) Probability-based energy reinforced management of electric vehicle aggregation in the electrical grid frequency regulation. IEEE Access 8:110598–110610. https://doi.org/10.1109/ACCESS.2020.3002693
17. Khooban MH, Gheisarnejad M (2020) A novel deep reinforcement learning controller based type-II fuzzy system: frequency regulation in microgrids. IEEE Trans Emerg Top Comput Intell 5(4):689–699. https://doi.org/10.1109/TETCI.2020.2964886
18. Zhao M, Zhang R, Lin C, Zhou H, Shi J (2020) Stochastic model predictive control for dual-motor battery electric bus based on signed Markov chain Monte Carlo method. IEEE Access 8:120785–120797. https://doi.org/10.1109/ACCESS.2020.3006285

19. Lei Z, Sun D, Liu J, Chen D, Liu Y, Chen Z (2019) Trip-oriented model predictive energy management strategy for plug-in hybrid electric vehicles. IEEE Access 7:113771–113785. https://doi.org/10.1109/ACCESS.2019.2933015
20. Shokri Gazafroudi A, Corchado JM, Keane A, Soroudi A (2019) Decentralised flexibility management for EVs. IET Renew Power Gener 13(6):952–960. https://doi.org/10.1049/iet-rpg.2018.6023
21. Ahmed EM, Mohamed EA, Elmelegi A, Aly M, Elbaksawi O (2021) Optimum modified fractional order controller for future electric vehicles and renewable energy-based interconnected power systems. IEEE Access 9:29993–30010. https://doi.org/10.1109/ACCESS.2021.3058521
22. Zhang S, Zhou Z, Luo R, Zhao R, Xiao Y, Xu Y (2023) A low-carbon, fixed-tour scheduling problem with time windows in a time-dependent traffic environment. Int J Prod Res 61(18):6177–6196. https://doi.org/10.1080/00207543.2022.2153940
23. Potter MA, Jong KAD (2000) Cooperative coevolution: An architecture for evolving coadapted subcomponents. Evol Comput 8(1):1–29. https://doi.org/10.1162/106365600568086
24. Battistelli C, Baringo L, Conejo AJ (2012) Optimal energy management of small electric energy systems including V2G facilities and renewable energy sources. Electr Power Syst Res 92(1):50–59. https://doi.org/10.1016/j.epsr.2012.06.002
25. Bahramirad S, Reder W, Khodaei A (2012) Reliability-constrained optimal sizing of energy storage system in a microgrid. IEEE Trans Smart Grid 3(4):2056–2062. https://doi.org/10.1109/PESMG.2013.6672068
26. Khokhar B, Dahiya S, Parmar KS (2021) A novel hybrid fuzzy PD-TID controller for load frequency control of a standalone microgrid. Arab J Sci Eng 46(1):1053–1065. https://doi.org/10.1007/s13369-020-04761-7
27. Mehrjerdi H, Hemmati R (2020) Coordination of vehicle-to-home and renewable capacity resources for energy management in resilience and self-healing building. Renew Energy 146(1):568–579. https://doi.org/10.1016/j.renene.2019.07.004
28. Sravanthi P, Sandeep V (2022) Solving realistic reactive power market clearing problem of wind-thermal power system with system security. Int J Emerg Electr Power Syst 23(2):125–144. https://doi.org/10.1515/ijeeps-2021-0060
29. Kumar JV, Sesham H, Salkuti SR (2024) Smart energy management model for electric vehicles. In: Salkuti SR (eds) Energy and environmental aspects of emerging technologies for smart grid. Green energy and technology. Springer, Cham. https://doi.org/10.1007/978-3-031-18389-8_9
30. Guo L, Liu W, Cai J, Hong B, Wang C (2013) A two-stage optimal planning and design method for combined cooling, heat and power microgrid system. Energy Convers Manage 74:433–445. https://doi.org/10.1016/j.enconman.2013.06.051
31. Deb K, Pratap A, Agarwal S, Meyarivan TAMT (2002) A fast and elitist multiobjective genetic algorithm: NSGA-II. IEEE Trans Evol Comput 6(2):182–197. https://doi.org/10.1109/4235.996017
32. Marler RT, Arora JS (2004) Survey of multi-objective optimization methods for engineering. Struct Multidiscip Optim 26(1):369–395. https://doi.org/10.1007/s00158-003-0368-6
33. Salkuti SR, Ray P, Pagidipala S (2022) Overview of next generation smart grids. In: Salkuti SR, Ray P (eds) Next generation smart grids: modeling, control and optimization. Lecture notes in electrical engineering, vol 824. Springer, Singapore. https://doi.org/10.1007/978-981-16-7794-6_1

Frequency Stability of Highly Ranged Renewable Energy Penetrated Multi-area Interconnected Power System Under a 3DOFPID Controller

Ch. Naga Sai Kalyan and Surender Reddy Salkuti

Abstract The load frequency control (LFC) of a multi-area interconnected power system (IPS) is presented in this chapter. The multi-area IPS in this work has two areas that incorporate power generation units like thermal and hydro with non-linear practical constraints. Moreover, area 1 of the multi-area IPS is integrated with renewable-based systems such as wind turbine generation, fuel cells, and aqua electrolyzers, and further diesel power generation is also penetrated. For better control over the stability of the IPS network, this work suggests the osprey optimization algorithm (OOA) tuned three-degree-of-freedom proportional-integral-derivative (3DOFPID) controller as a secondary controller, and the investigation is performed for a 10% step load perturbation (SLP) in area 1. Moreover, the communication time delays (CTDs) are deliberated with the investigation model, and their subsequent effect on the IPS performance is showcased. Later, the coordinated control approach of capacitive energy storage (CES) and a thyristor-controlled series compensator (TCSC) is employed for the investigative model in this work under the OOA-tuned 3DOFPID controller. The improvement in the dynamic behavior of the multi-area IPS is achieved with the deployment of the additional CES-TCSC control strategy. Finally, the multi-area IPS is targeted with a wide range of loads and parametric uncertainties to validate the robustness of the implemented control techniques.

Keywords Dynamic analysis · 3DOFPID controller · Osprey optimization algorithm · CES-TCSC strategy · Step load perturbation

---

Ch. Naga Sai Kalyan
Department of Electrical and Electronics Engineering, Vasireddy Venkatadri International Technological University, Nambur, Guntur 522508, India

S. R. Salkuti (✉)
Department of Railroad and Electrical Engineering, Woosong University, Daejeon, Republic of Korea
e-mail: surender@wsu.ac.kr

S. R. Salkuti (ed.), *Artificial Intelligence for Integrated Smart Energy Systems in Electric Vehicles*, Lecture Notes in Electrical Engineering 1427,
https://doi.org/10.1007/978-3-031-94276-1_23

Nomenclature

CES	Capacitive energy storage
DG	Distributed generation
DOF	Degree-of-freedom
ESDs	Energy storage devices
GDB	Governor dead band
GRC	Generation rate constraint
LFC	Load frequency control
OOA	Osprey optimization algorithm
PID	Proportional-Integral-Derivative
REP	Renewable energy plant
RPM	Real power imbalance
SLP	Step load perturbation

1 Introduction

The excessive penetration of renewable energy sources (RES) is going on in the power grid with an increase in power demands as well as environmental concerns. Moreover, the power generation through RES is very low, and to capture the benefit of this, RES penetration is increasing around the world. However, the uncertainty in power generation by the RES due to its intermittency is bringing hurdles to the electrical network operators. Of which, the real power mismatch (RPM) is the key indicator, which not only affects the grid performance but also the stability and resilience of the system. The RPM arises from the imbalance between available generation and demand. Moreover, it has a direct relationship with system frequency and hence affects the electrical system's optimal performance. The LFC has the potential to cancel out the negative effects that emerge in the system due to the development of the RPM [1]. Maintaining the electrical frequency to not deviate beyond the specified boundaries and to limit the tie-line power that flows unintended is the primary task of LFC.

The literature revealed a bulk quantity of studies on LFC, and the majority of them are in and around conventional LFC actions [2]. The design of PI using the age of information method to control an isolated thermal system is presented in [3], whose dominance is demonstrated by the traditional PI performance. For a single-area sine-source system, the PI controller was suggested based on the elephant herd optimization algorithm (EHO) [4], and further, the efficacy of EHO is revealed with the comparative performances of differential evolution, teacher-learner, and grey wolf algorithms. The PID controller using the quadratic regulator approach [5] and the particle swarm algorithm (PSA) [6] are suggested for multi-source, two-area IPS networks. In [7], the integrated multi-area IPS stability of the integrated multi-area IPS stability of the renewable energy plant (REP) is investigated with the enhanced whale optimizer-tuned PID controller, and further analysis is extended to the implementation of penetrating energy storage devices (ESD). The PI-PD controller is suggested in [8] for a single-area system whose parameters are obtained using the design synthesis method. The PID for the REP multi-area network is suggested using the artificial field algorithm [9], and later the analysis is stretched out to the integration of ESDs and the fact controllers. Apart from the above-mentioned optimization approaches, algorithms like the levy flight technique, slap swarm optimizer, artificial rabbit algorithm, falcon optimizer, symbiotic organism search technique, naked more rat algorithm, COOT optimizer, and dragon fly algorithm, etc. are implemented for conventional controller optimizations, and the researchers are tested on various IPS models with and without deliberating the constraints of practicality as discussed extensively in [10].

The inability of the classical regulators to control the dynamical characteristics of the non-linearized IPS network made the researchers look for alternate control structures. In this regard, modifications are made to the conventional PID controllers to enhance their performance and avoid the pitfalls in the conventional structure. The dual-mode PID [11] is studied on the three-area thermal network with generation rate constraints (GRC) carried out using the recent technique of ion motion optimization. The efficacy of the modified PID based on grasshopper optimization is tested on the REP-integrated multi-area system in [12]. The PID with the addition of a double derivative parameter was employed recently in [13] using the COOT algorithm and the efficacy of the controller was revealed with other reported conventional techniques. However, the modified conventional controllers are also inevitable, as their performance supremacy with the non-linearized plant models is restricted to a certain extent.

Fractional order (FO) controllers, intelligent-based fuzzy logic controllers (FLC), and tilt-integral-derivatives (TID) have been widely adopted in the literature over the past ten years. The FOPID [14] is implemented for the pumped storage unit integrated multi-area IPS and further integrated with the ESDs. The hybrid artificial bee colony and PSA-based FOPID [15] are suggested and tested on three different types of IPS models, and the efficacy of the suggested technique is showcased. Further, the combination of FO and FLC is also reported in the literature. The fuzzy TID optimized with the sailfish algorithm, Aquila optimization, and wild horse optimizations are presented in [16–18], respectively. The adaptive variable FLC is designed in [19] and reflects the improvement in LFC actions. However, the complexity of the optimization of the FO controllers and the assumptions being made in the development of the fuzzy controllers might not be efficient and robust enough to handle the system's non-linear behavior. Thus, this work implemented the 3DOFPID [20, 21] controller, which has multiple control loops, which helps it exhibit performance dominance. In the literature, a few works are available on the 3DOFPID controller based on different optimization techniques, but the OOA has not been reported in the LFC study so far to the best of the author's knowledge. Moreover, the OOA-tuned 3dOFPID is tested on MAPS with the integration of multiple REPs, and the investigation of such IPS models is very limited in the literature. The contributions of this work are:

- The LFC of renewable energy penetrated MAPS is investigated for 10% SLP in area 1.
- The OOA-tuned 3DOFPID is enforced as the regulator and its sovereignty is demonstrated with PID and FOPID performances.
- The CTDs are deliberated, and their subsequent effect on the IPS performance is showcased.
- The CES-TCSC strategy has been implemented and has achieved further performance improvement.
- The suggested control approaches' robustness is confirmed by conducting a robustness test.

In this chapter, the remaining sections are organized by presenting the test system model in Sect. 2 and the structure 3DOFPID controller along with the objective function in Sect. 3. The detailed mathematical modeling of the OOA and the CES-TCSC strategy is presented in Sects. 4 and 5, respectively. In Sect. 6 the simulation analysis is presented followed by the major conclusions of this work in Sect. 7.

2 Power System Model

The transfer function of the investigative model is shown in Fig. 1, with the two control regions having the non-linearity characteristics of the hydrothermal units in the two regions. Furthermore, in Area 1, renewable energy sources such as wind turbine generation, fuel cells, and aqua electrolyzer, along with the diesel power generation unit, are penetrated as distributed generation sources (DG). The analysis of the renewable-penetrated MAPS is performed under the conditions of 10% SLP on area 1. The DGs are modeled in Fig. 2, and the system dynamic behavior is analyzed by developing a model of renewable energy penetrated MAPS in MATLAB/Simulink (R2019a), and the parameters are made available in [22, 23]. However, the MAPS is considered the realistic constraint with the CTDs, which is the inherited behavior of the communication channels. In the realistic practice of the IPS network, data or signal transfer among several devices that are installed at numerous locations is done through these channels. Based on the information carried out by these channels from the terminal points to the control center, the area control error (ACE) signal is generated, which can then be used as input to the controller. The controller then took the necessary measures to regulate the system's behavior. Because the peripherals have their delays, it took longer for the regulator to make the necessary corrections. These effects on the IPS performance need to be studied and shown. The modeling of the CTDs is given in Eq. (1) [24]. Further, the hydrothermal units are deliberated with the GRC and GDB effects to investigate the real-time environment.

$$e^{-s\tau_d} = \frac{1 - s\frac{\tau_d}{2}}{1 + s\frac{\tau_d}{2}} \tag{1}$$

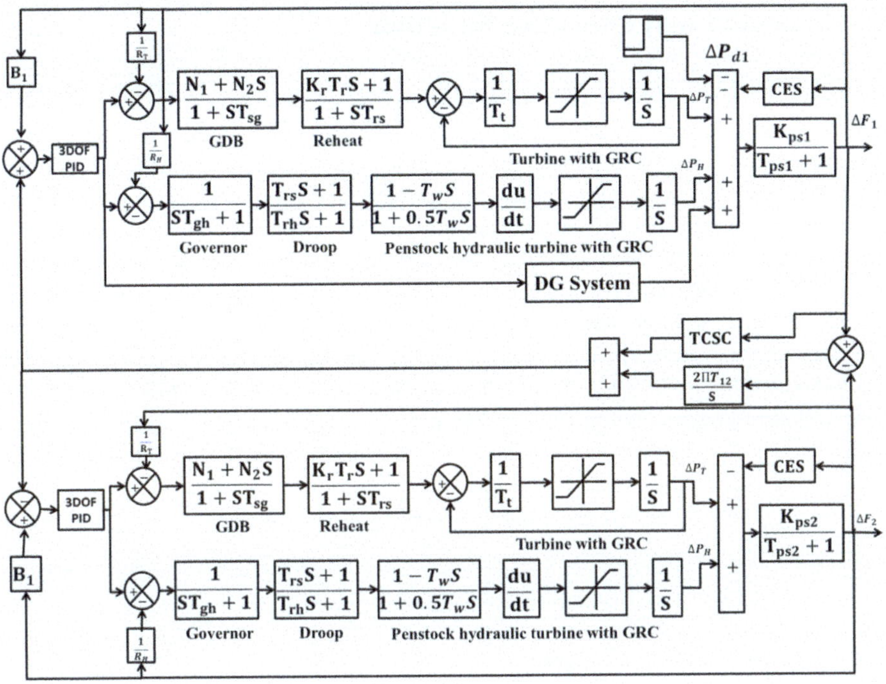

Fig. 1 Model of MAPS with DG integration

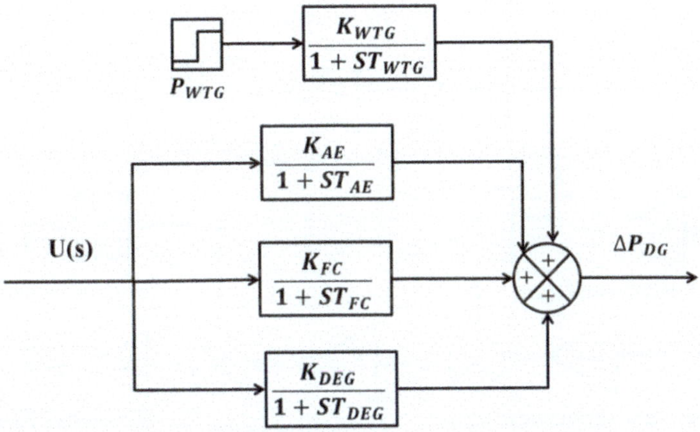

Fig. 2 Transfer function of DG sources

3 Controller and Objective Function

The literature review showcased the extensive implementation of the traditional PI and PID and their modified control structures in the LFC study. The test system model conceived in this work has the non-linearity characteristics of CTDs, GDB, and GRC, and hence the performance of conventional regulators is inevitable. The DOF type of controllers offers promising regulation over non-linear systems [25], and thus the 3DOFPID as a secondary regulator as shown in Fig. 3 is enforced. The K_P, K_I, K_D, K_1, K_2, and K_{ff} are the proportional, integral, derivative, proportional scaling, integral scaling, and feed-forward gains of the 3DOFPID on which it is based for optimal performance. Thus, the recent optimization methodology of OOA is selected for the 3DOFPID optimal tuning subjected to integral square error (ISE) time-domain indices given in Eq. (2) [26, 27].

$$J_{ISE} = \int_0^{T_{sim}} \left(\Delta f_1^2 + \Delta P_{tie12}^2 + \Delta f_2^2\right)dt \quad (2)$$

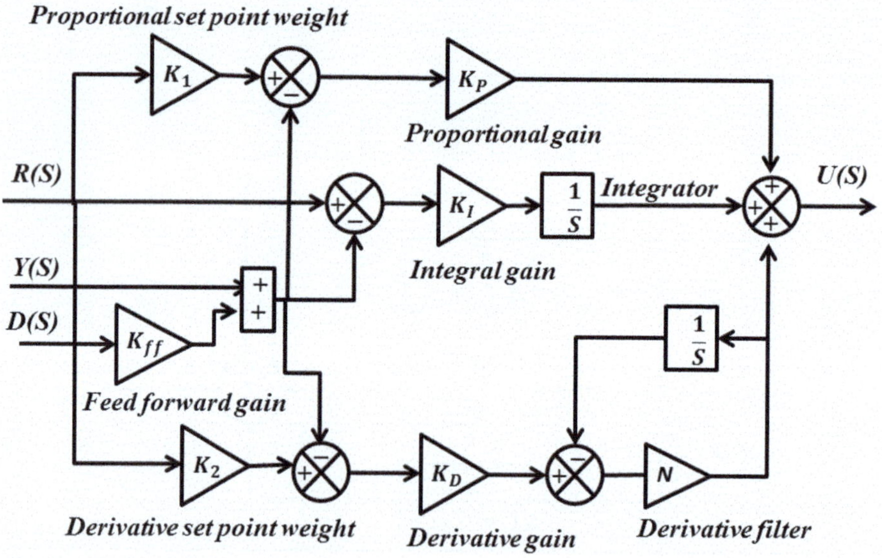

Fig. 3 Structure of 3DOFPID controller

4 Osprey Optimization Algorithm (OOA)

The osprey is a fish-eating bird that can also be called a fish hawk; it merely resembles the eagle. They weigh only about 2 kg and can catch underwater fish. The osprey has powerful vision and can locate the underwater fish even when it is flying a few meters away. They catch the fish with their legs and land on the nearby land or rock to eat the prey. These birds play a significant role in maintaining biodiversity and balancing aquatic creatures. The strategies of the osprey in hinting and catching the fish are intelligent, which facilitates the researchers [28] to propose the OOA. In OOA, Every osprey serves as a means of resolving an issue, and the initialization of the ospreys is performed in Eq. (3). At first, the osprey in OOA is initialized randomly using Eq. (4).

$$Z = \begin{bmatrix} Z_1 \\ \cdot \\ \cdot \\ Z_i \\ \cdot \\ Z_N \end{bmatrix} = \begin{bmatrix} Z_{1,1} & \cdots & Z_{1,m} \\ \vdots & \ddots & \vdots \\ Z_{N,1} & \cdots & Z_{N,m} \end{bmatrix} \tag{3}$$

$$Z_{i,j} = lb_j + r_{ij} \cdot \left(ub_j - lb_j\right), \quad i = 1, 2, \ldots N, j = 1, 2, \ldots m \tag{4}$$

where the matrix 'Z' represents the osprey's locations and the parameter, 'i' represents the position of the osprey for the jth problem variable. The parameters 'N' and 'm' demonstrate the number of ospreys and the problem variables. In this work, the total population, or the osprey, is considered to be 50 and the problem variables as 6. Because the 3DOFPID has six parameters that must be identified to best suit the investigative model. The lower and upper boundaries of the jth variable are indicated with lb_j and ub_j, and r_{ij} is the randomly generated value from [0–1]. For each osprey, the objective function (OF) can be evaluated using Eq. (2), and the respective matrix is provided in Eq. (5).

$$OF = \begin{bmatrix} OF_1 \\ \cdot \\ \cdot \\ OF_i \\ \cdot \\ OF_N \end{bmatrix} = \begin{bmatrix} OF(X_1) \\ \cdot \\ \cdot \\ OF(X_i) \\ \cdot \\ OF(X_N) \end{bmatrix} \tag{5}$$

Later, the exploration phase begins, which indicates the location and hunting of the fish by the ospreys. Before entering this phase, the location of the osprey with the best OF indicated with 'Z_{Best}' is identified, and the fish positions (FP_i) are initialized as given in Eq. (6).

$$FP_i = \{Z_k | k \in \{1, 2, \ldots N\} \wedge OF_k < OF_i\} \cup Z_{Best} \tag{6}$$

After the detection of the fish by the osprey randomly, the movement of the osprey is given in Eq. (7), and the update in the osprey position vector is done as in Eq. (8).

$$Z_{i,j}^{p1} = Z_{i,j} + r_{i,j} \cdot (SF_{i,j} - I_{i,j} \cdot Z_{i,j}) \tag{7}$$

$$Z_i = \{Z_i, \ OF_i^{p1} < OF_i; \ \text{else } Z_i\} \tag{8}$$

The parameter 'SF' indicates the selected fish by the osprey, and the superscript 'p1' represents phase one. After catching the fish, the osprey looks for a suitable place to eat the prey, and this stage is termed phase 2 (p2) and is part of exploitation. There, the position of the osprey is updated using Eqs. (9) and (10).

$$Z_{i,j}^{p2} = Z_{i,j} + \frac{lb_j + r \cdot (ub_j - lb_j)}{t} \tag{9}$$

$$Z_i = \{Z_i^{p2}, \ OF_i^{p2} < OF_i; \ \text{else } Z_i\} \tag{10}$$

The 't' represents the iteration count, and the process from Eqs. (6) to (10) will be repeated until the maximum number of iterations is reached.

5 CES-TCSC Strategy

The integration of the CES devices with the IPS network optimized energy utilization. These devices store the energy during peak generation periods and deliver it back to the grid during high-demand periods. Because of this, the CES can provide support for emergency or critical loads for short periods of time. Moreover, the CES integration with the grid facilitates the smoothening of the power output, especially when the grid is integrated with renewable sources. Overall, it has the capability of regulating frequency fluctuations to a certain extent, and hence the integration of CES [29] is considered in this work, whose modeling is placed in Eq. (11).

$$G_{CES} = \frac{K_{CES}}{1 + sT_{CES}} \tag{11}$$

For the planners and operators of the IPS network, the TCSC device is a very valuable thing in getting better control over the transmission line. In this work, the TCSC is placed with the time to dampen the power oscillations that emerge in the tie-line during load perturbations. The damping control structure of the TCSC [30, 31] is shown in Fig. 4.

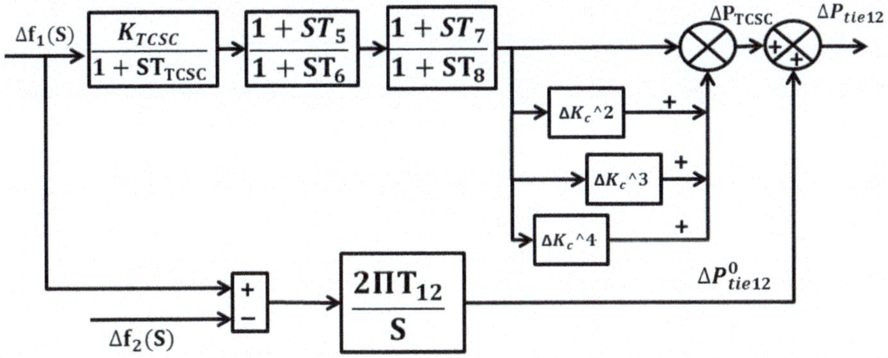

Fig. 4 TCSC as damping controller

6 Results and Discussion

In this section, the analysis is presented totally in four cases. In case 1 the performance of the MAPS I tested on various controllers to reveal the best control approach and from there onwards the analysis is continued under the same regulator. In case 2, the effect of CTDs is revealed and the employment of the CES-TCSC strategy is performed in case 3. In case 4, the robustness study is conducted in terms of wide variations in nominal loading and the system tie-line parameter.

6.1 Case 1: Analysis of Renewable Energy Penetrated MAPS Performance Under Various Controllers

In this case, the dynamical behavior of the renewable energy penetrated MAPS is investigated by enforcing various controllers at the secondary level: the PID, FOPID, and 3DOFPID individually. These controllers are fine-tuned with the OOA to reveal the controller that can exhibit the best performance. However, the CTDs of 0.25 s are taken with the IPS to assess the controller performance in a realistic environment. Area 1 of the system is subjected to 10% SLP, and the system performance under PID, FOPID, and 3DOFPID is compared in Fig. 5. The responses in Fig. 5 are assessed in terms of peak undershoot (PUS) and the time conceded by the responses in reaching stable positions, i.e., the settling time in seconds. The numerical data of the responses in Fig. 5 are placed in Table 1, and it is clear that the performance of 3DOFPID is dominating and outperforms the PID and FOPID in terms of regulating the PUS values, eliminating the oscillations, and settling time. Moreover, the 3DOFPID effectively handled the ISE index, and its value improved by 45.35 and 93.17% with FOPID and PID. The optimal parameters of PID, FOPID, and 3DOFPID using OOA are placed in Table 2.

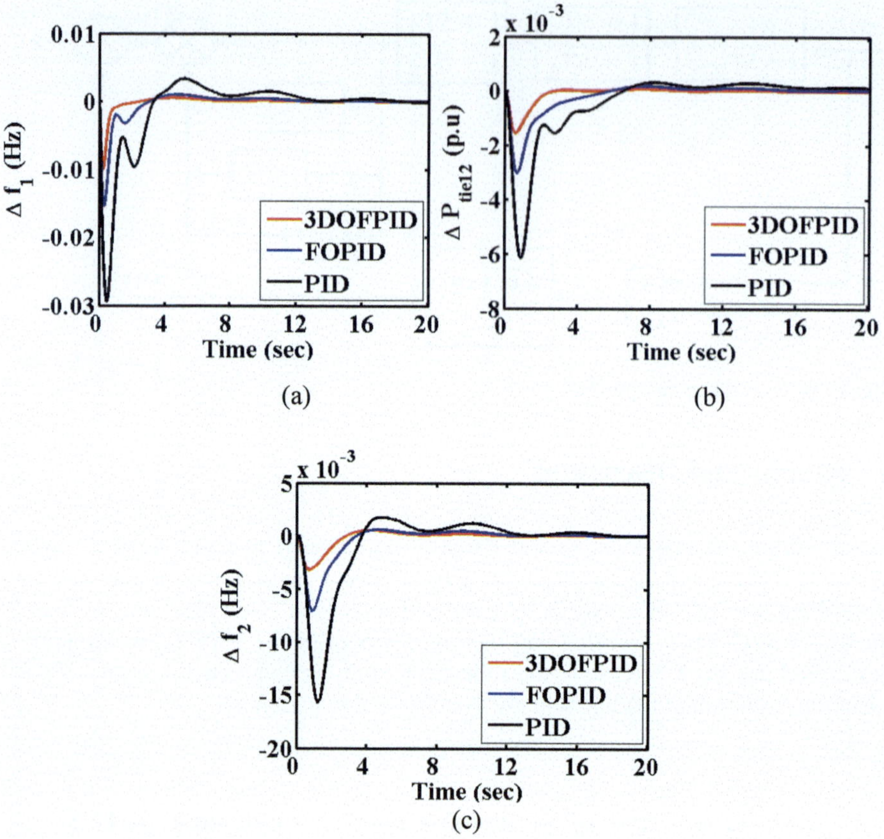

Fig. 5 Case 1 responses. **a** $\Delta f1$; **b** ΔP_{tie12}; **c** Δf_2

Table 1 Responses settling time

Settling time (s)	OOA optimization			OOA tuned 3DOFPID	
	PID	FOPID	3DOFPID	CES only	CES-TCSC
Δf_1	17.22	11.52	6.61	5.51	4.27
ΔP_{tie12}	18.76	14.86	9.8	9.13	7.61
Δf_2	17.66	11.99	9.89	8.81	6.11
ISE $\times 10^{-3}$	84.178	10.516	5.746	–	–

6.2 Case 2: Effect of CTDs on Renewable Energy Penetrated MAPS Performance

The renewable energy-permeated MAPS network is considered with a CTD of 0.25 s for realistic analysis and is required to demonstrate its effect on the system's dynamic

Table 2 Controller optimal gains

Parameters	Area 1			Area 2		
	PID	FOPID	3DOFPID	PID	FOPID	3DOFPID
K_P	1.943	2.476	1.578	1.524	2.628	1.371
K_I	0.235	1.018	0.0568	0.329	1.566	1.460
K_D	0.768	0.977	1.268	0.618	0.434	0.993
K_1	–	–	0.095	–	–	0.046
K_2	–	–	0.186	–	–	0.095
λ	–	0.067	–	–	0.015	–
μ	–	0.059	–	–	0.064	–
K_{ff}	–	–	0.039	–	–	0.125

behavior. Hence, under the regulation of the OOA-tuned 3DOFPID, the impact of CTDs on the system performance is showcased in this subsection by investigating the system without considering the CTDs. The system responses with and without CTDs under the OOA-tuned 3DOFPID for the unique disturbance conditions of 10% SLP on area 1 are compared in Fig. 6. Noticing that renewable energy penetrated MAPS behavior for the two conditions of neglecting and deliberating the CTDs, it is clear that the fluctuations in performance are more for the instance of considering the CTDs with the IPS network. A lag time will exist between the time the control signal is generated and the time it takes to modify the system's operational point. This delay might increase the RPM, and this reflects on the frequency of the IPS to deviate further. But, in realistic practice, the communication channels have a delayed nature, and hence this work recommends deliberating for realistic analysis. Suppose a secondary regulator is designed by neglecting the CTDs. That controller might be ineffective in dealing with the disturbances and unpredictable delays that arise in the network.

6.3 Case 3: Analysis of Renewable Energy Penetrated MAPS Performance with CES-TCSC Strategy Under 3DOFPID Controller

It is always encouraged to adopt the territory control approach for the better LFC of the IPS network. In this work, the CES-TCSC strategy is implemented to enhance the enrichment of renewable energy in MAPS. At first, the CES is integrated in both areas and under the OOA-tuned 3DOFPID controller, and for the 10% SLP in Area 1, the responses are shown in Fig. 7. Thereafter, placing the CES in both areas, the TCSC is placed with the tie-line, and for the same unique conditions, the system responses are compared in Fig. 7. Through observation of the system responses, it was revealed that an improvement in the dynamic behavior was reported with

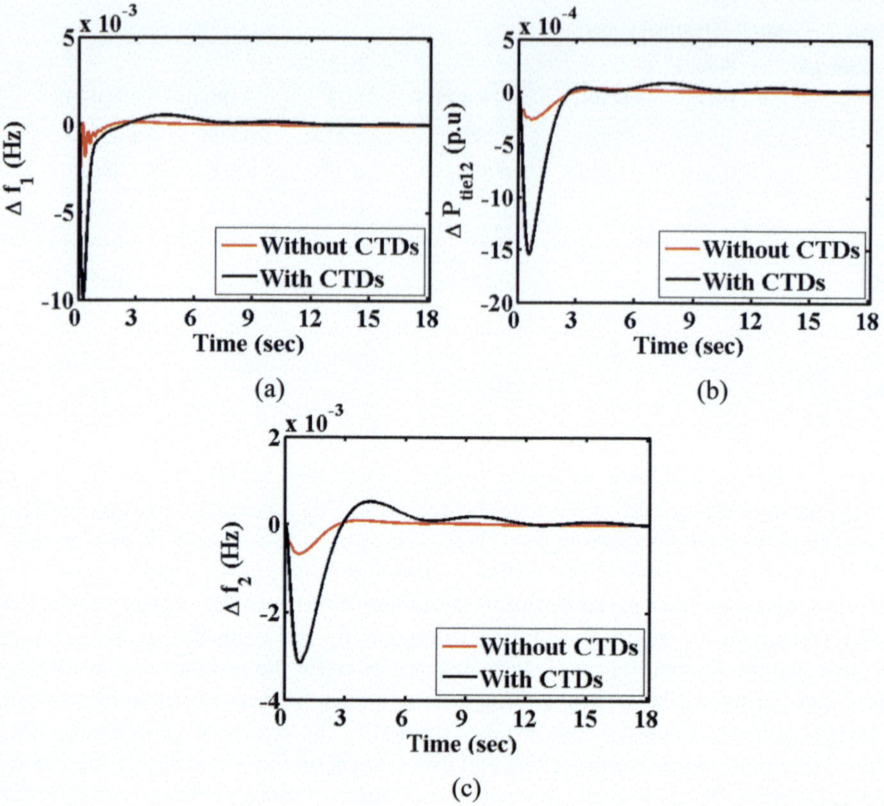

Fig. 6 Case 2 responses. **a** Δf_1; **b** ΔP_{tie12}; **c** Δf_2

the CES integration, and further improvement was achieved with the CES-TCSC strategy. With the CES and its quick response nature, it can effectively regulate the RPM in less time than the control actions taken by the regulator by injecting the stored energy into the network. The TCSC can alter the tie-line parameters, and hence the allowable line capability can regulate the inter-area power flow deviations. Overall, with the implementation of the CES-TCSC strategy, the dynamic behavior of the renewable energy-infused MAPS performance has improved considerably. For easy representation, the response settling time for the case of the CES-TCSC strategy is shown in Fig. 8 as a bar chart.

6.4 *Case 4: Robustness Test*

It is important to check the robustness of the suggested control approaches for stabilizing the IPS frequency during perturbations. This means that changes to both the

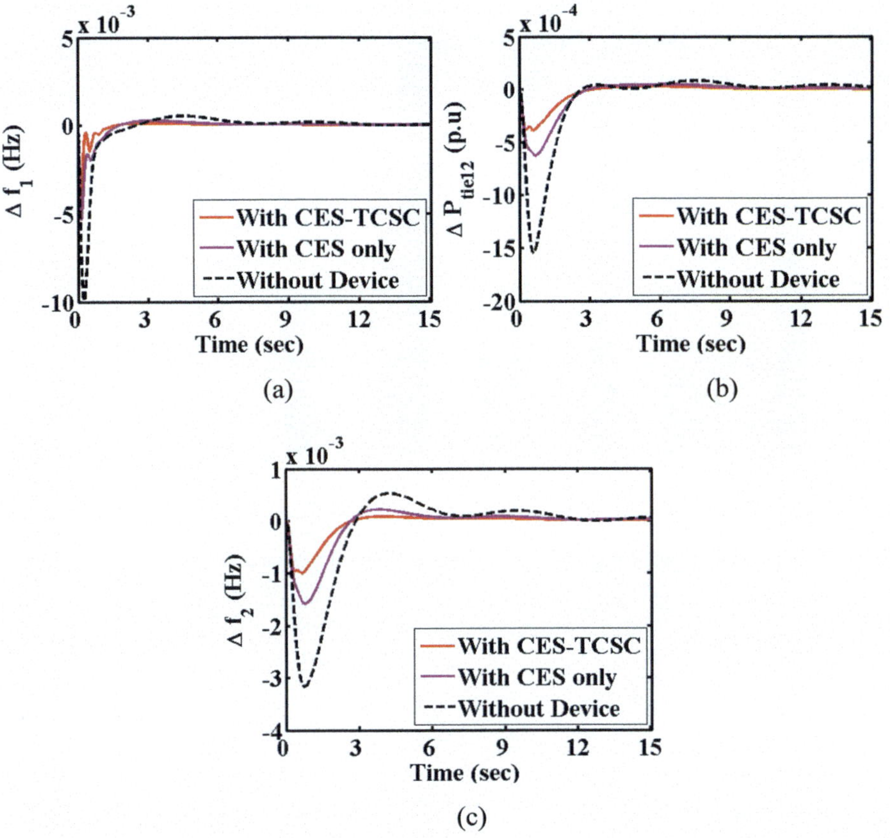

Fig. 7 Case 3 responses. **a** Δf_1; **b** ΔP_{tie12}; **c** Δf_2

load and the tie-line parameter are applied to the testing system to perform the robustness test. Figure 9 displays the results of the robustness test that was performed on the system using the OOA-tuned 3DOFPID and the CES-TCSC strategy, subjected to load fluctuations of $\pm$ 50% from the nominal load. The disturbances in system performance are noticed hardly for the instance of varying the load to the wide variations when compared to the case of the system with a nominal load. Further, the tie-line parameter value is varied in the range of $\pm$ 50% from the nominal value, and the deviations are lightly noticed as shown in Fig. 10. From this, it is concluded that the suggested controllers of the OOA-tuned 3DOFPID at the secondary and the CES-TCSC at the supplementary level are robust and effective enough to handle the IPS stability even under large load disturbances.

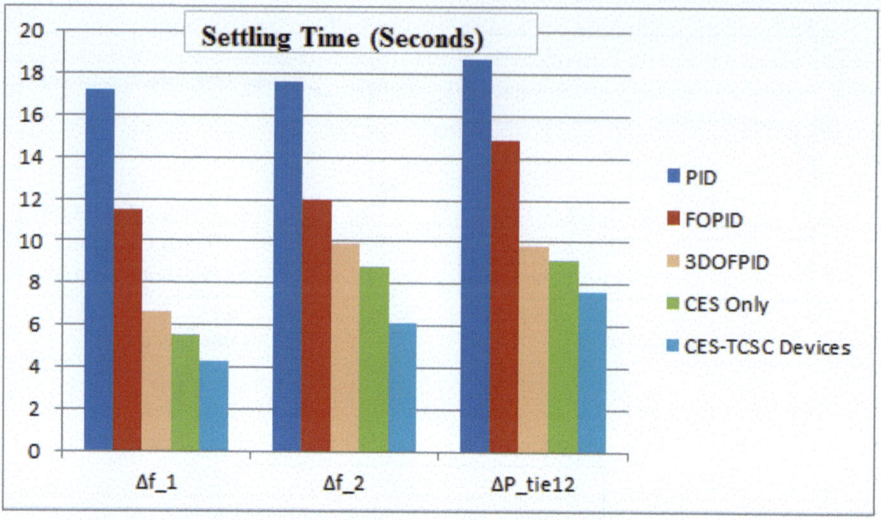

Fig. 8 Comparison of case 1 and case 3 responses settling time in seconds

7 Conclusions

The OOA-tuned 3DOFPID is suggested as the better frequency regulator for the renewable energy-penetrated two-area IPS in this work. On the other hand, the 3DOFPID's effectiveness is shown by the results of the PID and FOPID by subjugating the area 1 of the MAPS with 10% SLP. Moreover, the investigation of the MPS is carried out with the consideration of the CTDs to analyze a realistic environment. Further, the necessity of considering the CTDs with the MAPS network while designing the frequency regulator is discussed extensively, and their subsequent effect on the MAPS performance is demonstrated. The MAPS dynamic behavior is slightly more deviated when deliberating the network with CTDs, and to further compensate for these fluctuations, this work adopted the CES-TCSC strategy at the territory level. At first, the CES is integrated into areas 1 and 2, and later, the TCSC is placed with the tie-line. The simulation results confirmed the performance enhancement with the deployment of the CES-TCSC control strategy. The CES aids the MAPS in regulating the frequency fluctuations by delivering the stored energy to the grid to diminish the RPM, and the TCSC facilitates control over the tie-line behavior, resulting in a reduction in the inter-area oscillations. Thus, this work suggests the CES-TCSC strategy as a better control approach to obtain further improvement in the MAPS dynamic behavior. The last step is to run the robustness test, which shows how well the presented control method holds up.

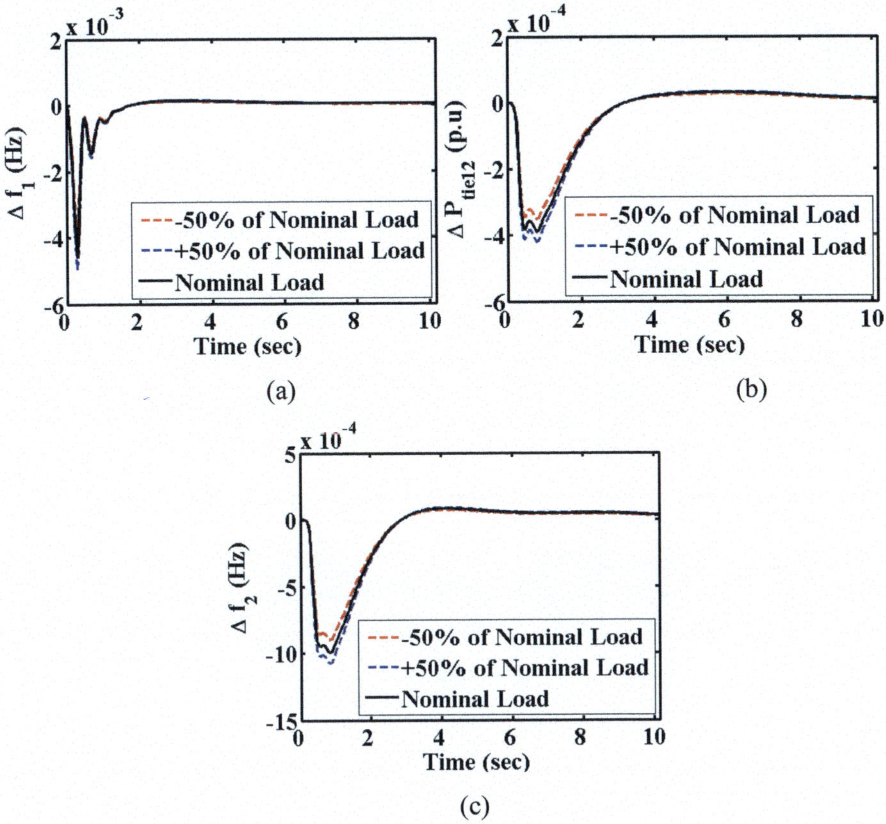

Fig. 9 Case 4 responses for loading uncertainty. **a** Δf_1; **b** ΔP_{tie12}; **c** Δf_2

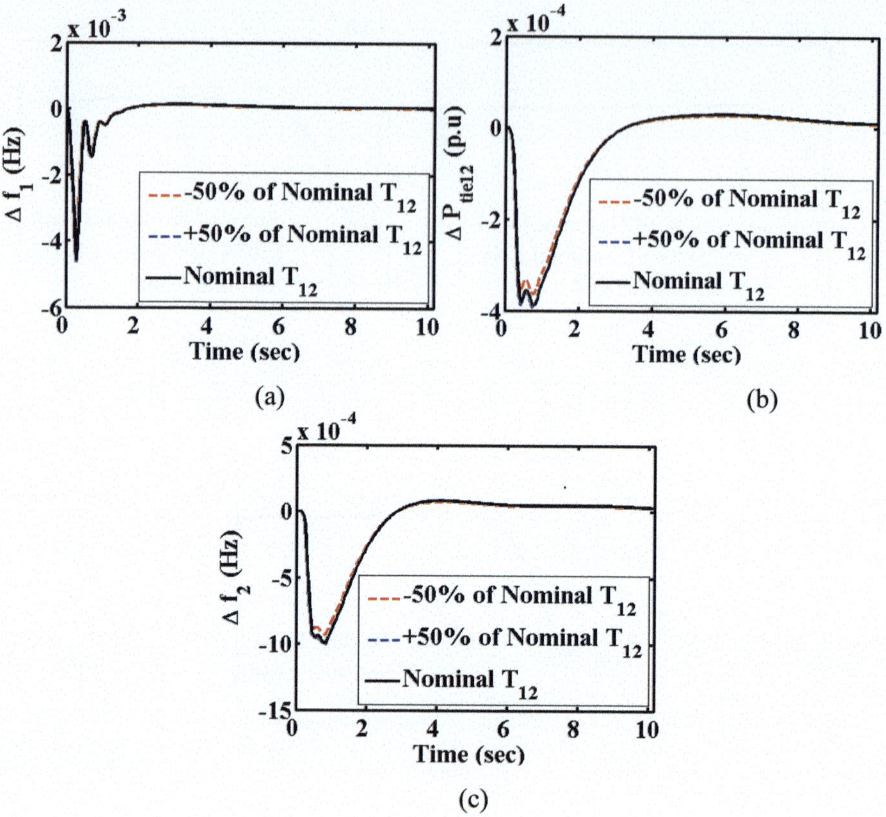

Fig. 10 Case 4 responses for tie-line parameter uncertainty. **a** Δf_1; **b** ΔP_{tie12}; **c** Δf_2

References

1. Kalyan C, Rao G, Bajaj M, Khan B, Zaidan M (2024) Frequency regulation of geothermal power plant-integrated realistic power system with 3DOFPID controller. Cogent Eng 11(01):2322820. https://doi.org/10.1080/23311916.2024.2322820
2. Khan I, Mokhlis H, Mansor N, Illias H, Awalin L, Wang L (2023) New trends and future directions in load frequency control and flexible power system: a comprehensive review. Alex Eng J 71:263–308. https://doi.org/10.1016/j.aej.2023.03.040
3. Jiao D, Shao C, Hu B, Xie K, Lin C, Ju Z (2023) Age-of-information aware PI controller for load frequency control. Prot Control Mod Power Syst 8:39. https://doi.org/10.1186/s41601-023-00311-z
4. Dewangan S, Prakash T, Singh V (2021) Design and performance analysis of elephant herding optimisation based controller for load frequency control in thermal interconnected power system. Optim Control Appl Methods 42(01):144–159. https://doi.org/10.1002/oca.2666
5. Hanwate S, Hote Y (2018) Optimal PID design for load frequency control using QRAWCP approach. IFAC-PapersOnline 51(04):651–656. https://doi.org/10.1016/j.ifacol.2018.06.170
6. Dhanaseakaran B, Kaliannan J, Baskaran A, Dey N, Taveras J (2023) Load frequency control assessment of a PSO-PID controller for a standalone multi-source power system. Technologies 11(01):22. https://doi.org/10.3390/technologies11010022

7. Gbadega P, Sun Y (2023) Multi-area load frequency regulation of a stochastic renewable energy based power system with SMES using enhanced-WOA-tuned PID controller. Heliyon 9:e19199. https://doi.org/10.1016/j.heliyon.2023.e19199
8. Guler Y, Kaya I (2023) Load frequency control of single-area power system with PI-PD controller design for performance improvement. J Electr Eng Technol 18:2633–2648. https://doi.org/10.1007/s42835-022-01371-1
9. Kalyan C, Rao G (2022) Coordinated control strategy for simultaneous frequency and voltage stabilisation of the multi-area interconnected system considering communication time delays. Int J Ambient Energy 43(01):5512–5524. https://doi.org/10.1080/01430750.2021.1967192
10. Kumari S, Pathak P (2024) A state-of-the-art review on recent load frequency control architectures of various power system configurations. Electr Power Compon Syst 52(5):722–765. https://doi.org/10.1080/15325008.2023.2234373
11. Mohapatra G, Debnath M, Mohapatra K (2020) IMO-based novel adaptive dual-mode controller design for AGC investigation in different types of systems. Cogent Eng 7(01):1711675. https://doi.org/10.1080/23311916.2020.1711675
12. Nosratabadi S, Bornapour M, Gharaei M (2019) Grasshopper optimisation algorithm for optimal load frequency control considering predictive functional modified PID controller in restructured multi-resource multi-area power system with redox flow battery units. Control Eng Pract 89:204–227. https://doi.org/10.1016/j.conengprac.2019.06.002
13. Shukla H, Raju M (2023) Application of COOT algorithm optimized PID plus D2 controller for combined control of frequency and voltage considering renewable energy sources. e-Prime Adv Electr Eng Electron Energy 6:100269. https://doi.org/10.1016/j.prime.2023.100269
14. Dong M, Liu K, He J, Li X, Qiao L, Zhao X (2023) Power grid load frequency control based on fractional order PID combined with pumped storage and battery energy storage. Energy Rep 9(10):1402–1411. https://doi.org/10.1016/j.egyr.2023.05.117
15. Iqbal M, Limon M, Kabir M, Rabby M, Soeb M, Jubayer M (2024) A hybrid optimization algorithm for improving load frequency control in interconnected power systems. Expert Syst Appl 249:123702. https://doi.org/10.1016/j.eswa.2024.123702
16. Rai A, Das D (2022) The development of a fuzzy tilt integral derivative controller based on the sailfish optimizer to solve load frequency control in a microgrid, incorporating energy storage systems. J Energy Storage 48:103887. https://doi.org/10.1016/j.est.2021.103887
17. Sahoo A, Sahoo A, Nayak S, Kar S (2024) Load frequency and voltage control of two area interconnected power system using covariance matrix adaptation evolution strategy based Aquila optimisation optimized fast fuzzy-fractional order tilt integral derivative controller. e-Prime Adv Electr Eng Electron Energy 8:100488. https://doi.org/10.1016/j.prime.2024.100488
18. Ali M, Kotb H, AboRas M, Abbasy H (2022) Frequency regulation of hybrid multi-area power system using wild horse optimizer based new combined fuzzy fractional-order PI and TID controllers. Alex Eng J 61(12):12187–21221. https://doi.org/10.1016/j.aej.2022.06.008
19. Ghany M, Mosa M, Ghany A, Yousef M (2024) Improving the frequency response of a multi-generation power system using adaptive variable structure fuzzy controller. Electr Power Syst Res 232:110368. https://doi.org/10.1016/j.epsr.2023.109511
20. Kalyan C, Srihari P, Nagalingachary K, Kumar A, Imran A, Goud B, Kumar M (2023) Path finder optimization algorithm tuned 3DOFPID controller for frequency stabilization in wind integrated realistic power system with HVDC line. Int J Renew Energy Res 13(01):208–220. https://doi.org/10.20508/ijrer.v13i1.13302.g8676
21. Kalyan C, Goud B, Reddy C, Bajaj M, Rao G (2023) SMES and TCSC coordinated strategy for multi-area multi-source system with water cycle algorithm based 3DOF-PID controller. Smart Sci 11(01):1–15. https://doi.org/10.1080/23080477.2022.2054199
22. Harideep P, Reddy V, Ganesh G, Prakash O, Kalyan C, Suresh C (2021) Seagull optimization tuned traditional PID controller with set point filter for LFC of hydro thermal system based on performance indices. In: 4th international conference on recent developments in control, automation & power engineering (RDCAPE), Noida, 7–8 Oct 2021, pp 85–89. https://doi.org/10.1109/RDCAPE52977.2021.9633523

23. Raju M, Saikia L, Sinha N (2017) Load frequency control of multi-area hybrid power system using symbiotic organisms search optimized two degree of freedom controller. Int J Renew Energy Res 7(04):1663–1674. https://doi.org/10.20508/ijrer.v7i4.6212.g7207
24. Kalyan C, Rao G (2021) Impact of communication time delays on combined LFC and AVR of a multi-area hybrid system with IPFC-RFBs coordinated control strategy. Prot Control Mod Power Syst 6(1):1–20. https://doi.org/10.1186/s41601-021-00185-z
25. Karnavas Y, Nivolianiti E (2023) Optimal load frequency control of a hybrid electric shipboard microgrid using jelly fish search optimization algorithm. Appl Sci 13(10):6128. https://doi.org/10.3390/app13106128
26. Delassi A, Arif S, Mokrani L (2018) Load frequency control problem in interconnected power systems using robust fractional PID controller. Ain Shams Eng J 9(1):77–88. https://doi.org/10.1016/j.asej.2015.10.004
27. Kalyan C, Rao G (2022) Combined frequency and voltage stabilization of multi-area multi-source system by DE-AEFA optimized PID controller with coordinated performance of IPFC and RFBs. Int J Ambient Energy 43(1):3815–3831. https://doi.org/10.1080/01430750.2020.1860130
28. Dehghani M, Trojovsky P (2023) Osprey optimisation algorithm: a new bio-inspired meta-heuristic algorithm for solving engineering optimisation problems. Front Mech Eng 8:1126450. https://doi.org/10.3389/fmech.2022.1126450
29. Kumar A, Shankar G (2020) Load frequency control assessment of tidal power plant and capacitive energy storage systems supported microgrid. IET Gener Transm Distrib 14(07):1279–1291. https://doi.org/10.1049/iet-gtd.2019.1249
30. Kalyan C, Rao G (2020) Coordinated SMES and TCSC damping controller for load frequency control of multi area power system with diverse sources. Int J Electr Eng Inform 12(04):747–769. https://doi.org/10.15676/ijeei.2020.12.4.4
31. Nandi M, Shiva C, Mukherjee V (2017) TCSC based automatic generation control of deregulated power system using quasi-oppositional harmony search algorithm. Eng Sci Technol Int J 20(04):1380–1395. https://doi.org/10.1016/j.jestch.2016.08.021

A Cascaded (1 + PI)-PID Controller for the Improvement in Load Frequency Control of the Diversified Interconnected Power System

Ch. Naga Sai Kalyan and Surender Reddy Salkuti

Abstract This work aims to develop a robust controller design for the optimal operation and stability of the diversified interconnected power system (IPS). With regards to this connection, a cascade one plus proportional, integral (1 + PI), and proportional-integral-derivative (PID) ((1 + PI)-PID) is presented to regulate the deviations in the behavior of the diversified IPS. However, a recent soft computing technique called the butterfly optimization algorithm (BFOA) is utilized for the parameter tuning of the cascade (1 + PI)-PID controller to obtain the most optimal performance. The diversified IPS examined in this work has two areas with four types of generation units, such as thermal, hydro, gas, and wind energy conversion systems, in each area. The performance of the presented cascade controller is tested on the two-area, four-source system, and its efficacy is showcased. Further, to obtain an improvement in the diversified IPS performance, a high-voltage DC (HVDC) line is operated as the tie-line. The simulation analysis confirmed the considerable enhancement in the IPS performance with the HVDC line incorporation. Finally, the cascade (1 + PI)-PID and the incorporation of the HVDC line robustness are validated through the sensitivity analysis.

Keywords Cascade (1 + PI)-PID controller · Butterfly optimization · HVDC line · Interconnected power system · Load frequency control

Nomenclature

BFOA Butterfly optimization algorithm

Ch. N. S. Kalyan
Department of Electrical and Electronics Engineering, Vasireddy Venkatadri International Technological University, Nambur, Guntur 522508, India

S. R. Salkuti (✉)
Department of Railroad and Electrical Engineering, Woosong University, Daejeon 34606, Republic of Korea
e-mail: surender@wsu.ac.kr

S. R. Salkuti (ed.), *Artificial Intelligence for Integrated Smart Energy Systems in Electric Vehicles*, Lecture Notes in Electrical Engineering 1427,
https://doi.org/10.1007/978-3-031-94276-1_24

HHO	Harris hawks optimizer
IPS	Interconnected power system
ISE	Integral square error
PID	Proportional-integral-derivative
PSO	Particle swarm optimization
PUS	Peak undershoot
RPB	Real power balance
SLP	Step load perturbation
WEC	Wind energy conversion

1 Introduction

The operators of the electrical network are tending towards the interconnection of the existing networks and further to the penetration of renewable generation units to cater to the requirements of the demand. These developments might succeed in meeting the increase in power demands and can affect the optimal operation, security, and reliability of the interconnected power system (IPS) network at the same time. Moreover, the establishment of equilibrium between demand and generation is a difficult task because the demand on the electrical network is continuously varying and unpredictable. Thus, the power system network operators continuously monitor the real power balance (RPB), which indicates the imbalance between generation and demand. If the mismatch in the RPB stands for a long time, then there will be an adverse effect on the system frequency [1]. Through the load frequency control (LFC) action, the RPB can be achieved and the regulation of the IPS frequency can be obtained. Moreover, the LFC approach also regulates the power flow via the tie-line when perturbations occur in the system. At present, the penetration of RES into the existing grid is taking place rapidly to meet the rise in electricity demands. The harnessing of power from renewable energy sources (RESs) at the cheapest prices made their integration into the grid more rigorous, and at the same time, it impacted the grid's stability and reliability. Out of all the RESs, harnessing power from the wind energy conversion (WEC) system using the wind is more likely to be integrated with the grids [2]. It is well known that the WEC system tends to affect the grid frequency characteristics. Hence, there is an increase in concern regarding the WEC systems and their penetration, and therefore, it is required to investigate the behavior of the WEC systems and their subsequent impact on the IPS stability. Thus, this work deliberated the test system model with the WEC system along with the other types of generation units.

So far, many control approaches have been invented to regulate the IPS frequency. Of all, the traditional PI/PID-based control mechanisms are widely spread, irrespective of the IPS models. The PI using Harris Hawks Optimizer (HHO) [3] is implemented in the multi-area IPS with renewable energy plants, and the dominance of the HHO is demonstrated with multi-verse optimization and the Ant Lion algorithm.

For the isolated system with multiple sources, the particle swarm optimization (PSO) tuned PID is adopted in [4], and the superiority of the PSO is revealed with the differential evolution (DE) and genetic algorithm. The fitness-dependent optimizer-tuned PID is developed in [5] for a two-area system, and the comparative performance assessment is performed with the PID, PI-tuned PSO, and firefly algorithm, respectively. The hybrid algorithm of DE and the artificial field algorithm is adopted in [6] for PID tuning and has been tested on the two-area diver source system. Further, in [7], the analysis is extended to the non-linear system with renewable energy plants, and the efficacy of the suggested hybrid algorithm is validated with other algorithms of DE and PSO. Apart from the above-discussed algorithms, the other algorithms that are adopted by the researchers in the LFC study for the optimal tuning of the PI/PID controllers include the sine–cosine algorithm, marine predators approach, doctor-patient technique, teaching–learning optimization, chimp optimization, mine blast optimization, simulated annealing, fruit fly algorithm, ant swarm algorithm, mountain gazelle optimization, sine–cosine technique, artificial rabbit approach, dragonfly approach, crow search mechanism, slap swarm technique, gorilla troops algorithm, lion optimization, dwarf mongoose approach, and bald eagle- manta ray technique, etc., as discussed in [8] recently. However, the PI/PID traditional controllers can offer the best control mechanism for linearized plant structures, and in realistic practice, such kinds of IPS networks are inevitable. Thus, researchers are looking for and making some alternative efforts for better LFC regulator design.

The implementation of deep learning [9], fuzzy logic regulators, artificial neural networks (ANN), and other fractional order (FO) [10] controllers has been encouraged in recent years. The Sooty Tern approach for model predictive control is applied in [11], and the efficacy of the algorithm is demonstrated with the Bird-Bearer algorithm. The Coronavirus immunity algorithm [12] and crow search technique [13] tuned fuzzy PID have been adapted recently to the LFC, and the performance superiority of the fuzzy controller is revealed with some other traditional approaches. On the other hand, FO-aided PI/PID and tilt-based PID are also reported using optimizations like COOT optimization [14], flower pollination algorithm [15], Manta-Ray optimizer [16], wild horse optimizer [17], satin bowerbird optimizer [18], seagull optimization [19], cohort intelligent optimization [20], and water cycle optimization [21], etc. However, the complexity of rendering the FO parameters via the optimization process and the assumptions in the fuzzy rule-based system, their efficacy, and their robustness to the non-linear practical system are not extensively validated.

Hence, this work focused on the development of cascade structures for LFC actions. The cascade controller has multiple stages, mostly two that are concentrated on eliminating disturbances in the input signal by the outer loop and then changing the plant's operating point smoothly by the inner loop. The literature survey revealed that, only to a limited extent, the research work was reported on the LFC with cascade controllers, and there is a lot more scope in developing the most efficient and robust cascade control structures for optimal LFC actions. The PDN-PI based on the coyote technique in [22], the fuzzy cascade regulator using the crow-search technique in [23], and the cascaded PIDA tuned with the marine predators' algorithm [24], etc. are recently reported. However, the BFOA for tuning the cascade (1 + PI)-PID controller is not available in the literature, and thus this work aims to investigate the LFC study with the motto of offering robust frequency control. The contributions of this work are:

- The dynamic stability of a two-area, four-source system is studied.
- The cascade (1 + PI)-PID controller is suggested as the secondary controller.
- The BFOA approach optimizes the suggested control operation.
- The suggested control technique's effectiveness is contrasted with that of other control approaches found in the literature.
- HVDC line as the tie-line is laid with the system and enrichment in the system's dynamic performance is achieved.
- Sensitivity analysis is carried out, confirming the enhanced control approach's resilience.

The remaining sections in this chapter are organized by demonstrating the investigated model in Sect. 2 and the cascade (1 + PI)-PID controller in Sect. 3. Section 4 presents the comprehensive modeling of the BFOA followed by the simulation studies in Sect. 5. Ultimately, Sect. 6 presents the study's main findings.

2 Power System Model

The two-area, four-source system taken for investigation in this work has two areas with thermal, hydro, wind, and gas-based power generation units in each area, and the two areas have a power generation capacity of 1:1. The disturbance of 10% step load perturbation (SLP) is laid on area-1 to analyze the dynamical behavior of the considered IPS network. The model of the two-area, four-source system shown in Fig. 1 is developed in the (R2019a) MATLAB/Simulink domain, and the required parameters are made available in [25]. Later, the analysis of the two-area, four-source system is extended to the incorporation of the HVDC line. The implementation of the HVDC line as the tie-line between the control regions is gaining momentum to ensure further improvement in the IPS network performance. The modeling of the HVDC line is given in Eq. (1) [26], and the gain (K_{HVDC}) and time (T_{HVDC}) constants are taken as 0.9832 and 0.8976 s, respectively.

$$G_{HVDC} = \frac{K_{DC}}{1 + sT_{DC}} \tag{1}$$

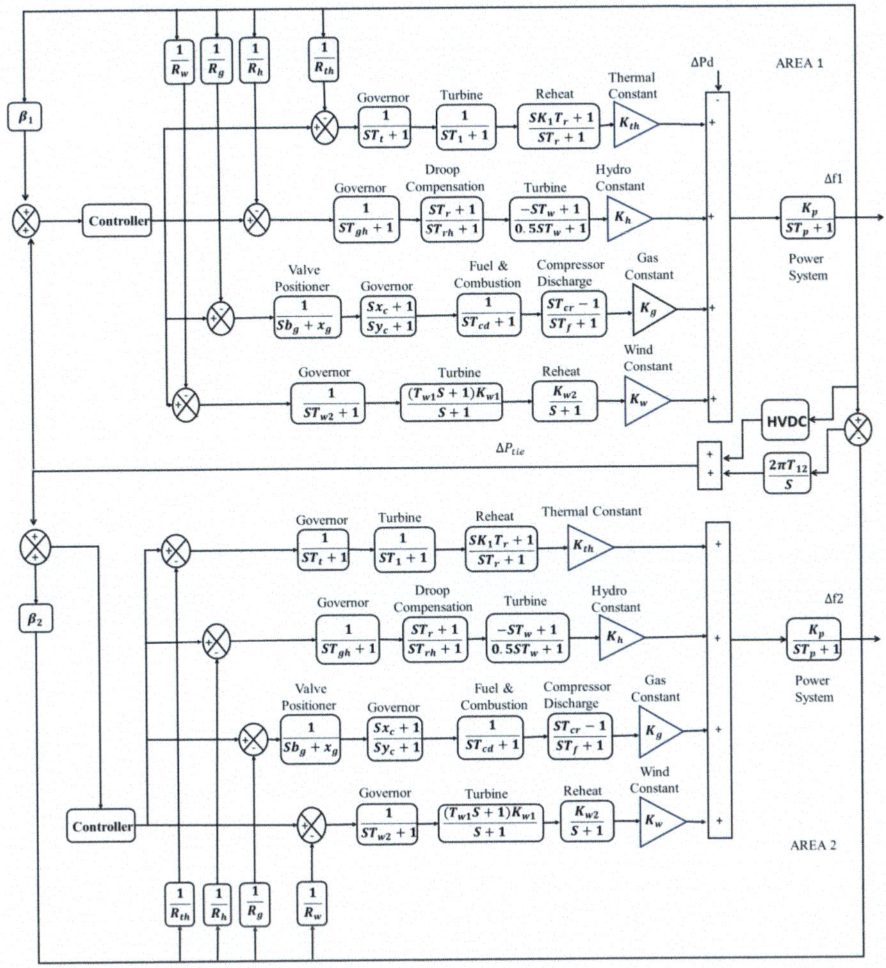

Fig. 1 Model of two-area four-source system with HVDC line

3 Cascade (1 + PI)-PID Controller

The variations in input to the controller lead to an abrupt change in the control signal delivered to the plant. The transient response of the plant might deteriorate with the changes involved in the control signal. To avoid such circumstances, a modified control structure is proposed for the PID controller and is termed a cascade (1 + PI)-PID controller in this work. This controller has two loops: the control part (1 + PI) works in the outer loop, and the PID acts in the inner loop [27]. The output of the outer loop is given to the inner loop as the input, which can be used as the set pin for the PID controller. The combination of frequency deviations in the control regions and the tie-line power is formulated as the area control error (ACE), which can be

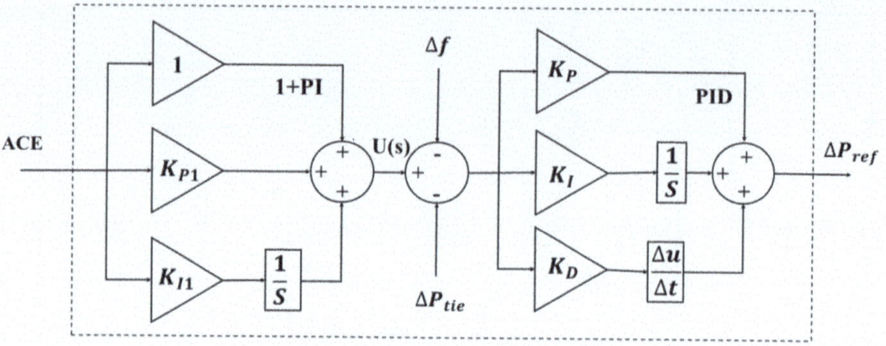

Fig. 2 Structure of cascade (1 + PI)-PID controller

fed to (1 + PI) as input. The structure of the (1 + PI)-PID controller has been shown in Fig. 2 and has multiple tuning hubs. This might enhance the response's dynamical characteristics by eliminating the sluggish nature through the integral action in the (1 + PI) loop. The parameters like K_P, K_I, and K_D represent the proportional, integral, and derivative gains of the PID controller, and the parameters K_{PI} and K_{I1} represent the proportional and integral gains in the (1 + PI) part. These parameters are tuned to the integral square error (ISE) [28, 29] placed in Eq. (2).

$$J_{ISE} = \int_0^{T_{sim}} (\Delta f_1^2 + \Delta P_{tie12}^2 + \Delta f_2^2)dt \quad (2)$$

4 Butterfly Optimization Algorithm

Being in the category of Lepidoptera in the family of insects, butterflies have survived for so many years and are of nearly twenty thousand species spread over the planet. The reason behind it is their sensory receptors, which facilitate their finding of nectar as food. Apart from the use of smell by the butterflies, they can also use their senses of taste, hearing, and touch to find a partner for mating. Moreover, these senses also help the butterflies move from one location to another. Out of all the senses, in search of nectar, the butterflies strongly rely on the sense receptors that are scattered over their body and can work like antennas. For mating, the chemoreceptors in the butterflies guide them in finding their best mating partner. The observations from the scientific investigations revealed that the butterflies can sense multiple fragrances and can separate them. Further, the intensity of detecting the fragrances will be different for the different butterflies. In this work, the BFOA is implemented, which emerged from the socio-behavioral behavior of the butterflies, to locate the cascade (1 + PI)-PID parameters. The butterflies are the search agents in the BFOA, and after initializing the initial parameters, the formulation of the fragrance (F) is evaluated using Eq. (3) [30].

$$F = C.\,I^{a} \tag{3}$$

Where the indicators C, I, and a represent the sensory fragrance, stimulus, and power exponent parameters, respectively. The values of C and a will be taken from [0–1]. Moreover, the stimulus 'I' is closely associated with the objective function, which can be calculated using Eq. (2). Overall, the initialization, iterative, and final stages are the phases that are operating the BFOA. In the initialization phase, the boundaries for the search space are defined within the range of [0–3] to find the (1 + PI)-PID parameters. The iterative phase deals with the movement of the butterflies in an iterative manner in search of the best solution. However, it has two movements: one is local and the other is global. The movement of the butterflies in the global as well as in the local searchers [31] is given in Eqs. (4) and (5), respectively.

$$P_j^{k+1} = P_j^k + \left(r()^2 .\, g^* - P_j^k\right).\, F_j \tag{4}$$

where the subscript 'j' indicates the butterfly in the kth iteration, whose solution vector is represented by the P_j^{k+1}. In the present iteration, the best solution is represented by g*, and r() indicates the generation of random values in the form [0–1].

$$P_j^{k+!} = P_j^k + \left(r()^2 .\, P_i^k - P_l^k\right).\, F_j \tag{5}$$

where the sub-scripts 'i' and 'l' represent the ith and lth butterflies in the solution vector, The movement of the butterflies can be decided by the parameter 'P' whether the searching can take place in the local or as in the global level. A comparison is done between the 'P' and the randomly generated value 'r()'. If r() is less than 'P' then the searching strategy will be done at the global level, and otherwise it will be performed at the local level. The steps will be carried out again and again until the program reaches its maximum number of trials and the solutions can be shown. The detailed flowchart is shown in Fig. 3, and in this work, the maximum iterations are taken as 50.

5 Simulation Results

The simulation study has been carried out in three cases for 10%SLP on the area-1 and the entire investigations are undergone in MATLAB/Simulink domain. In case 1, the performance comparison of various controllers on the test system model is performed to reveal the best one and further in case 2, under the best controller, the test system performance is investigated by incorporating the HVDC line. Finally, in case 3, the

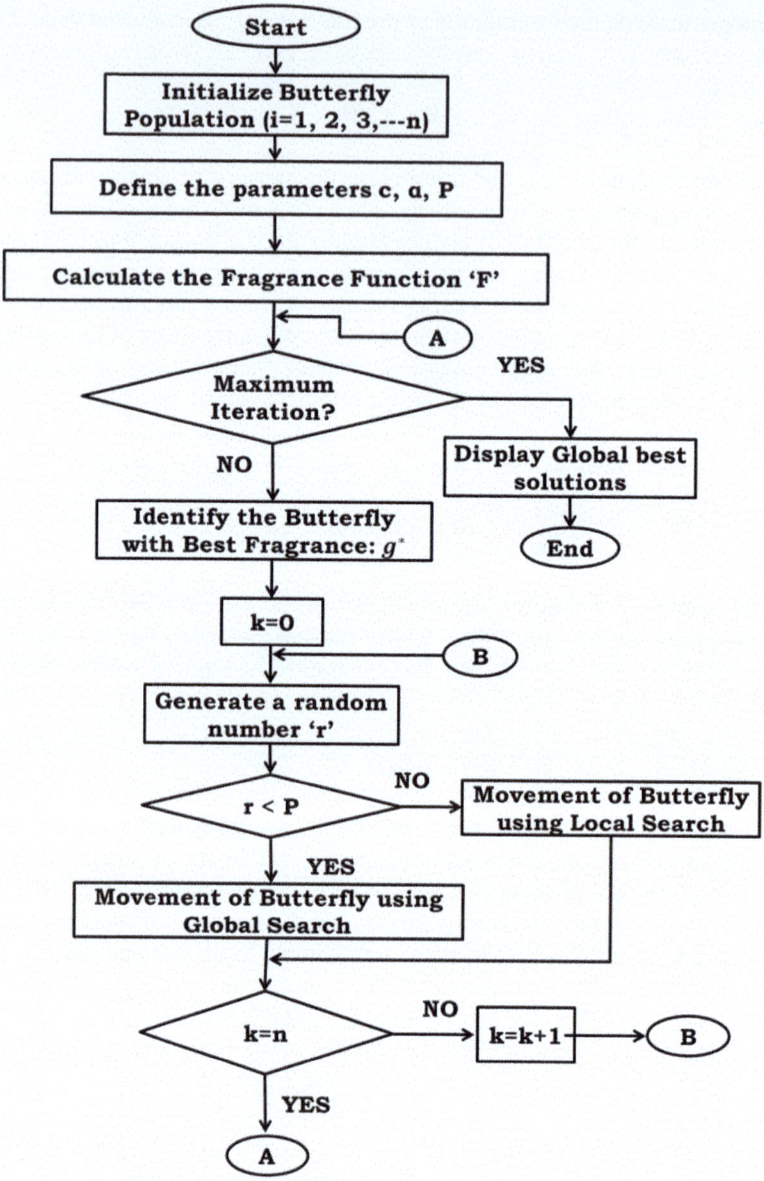

Fig. 3 BFOA flowchart

sensitivity test is conducted in terms of loading uncertainty and random loadings to reveal the robustness of the proposed cascade controller and the deployment of the HVDC line with the system.

5.1 *Case 1: Analysis of Two-Area Four Source System Performance Under Various Controllers*

The considered model of the two-area, four-source system for the investigation is analyzed for laying the area-1 with 10% SLP. To get the comparative performance assessment, the controllers like cascade (1 + PI)-PID, FOPID, 2DOFPID, and PID based on BFOA are enacted individually as the secondary controllers, and the operational performance is checked under the same load disturbance. The dynamic behavior of the investigative power system under these controllers is showcased in Fig. 4 and is interpreted numerically in terms of the peak undershoot (PUS) and the settling time in seconds, as placed in Table 1. Noticing Fig. 4 and Table 1, it is known that the proposed cascade (1 + PI)-PID controller outperforms the other controllers in diminishing the PUS, and the presented controller very effectively dampens the oscillations. Further, the objective function is very effectively minimized with the cascade (1 + PI)-PID controller and is enhanced by 55.84% with FOPID, 65.08% with 2DOFPID, and 73.08% with the PID controller. The parameters of the cascade (1 + PI)-PID, FOPID, 2DOFPID, and PID controllers that are found using the BFOA are placed in Table 2.

5.2 *Case 2: Analysis of Two-Area Four Source System Performance with HVDC Line Under the Cascaded Controller*

Later, the dynamic behavior of the investigated IPS network is analyzed by incorporating the HVDC line as the tie-line between control regions one and two. The analysis for this case is performed under the cascade (1 + PI)-PID controller as its performance efficacy is established in the above subsection, and further analysis is continued under the same regulator. The responses of the system without the HVDC tie-line and with the incorporation of the HVDC line under the same BFOA-tuned (1 + PI)-PID controller are shown in Fig. 5. After careful observation of the behavior depicted in Fig. 5, it is elucidated that the system responses are more promising with the employment of the HVDC, the oscillations are damped out further, and the responses concede very little time in attaining the steady state position. As you can see from the bar chart in Fig. 6, the settling time of the system reactions with the HVDC line is very clear. The ability of the HVDC line to deliver bulk power makes it feasible to eliminate power flow oscillations during load fluctuations. This results

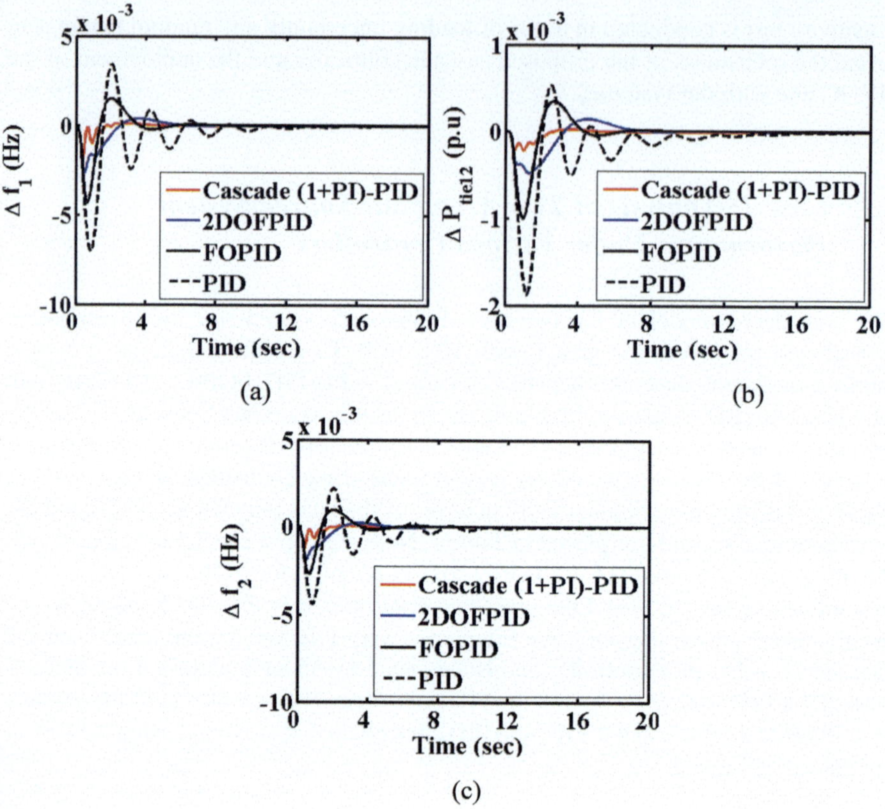

Fig. 4 Case 1 responses. **a** Δf_1, **b** ΔP_{tie12}, **c** Δf_2

Table 1 Responses settling time in seconds

Settling time (s)	BFOA tuning				BFOA tuned cascade (1 + PI)-PID
	PID	2DOFPID	FOPID	Cascade (1 + PI)-PID	HVDC line
Δf_1	17.41	10.57	7.71	5.73	4.77
ΔP_{tie12}	18.24	15.46	9.13	6.81	4.52
Δf_2	17.2	11.37	7.50	5.77	4.63
ISE * 10^{-3}	113.754	87.693	69.331	30.617	–

Table 2 Controller optimal parameters

Parameters	PID	2DOFPID	FOPID	Cascade (1 + PI)-PID
K_P	2.667	1.645	1.886	0.898
K_I	1.895	1.134	1.729	0.064
K_D	1.858	1.786	0.542	1.146
K_1	–	0.072	–	–
K_2	–	0.031	–	–
λ	–	–	0.045	–
μ	–	–	0.047	–
K_{P1}	–	–	–	0.525
K_{I1}	–	–	–	0.098

in the enhancement of the performance, and hence the incorporation of the HVDC line with the IPS networks is handy.

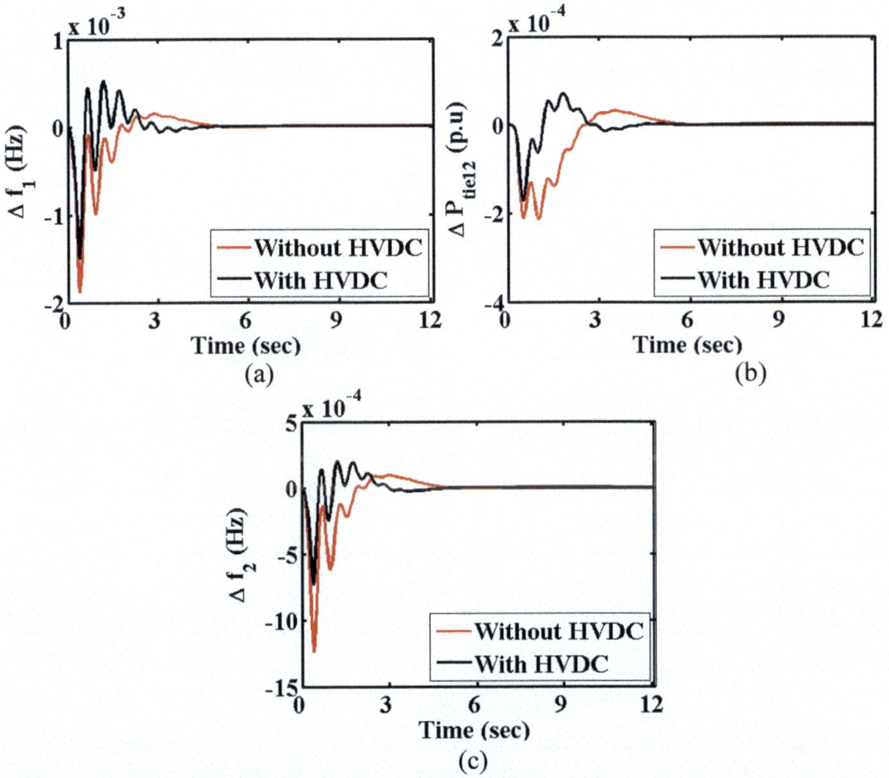

Fig. 5 Case-2 responses. a. Δf_1 b. ΔP_{tie12} c. Δf_2

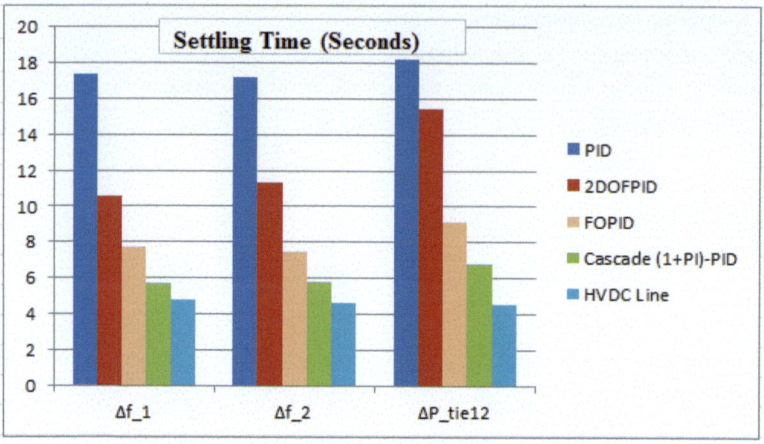

Fig. 6 Comparison of responses setting time in seconds for case 1 and case 2

5.3 *Case 3: Sensitivity Analysis*

The sensitivity analysis is further conducted about the subjection of different loadings and the random loadings on the two-area, four-source model of the IPS network. The BFOA-tuned cascade (1 + PI)-PID controller oversees the system's sensitivity analysis, which takes into account the use of the HVDC line. Initially, the system performance is tested on different loadings of ± 25% from the nominal load on the area. The system responses for different loadings are shown in Fig. 7, and it is clear that the deviations in the responses are hardly noticed even when the load is varied in the range of ± 25% from the nominal load. Thus, it is preliminarily concluded that the designed controller is robust and can withstand a wide range of load fluctuations without any change in the controller parameters. Later, the system is targeted with random loadings, and the system behavior with and without the HVDC line is analyzed as shown in Fig. 8. From Fig. 8, it is clear that the system with the HVDC tie-line under random loadings performs well in ensuring the system's dynamic behavior is stable. From the sensitivity analysis, it is concluded that the suggested BFOA-tuned cascade (1 + PI)-PID controller and the employment of the HVDC line with the system for attaining stable operation are robust.

6 Conclusions

For the dynamical stability of the diversified IPS network, this work suggested the cascade (1 + PI)-PID controller be fine-tuned with the BFOA technique. However, the operational efficacy of the suggested cascade (1 + PI)-PID controller is demonstrated by the performances of the PI, PID, and FOPID controllers in the literature.

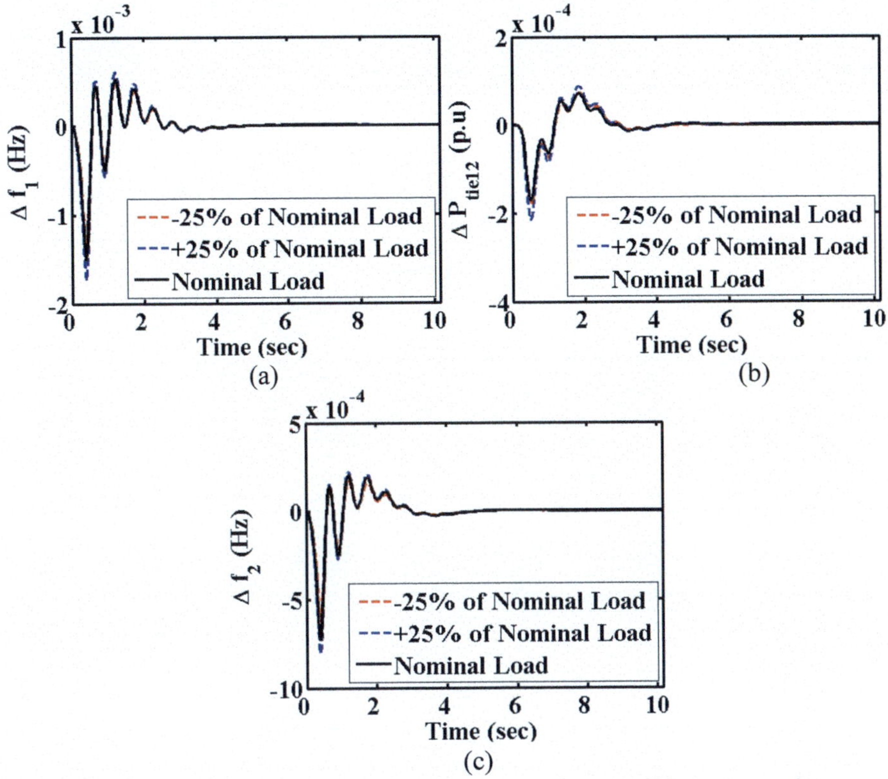

Fig. 7 Case 3 responses for loading uncertainty. **a** Δf_1, **b** ΔP_{tie12}, **c** Δf_2

The operational performances of these controllers are investigated by subjecting the considered power system model to a disturbance of 10% SLP in area 1. Further, the investigative system performance is assessed by laying the HVDC line as the tie-line, and the simulation analysis revealed further improvement in the performance. Later, the sensitivity analysis is conducted in terms of injecting the different loads and random loadings into the system under the supervision of the cascade controller and the HVDC line. The dynamic responses of the system under the sensitivity test revealed that the suggested control techniques for the stability of the diversified IPS network are robust, and the controller optimization need not be undergone again for the different load perturbations.

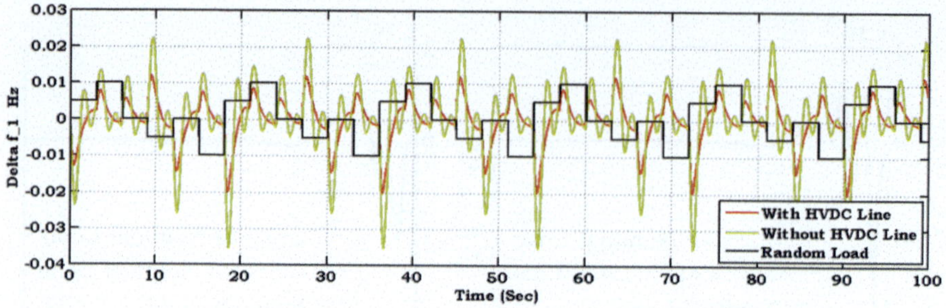

(a)

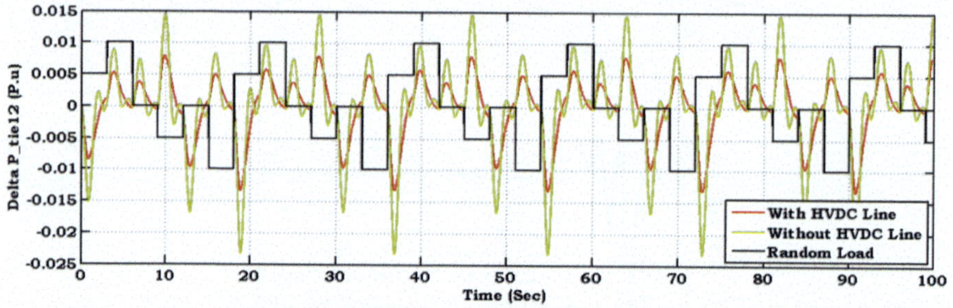

(b)

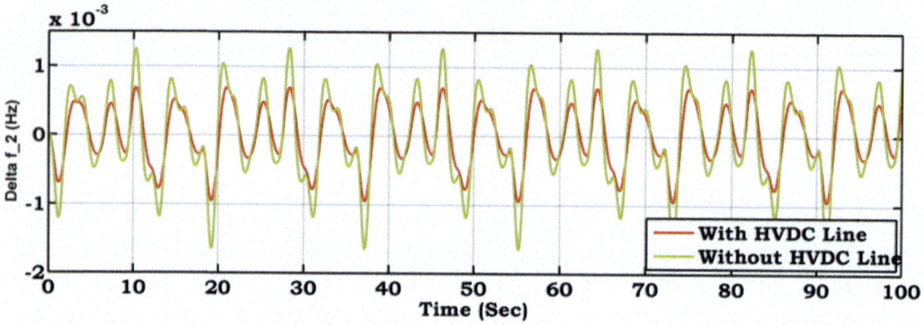

(c)

Fig. 8 Case 3 responses for random loading. **a** Δf_1, **b** ΔP_{tie12}, **c** Δf_2

References

1. Kalyan C, Rao G (2020) Frequency and voltage stabilisation in combined load frequency control and automatic voltage regulation of multiarea system with hybrid generation utilities by AC/DC links. Int J Sustain Energ 39(10):1009–1029. https://doi.org/10.1080/14786451.2020.1797740
2. Wang Z, Li D, Lyu X, Gao S, Fu C, Zhu S, Wang B (2023) Intelligent load frequency control for improving wind power penetration in power systems. Energy Rep 9:1225–1234. https://doi.org/10.1016/j.egyr.2023.04.091

3. Yousri D, Babu T, Fathy A (2020) Recent methodology based Harris hawks optimizer for designing load frequency control incorporated in multi-interconnected renewable energy plants. Sustain Energy Grids Netw 22:100352. https://doi.org/10.1016/j.segan.2020.100352
4. Dhanasekaran B, Kaliannan J, Baskaran A, Dey N, Tavares J (2023) Load frequency control assessment of a PSO-PID controller for a standalone multi-source power system. Technologies 11(01):22. https://doi.org/10.3390/technologies11010022
5. Daraz A, Malik S, MokhliS H, Haq I, Laghari G, Mansor N (2020) Fitness dependent optimizer-based automatic generation control of multi-source interconnected power system with non-linearities. IEEE Access 8:100989–101003. https://doi.org/10.1109/ACCESS.2020.2998127
6. Kalyan C, Rao G (2020) Coordinated SMES and TCSC damping controller for load frequency control of multi area power system with diverse sources. Int J Electr Eng Inform 12(04):747–769. https://doi.org/10.15676/ijeei.2020.12.4.4
7. Kalyan C, Rao G (2022) Combined frequency and voltage stabilisation of multi-area multi-source system by DE-AEFA optimised PID controller with coordinated performance of IPFC and RFBs. Int J Ambient Energy 43(01):3815–3831. https://doi.org/10.1080/01430750.2020.1860130
8. Wadi M, Shobole A, Elmasry W, Kucuk I (2024) Load frequency control in smart grids: a review of recent developments. Renew Sustain Energy Rev 189:114013. https://doi.org/10.1016/j.rser.2023.114013
9. Wu M, Ma D, Xiong K, Yuan L (2024) Deep reinforcement learning for load frequency control in isolated microgrids: a knowledge regression approach with emphasis on power symmetry and balance. Symmetry 16(03):322. https://doi.org/10.3390/sym16030322
10. Kumar A, Pan S (2022) Design of fractional order PID controller for load frequency control system with communication delay. ISA Trans 129:138–149. https://doi.org/10.1016/j.isatra.2021.12.033
11. Ali H, Fathy A, Kassem A (2020) Optimal model predictive control for LFC of multi-interconnected plants comprising renewable energy sources based on recent sooty tern approach. Sustain Energy Technol Assess 42:100844. https://doi.org/10.1016/j.seta.2020.100844
12. Biswas J, Bera P, Chakrabarty K (2023) Determination of control area and design of fuzzy-rule tuned PID controller for LFC of multimachine power system. Electr Power Syst Res 221:109411. https://doi.org/10.1016/j.epsr.2023.109411
13. Nayak J, Shaw B, Sahu B, Naidu K (2022) Application of optimized adaptive crow search algorithm based two degree of freedom optimal fuzzy PID controller for AGC system. Eng Sci Technol Int J 32:01061. https://doi.org/10.1016/j.jestch.2021.09.007
14. Moschos I, Parisses C (2023) Combined frequency and voltage control of two-area multi-source interconnected microgrids via the 2DOF-TID controller. e-Prime-Adv Electr Eng, Electron Energy 5:100268. https://doi.org/10.1016/j.prime.2023.100268
15. Hongesombut K, Keteruksa R (2023) Fractional order based on a flower pollination algorithm PID controller and virtual inertia control for microgrid frequency stabilisation. Electr Power Syst Res 220:109381. https://doi.org/10.1016/j.epsr.2023.109381
16. Ahmed E, Mohammad E, Selim A, Aly M, Alsadi A, Alhosaini W, Alnuman H, Ramadan H (2023) Improving load frequency control performance in interconnected power systems with a new optimal high degree of freedom cascaded FOTPID-TIDF controller. Ain Shams Eng J 14(10):102207. https://doi.org/10.1016/j.asej.2023.102207
17. Ali M, Kotb H, AboRas M, Abbasy H (2022) Frequency regulation of hybrid multi-area power system using wild horse optimizer based new combined fuzzy fractional order PI and TID controllers. Alex Eng J 61(12):12187–12210. https://doi.org/10.1016/j.aej.2022.06.008
18. Chinta D, Jatoth R, Biswal M (2023) Power system frequency profile improvement using satin bowerbird optimised tilt fractional order cascade controller with reduced data based cost function. J Eng Res 11(04):414–424. https://doi.org/10.1016/j.jer.2023.100116
19. Kalyan C, Venkateswarlu A, Reddy C, Goud B, Prasad A, Sriram C, Aymen F (2021) Frequency regulation of multi area renewable energy source system with practical constraints under fractional-order fuzzy controller. Int J Renew Energy Res 11(03):992–1002. https://doi.org/10.20508/ijrer.v11i3.12318.g8286

20. Murugesan D, Jagatheesan K, Shah P, Sekhar R (2023) Fractional order PID controller for microgrid power system using cohort intelligence optimization. Results Control Optim 11:100218. https://doi.org/10.1016/j.rico.2023.100218
21. Kalyan C, Goud B, Rajasekaran S, Swapna G, Venkateswarlu A, Reddy C (2022) Coordinated IPFC and SMES strategy for stability analysis of renewable energy based contemporary interconnected power system with FOPID controller. Int J Renew Energy Res 12(02):1052–1062. https://doi.org/10.20508/ijrer.v12i2.13012.g8493
22. Ela A, Sehiemy R, Shaheen A, Diab A (2022) Design of cascaded controller based on coyote optimiser for load frequency control in multi-area power systems with renewable sources. Control Eng Pract 121:105058. https://doi.org/10.1016/j.conengprac.2021.105058
23. Sahu P, Samantaray S (2023) Resilient frequency stability of a PV/Wind penetrated complex power system with CSA tuned robust type-2 fuzzy cascade PIF controller. Electr Power Syst Res 125:109815. https://doi.org/10.1016/j.epsr.2023.109815
24. Yakout A, Attia M, Kotb H (2021) Marine predator algorithm based cascaded PIDA load frequency controller for electric power systems with wave energy conversion systems. Alex Eng J 60(04):4213–4222. https://doi.org/10.1016/j.aej.2021.03.011
25. Hakimuddin N, Nasiruddin I, Bhatti T, Arya Y (2020) Optimal automatic generation control with Hydro, thermal, gas and wind power plants in 2-area interconnected power system. Electr Power Compon Syst 48(06):558–571. https://doi.org/10.1080/15325008.2020.1793829
26. Kalyan C, Srihari P, Nagalingachary K, Kumar A, Imran A, Goud B, Kumar M (2023) Path finder optimization algorithm tuned 3DOFPID controller for frequency stabilization in wind integrated realistic power system with HVDC line. Int J Renew Energy Res 13(01):208–220. https://doi.org/10.20508/ijrer.v13i1.13302.g8676
27. Arya Y (2019) AGC of PV-thermal and hydro-thermal power systems using CES and a new multi-stage FPIDF-(1+PI) controller. Renew Energy 134:796–806. https://doi.org/10.1016/j.renene.2018.11.071
28. Kalyan C, Rao G, Bajaj M, Khan B, Zaidan M (2024) Frequency regulation of geothermal power plant-integrated realistic power system with 3DOFPID controller. Cogent Eng 11(01):2322820. https://doi.org/10.1080/23311916.2024.2322820
29. Kalyan C, Rao G (2022) Coordinated control strategy for simultaneous frequency and voltage stabilisation of the multi-area interconnected system considering communication time delays. Int J Ambient Energy 43(01):5512–5524. https://doi.org/10.1080/01430750.2021.1967192
30. Arora S, Singh S (2019) Butterfly optimization algorithm: a novel approach for global optimization. Soft Comput 23:715–734. https://doi.org/10.1007/s00500-018-3102-4
31. Makhadmeh S, Betar M, Abasi A, Awadallah M, Doush I, Alyasseri Z, Alomari O (2023) Recent advances in butterfly optimization algorithm, its versions and applications. Arch Comput Methods Eng 30:1399–1420. https://doi.org/10.1007/s11831-022-09843-3

FOTIDF Controller for Stability of Renewable Integrated Multi-area Realistic Power System with AC-DC Links

Ch. Naga Sai Kalyan and Surender Reddy Salkuti

Abstract This work focuses on the design of a fractional order tilt-integral-derivative plus filter controller (FOTIDF) using the water cycle optimization algorithm (WCOA) as load frequency controller (LFC) for the dynamical stability of multi-area renewable integrated realistic power system (MARIRPS). The analysis on MARIRPS is carried out by injecting the step load disturbance (SLD) of 10% in area 1. Moreover, the investigative model is perpetuated with realistic constraints for obtaining the analysis near the real-time environment. The dominance of one such practical constraint as communication time delays (CTDs) on the investigative model operation is revealed in this work. Further, the MARIRPS operates with the AC-DC links to enhance dynamic behavior. Finally, the MARIRPS is impingement to a large range of load disturbances to test the robustness of the suggested control strategy.

Keywords Communication time delays · FOTIDF controller · Water cycle algorithm · Dynamical analysis · 10%SLD

Nomenclature

CTDs	Communication time delays
SLD	Step load disturbance
ISE	Integral square error
WCOA	Water cycle optimization algorithm
FLC	Fuzzy logic control

Ch. N. S. Kalyan
Department of Electrical and Electronics Engineering, Vasireddy Venkatadri International Technological University, Nambur, Guntur 522508, India

S. R. Salkuti (✉)
Department of Railroad and Electrical Engineering, Woosong University, Daejeon 34606, Republic of Korea
e-mail: surender@wsu.ac.kr

S. R. Salkuti (ed.), *Artificial Intelligence for Integrated Smart Energy Systems in Electric Vehicles*, Lecture Notes in Electrical Engineering 1427,
https://doi.org/10.1007/978-3-031-94276-1_25

LFC	Load frequency control
RPI	Real power imbalance
GRC	Generation rate constraint
PID	Proportional, integral, and derivative
TIDF	Tilt, integral, derivative and filter

1 Introduction

The increase in concern regarding the expenses incurred in electric power transmission and distribution, and the rapid diminution of conventional fuel reserves made a prominent entrance for the injection of renewable sources into the electric sector. It is well known that the generation of power through renewable energy conversion systems is highly intermittent and it can greatly influence the electrical network stability. In addition to this, the continuous changes in load demand greatly impacted the stability of the electrical network, which became the primary task for the power system technology. The power system operators have to change the system operating point continuously to keep the system stable. Thus, a highly developed regulatory mechanism is necessary to control the real power imbalance (RPI) as it can directly influence network stability. The RPI arises with the difference in generation and demand of real power. The RPI directly aligns with the key indicator of the power system, the control area frequency. The change in the system operating point about the momentary fluctuation in load cannot be performed manually. Hence, an automatic regulation is required and can be fulfilled by the LFC actions [1].

It is evident from the recent work listed in the literature that, rigorous investigations have been performed in the LFC for twenty years. Most of the work is mainly concentrated on developing conventional proportional-integer-derivative (PID) controllers based on meta-heuristic optimization methods [2]. However, the designed controllers are adapted to numerous models of conventional power systems, conventional with renewable penetration such as solar thermal, wind energy conversion with doubly fed induction generators, solar photo voltaic, geothermal plants [3], etc. The meta-heuristic algorithmic methods of electro-search optimization, crow search algorithm, modified JAYA algorithm [4], chaotic whale technique, bumble bees algorithm [5], stochastic fractal search, goat optimization technique, COOT optimizer [6], coyote optimization, vortex search algorithm, sine–cosine algorithm [7], pathfinder algorithm, spotted hyena optimization, school fishing algorithm, black widow algorithm, skill optimization algorithm, back tracking search algorithm [8], Ray optimization, culinary chef optimization, flower pollination algorithm [9], Harris hawks algorithm [10], population-extremal-algorithm [11], levy flight optimization, lightning search technique [12] etc. are reported.

The fractional order (FO) aided conventional control schemes are also reported extensively. The effective performance of FO-aided regulators also depends on the parametric gains that are intended to be found using optimization techniques. Algorithms of big-bang big-crunch [13], equilibrium optimization, volley-ball premier algorithm, chemical reaction algorithm, sooty tern algorithm, moth flame algorithm [14], water cycle algorithm, and may fly algorithm, etc. are implemented. Further, the fuzzy logic control (FLC) [15] aided controllers based on the algorithms of particle swarm optimization, sine–cosine, imperialistic competitive algorithm [16]; squirrel search algorithm [17], symbiotic organisms search mechanism [18], doctor-patient algorithm, bacterial foraging algorithm [19], artificial bee colony, honey bees algorithm, crow search algorithm [20], exchange market algorithm, harmony search technique [21], bibliography algorithm, and dragonfly algorithm etc. are reported.

Compared to FLC, the complexity involved in designing FO-aided controllers for real-time environments is hassle-free. Moreover, the approximations and the assumptions that are intervened in FLC design made their application inevitable for realistic practice [22, 23]. Thus, this paper focused on the development of FO-type controllers as the available traditional controllers are not robust enough for generation units with practical constraints. Assessing the overall factors from the careful literature survey, this paper put forward the possible implementation of FOTIDF based on the effective optimization of WCOA for LFC. The selection of WCOA is undergone by considering its ability to deal with the phases of exploration and exploitation. Moreover, it has been tested extensively for numerous electrical optimization problems which motivate this chapter to consider. The paper contributions are:

- Designed the FOTIDF controller using WCOA for the effective operation of a realistic power system model.
- The efficacy of FOTIDF performance is demonstrated with other controllers.
- The controller performance is tested on MARIRPS for 10%SLD.
- MARIRPS is considered with CTDs, GDB, and GRC for real-time analysis.
- Employed the AC-DC links for MARIRPS to obtain further performance improvement.

The remaining sections are organized by discussing the investigative model and the designed controller along with the deliberated objective function in Sects. 2 and 3 respectively. Section 4 gives insightful information and the appropriate modeling of the WCOA and the simulation results are presented in Sect. 5. The conclusions of this work are discussed in Sect. 6.

2 Power System Model

The MARIRPS deliberated to assess the dynamical behavior comprised of the traditional and renewable generation units like thermal, hydro, and wind in area-1 and area-2 as depicted in Fig. 1 [24]. The generation capacity in the two areas is unique and also the generation rate constraint (GRC) is believed for the traditional plants. Moreover, the thermal unit is considered with the generation dead band (GDB) as non-linearity behavior to put forward the analysis nearer to practicality. Further, the

CTDs are taken with the power system as one more realistic constraint. The power system in general is located in a wide region and the controlling of the widespread power system is done from the control center, depending on the data measured by the sensing units. The data recorded by the instruments and the sensing devices are transmitted to the control center via the communication channels and based on that the appropriate actions can be initiated. The data transmission from the location of the sensors or instruments cannot be delivered to the destination instantly and there will always exist a certain delay. This might affect the corrective actions taken by the regulator and to demonstrate its predominance on the regulator performance the CTDs are considered in this work. The modeling of CTDs [25] is presented in Eq. (1).

$$e^{-s\tau_d} = \frac{1 - s\frac{\tau_d}{2}}{1 + s\frac{\tau_d}{2}} \tag{1}$$

The AC-DC lines as additional control [26, 27] are implemented with the MARIRPS to further achieve better performance. The modeling of the HVDC line is given in Eq. (2).

$$G_{HVDC} = \frac{K_{DC}}{1 + sT_{DC}} \tag{2}$$

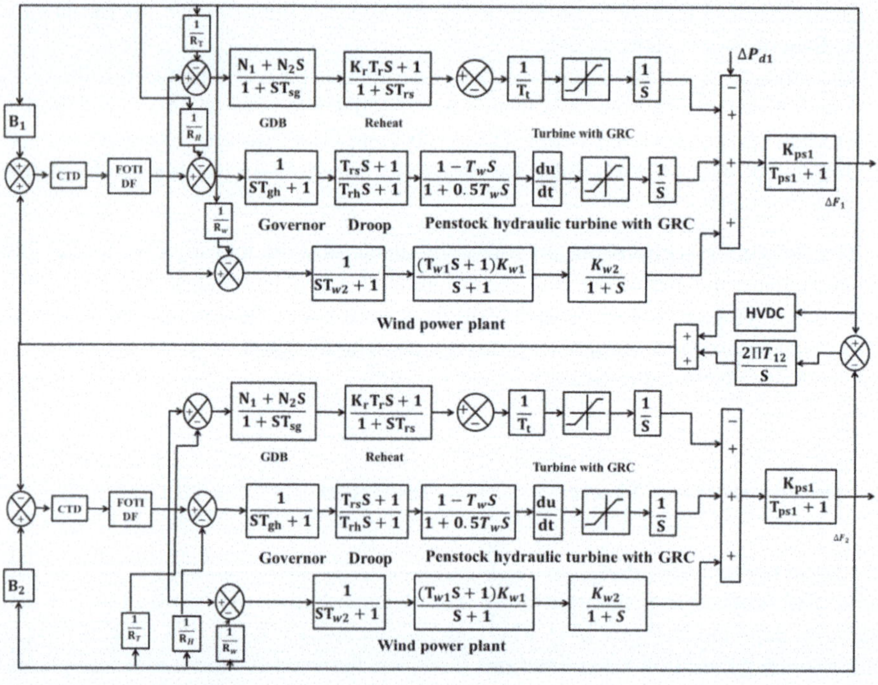

Fig. 1 Model of MARIRPS for dynamic analysis

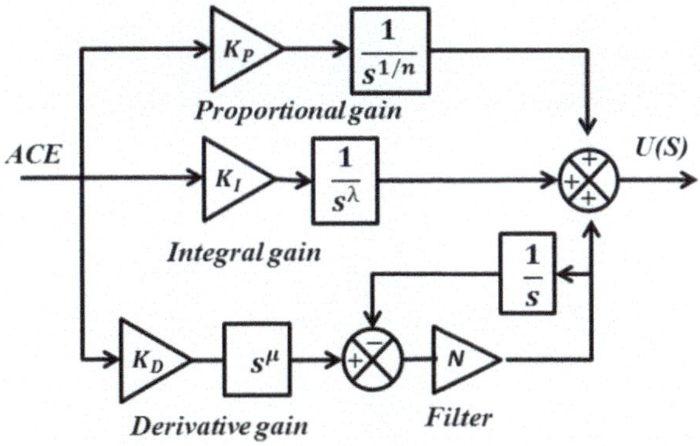

Fig. 2 Structure of FOTIDF controller

3 Controller and Objective Function

The FOTIDF regulator is adopted for the LFC of MARIRPS in this work to obtain better control over the dynamic performance. Being the category of FO controllers, they need better optimization methods for optimal performance. The FOTIDF includes the tunable parameters such as K_T, K_I, K_D, N, λ, and μ as shown in Fig. 2, and are named tilt, integral, derivative, filter, and the FO integral and derivative gains respectively. The optimization of these gains is performed with integral square error (ISE) [28] Eq. (3).

$$J_{ISE} = \int_0^{T_{sim}} (\Delta f_1^2 + \Delta P_{tie12}^2 + \Delta f_2^2)dt \tag{3}$$

4 Water Cycle Optimization Algorithm

The WCOA in recent times has been extensively adopted by researchers in finding out the solutions to various problems. Till now, the WCOA application has been reported in the fields of portfolio selection in finance, traveling salesman problems, and the other domains [29] in electrical engineering fields like LFC [30], optimal power flow, etc. Thus, this chapter motivates to implementation of this algorithm for the optimal dynamical performance of MARIRPS under the supervision of the FOTIDF controller. The operation of the WCOA begins with the initialization of water droplets and terminates systematically while the droplets merge to form the rivers/streams that finally meet the sea. The water droplets (WDs) initialization is presented in Eqs. (4) and (5) as an array of 'N' variables.

$$WD_i = [x_1, x_2, \ldots x_N] \tag{4}$$

$$WD = \begin{bmatrix} WD_1 \\ . \\ WD_i \\ . \\ WD_N \end{bmatrix} \quad (5)$$

These WD can collectively form the streams/rivers and the corresponding positions (P) are initialized as modeled in Eqs. (6) and (7).

$$P_{stream}^{new} = P_{stream} + r_1() * (P_{river} - P_{stream}) \quad (6)$$

$$P_{river}^{new} = P_{river} + r_2() * (P_{sea} - P_{river}) \quad (7)$$

where the r1() and r2() are randomly generated from [0–1]. River/stream locations are dependent on fitness values and are not set. The river receives the position with the highest fitness value, while the other positions go to the streams. The same Performa is applicable for rivers/sea. The loop of rain and precipitation emerges with the following Eqs. (8) and (9).

$$|P_{sea} - P_{river}| \leq d_{\max} \quad (8)$$

$$d_{\max}^{new} = d_{\max} - \left(d_{\max} / Iter.\mathrm{Max} \right) \quad (9)$$

After precipitation, the new WD joins the water cycle, and the positions of the streams are relocated as given in Eq. (10).

$$\boldsymbol{P_{stream}^{new}} = \boldsymbol{P_{sea}} + \sqrt{\boldsymbol{U}} \, \boldsymbol{*r}\,() \quad (10)$$

where the parameter d, indicates a value close to zero and ‘U’ denotes the search close to the sea.

5 Simulation Results

The simulation analysis is presented in a total of five cases. The investigations in case 1 and case 2 are studied by omitting and deliberating the CTDs under several controllers. The effect of CTDs under FOTIDF on MARIRPS performance is presented in case 3. The study of the system with AC-DC links, followed by the robustness analysis is analyzed in case 4 and case 5, respectively.

5.1 Case-1: Analysis of MARIRPS Without CTDs Under Various Controllers

In this subsection, the MARIRPS dynamical behavior is investigated and the constraint of CTDs is omitted. The analysis is conducted by laying 10%SLD and the controllers such as FOTIDF, TIDF, and FOPID are enacted as secondary regulators that are optimized with the WCOA to obtain the comparative results. For this particular case, MARIRPS responses are shown in Fig. 3, and are interpreted in the settling time aspect as tabulated in Table 1. Observing Fig. 3 it is clear that the MARIRPS have very little fluctuation under the supervision of the FOTIDF controller and from Table 1 it is elucidated that the responses provided under the FOTIDF concede a shorter amount of time. Further, the FOTIDF predominantly regulated the ISE and is improvised by 18.44 and 62.5% with TIDF and FOPID controllers. The FOTIDF, TIDF, and FOPID controller parameters found optimally using the WCOA are provided in Table 2.

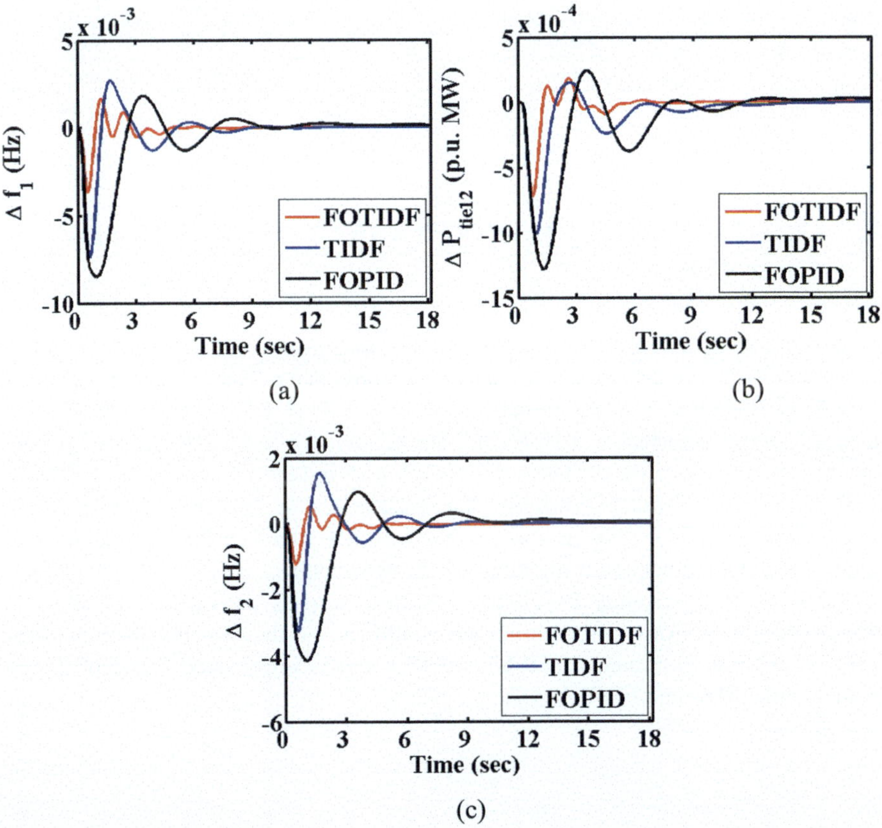

Fig. 3 Case-1 responses. **a** Δf_1, **b** ΔP_{tie12}, **c** Δf_2

Table 1 Responses settling time

Settling time (s)		Δf_1	Δf_2	ΔP_{tie12}	ISE*10^{-3}
Case-1	FOPID	12.71	11.72	13.92	31.95
	TIDF	10.34	9.937	13.17	14.69
	FOTIDF	6.35	8.68	9.43	11.98
Case-2	FOPID	16.84	18.58	18.45	90.49
	TIDF	16.15	15.77	16.05	46.72
	FOTIDF	13.97	12.33	15.29	25.39

Table 2 Controller optimal gains

Parameter	Area-1			Area-2		
	FOPID	TIDF	FOTIDF	FOPID	TIDF	FOTIDF
K_P/K_T	0.582	0.428	0.216	0.255	0.663	0.252
K_I	0.061	0.090	0.394	0.041	0.531	0.413
K_D	0.119	0.246	0.523	0.277	0.953	0.333
n	–	0.630	0.812	–	0.422	0.467
N	–	221.17	219.19	–	226.63	229.63
λ	0.019	–	0.056	0.033	–	0.089
μ	0.032	–	0.143	0.154	–	0.167

5.2 *Case-2: Analysis of MARIRPS with CTDs Under Various Controllers*

The MARIRPS in this case is investigated by considering the CTDs and the time delay indicator of 0.25 s is deliberated as the realistic parameter. The dynamical behavior of the MARIRPS under the individual supervision of FOTIDF, TIDF, and FOPID are analyzed for the disturbance of 10%SLD. The behavior of MARIRPS under different control structures with the CTDs is shown in Fig. 4 and their corresponding settling times are noted in Table 1. From Fig. 4 and Table 1, it is primarily concluded that the performance of FOTIDF is superior to the TIDF and FOPID performances in handling the deviations and mitigating the oscillations.

Moreover, the presented FOTIDF controllers surpass the other controllers in striving the fluctuations to the zero state condition. Further, in this case, also the FOTIDF exhibits dominance in ISE reduction and is improvised by 45.65% and 71.94% with TIDF and FOPID.

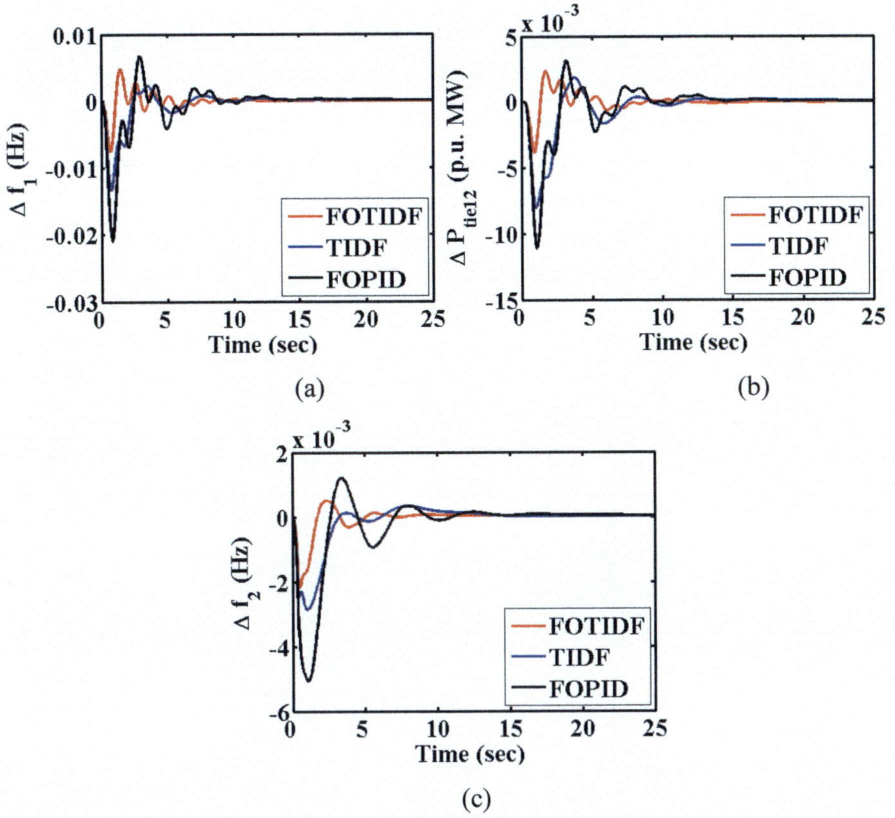

Fig. 4 Case-2 responses. **a** Δf_1, **b** ΔP_{tie12}, **c** Δf_2

5.3 *Case-3: Demonstrating the Impact of CTDs on MARIRPS Performance*

The dynamical behavior of the system, which is under the direction of the FOTIDF and is based on WCOA, is contrasted in Fig. 5 for instances with and without CTDs to demonstrate the impact of CTDs on the MARIRPS operation. From the analysis in the above cases, the FOTIDF is proven to be the better controller and thus further investigation on the MARIRPS is carried out with this controller. By looking at Fig. 5, it is clear that the MARIRPS performance when CTDs are taken into account deviates more from the norm than when CTDs are not taken into account. This means, with CTDs the sending of error signal to the regulator and the action taken by it to minimize the RPI is performed with the delay. The delay in the correction measures handled by the regulator might widen the real power gap and thereby more fluctuations in the power system stability indicators. This work recommends considering the CTDs with power system network models at the time of design, development, and performance

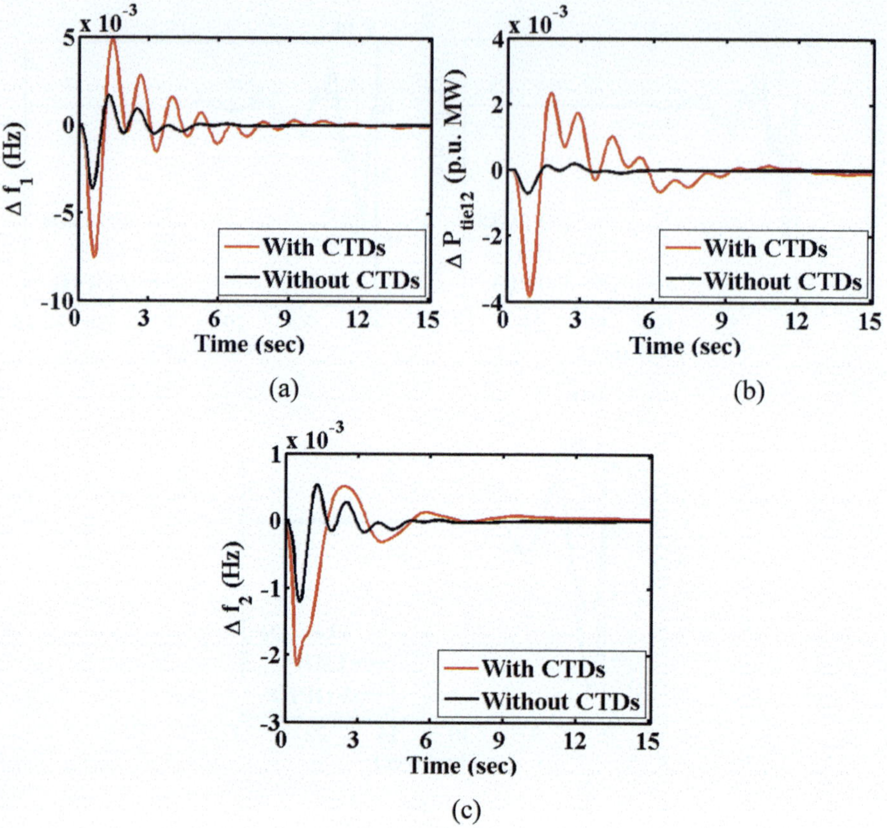

Fig. 5 Case-3 responses. **a** Δf_1, **b** ΔP_{tie12}, **c** Δf_2

assessment of the secondary regulator. Otherwise, the suggested regulator couldn't handle the deviations in the performance when unintended delays emerged with the network.

5.4 *Case-4: Performance of MARIRPS with AC-DC Links*

The MARIRPS is further analyzed by integrating the DC line and the tie-line operates as the AC-DC lines. For the disturbance of 10%SLD and under the supervision of the FOTIDF controller the performance of MARIRPS is investigated with the AC-DC links as shown in Fig. 6. Noticing the MARIRPS behavior in Fig. 6, it is identified that the considerable enhancement in the dynamical behavior is reported with the employment of AC-DC links. With the AC-DC links the allowable transfer capability of the tie-line is improved and there the power exchange might be performed more

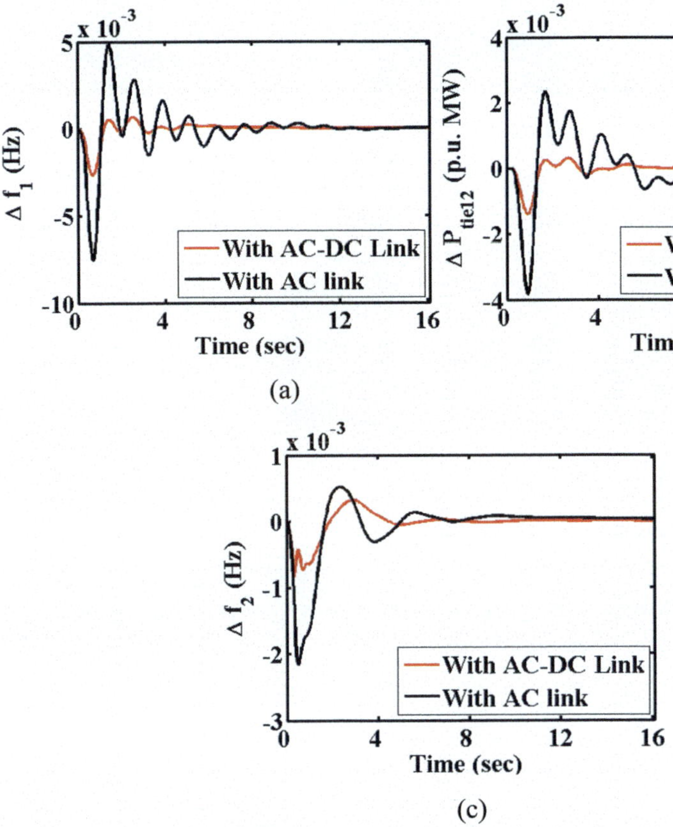

Fig. 6 Case-4 responses. **a** Δf_1, **b** ΔP_{tie12}, **c** Δf_2

comfortably from the surplus to deficit generation areas at the time of a sudden rise in large demands. Moreover, the responses with the employment of AC-DC links attain a stable condition in less time and the same is shown in Fig. 7.

5.5 Case-5: Robustness Analysis

To address the robustness of the suggested FOTIDF controller performance and the territory control approach of employing the AC-DC links with MARIRPS, the robustness analysis is conducted. In this test, the MARIRPS under the secondary control action of FOTIDF and the territory method of AC-DC links incorporation is laid with a large range in load of $\pm$ 50% from the normal load in the area-1. The results are presented in Fig. 8, and it is concluded that despite the power system being subjected to significant disturbances, any variations in the performances are

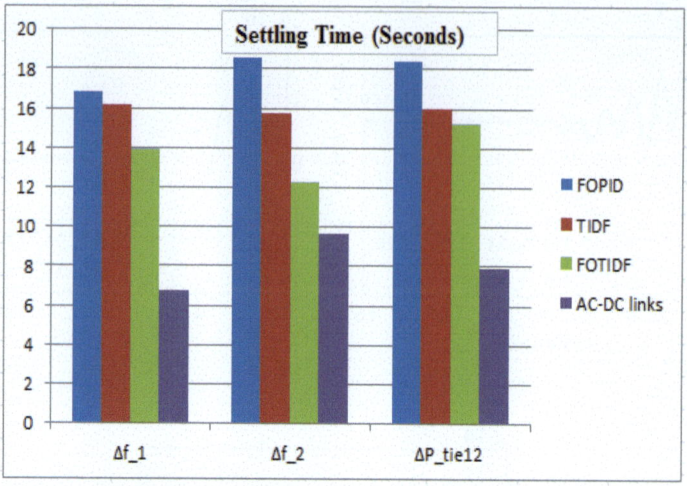

Fig. 7 Bar chart indicates responses settling time

hardly noticeable. This demonstrates the FOTIDF controller's reliability and the use of AC-DC connections as a territorial control method.

6 Conclusions

This work tested the performance efficacy of the WCOA-optimized FOTIDF controller for controlling the dynamical behavior of MARIRPS. Area-1 has first impinged with 10%SLD to begin the research under the supervision of FOTIDF, TIDF, and FOPID controllers for the instances of MARIRPS with and without considering the CTDs. The comparative investigation revealed the sovereignty of the FOTIDF controller. Additionally, by doing an individual study on the system for the scenarios of contemplating and not contemplating the CTDs, the relevance of the CTDs on the performance of the power system is proved and justified. Further, the MARIRPS is integrated with the AC-DC lines for power exchange. Simulation results revealed a considerable improvement in the dynamical behavior of MARIRPS with the AC-DC links. Finally, the MARIRPS under the WCOA-based FOTIDF controller with AC-DC links is subjugated with a wide range of load disturbances. Since the recommended control method hardly deviates from the system's dynamic behavior under a variety of load shocks, it is deemed to be resilient.

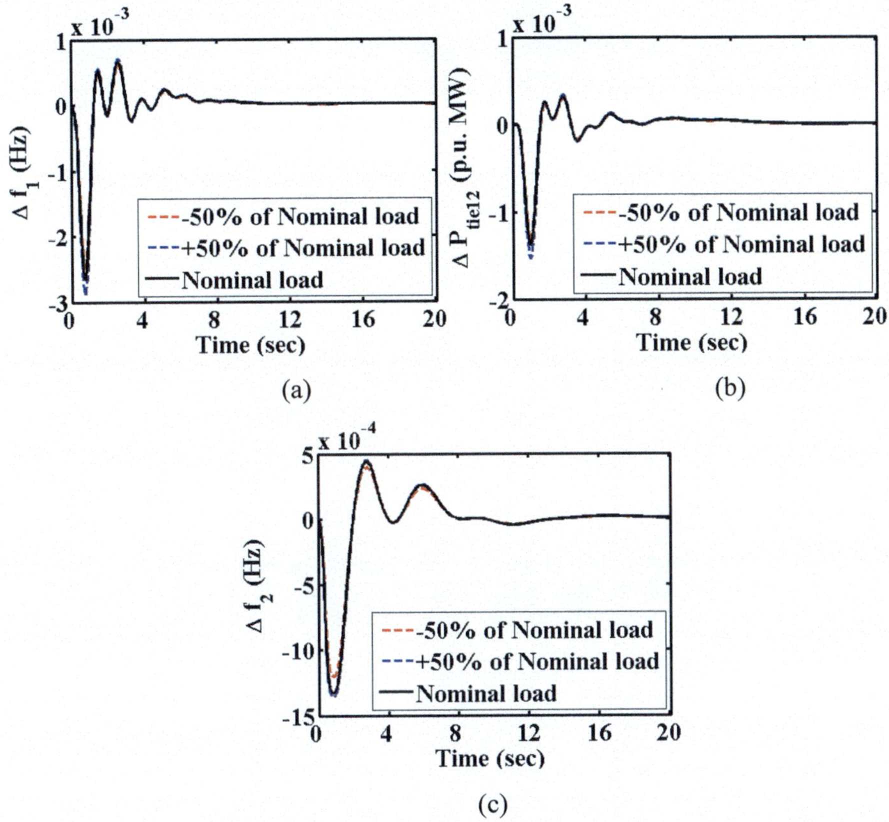

Fig. 8 Case-5 responses. **a** Δf_1, **b** ΔP_{tie12}, **c** Δf_2

References

1. Kalyan C, Rao G (2020) Frequency and voltage stabilization in combined load frequency control and automatic voltage regulation of multiarea system with hybrid generation utilities by AC/DC links. Int J Sustain Energ 39(10):1009–1029. https://doi.org/10.1080/14786451.2020.1797740
2. Kalyan C, Rao G (2022) Combined frequency and voltage stabilization of multi-area multi-source system by DE-AEFA optimized PID controller with coordinated performance of IPFC and RFBs. Int J Ambient Energy 43(1):3815–3831. https://doi.org/10.1080/01430750.2020.1860130
3. Ranjan M, Shankar R (2022) A literature survey on load frequency control considering renewable energy integration in power system: recent trends and future prospects. J Energy Storage 45:103717. https://doi.org/10.1016/j.est.2021.103717
4. Pradhan C, Bhende C (2019) Online load frequency control in wind integrated power systems using modified JAYA optimization. Eng Appl Artif Intell 77:212–228. https://doi.org/10.1016/j.engappai.2018.10.003

5. Shouran M, Anayi F, Packianather M (2021) The bees algorithm tuned sliding mode control for load frequency control in two area power system. Energies 14(18):5701. https://doi.org/10.3390/en14185701
6. El-Bahay M, Lotfy E, El-Hameed A (2023) Effective participation of wind turbines in frequency control of a two-area power system using coot optimization. Prot Control Mod Power Syst 8:14. https://doi.org/10.1186/s41601-023-00289-8
7. Rajbongshi R, Saikia L (2018) Combined voltage and frequency of a multi-area multisource system incorporating dish-Stirling solar thermal and HVDC link. IET Renew Power Gener 12(3):323–334. https://doi.org/10.1049/iet-rpg.2017.0121
8. Guha D, Roy P, Banerjee S (2018) Application of backtracking search algorithm in load frequency control of multi-area interconnected power system. Ain Shams Eng J 9(2):257–276. https://doi.org/10.1016/j.asej.2016.01.004
9. Madasu S, Kumar M, Singh A (2018) A flower pollination algorithm based automatic generation control of interconnected power system. Ain Shams Eng J 9(4):1215–1224. https://doi.org/10.1016/j.asej.2016.06.003
10. Yousri D, Babu T, Fathy A (2020) Recent methodology based Harris Hawks optimizer for designing load frequency control incorporated in multi-interconnected renewable energy plants. Sustain Energy, Grids Netw 22:100352. https://doi.org/10.1016/j.segan.2020.100352
11. Lu K, Zhou W, Zeng G, Zheng Y (2019) Constrained Population extremal optimization based robust load frequency control of multi-area interconnected power system. Electr Power Energy Syst 105:249–271. https://doi.org/10.1016/j.ijepes.2018.08.043
12. Rajbongshi R, Saikia L (2017) Combined control of voltage and frequency of multi-area multisource system incorporating solar thermal power plant using LSA optimized classical controllers. IET Gener Transm Distrib 11(10):2489–2498. https://doi.org/10.1049/iet-gtd.2016.1154
13. Jain S, Hote Y, Saxena S (2021) Fractional order PID design using big bang big crunch algorithm and order reduction: application to load frequency control. Electr Power Compon Syst 49(6–7):624–636. https://doi.org/10.1080/15325008.2021.2011482
14. Gope S, Hemakumar G, Singh K (2023) Frequency regulation analysis for renewable bio generated autonomous multi-microgrid using moth flame optimized fractional order controller. Mater Today Proc 80(2):753–761. https://doi.org/10.1016/j.matpr.2022.11.082
15. Kalyan C, Goud B, Bajaj M, Kumar K, Ahmed E, Kamel S (2022) Water-cycle-algorithm-tuned intelligent fuzzy controller for stability of multi-area multi-fuel power system with time delays. Mathematics 10(3):508. https://doi.org/10.3390/math10030508
16. Singh K, Dahiya M, Grover A, Adlakha R, Amir M (2023) An effective cascade control strategy for frequency regulation of renewable energy based hybrid power system with energy storage system. J Energy Storage 68:107804. https://doi.org/10.1016/j.est.2023.107804
17. Kalyan C, Goud B, Kumar M, Thakur P, Bajaj M, Bansal R (2023) Squirrel search algorithm based intelligent controller for interconnected power system. Int J Model Simul. https://doi.org/10.1080/02286203.2023.2205989
18. SinghvS PT, Singh V (2020) Coordinated tuning of controller parameters using symbiotic organisms search algorithm for frequency regulation of multi-area wind integrated power system. Eng Sci Technol Int J 23(1):240–252. https://doi.org/10.1016/j.jestch.2019.03.007
19. Ali E, Abd-Elazim S (2011) Bacteria foraging optimization algorithm based load frequency controller for interconnected power system. Int J Power Energy Syst 33(3):633–638. https://doi.org/10.1016/j.ijepes.2010.12.022
20. Nayak J, Shaw B, Sahu B, Naidu K (2022) Application of optimized adaptive crow search algorithm based two degree of freedom optimal fuzzy PID controller for AGC system. Eng Sci Technol Int J 32:101061. https://doi.org/10.1016/j.jestch.2021.09.007
21. Shivaie M, Kazemi M, Ameli M (2015) A modified harmony search algorithm for solving load frequency control of non-linear interconnected hydrothermal power systems. Sustain Energy Technol Assess 10:53–62. https://doi.org/10.1016/j.seta.2015.02.001
22. Kalyan C, Suresh C (2022) Higher order degree of freedom controller for load frequency control of multi area interconnected power system with time delays. Global Transition Proc 3(1):332–337. https://doi.org/10.1016/j.gltp.2022.03.020

23. Ali M, Kotb H, AboRas M, Abbasy N (2022) Frequency regulation of hybrid multi-area power system using wild horse optimizer based new combined fuzzy fractional order PI and TID controllers. Alex Eng J 61(12):12187–12210. https://doi.org/10.1016/j.aej.2022.06.008
24. Kalyan C, Srihari P, Nagalingachary K, Kumar R, Imran A, Goud B, Kumar M (2023) Path finder optimization algorithm tuned 3DOFPID controller for frequency stabilization in wind integrated realistic power system with HVDC line. Int J Renew Energy Res 13(1):208–220. https://doi.org/10.20508/ijrer.v13i1.13302.g8676
25. Kalyan C, Rao G (2021) Impact of communication time delays on combined LFC and AVR of a multi-area hybrid system with IPFC-RFBs coordinated control strategy. Prot Control Mod Power Syst 6:1–20. https://doi.org/10.1186/s41601-021-00185-z
26. Khanjanzadeh A, Soleymani S, Mozafari B (2020) Integrated multi-area power system with HVDC tie-line to enhance load frequency control and automatic generation control. Electr Eng 102:1223–1239. https://doi.org/10.1007/s00202-020-00944-5
27. Rajamand S (2021) Load frequency control and dynamic response improvement using energy storage and modeling of uncertainty in renewable distributed generators. J Energy Storage 37:102467. https://doi.org/10.1016/j.est.2021.102467
28. Shanngguan X, He Y, Zhang C, Jiang L, Wu M (2022) Load frequency control of time-delayed power system based on event-triggered communication scheme. Appl Energy 308:118294. https://doi.org/10.1016/j.apenergy.2021.118294
29. Nasir M, Sadollah A, Choi Y, Kim J (2020) A comprehensive review on water cycle algorithm and its applications. Neural Comput Appl 32:17433–17488. https://doi.org/10.1007/s00521-020-05112-1
30. Kalyan C, Goud B, Reddy C, Bajaj M, Rao G (2023) SMES and TCSC coordinated strategy for multi-area multi-source system with water cycle algorithm based 3DOF-PID controller. Smart Sci 11(1):1–15. https://doi.org/10.1080/23080477.2022.2054199

Simultaneous Optimal Allocation and Sizing of Distributed Generations and Shunt Capacitors in Distribution Network for Voltage Profile Improvement and Loss Reduction Using IMDE Algorithm

Mohammad Yousefzadeh, Ehsan Monfared, Mehrdad Ahmadi Kamarposhti, El Manaa Barhoumi, and Surender Reddy Salkuti

Abstract The development of smart grid technology has received a lot of attention in recent years to enhance the conventional power distribution system. This includes incorporating distributed generations (DGs) to meet growing demand and enhance network efficiency. Alongside DGs, the integration of shunt capacitors (SCs) and network reconfiguration can also contribute significantly to improving network performance. Distribution companies must deploy DGs and SCs in the best possible locations to operate the distribution network in optimal circumstances. The intersect mutation differential evolution (IMDE) method is a new differential evolution (DE) algorithm with intersect mutation operation that is proposed in this research for simultaneous optimal size and allocation of DGs and SCs in distribution networks. As

---

M. Yousefzadeh
Faculty of Electrical and Computer Engineering, University of Birjand, Birjand, Iran

E. Monfared
Deputy of Engineering, Electricity Distribution Company, Neyriz, Iran

M. A. Kamarposhti (✉)
Department of Electrical Engineering, Jo.C., Islamic Azad University, Jouybar, Iran
e-mail: mehrdad.ahmadi@iau.ac.ir

E. M. Barhoumi
Department of Electrical and Computer Engineering, College of Engineering, Dhofar University, Salalah, Oman
e-mail: ebarhoumi@du.edu.om

S. R. Salkuti
Department of Railroad and Electrical Engineering, Woosong University, Daejeon, Republic of Korea
e-mail: surender@wsu.ac.kr

S. R. Salkuti (ed.), *Artificial Intelligence for Integrated Smart Energy Systems in Electric Vehicles*, Lecture Notes in Electrical Engineering 1427,
https://doi.org/10.1007/978-3-031-94276-1_26

long as the bus voltage and branch current do not exceed their restrictions, the objective function is thought to be reducing power loss and improving the voltage profile. IEEE 33 and 69-bus distribution networks to test and evaluate the performance of the suggested technique in the MATLAB environment. When the suggested method is compared with other strategies described in the literature, results from the simulations indicate that this algorithm not only achieves faster convergence but also performs better than previous methods in this field. The findings demonstrate a significant reduction loss reduction of 83.05 and 89.716% as well as the worst bus voltage of 0.9804 and 0.9903pu in the simultaneous presence of DGs and SCs while voltages of bus voltages and line currents keep within their limits.

Keywords Distribution network · Distributed generation · Shunt capacitors · Power loss · Voltage profile · IMDE algorithm

Nomenclature

ABC	Artificial bee colony
ANN	Artificial neural network
BFOA	Bacterial foraging optimization algorithm
BPSO	Binary particle swarm optimization
BSO	Bee swarm optimization
DE	Differential evolution
DG	Distributed generation
DPSO	Discrete particle swarm optimization
FGA	Fuzzy genetic algorithm
GA	Genetic algorithm
ICA	Imperialist competitive algorithm
IMDE	Intersect mutation differential evolution
IPSO	Improved particle swarm optimization
SC	Shunt capacitor
TLBO	Teaching–learning-based optimization
VDI	Voltage deviation index
UPQC	Unified power quality conditioner
C	Crossover parameter
D	Number of variables
F	Mutation parameter
G	Number of generations
I_{kj}	Current flows from bus k to bus j
NP	Number of population
P_{DG}	Active power of each DG
P_k	Active power of bus k
Q_{SC}	Reactive power of each capacitor
Q_k	Reactive power of bus k

r(j)	Random number between [0,1]
V_k	Voltage of bus k

1 Introduction

The present emphasis on smart grid technology has benefited both the expansion of the traditional distribution network and the addition of DG units to meet increasing required loads and enhance network performance. SCs which are combined with network reconfiguration in addition to DGs, can significantly contribute to improving network performance [1]. The two significant challenges facing the electrical system are the growing consumers and how to supply the loads. Furthermore, distribution networks encompass the greatest amount of power losses within the power system due to its low-level voltage and high current [2]. Network reorganization, the use of DGs and capacitor banks, and relocating the system's feed point are some of the most affordable and practical ways to increase the efficiency of the distribution network [3].

Both active and reactive power have a significant role in distribution networks. While reactive power supports the voltage that requires management from a network reliability perspective, active power does the beneficial work, as the variation of voltage from the normal range may cause unintentional operation and early component failure. This is why it's crucial to develop and operate DGs and CSs as efficiently as possible while taking power system uncertainties into account, especially in today's smart distribution networks.

To increase voltage stability, network capacity release, power loss reduction, and reliability enhancement, several researchers have lately developed various ways to concurrently estimate the positions and sizes of DGs and SCs. Aman et al. [4] uses the PSO approach to determine the appropriate size and location for DG and SC in IEEE 12, 30, 33, and 69-bus test networks for loss minimization. In Esmaeilian et al. [5], the IEEE 33-bus network is used to demonstrate the benefits of adopting an enhanced evolutionary algorithm for DG and SC location. A similar goal was pursued in [6] by utilizing bacterial foraging optimization algorithm (BFOA) on the 33-bus network, and numerical results are contrasted for three distinct scenarios: when just DG, when DG and SC are used in tandem, and when none of them are utilized at all. In Yousefzadeh et al. [7] and Yousefzadeh et al. [8], the problem of allocating both DGs and SCs is resolved by employing the artificial bee colony (ABC) algorithm in both balanced and unbalanced distribution networks, whereby, in addition to the primary goals of loss reduction, reliability indices are employed.

To reduce distribution network losses, the optimal locating of both DGs and CSs is presented in [9–12]. In Zeinalzadeh et al. [13], it was determined where and how size to make DGs and CSs in the network of distribution while taking into consideration the uncertainty caused by the load, which was described using fuzzy logic.

To lower yearly power losses and boost bus voltage profiles, the optimal simultaneous allotment of DGs and SCs in transmission networks was studied in [14] using advanced meta-heuristic optimization approaches that took demand load fluctuation into account. To achieve the goals of reducing power losses and enhancing the voltage profile, the authors in [15] used the binary collective behavior of animal optimization approach for the best simultaneous position and size of DGs and SCs in distribution networks. In Ghaffarzadeh and Sadeghi [16], DGs and CSs optimal allocation problems were addressed using the biogeography-based optimization technique while taking into consideration the optimization goals of cutting power losses, lowering overall harmonic distortion, and optimizing the voltage profile.

To lower the expenses of both the initial purchase and continuing maintenance, DGs and SCs were concurrently properly scaled in reference [17] using a GA optimization approach. For the concurrently ideal placement of renewable DGs and CSs, the authors in reference [18] introduced and utilized a sorting-based, multi-objective PSO optimization method that considered fuzzy choice principles. Numerous optimization strategies were used in [19–21] to choose the best position and size of DGs and SCs in distribution networks to minimize power losses. A probabilistic approach utilizing a method of point estimation was used in [22] to allocate optimally CSs in a distribution network integrated with wind energy-based DGs. To find the optimal DGs and CSs allocation and sizing, a hybrid optimization strategy combining weight-enhanced PSO and the gravitational search algorithm (GSA) was used [23]. In Pereira et al. [24], the irregular and unpredictable behavior of DGs was taken into account while determining the simultaneous ideal location of DGs and CSs. Technical–economic analysis and optimal sizing of battery storage for a hybrid rural electrification system based on renewable energy for selected regions of Indian ecology by using HOMER software is conducted [25]. For use in the shunt active power filter of UPQC, a hybrid controller trained by an artificial neural network controller (SANNC) and based on the soccer league algorithm was introduced in [26].

This research suggests a new DE algorithm with intersect mutation operation called IMDE algorithm to find the ideal placement for DGs and CSs in radial distribution networks as well as determine their sizes. Instead of focusing on setting appropriate parameters, the IMDE algorithm divides all individuals into the worst and better parts according to their fitness. Subsequently, novel crossover and mutation operations were developed to create new individuals. This method offers a significant benefit due to less probability of becoming trapped in a local optimal solution, its superior performance in terms of convergence speed, and the ability to find more optimal solutions than with the conventional DE and several other optimization algorithms. The main innovations and paper's contributions are listed in the following:

- Applying a powerful meta-heuristic algorithm based on population called the IMDE optimization for solving the problem of optimal allocation of DG units and SCs in distribution networks and results comparison with the results obtained in the literature.

- Satisfaction of two technical objectives: voltage deviation improvement and power loss reduction.
- Calculation of the bus voltage profile and worst voltage for the performance evaluation of algorithms in each scenario.
- Raising awareness about the significance of integrating DGs and SCs to improve the functionality of electrical grids.

The rest of the chapter is organized as follows: Sect. 2 discusses the optimization problem's mathematical formulation, including the description of the load flow, the objective function, and any limits that may apply to the power flow in the distribution network. Section 3, elaborates on the description and implementation of the IMDE algorithm. On 33 and 64 bus test networks, case studies are conducted in Sect. 3 and the discussion and comparison of numerical results with several other algorithms are presented. Finally, the conclusion will be discussed in Sect. 5.

2 Problem Formulation

Determining the proper size and discrete location of DGs and SCs in distribution networks is a challenging discrete optimization problem that calls for an effective approach. As will be shown in the following subsections, the purpose of this study is to resolve this problem by employing the IMDE method.

2.1 Power Flow Equations

Figure 1 displays a single-line diagram of a basic radial feeder. Power flow can be calculated employing a series of recursive equations given in [27]:

$$P_{k+1} = P_k - P_{Lk+1} - \left(R_{j,k+1} \cdot \frac{P_j^2 + Q_k^2}{|V_k|^2} \right) \tag{1}$$

$$Q_{k+1} = Q_k - Q_{Lk+1} - \left(X_{j,k+1} \cdot \frac{P_j^2 + Q_k^2}{|V_k|^2} \right) \tag{2}$$

$$V_{k+1}^2 = V_k^2 - 2\left[P_k.R_{j,k+1} + Q_k.X_{j,k+1}\right] + \left[R_{k,k+1}^2 + X_{k,k+1}^2\right] \cdot \frac{P_j^2 + Q_k^2}{|V_k|^2} \tag{3}$$

where the power going out of bus k is represented by P_k and Q_k. The active and reactive load power at bus k + 1 is denoted by P_{Lk+1} and Q_{Lk+1} respectively. Buses k and $k+1$ have a line segment with resistance $R_{k,k+1}$ and reactance $X_{k,k+1}$ between them. $|V_k|$ is the bus k voltage magnitude. Equations (1) and (2) of the power balance must be met for the power flow to converge. Additionally, transmitting and receiving

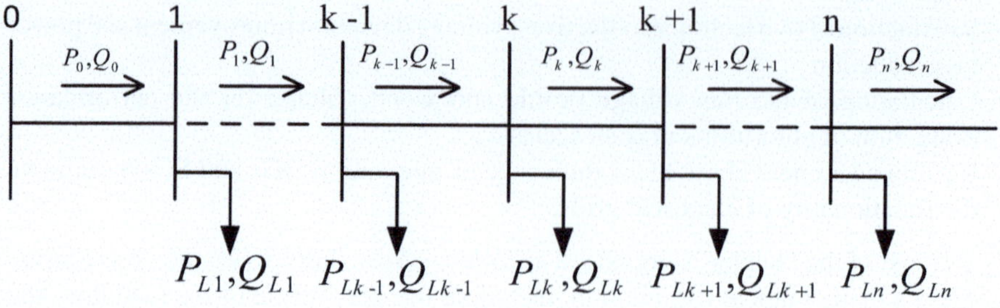

Fig. 1 Single-line diagram of a radial feeder

end-bus voltage magnitudes need to match up with Eq. (3). Using the following two equations, the power loss in the portion of the line linking buses k and $k+1$ can be determined:

$$P_{Loss}(k, k+1) = R_{k,k+1} \cdot \left(\frac{P_k^2 + Q_k^2}{|V_k|^2} \right) \tag{4}$$

$$Q_{Loss}(k, k+1) = X_{k,k+1} \cdot \left(\frac{P_k^2 + Q_k^2}{|V_k|^2} \right) \tag{5}$$

where P_{Loss} and Q_{Loss} represent the active and reactive power loss, respectively. By adding together all of the line losses, the network's overall power loss can be determined:

$$P_{TLoss} = \sum_{k=0}^{n-1} P_{Loss}(k, k+1) \tag{6}$$

$$Q_{TLoss} = \sum_{k=0}^{n-1} Q_{Loss}(k, k+1) \tag{7}$$

where P_{TLoss} denotes the overall loss of active power and Q_{TLoss} denotes the loss brought on by reactive power across the network. The recursive equations provided in (1) and (2) need to be adjusted to employ the IMDE approach.

$$P_{k+1} = P_k - P_{Lk+1} - R_{j,k+1} \cdot \frac{P_k^2 + Q_k^2}{|V_k|^2} + \phi_p . AP_{k+1} \tag{8}$$

$$Q_{k+1} = Q_k - Q_{Lk+1} - X_{j,k+1} \cdot \frac{P_k^2 + Q_k^2}{|V_k|^2} + \phi_q . RQ_{k+1} \tag{9}$$

where the active power coefficient is denoted by ϕ_p. ϕ_p is equal to one when there is an active power source; otherwise, it is equal to zero. Similar to this, the reactive power coefficient ϕ_q, is equal to one in the presence of a reactive power source and zero in the absence of one. Additionally, AP_{k+1} and RQ_{k+1} are the amount of active and reactive power that were injected into bus $k + 1$, respectively.

2.2 Objective Function

This study aims to minimize the overall loss and the voltage deviation index (VDI). The objective function is formulated as Eq. (10). VDI is defined as the difference between the nominal reference voltage and the actual voltage in each bus.

$$\begin{aligned}
\text{Minimize } f(x) &= \{f_1(x) + f_2(x)\}a \\
\text{Max } f_1(x) &= \sum\left[\Delta P^{DG}_{Loss} + \Delta P^{CS}_{Loss}\right] \\
\text{Min } f_2(x) &= \sum_{k=1}^{n} |1 - V_k|^2 \\
\Delta P^{DG}_{Loss} &= P_{Loss0} - P^{DG}_{Loss} \\
\Delta Q^{CS}_{Loss} &= Q_{Loss0} - Q^{SC}_{Loss}
\end{aligned} \tag{10}$$

where $f_1(x)$ and $f_2(x)$ denote power loss reduction and voltage deviation vector respectively. The network loss before DGs and SCs installation. ΔP^{DG}_{Loss} and ΔP^{CS}_{Loss} are the network losses after DGs and SCs installation respectively.

2.3 Operational Constraints

The voltage restrictions of the buses, the current flow across the lines, and the lower limit and upper limit of capacity allowed for the setting up of DGs and SCs are all operating constraints that are included in this optimizing issue. Any bus voltage must fall between the defined maximum and lowest voltage ranges in terms of magnitude. Any branch's current cannot be greater than its rated capacity. These limitations can be expressed mathematically as:

$$V_k^{\min} \le V_k \le V_k^{\max} \tag{11}$$

$$I_{kj} \le I_{kj}^{\max} \tag{12}$$

$$P_{DG}^{\min} \le P_{DG} \le P_{DG}^{\max} \tag{13}$$

$$Q_{SC}^{\min} \le Q_{SC} \le Q_{SC}^{\max} \tag{14}$$

When the permitted voltages for each bus in the network, $V_k^{\min}$ and $V_k^{\max}$, are given, between 97 and 103% of each bus's rated voltage are considered the voltage limitations. I_{kj} measures how much current flows from bus k to bus j. The maximum current of the lines between k and j buses that can pass through the same branch while primarily taking into account the line's thermal capacity restriction is called $I_{kj}^{\max}$ [28, 29]. P_{DG} and Q_{SC} represent the range of active and reactive power for each DG and SC.

3 Description and Implementation of the Intersect Mutation Differential Evolution Algorithm

One of the metaheuristic algorithms that is widely employed is differential evolution (DE), which is known for its unique character and quick convergence. The phases of mutation and recombination in DE are different from those in other evolutionary algorithms. In contrast to other stochastic approaches like GA and expert systems, which disturb the population according to a random quantity, it employs solution vector differences to modify the population. Additionally, this algorithm offers a quick and straightforward method for computing the global optimum solutions in two discrete and continuous areas [30].

The behavior of this algorithm is depicted in broad strokes in Fig. 2. Traditional DE, like the majority of evolutionary algorithms, uses three operations consist of selection, crossover, and mutation. It also uses three crucial parameters: NP (number of populations), CR (crossover parameter), and F (mutation parameter). One population consists of NP individuals:

$$X_{i,G}, i = (1, 2, 3, 4, ..., NP) \tag{15}$$

where G gives a generation in this case, one individual, $X_{i,} G, i_{,,}$ is made up of D variables that are limited by the search range of [Xmin, j Xmax, j], where j = (1, 2, 3, …, D). Individuals in the initial population are chosen at random. New individuals are then created through mutation and crossover. The selection operation is then used to decide whether the new individuals or the original Individuals will make up the future generation. This tactic's key benefit is that it avoids population diversity loss by avoiding the transfer of target and test vectors to the following generation.

This part introduces a brand-new technique dubbed IMDE that [31] has proposed to enhance the DE's global search functionality. The key benefit of this approach is that it performs better in terms of finding solutions and convergence speed. Compared to conventional DE, there is a lower chance of being trapped in a local optimum solution and a higher chance of obtaining an ideal solution. In conventional DE, some random individuals are chosen to produce new individuals, however, this slows down convergence and makes it less effective at finding the global optimum for some multi-modal functions. Thus, a new method is developed by sorting the individuals into

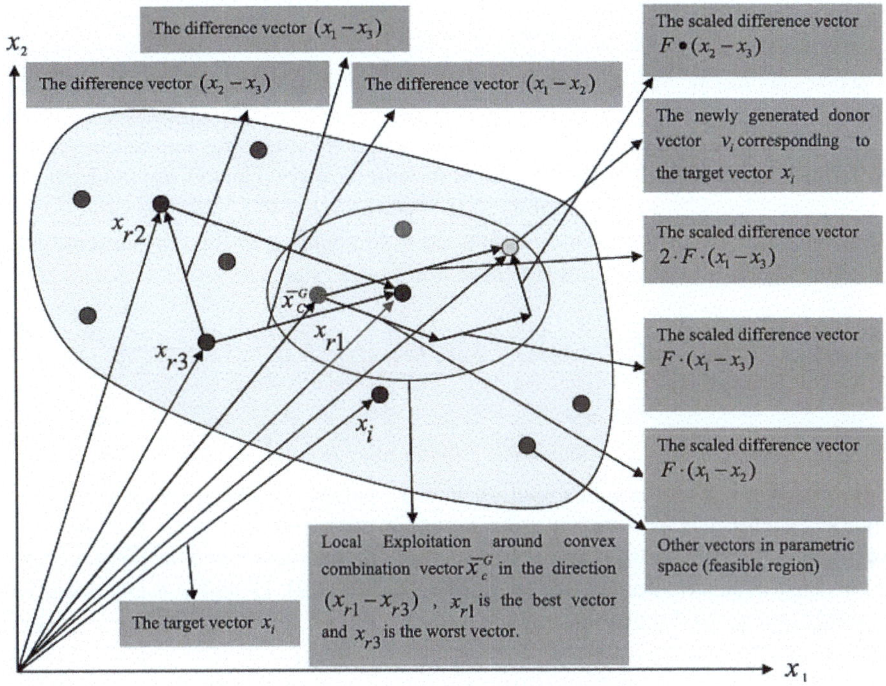

Fig. 2 Illustrating behavior overview of the DE algorithm in 2-D parametric space [30]

the worse and better parts using a constant number and ranking them in ascending order of fitness. Instead of randomly choosing, an intersection operation between the worst and better parts is used to generate new mutate vectors. To generate new individuals, an improved crossover operation is developed.

The first thing that happens in IMDE is that the population is ranked in order of their fitness levels, from worst to best. Then, the worst and better parts are separated from one another. To produce the following generation, new mutation, crossover, and selection processes are finally utilized. Two distinct processes need to be carried out to produce the following generation. For obtaining the following generation, two distinct methods have been proposed. Since most activities in these two systems are similar, there are just a few differences in how mutations are carried out.

3.1 *The First Process*

This element of the approach is split into two sections due to the separate new crossover and mutation procedures for the poorer and better halves. With two individuals (*br*1 and *br*2) chosen from the better half and one individual (*wr*1) chosen from the worst, the formula below demonstrates how the vectors are adjusted for the better portion:

$$V_{i,G+1} = X_{wr1,G} + F\left(X_{br1,G} - X_{br2,G}\right)\ wr1 \neq br1 \neq br2 \neq j \tag{16}$$

The objective, as previously said, is to enhance searching capabilities. The crossover procedure is therefore modified in various ways. Checks are made to see if $V_{i,G}$ fitness level exceeds that of $X_{i,G}$. If $V_{i,G}$ triumphs, $V^j_{i,G+1}$ is given to $U^j_{i,G+1}$; otherwise, the crossover operation is applied in the manner of the conventional DE algorithm. The following is how this operator may be stated,

$$U^j_{i,G+1} = \begin{cases} V^j_{i,G+1} F\left(V_{i,G+1}\right) \leq F\left(X_{i,G}\right)\ or \quad r(j) \geq CR \quad or \quad j = n_j \\ X^j_{i,G} \quad otherwise \end{cases} \tag{17}$$

In this scenario, $U^j_{i,G+1}$ means the j th number of the i th trial vector, $V^j_{i,G+1}$ is the j th number of the $\boldsymbol{i}$ th mutation vector, and $X^j_{i,G}$ is the jth member of the ith target vector. CR demonstrates the binomial crossover parameter that controls exchanging the dimensions between the offspring mutant vector and the target vector. $r(j)$ is considered a random number within the range of [0, 1], and n_j is considered a randomly generated dimension to ensure that the trial vector's at least one dimension is closed from the mutant vector. One vector ($br1$) and two vectors ($wr1$ and $wr2$) retrieved from the worse portion, respectively, are used to change the vectors in the worse section, as shown in Eq. (18).

$$V_{i,G+1} = X_{br1,G} + F\left(X_{wr1,G} - X_{wr2,G}\right)\ br1 \neq wr1 \neq wr2 \neq j \tag{18}$$

The crossover operation is the same as the other for the DE algorithm and this process. therefore, $U_{i,G+1}$ can be calculated as,

$$U^j_{i,G+1} = \begin{cases} V^j_{i,G+1} \quad r(j) \leq CR\ or \quad j = n_j \\ X^j_{i,G} \quad otherwise \end{cases} \tag{19}$$

The selection process remains unchanged. The selection process for IMDE is identical to that for DE. Each number's fitness function from the population is calculated by Eq. (20). The trial vector's fitness function will be selected instead of the prior generation if it is more favorable, otherwise the prior individuals will also be in the next generation.

$$x_i^{G+1} = \begin{cases} u_i\ f(u_i) \geq f\left(x_i^G\right) \\ x_i^G\ otherwise \end{cases} \tag{20}$$

where x_i^{G+1}, u_i, x_i^G are the next generation of the i th vector, trial vector, and prior generation's target vector respectively.

3.2 The Second Process

To strengthen the capability of global search, the first process uses the individuals of the worst part to search for larger regions. However, it does not release from the weak point of lacking the tendency to approach better solutions in the DE/rand/1/bin strategy. In this strategy, "rand" denotes that base vectors are selected at random in this strategy, "1" denotes that the mutated population is formed using only one vector difference, and "bin" (from the binomial distribution) denotes that uniform crossover is used during formation of the trial population. Therefore, the second process uses another new mutation operation to avoid the shortcomings mentioned above. The procedure is identical to the DE/current-best/1/bin strategy [32]. The strategy DE/current-best/1/bin is a more avaricious version where "best" denotes that the base vector being used is the population's best vector at the moment. However, this tendency could lessen the search ability for multi-modal problems globally. The formulation in the second procedure changes for the better part and the worse part as (21–22) and (23–24) respectively:

$$V_{i,G+1} = X_{wr1,G} + F\left(X_{wr1,G} - X_{br2,G}\right) \; br1 \neq wr1 \neq wr2 \neq j \tag{21}$$

$$U^{j}_{i,G+1} = \begin{cases} V^{j}_{i,G+1} F\left(V_{i,G+1}\right) \leq F\left(X_{i,G}\right) \; or \;\; r(j) \geq CR \;\; or \;\; j = n_j \\ X^{j}_{i,G} \qquad\qquad\qquad\qquad otherwise \end{cases} \tag{22}$$

$$V_{i,G+1} = X_{br1,G} + F\left(X_{wr1,G} - X_{wr2,G}\right) \; br1 \neq wr1 \neq wr2 \neq j \tag{23}$$

$$U^{j}_{i,G+1} = \begin{cases} V^{j}_{i,G+1} & r(j) \leq CR \; or \;\; j = n_j \\ X^{j}_{i,G} & otherwise \end{cases} \tag{24}$$

According to the problem and algorithm formulation presented in Sects. 2 and 3, the working steps of the proposed IMDE optimization for solving the optimal allocation problem of DGs and SCs are implemented in the flowchart given in Fig. 3.

4 Case Studies: Simulation and Numerical Results

Two test systems the IEEE 33-bus and 69-bus distribution networks are used in the optimization study to assess the effectiveness of the recommended method for the concomitant location and size of DGs and SCs.

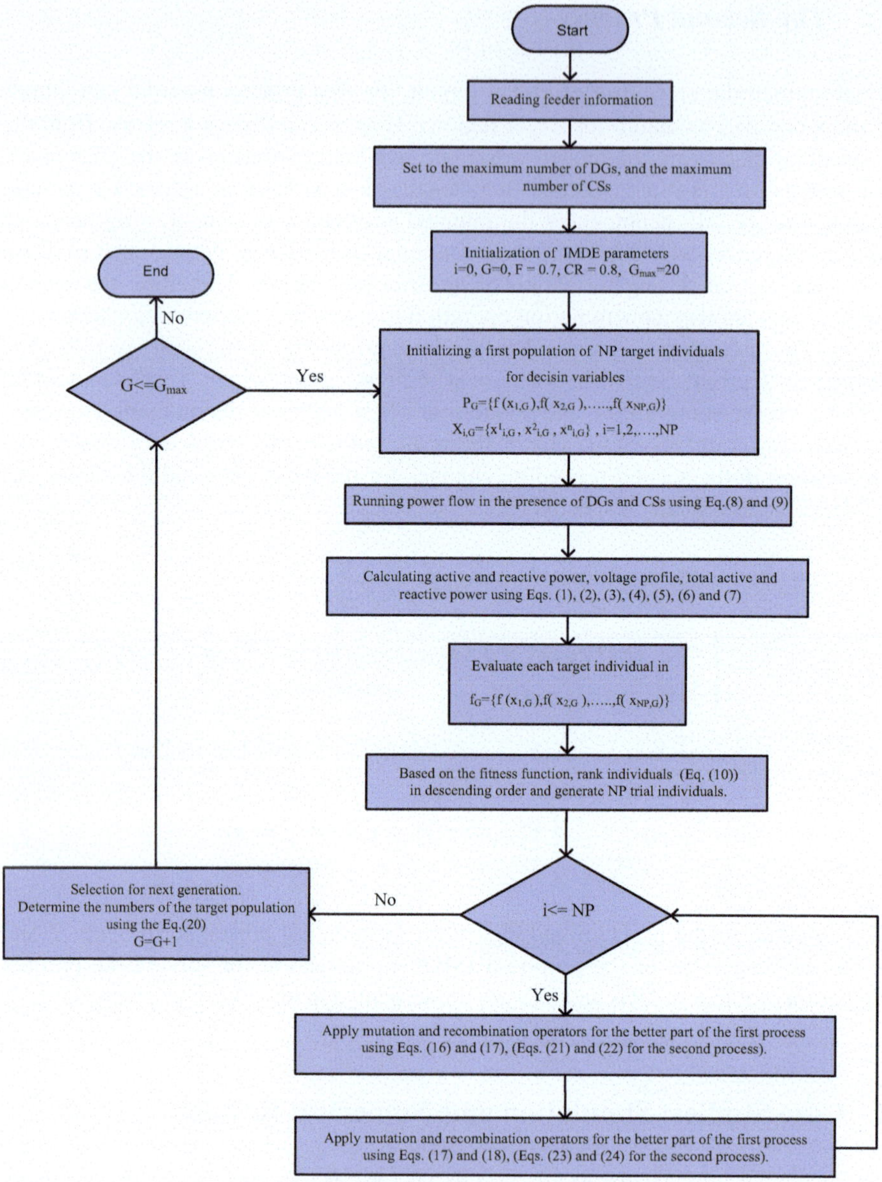

Fig. 3 Flowchart of IMDE algorithm for optimal allocation of DGs and SCs

4.1 System 1: IEEE-33 Bus Distribution Network

Figure 4 shows the typical 33-bus network with 32 branches. The network's total active and reactive load power is 3.72 MW and 2.3 MVAr, respectively. The network's base voltage and power are considered 12.66 kV and 10 MVA, respectively. Mishra et al. [33] shows statistics for lines and loads.

In this study, many eventualities are looked into. There are no DGs or SCs in the test network in the first scenario. It should be noted that the network is unable to support the installation of any number of DGs and SCs because of the substantial upfront expenses of this equipment [34, 35]. The second scenario just uses SCs, and the third scenario only uses DGs [36]. To analyze and compare the consequences of situations, the final scenario employs two reactive and two active sources simultaneously.

Table 1 displays the numerical results for the 33-bus network under various optimization techniques and scenarios that have been discussed in the literature. Also, it should be mentioned that the objective function and operating constraints applied in this study are identical to those used in [4, 6, 11, 34], except a minor difference.

Table 1 shows that the IMDE algorithm outperforms other techniques in terms of increasing bus voltage and decreasing network losses. In most situations, the installed values of active and reactive power are also lower than those of other algorithms.

For instance, when only SCs are allocated by utilizing the IMDE algorithm, a decrease of 31.40% in loss is achieved by utilizing two capacitors of 0.40 MVAr at bus 13 and 1.061 MVAr at bus 29 (total, 1.461 MVAr). The minimum bus voltage becomes visible with the value of 0.9334 pu at bus 18. In scenario 3, using IMDE, it is possible to achieve a decrease of 58.44% in loss and a minimum voltage of 0.9647 pu at bus 33, which only has two DGs (one DG of 0.778 MW at bus 14 and another of 1.193 MW at bus 29) situated. The results from BFOA [6], although still inferior to the approach suggested in this study, are the best for scenario 3 among

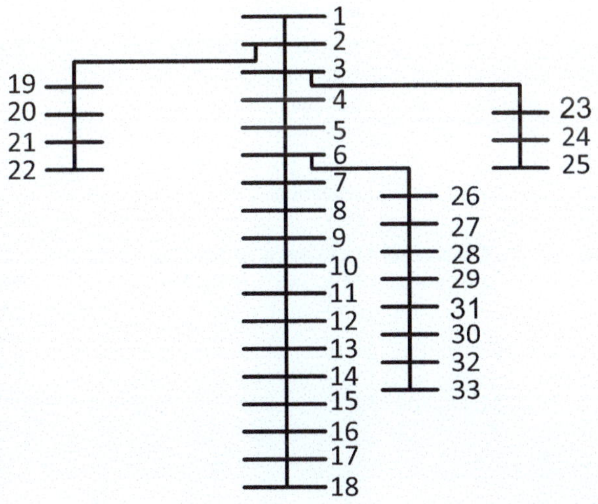

Fig. 4 IEEE-33 bus test distribution network

Table 1 The comparison of numerical results on the 33-bus network

Scenario	Description parameter		Various optimization methods				
			Analytical [11]	PSO [4]	BFOA [6]	FGA [34]	IMDE
1	Original network	Real power loss (KW)	210.8433	210.8433	210.8433	210.8433	210.8433
		Min bus voltage in p.u. (bus no.)	0.90394 (18)	0.90394 (18)	0.90394 (18)	0.90394 (18)	0.90394 (18)
2	SC installation	SC size in MVAr (bus no.)	1 (33)	–	0.35 (18) 0.82 (30) 0.28 (33)	0.95 (18) 0.7 (30)	0.40 (13) 1.061 (29)
		Real power loss (KW)	164.6	–	144.05	140.3	144.6204
		Min bus voltage in p.u. (bus no.)	0.916 (18)	–	0.9361	0.929 (18)	0.9334 (18)
		Loss reduction (%)	22.83	–	31.72	33.03	31.4086
3	DG installation	DG size in MW (bus no.)	1 (18)	–	0.633 (7) 0.090 (18) 0.947 (33)	0.6 (7) 1.1 (32)	0.778 (14) 1.193 (29)
		Real power loss (KW)	142.34	–	98.3	119.7	87.63
		Min bus voltage in p.u. (bus no.)	0.931 (33)	–	0.9645	0.936 (18)	0.9647 (33)
		Loss reduction (%)	33.29	–	53.41	43.27	58.44
4	DG and SC allocation	DG size in MW (bus no.)	0.447 (18) 0.56 (17)	2.51 (6)	0.542 (17) 0.160 (18) 0.895 (33)	0.6 (7) 1.1 (32)	0.952 (30) 1.095 (11)

(continued)

Table 1 (continued)

Scenario	Description parameter		Various optimization methods				
			Analytical [11]	PSO [4]	BFOA [6]	FGA [34]	IMDE
		SC size in MVAr (bus no.)	0.4 (33) 0.5 (32)	1.45 (30)	0.163 (18) 0.541 (30) 0.338 (33)	0.8 (33) 0.65 (16)	0.304 (10) 1.185 (28)
		Real power loss (KW)	84.28	59.7	41.42	59.5	35.74
		Min bus voltage in p.u. (bus no.)	0.961 (30)	0.955 (18)	0.9783	0.96 (18)	0.9804 (25)
		Loss reduction (%)	60.05	71. 7	80.37	71.8	83.05

the other methods, as can also be shown in Table 1. Although the FGA algorithm used in [36] only requires a DG of 1.7 MW, with a reduction of 43.27% in losses and the lowest voltage was 0.936 pu at bus 18. This demonstrates unequivocally that the performance of the IMDE algorithm is superior. IMDE displays a loss reduction of 83.05% for the situation of simultaneous DGs and CSs location and size, and its minimum bus voltage is 0.9804 pu. Both of these data clearly show that a considerable improvement has been made.

The bus voltage profiles for several scenarios for the 33-bus network are shown in Fig. 5. When their positions and sizes are chosen using the suggested optimization approach, it can be observed that the voltage profile at buses has been significantly improved by employing both DGs and SCs. Additionally, the loss reduction for the 33-bus network is shown in Fig. 6 using various optimization techniques. This graph shows unequivocally that the IMDE technique outperforms other methods. This figure implies that the performance of the IMDE method is superior to that of other algorithms. Loss reduction by any amount is of economic and technical advantage. Energy represents a cost to both the utility and the customer, and any energy savings, when added up annually, in the end, we discover a remarkable amount. In addition, Reduced heat production is a direct result of less real power loss. Therefore, the long-term benefits of reduced heat emission also improve the reliability of system components.

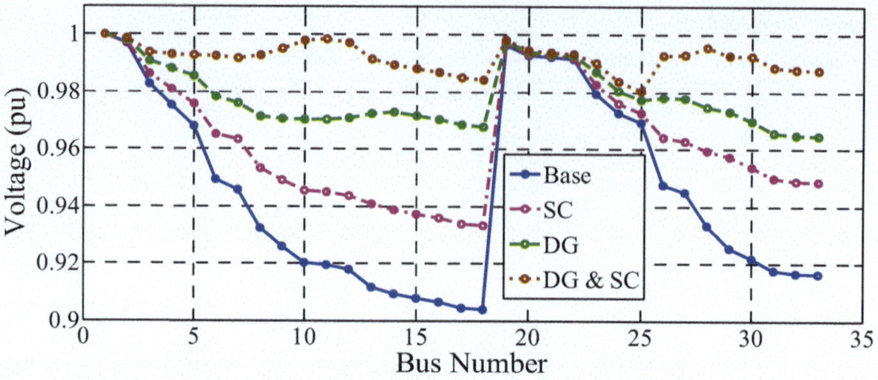

Fig. 5 The voltage profile of the 33-bus network in the different scenarios

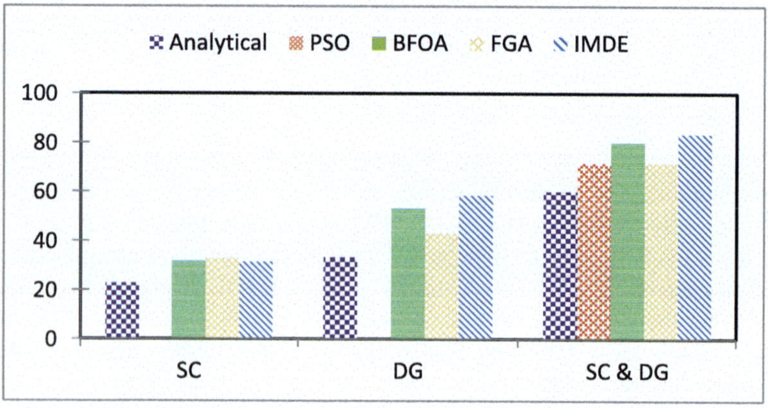

Fig. 6 The loss reduction of the 33-bus network with various methods in the different scenarios

4.2 System 2: IEEE-69 Bus Distribution Network

The typical 69-bus network with 68 branches is shown in Fig. 7. The network's overall load power is 2.69 MVAr and 3.8 MW. Line and load information may be found in [30]. The system's base voltage and base power are 12.66 kV and 10 MVA, respectively. Several situations are examined in this study. The test network in the first scenario is devoid of SCs and DGs. To examine the effects of each situation independently and to contrast with the final scenario, which employs both two active and two reactive sources, only SCs are utilized for the second scenario, and just DGs are used for the third scenario.

Table 2 shows that the IMDE algorithm outperforms other methods for reducing system losses and bus voltage deviation in a 69-bus test network. When DGs and SCs are allocated at the same time, it results in a minimum voltage of 0.9903pu at

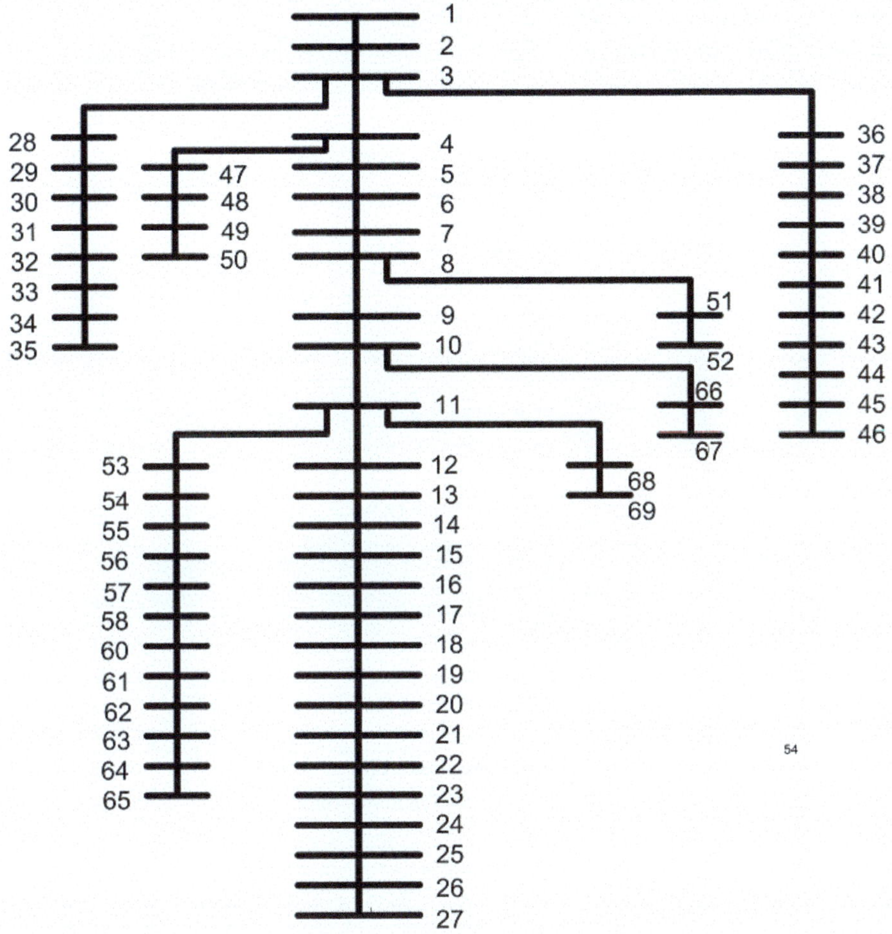

Fig. 7 Standard IEEE-69 bus distribution network

bus 27 and a loss reduction of 89.71%. The PSO algorithm's minimum voltage for this scenario was 0.97pu, resulting in a loss reduction of 88.4%, demonstrating that the technique suggested in this research produces better outcomes.

Figure 8 displays the bus voltage profile for various scenarios on the 69-bus network using the IMDE approach. It can be shown that utilizing both DGs and SCs has greatly boosted the bus voltage when their placements and capacities are set using the suggested optimization technique. Figure 9 also displays a decrease in loss for the 69-bus network obtained using different methods. This figure indisputably demonstrates that the IMDE methodology works better than other techniques.

Table 2 The comparison of numerical results on 69-bus network

Scenario	Description parameter		Various optimization methods	
			PSO	IMDE
1	Original network	Real power loss (KW)	–	224.9387
		Min bus voltage in p.u. (bus no.)	–	0.90919 (65)
2	SC installation	SC size in MVAr (bus no.)	–	1.142 (61) 0.905 (10)
		Real power loss (KW)	–	148.1138
		Min bus voltage in p.u. (bus no.)	–	0.93091 (65)
		Loss reduction (%)	–	34.1537
3	DG installation	DG size in MW (bus no.)	–	0.709 (17) 1.693 (61)
		Real power loss (KW)	–	72.9864
		Min bus voltage in p.u. (bus no.)	–	0.9769 (65)
		Loss reduction (%)	–	67.5528
4	DG and SC installation	DG size in MW (bus no.)	1.566 (61)	0.259 (39) 1.736 (63)
		SC size in MVAr (bus no.)	1.4013 (61)	0.3 (36) 1.124 (62)
		Real power loss (KW)	26	23.1328
		Min bus voltage in p.u. (bus no.)	0.97 (27)	0.9903(27)
		Loss reduction (%)	88.4	89.716

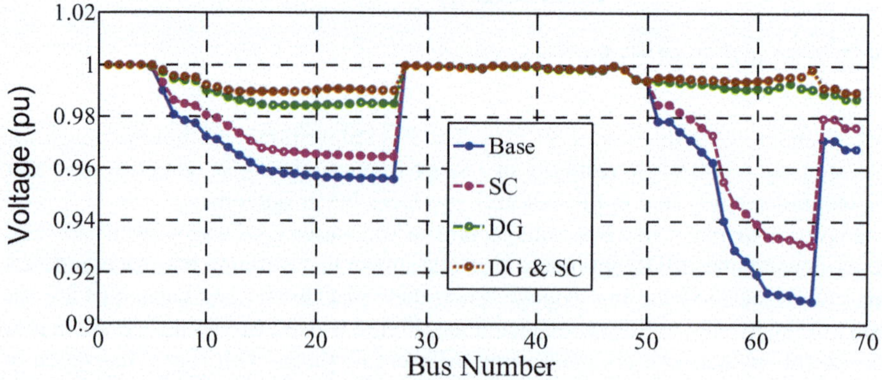

Fig. 8 The voltage profile of the 69-bus network in the different scenarios

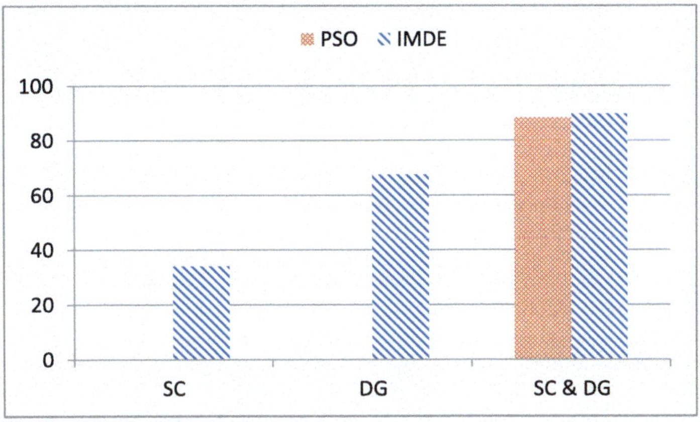

Fig. 9 The loss reduction of the 69-bus network with various methods in the different scenarios

5 Conclusions

The IMDE optimization method has been used in this research to optimize renewable DGs and SCs integration in the distribution networks under consideration, to reduce power loss, voltage deviation, and to enhance the network voltage profile while meeting a variety of operational restrictions. Comparing the simulation results of the suggested algorithm on two IEEE 33-bus and 69-bus test networks with recently published methods in this field. The simulation results demonstrated that IMDE optimization has performed better and more efficiently than previous research, for reducing power losses and voltage variation while maintaining bus voltages and branch currents within predetermined limits. For the case of simultaneous DGs and SCs allocating in the 33-bus test network, IMDE has shown a minimum bus voltage of 0.9804pu and a loss reduction of 83.05%. Both of these results clearly show that there has been a considerable improvement. Additionally, the IMDE algorithm performs better in the 69-bus network where it has shown and a minimum voltage of 0.9903pu and a loss reduction of 89.71% when DGs and SCs allocating has been done simultaneously.

Despite the proper performance and contribution of the IMDE algorithm, many challenges still need to be addressed in future research. First off, since we used size constant values and constant population size in our experiments, it might be intriguing to combine the IMDE algorithm with other enhanced DE algorithms that use parameter adaptive. Secondly, Nowadays, with the integration of novel technologies like energy storage systems and electric vehicles into the local distribution network, employing the proposed algorithm to solve optimization problems including more objectives and constraints will be challenging work.

References

1. Sayed MM, Mahdy MY, Abdel Aleem SH, Youssef HK, Boghdady TA (2022) Simultaneous distribution network reconfiguration and optimal allocation of renewable-based distributed generators and shunt capacitors under uncertain conditions. Energies 15(6):22–99. https://doi.org/10.3390/en15062299
2. Yousefzadeh M, Kamarposhti MA, Monfared E (2021) Enhancement of monitoring and non-technical power loss detection in power distribution networks using WSN state estimation. Elektrotehniski Vestnik 88(4):210–217. https://doi.org/10.1109/EV.2021.2853192
3. Narimani MR, Azizi Vahed A, Azizipanah-Abarghooee R, Javidsharifi M (2014) Enhanced gravitational search algorithm for multi-objective distribution feeder reconfiguration considering reliability, loss and operational cost. IET Gener Trans Distrib 8(1):55–69. https://doi.org/10.1049/iet-gtd.2013.0117
4. Aman MM, Jasmon GB, Solangi KH, Bakar AH, Mokhlis H (2013) Optimum simultaneous DG and capacitor placement on the basis of minimization of power losses. Int J Comput Electr Eng 5(5):516. https://doi.org/10.7763/IJCEE.2013.V5.764
5. Esmaeilian HR, Darijany O, Mohammadian M (2014) Optimal placement and sizing of DG units and capacitors simultaneously in radial distribution networks based on the voltage stability security margin. Turkish J Elect Eng Comput Sci 1–14. https://doi.org/10.3906/elk-1203-7
6. Kowsalya M (2014) Optimal distributed generation and capacitor placement in power distribution networks for power loss minimization. In: 2014 international conference on advances in electrical engineering (ICAEE), pp 1–6. https://doi.org/10.1109/ICAEE.2014.6838519
7. Yousefzadeh M, Najafi HR, Monfared E (2020) Simultaneous optimal placement and sizing of distributed generation resources and capacitor banks in neyriz distribution systems for reliability improvement and losses reduction using artificial bee colony algorithm. In: Proceedings of the first national conference on applied research in electrical engineering, Tehran, Iran, pp 1–8. NCAREE.2020/41.793345
8. Yousefzadeh M, Najafi HR, Monfared E (2023) Simultaneous optimal placement and sizing of distributed generations and capacitor banks in unbalanced distribution networks for reliability improvement and losses reduction. Int J Adv Res Electr, Electron Instrum Eng 12(3):501–516. https://doi.org/10.15662/IJAREEIE.2022.1203001
9. Moradi MH, Zeinalzadeh A, Mohammadi Y, Abedini M (2014) An efficient hybrid method for solving the optimal sitting and sizing problem of DG and shunt capacitor banks simultaneously based on imperialist competitive algorithm and genetic algorithm. Int J Electr Power Energy Syst 54:101–111. https://doi.org/10.1016/j.ijepes.2013.06.023
10. Jain N, Singh SN, Srivastava SC (2014) PSO based placement of multiple wind DGs and capacitors utilizing probabilistic load flow model. Swarm Evol Comput 19:15–24. https://doi.org/10.1016/j.swevo.2014.08.001
11. Naik SG, Khatod DK, Sharma MP (2013) Optimal allocation of combined DG and capacitor for real power loss minimization in distribution networks. Int J Electr Power Energy Syst 53:967–973. https://doi.org/10.1016/j.ijepes.2013.06.008
12. Ahmed MM, Rawa M, Ibrahim AM, Abdel Mageed HM, Abdel Aleem SH (2023) Optimal coordination of distributed generation units and shunt capacitors in Egyptian distribution system using sine-cosine optimization algorithm. In: Modernization of electric power systems: energy efficiency and power quality. Springer International Publishing, Cham, pp 205–233. https://doi.org/10.1007/978-3-031-18996-8_8
13. Zeinalzadeh A, Mohammadi Y, Moradi MH (2015) Optimal multi objective placement and sizing of multiple DGs and shunt capacitor banks simultaneously considering load uncertainty via MOPSO approach. Int J Electr Power Energy Syst 67:336–349. https://doi.org/10.1016/j.ijepes.2014.12.010
14. Kanwar N, Gupta N, Niazi KR, Swarnkar A (2015) Improved meta-heuristic techniques for simultaneous capacitor and DG allocation in radial distribution networks. Int J Electr Power Energy Syst 73:653–664. https://doi.org/10.1016/j.ijepes.2015.05.049

15. Khan NA, Ghoshal SP, Ghosh S (2015) Optimal allocation of distributed generation and shunt capacitors for the reduction of total voltage deviation and total line loss in radial distribution systems using binary collective animal behavior optimization algorithm. Electr Power Compon Syst 43(2):119–133. https://doi.org/10.1080/15325008.2014.975384
16. Ghaffarzadeh N, Sadeghi H (2016) A new efficient BBO based method for simultaneous placement of inverter-based DG units and capacitors considering harmonic limits. Int J Electr Power Energy Syst 80:37–45
17. Rahmani-Andebili M (2016) Simultaneous placement of DG and capacitor in distribution network. Electr Power Syst Res 131:1–10
18. Kayal P, Chanda CK (2016) Strategic approach for reinforcement of intermittent renewable energy sources and capacitor bank for sustainable electric power distribution system. Int J Electr Power Energy Syst 83:335–351
19. Khan NA, Ghosh S, Ghoshal SP (2013) Optimal allocation and sizing of DG and shunt capacitors using differential evolutionary algorithm. Int J Power Energy Convers 4(3):278–292
20. Dixit M, Kundu P, Jariwala HR (2017) Incorporation of distributed generation and shunt capacitor in radial distribution system for techno-economic benefits. Eng Sci Technol Int J 20(2):482–493. https://doi.org/10.1016/j.jestch.2017.01.003
21. Mohamed EA, Mohamed AAA, Mitani Y (2018) Hybrid GMSA for optimal placement and sizing of distributed generation and shunt capacitors. J Eng Sci Technol Rev 11(1):55–65
22. Baziareh A, Kavousi-Fard F, Zare A, Abasizade A, Saleh S (2015) Stochastic reactive power planning in distribution systems considering wind turbines electric power variations. J Intell Fuzzy Syst 28(3):1081–1087
23. Rajendran A, Narayanan K (2020) Optimal multiple installation of DG and capacitor for energy loss reduction and loadability enhancement in the radial distribution network using the hybrid WIPSO–GSA algorithm. Int J Ambient Energy 41(2):129–141. https://doi.org/10.1080/01430750.2018.1451371
24. Pereira BR, Da Costa GRM, Contreras J, Mantovani JRS (2016) Optimal distributed generation and reactive power allocation in electrical distribution systems. IEEE Trans Sustain Energy 7(3):975–984
25. Krishnamoorthy M, Periyanayagam ADVR, Santhan Kumar C, Praveen Kumar B, Srinivasan, S, Kathiravan P (2022) Optimal sizing, selection, and techno-economic analysis of battery storage for PV/BG-based hybrid rural electrification system. IETE J Res 68(6):4061–4076
26. Srilakshmi K, Sujatha CN, Balachandran PK, Mihet-Popa L, Kumar, NU (2022) Optimal design of an artificial intelligence controller for solar-battery integrated UPQC in three phase distribution networks. Sustainability 14(21):13992
27. Khan NA, Ghosh S, Ghoshal SP (2013) Optimal allocation and sizing of DG and shunt capacitors using differential evolutionary algorithm. Int J Power Energy Convers 4(3):278–292. https://doi.org/10.3923/ijscomp.2011.128.135
28. Duan Y, Zhao Y, Hu J (2023) An initialization-free distributed algorithm for dynamic economic dispatch problems in microgrid: modeling, optimization and analysis. Sustain Energy Grids Netw 34(2):101004. https://doi.org/10.1016/j.segan.2023.101004
29. Xu S, Huang W, Huang D, Chen H, Chai Y, Ma M, Zheng WX (2023) A reduced-order observer-based method for simultaneous diagnosis of open-switch and current sensor faults of a grid-tied NPC inverter. IEEE Trans Power Electron 38(7):9019–9032. https://doi.org/10.1109/TPEL.2023.3268666
30. Eltaeib T, Mahmood A (2018) Differential evolution: a survey and analysis. Appl Sci 8(10):19–45. https://doi.org/10.3390/app8101945
31. Zhou Y, Li X, Gao L (2013) A differential evolution algorithm with intersect mutation operator. Appl Soft Comput 13(1):390–401. https://doi.org/10.1016/j.asoc.2012.08.014
32. Price K, Storn, RM, Lampinen JA (2006) Differential evolution: a practical approach to global optimization. Springer Science & Business Media. https://doi.org/10.1007/3-540-31306-0
33. Mishra S, Das D, Paul S (2017) A comprehensive review on power distribution network reconfiguration. Energy Syst 8:227–284. https://doi.org/10.1007/s12667-016-0195-7

34. Reddy SC, Prasad PVN, Laxmi AJ (2013) Placement of distributed generator, capacitor and DG and capacitor in distribution system for loss reduction and reliability improvement. J Electr Eng 13(4):329–337. https://doi.org/10.1016/jee.2013.08.014
35. Sravanthi P, Sandeep V (2022) Solving realistic reactive power market clearing problem of wind-thermal power system with system security. Int J Emerg Electr Power Syst 23(2):125–144. https://doi.org/10.1515/ijeeps-2021-0060
36. Salkuti SR (2022) Modeling of various renewable energy resources for smart electrical power systems. In: Salkuti SR, Ray P (eds) Next generation smart grids: modeling, control and optimization. lecture notes in electrical engineering, vol 824. Springer, Singapore. https://doi.org/10.1007/978-981-16-7794-6_2

Application of Advanced Photovoltaic Control Algorithms for Stability of the Power Grid

Saroj Kumar Panda, Surender Reddy Salkuti, and Manoj Kumar Panda

Abstract The demand for sustainable development and renewable energy has resulted in a growing integration of grid-tied systems of photovoltaic (PV). This resource typically runs in the mode of maximum power point trackers (MPPT) to maximize income. However, system operators will keep adding new regulations to guarantee grid stability and dependability, eventually compelling PV plants to follow primary frequency control. PV plants must be able to operate outside of the maximum power point and adjust their power output to maintain the reserve of active power in response to grid requests to fulfill this mission. A new algorithm of model-based (MB) tracking for the regulation of power networks is presented in this research. The algorithm can now change the curtailment of power based on the factor of reduction provided by the operators of the power system, thanks to the improved formulation. The new algorithm has outstanding performance and precision, as demonstrated by the results. It can regulate power between 10 and 100% of the maximum power available. Furthermore, an assessment has been conducted regarding the effects of utilizing constant reduction coefficients on algorithmic performance. Based on information acquired during a longer than month-long period of observation of six, validation has been carried out. This could serve as the foundation for PV systems' engagement in frequency regulation because of its versatility.

Keywords Photovoltaic · Maximum power point tracker · Model-based tracking · Renewable energy sources

---

S. K. Panda
Department of Electrical Engineering, VSSUT, Burla, Odisha, India

S. R. Salkuti (✉)
Department of Railroad and Electrical Engineering, Woosong University, Daejeon, Republic of Korea
e-mail: surender@wsu.ac.kr

M. K. Panda
Department of Electronics and Communication Engineering, GIET University, Gunupur, Odisha, India
e-mail: manojkumarpanda@giet.edu

S. R. Salkuti (ed.), *Artificial Intelligence for Integrated Smart Energy Systems in Electric Vehicles*, Lecture Notes in Electrical Engineering 1427,
https://doi.org/10.1007/978-3-031-94276-1_27

Nomenclature

PV	Photovoltaic
MPPT	Maximum power point trackers
MB	Model-based
RES	Renewable energy sources
GW	Gigawatt
CPG	Continuous power generation
TSO	Transmission system operator
RP	Reduced power
RPP	Reduced power point
P&O	Perturb and observe
C_V	Voltage ratio coefficient
C_I	Current ratio coefficient
EU	European Union
P_n	Nominal power
P_r	Rated power
V_0	Open circuit voltage
I_{sc}	Short-circuit current
LL	Linear load
PC	Personal computer
R	Radiation
T	Temperature
WLS	Weighted least square

1 Introduction

The demand for renewable energy sources (RESs) has surged due to growing concerns about global warming and rising petrol prices brought on by the conflict in Ukraine. According to estimates, the gigawatt (GW) of 314 of additional capacity of renewable power will be added globally in 2021 [1]. Out of that, the PV contributes a GW of 175, surpassing hydropower's 26 GW and wind capacity addition's 105 GW. For the last five years, the annual growth in PV capacity has outpaced the growth in wind power generation.

According to the "Scenario of Medium" outlined in the Europe solar power report [2], it is projected that the photovoltaic sector will grow steadily in the upcoming years, adding 347 GW of installed capacity annually by 2026. Without a doubt, PV systems are anticipated to be crucial to the global energy shift. This transformation will present a number of significant obstacles to the usual operation, control, and planning of the power system. The primary duty of power system operators, aside from supplying energy to final users, is to ensure the security and dependability of the supply. Grid instability is caused by any discrepancy between the active power utilized by consumers and the power provided by the generators. This causes gearbox line stress, equipment damage, and early aging of producing units. Primary frequency regulation in conventional power plants is accomplished by controlling the droop and inertial response of rotating generators of electric. Nevertheless, PVs grid-tied that are operating at MPP lack the moving parts and stored energy necessary to take part in frequency regulation. The intrinsic energy buffer is reduced as a result, endangering the system's overall dependability and functionality. Thus, the importance of PVs in grid control grows together with their grid penetration, leading to the implementation of particular grid regulations [3, 4]. Emulating the response of inertia of conventional generators by enabling systems of PV to adjust their power output in response to grid demands is the way to solve the problem.

Various methods have been proposed by scientific literature to accomplish this task: putting in place energy storage systems such as batteries, PVs can be operated in power curtailment mode by modifying MPPT algorithms to provide some frequency management of output power, or they can be dump loaded to dissipate surplus power [5–9]. Increasing the storage of battery energy is the most widely applied fix. The photovoltaic system can run in maximum power point tracker mode if there is excess PV power that can be stored in the energy storage device. Because battery systems have a lower lifespan than photovoltaic systems and a greater initial cost, the solution for the storage of batteries is also the priciest [10, 11]. Theoretically, the most economical strategy is power reduction. Frequency control is often divided into two subcategories by the electric power industry: primary (which operates on a few seconds and lower time scales) and secondary (which operates on a few seconds to several minutes of time scales). Numerous studies demonstrate the viability of photovoltaic plants' involvement in both kinds of the regulation of grid frequency and the ideal degree of the curtailment of power, but they do not concentrate on power reduction techniques [12–15].

Active power control from PV plants will also be used in the markets for regulation and islanded microgrids [16]. Particularly difficult with microgrid power systems of the island is the fluctuation of renewable energy. For instance, to offer the regulation of primary frequency and the ramp rates of controlled power, the island of Puerto Rico needs substantial additional photovoltaic facilities of battery, which will greatly raise the costs of capital and maintenance [17]. On the other hand, privately owned PV plants will find it more profitable to offer the services of frequency regulation rather than using maximum power point trackers as the necessity of frequency control increases in liberalized markets of power. PV plants must contribute to the primary frequency regulation effort in the countries where grid laws have been amended by lowering their production of power when the frequency of the grid is larger than the value of nominal. Using conventional climbing of hill techniques, this can be accomplished by setting a maximum power output that, when reached, disables further MPPT. Working at reduced power (RP) is made possible by these and numerous other strategies of continuous power generation (CPG) that were suggested in [18–20]. These strategies do not require knowledge of the power of maximum available or the reserve of available power at any particular time. All of these tactics, however, have one drawback: while the transmission system operator (TSO) can use them to lower frequency to the value of nominal, they cannot be applied to the opposing task of raising the frequency of the grid, which necessitates knowledge of the available reserve of the power. The algorithm of model-based (MB) MPP tracking that the authors of the earlier works [21, 22] published calls for the first collection of a database of parameters of MPP to determine the coefficients of the model.

This chapter is a technological continuation of a previously published study that modified the model of maximum power point to follow a predetermined RP [23]. Working at reduced power as a proportion of the existing power of maximum is made possible by the suggested MB technique in this article, which permits the estimation of accurate maximum power point even when operating at RP point (RPP). Because of this, the reserve of power that is available for frequency management at any given working instance is known. The accuracy of the algorithm's operation at 80 and 90% of the maximum power available was demonstrated by an evaluation of its performance using a dataset of curves obtained over more than six months on a real panel. Furthermore, [24] demonstrated via simulation that the approach can give the module of photovoltaic the capacity to run at any level of power level percentages of 80 and 100 of its maximum possible power under a variety of temperature and irradiance scenarios. A thorough validation of the approach will be presented in this article, expanding its range of operating to percentages of 10–100 of the existing power. Additionally, three methods for figuring out the model's required coefficients will be discussed and contrasted. To assess the effects of ignoring the factors of reduction dependence on the dataset, the involvement of the uncertain evaluation of the factors of reduction has finally been examined.

2 MPPT Algorithm

2.1 Conventional Algorithms

In the past, the regulation of the work of inverters attached to the photovoltaic sheets has been designed to higher the power formed by making use of the panels' usual feature of P–V, which is defined by a single MP point and a hill shape. The approach of perturb and observe (P&O) is a popular algorithm that yields good results with little processing overhead. It determines MPP by disturbing the situation and evaluating the energy generated by the photovoltaic sheet to its initial value. The voltage perturbation selected has a strict bearing on the algorithm's efficiency. The method's performance is constrained by the large number of iterations needed for convergence if the step of voltage (ΔV) is tiny. Furthermore, it may be unstable if it is likewise less than the measurement noise. Conversely, if a large ΔV is selected, the attained PP may theoretically be distant from the theoretical one (by at least ΔV), but its velocity will rise. The techniques of changing steps have been used to reduce the energy that the panel does not produce as a result of the algorithm to get around this problem.

Similar ideas form the basis of other algorithms. For example, the incremental conductance approach calculates the ratio of $\Delta I/\Delta V$, considering that it has a value of zero in the MPP and in the area to the right of the curve, it changes sign. Once more, this approach is predicated on the requirement that the photovoltaic module avoid the local maximum. When there are abrupt fluctuations in the amount of radiation during the day, such as when there are clouds and partial shadows, these algorithms are ineffective. In the first scenario, the algorithm's dynamic is constrained by the selection of ΔV; in the scenario of the second, the curve of P–V has distinct local maxima, which means that the algorithm may converge to local maximum power points that diverge from the global maximum power point.

2.2 *Algorithms of MB*

The application of MB algorithms has improved MPP tracking even further. These algorithms are based on the panel's electrical behavior rather than its PV characteristics. Owing to their existing expertise in photoconversion, they may be able to improve the MPPT algorithms' dynamic performance and tracking speed. Naturally, those techniques are intricate and could involve a heavy computational load. Two distinct approaches could be used to prevent this. One might combine MB models with conventional tracking methods in one way, or one could simplify the models to lessen the computing load in terms of the parameters of physical and required measurements.

The single-diode models of the panel's equations were solved by the authors to create a simplified MB MPPT technique. The panel's electrical characteristics and the weather can be used to determine the MPP voltage and current.

$$V_{MP} = V_{MP0}\left[1 + \alpha(T - T_0) + \beta_1 \ln\left(\frac{R}{R_0}\right) + \beta_2 \ln^2\left(\frac{R}{R_0}\right)\right] \quad (1)$$

$$I_{MP} = I_{MP0}\frac{R}{R_0}[1 + \gamma(T - T_0)] \quad (2)$$

where V_{MP0} and I_{MP0} are the MPP voltage and current at reference operating circumstances R_0 and T_0, R is the value of radiation, T is the value of temperature of panel, and α, β_1, β_2 and γ are the coefficients of correction based on the panel. Equations (1) and (2) can be rearranged in the well-known four equations of constraints with a few straightforward considerations [21].

$$V_{MP} = B_0 + B_1 T + B_2 \ln(I_{MP}) + B_3 \ln^2(I_{MP}) \quad (3)$$

Since the MP voltages and currents in real-world applications are unknown, Eq. (3) can be applied repeatedly to get to the MPP. In practical working settings, this method requires fewer iteration steps than five, and on bright, sunny days, it produces results comparable to those obtained by the hill-climbing methods [25]. In contrast to hill-climbing methods, MB MPPT dynamic responsiveness improves unpredictable weather days and is unaffected by abrupt changes in radiation [22]. However, the identification of the alpha parameters requires special attention because they vary significantly over time and require the implementation of an automatic update. When poor data, like partially shaded curves, are available in the estimation database, it has been demonstrated that the approach of weighted least squares increases the robustness of the estimation of parameters [26].

3 Algorithm for Energy Management

3.1 State of Art

The literature contains a variety of algorithms that allow a photovoltaic (PV) plant to operate at reduced power. The most straightforward approach uses a ratio between the open-circuit voltage and the RPP voltage to implement a modified method of fractional open-circuit voltage maximum power point trackers [27]. The fact that this method occasionally necessitates disconnecting the plant to test the voltage of the open circuit is an evident disadvantage. Similar to this, Reference [28] uses a rapid maximum power point tracker procedure and then limits the required power after determining the maximum power. Although it is a sensorless approach that is reasonably priced, it necessitates constant oscillation of power between the maximum and desired.

To adopt a power reduction strategy, many authors [29–31] propose techniques to estimate the maximum power point trackers using measured irradiance and photovoltaic cell temperature. Curve fitting is the foundation of other active power control techniques [32–35]. The most current one [34] uses the fitting of a curve technique derived from a single-diode photovoltaic model to achieve a real-time estimation of the maximum power point and parameters of the model. The lack of need for temperature and irradiance sensors reduces the cost and complexity of the system. The earlier work is improved upon in reference [35], which is successful in keeping functioning reserves under partial shading conditions. Based on the management of power, current, or the algorithm of P&O, three control of CPG strategies for the systems of single-phase grid-connected photovoltaic systems have been compared in [18]. The most reliable option was found to be P&O-CPG, which is also covered in separate sections in [19, 20].

Without revealing the maximum power available or the power reserve that is accessible at any given time, all CPG techniques allow for working at RP. This restriction is addressed in [36, 37] by taking into account the system of multi-string and applying the slave of master strategy. In [36], the main photovoltaic sequences operate in maximum power point trackers to calculate the whole existing photovoltaic energy. But if this method is used, there will always be some estimation inaccuracy due to the inherent variances that exist even amongst "identical" panels, and the master PV panels in [37] are both in the circuits of short and open and supply the amounts for the MPP evaluation, in contrast to the other slave photovoltaic strings that use the master photovoltaic string's estimated available power to determine their point of operation in the PV's characteristic of the P–V curve. Algorithms for hill climbing have been designed to operate at maximum power and are not ideally effective beyond this point. As a result, using them will produce an oscillating tracker that alternates between the intended and maximum RPP.

In [38], the majority of these algorithms were compared to their capabilities, steady state, and dynamic mistakes. They are all predicated on either direct PV voltage computation using modified hill climbing techniques or power generation regulation. The chapter's conclusion highlighted the fact that high-performance the algorithms of RPP for the operations of steady-state significantly deteriorate under dynamic conditions; on the other hand, a modified algorithm of MB, which exhibits the performances of large energetic, is intended to fulfill both dynamic and the functions of a steady state with a lower error rate.

3.2 *Tracker of MB-RPP*

Since the algorithms of MB are based on the panel's physical behavior, they can theoretically predict a generic point outside the maximum in addition to the MPP, provided that the simplification hypothesis is satisfied. Specifically, the photovoltaic current needs to be close to the maximum power current for the algorithm of four parameters to work. The photovoltaic characteristic's flatness allows for the fulfillment of this restriction at the leftmost region of the curve of V–I.

Taking Eq. (3) into consideration, it has already been shown that the algorithm converges to the MP voltage in a few steps when a current other than IMP is used. Thus, by modifying the original equation, it would be able to work at any specified power if the fraction between the current and voltage at the required reduced power and the current and voltage at the maximum power point is identified. If the subsequent coefficients are added:

$$C_V = \frac{1}{\left(\frac{V_{MP}}{V_{RP}}\right)} \tag{4}$$

$$C_I = \frac{1}{\left(\frac{I_{MP}}{I_{RP}}\right)} \tag{5}$$

The equation of RPP tracking can be calculated by putting Eqs. (4) and (5) in the MB MPP tracker Eq. (3) as the current and voltage ratio coefficients.

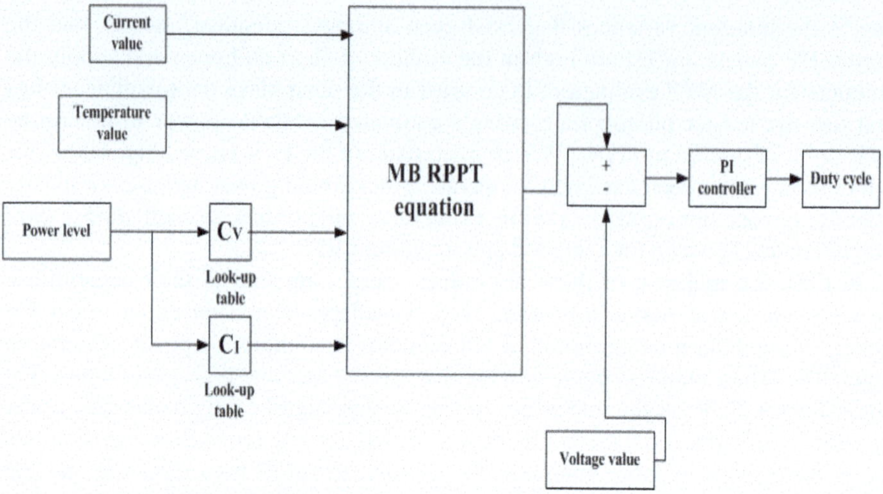

Fig. 1 Control strategy of MB RPPT

$$V_{RP} = C_V \left[B_0 + B_1 T + B_2 \ln\left(\frac{1}{\frac{C_I}{I_{RP}}}\right) + B_3 \ln^2\left(\frac{1}{\frac{C_V}{I_{RP}}}\right)\right] \quad (6)$$

Theoretically, Eq. (6) can be implemented in such a way that it converges to any required level of power, just as the conventional Eq. (3) would converge to the point of maximum power. It is important to note that the MB coefficients are stable about the intended level of power because they have already been evaluated for the MPP tracker. It is important to emphasize that the factor of voltage reduction C_V cannot be assumed to be equal to the factor of power reduction; hence, the C_V value needs to be determined numerically. The logic depicted in Fig. 1 would be used to implement a control strategy based on the reduced power point tracking equation onto an inverter or the boost converter of dc/dc (in the event of a two-stage arrangement).

The algorithm receives the desired power level as an input. The reference voltage is the output of the MB equation, which also requires the panel's temperature and current to be measured. A PI controller receives the resultant voltage, which is subtracted from the voltage of reference and specifies the necessary duty cycle.

3.3 Identification of Parameters

The parameters for the conventional MB algorithm, in this instance as well, need to be found on an actual curve database. First, a method of weighted least squares can be used to determine the B_0–B_3 parameters. Once VMP and IMP have been calculated from the N obtained the characteristics of V–I and the associated value of temperature TP, the problem can be expressed as follows:

$$\begin{bmatrix} V_{MP,1} \\ V_{MP,2} \\ V_{MP,3} \\ \vdots \\ V_{MP,N} \end{bmatrix} = \begin{bmatrix} B_0 \\ B_1 \\ B_2 \\ B_3 \end{bmatrix} \begin{bmatrix} 1 \; T_{P,1} \; ln(I_{MP,1}) \; ln^2(I_{MP,1}) \\ 1 \; T_{P,2} \; ln(I_{MP,2}) \; ln^2(I_{MP,2}) \\ 1 \; T_{P,3} \; ln(I_{MP,3}) \; ln^2(I_{MP,3}) \\ \vdots \; \vdots \; \vdots \; \vdots \\ 1 \; T_{P,N} \; ln(I_{MP,N}) \; ln^2(I_{MP,N}) \end{bmatrix} \tag{7}$$

$$V_{MP} = BF \tag{8}$$

$$P_{MP} = \begin{bmatrix} P_{MP}(1) \; 0 \; 0 \; \cdots \; \cdots \cdots \; 0 \\ 0 \; P_{MP}(2) \; 0 \; \cdots \; \cdots \cdots \; 0 \\ 0 \; 0 \; P_{MP}(3) \; \cdots \; \cdots \cdots \; 0 \\ \vdots \; \vdots \; \vdots \; \vdots \; \vdots \; \vdots \\ 0 \; 0 \; 0 \; 0 \; 0 \; 0 \; P_{MP}(N) \end{bmatrix} \tag{9}$$

$$B_{WLS} = V_{MP} F^T P_{MP} \left(F^T P_{MP} F \right)^{-1} \tag{10}$$

When employing the approach of squares of weighted least, which minimizes the produced power in this scenario, the maximum value of power supplies the matrix of weight, providing a good balance between precision and intricacy when reversing the problem. Since a complete collection of the curves of I-V has been produced, Eqs. (4) and (5) can be used to find the values of C_V and C_I at any level of power.

4 Collection of Database

An actual set of photovoltaic curves from a pane of real test 1 with 72 cells has been obtained in order to evaluate the suggested algorithm and defined as having a rated power $P_r = 160$ W, an open voltage $V_O = 40.12$ V, and a short-circuit current $I_{sc} = 5.26$ A. To obtain the curves, a linear load (LL) that is associated with the panel and can switch the potential at its end between short- and open-circuit voltages was employed [41]. In order to acquire PV current and voltage, a linear change in the two-panel conditions has been used, with a ramp duration of 100 ms. A resistive divider was used to monitor the voltage, and the sensor of the closed-loop Hall effect (LA25NP) was used to detect the current. The personal computer (PC), which also regulates the slope of the voltage of the linear LL, received data from the transducers' outputs after they were linked to a DAQ board. Every curve was recorded using a 100 kHz sampling rate. There is a one-minute pause in between every two acquisitions.

Since temperature and irradiation are known to have a stringent influence on panel V–I properties, it is also vital to keep an eye on their values. The sun radiation has

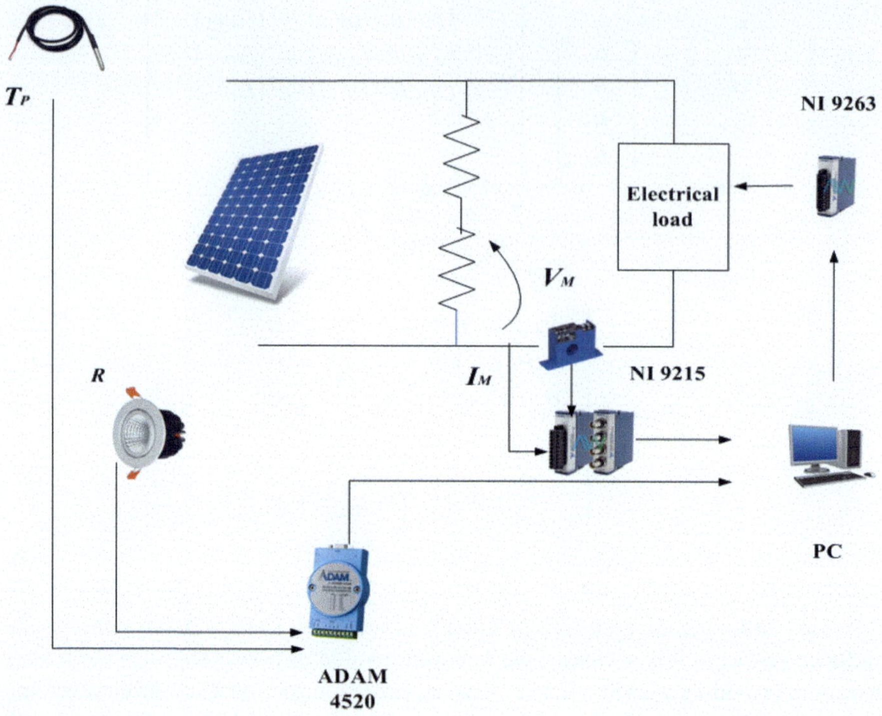

Fig. 2 Laboratory diagram

been measured using a CMP21 pyranometer, and the panel temperature has been recorded using a PT100 RTD. The entire configuration is shown in Fig. 2. 113 days of actual measurement data, including cloudy and partially shaded days, were used throughout six months. A total of 55,613 curves have been considered.

5 Validation of Algorithm

The suggested algorithm has been verified using the gathered database. When power is required, ranging from percentages of 10–100 of its maximum value of power, its performance is assessed.

5.1 Identification of Parameters

Based Since it is known that the AWLS coefficients change over time, attention has been directed to the power control method regularly [39]. Specifically, data gathered on the first sunny day has been taken into consideration while selecting a monthly update. The coefficients have been computed using Eqs. (9) and (10), and Table 1 presents the findings.

For all the curves of V-I with produced power between 10 and 100% of its maximal power at a 5% step, C_V and C_I have first been computed. The computed C_V and C_I distributions for producing percentages of 10, 30, 50, and 70 of the maximum possible power are displayed in Figs. 3 and 4.

The figures make it evident that the C_V and C_I distributions behave differently from one another. In the first scenario, there is a discernible difference in the C_V values, and their distribution width grows as the power value increases. Furthermore, there is a substantial connection seen even though each distribution's mean value is adversely biased in proportion to the RP value. However, C_I distributions are nearly overlapping in any scenario; this can be explained by the low voltage levels and flat V-I curve feature. Furthermore, the standard deviation of the various C_I values is comparable, indicating a limited link between C_I and power decrease. As a result,

Table 1 Estimated parameter values

Estimation date	B_0	B_1(VC-1)	B_2(V)	B_3(V)
26/01	37.123	– 0.143	– 1.134	– 0.345
01/03	37.056	– 0.134	– 1.234	– 1.123
03/04	37.465	– 0.145	– 0.023	– 1.345
13/05	37.058	– 0.146	– 0.245	– 2.365
18/06	37.864	– 0.156	– 0.345	– 1.145

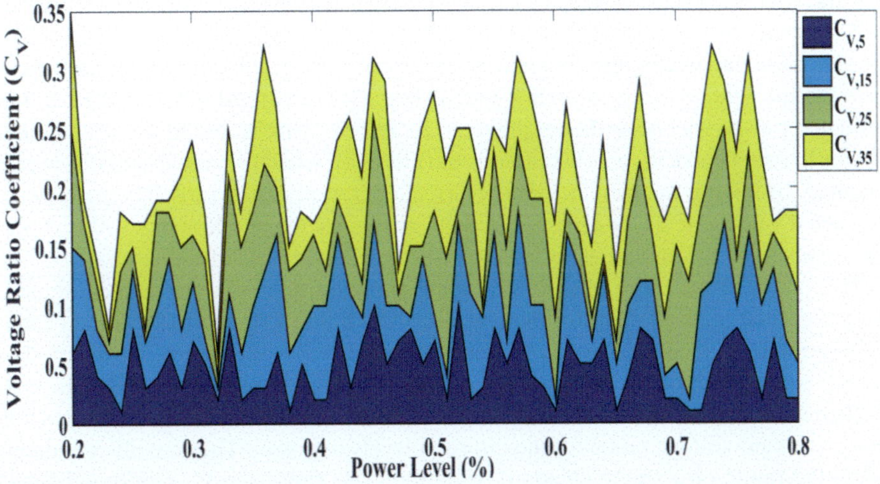

Fig. 3 Comparison between C_V and power level

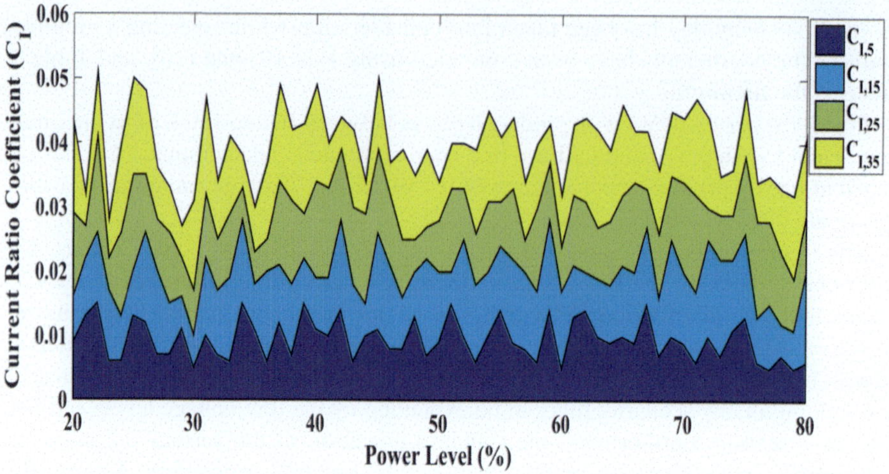

Fig. 4 Comparison between C_I and power level

the MPP current can be accurately estimated in a single cycle, starting from any position along the characteristic.

From the perspective of actual implementation, the tracker must select a single value for C_V and C_I. Using the information at hand, three methods for obtaining C_V and C_I have been examined in this regard. The first involved using all 55,613 of the available data to determine the mean value of C_V and C_I. The second method involved figuring out a C_V and C_I based on a single observation day. To that end, May 13 has been chosen as the midpoint of the six-month observation period. The method uses the computed mean values of C_V and C_I as its parameters. Utilizing simulated data to compute the coefficients was the third tactic. To achieve this, a Simulink panel was created with extremely comparable specs to the actual panel, including an open-circuit voltage of $V_0 = 40.5$ V and a short-circuit current of $I_{sc} = 4.12$ A. Its evaluated power was nevertheless lowered to 160 W. This is due to a 0.5% annual decline in panel capacity was taken into consideration when the actual statistics were obtained, following the panel's seven years of operation on the roof. The panel was connected to a perfect voltage source with variable voltage, allowing 1000 whole I-V curves to be calculated at various temperatures and levels of radiation.

The values of the mean of the current and voltage fractions were then computed using the same methodology for the real data. The limit of standard deviation of the distribution of current and voltage fractions derived from all data is shown by the lines of blue dotted in Figs. 5 and 6. The same two figures show that using data from a single measurement day or even simulation results in remarkably similar numbers for the current and voltage ratios for the different levels of power. They are all contained inside the range of a single standard deviation. The simulated C_I ratio coefficient at lower power levels deviates from the other two but stays within one standard deviation. It also results in an extremely comparable RP operation, as the following example will demonstrate. Consequently, when analyzing the data simulated, the value of the average over a day of single or the curves of the total number shouldn't produce significantly different outcomes. Figures 4 and 6 demonstrate that the power reduction has little effect on the coefficient C_I. As previously mentioned, this property results from the panel's V-I curve, which has a low current slope between the short circuit and maximum power point. Conversely, several factors can be taken into account for C_V as given in Figs. 3 and 5. Its value falls when the power requirement is reduced, and it is highly dependent on the selected power reduction. It must be emphasized once more that this value cannot be regarded as constant throughout the entire set of data or as equal to the power decrease.

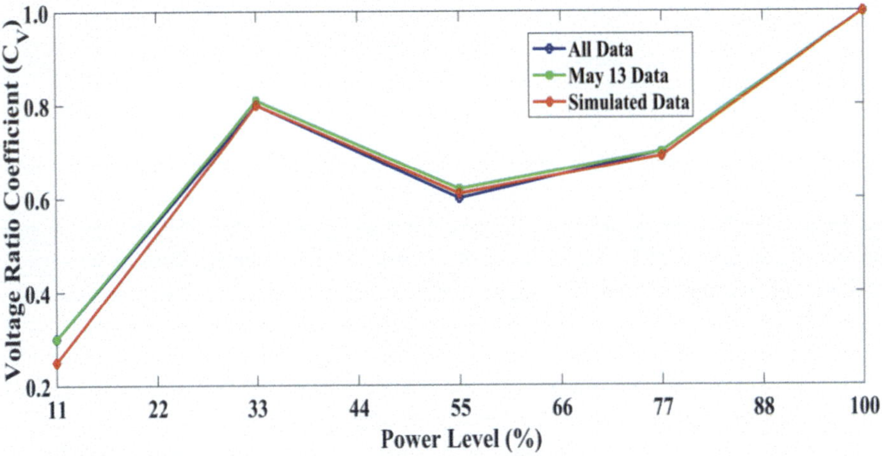

Fig. 5 Comparison of all datasets between C_V and power level

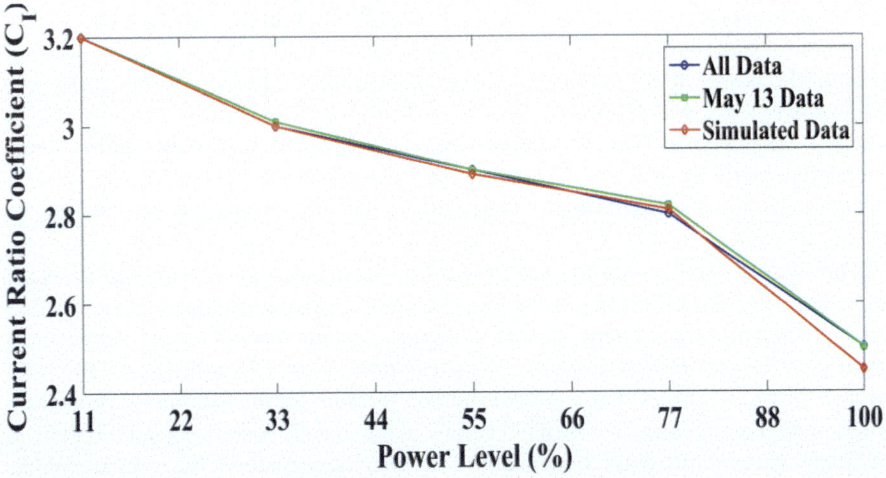

Fig. 6 Comparison of all datasets between C_I and power level

5.2 Performance Index

As previously noted, the assessment of the algorithm's correctness within the 10% to 100% power range, utilizing several methods for evaluating the reduction coefficients, is the focus of this paper. Because of this, the total set of 55 613 curves has been used to compute the error in the RP estimation. All three C_V and C_I estimate techniques have been examined, and the outcomes have been contrasted. The RPPT algorithm's performance has been assessed using the relative error. Since each analysis point's maximum power is known beforehand, it is possible to calculate the relative inaccuracy.

$$\delta_{RP}(c) = \frac{-C_R P_{MP}(c) + P_{RP}(c)}{P_n} \tag{11}$$

where P_{MP} is the theoretical value of MPP, C_R is the factor of power reduction, and P_{RP} is the value of the power that the predicts of the algorithm. The power plant operating in the mode of power curtailment shall operate with an error smaller than the value of $\pm$ 2% with respect to the value of the nominal value of power $P_{n,}$ as per the European Union (EU) regulation [40]. Table 2 shows the fraction of the curve inside such bounds if the $\pm$ 1% value of the limit is considered into account. The algorithm functions sufficiently for (90–99)% of the information, depending on the set of coefficients utilized. The lowest results are produced when operating at 85% of the maximum available power. The limits specified by the standard indicate that the algorithm's accuracy is even higher for all other power levels.

Using the three sets of coefficients, the relative error distribution when operating at 85% of maximum power is displayed in Fig. 7. All other power levels provide distributions that are likewise similar or narrower. C_V and C_I values can be produced for any required power level by linear interpolation. Based on the calculated values of the fractions for percentages of 80 and 85 of the maximum value of the power, C_V and C_I are estimated for the intended point of percentage of 82.5 of the maximum value of the power in this study using linear interpolation. The algorithm works best at this power level, which is the midway point between the two levels of power, which is why it was chosen. Table 3 illustrates how accurately the program can track this intended power with 99–92% of the data falling within the $\pm$ 1% range of about 80% of maximum power. Additionally, the relative error distribution, as seen in Fig. 8, does not change.

The figures demonstrate the exceptional performance of the proposed method. To achieve the TSO limit in almost all panel working circumstances, it seems that the method employs a coefficient of a single constant in both cases. However, a small positive bias may be seen in the distribution of error in both cases. The total number of faulty curves that were overlooked throughout the estimating procedure or a potential association between the values of C_V and C_I because of solar radiation and panel temperature could both account for bias. Nevertheless, the error is smaller than $\pm$ 1% in all scenarios when accounting for the 90th percentile of the symmetric

Table 2 Comparison of data

Level of power (%)	All actual data	13th May real data	Simulated data
20	99.998	99.998	99.997
25	99.992	99.993	99.923
30	99.985	99.985	99.945
35	99.974	99.976	99.956
40	99.931	99.932	99.982
45	99.871	99.878	99.828
50	99.801	99.801	99.831
55	99.670	99.666	99.606
60	99.458	99.456	99.456
65	99.172	99.170	99.172
70	98.840	98.823	98.803
75	98.433	98.408	98.428
80	98.068	98.046	98.046
85	97.961	97.928	97.918
90	98.115	98.110	98.110
95	98.583	98.583	98.583

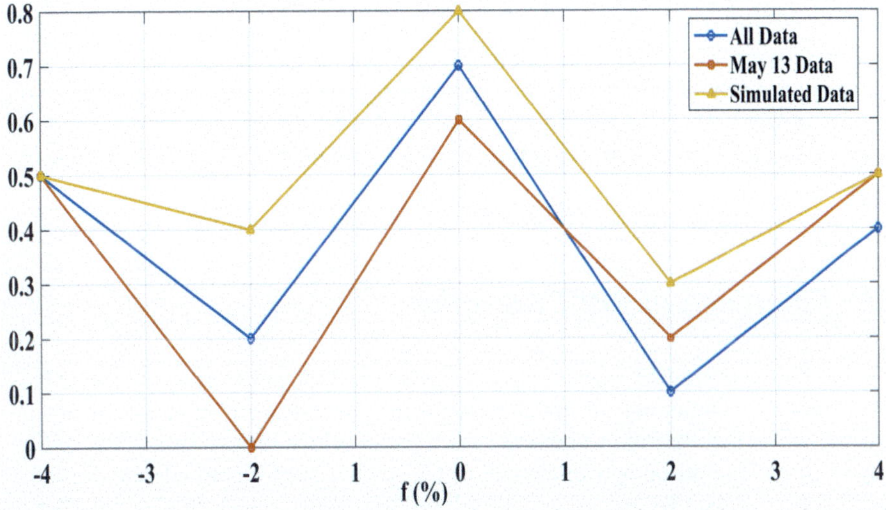

Fig. 7 Comparison of all datasets with respect to error distribution (f %) for the requirement of the maximum power of 80%

Table 3 Estimated dataset

Level of power (%)	All actual data	13th May real data	Simulated data
82.5	97.9257	97.9501	99.0345

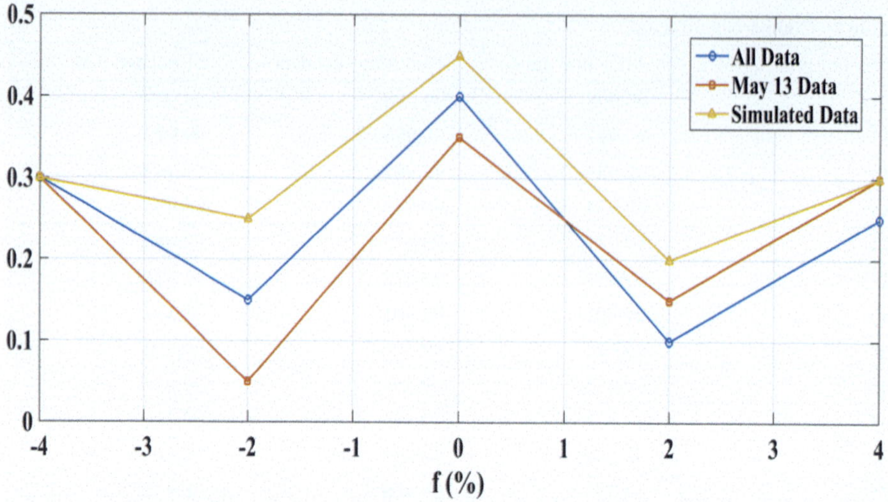

Fig. 8 Comparison of all datasets with respect to error distribution (f %) for the requirement of the maximum power of 82.5%

of the distributions. The fact that the V-I database has not been filtered needs to be emphasized. The errors of the proposed MB RPPT algorithm are comparable to those of the other algorithms when it is asked to operate in two steady-state conditions: a steady-state value of MPP situation [25] (a value of the mean of percentages of 0.1 versus 0.1–9 reached by the other approaches) and the condition of steady-state reduced power point (a value of the mean of 1% for the model-based reduced power point against 0.1–7%). Comparing the outcomes with those found in [38, 40] makes this clear.

5.3 *Uncertainty of RPPT Algorithm*

It is helpful to assess the inaccuracy caused by the lack of certainty on the values of C_I and C_V. Obtaining an updated value for the reduction factors will not be feasible when the algorithm operates under real-world settings. This is the reason why the impact of utilizing a single pair of C_V–C_I has been identified, and the value of accuracy reached by the approach utilizing the exact values of C_V and CI has been compared to the inaccuracy caused by using the values of C_V and C_I from the suggested strategies. If the error values overlap with the ones found using the three procedures for evaluating the reduction factors that have been explained, it indicates that the mistake in the RPPT that results from incorrect reduction factor identification is not very significant.

Because of this, the algorithm has been obliged to operate with the real C_V and C_I values that have been determined for each curve. The error was then calculated once more using the method from Eq. (7) and contrasted with the results found in 5.2 of Sect. 5. Figure 8 displays the distribution of error, for instance, for a generation that uses 80% of the available power. When a single C_V–C_I couple is used, the error distribution increases as predicted; nevertheless, the distribution has a reduced moment of second-order when the technique is applied to the real values, which in the example under study have a standard deviation of 0.82% as opposed to 0.93%. Nevertheless, neither the 1% standard limit nor the distribution's 97% is much impacted by the usage of a single extraction of C_V and C_I. Although the error's standard deviation is higher, its percentile does not reflect this because the error is not Gaussian. Thus, using just a few numbers for C_V and C_I raises the overall uncertainty slightly from a metrological perspective, but for more than 99% of the observations, it still satisfies all TSO requirements.

6 Conclusions

Since grid-tied photovoltaic (PV) systems don't have any energy stored and are made to run at maximum power, they don't take part in power network frequency control. It is becoming more and more strategic to give these plants the flexibility to adjust their power output in response to requests from the operators of the grid as PVs become more and more integrated into the system. Starting from the approach of model-based to the approach of maximum power point tracker, the authors have suggested an algorithm of model-based curtailment of power intended to function at percentages of 10–100 of maximum available power in the optics of the control of any panel alone. A model estimation process has been conducted, and 113 days of data measurement gathered over a six-month period have been used to validate the technique for various power levels. It has been shown that replicating a single standard curve can yield values for the parameters C_V and C_I, leading to comparable results. The algorithm's effectiveness is validated by the results, which show that over 99% of the measurable information across the whole observed the range of power. Additionally, it has been shown that, from a metrological perspective of perspective, the overall uncertainty is slightly increased by using a single set of C_V and C_I values instead of the actual values computed on each curve, but it still complies with all TSO criteria. Since the approach is based on the behavior of electrical of a single cell, similar outcomes should be expected when a string is taken into consideration. This could be the first step towards the actual adoption of photovoltaic plants' participation in the network of power control because of its robustness and flexibility.

Acknowledgements This research work was supported by "Woosong University's Academic Research Funding—2025".

References

1. Salkuti SR (2022) Emerging and advanced green energy technologies for sustainable and resilient future grid. Energies 15(18):6667. https://doi.org/10.3390/en15186667
2. Salkuti SR (2023) Advanced technologies for energy storage and electric vehicles. Energies 16(5):2312. https://doi.org/10.3390/en16052312
3. Park S, Salkuti SR (2019) Optimal energy management of railroad electrical systems with renewable energy and energy storage systems. Sustainability 11(22):6293. https://doi.org/10.3390/su11226293
4. Jung CM, Pagidipala S, Salkuti SR (2024) Emerging technologies for the integration of renewable energy, energy storage and electric vehicles. In: Salkuti SR (eds) Energy and environmental aspects of emerging technologies for smart grid. Green energy and technology. Springer, Cham. https://doi.org/10.1007/978-3-031-18389-8_5
5. Hill CA, Such MC, Chen D, Gonzalez J, Grady WM (2012) Battery energy storage for enabling integration of distributed solar power generation. IEEE Trans Smart Grid 3(2):850–857. https://doi.org/10.1109/TSG.2012.2190113H
6. Ma W et al (2019) Optimal allocation of hybrid energy storage systems for smoothing photovoltaic power fluctuations considering the active power curtailment of photovoltaic. IEEE Access 7:74787–74799. https://doi.org/10.1109/ACCESS.2019.2921316
7. Rajan R, Fernandez FM, Yang Y (2021) Primary frequency control techniques for large-scale PV-integrated power systems: a review. Renew Sustain Energy Rev 144:110998. https://doi.org/10.1016/j.rser.2021.110998
8. Shivashankar S, Mekhilef S, Mokhlis H, Karimi M (2016) Mitigating methods of power fluctuation of photovoltaic (PV) sources—a review. Renew Sustain Energy Rev 59:1170–1184. https://doi.org/10.1016/j.rser.2016.01.059
9. Aghamohammadi MR, Abdolahinia H (2014) A new approach for optimal sizing of battery energy storage system for primary frequency control of islanded microgrid. Int J Electr Power Energy Syst 54:325–333. https://doi.org/10.1016/j.ijepes.2013.07.005
10. Omran WA, Kazerani M, Salama MMA (2011) Investigation of methods for reduction of power fluctuations generated from large gridconnected photovoltaic systems. IEEE Trans Energy Convers 26(1):318–327. https://doi.org/10.1109/TEC.2010.2062515
11. Liao S, Xu J, Sun Y, Bao Y, Tang B (2018) Wide-area measurement system-based online calculation method of PV systems de-loaded margin for frequency regulation in isolated power systems. IET Renew Power Gener 12(3):335–341. https://doi.org/10.1049/ietrpg.2017.0272
12. Yang B, Wang X, Xie D (2019) Guo Y (2019) Novel control strategy of grid-connected photovoltaic power supply for frequency regulation. J Eng 16:1488–1491. https://doi.org/10.1049/joe.2018.8590
13. Yuan H et al (2020) Machine learning-based PV reserve determination strategy for frequency control on the WECC system. In: Proceedings of IEEE power energy society innovative smart grid technologies conference (ISGT), pp 1–5. https://doi.org/10.1109/ISGT45199.2020.9087744
14. Dajun J, Lei Z, Dawei Z, Ning C, Minhui Q (2018) Research on PV generation participating in power grid frequency regulation. In Proceedings China international conference on electricity distribution (CICED), pp 1713–1717. https://doi.org/10.1109/CICED.2018.8592122
15. Rajan R, Fernandez FM (2019) Power control strategy of photovoltaic plants for frequency regulation in a hybrid power system. Int J Electr Power Energy Syst 110:171–183
16. Hoke A, Maksimovic D (2013) Active power control of photovoltaic power systems. In: Proceedings of 1st IEEE conference technologies sustainability (SusTech), pp 70–77. https://doi.org/10.1109/SusTech.2013.6617300
17. Minimum Technical Requirements for Photovoltaic Generation (PV) Projects, Puerto Rico Electric Power Authority, San Juan, Puerto Rico, Nov. 2011
18. Sangwongwanich A, Yang Y, Blaabjerg F, Wang H (2018) Benchmarking of constant power generation strategies for single-phase grid connected photovoltaic systems. IEEE Ind Appl Mag 54(1):447–457. https://doi.org/10.1109/TIA.2017.2740380

19. Sangwongwanich YY, Blaabjerg F (2016) High-performance constant power generation in grid-connected PV systems. IEEE Trans Power Electron 31(3):1822–1825
20. Tafti HD, Sangwongwanich A, Yang Y, Konstantinou G, Pou J, Blaabjerg F (2018) A general algorithm for flexible active power control of photovoltaic systems. In: Proceedings of IEEE applied power electronics conference and exposition (APEC), pp 1115–1121. https://doi.org/10.1109/APEC.2018.8341156
21. Faifer M, Cristaldi L, Toscani S, Soulantiantork P, Rossi M (2015) Iterative model-based maximum power point tracker for photovoltaic panels. In: Proceedings of IEEE international instrumentation and measurement technology conference (I2MTC), pp 1273–1278. https://doi.org/10.1109/I2MTC.2015.7151456
22. Soulatiantork P, Cristaldi L, Faifer M, Laurano C, Ottoboni R, Toscani S (2018) A tool for performance evaluation of MPPT algorithms for photovoltaic systems. Measurement 128:537–544. https://doi.org/10.1016/j.measurement.2018.07.005
23. Cristaldi L, Faifer M, Laurano C, Petkovski E, Toscani S, Ottoboni R (2022) An innovative model-based algorithm for power control strategy of photovoltaic panels. In: Proceedings of IEEE international instrumentation and measurement technology conference (I2MTC), pp 1–6. https://doi.org/10.1109/I2MTC48687.2022.9806605
24. Cristaldi L, Faifer M, Laurano C, Ottoboni R, Petkovski E, Toscani S (2022) Reduced power model-based tracker for photovoltaic panels. In: Proceedings of IEEE 16th international conference on compatibility, power electronics, and power engineering (CPE-POWERENG), pp 1–6
25. Cristaldi L, Faifer M, Laurano C, Ottoboni R, Toscani S (2016) Experimental comparison of MPPT algorithms. In: Proceedings of IEEE international instrumentation and measurement technology conference proceedings, pp 1–6. https://doi.org/10.1109/I2MTC.2016.7520569
26. Cristaldi L, Faifer M, Laurano C, Ottoboni R, Toscani S, Zanoni M (2020) Model-based maximum power point tracking for photovoltaic panels: parameters identification and training database collection. IET Renew Power Gener 14(15):2876–2884. https://doi.org/10.1049/iet-rpg.2020.0101
27. Pappu VAK, Chowdhury B, Bhatt R (2010) Implementing frequency regulation capability in a solar photovoltaic power plant. In: Proceedings of North American power symposium, pp 1–6. https://doi.org/10.1109/NAPS.2010.5618965
28. Sangwongwanich A, Yang Y, Blaabjerg F (2016) Sensorless reserved power control strategy for two-stage grid-connected photovoltaic systems. In: Proceedings of IEEE 7th international symposium on power electronics for distributed generation systems (PEDG), pp 1–8. https://doi.org/10.1109/PEDG.2016.7527032
29. Hoke, Muljadi E, Maksimovic D (2015) Real-time photovoltaic plant maximum power point estimation for use in grid frequency stabilization. In: Proceedings of IEEE 16th workshop control modeling power electronics (COMPEL), pp 1–7
30. Li Q, Baran ME (2020) A novel frequency support control method for PV plants using tracking LQR. IEEE Trans Sustain Energy 11(4):2263–2273. https://doi.org/10.1109/TSTE.2019.2953684
31. Verma P, Kaur T, Kaur R (2021) Power control strategy of an integrated PV system for active power reserve under dynamic operating conditions. Sustain Energy Technol Assessments 45:101066. https://doi.org/10.1016/j.seta.2021.101066
32. Xin H, Liu Y, Wang Z, Gan D, Yang T (2013) A new frequency regulation strategy for photovoltaic systems without energy storage. IEEE Trans Sustain Energy 4(4):985–993. https://doi.org/10.1109/TSTE.2013.2261567
33. Wang Z et al (2022) Active power control of voltage-controlled photovoltaic inverter in supporting islanded microgrid without other energy sources IEEE. J Emerg Sel Top Power Electron 10(1):424–435. https://doi.org/10.1109/JESTPE.2021.3069700
34. Batzelis E, Kampitsis G, Papathanassiou S (2017) Power reserves control for PV systems with real-time MPP estimation via curve fitting IEEE Trans. Sustain Energy 8(3):1269–1280. https://doi.org/10.1109/TSTE.2017.2674693

35. Batzelis EI, Papathanassiou SA, Pal BC (2018) PV system control to provide active power reserves under partial shading conditions. IEEE Trans Power Electron 33(11):9163–9175. https://doi.org/10.1109/TPEL.2018.2823426
36. Sangwongwanich A, Yang Y, Blaabjerg F, Sera D (2017) Delta power control strategy for multistring grid-connected PV inverters. IEEE Trans Ind Appl 53(4):3862–3870
37. Hua T, Yan X, Fan W (2017) Research on power point tracking algorithm considered spinning reserve capacity in gird-connected photovoltaic system based on VSG control strategy. In: Proceedings of IEEE 3rd international future energy electronics conference ECCE Asia (IFEEC ECCE Asia), pp 2059–2063. https://doi.org/10.1109/IFEEC.2017.7992368
38. Tafti HD et al (2020) Extended functionalities of photovoltaic systems with flexible power point tracking: recent advances. IEEE Trans Power Electron 35(9):9342–9356. https://doi.org/10.1109/TPEL.2020.2970447
39. Cristaldi L, Faifer M, Laurano C, Leone G, Toscani S (2017) Model based maximum power point tracker stability over time. In: Proceedings of 6th international conference on clean electrical power (ICCEP), pp 591–595. https://doi.org/10.1109/ICCEP.2017.8004748
40. Commission Regulation (EU) 2016/631 (2016) Establishing a network code on requirements for grid connection of generators. [Online]. Available: https://eur-lex.europa.eu/eli/reg/2016/631/oj
41. Cristaldi L, Faifer M, Laurano C, Ottoboni R, Petkovski E, Toscani S (2023) Power generation control algorithm for the participation of photovoltaic panels in network stability. IEEE Trans Instrum Meas 72(1–9):9000809. https://doi.org/10.1109/TIM.2023.3238745

Performance Evaluation of Quasi Z-Source Inverter Utilize for Grid Connected with Wind-PV Co-generation Systems

Ch. Sreenu, G. Mallesham, T. Chandra Shekar, and Surender Reddy Salkuti

Abstract This chapter provides a thorough performance evaluation of a Quasi Z-Source Inverter (qZSI) used in grid-connected wind-photovoltaic (PV) co-generation systems. Reducing dependency on fossil fuels and attaining sustainable energy objectives need the integration of clean energy sources, especially solar and wind. The qZSI topology offers several advantages, including improved voltage-boosting capabilities, reduced component count, and enhanced system reliability. This study examines the operational characteristics of the qZSI in combination with wind and PV systems, focusing on efficiency, power quality, and dynamic response under varying load conditions and environmental factors. Through MATLAB simulation validation, the study analyzes the qZSI's effectiveness concerning its ability to maintain grid stability and optimize the energy output from both wind and solar sources. The findings show how the qZSI may improve grid-connected cogeneration systems' overall performance and highlight its potential for use in future integration of green energy applications.

Keywords Quasi Z-source inverter · Wind-PV Co-generation · Grid integration · Renewable energy · Permanent magnet synchronous generator

---

Ch. Sreenu · G. Mallesham
Department of Electrical Engineering, University College of Engineering, Osmania University, Hyderabad, Telangana, India
e-mail: drcnu@osmania.ac.in

G. Mallesham
e-mail: drgm@osmania.ac.in

T. Chandra Shekar
Department of Electrical Engineering, Maturi Venkata Subba Rao Engineering College, Hyderabad, India

S. R. Salkuti (✉)
Department of Railroad and Electrical Engineering, Woosong University, Daejeon, Republic of Korea
e-mail: surender@wsu.ac.kr

S. R. Salkuti (ed.), *Artificial Intelligence for Integrated Smart Energy Systems in Electric Vehicles*, Lecture Notes in Electrical Engineering 1427,
https://doi.org/10.1007/978-3-031-94276-1_28

Nomenclature

QZSI	Quasi Z source inverter
ZSI	Z source inverter
CSI	Current source inverter
VSI	Voltage source inverter
B	Boost factor
M	Modulation index
PCC	Point of common coupling
V_{link}	DC-link voltage
PMSG	Permanent magnet synchronous generator
PWM	Pulse width modulation
WECS	Wind energy conversion system
β	Pitch angle
P_w	Wind power
V_{out}	Output voltage
v_{ds} and v_{qs}	D-q axes in stator voltages
i_{ds} and i_{qs}	D-q-axes in stator current
ψ_d and ψ_q	D-q-axes in flux linkage
I_D	Diode current
I_{sh}	Shunt current
p_n	Pole pairs
V_{sh}	Voltage across the shunt resistor
I_0	Diode reverse saturation current
V_D	Voltage across the diode
V_T	Thermal voltage
STC	Standard test condition

1 Introduction

Clean sources of energy may meet the world's current energy needs and provide several benefits. They can provide a long-term, environmentally friendly power supply, lower worldwide air emissions, and diversify the markets for the production of energy [1]. Traditional converters like VSI [2], CSI [3], and ZSI [4, 5] have specific limits in terms of their planning and functioning. A high-value capacitor in the VSI is linked to the source end to provide a steady DC input voltage. However, it has drawbacks: (a) Additional devices are required for voltage improvement, leading to increased costs and reduced overall system efficiency. (b) There is a risk of short-circuit occurrences that can damage devices. (c) The reliability of the VSI is limited. For the CSI, a high inductance linked to the source's end makes it easier to maintain a steady DC input current. The discussed configuration involves six switches similar to a VSI, with IGBTs instead of MOSFETs and diodes in series. Six of the nine switching states that the CSI uses for operation are active. However, the CSI has drawbacks, including being a type of boost inverter, high cost, low power factor, vulnerability to electromagnetic interference, and slow dynamic response [6–8]. Integrating boost converters into the system is an interface between lower DC voltages and a more DC voltage power supply when needed. However, this increases the system's cost and size while reducing performance. This highlights the limitations of conventional inverters, necessitating multiple operations to enhance the system's performance within acceptable limits [9–11].

Usually, a normal VSI is used; however, to provide the voltage required for the network circumstances, an additional boost stage is needed. Nevertheless, this method reduces system efficiency and dependability while increasing the price and overall dimensions of the converter [12]. A qZSC is employed to solve this problem in renewable hybrid microgrids. It is better suited for PV systems because of its advantages, which include low voltage stress, steady incoming currents, and general grounding of the DC source [13]. The ZSC is operated, modeled, and controlled by the qZSC, which is a component of the ZSC [14, 15]. The qZSC has been used in systems for clean energy in earlier research [16–18]. However, its possible use in combination microgrid systems hasn't been investigated yet. An experimental examination of a combined PV-wind standalone system founded on the qZSC is shown in Reference [19]. With an emphasis on energy storage control, the qZSC was used in reference [20] as an Inverters Controller (IC) in a hybrid AC or DC microgrid. The conventional inverters listed above require more reliable operations while maintaining acceptable efficiency. To improve system performance within reasonable bounds, it is important to execute numerous operations inside a single job. An innovative single-stage buck-boost electrical power conversion device for green energy applications is ZSI [21–26]. Rearranging the supply side components results in a quasi-impedance source inverter (QZSI), which is designed to achieve low voltage stress and continuous input current, starting from the fundamental ZSI impedance. With the use of the modulation index, this inverter can perform two main functions: boost operation when the shoot-through condition is reached and buck operation when it is not. Compared to QZSI, ZSI provides less DC voltage gain and an interrupted input current [27–31]. The stable operation of energy from green sources, such as solar panels and uninsulated wind inverters with large direct current gain for grid injection, is severely hampered by wide voltage fluctuations. This issue may be resolved with the aid of the QZSI, which also contributes to more accurate and reliable functioning. Furthermore, the QZSI's straightforward design enables low-cost and efficient operations for variable-speed drive applications. The study aims to optimize Quasi-Z-Source Inverters (qZSI) for integrating wind and PV systems into the power grid by developing efficient qZSI topology and advanced control strategies. The goal is to enhance voltage gain, power quality, and overall efficiency of renewable energy systems to improve grid integration of wind and PV technologies.

This research contributes to the integration of renewable energy by comprehensively analyzing Quasi-Z-Source Inverters (qZSI) for wind-PV co-generation systems. It presents a detailed examination of qZSI design principles, including optimizing critical parameters such as inductance and capacitance to enhance system performance. By demonstrating the practical applications and benefits of qZSI through simulation results, the study highlights its effectiveness in improving power quality, increasing energy yield, and supporting grid stability. This work advances the understanding of qZSI technology and provides valuable insights for designing and implementing robust and efficient grid-connected renewable energy systems.

After the introduction, this chapter is organized as follows: Sect. 2 covers mathematical modeling and provides an overview of hybrid wind-solar generators. Section 3 presents the three-phase qZSI. A discussion of the results follows in Sect. 4. Finally, the chapter concludes with a summary and recommendations in Sect. 5.

2 Modeling and Overview of Hybrid Wind-Solar Generators

Figure 1 shows the hybrid wind-solar generators combine wind turbines and solar photovoltaic (PV) panels to capture renewable energy from various sources. This setup optimizes electricity production even in changing environmental conditions. By merging these two complementary energy systems, hybrid setups can offer a more consistent and dependable power output. Solar panels generate energy during the day, while wind turbines often produce more power during evening and night hours [32]. These systems are well-suited for off-grid and grid-connected applications, as they enhance efficiency, reduce reliance on fossil fuels, and improve energy security. To model these systems effectively, it is crucial to consider factors such as resource availability, power electronics, control strategies, and energy storage to achieve optimal performance and grid stability.

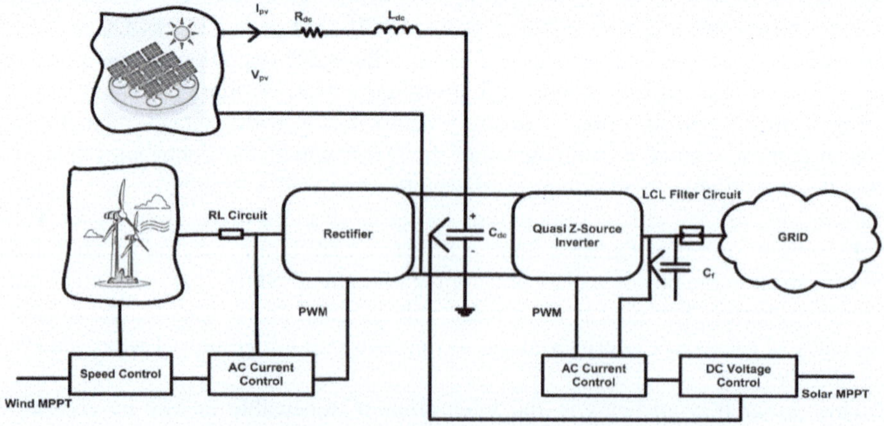

Fig. 1 The proposed wind-PV cogeneration system

2.1 Modeling of Wind Power Generation System

Wind power technologies have been around for centuries. The earliest use of wind power was found in ancient civilizations like Persia and China, where windmills were used to pump water and grind grain. Modern wind energy systems began in the late nineteenth century when scientists like Charles F. Brush and Poul la Cour experimented with wind-driven electricity generators. The 1970s oil crisis fueled demand for wind energy as a viable alternative to energy from fossil fuels, resulting in the construction of more powerful and efficient wind turbines. With advancements in materials, aerodynamics, and control systems, wind has become one of the fastest-growing green energy sources worldwide, contributing significantly to the global transition towards sustainable energy [33]. Today, wind farms are installed onshore and offshore, delivering clean electricity to millions of households, with continuous innovations driving cost reductions and performance improvements. The main components of a wind energy conversion system are shown in Fig. 2a as the wind turbine, generator, control system, and support structure. To develop the mathematical modeling of PMSG-based WECS is essential to predict the system behavior and response for grid interconnection during normal and abnormal conditions. Figure 2b displays the electrical power characteristics associated with wind turbines. For the sake of this study, β is fixed to 0° and the base velocity is set at 12 m/s.

2.2 Wind Energy Conversion

If an object has mass m and speed v and stays at speed v, its kinetic energy E is the same as the work done W to move it from rest to a location s by a force F, i.e., $E = W = F \times s$. The 2nd rule for motion by Newton says $F = ma$. Using air mass m and wind speed v, we can find the wind's kinetic energy (K.E). Thus, the K.E becomes $E = mas$. Using solid motion kinematics, $v^2 = u^2 + 2as$ here u represents the object's starting velocity. This suggests that. $a = \frac{v^2-u^2}{2s}$ Assume the object's starting velocity is zero. We possess that $a = \frac{v^2}{2s}$. we have that $E = \frac{1}{2}mv^2$. Kinetic energy is calculated by assuming that the solid's mass stays the same. Moving air, like wind, is a fluid, so its density and speed can change, so it doesn't have a steady mass. So, we can figure out the K.E (in joules) of air with mass m moving at speed v_w (waving wind). We can figure out wind power P by looking at how fast kinetic energy changes.

$$P_w = \frac{dE}{dt} = \frac{1}{2}\frac{dm}{dt}v_w^2 \tag{1}$$

However, the mass flow rate is $\frac{dm}{dt}$ is given by $\frac{dm}{dt} = \rho A v_w$. Where A represents the region that the wind is now blowing across and ρ denotes the air density. Equation (1) is transformed into, using this expression.

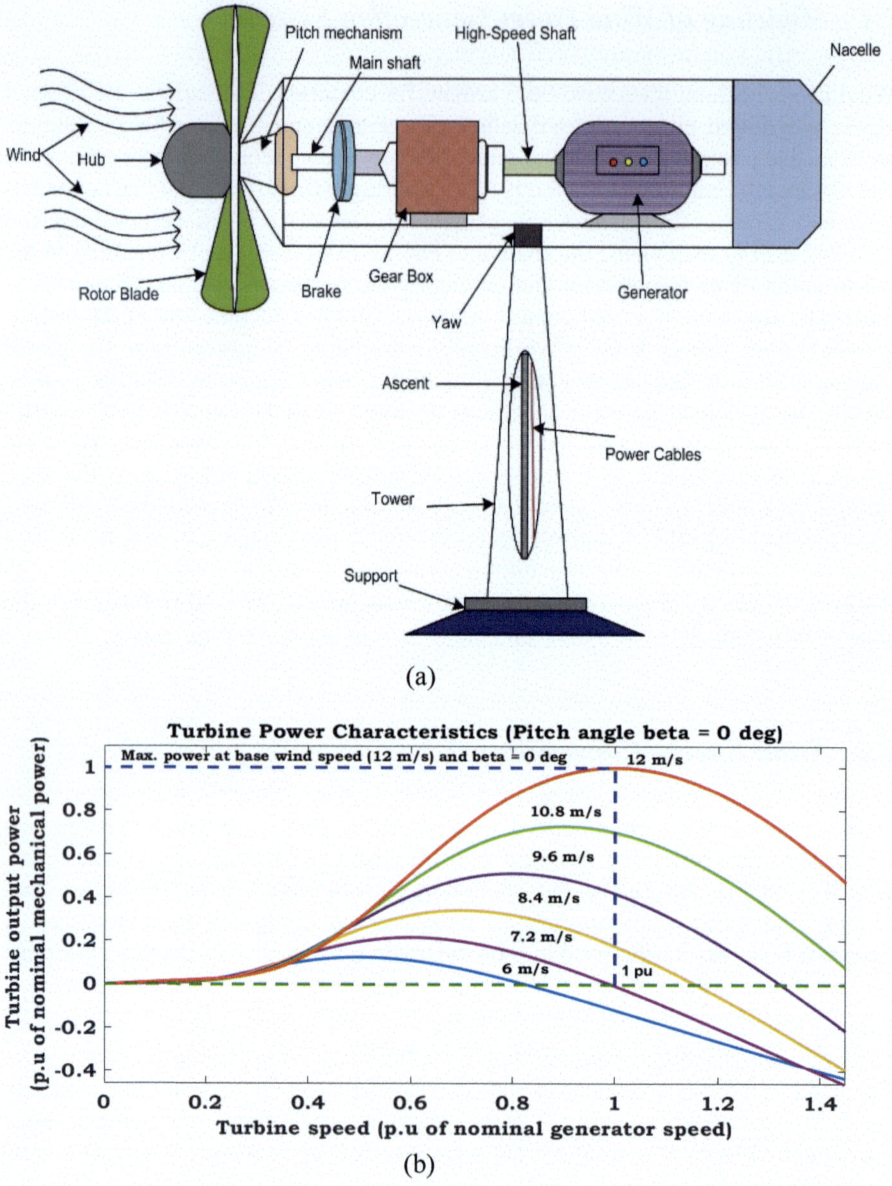

Fig. 2 **a** Essential parts of a system for converting wind energy. **b** The wind turbine's mechanical attributes with varying wind speeds

$$P_w = \frac{1}{2}\rho A v_w^3 \tag{2}$$

The actual mechanical energy Pw that the blades of the rotor extract is represented by the difference between the downstream and upstream wind powers, or watts.

$$\rho A v_w = \frac{\rho A(v_u - v_d)}{2} \tag{3}$$

where v_u and v_d are the upstream and downstream wind speeds at the rotor blades' entry and departure, respectively, in m/s. We'll find subsequently that these two velocities determine the blade-tip speed ratio. In terms of mass flow rate,

$$\rho A v_w = \frac{\rho A(v_u - v_d)}{2} \tag{4}$$

The mean of the velocity at the turbine rotor blades' entrance and exit is called v_w. This expression transforms Eq. (3) into $P_w = \frac{1}{2}\rho A\left(v_u^2 - v_d^2\right)\frac{(v_u - v_d)}{2}$ which may be simplified as follows:

$$\begin{aligned} P_w &= \frac{1}{2}\left[\rho A\left\{\frac{v_u}{2}\left(v_u^2 - v_d^2\right) + \frac{v_d}{2}\left(v_u^2 - v_d^2\right)\right\}\right] \\ &= \frac{1}{2}\left[\rho A\left\{\frac{v_u^3}{2} - \frac{v_u v_d^2}{2} + \frac{v_d v_u^2}{2} - \frac{v_d^3}{2}\right\}\right] \\ &= \frac{1}{2}\left[\rho A v_u^3\left\{\frac{1 - \left(\frac{v_d}{v_u}\right)^2 + \left(\frac{v_d}{v_u}\right) - \left(\frac{v_d}{v_u}\right)^3}{2}\right\}\right] = \frac{1}{2}\rho A v_u^3 C_p \end{aligned} \tag{5}$$

$$Where,\ C_p = \frac{1 - \left(\frac{v_d}{v_u}\right)^2 + \left(\frac{v_d}{v_u}\right) - \left(\frac{v_d}{v_u}\right)^3}{2};\ \text{or}$$

$$C_p = \frac{\left(1 + \frac{v_d}{v_u}\right) + \left(1 - \left(\frac{v_d}{v_u}\right)^2\right)}{2} \tag{6}$$

The formula for Cp in Eq. (6) reflects the fraction of upstream wind power collected by the rotor blades. Since its discovery in 1919 by German scientist Albert Betz, Cp is often referred to as the Betz limit. This value is often known as the rotor's coefficient of power or efficiency. The energy coefficient is not equivalent to a static value. It fluctuates depending on the wind turbine's tip-speed ratio. The wind speed v_d downstream to the wind speed v_u upstream of the turbine is represented by λ, i.e. $\lambda = \frac{v_d}{v_u}$ or $\lambda = \frac{BladeTipSpeed}{WindSpeed}$. Where λ is called the tip speed ratio of the wind turbine. The turbine's rotating speed and the turbine's blade length may be used to compute the tips of the blade's speeds in meters per second, i.e.,

$$BladeTipSpeed = \frac{Angular\ Speed\ of\ Turbine\ (\omega) \times R}{Wind\ Speed} \tag{7}$$

where R represents the turbine's radius, while ω is expressed in radians per second.

2.3 Mathematical Modelling of Wind Turbine

A wind turbine's aerodynamic, mechanical, and electrical components are represented mathematically in order to forecast how the machine would behave under different operating scenarios. The power that is taken from the wind and is controlled by Eq. (8) forms the basis of the model. The power coefficient, or Cp, is a measure of how well wind energy is converted to mechanical energy and is influenced by turbine design and operating parameters. This power is converted into rotational motion by the mechanical model, which also takes the turbine's inertia and damping into consideration. Electrical power is produced by means of a generator from mechanical energy. After that, the generator is connected to the grid, generally with the help of power electronics, enabling variable-speed operation and grid integration. The definition of wind power is given by,

$$P_w = \frac{1}{2}\rho A \upsilon_w^3 \times C_p(\lambda, \beta) \tag{8}$$

where A is the swept area covered by turbine blades (in m^2), C_p is the coefficient of wind power, ρ is the air density factor in (kg/m^3), and v_ω is wind speed (in m/s).

$$P_w = \frac{1}{2}\rho \pi r^2 \upsilon_w^3 \times C_p(\lambda, \beta) \tag{9}$$

The aerodynamic torque produced by the aerodynamic force is $T_t = \frac{P_w}{\omega_m}$. For a wind turbine, the ratio of the turbine's tip speed to wind velocity is known as the Tip Speed Ratio (TSR). $TSR = \lambda_{opt} = \frac{\omega_m R}{\upsilon_\omega}$, where ω_m is the rotor's angular speed (in rad/s), and R is the turbine's radius (in meters). Therefore, wind velocity is given by $\upsilon_\omega = \frac{\omega_m R}{\lambda_{opt}}$. From Eq. (8), the wind power is given by,

$$P_w = 0.5\rho A C_p \left(\frac{\omega_m R}{\lambda_{opt}}\right)^3 \tag{10}$$

To demonstrate its effectiveness, the proposed controller is implemented in MATLAB/Simulink using a 3 MW full-scale wind turbine as described in Table 1.

Table 1 Wind modeling simulation parameters

Wind parameters	
Mechanical output power (W)	3×10^6
Basic power of the electrical generator (VA)	$3 \times 10^6/0.9$
Basic wind speed (m/s)	12
Maximum power at base wind speed (p.u. of nominal mechanical power)	1
Base rotational speed (p.u. of the base generator speed)	1
Wind-turbine power shown by pitch angle beta (>0) (deg	0°
PMSG details Specifications	
Resistance (R) (ph–ph)	0.36 Ω
Inductance (L_{ab}) (ph–ph)	1.67 mH
D axis Inductance (L_d) (ph)	0
Q axis inductance (L_q) (ph)	0
Constant of voltage (K_e)	0
Constant of torque (k_t)	60.7
Inertia (J)	5.5 E^{-3}
Damping viscosity (F)	4.5
Pole pairs (P)	4

2.4 *Mathematical Modelling of PMSG*

The PMSG is a power conversion machinery that converts energy from the mechanical form to the electrical form. The following assumptions are assumed to shorten the PMSG mathematical modeling:

- The stator core is saturated and the linear magnetic circuit is assumed to be linear.
- Parameters of the inductance are not changed.
- Damping winding is absent on the rotor side.

The 3-phase PMSG voltage is given by,

$$v_a = v_m \sin(\omega t);\ v_b = v_m \sin\left(\omega t - \frac{2\pi}{3}\right);\ v_c = v_m \sin\left(\omega t + \frac{2\pi}{3}\right) \tag{11}$$

The time-varying quantities are transformed into time-invariant quantities using Park transformation.

$$\begin{bmatrix} v_d \\ v_q \\ v_0 \end{bmatrix} = \sqrt{\frac{2}{3}} \begin{bmatrix} \cos\theta_r & \cos\left(\theta_r - \frac{2\pi}{3}\right) & \cos\left(\theta_r + \frac{2\pi}{3}\right) \\ -\sin\theta_r & -\sin\left(\theta_r - \frac{2\pi}{3}\right) & -\sin\left(\theta_r + \frac{2\pi}{3}\right) \\ \frac{\sqrt{2}}{2} & \frac{\sqrt{2}}{2} & \frac{\sqrt{2}}{2} \end{bmatrix} \begin{bmatrix} v_a \\ v_b \\ v_c \end{bmatrix} \tag{12}$$

The PMSG three-phase stator voltage in the d-q reference frame is represented by,

$$v_{ds} = i_{ds}R_s + L_d\frac{di_{ds}}{dt} - \omega_e\psi_q;\ v_{qs} = i_{qs}R_s + L_q\frac{di_{qs}}{dt} + \omega_e\psi_d \tag{13}$$

R_s represents the PMSG stator's resistance., v_{ds} and v_{qs} is the stator voltages in d and q axes, i_{ds} and i_{qs} are stator current in d and q-axes, L_d and L_q are components of stator inductance in dq-axes, ω_e is the velocity of rotation of the generator, and ψ_d and ψ_q are linked by stator flow in dq-axes. The electrical and mechanical rotational speed is related by $\omega_e = p_n\omega_m$, where p_n is pole pairs. The PMSG stator flux linkages in dq-axes are given by,

$$\psi_d = L_d i_{ds} + \psi_m;\ \psi_q = L_q i_{qs} \tag{14}$$

where ψ_m is the flux linkage generated by permanent magnets. Then, Eq. (13) can be expressed by,

$$v_{ds} = i_{ds}R_s + L_d\frac{di_{ds}}{dt} - \omega_e L_q i_{qs};\ v_{qs} = i_{qs}R_s + L_q\frac{di_{qs}}{dt} + \omega_e L_d i_{ds} + \omega_e\psi_m \tag{15}$$

Figure 3 shows PMSG's dq-axes equivalent circuit.

The source of the PMSG stator's actual and reactive power is,

$$P_g = \frac{3}{2}\left(v_{ds}i_{ds} + v_{qs}i_{qs}\right);\ Q_g = \frac{3}{2}\left(v_{qs}i_d - v_{qs}i_q\right) \tag{16}$$

The PMSG's electromagnetic torque is given by,

$$T_e = \frac{3}{2}P_n\left(\psi_d i_{qs} - \psi_q i_{ds}\right) = \frac{3}{2}P_n\left(\psi_m i_{qs} + \left(L_d - L_q\right)i_{ds}i_{qs}\right) \tag{17}$$

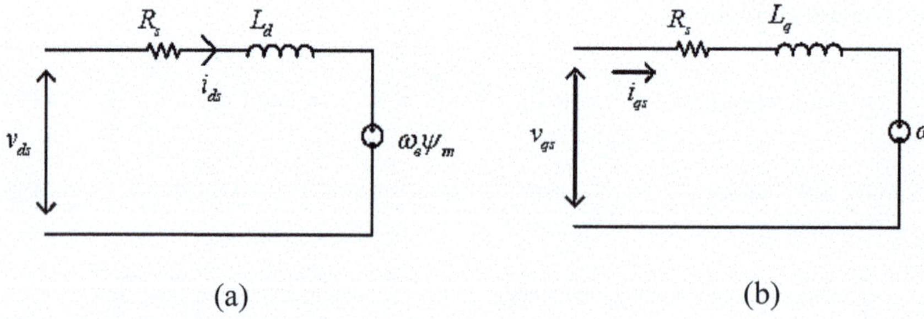

Fig. 3 PMSG equivalent dynamic model in **a** d-axis and **b** q-axis

By taking L_d and L_q equal for a non-saliency, the PMSG electromagnetic torque is given as $T_e = \frac{3}{2} P_n \psi_m i_{qs}$. The generator's mechanical rotational speed is defined by $J_m \frac{d\omega_m}{dt} + B_m \omega_m = T_m - T_e$. The generator's mechanical rotational velocity is defined by $\frac{d\omega_m}{dt} = \omega = \omega_r = \frac{(T_m - T_e - \omega_m B_m)}{J_m}$. Where J_m is the rotor moment of inertia, B_m is the viscous friction coefficient, T_m and T_e are mechanical and electrical torque.

2.5 *Photovoltaic Energy System*

A French scientist named Edmund Becquerel postulated in 1839 that not many materials could generate electricity while placed in solar. Albert Einstein, however, provided clarification on the fundamental properties of light and the photoelectric impact in 1905. According to the photoelectric effect, electrons will flow across a metal surface when photons or sunshine reach it. Eventually, the fundamental idea behind solar power-generating electricity technology was the photoelectric effect. In 1954, Bell Laboratories manufactured the first solar module.

2.6 *Types of PV Cells*

Solar cells are classified into three types: monocrystalline, polycrystalline, and thin film, depending on their installation. Silicon wafers joined in both directions are used to create monocrystalline cells, which are then covered with a glass sheet from the top. It goes by the name of the one crystalline panel as well. It has a premium price tag and provides excellent performance. One silicon crystal is used to create polycrystalline cells, which then melt together to make wafers. Another name for it is multi-crystalline silicon. It is less expensive and has a lower efficiency than mono-crystalline cells. The amorphous forms of silicon and cadmium are used to create thin-film cells, which are then applied on plastic surfaces. It is not as thick as other cells. It is lighter and less efficient. It is also versatile and portable.

2.7 *Mathematical Description of Solar PV Cell/Module/Array*

Solar panels can generate higher voltages, currents, and power levels by connecting photovoltaic cells in series and parallel circuits. These panels comprise one or more photovoltaic modules, which are PV cell circuits enclosed in a protective laminate. The total power generation equipment is a collection of sunlight cells and panels. Photovoltaic modules and arrays are often assessed for performance using their maximum DC power output measured under STC, which includes a 25 °C module operating temperature, 1000 W/m^2 of incoming solar radiation, and a 1.5

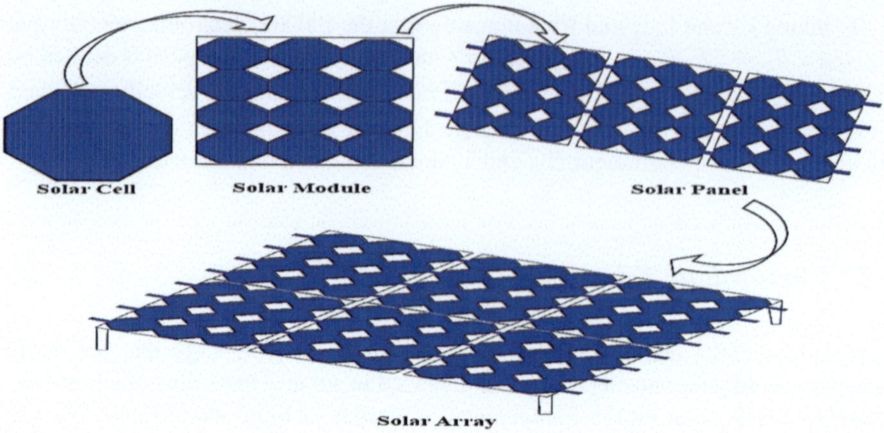

Fig. 4 Solar cells are arranged in a module to create the solar array

spectral distribution for Air Mass. However, since these conditions do not fully reflect real-world performance, the actual output might be between 85 and 90% of the STC rating. PV cells are used to convert photon energy into electrical power. A single PV cell can only provide a small amount of electricity. Multiple solar cells are connected in series to create a PV module to increase power output. These PV modules may then be connected in series and parallel to construct PV arrays that can handle enormous power needs. The solar array created by the module made up of solar cells is shown in Fig. 4.

2.8 Characteristics of Solar PV Module

When a PV cell is disconnected from any external load, its maximum voltage is referred to as the open circuit voltage. The maximum current that a solar cell can produce when its terminals are shorted together is known as the short circuit current. When the photovoltaic cell operates at the MPP, or peak power operational point, on the I-V curve, its voltage multiplied by current reaches its maximum. Figure 5 depicts the PV cell's I-V and P–V curves.

Radiation has a considerable impact on the current flowing through a PV module, resulting in a current-regulated device. Similarly, temperature affects the voltage of a PV module. The Mitsubishi Electric PV-UD190MF5PV module is being used to study the impact of variations in irradiation. In Fig. 6, characteristic curves for various irradiation levels are displayed. The P–V characteristics further illustrate how changes in irradiance affect the output power of the solar PV system. Figure 7 shows the characteristic curve of the PV module to study the impact of temperature variations. This chart demonstrates that changes in temperature cause voltage to fluctuate while the current remains relatively constant with minimal variation.

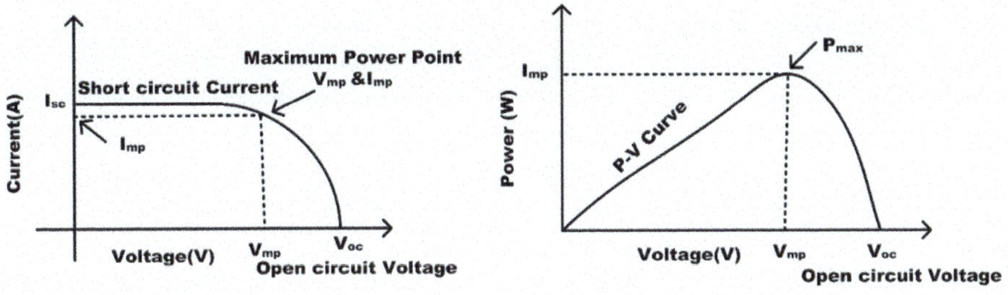

Fig. 5 I-V and P–V curve of the PV cell

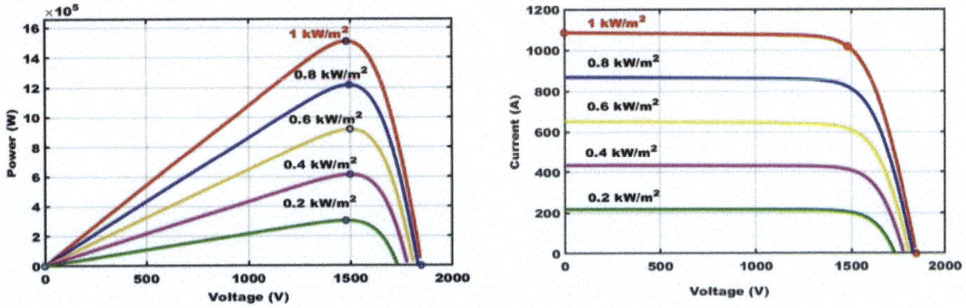

Fig. 6 I-V and P–V characteristics of PV module for different irradiations

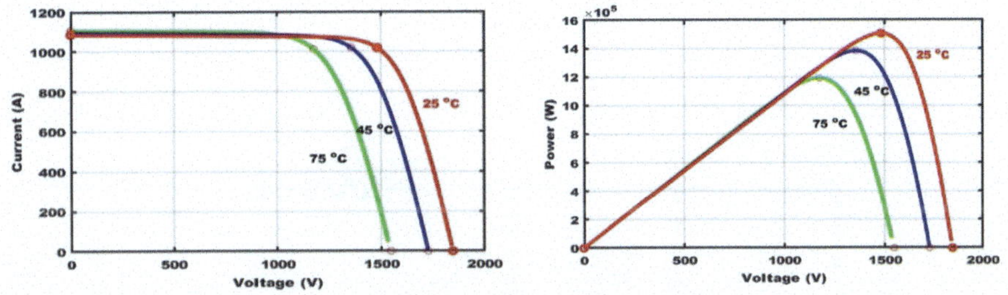

Fig. 7 I-V and P–V characteristics of PV module for different temperatures

2.9 *Modeling of PV Module*

PV modeling is a useful tool for evaluating solar module operation in various climatic scenarios. The literature presents various circuit models, including modified, ideal, single, and two-diode versions. The one-diode model is the most straightforward and accurate.

2.9.1 Single Diode Model

Photovoltaic (PV) modeling commonly employs the one-diode model due to its simple implementation and precise methodology. The ideal model is primarily used for research and does not depict controlling components such as resistors. However, a solar cell's behavior changes in real time in response to its surroundings. Consequently, the ideal model with a one-diode, known as the one-diode model, includes series and parallel resistors. Figure 8 illustrates an analogous electrical circuit for a one-diode model.

By using the KCL rule, the output current equation is created as follows:

$$I = I_{ph} - I_D - I_{sh} \tag{18}$$

$$I_D = I_0\left[\left(\frac{V_D + IR_{se}}{\eta V_T}\right) - 1\right] \tag{19}$$

$$V_T = \frac{kT_c}{q} \tag{20}$$

$$I = I_{ph} - I_0\left[\left(\frac{V_D + IR_{se}}{\eta V_T}\right) - 1\right] - \left[\frac{V_D + IR_{se}}{R_{sh}}\right] \tag{21}$$

where I_{pv} is photovoltaic current, I_0 is reverse saturation current, V_D is voltage across the diode, η is ideality factor, V_T is thermal voltage, R_s is series resistance, and R_{sh} is shunt resistance.

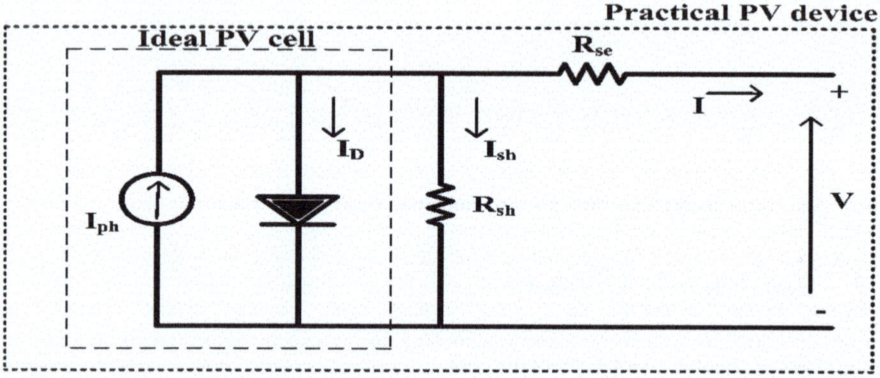

Fig. 8 Electrical circuit equivalent to the one diode model

2.9.2 Mathematical Descriptions of Single Diode Model

The manufacturer's datasheet only provides standard test conditions (STC) for limiting parameter values. The datasheet does not include the values for the five unknown parameters in the single-diode variant: Series resistance, parallel resistance, diode ideality factor, photocurrent, and saturation current. As a result, an iterative process is necessary to extract these characteristics.

At short-circuit (S.C)point ($I = I_{sc}$, $V = 0$):

$$I_{SC} = I_{ph} - I_0\left[exp\frac{I_{SC}R_{se}}{\eta V_T N_{se}}\right] - \frac{I_{SC}R_{se}}{R_{sh}} \tag{22}$$

At open-circuit (O.C) point ($I = 0$, $V = V_{oc}$):

$$0 = I_{ph} - I_0\left[exp\frac{V_{OC}}{\eta V_T N_{se}} - 1\right] - \frac{V_{OC}}{R_{sh}} \tag{23}$$

At the MPP ($I = I_{mpp}$, $V = V_{mpp}$):

$$I_{mpp} = I_{ph} - I_0\left[exp\frac{V_{mpp} + R_{se}I_{mpp}}{\eta V_T N_{se}} - 1\right] - \frac{V_{mpp} + R_{se}I_{mpp}}{R_{Sh}} \tag{24}$$

The power derivative with respect to voltage,

$$\left.\frac{dP}{dV}\right|_{V=V_{mpp}} = 0, \; I = I_{mpp} \tag{25}$$

The current's derivative in relation to voltage,

$$\left.\frac{dI}{dV}\right|_{V=I_{SC}} = -\left(\frac{1}{R_{sh}}\right) \tag{26}$$

The maximum power can be stated as,

$$P_{max} = V_{max}I_{max} \tag{27}$$

Temperature and radiation levels determine the photocurrent in PV cells, which is represented as,

$$I_{ph} = \left(I_{ph\_STC} + K_1\Delta T\right)\frac{G}{G_{STC}} \tag{28}$$

where K_1 is cell's short circuit current temperature coefficient, G is solar irradiation in W/m^2, G_{STC} is nominal solar irradiation in W/m^2, and I_{PV_STC} is light-generated current under standard test conditions. The temperature-dependent reverse saturation current is expressed as a cubic coefficient of temperature.

$$I_0 = I_{0\_STC}\left(\frac{T_{STC}}{T}\right)^3 exp\left[\frac{qE_g}{aK}\left(\frac{1}{T_{STC}} - \frac{1}{T}\right)\right] \tag{29}$$

where I_{0_STC} is nominal saturation current, E_g is the energy band gap of the semiconductor, T_{STC} is the temperature at standard test conduction, and q is the charge of the electron. The reverse saturation current can be further improved as a function of temperature as follows.

$$I_0 = \frac{(I_{SC\_STC} + K_1\Delta T)}{exp\left[\frac{V_{OC\_STC}+K_2\Delta T}{\eta V_T}\right] - 1} \tag{30}$$

where I_{SC_STC} is S.C current at standard test conduction, V_{OC_STC} is O.C voltage at standard test conduction, K_2 is the temperature coefficient of open circuit voltage. Since an individual PV cell generates relatively little power overall, we combined many PV cells to provide the required amount, as seen in Fig. 9. An array of photovoltaic cells is called this grid. The PV array's equations may be expressed as,

$$I = I_{ph}N_{sh} - I_0N_{sh}\left[exp\left(\frac{V + IR_{se}\left(\frac{N_{se}}{N_{sh}}\right)}{\eta V_T N_{se}}\right) - \left(\frac{V + IR_{se}\left(\frac{N_{se}}{N_{sh}}\right)}{R_{sh}\left(\frac{N_{se}}{N_{sh}}\right)}\right)\right] \tag{31}$$

where N_{se} is the number of series cells and N_{sh} is the number of parallel cells.

The efficiency of PV cells may be more affected by a little difference in series resistance, but not more by variation in shunt resistance. Shunt resistance may be seen as open when there is very little current leakage to the ground since it is believed to be infinite. The model's mathematical equation may be written as follows after

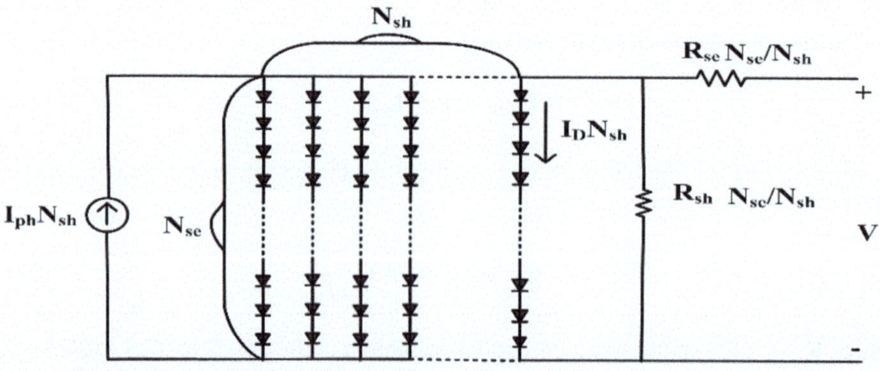

Fig. 9 PV module representation

taking into account shunt resistance infinity:

$$I = I_{ph}N_{se} - I_0 N_{sh}\left[exp\left(\frac{V + IR_{se}\left(\frac{N_{se}}{N_{sh}}\right)}{\eta V_T N_{se}}\right) - 1\right] \quad (32)$$

To construct a PV panel in the MATLAB/Simulink software environment, utilize the equations shown below. The datasheet of the Mitsubishi Electric PV-UD190MF5 PV module is tabulated in Table 2 under STC conditions. In STC conditions, the temperature is 25 degrees Celsius and irradiation is 1000 Watt per meter square. The extracted parameters are applied in a single-diode model. The output characteristics of the Mitsubishi Electric PV-UD190MF5 PV module are examined using P–V and I-V characteristic curves under various environmental situations. Figures 6 and 7 illustrate the P–V and I-V characteristic curves of PV modules under varied irradiations at constant temperatures. The single-diode extracted parameters are shown in Table 3.

Table 2 The specification of Mitsubishi Electric PV-UD190MF5 solar PV module

Solar/PV parameters	
Strings in parallel	132
Modules linked in series for each string	60
Module-specific cells (N_{cell})	50
Resistance (Rdc)	0.2e-3 Ω
Inductance (Ldc)	0.5e-6 H
Voltage in the open circuit (V_{oc})	30.8 V
Current during a short circuit (I_{SC})	8.23 A
Temperature of operation (T)	25 °C
Irradiation (S)	1000 W/m^2
Power at maximum (P_{ma}x)	190.437W
Maximum power point voltage (V_{mp})	24.7 V
Maximum power point current (I_{mp})	7.71A
Coefficient of temperature (V_{oc})	−0.3257 (%/°C)
Coefficient of temperature I_{sc}	0.038202 (%/°C)

Table 3 Estimated value for single diode PV module at STC condition

PV single diode model parameters	
Current from light I_L	8.2413 A
Current diode saturation I_0	7.6947e−11 A
Ideality factor of diodes	0.94464
Shunt resistance R_{sh}	236.4306 ohms
Series resistance R_s	0.32377 ohms

3 Operational Modes of Quasi Z-Source Inverter (QZSI)

To convert DC to AC, Z-source inverters boost voltage. Unlike the ZSI, the QZSI boosts and inverts voltage via an impedance network. zsi and qzsi enhance output voltage without power interruption. The dependability, reduced component ratings, and continuous input current of qzsi make them better than conventional ones. They aid renewable energy power conditioning. QZSI control via pulse-width modulation (PWM) is common. Sinusoidal PWM using a simple boost approach for shoot-through control improves triangular carrier PWM output voltage gain, THD, and bridge voltage stress. The Quasi-Z-supply Inverter (QZSI) has an LC impedance circuit between the power supply and the inverter, unlike the ZSI. ZSIs and QZSIs function in shoot-through and non-shoot-through modes. In non-shoot-through mode, the QZSI switches like a Voltage-Source Inverter. The output capacitor sees the inverter bridge as a current source. The shoot-through condition temporarily turns on switching in the same inverter bridge phase. The technology increases output voltage because the LC network avoids shorting the source. The QZSI behaves like a VSI in active mode since the input DC supply voltage is the inverter's DC link voltage. The voltage on the capacitor is increased during shoot-through intervals by a boosting factor that is dependent on the duty cycle and a specific modulation index. Figure 10 depicts QZSI's fundamental architecture.

To manage both the active and shoot-through states, a modified carrier signal is used in a PWM control approach for the QZSI. This method replaces the zero states seen in traditional VSIs with shoot-through states. To preserve the integrity of the output voltage waveform, on the other hand, the active states don't change. Sine waves are used in the procedure as the carrier and reference signals. In this control technique, a sine carrier wave is compared with two constant voltage envelopes. The sine carrier's amplitude rises or falls over the positive or negative constant voltage envelope, respectively, causing pulses to be produced. The shoot-through duty ratio denoted as D_0, is governed by these pulses. When mixed with the sinusoidal PWM

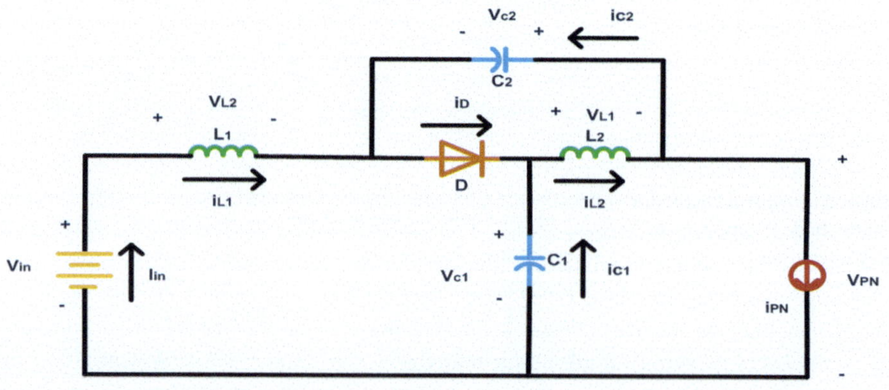

Fig. 10 Quasi z-source inverter model

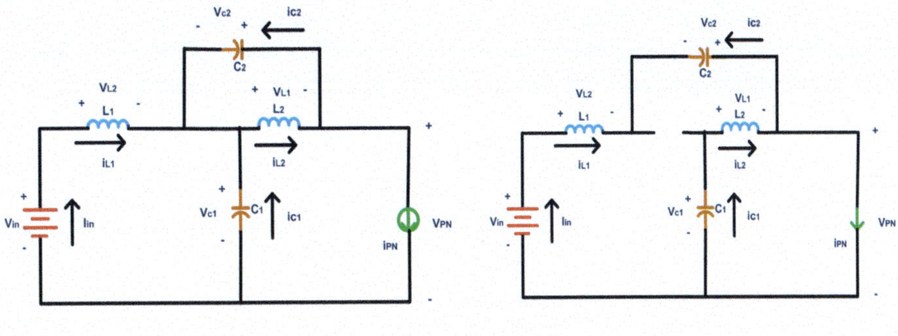

Fig. 11 **a** Non-shoot-through mode, **b** Shoot-through mode

Table 4 QZSI Parameters

Parameters	Symbols	Values
Inductor	$L_1 = L_2$	150 mH
Capacitors	$C_1 = C_2$	1200 μF
QZSI Switching frequency	f_{sw}	3420 Hz

pulses, they provide the gating signals for the inverter switches. In the shoot-through state, the peak voltage across the inverter bridge is increased by a boost factor, B. Figure 11 depicts the setup for the shoot-through and noon shoot-through modes.

MATLAB/Simulink simulation experiments were conducted to assess the three-phase QZSI's performance. MATLAB/Simulink was used in the whole design of the QZSI's control architecture. Table 4 lists the parameters that were used in the study.

Therefore, the voltage gain G of QZSI is given by,

$$G = \frac{Output\ peak\ AC\ voltage}{DC - Link\ voltage} \tag{33}$$

$$V_{Link} = \frac{V_s}{2} \tag{34}$$

$$V_{ac} = MB\frac{V_s}{2} \tag{35}$$

where V_{ac} is the peak AC output voltage, B is the boost factor, M is the modulation index, and V_{link} is the inverter's DC link voltage. The boost factor comes from,

$$B = \frac{T}{T_1 - T_0} = \frac{1}{1 - 2\frac{T_0}{T}} = \frac{1}{1 - 2D_0} \tag{36}$$

The shoot-through interval T0 denotes the shoot-through ratio of QZSI, while the switching cycle is denoted by the symbol T. The shoot-through duty ratio, boost factor, and voltage gain are specified as follows when using a triangular carrier:

$$D_0 = 1 - M \tag{37}$$

$$B = \frac{1}{2M - 1} \tag{38}$$

$$G = \frac{M}{2M - 1} \tag{39}$$

Reducing M limits the amplification and increases the voltage stress on the switch, making it impossible to increase the duty ratio. We can compute the voltage gain, boost factor, and shoot-through ratio of QZSI using sine carrier PWM.

$$D_0 = \frac{T_0}{T} = 1 - \frac{2}{\pi} \sin^{-1} M \tag{40}$$

$$B = \frac{\pi}{4 \sin^{-1} M - \pi} \tag{41}$$

$$G = \frac{\pi M}{4 \sin^{-1} M - \pi} \tag{42}$$

The output is devoid of low-frequency ripples because of the straightforward boost control method's equal distribution of shoot-through states. In addition to improving fundamental components, using a sine carrier wave has reduced THD in the output voltage.

4 Results and Discussion

The efficiency of the suggested control technique is shown by simulating using idealized model components created with the MATLAB simulation software. Wind-PV co-generation energy is the simultaneous generation of electricity from wind and solar PV sources. This hybrid system makes use of solar and wind energy's complementing qualities. Solar energy works best during the day and in bright seasons, but wind energy is usually more plentiful at nighttime and in the winter. By integrating both systems, the co-generation approach ensures a highly consistent and reliable supply of green energy, reducing the intermittency that can occur when relying on a single energy source. Additionally, wind-PV co-generation enhances the overall efficiency of renewable energy systems, making better use of available land and infrastructure. It also reduces dependence on fossil fuels, contributing to sustainability and reducing carbon emissions. This combined approach is precious for grid-connected systems, where maintaining a stable and constant energy supply is critical for meeting fluctuating demand while supporting clean energy goals. Under various meteorological circumstances, the cogeneration of solar and wind power is studied. The wind speeds rise from 6.5 to 8.3, fall to 4.9, and then rise to 12 m/s at t = 2, 4, and 6 s, respectively, as Fig. 12a illustrates. When wind speed varies, the sun irradiance level also changes, falling from 1 to 0.5, then to 0.3, and ultimately rising to 0.8 kW/m^2 at t = 1, 3, and 5 s, in that order. The MPPTs result in the ideal ω^*_r and V^*_{dc} when compared to Fig. 2b and 6. Figures 12c and d, which illustrate the changing frequencies working of the VSR, show that the controllers designed for the combination wind-solar system do not fully saturate the pulse-width modulating of the BtoB QZSCs. Both ωr and Vdc exhibit well-damped performance in Figs. 12b and 14a, which is mirrored in the voltage and current produced by solar and wind energy's active and reactive power, as shown in Figs. 13 and 14. For further analysis, the maximum solar power of 1.239 MW and wind power of 2.354 MW are produced at t = 7 s. A maximum overrun of 0.061 p.u. is necessary to preserve DC-link stability. Figure 15 illustrates how the injected reactive and active power maintains a unity PCC voltage and current under all circumstances. The utility grid's voltage and current are shown in Fig. 16, which is maintained by producing 3.049 MW of total active power and 1.044 MW of reactive power at the grid side.

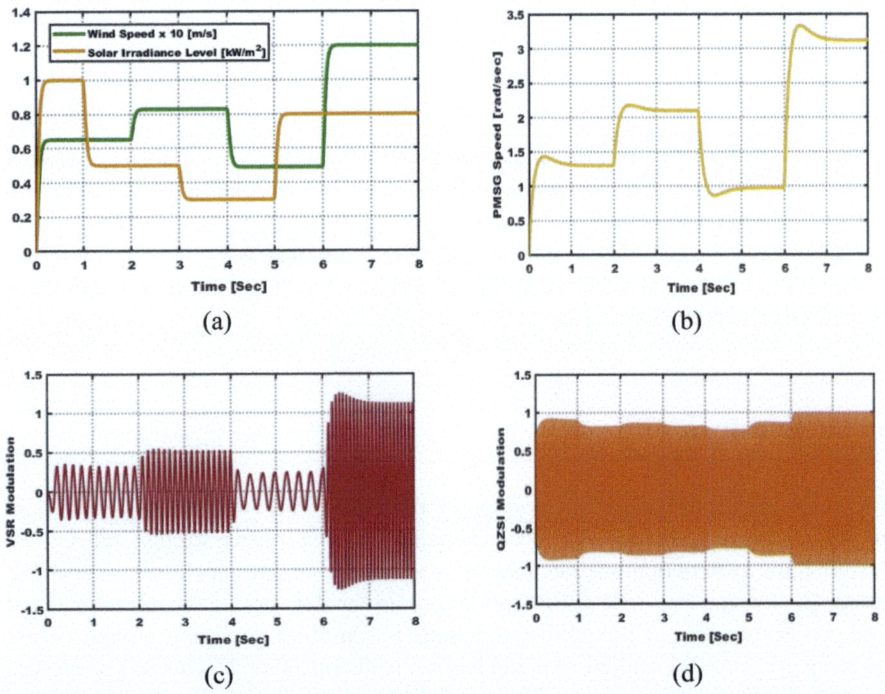

Fig. 12 The solar and wind generators' performance. **a** The solar radiation level and wind speed. **b** PMSG Speed. **c** VSR Modulation. **d** QZSI Modulation

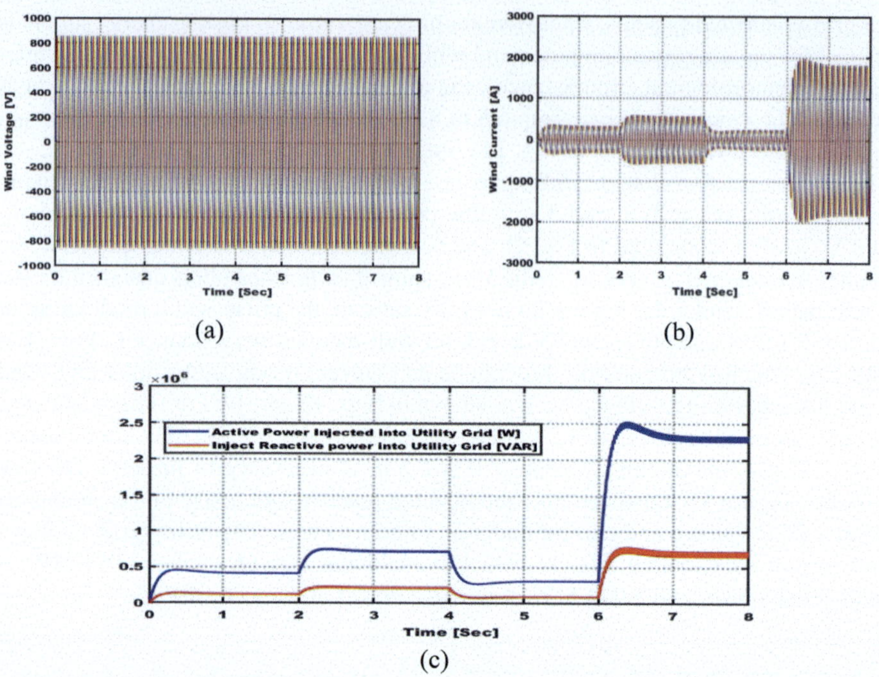

Fig. 13 **a** Wind generation voltage. **b** Wind generation current. **c** Injected active and reactive power to the wind

Both control strategies must be integrated into a combined wind-PV co-generation system to optimize overall performance. This involves coordinating between the two sources to balance power output, managing energy storage solutions, and ensuring a stable supply to the grid. The system can adapt to changing environmental conditions and continuously optimize energy production, enhancing the reliability and efficiency of renewable energy generation.

5 Conclusions

The study showcased the effectiveness of the Quasi Z-Source Inverter (qZSI) in improving the performance of grid-connected wind-PV co-generation systems. With the qZSI, the system demonstrated superior voltage-boosting capabilities, enhanced power quality, and a more stable dynamic response under varying environmental and load conditions. The qZSI's ability to handle intermittent renewable sources while ensuring grid stability and minimizing harmonic distortion highlights its importance in integrating renewable energy. The simulation results confirm the system's operational efficiency, indicating its potential to enhance the reliability and performance of hybrid energy systems, particularly for sustainable power generation. Future research

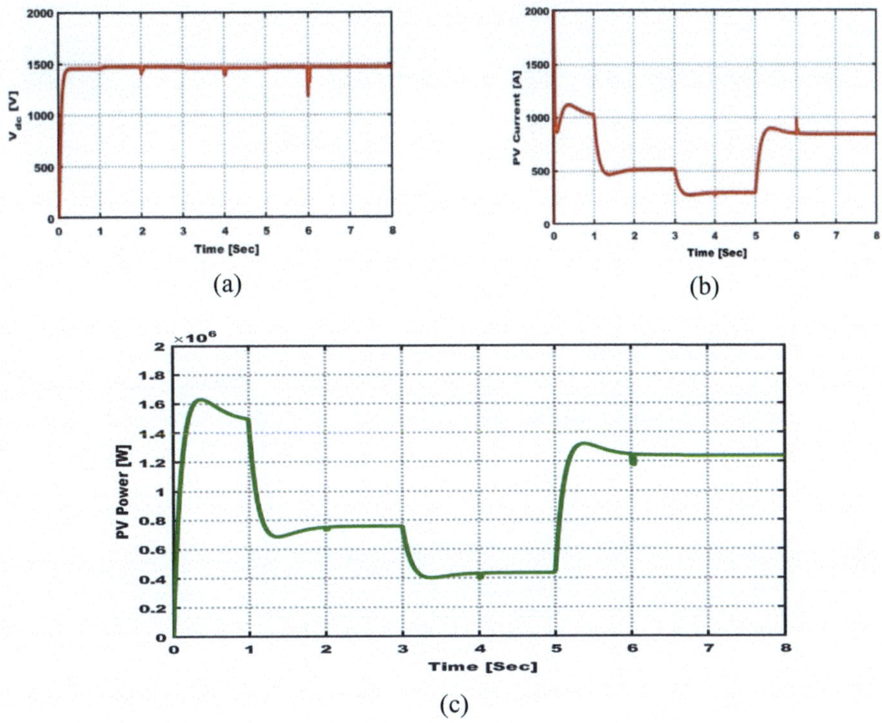

Fig.14 **a** DC-link voltage. **b** PV current. **c** PV power

will explore advanced control algorithms, such as artificial intelligence-based optimization techniques, to further improve the efficiency of qZSI in hybrid renewable systems. Additionally, investigating the integration of energy storage systems with qZSI for better energy management and load balancing will be considered. Further studies will also examine the scalability of this system for large-scale renewable energy projects, taking into account economic feasibility, grid code compliance, and real-time monitoring solutions for long-term deployment. Furthermore, optimizing the converter design for higher power ratings and examining its potential in multi-source grid applications, such as electric vehicle (EV) charging stations and microgrids, presents promising opportunities.

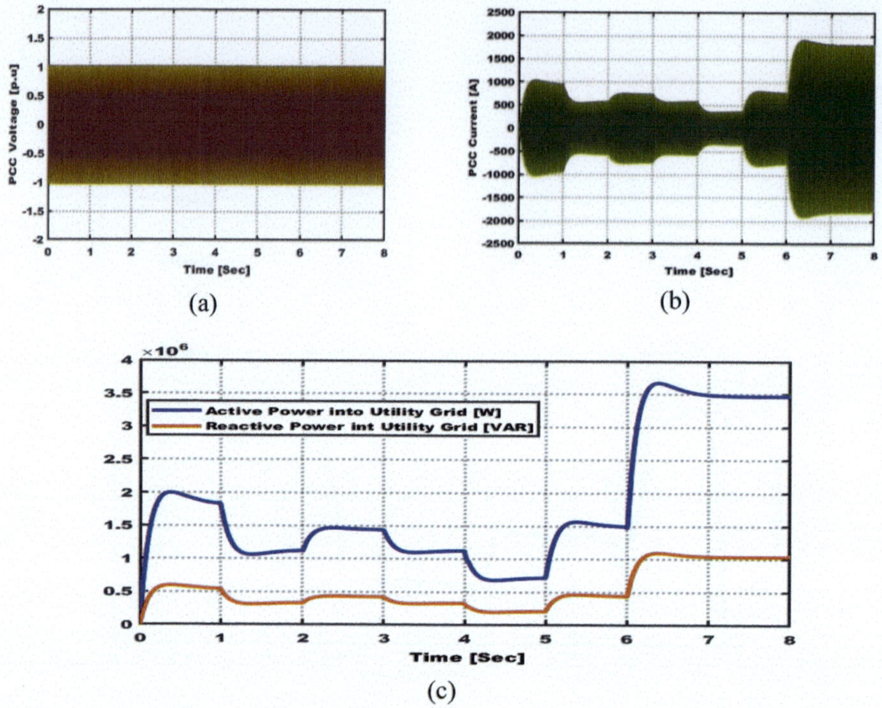

Fig. 15 **a** PCC Voltage. **b** PCC current. **c** Injected active and reactive power to the PCC

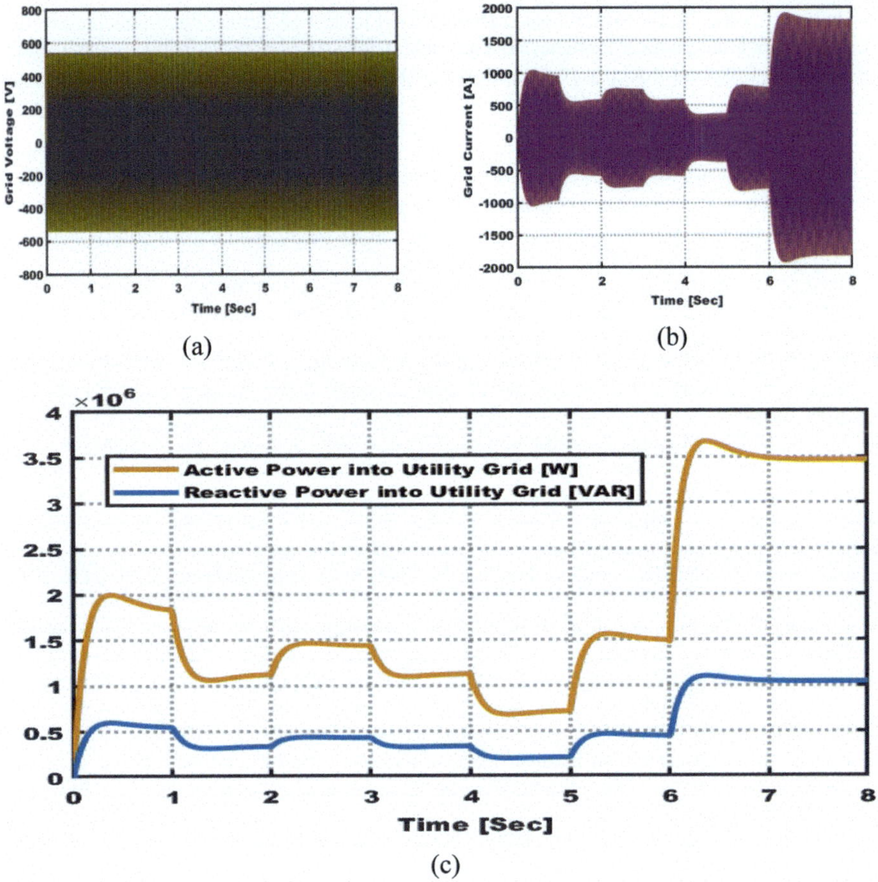

Fig. 16 **a** Grid voltage. **b** Grid current. **c** Injected active and reactive power into the utility grid

Acknowledgements This research work was supported by "Woosong University's Academic Research Funding—2025".

References

1. Kaviani A, Riahy A, Kouhsari GH, SH M (2009) Optimal design of a reliable hydrogen-based stand-alone wind/PV generating system, considering component outages. Renew Energy Elsev 34:2380-2390. https://doi.org/10.1016/j.renene.2009.03.020
2. Zende RM, Bhalkar R (2018) Additional arm for voltage boost operation in three phase voltage source inverter. Int Conf Inf Commun Eng Technol (ICICET) 1–4. https://doi.org/10.1109/ICICET.2018.8533688

3. Paikray P, Swain S C, Dash R, Panda PC (2017) A review on current control techniques for inverter for three-phase grid-connected renewable sources. Innovat Power Adv Comput Technol (i-PACT) 1–6. https://doi.org/10.1109/IPACT.2017.8245203
4. Loh PC, Li D, Blaabjerg F (2013) Γ-Z-source inverters. IEEE Trans Power Electron 28:4880–4884. https://doi.org/10.1109/TPEL.2013.2243755
5. Tang Y, Xie S, Zhang C (2011) Single-phase Z-source inverter. IEEE Trans Power Electron 26:3869–3873. https://doi.org/10.1109/TPEL.2009.2039955
6. Tran Q, Chun T, Ahn J, Lee H (2007) Algorithms for controlling both the DC boost and AC output voltage of Z-source inverter. IEEE Trans Ind Electron 54:2745–2750. https://doi.org/10.1109/TIE.2007.895146
7. Tang Y, Xie S, Zhang C, Xu Z (2009) Improved Z-source inverter with reduced Z source capacitor voltage stress and soft-start capability. IEEE Trans Power Electron 24:409–415. https://doi.org/10.1109/TPEL.2008.2006173
8. Jagan V, Kotturu J, Das S (2017) Enhanced-boost quasi-Z-source inverters with two switched impedance networks. IEEE Trans Ind Electron 64:6885–6897. https://doi.org/10.1109/TIE.2017.2688964
9. Mohammadi M, Moghani JS, Milimonfared J (2018) A novel dual switching frequency modulation for Z-source and quasi-Z-source inverters. IEEE Trans Ind Electron 65:5167–5176. https://doi.org/10.1109/TIE.2017.2784346
10. Hasan Babayi Nozadian M, Babaei E, Shokati HSH, Asl E (2017) Steady-state analysis and design considerations of high voltage gain switched Z-source inverter with continuous input current. IEEE Trans Industr Electron 64:5342–5350. https://doi.org/10.1109/TIE.2017.2677315
11. Ahmed HF, Cha H, Kim S, Kim H (2016) Switched-coupled-inductor quasi-Z-source inverter. IEEE Trans Power Electron 31:1241–1254. https://doi.org/10.1109/TPEL.2015.2414971
12. Siwakoti YP, Peng FZ, Blaabjerg F, Loh PC, Town GE (2015) Impedance-source networks for electric power conversion part I: a topological review. IEEE Trans Power Electron 30:699–716. https://doi.org/10.1109/TPEL.2014.2313746
13. Zhu X, Bo Z, Qiu D (2018) Enhanced boost quasi-Z-source inverters with active switched-inductor boost network. IET Power Electron 11:1774–1787. https://doi.org/10.1049/iet-pel.2017.0844
14. Li Y, Jiang S, Cintron-Rivera J G, Peng F Z (2013) Modeling and control of quasi-Z-source inverter for distributed generation applications. IEEE Trans Ind Electron 60:1532–1541. https://doi.org/10.1109/TIE.2012.2213551
15. AsSakka AO, ElShahed MA, Moustafa Hassan MA, Senjyu T (2017) Modeling and control of a PV based QZSI for grid connected applications. Int Conf Cont Automat Inf Sci (ICCAIS) 67–72. https://doi.org/10.1109/ICCAIS.2017.8217595
16. Monjo L, Sainz L, Mesas JJ, Pedra J (2021) Quasi-Z-source inverter-based photovoltaic power system modeling for grid stability studies. Energies 14:508. https://doi.org/10.3390/en14020508
17. Grgić I, Bašić M, Vukadinović D, Bubalo M (2020) Optimal control of a standalone wind-solar-battery power system with a quasi-Z-source inverter. In: 9th International conference on renewable energy research and application (ICRERA) pp 61–66. https://doi.org/10.1109/ICRERA49962.2020.9242854
18. Raja Nayak M, Tulasi VVK, Divya Teja K, Koushic K, Suresh Naik B (2022) Implementation of quasi Z-source inverter for renewable energy applications. Mater Today Proc 80:2458–2463. https://doi.org/10.1016/j.matpr.2021.06.383
19. Priyadarshi N, Padmanaban S, Ionel DM, Mihet-Popa L, Azam F (2018) Hybrid PV-wind, micro-grid development using quasi-Z-source inverter modeling and control—experimental investigation. Energies 11:2277. https://doi.org/10.3390/en11092277
20. Sun D, Du L, Lu X, He L (2018) An energy-stored quasi-z source converter based interlinking converter for hybrid AC/DC microgrids. In: IECON-44th annual conference of the IEEE industrial electronics society, pp 3821–3826. https://doi.org/10.1109/IECON.2018.8591538

21. Singh SA, Carli G, Azeez NA, Williamson SS (2018) Modeling, design, control, and implementation of a modified Z-source integrated PV/Grid/EV DC charger/ inverter. IEEE Trans Ind Electron 65:5213–5220. https://doi.org/10.1109/TIE.2017.2784396
22. Kumar TS, Nayak MR, Krishna RV, Rao KP (2020) Enhanced performance of solar PV array-based machine drives using zeta converter. In: IEEE international conference on advances and developments in electrical and electronics engineering (ICADEE), pp 1–5. https://doi.org/10.1109/ICADEE51157.2020.9368937
23. Nayak MR, Mujeer SA (2020) New computational method for study of ionic current environment of HVDC transmission lines. In: IEEE international conference on advances and developments in electrical and electronics engineering (ICADEE), pp 1–5. https://doi.org/10.1109/ICADEE51157.2020.9368934
24. Sreenu C, Mallesham G, Chandra Shekar T, Salkuti SR (2024) Pairing voltage-source converters with PI tuning controller based on PSO for grid-connected wind-solar cogeneration. Franklin Open 8:2773–1863. https://doi.org/10.1016/j.fraope.2024.100138
25. Tang Y, Xie S, Ding J (2014) Pulse width modulation of Z-source inverters with minimum inductor current ripple. IEEE Trans Ind Electron 61:98–106. https://doi.org/10.1109/TIE.2013.2240632
26. Fang XP, Chen ZQ, Wang XG (2014) Current-fed quasi-Z-source inverter-based adjustable speed drive system with bidirectional power flow. IEEE Trans Appl Superconductivity 24:1–6. https://doi.org/10.1109/TASC.2014.2333531
27. Deepak Reddy P, Raja Nayak M (2021) A novel review on modular multilevel converter for induction motor drive applications. Turkish J Comput Math Educ (TURCOMAT) 12:6303–6309. https://doi.org/10.17762/turcomat.v12i10.5471
28. Sreenu C, Mallesham G, Chandra Shekar T (2023) Control of quasi Z-source converter in a microgrid using modified power ratio P&O MPPT. second international conference on emerging trends in engineering ICETE Atlantis Press, vol 9, pp 673–686. https://doi.org/10.2991/978-94-6463-252-168
29. Raja M, Nayak, Saritha M, Abdul Mujeer S, Devulal S, Santhosh Kumar T (2021) A photovoltaic based multilevel inverter fed induction motor drive. Turkish J Comput Math Educ 12:6196–6212. https://doi.org/10.17762/turcomat.v12i3.1647
30. Ho A V, Chun T W (2018) Single-phase modified quasi-Z-source cascaded hybrid five-level inverter. IEEE Trans Ind Electron 65:5125–5134. https://doi.org/10.1109/TIE.2017.2779419
31. Liu Y, Ge B, Abu-Rub H, Peng FZ (2014) Overview of space vector modulations for three-phase Z-Source/Quasi-Z-Source inverters. IEEE Trans Power Electron 29:2098–2108. https://doi.org/10.1109/TPEL.2013.2269539
32. Sravanthi P, Sandeep V (2022) Solving realistic reactive power market clearing problem of wind-thermal power system with system security. Int J Emerg Electr Power Syst 23(2):125–144. https://doi.org/10.1515/ijeeps-2021-0060
33. Salkuti SR (2023) Advanced technologies for energy storage and electric vehicles. Energies 16(5):2312. https://doi.org/10.3390/en16052312

Application of Swarm Intelligence Techniques for Quasi-Z-Source Inverters in Renewable Energy Systems

Ch. Sreenu, G. Mallesham, T. Chandra Shekar, and Surender Reddy Salkuti

Abstract Swarm intelligence (SI) techniques have garnered significant attention for optimizing complex, nonlinear systems due to their adaptive, decentralized, and self-organizing nature. Particle Swarm Optimization (PSO) and Artificial Bee Colony (ABC) optimization are two swarm intelligence techniques discussed in this chapter that aim to enhance the performance of quasi-source inverters (QZSI) in grid-connected renewable energy systems. QZSI is well-known for its ability to increase voltage and perform inversion in a single stage, which results in improved power density, reduced switching stress, and excellent reliability. This makes it suitable for integrating wind and solar photovoltaic (PV) systems. However, these hybrid systems' nonlinear characteristics and dynamic behaviour challenge conventional control methods. This study uses SI-based algorithms to optimize the parameters of PI controllers for QZSI to ensure optimal power flow, fast dynamic response and minimized total harmonic distortion (THD). The PSO and ABC algorithms dynamically fine-tune the control parameters by searching for optimal solutions in a multidimensional space. This results in higher tracking accuracy and system stability under varying grid conditions. The comparative analysis highlights the two algorithms' strengths: faster PSO convergence and greater ABC flexibility for local

Ch. Sreenu · G. Mallesham
Department of Electrical Engineering, University College of Engineering, Osmania University, Hyderabad, Telangana, India
e-mail: drcnu@osmania.ac.in

G. Mallesham
e-mail: drgm@osmania.ac.in

T. Chandra Shekar
Department of Electrical Engineering, Maturi Venkata Subba Rao Engineering College, Hyderabad, India

S. R. Salkuti (✉)
Department of Railroad and Electrical Engineering, Woosong University, Daejeon, Republic of Korea
e-mail: surender@wsu.ac.kr

S. R. Salkuti (ed.), *Artificial Intelligence for Integrated Smart Energy Systems in Electric Vehicles*, Lecture Notes in Electrical Engineering 1427,
https://doi.org/10.1007/978-3-031-94276-1_29

optima avoidance. Simulation results demonstrate significant improvements in efficiency, voltage regulation, and transient response, affirming the effectiveness of SI-based techniques in renewable energy integration. This research provides a scalable framework for optimizing QZSI-based wind-PV cogeneration systems, contributing to sustainable and reliable energy networks.

Keywords Swarm intelligence · Particle swarm optimization (PSO) · Artificial bee colony (ABC) · Quasi Z source inverter (QZSI) · Renewable Energy Source · PI controller · Grid-connected systems

Nomenclature

QZSI	Quasi Z-source inverter
ABC	Artificial bee colony
PSO	Particle swarm optimization
PV	Photovoltaic system
ABC	Artificial bee colony
ANN	Artificial nural network
SI	Swarm intelligence
V_{link}	DC-link voltage
PMSG	Permanent magnet synchronous generator
P_m	The mechanical input power
P_e	The electrical output power
WECS	Wind energy conversion system
AI	Artificial intelligence
AFS	Artificial fish swarms
V_{out}	Output voltage
BFO	Bacterial foraging optimization
ACO	Ant colony optimization
THD	Total harmonic distortion
I_D	Diode current
I_{sh}	Shunt current
p_n	Pole pairs
V_{sh}	Voltage across the shunt resistor and diode
I_0	Diode reverse saturation current
V_D	Voltage across the diode
V_T	Thermal voltage
MPP	Maxmim power point
VSI	Voltage source inverters
CSI	Current source inverters
p_{Best}	Personal best
g_{Best}	Global best
C_1 and C_2	Acceleration coefficient

1 Introduction

The global energy landscape is transforming significantly with the increasing integration of renewable energy sources into the electrical grid. Quasi-Z-Source Inverters (QZSI) play a crucial role in this transition because they simultaneously handle voltage boosting and inversion, improving reliability and efficiency compared to traditional inverter topologies. However, the dynamic behaviour of QZSI requires sophisticated control strategies to optimize performance. Drawing inspiration from the collective behaviour of social organisms, swarm intelligence techniques offer a promising approach for optimizing complex control systems. This research explores the application of PSO and ABC algorithms for tuning PI controllers in QZSI to enhance system performance and stability in renewable energy. Diesel power production is not always the ideal option owing to practical issues such as high running costs, fuel transportation issues, and complex maintenance. Rising sustainability issues and advancements in renewable energy technology make green energy sources an effective answer for sustainable energy development and pollution control. Diversifying energy sources enhances system efficiency and dependability. Combination solar-wind power generating systems with storage banks provide dependable electricity due to the complimentary properties of solar and wind energy resources in particular places [1]. Swarm intelligence systems are the result of research on individual characteristics and how they relate to one another in a group. For artificial intelligence (AI), this new area of computing influenced by biology is essential. Fuzzy systems, ANNs, and evolutionary computing are all included in swarm intelligence techniques, a subset of computational intelligence (CI) approaches. These naturalistic approaches resolve challenging optimization issues that conventional mathematical approaches cannot handle [2].

In their research on cellular robotics, Beni and Wang introduced the term "swarm intelligence" [3]. Modeling self-organizing agent populations to address various issues is the primary goal of this branch of artificial intelligence. Based on biological principles, swarm intelligence algorithms may be divided into six categories: artificial fish swarms (AFS), bacterial foraging optimization (BFO), ABC, PSO, ant colony optimization (ACO), and others. Due to its robust and self-organizing properties, swarm intelligence has widespread use in scheduling, robotics, power systems, and optimization problems. It is beneficial to comprehend these algorithms from an academic and practical standpoint. Alternative techniques include probabilistic techniques like simulated annealing and hill climbing. EC methods include genetic algorithm (GA), evolutionary programming, genetic programming, and SI algorithms like optimization of PSO, ABC, ACO, bacterial foraging, etc. Inspired by nature, EC optimization uses mutation, crossover, natural selection, and survival of the fittest. SI has various excellent optimization methods based on swarm social behavior research. Studying ants, bees, termites, or social creature swarms like birds and fish led to SI. SI has various novel and effective optimization techniques. Swarms are groups of agents that interact with their immediate surroundings to communicate. Agent interactions create distributive collective problem-solving solutions called SI.

Swarm intelligence (SI) is a research field inspired by the behaviour of social insect colonies and other animal societies. This study investigates the principles and models of behavior that arise from group interactions. SI is connected to optimization using natural swarm behavior to create practical algorithms and methods. Bonabeau et al. [4] propose that SI utilizes social insect behaviour to develop algorithms and distributed problem-solving tools. By using a unique impedance network to connect the inverter's primary circuit to the power source, the impedance source inverter provides a novel concept for energy conversion [5]. Several unique characteristics of the Z-source inverter improve performance. It may increase the output voltage range, reduce route harmonics, increase power factor, and withstand voltage swings. The modulation index and shoot-through duty cycle are controlled to accomplish this. As a result, by using passive components, the inverter may either raise or decrease the input supply voltage (buck). In addition to increasing dependability, this lowers overall expenses [6]. All of the benefits of the conventional ZSI are carried over into the newly suggested quasi-Z-source inverter (qZSI), which also offers variable input profiles and lower passive circuit strain [7]. The qZSI architecture is very appealing for applications for sustainable energy including solar, wind, and fuel cell systems because of the aforementioned characteristics. When the load is low or the voltage source is sufficient, such clean energy power systems often need a battery for energy storage to provide the load power during times when the voltage source is either insufficient or non-existent [6, 8]. Algorithms for a similar process referred to as "swarm intelligence," have been established by studying individual traits and their interactions with groups. These are considered a developing topic and a vital component of artificial intelligence (AI), which consists of calculations influenced by biology. Several computational intelligence techniques, including adaptive processing, artificial neural networks (ANNs), and fuzzy systems, may naturally combine to form intelligence swarm algorithms [9].

ABC is a newly created algorithm that draws inspiration from honey bees' clever behavior. Food supplies, employed bees, and unemployed bees are components of ABC that are crucial to the behavior that results in the development of collective intelligence. When a swarm of produced bees discovers a goal solution with the best quality, they communicate and travel randomly around an n-dimensional search space. The bees leave the solution and start again with different food sources when the solution is of poor quality. The method is based on the number of cycles and colony size of bees [10]. PSO is an additional population-based stochastic search technique that employs a swarm of particles, each of which stands for an alternative to the optimization problem. Finding the patterns that control birds' synchronized flight is the goal of PSO. As they go through a multifaceted search space, particles modify their location based on their own experiences as well as those of their neighbors. Each particle's inertia, cognition, and the information it learns from its peers via social contact all have an impact. Particles' collective activity finds answers in the best areas of a level search area [11]. GA and PSO have been used to determine the ideal size of a wind, solar, and battery power system to illustrate the dependability index of the Loss of Energy Supply Probability. However, it is necessary to take into account the possibility of interruptions for system elements like solar arrays and wind turbines.

In actuality, dependability concerns are not considered while designing the hybrid power system optimally. For example, certain extremely common occurrences, such as producing unit failures and outages, are often disregarded even though they might significantly impact the system's cost and dependability [12, 13].

Utilizing the data from the colony's prior solutions, the ABC technique search equation provides an innovative solution that excels at exploration but struggles at exploitation. These circumstances generate a very small balance between the two processes, which traps the algorithm in positional optima [14]. In a microgrid, a Modified Power Ratio Perturb and Observing (P&O) Maximum Power Point Tracking (MPPT) method may be used to improve the management of a Quasi-Z-Source Converter (QZSC). The QZSC is appropriate for renewable energy sources because of its single-stage power conversion and voltage-boosting capabilities. In situations that change quickly, traditional P&O MPPT algorithms often exhibit decreased efficiency and oscillations at the maximum power point. To maximize energy extraction from solar and wind sources, the improved algorithm improves convergence speed and accuracy by adjusting its perturbation step size depending on real-time power changes. Through the control of voltage and current, this technique also stabilizes the microgrid. The QZSC and the improved MPPT work together to improve the microgrid's dependability, minimize harmonic distortion, and facilitate the efficient integration of renewable energy sources [15]. Voltage-source converters (VSCs) are essential for efficiently integrating wind-solar cogeneration systems into the grid, enabling DC-AC conversion and power flow regulation. When combined with a Proportional-Integral (PI) controller optimized through Particle Swarm Optimization (PSO), VSCs exhibit improved control accuracy and dynamic response under varying conditions. Traditional tuning methods often need to be revised due to the nonlinear nature of renewable energy systems. In contrast, the PSO approach iteratively finds optimal PI parameters, minimizing control errors and enhancing stability amid disturbances [16].

The investigation of a grid-connected mixed PV/WIND network with a minimal number of converters is the focus of this work. Maximum PV/wind power may be extracted and transferred thanks to the construction under study. The amalgamation of Solar and WIND sources has been covered in a number of literary publications. In [17], a hybrid system using PV, wind, and batteries is proposed. A shared DC bus and individual adaption circuits for each source define the system under study. An inverter connects the general DC Bus to the grid. A hybrid system that uses PV/PMSG sources and is linked to the grid is studied by Bae and Kwasinski [18]. However, this work does not use the MPPT control. A number of PV/Wind setups are documented. To increase system performance even further, the research suggests an exact sizing technique. The research presents findings for twelve distinct days with diverse weather circumstances to show how efficient the suggested technique in references [19, 20]. A PV/wind hybrid structure linked to a one-phase electrical grid is described in the paper [21]. The author of this paper suggests research in which the solar power plant is linked straight to the standard DC bus. The use of a five-phase voltage source inverter (VSI) fed motor drive [23] has grown significantly over the last four decades. The necessity to analyze the topology, driving scheme, control method, and application has increased due to recent advancements in VSI-fed motor drives. The Z-source inverter (ZSI), originally suggested by Peng [24], is one kind of VSI-fed motor drive. Further studies discovered [25] that ZSI has made significant strides in the electric car, where the driving voltage is under a lot of stress. As a result, their study has become an ongoing worry in the context of multiphase systems. Although there are many published papers outlining the functions of ZSI in multiphase systems [26–28].

This combined strategy enhances power quality by reducing harmonics and ensuring voltage and frequency stability while improving system reliability through efficient power dispatch and grid code compliance. Ultimately, it is vital for seamless grid integration and maximizing renewable energy utilization.

1.1 Contributions

This study combines swarm intelligence methods with Quasi-Z-Source Inverters (QZSI) in power electronics and renewable energy systems. The following summarizes the primary contributions: Improved QZSI Performance: Swarm intelligence algorithms maximize QZSI performance, lowering total harmonic distortion (THD) and enhancing efficiency and voltage control. Adaptive Control Techniques: Swarm intelligence-based advanced control techniques improve system responsiveness to energy variations, guaranteeing steady operation under various grid situations. Comparative Analysis: The paper examines the PSO and ABC algorithms for QZSI parameter optimization to help choose the best optimization method. Flexibility and Scalability: The framework encourages broad adoption of QZSI technology by being flexible and scalable to different renewable energy combinations. Support for Sustainable Energy Systems: In line with international initiatives to lower carbon emissions, the study encourages the integration of renewable energy sources using sophisticated control mechanisms. Basis for Future Research: The study lays the groundwork for future research in swarm intelligence applications in power electronics by directing more studies and encouraging creativity.

1.2 Objectives

The goals of this research on applying swarm intelligence techniques for Quasi-Z-Source Inverters (QZSI) in renewable energy systems are to optimize Control Parameters using swarm intelligence algorithms. Enhance System Performance through optimized control strategies. Compare and analyze intelligent swarm techniques. Analyze the System Dynamics of QZSI integrated with renewable energy sources. Develop a Scalable Framework for applying swarm intelligence techniques. Contribute to Sustainable Energy Solutions by promoting efficient integration of renewable energy sources. Establish a Foundation for Future Research in swarm intelligence applications within power electronics.

A new method enhances system efficiency, stability, and flexibility by applying swarm intelligence techniques, such as Particle Swarm Optimization (PSO) and Artificial Bee Colony (ABC), to Quasi-Z-Source Inverters (QZSI) in renewable energy systems. This approach relies on collective behaviour and decentralized problem-solving to fine-tune control parameters for QZSI's unique properties. It ensures strong grid integration and optimized performance in managing variable inputs from wind and solar sources, addressing issues like dynamic stability, power imbalance, and harmonic distortion. This innovative method presents a significant advancement for sustainable energy production and distribution.

After the introduction will discuss objectives, contribution, and novality, this chapter is organized as follows: Sect. 2 describes the hybrid system, which incorporates wind and solar energy generation technologies. Section 3 presents the mathematical model and analysis of the three-phase qZSI, focusing on both shoot-through and non-shoot-through modes. Section 4 provides a detailed explanation of ABC and PSO optimization approaches and discusses various soft computing techniques. Section 5 covers the discussion of the results. Finally, Sect. 6 concludes the chapter with a summary and suggestions.

2 Description of the Hybrid System

A hybrid wind-PV co-generation system combines wind turbines and solar photovoltaic (PV) panels to generate electricity in an integrated way, as illustrated in Fig. 1. With this hybrid arrangement, electricity production is more dependable throughout the day and in different seasons because it uses the complementary capabilities of solar and wind energy. The PV panels generate their highest power during the daytime when sunshine is plentiful. At the same time, wind turbines provide electricity when wind speeds are more significant, often at night or on sunny days. The synergy between these energy sources helps reduce the variability and intermittency typically associated with standalone renewable systems, thereby improving energy reliability and making the system more effective for grid integration and off-grid applications. These systems are widely used for residential and industrial energy, especially in remote areas with limited grid access. A wind-PV hybrid system usually includes essential components such as inverters, controllers, and power management strategies to balance supply and demand. Advanced control techniques, like ABC-PSO-based PI controllers, enhance system efficiency by optimizing the operational performance of converters, such as high-gain Quasi-Z-Source Inverters, ensuring stable grid integration. These systems are environmentally friendly, reduce dependency on fossil fuels, and offer a sustainable energy solution by efficiently using local renewable resources.

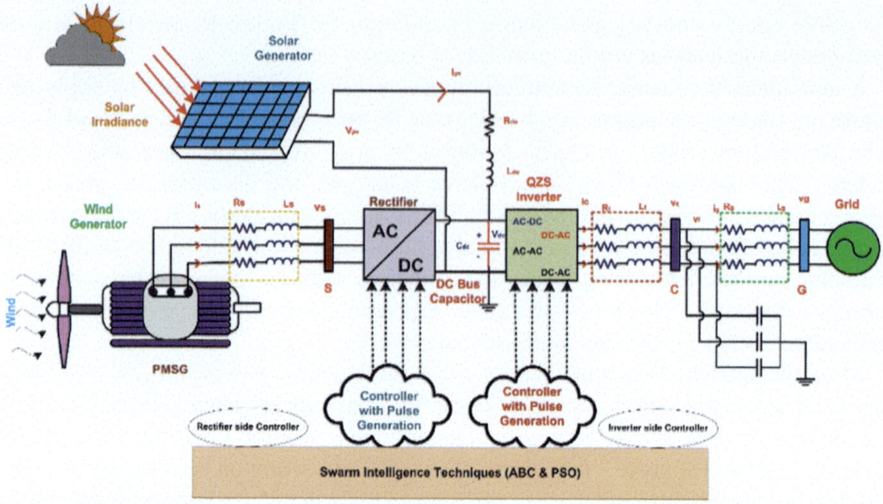

Fig. 1 Block diagram of Proposed wind-solar cogeneration system

2.1 *Wind Energy Generation*

One renewable energy source that is gaining increasing importance is wind energy generation. This process generates electricity by harnessing the kinetic energy of the wind. Wind turbines play a crucial role in this conversion, transforming kinetic energy into mechanical energy, which a generator converts into electrical energy. A straightforward formula can calculate the amount of energy a wind turbine produces. The Permanent Magnet Synchronous Generator (PMSG) is a popular choice for modern wind turbines due to its efficiency, reliability, and capability to operate at variable speeds. It enables efficient energy conversion from the wind, especially in direct-drive systems (which avoid gearboxes). Its design makes it suitable for offshore and onshore wind applications, contributing significantly to the growth of renewable energy technologies.

For wind turbines, the mechanical input power P_m is given by,

$$P_m = \frac{1}{2}.\rho.v^3.C_p \tag{1}$$

The electrical output power from the PMSG is,

$$P_e = T_e.\omega_r \tag{2}$$

where P_m is power output (W), ρ is air density (kg/m^3), A is swept area of the turbine blades (m^2), v is wind speed (m/s), Cp is power coefficient (typically ≤ 0.59, known as Betz's limit), ω_r is rotor mechanical angular speed (rad/s).

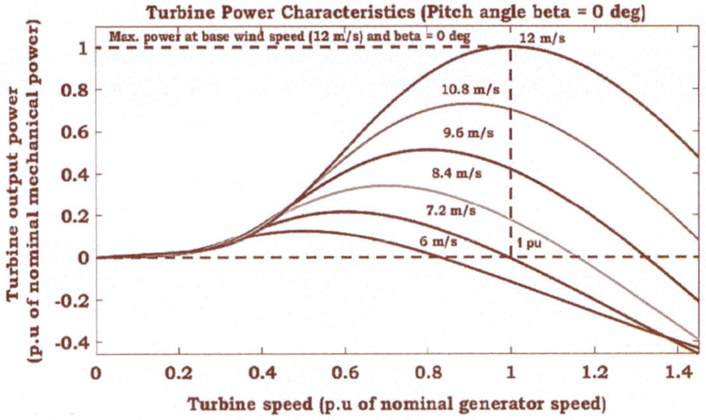

Fig. 2 Maximum power characteristics for PMSG wind turbine

Figure 2 shows the mechanical power characteristics of wind energy highlighting the importance of wind speed, rotor size, and air density in determining a turbine output. By maximizing the swept area and optimizing the turbine design, engineers can improve efficiency and achieve better energy yields. Understanding these power characteristics is essential for site selection, turbine design, and predicting energy production for wind farms.

2.2 Photovoltaic System

A photovoltaic (PV) model mathematically represents the behaviour of a solar cell or PV module. The single-diode PV model, which represents the electrical properties of a solar cell under various environmental variables, including temperature and irradiance, is the most often used. This paradigm aids PV system analysis, design, and simulation. A simplified model known as a single-diode solar PV system represents the electrical activity of a solar cell. This model aids in analyzing the solar cell's performance parameters, such as the relationship between current and voltage (I-V). As shown in Fig. 3, the single-diode model considers the solar cell's internal resistance by including a resistive element, a diode, and a current source.

The following equation can express the current–voltage relationship for the single-diode model:

$$I = I_{ph} - I_d - \frac{V}{R_s} \tag{3}$$

The Shockley equation gives the diode current (I_d),

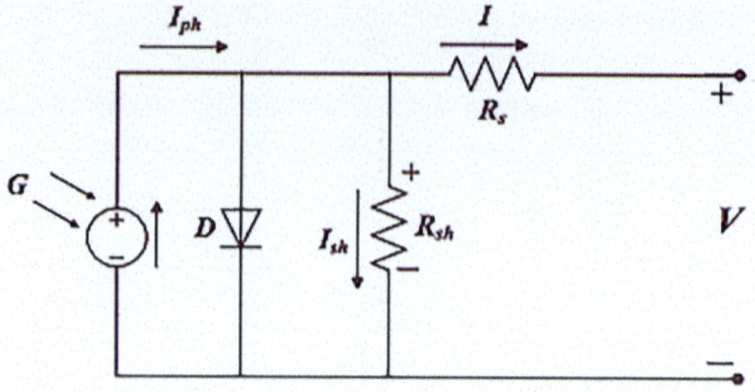

Fig. 3 Simplest single diode model of solar cell equivalent circuit

$$I_d = I_0\left(e^{\frac{qV}{\eta kT}} - 1\right) \tag{4}$$

where I is the output current of the solar cell (A), I_{ph} is the photogenerated current (A), I_d is diode current (A), V is output voltage of the solar cell (V), R_s is series resistance (Ω), I_0 is reverse saturation current (A), q is charge of an electron (1.602 × 10^{-19} coulombs), n is ideality factor of the diode (typically between 1 and 2), k is Boltzmann's constant (1.38 × 10^{-23} J/K), and T is the temperature (K). By substituting the expression for I_d into the original current equation, we obtain,

$$I = I_{ph} - I_0\left(e^{\frac{qV}{\eta kT}} - 1\right) - \frac{V}{R_s} \tag{5}$$

This formula thoroughly explains how the solar cell behaves in different lighting and temperature scenarios. A PV cell's I–V curve illustrates the connection between output voltage and current under certain circumstances. The P–V curve shows the power output's relationship to voltage. The highest Power Point (MPP) is when the power reaches its highest magnitude. The PV system will run as efficiently as possible if this point is tracked using Maximum PowerPoint Tracking (MPPT) algorithms. Grading how solar cells and PV modules work requires grasping the photovoltaic model. The single-diode model is commonly used in designing and modeling photovoltaic (PV) systems due to its balance of simplicity and accuracy. By analyzing the current–voltage (P–V) and power-voltage (I-V) characteristics, engineers can optimize the energy production of photovoltaic systems under different climatic conditions, as illustrated in Fig. 4a, b.

3 Basic Analysis of the Quasi-Z-Source Inverter (QZSI)

A qZSI is an advanced type ZSI that overcomes the limitations of traditional voltage-source and current-source inverters. Unlike conventional inverters, which can only step down or step up voltage, the qZSI achieves both functions in one power conversion stage. It features a unique impedance network of inductors, capacitors, and a

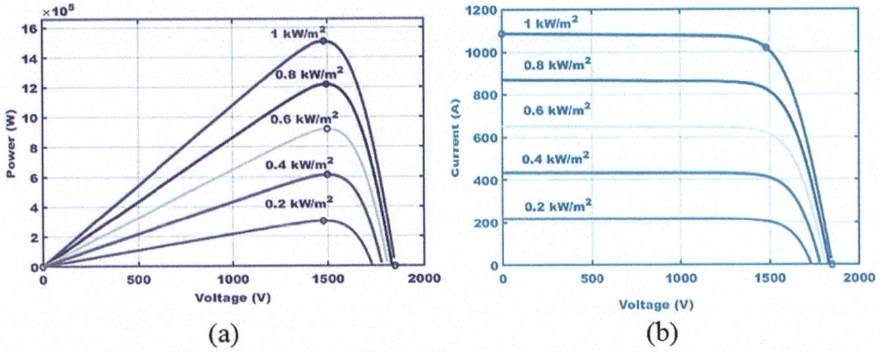

Fig. 4 **a** P–V characteristics. **b** I–V characteristics

diode, which allows for a comprehensive voltage gain, eliminates dead time, and enhances system reliability by protecting against short circuits and voltage sags. The qZSI produces continuous input current, which is particularly advantageous for renewable energy sources like solar PV and wind, as it reduces ripple and improves energy harvesting. This design also facilitates easier integration of energy storage devices, improving the stability of grid-tied systems. Because it operates efficiently under various input conditions, the qZSI is widely used in EVs, renewable energy systems, and motor drives. It is ideal for applications requiring dynamic voltage adjustments. The QZSI's comparable circuit as seen from the DC link is shown in Fig. 5.

The figure gives an equivalent circuit of the impedance source network, a two-port network that is used to provide mathematical analysis, Assume the inductor (L_1 & L_2) and capacitor (C_1 & C_2) have the same inductance and capacitance values respectively.

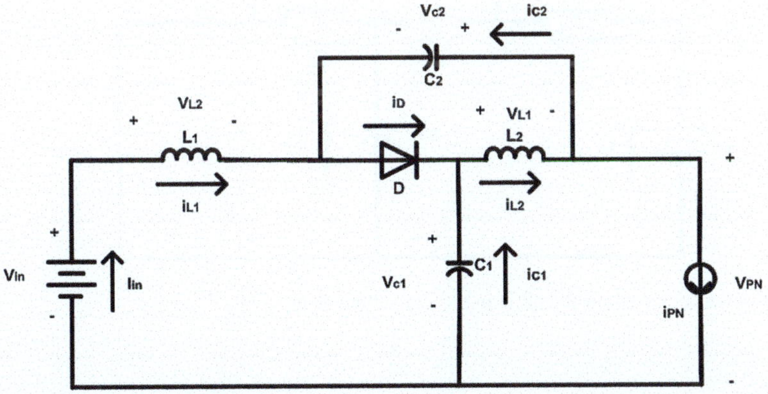

Fig. 5 Equivalent circuit of QZSI

Where L_1 and L_2 are series arm inductors, C_1 and C_2 are parallel arm capacitors, D Diode switch, I_D Diode current, V_D The voltage across the diode, V_{in} and I_{in} Input voltage and current, V_{PN} and I_{PN} DC link voltage and current, V_{L1} and V_{L2} The voltage across the inductors L_1 and $L_{2,}$ I_{L1} and I_{L2}. The current through inductors L_1 and L_2, V_{C1} and V_{C2} The voltage across the capacitors C_1 and C_2, I_{C1} and I_{C2} The current through the capacitors C_1 and $C_{2.}$ Similar to the ZSI, the QZSI offers several benefits, including in certain combinations. Reduced component ratings, less source stress, fewer components, and simple control techniques.

The operation of the QZSI can be broken down into two types:

- Shoot-Through (or) Short-Circuit (or) Zero-State, if the Shoot-Through interval is T_0 (or) T_{sh}.
- Non-Shoot-Through (or) Non-Short-Circuit (or) Active state, if the Non-Shoot-Through interval is T_1 (or) T_{nsh}.

If the shot-through interval is T_{sh} during a switching cycle of period T then the shoot-through duty ratio is $D = T_{sh}/T$. Similarly, the non-shoot-through interval is $T_{nsh} = T\text{-}T_{sh.}$ Therefore, the total Period $T = T_{sh} + T_{nsh}$.

The new power converter design known as the qZSI combines the characteristics of traditional Voltage Source Inverters (VSI) and Current Source Inverters (CSI). Its special qualities allow boost and buck operations to be completed in a single conversion phase. Its shoot-through and non-shoot-through modes make the qzsi particularly suitable for integrating energy from renewable sources like wind and solar (PV) systems. Figures 6 and 7 depict the circuits that correspond to the two states.

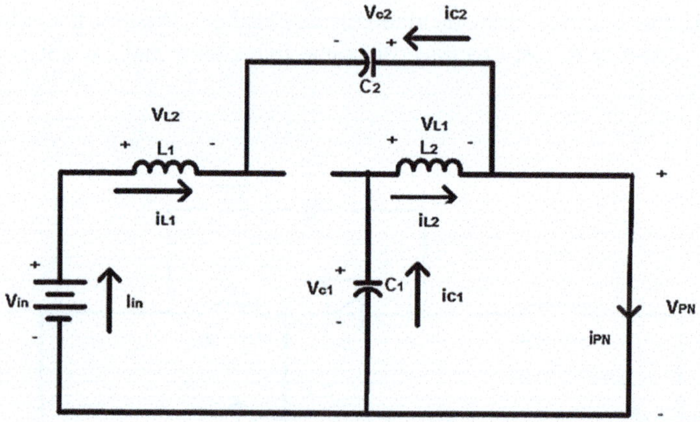

Fig. 6 Equivalent circuit of the qZSI in the Zero-State state

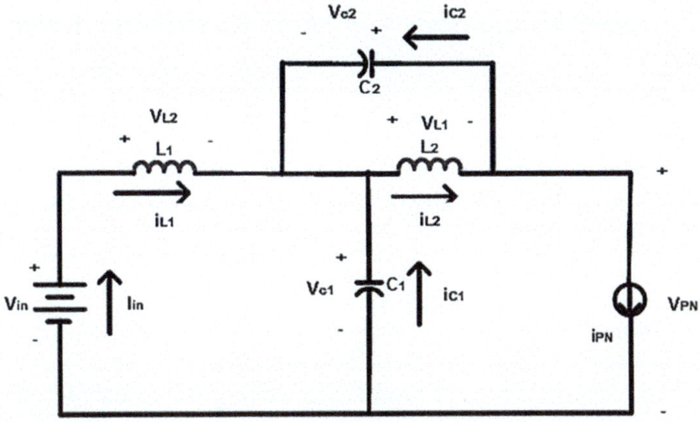

Fig. 7 The qZSI's equivalent circuit in an active state

3.1 Shoot-Through (or) Short-Circuit (or) Zero-State

Shoot-through occurs when two switches in the same leg conduct simultaneously, leading to a short circuit in conventional voltage-source inverters (VSIs). In contrast, the shoot-through condition in the quasi-Z-source inverter (qZSI) is intentional and critical for enhancing the input voltage.

3.1.1 Operation

The switches in one or more inverter legs are switched on at the same time. A shoot-through state is produced by this circumstance, which applies a short circuit across the inverter bridge. qZSI inductors and the input DC source provide energy that is stored in the network's capacitors. This mode enhances the voltage across the inverter by increasing the stored energy (boosting the voltage through the passive elements).

While the QZSI is in a shoot-through state for T_0 seconds after the switching cycle of T. During the shoot-through state, the QZSI mode is represented by short-circuited as shown in Fig. 6. By applying KVL to Fig. 6, the inductor voltage (V_{L1} and V_{L2}), DC link voltage (V_{PN}), out pot voltage (V_{out}), and diode voltage (V_{diode})

$$V_{L1} = V_{in} + V_{C2};\ V_{L2} = V_{C1};\ V_{out} = 0;\ V_{PN} = 0;\ V_{diode} = V_{C1} + V_{C2} \quad (6)$$

3.2 Non-shoot-Through (or) Non-short-Circuit (or) Active State

This is the normal operating mode, where the inverter switches follow the desired modulation scheme (like SPWM or SVPWM) to generate the AC output. In this mode, no shoot-through occurs, and the power is transferred from the DC input to the AC load or the grid.

3.2.1 Operation

The inverter operates in a typical fashion, generating sinusoidal AC output. The inverter releases the energy that was stored in the inductors during the shoot-through condition to the load during this mode. The output voltage is stepped down (if needed) based on the required AC output, but the DC-link voltage remains boosted due to the earlier shoot-through state.

3.2.2 Energy Flow

The inductor discharges and the capacitor supplies power to maintain a stable output voltage. The non-shoot-through operation ensures continuous power flow to the grid or load.

During the period T_1 from the switching cycle of T when the QZSI is not in a shoot-through state Figure 7 illustrates how a steady current source in the non-shoot-through condition represents the QZSI mode. By applying KVL to Fig. 7, the inductor voltage (V_{L1} and V_{L2}), DC link voltage (V_{PN}), out pot voltage (V_{out}), and diode voltage (V_{diode}).

$$V_{L1} = V_{in} - V_{C1} = V_{in} - V_{out} + V_{C2} \tag{7}$$

$$V_{L2} = V_{C1} - V_{out} = -V_{C2} \tag{8}$$

$$V_{PN} = V_{C1} - V_{L1} = V_{C1} + V_{C2};\ V_{diode} = 0 \tag{9}$$

Equations (6), (7), and (8) may be put together to create the following equation throughout the whole switching time T, assuming that the average voltage across inductors L1 and L2 is zero in a steady state.

$$V_{L2} = \frac{T_0 V_{C1} + T_1(-V_{C1})}{T} = \frac{T_0 V_{C1} + T_1(V_{C1} - V_{out})}{T} = 0 \tag{10}$$

$$T_0 V_{C1} = T_1 V_{C2}\ V_{C1} = \frac{T_1}{T} V_{out} \tag{11}$$

Based on equations in (8), (10), and (11),

$$V_{C2} = \frac{T_0}{T} V_{out}\ V_{out} = V_{C1} + V_{C2} \tag{12}$$

This also applies to the average voltage between the L1 and VL1 during a single switching period, which has the following definition:

$$V_{L1} = \frac{T_0(V_{C2} + V_{in}) + T_1(V_{in} - V_{C1})}{T} = \frac{T_0(V_{C2} + V_{in}) + T_1(V_{in} - V_{out} + V_{C2})}{T} = 0 \tag{13}$$

$$T_0 V_{out} + V_{in} T - T_1 V_{out} = 0 \tag{14}$$

$$V_{out} = \frac{T}{T_1 - T_0} V_{in}\ V_{in} = V_{C1} - V_{C2} \tag{15}$$

Next, the following formula may be used to get the average DC link voltage across the inverter bridge:

$$V_{out} = \frac{T_0(0) + T_1(V_{C1} - V_{L2})}{T} = \frac{T_1(V_{C1} + V_{C2})}{T} = V_{C1} \Rightarrow V_{out} = V_{C1}; \\ V_{C1} = V_{L1};\ V_{C1} = -V_{C2} \tag{16}$$

From Eqs. (11) and (14), the average voltage across the bridge can be defined as follows:

From Eq. (11),

$$V_{C1} = \frac{T_1}{T} V_{out} \Rightarrow \quad V_{out} = V_{C1} = \frac{T_1}{T_1 - T_0} V_{in} \tag{17}$$

$$V_{C2} = \frac{T_0}{T_1 - T_0} V_{in} \tag{18}$$

Defining shoot-through duty ratio $D = \frac{T_0}{T}$ and $T = T_0 + T$.

$$V_{out} = V_{C1} = \frac{T_1}{T_1 - T_0} V_{in} = \frac{1 - \frac{T_0}{T}}{1 - 2\frac{T_0}{T}} V_{in} \Rightarrow V_{C1} = \frac{1 - D}{1 - 2D} V_{in} \tag{19}$$

Adjusting the Shoot-Through time within a single switching cycle can increase the input voltage. In this scenario, the output of the PV array serves as the input voltage ($V_{in} = V_{pv}$). According to Eq. (19), the input voltage Vpv increases as the shoot-through time decreases and decreases as the shoot-through time increases if the capacitor voltage VC1 remains constant.

$$V_{in} = \frac{1-2D}{1-D} V_{C1} (OR) V_{out} \Rightarrow V_{out} = \left[\frac{1}{1-2D}\right] V_{in} \Rightarrow V_{out} = BV_{in} \Rightarrow$$
$$B = \frac{V_{out}}{V_{in}} \tag{20}$$

$$Where V_{in} = V_{pv}, B = BoostFactor = \frac{1}{1-2D}$$

Now if a simple boost control method is used in the QZSI topology then,

$$B = \frac{1}{1-2D} \tag{21}$$

where D is the shoot-through duty ratio. From Eq. (12),

$$V_{C1} = \frac{T_1}{T_1 - T_0} V_{in} = \left(\frac{1 - \frac{T_0}{T}}{1 - \frac{2T_0}{T}}\right) V_{in} \Rightarrow V_{C1} = \frac{1-D}{1-2D} V_{in} \tag{22}$$

$$V_{C2} = \frac{D}{1-2D} \tag{23}$$

From Eqs. (22) and (23),

$$V_{out} = \frac{1}{1-2D} V_{in} \tag{24}$$

$$V_{out} = BV_{in} \Rightarrow B = \frac{V_{out}}{V_{in}} = \frac{1}{1-2D} \tag{25}$$

$where, B = BoostFactor = 1 - M$, and M is the modulation index.

$$\text{Swithching stress } V_s = BV_{dc} \tag{26}$$

$$V_s = \frac{V_{dc}}{2M - 1} \tag{27}$$

Peak of the output voltage:

$$\text{For traditional source } V_{ph} = M \frac{V_i}{2} \tag{28}$$

$$\text{For Quasi Z} - \text{Source } V_{ph} = MB \frac{V_{dc}}{2} \tag{29}$$

$$V_{ph} = M \frac{V_{dc}}{2(2M - 1)} \tag{30}$$

3.3 Switching Sequence

Shoot-through states are applied strategically during the zero states of a modulation scheme (SPWM or SVPWM) to avoid interference with the output waveform. This technique allows the qZSI to boost voltage while maintaining a high-quality AC output, ensuring better performance for grid integration.

This dual-mode operation makes qZSI robust and efficient for renewable energy systems, such as wind-PV cogeneration, by providing voltage flexibility and enhanced reliability.

4 Soft Computing Techniques

Soft computing is a group of computer methods used to represent and resolve complicated issues marked by imprecision, ambiguity, and partial truth. Soft computing uses Approximation and heuristic approaches to provide reliable and effective answers, in contrast to classic complex computing techniques that need precise solutions and deterministic algorithms. Different types of optimization techniques are depicted in Fig. 8. Natural or biological processes serve as the inspiration for techniques like fuzzy logic, neural networks, genetic algorithms, and swarm intelligence. These methods are especially well-suited when it comes to issues like optimization, pattern identification, and decision-making in the face of ambiguity. In various fields, such as robotics, artificial intelligence, and renewable energy, soft computing offers high-performance, flexible, and economical solutions using adaptive learning, parallelism, and tolerance for imprecision. Soft computing can handle uncertainty and imprecision; it provides adaptable solutions to challenging real-world issues. Traditional computing, on the other hand, depends on exact models and inputs. Soft computing approaches, which draw inspiration from biological processes, mimic human problem-solving methods. The collective behaviour of social species like ants, bees, and birds serves as the foundation for swarm intelligence systems. To tackle complicated optimization issues, these algorithms simulate individual interactions within a population, or "swarm". The primary categories of swarm intelligence algorithms, each influenced by distinct social behaviours, are as follows:

4.1 Artificial Bee Colony (ABC) Algorithm

Inspired by the foraging behavior of honeybees, Karaboga created the Artificial Bee Colony (ABC) algorithm in 2005. The algorithm incorporates three different kinds of artificial bees: scout, worker, and spectator bees. Worker bees, who stand in for potential solutions, investigate food sources in their assigned regions. They then share their findings with spectator bees, which evaluate potential sources based on their

Fig. 8 Different types of optimization techniques

quality. When certain food sources stop producing high-quality results, scout bees randomly explore different places to guarantee complete exploration of the search area. This division of labor makes the ABC algorithm highly effective for solving continuous, nonlinear, and multidimensional optimization problems. It successfully balances exploring new possibilities with exploiting known high-quality solutions.

4.1.1 ABC Algorithm Basics

Bees as Agents: Every bee in the ABC algorithm stands for a different way to solve the optimization problem. The three types of bees—observer, scout, and employed bees—each have different roles to play. Food Sources as Solutions: Every food supply is a potential remedy, and a food source's "quality" correlates with the solution's objective value or fitness.

4.1.2 Types of Bees and Their Roles

Employed Bees: Every bee in use is linked to a particular food source (solution), and other bees are informed about it. To improve their existing solution, they investigate nearby options. Onlooker Bees: Based on their observations of the engaged bees, they choose food sources to take advantage of according to their suitability. A given source's likelihood of being selected by an observer directly correlates with its quality. Scout Bees: A recruited bee will become a scout if a food source does not improve after a certain number of attempts, known as the "limit." Scout bees then search randomly for new food sources in the area while ignoring low-quality ones.

4.1.3 Algorithm Steps

The ABC algorithm is iterative and consists of the following steps:

Initialization: Create the first food sources (solutions) in the search area at random. The quality of each solution is assessed based on its fitness.

Employed Bee Phase: Each employed bee explores a new solution by modifying its current food source. This is done by adjusting one of the parameters (dimensions) in the solution:

$$x_{new} = x_{old} + \phi \times \left(x_{old} - x_{neighbor}\right) \tag{31}$$

where ϕ is a random number between -1 and 1, and $x_{neighbor}$ is a randomly selected neighbor solution.

If the new solution is better than the previous one, the employed bee adjusts its food supply accordingly.

Onlooker Bee Phase: Every observer bee chooses a food source according to the possibility associated with each source's quality:

$$P_i = \frac{fitness\ of\ solution\ i}{\sum fitness\ of\ all\ solution} \tag{32}$$

Like the employed bee phase, onlooker bees also investigate and refine chosen food sources.

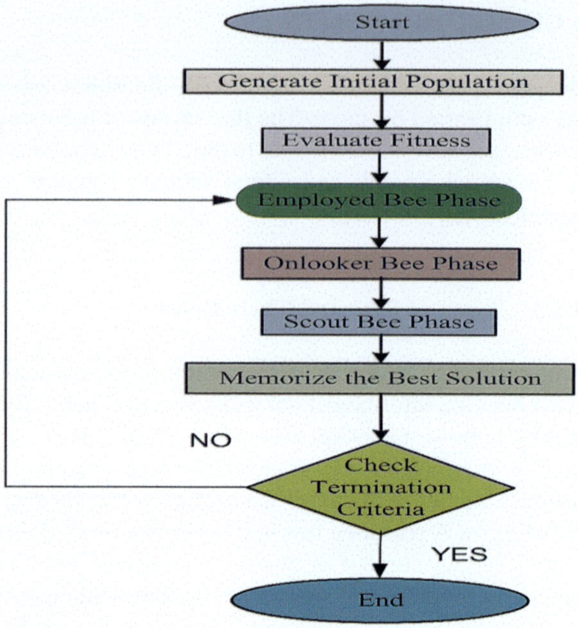

Fig. 9 Artificial bee colony flow chart

Scout Bee Phase: The associated worker bee becomes a scout if a food source is abandoned after several attempts without improvement. The scout then randomly searches for a new food source within the search space.

Termination: Iterate through the stages until a termination condition is satisfied, such as attaining a predetermined solution quality or a limited number of iterations. The flow chart of Artificial Bee Colony algorithm is depicted in Fig. 9.

4.1.4 Parameter Roles

Population Size: Total number of bees (employed and onlookers), influencing the search scope. Limit: The maximum number of iterations a food source can remain without improvement before it is abandoned. Probability Mechanism: Determines how likely an onlooker bee is to choose a food source based on its quality, balancing exploration and exploitation.

4.1.5 Applications of ABC Optimization

ABC is quite flexible and has been used well in fields like Power Systems: Used for optimization tasks in renewable energy, load dispatch, and power flow management. Renewable Energy: Enhancing efficiency in systems like solar and wind power converters, often by tuning controllers or optimizing system parameters. Machine Learning: Optimizing hyperparameters and neural network structures.

4.1.6 Advantages and Limitations

Advantages

Simple to implement and flexible. Good at balancing exploration (global search) and exploitation (local search). Effective in multi-objective and high-dimensional optimization.

Limitations

May require a high number of iterations for complex problems. Sensitive to parameter settings, especially in handling local optima.

The ABC algorithm's adaptability to different optimization landscapes, along with its simplicity, makes it a powerful tool in both theoretical and applied optimization contexts. The efficiency of ABC in resolving discrete and continuous optimization problems has enhanced the functionality of sustainable power systems, particularly when paired with other tactics like PSO for hybrid methods.

4.2 Particle Swarm Optimization (PSO)

Eberhart and Kennedy developed the PSO computer approach in 1995, drawing inspiration from the social behavior of flocks of birds and schools of fish. Modeling a collection of particles representing possible solutions traveling across a multi-dimensional search space is intended to tackle challenging optimization issues. To balance the exploration of new regions with the exploitation of known excellent solutions, each particle modifies its location depending on its personal experience (the best solution it has discovered) and the collective experience of the swarm (the best solution found by any particle). With its simplicity and efficiency, PSO has become popular in various applications, particularly in engineering and computer science, including tuning control parameters in renewable energy systems and optimizing complex functions in various domains.

4.2.1 Key Concepts in PSO

Particles: Every particle is a possible way to solve the optimization issue. Particles contain: Position: Indicates an alternative position in the search area. Velocity: Determines the direction and distance the particle will move in the next step. Personal Best (pBest): The optimal location or solution a particle has discovered so far. Global Best (gBest): The swarm as a whole came up with the greatest answer. Search Space:

The area or multi-dimensional space in which particles search for the optimal solution. Every aspect of the optimization issue is associated with a variable. Objective Function: The function that is being optimized. This function is used to assess the location of each particle, and the objective is to determine which position minimizes or maximizes this function.

4.2.2 PSO Mechanism

Each particle in PSO adjusts its location iteratively depending on two primary factors: The most well-known position it holds (based on personal experience). The swarm's most well-known position (social impact).

The steps are as follows:

Initialization: The location and velocity of the particles in the search space are initialized at random.

Update Velocity: The velocity of each particle is modified according to:

Its current velocity. The distinction between its present posture and its ideal position (attraction to personal experience). The distinction between its present location and the optimal place in the global (attraction to the optimum solution of the swarm). The velocity update formula for each particle is:

$$v_i = w.v_i + c_1.r_1.(pBest_i - x_i) + c_2.r_{2.}(gBest - x_i) \tag{33}$$

where: v_i: Current velocity of the particle. w: Inertia weight, controlling the influence of the previous velocity. c_1 and c_2: Acceleration coefficients, controlling the influence of personal and social components. r_1 and r_2: Random numbers between 0 and 1, adding stochasticity.

Update Position: Each particle's position is updated based on its new velocity:

$$x_i = x_i + v_i \tag{34}$$

In this case, xi represents the particle's current location.

Evaluate Fitness: At each particle's new location, the goal function is assessed. The particle updates the personal best if this location surpasses the particle's prior best. It updates the global best if it is the best of the whole swarm.

Repeat: The process continues for a fixed number of iterations or until convergence criteria (like a sufficiently small change in the global best value) are met. The flow chart of Practical Swarm Optimization algorithm is depicrted in Fig. 10.

4.2.3 Parameters in PSO

Inertia Weight (w): Controls the impact of the previous velocity on the new velocity, balancing exploration (higher w) and exploitation (lower w). Acceleration Coefficients (c_1 and c_2): Determine the influence of personal and global best positions.

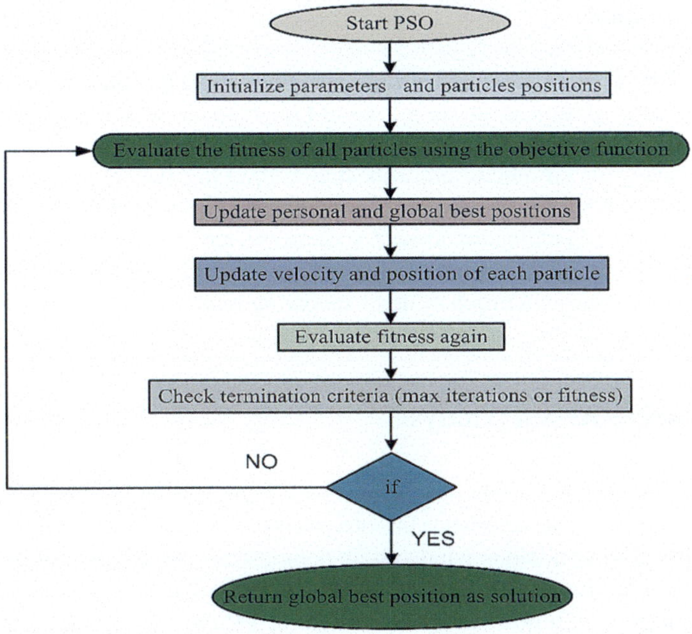

Fig. 10 Practical swarm optimization flow chart

Typical values are in the range [0, 2]. Population Size: The number of particles. A larger population can improve search quality but requires more computation.

4.2.4 Advantages of PSO

Simple to Implement: PSO has a straightforward set of equations and fewer parameters than many other optimization algorithms. Fast Convergence: PSO often converges quickly, particularly for simpler problems or lower-dimensional search spaces. Flexibility: This can be easily adapted to various types of problems and is widely used in continuous optimization.

4.2.5 Limitations of PSO

Local Optima: PSO sometimes converges too soon, particularly in multi-modal or complicated environments. Parameter Sensitivity: Parameter settings may have a significant impact on PSO performance and may need to be adjusted.

4.2.6 Applications

PSO is widely applied in engineering, machine learning (like neural network training), energy optimization (e.g., wind-PV cogeneration), and other fields requiring efficient solution searches for complex, non-linear, and high-dimensional problems. Due to its simplicity and robustness, PSO has become popular in optimizing control strategies for renewable energy integration, especially when combined with techniques like PI tuning for quasi-Z-source inverters. Its ease of implementation and ability to find optimal solutions in non-convex spaces make it particularly suited for grid-connected co-generation systems involving multiple renewable sources.

5 Simulation Result Analysis

In simulations comparing wind-only, PV-only, and co-generation wind-PV systems, distinct performance characteristics are observed in terms of power output stability and reliability. The standalone wind system demonstrates high power output during optimal wind conditions but shows significant variability when wind speeds fluctuate, leading to intermittent power supply. Similarly, the PV-only system provides consistent output during peak sunlight hours but underperforms during cloudy weather or nighttime, resulting in reduced reliability. On the other hand, wind-photovoltaic cogeneration demonstrates a synergistic impact in which the complementary qualities of solar and wind resources reduce the inherent unpredictability of each system. This combined approach ensures more stable power output throughout different weather conditions, harnessing solar energy during the day and wind energy during periods of low sunlight, which enhances grid reliability and maximizes renewable energy utilization. The co-generation system thus offers superior overall performance by balancing power production and addressing fluctuations, leading to a more consistent and sustainable energy supply for grid integration. Table 2 displays the suggested system's simulation parameters.

A simulation system analyzing wind power alone, photovoltaic (PV) power alone, and the cogeneration of wind-PV can provide critical insights into the operational efficiency and energy output of different configurations. Wind energy systems harness kinetic energy from wind using turbines, which can operate effectively in areas with steady wind patterns, producing clean electricity but with variability depending

Table 1 Summary of shoot-through vs. non-shoot-through modes

Mode	Purpose	Switching condition	Voltage effect
Shoot-through mode	Boosts the DC-link voltage	Both switches in one leg ON	Boosts input voltage
Non-shoot-through mode	AC power generation	Normal PWM operation	Delivers power to the load

Table 2 Simulation parameters

Wind parameters		
Mechanical output power (W)	3×10^6	
Basic power of the electrical generator (VA)	$3 \times 10^6/0.9$	
Basic wind speed (m/s)	12	
Maximum power at base wind speed (p.u. of nominal mechanical power)	1	
Wind-turbine power shown by pitch angle beta (>0) (deg	0°	
PMSG details specifications		
Resistance (R) (ph–ph)	0.36 Ω	
Inductance (L_{ab}) (ph–ph)	1.67mH	
Constant of torque (k_t)	60.7	
Inertia (J)	$5.5E^{-3}$	
Damping viscosity (F)	4.5	
Parameters	**Symbols**	**Values**
Inductor	$L_1 = L_2$	150 mH
Capacitors	$C_1 = C_2$	1200 μF
QZSI Switching frequency	f_{sw}	3420 Hz
Solar/PV parameters		
Strings in parallel	132	
Modules linked in series for each string	60	
Module-specific cells (N_{cell})	50	
Resistance (Rdc)	0.2e−3 Ω	
Inductance (Ldc)	0.5e−6 H	
voltage in the open circuit (V_{oc})	30.8 V	
current during a short circuit (I_{SC})	8.23 A	
Temperature of operation (T)	25 °C	
Irradiation (S)	1000 W/m^2	
Power at maximum (P_{ma}x)	190.437 W	
Maximum power point voltage (V_{mp})	24.7 V	
Maximum power point current (I_{mp})	7.71A	
Coefficient of temperature (V_{oc})	−0.3257 (%/°C)	
Coefficient of temperature I_{sc}	0.038202 (%/°C)	

on weather conditions. Similar to this, standalone PV systems use solar panels to directly convert sunlight into energy; they are most effective in bright weather but are limited by overcast or nocturnal situations. By integrating these two renewable sources into a wind-PV cogeneration system, the strengths of both can be synergized, enhancing energy reliability and grid stability. Since solar and wind resources often complement one another—solar production is usually greater during the day, while wind energy may peak at various times, including at night or during stormy conditions—the combined system may provide a more constant power output. Simulating these scenarios helps optimize control strategies, understand power management dynamics, and achieve more sustainable energy solutions through hybrid systems. The wind speeds rise from 8.4 m/s at t = 0–2 s, at t = 2–4 s fall to zero m/s then rise to 8.4 m/s at t = 4–6 s, and then rise to a maximum constant wind speed of 12 m/s after 6 s, as shown in Fig. 11a illustrates. When wind speed varies, the sun irradiance level also changes, falling from 1 to 0.5, then to 0.8, and ultimately rising to 1 kW/m^2 at t = 0−4, 4–6, and 6–8 s, in that order. The maximum power point tracking of wind and solar results in the ideal reference wind speed and DC link voltage compared with basic wind-solar generation characteristics when compared to Figs. 2 and 4. Both ω_r and V_{dc} exhibit well-damped performance in Fig. 11b, c, which is mirrored in the voltage and current produced by wind and solar energy's active power, as shown in Fig. 11e, d. For further analysis, the maximum solar power of 1.239 MW and wind power of 2.354 MW are produced at t = 8 s and t = 6 s. Figure 11f illustrates how the injected active power maintains a unity PCC voltage and current under all circumstances. As shown in Fig. 11g–i the utility grid's voltage and current are shown in Fig. 11, which is maintained by producing grid-side 3.49 MW of total active power and 3.54 MW at the point of common coupling.

6 Conclusions

This work explores how swarm intelligence techniques, namely Particle Swarm Optimization (PSO) and Artificial Bee Colony (ABC) algorithms, might improve quasi-Z-source inverters' (QZSI) efficiency for grid-connected renewable energy systems. We found that using these optimization algorithms to adjust the parameters of PI controllers leads to significant improvements in system efficiency, voltage regulation, and stability under various operational conditions. Our comparison of PSO and ABC shows that PSO has faster convergence, while ABC offers robustness in avoiding local optima. This demonstrates the effectiveness of swarm intelligence approaches in addressing the challenges of dynamic and nonlinear renewable energy systems. By making it easier to integrate renewable energy sources like solar and wind into the electrical grid, this study also advances the more general objective of encouraging sustainable energy alternatives. By providing a scalable framework for QZSI optimization, our research opens the door for further developments in renewable energy and power electronics. QZSI's performance is improved and the shift to cleaner and more sustainable energy systems is supported by the use of swarm intelligence methods. The knowledge gathered from this study offers a strong basis for further investigation and real-world applications in the dynamic field of renewable energy integration.

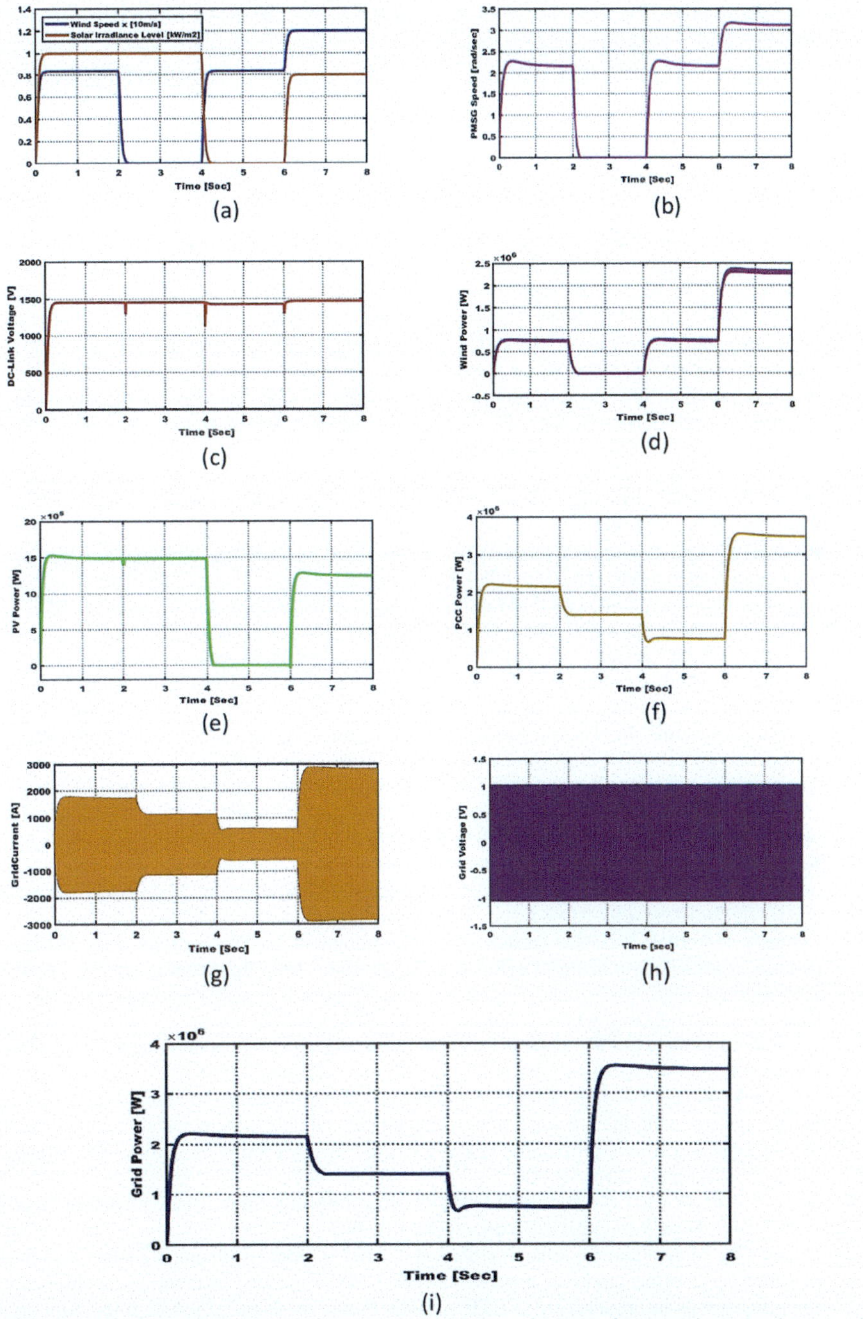

Fig. 11 **a** Wind speed and solar irradiance level, **b** PMSG speed, **c** DC-Link voltage, **d** Wind generated power, **e** Solar generated power, **f** Power at the point of common coupling, **g** Utility grid current, **h** Grid voltage, and **i** Utility grid power

Acknowledgements This research work was supported by "Woosong University's Academic Research Funding—2025".

References

1. Diaf S, Diaf D, Belhamel M, Haddadi M, Louche A (2007) A methodology for optimal sizing of autonomous hybrid PV/wind system. Elsevier Int J Energy Pol 35:5708–5718. https://doi.org/10.1016/j.enpol.2007.06.020
2. Chakraborty A, Kar AK (2017) Swarm intelligence: a review of algorithms. nature-inspired computing and optimization. Model Optim Sci Technol Springer Int Publishing 10:475–494. https://doi.org/10.1007/978-3-319-50920-4_19
3. Beni G, Jing W (1993) Swarm intelligence in cellular robotic systems. Robots and biological systems: towards new bionics. Berlin, Heidelberg: Springer, Berlin Heidelberg. 3342:703–712. https://doi.org/10.1007/978-3-540-30552-1_1
4. Bonabeau E, Marco D, Guy T (2024) Swarm intelligence: from natural to artificial systems (New York, 1999; online edn, Oxford Academic, 12 Nov. 2020), https://doi.org/10.1093/oso/9780195131581.001.0001.
5. Peng FZ (2003) Z-source inverter IEEE transactions on industry applications, 39:504–510 https://doi.org/10.1109/TIA.2003.808920
6. Peng FZ, Zhaoming SMQ (2005) Maximum boost control of the Z-source inverter. IEEE Trans Power Electron 20:833–838. https://doi.org/10.1109/TPEL.2005.850927
7. Anderson J Peng FZ (2008) Four quasi-Z-source inverters. IEEE power electronics specialists conference, Rhodes, Greece, pp 2743–2749. https://doi.org/10.1109/PESC.2008.4592360
8. Li Y, Anderson J, Peng FZ, Liu D (2009) Quasi-Z-source inverter for photovoltaic power generation systems. In: Twenty-fourth annual ieee applied power electronics conference and exposition, pp 918–924. https://doi.org/10.1109/APEC.2009.4802772
9. Chakraborty A, Kumar Kar A (2017) A review of algorithms. In: Nature-inspired computing and optimization: theory and applications, modeling and optimization, vol 10. Springer, Cham, pp 475–494. https://doi.org/10.1007/978-3-319-50920-4_19
10. Karaboga D, Basturk B (2007) A powerful and efficient algorithm for numerical function optimization: artificial bee colony (ABC) algorithm. J Global Optim 39:459–471. https://doi.org/10.1007/s10898-007-9149-x
11. Kennedy J Eberhart R (1995) Particle swarm optimization. In: Proceedings of the IEEE international conference on neural networks, vol 4, pp 1942–1948. https://doi.org/10.1109/ICNN.1995.488968
12. Kashefi Kaviani A, Riahy GH, Kouhsari SHM (2009) Optimal design of a reliable hydrogen-based stand-alone wind/PV generating system, considering component outages. Elsev J Renew Energy 34:2380–2390. https://doi.org/10.1016/j.renene.2009.03.020
13. Sambou V, Ndiaye PA, Kébé CMF, Ndongo M (2010) Optimal design of a hybrid solar–wind-battery system using the minimization of the annualized cost system and the minimization of the loss of power supply probability (LPSP). Elsev J Renew Energy 35:2388–2390. https://doi.org/10.1016/j.renene.2010.03.004
14. Gao WF, Liu SY (2012) A modified artificial bee colony algorithm. Elsev J Comput Oper Res 39:687–697. https://doi.org/10.1016/j.cor.2011.06.007
15. Sreenu Ch, Mallesham G, Chandra Shekar T (2023) Control of quasi Z-source converter in a microgrid using modified power ratio P&O MPPT. second international conference on emerging trends in engineering ICETE Atlantis Press, vol 9, pp 673–686. https://doi.org/10.2991/978-94-6463-252-168
16. Sreenu C, Mallesham G, Chandra Shekar T, Salkuti SR (2024) Pairing voltage-source converters with PI tuning controller based on PSO for grid-connected wind-solar cogeneration. Franklin Open 8:2773–1863. https://doi.org/10.1016/j.fraope.2024.100138

17. Chiang HC, Ma TT, Cheng YH, Chang JM, Chang WN (2010) Design and implementation of a hybrid regenerative power system combining grid-tie and uninterruptible power supply functions. Renew Power Generation, IET. 4:85–99. https://doi.org/10.1049/iet-rpg.2009.0033
18. Bae S, Kwasinski A (2012) Dynamic modeling and operation strategy for a microgrid with wind and photovoltaic resources. Smart Grid, IEEE Trans 3:1867–1876. https://doi.org/10.1109/TSG.2012.2198498
19. Hu K, Liaw C (2016) Incorporated operation control of DC microgrid and electric vehicle. IEEE Trans Industr Electron 63(1):202–215
20. Hu K-W, Liaw C-M (2015) Incorporated operation control of DC microgrid and electric vehicle. IEEE Trans Industr Electron 63:202–215. https://doi.org/10.1109/TIE.2015.2480750
21. Rekioua D, Rekioua T, Elsanabary A, Mekhilef S (2023) Power management control of an autonomous photovoltaic/wind turbine/battery system. Energies 16:2286. https://doi.org/10.3390/en16052286
22. Rajan MM, Daniel SA (2015) MPPT with single DC-DC converter and inverter for grid-connected hybrid wind-driven PMSG-PV system. IEEE Trans Industr Electron 62(4849):4857. https://doi.org/10.1109/TIE.2015.2399277
23. Levi E, Bojoi R, Profumo F, Toliyat H, Williamson S (2007) Multiphase induction motor drives - A technology status review. Electric Power Appl IET 1:489–516. https://doi.org/10.1049/iet-epa:20060342
24. Peng Fang Zheng (2003) Z-source inverter. IEEE Trans Ind Appl 39:504–510. https://doi.org/10.1109/TIA.2003.808920
25. Vijayaraghavan R, Sekhar CS, Ravi HK, Sasikumar M (2014) Comparative evaluation of generalized multicell impedance source inverter for drives. Int J Power Electron Drive Syst (IJPEDS). 4:61–68. https://doi.org/10.11591/ijpeds.v4i2.5128
26. Kouzou A, Abu-Rub H (2013) Multiphase Z-source inverter using maximum constant boost control. Arch Cont Sci 23:107–126. https://doi.org/10.2478/v10170-011-0045-8
27. Gao F, Loh PC, Blaabjerg F, Teodorescu R (2007) Modulation schemes of multi-phase three-level Z-source inverters, pp 1905–1911. https://doi.org/10.1109/PESC.2007.4342293
28. Barati AR, Moslehi M, Arab Khaburi D (2013) Z-source five leg inverter. In: 4th annual international power electronics, drive systems and technologies conference, vol 2, pp 205–211. https://doi.org/10.1109/PEDSTC.2013.6506705

Grey Wolf Optimization Based Optimal Charging Scheduling for Electric Vehicles with Real-Time Traffic Congestion Monitoring

K. Selvakumar, D. Selvabharathi, R. Palanisamy, T. M. Thamizh Thentral, and Surender Reddy Salkuti

Abstract Electric vehicles (EVs) are gaining popularity due to their environmental advantages over traditional vehicles. While considerable research has addressed EV performance, safety, and efficient grid power utilization, there is a noticeable lack of studies focusing on charging station capacity and Time of Use (TOU) power grids in relation to transport routes. This chapter seeks to fill this gap by introducing a novel approach for optimizing EV charging and scheduling using Grey Wolf Optimization (GWO). The proposed method accounts for various factors including route length, vehicle capacity, traffic congestion, charging time, travel time, backup requirements, and cost considerations. A comprehensive simulation platform has been created to facilitate the management of large-scale EV deployments under real-world charging conditions. This platform is instrumental in designing optimal strategies to alleviate congestion at charging stations and improve overall EV usage efficiency. Simulation results demonstrate significant enhancements in the utilization of EVs, optimization of TOU grid power, and reduction of charging station congestion. Our approach ensures the economical use of grid power and offers a structured strategy

K. Selvakumar · D. Selvabharathi · R. Palanisamy · T. M. T. Thentral
Department of Electrical and Electronics Engineering, College of Engineering and Technology, SRM Institute of Science and Technology, Kattankulathur, Chengalpattu 603203, Tamil Nadu, India
e-mail: selvakuk@srmist.edu.in

D. Selvabharathi
e-mail: selvabhd@srmist.edu.in

R. Palanisamy
e-mail: palanisr@srmist.edu.in

T. M. T. Thentral
e-mail: thamizht@srmist.edu.in

S. R. Salkuti (✉)
Department of Railroad and Electrical Engineering, Woosong University, Daejeon 34606, Republic of Korea
e-mail: surender@wsu.ac.kr

S. R. Salkuti (ed.), *Artificial Intelligence for Integrated Smart Energy Systems in Electric Vehicles*, Lecture Notes in Electrical Engineering 1427,
https://doi.org/10.1007/978-3-031-94276-1_30

for scheduling EVs based on real-time data and route-specific parameters. By effectively balancing charging demand with grid supply, the method reduces waiting times at charging stations and mitigates traffic congestion caused by multiple EVs arriving simultaneously. The integration of backup plans and cost considerations provides a thorough strategy for efficient EV deployment, leading to lower operational costs and improved environmental sustainability.

Keywords Electric vehicle · Optimal charge scheduling · Traffic congestion · SOC · Sustainable urban transportation

Nomenclature

EV	Electric vehicle
TOU	Time of use
GWO	Grey wolf optimization
SOC	State of charge
PSO	Particle swarm optimization
ATP	Arrival time-based priority
FCFS	First come first serve
C_{cost}	Total charging cost an electric vehicle
P_i^t	Required power needs to charge
Hp^t	Hourly electricity price at time 't'
$P_{capacity}$	Rated battery power capacity of an electric vehicle
P_{avail}	Available power capacity of an electric vehicle
C_{time}	Overall charging time of an electric vehicle
$C_{capacity}$	Charger capacity
T_{time}	Travelling time of an electric vehicle
D_i	Distance travelled by an electric vehicle
N_{max}	Maximum speed travelled by an electric vehicle
T_{time}^{index}	Travelling time index
$Tr_{congest}^{index}$	Traffic congestion index
R_{time}	Destination reaching time
S_{time}	Starting time of an electric vehicle
R_{time}^{index}	Reaching time with index
SOC_{min}	Minimum SOC of an electric vehicle
SOC_{max}	Maximum SOC of an electric vehicle
SOC_t	Available SOC of an electric vehicle at time 't'
P_{Charge}^{time}	Required charging time to reach maximum rated capacity

1 Introduction

The use of electric vehicles (EVs) has grown rapidly in recent years due to increasing concerns over climate change and air pollution, as well as advances in battery technology and charging infrastructure [1]. EVs are considered a crucial solution for reducing greenhouse gas emissions and improving air quality in urban areas. They offer several advantages over traditional gasoline-powered vehicles, including lower operating costs, reduced emissions, and quieter operation. However, EVs also face challenges such as limited driving range, longer charging times, and the need for extensive charging infrastructure [2]. Despite these challenges, the market for electric vehicles is growing rapidly, with many major automakers heavily investing in the development of new electric models. As battery technology continues to improve and charging infrastructure becomes more widespread, the use of EVs is expected to increase significantly in the coming years [3]. There are several types of charging options for EVs, each with its own charging time and power level [4]. The most common types of charging for EVs are:

- Level 1 Charging: This is the most basic and slowest method, involving the use of a standard 120 V household outlet. It delivers a charging rate of approximately 4–5 miles of range per hour, making it suitable for overnight charging at home or work.
- Level 2 Charging: This method employs a dedicated 240-V charging station, providing a charging rate of around 15–30 miles of range per hour. Level 2 charging is commonly used at public charging stations or home with a specialized charging station.
- DC Fast Charging: This offers the fastest charging time for electric vehicles, typically reaching 80% charge in about 30–40 min. DC fast charging requires specialized equipment and is frequently found at public charging stations along major highways.

To enhance the adoption of EVs, agreements have been established between charging infrastructure providers and regional authorities to promote the installation of more public charging stations [5, 6]. Based on the agreements most of the area is occupied or rented by the ventures for constructing the charging station [7]. Some countries are swapping batteries at stations to eliminate the delay in their movement. In the swapping technology, battery tracking is challenging [8, 9]. Optimal charge scheduling of EVs with real traffic congestion monitoring requires real-time data analysis. Vehicle data, recharge station data, traffic congestion data, and a route map have been gathered. Additionally, a travel plan has been established, which specifies the departure point, destination point, and recharge station for each vehicle type on these routes. Real-time charging station information was incorporated with transport information to minimize the waiting time for vehicles at charging stations [10]. This study presented various types of vehicles running through different routes with a basic State of Charge (SOC), which allows the vehicle to run to a further point and charge for further movement. When starting the travel plan, each vehicle must select the route and find a charging station for uninterrupted operation [11].

Over the past two decades, researchers have been working to address manufacturing issues related to EVs. While the trend is shifting towards the use of EVs, various operational issues are still being mitigated by researchers. In reference [12], an optimal static charging scheduling scheme is discussed to minimize charging costs. Algorithms such as Particle Swarm Optimization (PSO), Arrival Time Priority based (ATP) algorithm, and SOC-Based Priority (SBP) algorithm have been proposed to optimize and reduce energy consumption from the grid by properly shifting EV loads. IoT-based communication between end-users and charging stations is also recommended for effective operation. In reference [13], a strategy for charging PEV through Nash equilibrium is proposed to minimize generation costs by scheduling PEV demand electricity to obtain optimal and near-optimal charging patterns for operation. The strategy involves decentralizing charging for plug-in EVs. In reference [14], an intelligent method for controlling EV charging loads in TOU slots is discussed. It compares the cost of charging and energy demand in distinctive intervals related to the acceptable range and state of charge (SOC). In reference [15], changes in price in IEX on an hourly basis permit trading to access that cost and reduce peak time charging. This intelligent scheduling is based on grid demand and data optimization, and various scheduling schemes minimize total costs for future performance in the network [16, 17].

The chapter is organized as follows: Sect. 2 compares electric and conventional vehicles and identifies problems with charging schedules. Section 3 describes the methodology used to overcome these problems. Section 4 discusses the input data considered for the analysis, and Sect. 5 presents the results. Finally, Sect. 6 provides the conclusion.

2 Problem Formulation

In the future, EVs will overcome conventional vehicles, charging stations for EVs as low as compared to conventional vehicle fuel filling stations. If the future is running with EV vehicles there may be vast changes in CS and more charging stations to be installed in most of the areas to avoid congestion and to minimize cost [18]. Vehicles have been selected for a study focused on improving route selection and charging point allocation to avoid traffic congestion on roads and at charging stations. In the charging station, the FCFS procedure has been followed for charging the vehicle with various types of charging technology [19]. A further study was done for time minimization and avoiding overcrowding of vehicles near to charging station, whereas charging cost is dynamic due to changes in grid price. Online information on charging station data is not available in the present scenario.

In the future, efficient scheduling and charging for EVs may be implemented. To minimize time, appropriate charging points were selected to ensure vehicles can recharge as quickly as possible before continuing their route [20]. This approach aims to avoid long waiting times for trucks by ensuring that the appropriate chargers are readily available at the charging station when they arrive. To minimize waiting time, the First-Come-First-Served (FCFS) technique is used for charging vehicles. This technique helps to charge first-in-first-serve vehicles at any capacity of vehicle [21].

Additionally, congestion control is considered by selecting the optimal route for EVs. The selection of the optimal route is based on the charging station location and the available charging points. This helps in avoiding congestion near the charging stations, which in turn reduces the waiting time for the vehicles to get charged. To achieve this, the Shuffled Frog Leaping Algorithm is used, considering dynamic changes in grid prices and charging costs while optimizing the charging and routing schedule. However, one of the challenges in this approach is the lack of online information about charging station availability and occupancy. In the future, Efficient scheduling and charging for EVs can be implemented by using real-time charging station data and developing more advanced algorithms for congestion control and optimal routing [22]. The optimal charge scheduling problem for electric vehicles aims to minimize charging costs by factoring in variations in electricity prices, traffic congestion for different EVs, and the specific charging limits of each station. In this dynamic scheme, scheduling is carried out for each 60-min time slot to account for the random arrival of EVs and the hourly fluctuations in electricity prices.

2.1 Objective Function and Constraints

The objective function aims to reduce the overall cost, optimizing the economic operation of EVs by scheduling chargers efficiently, and taking into account electricity prices and charging limits [22]. The objective function to curtail the charging cost is given by:

$$C(t) = \min \sum_{t=1}^{T} C_{cost} + \sum_{v=1}^{V} R_{time}^{index} \tag{1}$$

Various calculations have been carried out to propose a method on charge scheduling, the formulae used are given below.

Cost: The cost of charging an EV during a particular hour is calculated by multiplying the actual power needed to charge the vehicle with the price during that hour. Therefore, the equation is,

$$C_{cost} = P_i^t * Hp^t \tag{2}$$

Power: The power needed to charge is calculated by multiplying the rated capacity of the vehicle battery with the available power to run, assuming the vehicle is running at full capacity [23]. Therefore, the formula is,

$$P_i^t = P_{capacity} - P_{avail} \tag{3}$$

Time: The time required for charging is calculated by dividing the power needed to charge by the charger capacity.

$$C_{time} = \frac{P_i^t}{C_{capacity}} \tag{4}$$

Travelling Time: The traveling time is calculated by dividing the distance by the maximum speed of the vehicle.

$$T_{time} = \frac{D_i}{N_{max}} \tag{5}$$

Travelling Time with Index: It is calculated by multiplying the traveling time with the traffic congestion index.

$$T_{time}^{index} = T_{time} * Tr_{congest}^{index} \tag{6}$$

Reaching Time: The reaching time is calculated by adding the starting time to the traveling time.

$$R_{time} = S_{time} + T_{time} \tag{7}$$

Reaching Time with Index: The reaching time with the index is calculated by adding the starting time to the traveling time with the index.

$$R_{time}^{index} = S_{time} + T_{time}^{index} \tag{8}$$

Distance to Run: The index for available power, maximum speed, and travel time is used to calculate the distance that can be covered, as follows:

$$D_i = P_{avail} - \left(N_{\max} * T_{time}^{index}\right) \tag{9}$$

The different limitations studied in the problem are given below.

3 Minimum SOC Constraint:

$$SOC_t \geq SOC_{min} \tag{10}$$

4 Maximum SOC Constraint:

$$SOC_t \leq SOC_{\max} \tag{11}$$

5 Maximum Charging Power Constraint:

$$P^{time}_{Charge} \leq P_{\max} \tag{12}$$

6 Operating Parameters in the Proposed Method

EVs face numerous challenges at every stage, from manufacturing to end-user adoption. While many improvements have been made in the manufacturing process, significant changes are still needed in EV operation. One critical aspect is the scheduling of charging for large-scale EVs, which presents a challenge for efficient charging. Another important consideration is congestion control at charging stations to prevent overcrowding [24]. To avoid rush, common techniques must be implemented at all charging stations. To facilitate this, the following inputs must be collected:

- Route map
- Vehicles data
- Charging station data
- Traffic congestion data
- scheduling of large-scale EVs
- Congestion control in EV station.

The various routes that are available on Google Maps for traveling from the departure point to the same destination are illustrated in Fig. 1. The specified route was plotted to ascertain the precise distance of the vehicle's voyage. A distinct route was chosen to travel from Chennai to Karaikal in this investigation.

Data from the manufacturer was collected for a variety of vehicles, as shown in Fig. 2. The data includes the motor capacity, speed, battery capacity, manufacturer year, and state of charge (SOC) for each vehicle [25]. This initial data was used to formulate the problem. The vehicles were then sorted using the GWO algorithm. Table 1 displays the different types of charging sockets available for rapid, fast, and slow chargers. Chargers are chosen based on their capacity and design for connecting to the vehicle. Further details are provided below.

Traffic data was collected on an hourly basis from Google Maps to determine the fastest route and time taken to travel on each route. The typical traffic congestion data is shown in Fig. 3. To reduce congestion at charging stations, each vehicle has been

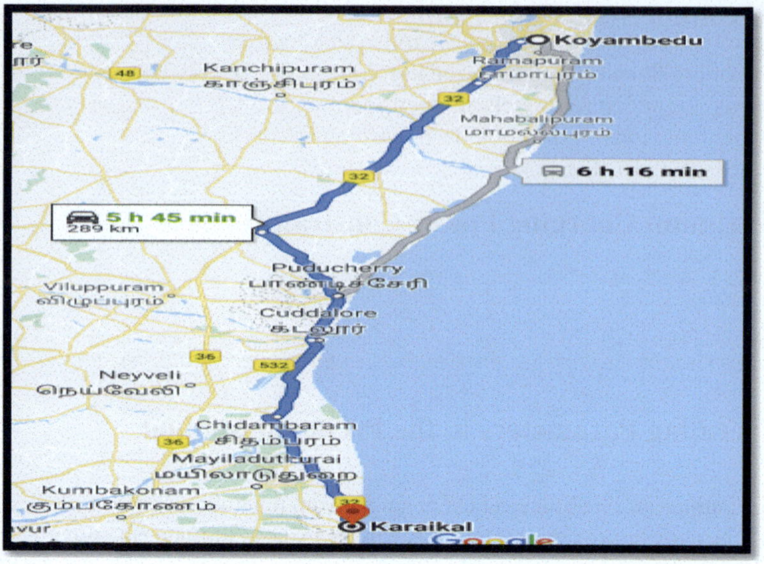

Fig. 1 Route map

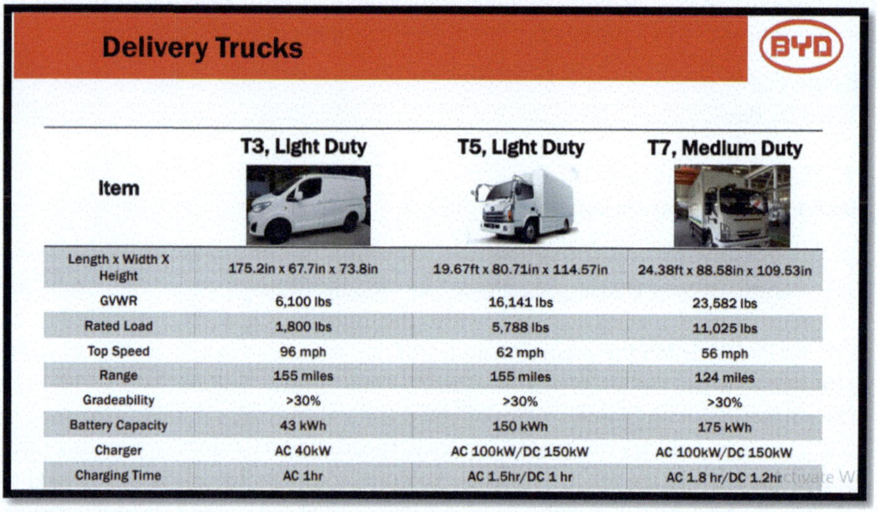

Item	T3, Light Duty	T5, Light Duty	T7, Medium Duty
Length x Width X Height	175.2in x 67.7in x 73.8in	19.67ft x 80.71in x 114.57in	24.38ft x 88.58in x 109.53in
GVWR	6,100 lbs	16,141 lbs	23,582 lbs
Rated Load	1,800 lbs	5,788 lbs	11,025 lbs
Top Speed	96 mph	62 mph	56 mph
Range	155 miles	155 miles	124 miles
Gradeability	>30%	>30%	>30%
Battery Capacity	43 kWh	150 kWh	175 kWh
Charger	AC 40kW	AC 100kW/DC 150kW	AC 100kW/DC 150kW
Charging Time	AC 1hr	AC 1.5hr/DC 1 hr	AC 1.8 hr/DC 1.2hr

Fig. 2 Vehicle data

allotted a designated charging station at the starting point. Additionally, a first-come-first-served (FCFS) method has been implemented to further reduce congestion. The illustration representing the FCFS method is shown in Fig. 4.

Table 1 Charging station data (charging time for 100 km of BEV range)

Power supply	Power (W)	Voltage (V AC)	Maximum current (A)	Charging time
One phase	33,000	230	16.00	5–6 h
One phase	74,000	230	32.00	2–2.5 h
Three phase	11,000	400	16.00	1.5–2 h
Three phase	22,000	400	32.00	44–55 min
Three phase	43,000	400	63.00	22–28 min
Direct current fast charge (DCFC)	50,000	400–500	100–125.00	19–24 min
Direct current fast charge (DCFC)	120,000	300-500	300–350.00	8–10 min

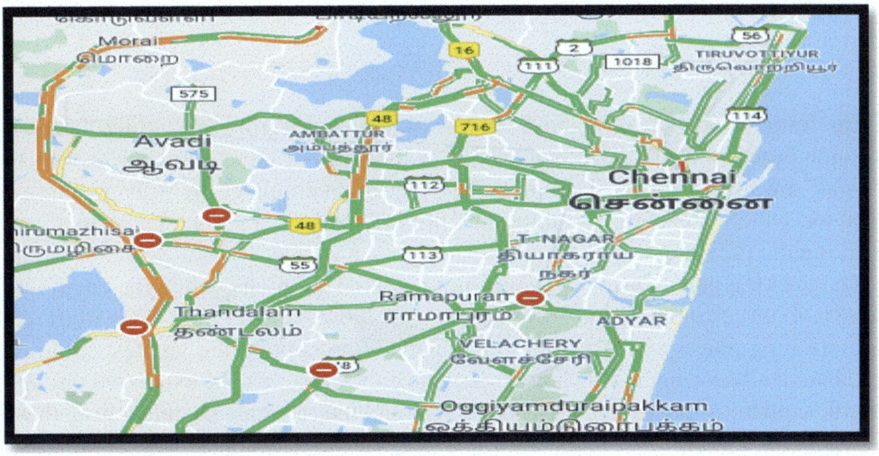

Fig. 3 Typical traffic congestion

7 Grey Wolf Optimization

In the detection of more effective computational problem-solving methods, researchers have increasingly drawn inspiration from nature, leading to the development of algorithms like Grey Wolf Optimizer (GWO) and Shuffled Frog Leaping Algorithm (SFLA) [26–29]. This approach has led to the development of nature-inspired algorithms, including a notable technique known as the Grey Wolf Optimizer (GWO), introduced in 2014 [30]. The GWO algorithm is inspired by the social hierarchy and hunting behavior of grey wolves. It mimics the wolves' leadership dynamics, where the finest solution is denoted by the alpha (α) wolf and the subsequent and 3rd best solutions are denoted by the beta (β) and delta (δ) wolves, correspondingly. The left applicant solutions are symbolized by omega (ω) wolves.

Fig. 4 Congestion control

In this framework, the α, β, and δ wolves play a central role in guiding the optimization process, while the ω wolves follow the leads of these dominant wolves to navigate towards the optimal solution. This algorithm emulates the wolves' hunting behavior, which is categorized into three primary phases: tracking, encircling, and attacking the prey.

The tracking phase involves identifying potential solutions, while the encircling phase is about surrounding the prey (or optimal solution) and refining the search area. The attacking phase represents the final approach to capture the greatest solution. Mathematically, the encircling activities of the grey wolves can be expressed through specific equations that model their movement and interactions. By integrating these natural behaviors into the algorithm, GWO effectively leverages the social structure and cooperative strategies of grey wolves to optimize complex problems. This nature-inspired methodology not only enhances the algorithm's performance but also provides a robust framework for exploring and solving a wide range of optimization challenges, making it a valuable tool in computational research and applications. As previously discussed, grey wolves exhibit a behavior of encircling their prey during hunts. This encircling activity can be mathematically developed using the following equation:

$$D = |R * Xprey(t) - X(t)| \tag{13}$$

$$X(t+1) = Xprey(t) - Q * D \tag{14}$$

Q and R are coefficient vectors used to adjust the wolves' positions relative to the prey. The calculation of these coefficient vectors is as follows:

$$Q = 2(a * ran_1) - a \tag{15}$$

$$R = 2 * ran_2 \tag{16}$$

The vector A modulates the encouragement of the prey's location on the grey wolf's position, diminishing over time to facilitate the wolves' gradual convergence on the target. This adjustment is represented by equations that model the wolves' movements relative to the prey, realistically mimicking the actual encircling behavior perceived in grey wolves throughout a hunt. By constantly revising their locations in relation to the prey, the wolves in the Grey Wolf Optimizer (GWO) algorithm can efficiently explore and navigate the search space, ultimately converging toward optimal solutions.

Grey wolves excel at recognizing and encircling their prey, typically guided by the alpha wolf during hunts, with the beta and delta wolves occasionally participating. To arithmetically reproduce the hunting performance of grey wolves, it is assumed that the alpha (best nominee result), beta, and delta wolves possess superior knowledge of the prey's potential location. Consequently, the algorithm retains the top three best answers identified thus far and requires the remaining explore representatives, including the omega wolves, to renew their locations based on these best solutions. The following equations are formulated to represent this process.

$$D_{alpha} = \left|R_1 * X_{alpha} - X\right| \tag{17}$$

$$D_{beta} = |R_2 * X_{beta} - X| \tag{18}$$

$$D_{delta} = |R_3 * X_{delta} - X| \tag{19}$$

$$X_1 = \left|X_{alpha} - Q_1 * D_{alpha}\right| \tag{20}$$

$$X_2 = |X_{beta} - Q_2 * D_{beta}| \tag{21}$$

$$X_3 = |X_{delta} - Q_3 * D_{delta}| \tag{22}$$

$$X_{(t+1)} = \frac{X_1 + X_2 + X_3}{3} \tag{23}$$

The proposed Grey Wolf Optimization (GWO) algorithm mimics the cooperative hunting behavior of grey wolves in nature to locate food. The steps for implementing the GWO technique are as follows:

- Step 1: Set the initial parameters, including the grey wolf population size, their positions, the number of repetitions, and the acceptance criteria.
- Step 2: Calculate the fitness value for each individual in the residents using the calculation

$$fitness = \min\left\{\sum_{i=1}^{k} w_i \frac{fi(x)}{f_i^{\max}}\right\} with\ w_i \geq 0\ and\ \sum_{i=1}^{k} w_i = 1 \quad (24)$$

- Step 3: Identify the top three wolves based on their fitness values, denoted as Alpha (best), Beta (second best), and Delta (3rd best).
- Step 4: Evaluate the stop criteria (number of repetitions and acceptance value). If the criteria are met, proceed to the end; otherwise, continue to the next step.
- Step 5: Generate new positions for Alpha, Beta, and Delta using the above equations and update them. The position was updated by using,

$$X_{(t+1)} = \frac{X_1 + X_2 + X_3}{3} \quad (25)$$

- Step 6: Choose the Universal-Greatest (Alpha) and other locations based on the renewed objective function.
- Step 7: Reassess the positions of the wolves against the termination criteria. If not satisfied, repeat steps 5 to 7 until the criteria are met.
- Step 8: Conclude the algorithm and conclude the optimal answer to the problem.

These steps collectively enable the GWO algorithm to actually navigate the search space and come together towards optimal solutions by leveraging the hierarchical and cooperative hunting strategies of grey wolves.

8 Result and Discussion

8.1 Input Data for Optimization

The proposed charging schedule problem in EVs needs a route map, vehicle data, charging station data, and traffic congestion data [31, 32]. The route from Chennai to Karaikal was selected using Google Maps, which provided three possible routes. The availability of charging stations along these routes was then plotted, as shown in Fig. 5.

Data was collected from a major EV manufacturer, as provided in Table 2. The data collected from EV manufacturers can include information on vehicle specifications, charging times and patterns, battery performance, and other related data points. Different types of charger points and charging stations were selected for charging the vehicle with appropriate connectors given in Table 3.

Table 4 presents the actual travel time of each vehicle, including charging time, and the traffic congestion index for the different routes from Chennai to Karaikal. The results show that the lowest possible travel time is 5.45 h, while the highest travel time is 7.61 h. The cost range of the routes varies from Rs. 1,116 to Rs. 1,395.

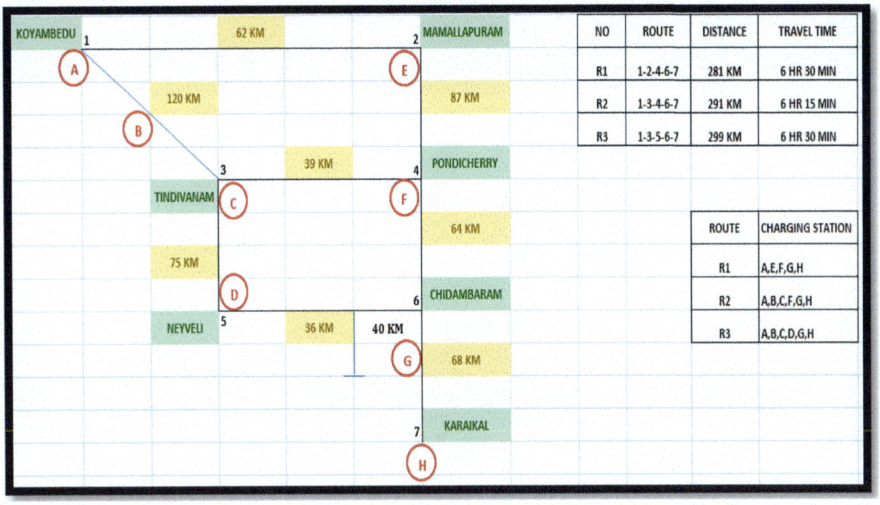

Fig. 5 Typical route map

The study identified the best route for travel from Chennai to Karaikal to be route 1–3–4–6–7 (Chennai-Thindivanam-Pondicherry-Chidambaram-Karaikal), which is both cost-effective and timesaving.

9 Conclusions

This work presented an optimization approach for the scheduling of electric vehicle (EV) charging and the management of traffic congestion, utilizing the Grey Wolf Optimization (GWO) procedure. The recommended method includes several modifications designed to enhance performance and minimize operational costs for EVs. Comprehensive simulation results, alongside theoretical calculations, have been rigorously validated, demonstrating the algorithm's effectiveness. Our findings reveal that the proposed approach can significantly improve the efficiency of EV usage by identifying optimal routes with available charging stations, thus saving time and reducing costs. The simulation results highlight the algorithm's ability to optimize the Time of Use (TOU) of grid power and alleviate congestion at charging stations, underscoring its practical benefits. The integration of factors such as route length, vehicle capacity, and traffic conditions has proven to be effective in balancing charging demand with grid supply. Looking ahead, advancements such as wireless charging and vehicle-to-grid (V2G) technology offer promising avenues for further enhancing the efficiency of electric vehicle transportation. These future improvements could build upon the current optimization framework, providing additional benefits and efficiencies. Overall, this study contributes to the improvement of actual approaches for optimal charge scheduling of EVs while monitoring traffic congestion.

Table 2 Vehicle data

S. No.	Type	Year of manufacturing	Make	Motor rating	Max speed	Running at 100% charge	Capacity	SOC in %
1	Truck	2018	BMW	125	93	230	42 kWH/ 200 AH	90
2	Truck	2019	VOLVO	150	90	212	23 kWH/ 180 AH	92
3	Truck	2017	FORD	107	84	195	23 kWH/ 175 AH	70
4	Truck	2015	MITSHUBISHI	67	100	180	24 kWH/ 180 AH	49
5	Truck	2016	VOLKSWAGEN	100	85	185	20 kWH/ 120 AH	63
6	Truck	2015	CHEVROLET	150	90	175	22 kWH/ 180 AH	55
7	Truck	2012	NISSAN	80	97	185	40 kWH/ 75 AH	45
8	Truck	2010	HYUNDAI	100	100	167	38.3 kWH/75 AH	70
9	Truck	2020	KARMA	200	125	200	26 kWH/ 300 AH	98
10	Truck	2009	GE	102	80	160	26.4 kWH/80 AH	45
11	Truck	2015	BMW	125	93	190	42 kWH /120 AH	60
12	Truck	2016	VOLVO	160	75	195	19 kWH/ 150 AH	70

Table 3 Charging station data

Charging time for 100 km of BEV range	Power supply	Power (W)	Voltage (V AC)	Maximum current (A)	Number of charging points							
					A	B	C	D	E	F	G	H
3–4 h	One phase	7400	230	32.00	0	0	1	1	0	0	0	0
2–3 h	3 Phase	11,000	400	16.00	0	1	1	1	1	2	1	1
1–2 h	3 Phase	22,000	400	32.00	1	1	1	2	1	0	1	1
20–30 min	DC	50,000	400–500	100–125.00	1	1	1	0	1	1	1	1
10 min	Direct current	120000	300–500	300–350.00	1	1	1	0	2	2	1	0
Total charging points					3	4	5	4	5	5	4	3

Table 4 Comparison of results

Route	Actual (km)	Travelling time	Cost (Rs.)
1–2–4–6–7	281	5.67	1170
1–3–4–6–7	291	5.45	1116
1–3–5–6–7	299	7.61	1395

References

1. Fernandez GS, Krishnasamy V, Kuppusamy S, Ali JS, Ali ZM, El-Shahat A, Abdel Aleem SHE (2020) Optimal dynamic scheduling of electric vehicles in a parking lot using particle swarm optimization and shuffled frog leaping algorithm. Energies 13:6384. https://doi.org/10.3390/en13236384
2. Albert B, Andrew C, Maxx S, David W (2011) Vehicle electrification: status and issues. Proc IEEE 99:1116–1138. https://doi.org/10.1109/JPROC.2011.2112750
3. Un-Noor F, Padmanaban S, Mihet-Popa L, Mollah MN, Hossain E (2017) A comprehensive study of key electric vehicle (EV) components, technologies, challenges, impacts, and future direction of development. Energies 10:1217. https://doi.org/10.3390/en10081217
4. Salkuti SR (2023) Advanced technologies for energy storage and electric vehicles. Energies 16(5):2312. https://doi.org/10.3390/en16052312
5. Sergejus M, Katarina K, Mattia M (2016) Management of power quality issues in low voltage networks using electric vehicles: experimental validation. IEEE Trans Power Delivery 32(2):971–979. https://doi.org/10.1109/TPWRD.2016.2614582
6. Ma Z, Callaway DS, Hiskens IA (2013) Decentralized charging control of large populations of plug-in electric vehicles. IEEE Trans Control Syst Technol 21(1):67–78. https://doi.org/10.1109/TCST.2011.2174059
7. Hui L, Zechun H, Yonghua S, Jianhui W, Xu X (2015) Vehicle-to-grid control for supplementary frequency regulation considering charging demands. IEEE Trans Power Syst 30:3110–3119. https://doi.org/10.1109/TPWRS.2014.2382979
8. Vagropoulos SI, Kyriazidis DK, Bakirtzis AG (2016) Real-time charging management framework for electric vehicle aggregators in a market environment. IEEE Trans Smart Grid 7(2):948–957. https://doi.org/10.1109/TSG.2015.2421299
9. Yang J, He L, Fu S (2014) An improved PSO-based charging strategy of electric vehicles in electrical distribution grid. Appl Energy 128:82–92. https://doi.org/10.1016/j.apenergy.2014.04.047
10. Polat H, Hosseinabadi F, Hasan MM, Chakraborty S, Geury T, El Baghdadi M, Wilkins S, Hegazy O (2023) A review of DC fast chargers with BESS for electric vehicles: topology, battery, reliability oriented control and cooling perspectives. Batteries 9:121. https://doi.org/10.3390/batteries9020121
11. Boomsma T, Capion K, Meibom P (2011) Optimal charging of electric drive vehicles in a market environment. Appl Energy 88:1940–1948. https://doi.org/10.1016/j.apenergy.2010.12.015
12. Moeini-Aghtaie M, Abbaspour A, Fotuhi-Firuzabad M (2014) Online multicriteria framework for charging management of PHEVs. Veh Technol IEEE Trans 63:3028–3037. https://doi.org/10.1109/TVT.2014.2320963
13. Yijia C, Shengwei T, Canbing L, Peng Z, Yi T, Zhikun Z, Junxiong L (2012) An optimized EV charging model considering TOU price and SOC curve. IEEE Trans. Smart Grid 3:388–393. https://doi.org/10.1109/TSG.2011.2159630
14. Kumar JV, Sesham H, Salkuti SR (2024) Smart energy management model for electric vehicles. In: Salkuti SR (eds) Energy and environmental aspects of emerging technologies for smart grid. green energy and technology. Springer, Cham. https://doi.org/10.1007/978-3-031-18389-8_9
15. El-Bayeh CZ, Alzaareer K, Aldaoudeyeh A-MI, Brahmi B, Zellagui M (2021) Charging and discharging strategies of electric vehicles: a survey. World Electr Veh J 12:11. https://doi.org/10.3390/wevj1201001116

16. Selvabharathi D, Muruganantham N (2022) Estimating the state of health of lead-acid battery using feed-forward neural network. J Circ Syst Comput 2250081(31):05. https://doi.org/10.1142/S0218126622500815
17. Linni J, Honghong X, Guoqing X, Xinyu Z, Dongfang Z, Shao ZY (2013) Regulated charging of plug-in hybrid electric vehicles for minimizing load variance in household smart microgrid. IEEE Trans Industr Electron 60:3218–3226. https://doi.org/10.1109/TIE.2012.2198037
18. Derakhshandeh SY, Amir M, Sara D, Mohammad M, Golshan H, Esmail M (2013) Coordination of generation scheduling with PEVs charging in industrial microgrids. IEEE Trans Power Syst 28:3451–3461. https://doi.org/10.1109/TPWRS.2013.2257184
19. Salkuti SR, Ray P, Pagidipala S (2022) Overview of next generation smart grids. In: Salkuti SR, Ray P (eds) Next generation smart grids: modeling, control and optimization. Lecture notes in electrical engineering, vol 824. Springer, Singapore. https://doi.org/10.1007/978-981-16-7794-6_1
20. Sharaf AM, Hani M, Abdollah A, Foad HG, Omid H, Heidarali S (2018) Application of new fast. Efficient-Self Adjusting PSO-Search Algo Power Syst. https://doi.org/10.1016/B978-0-12-812441-3.00003-3
21. Wenhui Z, Jie W, Weifeng Z, Haochuan Z, Lei S (2019) Optimal and elastic energy trading for green microgrids: a two-layer game approach. Mob Netw Appl24. https://doi.org/10.1007/s11036-018-1027-x
22. Jinghong Z, Xiaoyu W, Kun M, Chun Z, Shouzhen Z (2013) Aggregation model-based optimization for electric vehicle charging strategy. IEEE Trans Smart Grid 4:1058–1066. https://doi.org/10.1109/TSG.2013.2242207
23. Selvakumar K, Sathya Vignesh R, Vijayabalan R(2012) Shuffled frog leaping algorithm for solving profit based unit commitment problem. In: International conference on computing, communication and applications, pp 1–6. https://doi.org/10.1109/ICCCA.2012.6179234
24. Mirjalili S, Mirjalili SM, Lewis A (2014) Grey wolf optimizer. Adv Eng Softw 69:46–61. https://doi.org/10.1016/j.advengsoft.2013.12.007
25. Govinda CV, Udhay SV, Rani C, Wang Y, Busawon (2018) A review on various MPPT techniques for wind energy conversion system. In: 7th IEEE international conference on computation of power, energy, information and communication, Chennai, India. https://doi.org/10.1109/ICCPEIC.2018.8525219
26. Wu A, Zhao B, Mao J, Wu B, Yu F (2019) Adaptive active fault-tolerant MPPT control for wind power generation system under partial loss of actuator effectiveness. Int J Electr Power Energy Syst 105:660–670. https://doi.org/10.1016/j.ijepes.2018.09.015
27. Kucuk TV, Oncu S (2021) Wind energy conversion system With PDM controlled converter. In: International conference on renewable energy research and applications. IEEE, Istanbul, Turkey, pp 136–140. https://doi.org/10.1109/icrera52334.2021.9598618
28. Mousa HH, Youssef AR, Hamdan I, Ahamed M, Mohamed EE (2021) Performance assessment of robust PO algorithm using optimal hypothetical position of generator speed. IEEE Access 9:30469–30485. https://doi.org/10.1109/ACCESS.2021.3059884
29. Mohamed RG, Ebrahim MA, Alaas ZM, Ahmed MM (2022) Optimal energy harvesting of large-scale wind farm using marine predators algorithm. IEEE Access 10:24995–25004. https://doi.org/10.1109/ACCESS.2022.3156084
30. Yang B, Yu T, Shu H, Zhang X, Qu K, Jiang L (2018) Democratic joint operations algorithm for optimal power extraction of PMSG based wind energy conversion system. Energy Convers Manage 159:312–326. https://doi.org/10.1016/j.enconman.2017.12.090
31. Sravanthi P, Sandeep V (2022) Solving realistic reactive power market clearing problem of wind-thermal power system with system security. Int J Emerg Electr Power Syst 23(2):125–144. https://doi.org/10.1515/ijeeps-2021-0060
32. Salkuti SR (2022) Modeling of various renewable energy resources for smart electrical power systems. In: Salkuti SR, Ray P (eds) Next generation smart grids: modeling, control and optimization. Lecture notes in electrical engineering, vol 824. Springer, Singapore. https://doi.org/10.1007/978-981-16-7794-6_2

Unit Template-Based Control Algorithm for Control of Voltage and Frequency in Solar-Hydro System

Rajagopal Veramalla, Divya Gundeboina, Subhash Bochu, and Surender Reddy Salkuti

Abstract This chapter introduces a unit template control algorithm that uses less number of sensors as compared to various control algorithms of variable frequency control (VFC) off-grid asynchronous generators (IAG). The IAG is run by an uncontrolled hydro turbine for power generation integrated with a solar photovoltaic PV system with consumer loads. Solar PV system consists of a DC-DC converter, and solar panels with a maximum power point tracking (MPPT) controller. The VFC with a four-leg IGBT converter and storage battery on a DC link with a connected point of common coupling (PCC). The four-leg IGBT converter is utilized for suppressing neutral current. The solar PV panels with DC-DC converter operating with MPPT algorithm charged battery on DC link of VFC. The control algorithm computes reference source currents swiftly based on controller gains. Unit templates require less execution and delay time of reference source currents. The proposed system is modeled and simulated in MATLAB/Simulink using the Simpower system toolbox. The simulation results show an off-grid asynchronous generator with a solar PV system performs neutral current compensation, harmonics diminution, load balancing, frequency, and voltage control.

Keywords Hydro-solar PV system · Asynchronous generator · Point of common coupling · Battery energy storage system · Maximum power point tracking · Voltage and frequency controller · Maximum power point tracking

---

R. Veramalla · D. Gundeboina · S. Bochu
Department of Electrical and Electronics Engineering, Kakatiya Institute of Technology and Science, Warangal, India
e-mail: vrg.eee@kitsw.ac.in

D. Gundeboina
e-mail: m20pe004@kitsw.ac.in

S. Bochu
e-mail: subhash.eee@kitsw.ac.in

S. R. Salkuti (✉)
Department of Railroad and Electrical Engineering, Woosong University, Daejeon, Republic of Korea
e-mail: surender@wsu.ac.kr

S. R. Salkuti (ed.), *Artificial Intelligence for Integrated Smart Energy Systems in Electric Vehicles*, Lecture Notes in Electrical Engineering 1427,
https://doi.org/10.1007/978-3-031-94276-1_31

1 Introduction

With rising global energy demand, high prices for fossil fuels, and the inability to reach consumers in remote and isolated areas, small isolated producing units are becoming increasingly necessary. Non-conventional technological innovations are increasingly popular in this area since they provide highly economical, reliable, and clean energy to various industries using next-generation technologies [1]. A self-contained single-stage PV-hydro-battery microgrid MG is used to meet the demand for power in rural areas using renewable energy sources [2]. Tischer et al. highlighted the importance of the hybrid hydro-solar PV system power company will create systems with higher power ratings, ranging from 1 to 10 MW, managed at very low unit costs per generated kWh and require less overall maintenance over the next ten years. An advanced new hybrid topology for voltage regulating in micro hydropower plants has been created with self-excited IG in 3-phase four-wire systems [3–5]. Reference [4] describes about asynchronous generators which are more durable, brushless systems than synchronous generators, thus they are smaller and require less maintenance. In the generation of isolated power from uncontrolled hydro turbines, inaccessible asynchronous generators (IAG) have emerged as prominent competitors. IAGs provide many benefits, but they also have challenges with voltage and frequency profiles [6]. The majority of RES standards were created to ensure that these resources performed effectively in grid and islanded modes, as well as compliance with power quality standards [7, 8]. PV solar systems, diesel generators, small, Pico hydro model, and battery model banks with bidirectional inverter feed 3-phase 4-wire critical, non-critical load makeup microgrid [9–11]. Reference [12] described banks of capacitors as an asynchronous generator with a capacitor bank utilized in hydro solar PV system-producing plants that might combine to fulfill the reactive power requirement. The purpose of integrating a delta capacitor-connected bank in the distribution system is to progress system performance; the accuracy of the procedure is unaffected by the size of the capacitor bank. Reference [13] explained how to use solar energy to improve power quality and control a microgrid's voltage and frequency Solar with three-leg VSC, interface inductor, ripple filter at PCC, and maximum power point tracking MPPT controller. The microgrid's energy storage system that uses batteries, the BESS is now one of the most affordable available technologies for storing less energy [14, 15].

In the standalone mode for power system applications, a DC link connects a battery to a buck-boost converter to maintain power balancing and a VSC is to maintain the system's terminal voltage and frequency optimal values [16, 17]. Reference [18] state that this study uses the DC-DC bidirectional converter control technique for non-linear loading, MPPT of solar photovoltaic arrays, and DC voltage management. The control method analyzes distorted active, reactive component currents at full load current with effectively decreasing harmonics, compensating for reactive power, and regulating the voltage at PCC [19, 20]. The waveform qualities for three-phase 4-leg VSI output voltages are investigated beneath an unbalanced non-linear load. IGBTs are used to provide a neutral leg that is beautifully combined with the usual phase legs [21, 22]. A boost converter connects the solar photovoltaic array radiation to an innovative H-Bridge voltage source converter with AC-DC-AC conversion using IGBTs as the operating principle, control mechanisms investigated with a simplified PWM approach with redundant switching combination processor, and regulation and balancing of the capacitors [23, 24]. Reference [25] explains hybrid systems promote the value of current electricity by utilizing renewable energy sources; nevertheless, due to the hybrid hydro and solar energy's extremely intermittent nature, a reliable controller must be connected to the distant system. Reference [26] proposed a description of power converters that also feature filters for power buffering and filtering, and the DC-DC converter is used for transformer output. Reference [27] passed on the hybrid system's technology including a hierarchical grid connection electro-chemical storage subsystem, local network loads, and low-power systems. People were urged to express their concerns regarding the expenses, electricity supplied to local consumers by solar, and hybrid hydro systems, as well as a sub-criteria related to battery lifetime. Reference [28] described based on the literature future Hydro, wind, and solar energy conversion systems that were employed to electrify rural and railway transport and for a study of different types of energy storage elements used in remote electricity initiatives. Reference [29] gave the future scope of this proposed framework inter-connecting pico-hydro driven synchronous generators with a solar PV-based energy conversion system with bidirectional flow DC-DC converters with good MPPT, which is implemented due to its cost-effectiveness and reliability. Chaplin et al. gave the knowledgeable technical issues for future energy gain through the case study which investigates the various methods for achieving the green energy target set forth in the Environment, Climate Emergency Policy Statement which states that all sector operations must have carbon net zero influence in environmental gain by 2030 with a goal of being carbon net zero by 2050 [30].

2 System Configurations and Its Operation Principle

The proposed hybrid hydro-solar generating system (HSGS) connected to the consumer loads employing four-leg VSC for PQ improvement is shown in Fig. 1. The proposed HSGS contains an asynchronous generator with a Δ-connected excited system, VFC along with the battery on the DC bus, solar PV panel, DC-DC converter, interface inductors, and consumer loads. The VFC used a unit template algorithm to estimate the reference currents. The major goal of the MPPT algorithm for the converter is to keep output voltage optimal at 800V at all times on the DC link. The VFC's battery absorbs active power when the load is less than the generated power and delivers active power when the load is greater than the generated power, thereby controlling frequency. Four legs VSC compensates load neutral current. The rating of the excitation capacitor bank connected in delta is designed to create rated voltage during the no-load situation, while VFC regulates reactive power required by loads, thereby controlling terminal voltage.

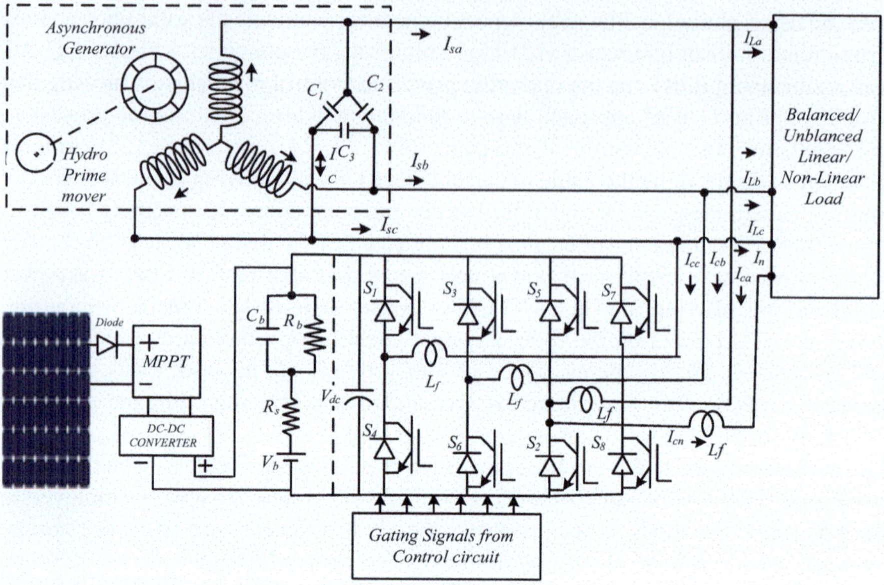

Fig. 1 Schematic four-leg VSC for VFC Hydro-Solar system

The main objectives of VFC are load balancing, harmonic elimination, reactive power generation, and integration of solar power into hydropower. The appendix contains the data for the hydro-solar PV system, and the next sections contain information on choosing the settings for the DC-DC converter, interface inductors, DC bus capacitors, and solar PV panels.

3 Control Algorithms

The next sections explain how the algorithm derived mathematical equations to estimate reference source currents.

3.1 Fundamental In-Phase Active, Reactive Components

Figure 2 shows the unit template algorithm for estimating and extracting the active and reactive fundamental source currents. The extraction of fundamental reference source currents needs in-phase components and quadrature components. 3-phase voltages v_a, v_b, v_c are sensed terminal voltage magnitude (V_t) is given as,

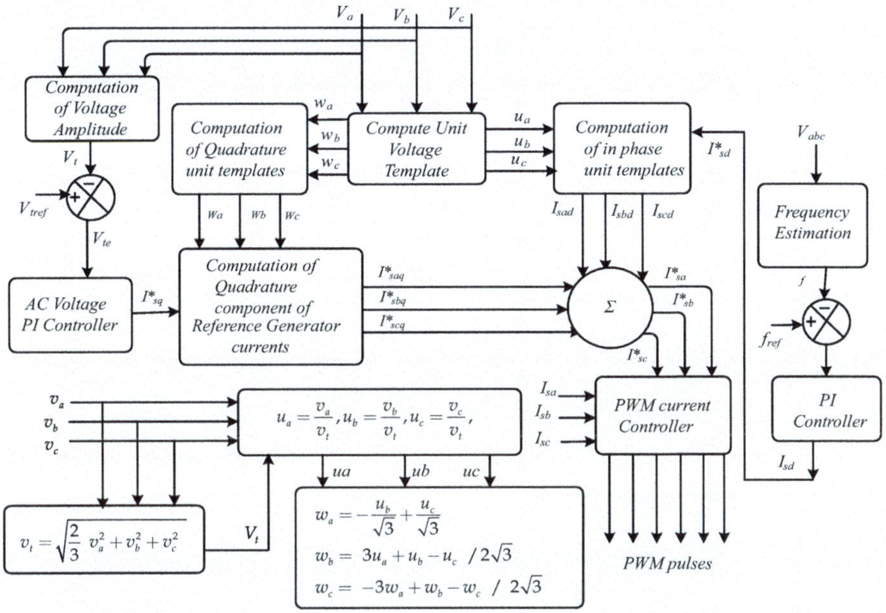

Fig. 2 Unit template control algorithm

$$V_t = \sqrt{\frac{2}{3}V_a^2 + V_b^2 + V_c^2} \tag{1}$$

The in-phase unit-templates (u_a, u_b, u_c) are derived as,

$$\{W_a\} = \left\{\frac{V_a}{V_t}\right\}, \{W_b\} = \left\{\frac{V_b}{V_t}\right\}, \{W_c\} = \left\{\frac{V_c}{V_t}\right\} \tag{2}$$

where quadrature unit templates (w_{qa}, w_{qb}, w_{qc}) are derived from in-phase unit templates (u_a, u_b, u_c) and calculated by the following expressions.

3.2 *Estimation of Frequency and Direct Axis Reference Source Currents*

The suggested algorithm is utilized to estimate the system frequency and individual phase voltage patterns online of HSPVS using every phase sine and cosine unit templates and estimated frequency. In the equation, quadrature, and in-phase unit patterns are used to determine the voltage frequency as,

$$[\omega_s] = \left[\text{Cos}\theta \times \frac{d}{dt}(\text{Sin}\theta) - \text{Sin}\theta\frac{d}{dt}(\text{Cos}\theta)\right] \quad (3)$$

where ω_s angular frequency rad/s. The cosθ, sinθ templates, system frequency f_s is selected as,

$$\left\{ \begin{array}{l} \text{Cos}\theta = u_{laq} \\ \text{Sin}\theta = u_{lad} \end{array} \right\} f_s = \frac{\omega_s}{2\pi} \quad (4)$$

The frequency error 'f_{err}' is as difference between reference frequency (f_{ref}) and normal frequency (f) written by,

$$[f_{err}(n)] = \{[f_{ref}(n)] - f(n)\} \quad (5)$$

The frequency PI controller output is determined on the nth sample instant and given in the following equation.

$$[i_{sd}(n)] = \{i_{sd}(n-1)\} + k_{pf}\{f_{ref}(n) - f_{ref}(n-1)\} + \{k_{if}\{f_{ref}(n)\}\} \quad (6)$$

where k_{pf}, k_{if} as proportional and integral gains of the frequency PI controller. The in-phase reference source currents are calculated below.

$$i^*_{sad} = i^*_{sd}u_a,\, i^*_{sbd} = i^*_{sd}u_b,\, i^*_{scd} = i^*_{sd}u_c \quad (7)$$

3.3 Estimation of Quadrature Axis and Reactive Power of Reference Source Currents

Voltage error (v_{te}) is the difference of V_t and V_{tr} and is given as,

$$[V_{te}(n)] = [V_{tr}(n) - V_t(n)] \quad (8)$$

The voltage error v_{te} given to PI controller (frequency), output I_{vq} at nth sampling instant is written as,

$$\left[i^*_{sq}(n)\right] = [\left\{i^*_{sq}(n-1)\right\} + k_{pv}\{V_{te}(n) - V_{te}(n-1)\} + \{k_{tv}V_{te}\{f_{ref}(n)\}\} \quad (9)$$

where k_{pv}, k_{iv} are proportional and integral gains of the voltage PI controller. The quadrature reference source currents are calculated as,

$$i^*_{saq} = i^*_{sq}w_a,\, i^*_{sbq} = i^*_{sq}w,\, i^*_{scq} = i^*_{sd}w_c \quad (10)$$

3.4 Estimation of Reference Source Currents

The reference source currents are a combination of direct axis and quadrature axis reference source currents which are estimated as,

$$\left(i^{*}_{saq}\right) = \left[i_{sap} + i_{saq}\right], \left(i^{*}_{sbq}\right) = \left[i_{sbp} + i_{sbq}\right], \left(i^{*}_{scq}\right) = \left[i_{scp} + i_{scq}\right] \quad (11)$$

3.5 PWM Signal Generator

From Eq. (11), we acquire 3-phase reference source currents (i^{*}_{sa}, i^{*}_{sb} and i^{*}_{sc}), which are compared with sensed source currents (i_{sa}, i_{sb} and i_{sc}), and errors compared with 10 kHz frequency triangular carrier to provide gate switching pulses to IGBT switches of VSC.

3.6 The Neutral Current Compensation

Four-leg of VFC is used to suppress and compensate neutral wire current. In order to realize the source neutral path current is sensed and compared reference neutral current ($i^{*}_{sn} = 0$). The error-neutral current is utilized to generate switching pulses for the fourth leg of VSC.

4 System Design

An asynchronous generator, capacitor bank, four-leg VSC, interface inductors, DC bus capacitor, MPPT-based DC-DC converter, and battery at the DC link of VSC are part of the proposed HSPVS. This section covers the design part with the selection of such components. The battery, DC bus capacitors, filtering inductors, and VSC design are in the following sections.

4.1 The Design of Electrical System

The electrical system is as 7.5 kW, 415 V, 50 Hz, star (Y) connected 4-Pole IG with Δ-connected excited capacitor to generate an optimal rated voltage on no load condition. In order to meet max rated voltage at 4.6 kVAR capacitor banks are utilized.

4.2 The Design of Battery

The capacitor is used to mimic the kilowatt-hours (kWh) energy storage element. Because battery capacity is reliant on the load-connected type, a low-capacity battery bank may have an impact on HSPVS reliability. We chose a big storage capacity element to enable greater contingency handling despite higher initial investment, space, and maintenance issues. The battery storage capacity determined by the capacitance C_b, is the same as the battery storage level, mini VSC DC bus link VFC voltage must be greater than twice the phase voltage peak value for appropriate PWM control, and derived the capacitance can be found out using the following equation and battery DC bus voltage determined using the given equation.

$$V_b = \left\{ \frac{\left(\sqrt[2]{2}\right) V_{ph}}{m} \right\} \tag{12}$$

$$C_b = \left\{ \frac{\text{kWH} * 3600 * 10^3}{0.5\left(V_{\text{Max}}^2 - V_{\text{Min}}^2\right)} \right\} \tag{13}$$

The battery model is the combination of the R_B and C_B in parallel and is used to identify the stored energy capacity of 700 V and to define the open circuit voltage when the battery is completely charged or entirely discharged.

4.3 Filtering AC Interface Inductor and RC Filter Design Model

The developed model and the values of the rating AC interface inductor coil and RC Filter are determined by the allowable ripple current I_{crpp}, the DC bus voltage V_{dc}, and VSC switched frequency fs. Based on these parameters, a tuned LPF on half switched frequency was used to filter a VFC higher frequency switched ripples components on PCC by selecting a VSC of VFC switching frequency 10 kHz, the RC ripple filter capacitor was found to have 5μF in series connected 5Ω resistance to have fundamental voltage. The current through the inductive filter for 5% ripple, the inductive filter inductance (L_f) can be written as,

$$L_f = \left\{ \frac{\left(\sqrt{3}\right) m V_b}{\left(12 a f_m i_{pp}\right) I_{sc}} \right\} \tag{14}$$

where m-modulation index, overload factor 'a' considered m = 1 beneath lowliest incident, designated dc bus voltage V_{dc}, $V_b = 700$V, 120% of current rating a = 1.0 or 1.2 steady-state values taken, VSC ripple current peak–peak i_{pp} = 5% or 10%

VSC currents, switching frequency, $f_m = 10$ kHz, $L_f = 3.4$mH. Round-off value of $L_f = 3.5$mH used for investigation.

4.4 Design DC Bus Capacitors

In a VFC VSC DC capacitor (C_{dc}) value is calculated by the amount of instant energy accessible to VSC throughout transients and the principle of energy conservation used as follows:

$$\frac{1}{2}C_{dc}\left[V_{dc}^2 - V_{dc1}^2\right] = \left\{3V_{ph}(aLI)t\right\} \quad (15)$$

where reference dc voltage V_{dc}, mini voltage dc bus V_{dc1}, an L overload factor, Vph phase voltage, I-phase current, 't' time. Consider mini voltage levels of dc bus, $V_{dc} = 700$ V, I = 14.5A, $V_{dc1} = 680$ V, $V_{ph} = 239.6$ V, t = 350 μs, $a_L = 1.2$, the calculated value $C_{dc} = 2777$ μF and designated 3000μF.

4.5 *Modeling of VFC*

VSC voltage grade (V_{sw}) solid state device (IGBT) could calculated beneath dynamic condition: $V_{sw} = (V_{dc} + V_d)$, where V_d 10% shoot in dc link voltage beneath condition. By using Eq. (16) switched operated voltage rating are Vsw = $415\sqrt{2} + 10\%$ of rated 586 = 645 V. Optimal rated current flows to 4-leg VSC 'I' peak current $\sqrt{2}$I where 'I' flow through VSC considered with safety factor 1.25. The maxi device current may be calculated by using,

$$(I_{sw}) = 1.25\left\{i_{LripplePP} + I_{sPeak}\right\} \quad (16)$$

Therefore, the Solid state Switch (IGBT) 1200 V and 60A are designated to form VSC of VFC. DC bus linkages, proposed VFC controller for a 9 kW 415 V IG includes VSC with BESS. Each phase generator bus connected four leg half-bridge midpoints individually through an AC interface inductor filter to eliminate triplen harmonics present in the proposed system with a zero-sequence current component that are existing loads flows and load neutral path terminal connected to neutral path four leg VSC system, neutral wire path with returned load currents, and neutral current compensation under unbalanced/balanced conditions. The designated AIG data, prime-mover data, four-leg VSC of VFC parameters, battery storage data, Solar panel, MPPT, DC-DC converter, and all other specifications are given in the appendix.

5 Results and Discussion

The recommended and planned designed VFC HSPVS AIG investigated and assessed its performance in terms of frequency, terminal voltage control, neutral path current compensation, and power factor correction in balanced, unbalanced 3Ø 4W loads. The performance HSPVS UT-based VSC VFC investigated and operated for various consumer loads applications such as domestic lighting, water pumping, and so on, of remote area power supply with good energy efficiency and load as well as the generator reactive power demand generated by VSC with reduced VSC rating by using delta connected capacitor bank at load bus. For the modeled manufactured 7.5 kW capacity HSPVS, 50Hz prime mover turbine driving AIG set, the control algorithm's optimum gains were discovered, used, and examined. The simulated investigation results were validated, verified, and tabularized for 3Ø 4W variable linear, non-linear loads. For balanced loads, the performance of AIG was evaluated. Figures 3 and 4 demonstrate the numerous linear, non-linear load waveforms for 1.72 s to 1.92 s, and the results are documented and these waveform graphs are described as a v_s and i_s source generator voltage, current respectively, different load current i_L, controller current i_c, instantaneous active power demand Pg, battery power Pb, load power PL, terminal, voltage Vt and v_b and i_b battery voltage, currents, neutral path terminal source current i_{sn}, load neutral wire terminal current i_{Ln} and system frequency cost providing and enlightened in further sections.

5.1 Performance Unit Template Synchronous Reference Frame Theory for Linear Loads

Balanced/unbalanced linear loads as Fig. 3 illustrates how the VF controller performs on IAG powered with Pico hydro turbine. A 3.5 kW IAG feeds each of the three single-phase loads. The battery bridges the power supply gap to load. 1.7 s load is switched off then IAG starts to supply surplus power to charge the battery. The second phase load is released at 1.72 s and IAG uses additional power to charge the battery. All 3 phase loads were detached at 1.9 s then IAG fed all power to the battery. The transformer with unbalanced currents, compensates for the load's neutral current and then the VFC maintains source generator 3Ø currents balanced with detected VSC switching ripples free by excitation capacitor currents, and the load neutral path current is connected through four leg system neutral path terminal, detected voltage, and frequency is exactly maintained fixed constant while load switching THD's percentage is within IEEE-519 limitations and developed load power is transferred to the battery to keep supply frequency constant.

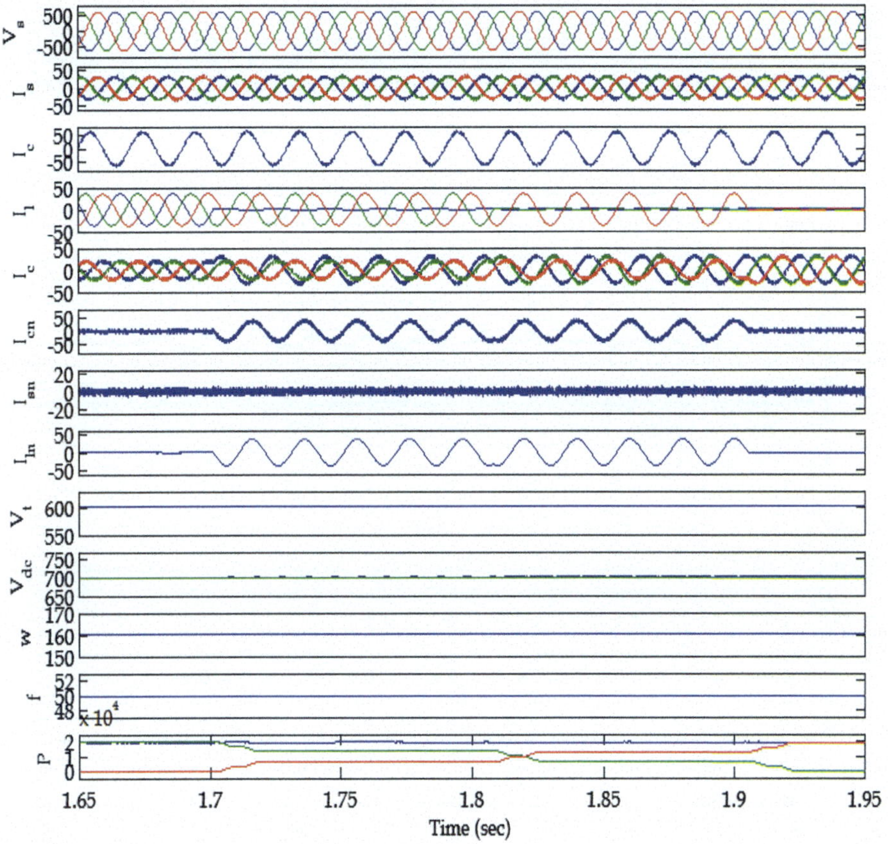

Fig. 3 Performance of Linear Load waveforms

5.2 *Performance Unit Template Synchronous Reference Frame Theory for Non-Linear Loads*

Today's nonlinear loads like power electronics devices, drives, changing speed drives, industrial drives, energy-efficient lighting loads, pull nonlinear currents from the source, which are verified by VFC when used with nonlinear loads. In the proposed HSPVS for voltage regulation, harmonics eradication, load leveling, VFC evaluated for 8.7 kW diode bridge rectifiers as non-resistive inductive load with circuit breaker and DC bus link. The AIG is driven by constant power and it performance of the VF controller when balanced or unbalanced nonlinear loads are applied. All 3 single phase loads are motivated at 1.95 s.

Figure 4 illustrates VFC performance balanced/ unbalanced non-linear loads with 3diode bridge rectifier R-L loads and 3 single phase LC filter diode rectifiers as non-linear loads with parameter values L = 1 mH, C = 470 F, and Rl = 25 connected

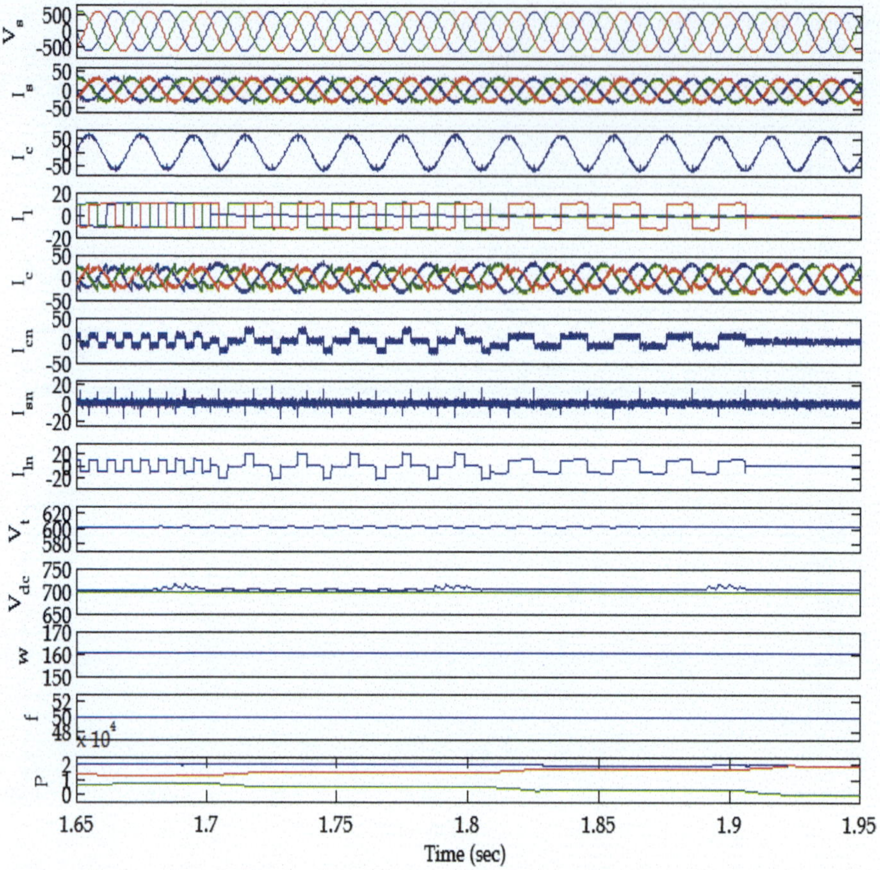

Fig. 4 Performance waveforms with non-linear loads

among every phase, neutral path wire. When each load is unplugged for 1.7s, the loads turn deranged, but terminal voltage V_t, and f-frequency remain closely constant just like in the linear load. The four-leg VSC arrangement provides the neutral path wire current. Consumer load disconnects the generator which produces a balancing current and starts charging the battery after 1.75 s.

In accordance with the IEEE 519 standard's 5 percent harmonic reduction requirements, the generator voltage THD at PCC is 0.70 percent, the current THD is 4.81 percent, and the load current THD is 46.05 percent. During an unbalanced load state, neutral load current i_{Ln} is recompensed through 4-leg VSC VFC, neutral current i_{sn} zero. In Fig. 5, the generator current is THD 0.70% and the load is THD 46.05%. DC bus link voltage v_{dc} is regulating reference value 800 V for all load turbulences. For load turbulences, the PCC voltage remained close to the base value of 339 V. The unit template method with generation reduces both execution time and phase angle calculation time un-uniformity, which are two of the standard PLL disadvantages.

However, these two can be eliminated, and the estimated angle between voltage and current is maintained uniformly throughout the system's operation.

6 Conclusions

This chapter constructed and modeled a simple standing uncontrolled hydro Solar PV system with MPPT powered generator IAG along with VF controller and Its performance under various loading situations like balanced/unbalanced, linear/nonlinear were simulated and demonstrated. VF controller has been found to provide satisfactory performance under a variety of loading scenarios. The four-legged VSC system's performance is proven as acceptable with 3phase 4wire consumer loads and enhanced system power quality problems like voltage and frequency sustained constant by PI controller's ideal two gains Ki, Kp which are obtained and neutral path current compensation. The recommended proposed unit templates sinθ and cosθ generator are very basic with fewer execution times, small phase angle calculation times, and shorter algorithm time delays than normal PLL obstacles. UT battery-based VSC VFC system performance is well-organized and generous delivering the best operational function parameters like load level, load balance, harmonic elimination, and neutral path currents compensators are done based on extended test results. The simulated results are shown, evaluated, documented, and realized in entire system operation modes without any delay time.

Fig. 5 System waveforms and harmonic spectrum

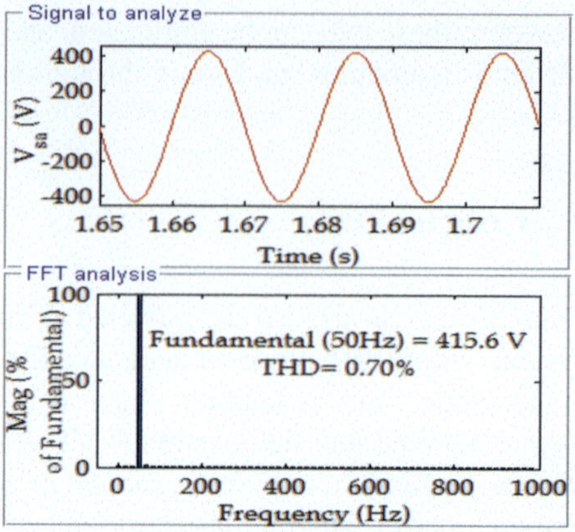

(a) %THD Source Voltage

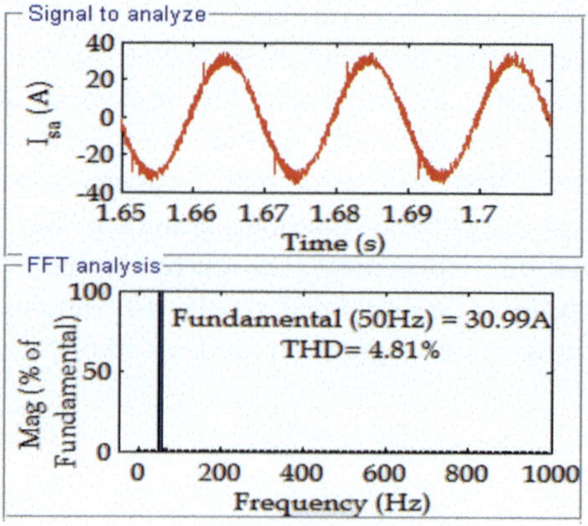

(b) %THD Source Current

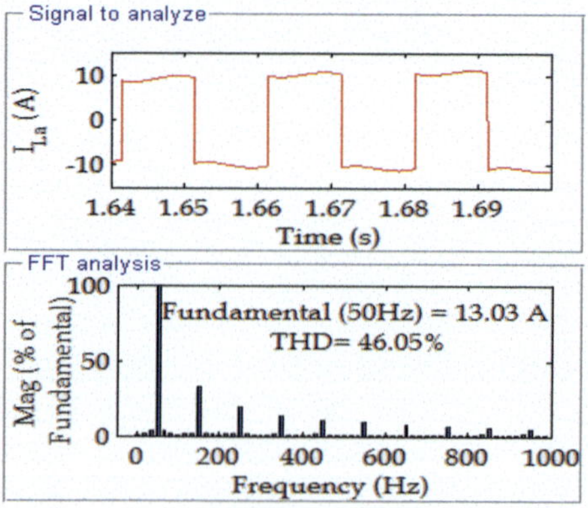

(c) %THD Load current

Acknowledgements This research work was supported by "Woosong University's Academic Research Funding – 2025".

Appendix

Parameters 7.5 kW Rating, 415 V, 50 Hz, Y Connected, 4 Pole Asynchronous Generator

$R_s = 1\ \Omega$, $R_r = 0.77\ \Omega$, $X_{lr} = X_{ls} = 1.5\Omega$, $J = 0.1384\ kg - m^2$, $L_m = 0.134$ H ($I_m <$ 3.16),

$L_m = 0.068$ H ($I_m > 12.72$), $L_m = 9e^{-5}\ I_m{}^2 - 0.0087\ I_m + 0.1643$ ($3.16 < I_m <$ 12.72).

Controller Parameters

$C_{dc} = 3000\ \mu F$, $L_f = 3.5$ mH, $R_f = 0.1\ \Omega$, Frequency PI controller gains $K_{pd} = 0.01$, $K_{id} = 0.01$, PI controller gains $K_{pa} = 0.02$, $K_{ia} = 0.02$.

Prime Mover and Its Physical Characteristics

$T_{shaft} = (K_1 - K_2\ \omega_m)$, where $K_1 = 1465$, $K_2 = 8.8$

The Battery Description

$R_{sb} = 5\ k\Omega$, $R_b = 0.1\ \Omega$, $C_b = 41326$ F, finally the $V_{oc} = 700$ V.

References

1. Tephiruk N, Kanokbannakorn W, Kerdphol T, Mitani Y, Hongesombut K (2018) Fuzzy logic control of a battery energy storage system for stability improvement in an islanded microgrid. Sustainability 10(5):1645. https://doi.org/10.3390/su10051645
2. Singh B (2010) Isolated wind energy conversion with asynchronous generator for rural electrification. IEEE Joint Int Conf Power Electron Drives Energy Syst, pp 1–7. https://doi.org/10.1109/PEDES.2010.5712389

3. Singh BN, Rastgoufard P, Singh B, Chandra A, Al-Haddad K (2004) Design, simulation and implementation of three pole/four pole topologies for active power filters. Proc IEE- Electr Power Appl 151(4):467–476. https://doi.org/10.1049/ip-epa:20040209
4. Rezkallah M, Singh B, Kasal GK (2008) Solid state voltage and frequency controller for a stand-alone wind power generating system. IEEE Trans Power Electron 23(3):1170–1177. https://doi.org/10.1109/TPEL.2008.921190
5. Rahman MM, Shakeri M, Tiong SK, Khatun F, Amin N, Pasupuleti J, Hasan MK (2021) Prospective methodologies in hybrid renewable energy systems for energy prediction using artificial neural networks. Sustainability 13(4):2393. https://doi.org/10.3390/su13042393
6. Salameh ZM, Casacca MA, Lynch WA (1992) A mathematical model for lead-acid batteries. IEEE Trans Energy Convers 7(1):93–98. https://doi.org/10.1109/60.124547
7. Kumar JV, Sesham H, Salkuti SR (2024) Smart energy management model for electric vehicles. In: Salkuti SR (eds) Energy and environmental aspects of emerging technologies for smart grid. Green Energy and Technology. Springer, Cham. https://doi.org/10.1007/978-3-031-18389-8_9
8. Goswami AK, Goswami DJ, Behera C, Reddy GH, Chakrapaani P (2017) System voltage and frequency control using DFIG based wind energy conversion system. IEEE IEEE region 10 annual international conference proceedings (TENCON), pp 766–771. https://doi.org/10.1109/TENCON.2016.7848107
9. Sravanthi P, Sandeep V (2022) Solving realistic reactive power market clearing problem of wind-thermal power system with system security. Int J Emerg Electr Power Syst 23(2):125–144. https://doi.org/10.1515/ijeeps-2021-0060
10. Salkuti SR (2022) Modeling of various renewable energy resources for smart electrical power systems. In: Salkuti SR, Ray P (eds) Next generation smart grids: modeling, control and optimization. Lecture Notes in Electrical Engineering, vol 824. Springer, Singapore. https://doi.org/10.1007/978-981-16-7794-6_2
11. Nirmal MCM, Sruthi M, Jayaprakash P (2021) Control of a pumped hydro storage power plant supported solar PV generation system for grid-side energy management. In: IEEE 4th international conference on computing, power and communication technologies (GUCON), Kuala Lumpur, Malaysia, pp 1–6. https://doi.org/10.1109/GUCON50781.2021.9573546
12. El-Ela AAA, El-Sehiemy RA, Abbas AS (2018) Optimal placement and sizing of distributed generation and capacitor banks in distribution systems using water cycle algorithm. IEEE Syst J 12(4):3629–3636. https://doi.org/10.1109/JSYST.2018.2796847
13. Sharma R, Seema, Singh B (2018) Riemann-liouville operator based FNLMS algorithm for 3 phase- 4 Wire SyRG-PV-BES based microgri. In: 8th IEEE India international conference on power electronics (IICPE), Jaipur, India, pp 1–6. https://doi.org/10.1109/IICPE.2018.8709420
14. Howlader AM, Matayoshi H, Sepasi S, Senjyu T (2018) Design and line fault protection scheme of a DC microgrid based on battery energy storage system. Energies 11(7):1823. https://doi.org/10.3390/en11071823
15. Salkuti SR, Ray P, Pagidipala S (2022) Overview of next generation smart grids. In: Salkuti SR, Ray P (eds) Next generation smart grids: modeling, control and optimization. Lecture notes in electrical engineering, vol 824. Springer, Singapore. https://doi.org/10.1007/978-981-16-7794-6_1
16. Jung CM, Pagidipala S, Salkuti SR (2024) Emerging technologies for the integration of renewable energy, energy storage and electric vehicles. In: Salkuti SR (eds) Energy and environmental aspects of emerging technologies for smart grid. green energy and technology. Springer, Cham. https://doi.org/10.1007/978-3-031-18389-8_5
17. Kewat S, Singh B (2020) Improved reweighted zero-attracting quaternion-valued LMS algorithm for islanded distributed generation system at IM load. IEEE Trans Industr Electron 67(5):3705–3716. https://doi.org/10.1109/TIE.2019.2916301
18. Kalla UK, Singh B, Murthy SS, Jain C, Kant K (2018) Adaptive sliding mode control of standalone single-phase microgrid using hydro, wind, and solar PV array-based generation. IEEE Trans Smart Grid 9(6):6806–6814. https://doi.org/10.1109/TSG.2017.2723845

19. Park S, Salkuti SR (2019) Optimal energy management of railroad electrical systems with renewable energy and energy storage systems. Sustainability 11(22):6293. https://doi.org/10.3390/su11226293
20. Luo Y, Liu C, Yu F, Lee CHT (2017) Design and evaluation of an efficient three-phase four-leg voltage source inverter with reduced IGBTs. Energies 10(4):530. https://doi.org/10.3390/en10040530
21. Luís R, Silva JF, Quadrado JC (2017) Output voltage quality evaluation of stand-alone four-leg inverters using linear and non-linear controllers. Energies 10(4):504. https://doi.org/10.3390/en10040504
22. Ye Z, Chen A, Mao S, Wang T, Yu D, Deng X (2018) A novel three-level voltage source converter for AC–DC–AC conversion. Energies 11(5):1147. https://doi.org/10.3390/en11051147
23. Tiwari S, Kewat S, Singh B (2020) UPQC controlled solar PV-hydro battery microgrid. In: IEEE international conference on power electronics, drives and energy systems (PEDES), Jaipur, India, pp 1–5. https://doi.org/10.1109/PEDES49360.2020.9379764
24. Aluko AO, Dorrell DG, Pillay Carpanen R, Ojo EE (2020) Heuristic optimization of virtual inertia control in grid-connected wind energy conversion systems for frequency support in a restructured environment. Energies 13(3):564. https://doi.org/10.3390/en13030564
25. Chen MH (2015) Use of three-level power converters in wind-driven permanent-magnet synchronous generators with unbalanced loads. Electronics 4(2):339–358. https://doi.org/10.3390/electronics4020339
26. Vlad C, Barbu M, Vilanova R (2016) Intelligent control of a distributed energy generation system based on renewable sources. Sustainability 8(8):748. https://doi.org/10.3390/su8080748
27. Mitrofanov SV, Kiryanova NG, Gorlova AM (2021) Stationary hybrid renewable energy systems for railway electrification: a review. Energies 14(18):5946. https://doi.org/10.3390/en14185946
28. Salkuti SR (2023) Advanced technologies for energy storage and electric vehicles. Energies 16(5):2312. https://doi.org/10.3390/en16052312
29. Chaplin G, Dibaj M, Akrami M (2022) Decarbonising universities: case study of the university of exeter's green strategy plans based on analysing its energy demand in 2012–2020. Sustainability 14(7):4085. https://doi.org/10.3390/su14074085
30. Salkuti SR (2022) Emerging and advanced green energy technologies for sustainable and resilient future grid. Energies 15(18):6667. https://doi.org/10.3390/en15186667

Overview of Electric Vehicle Management Using Artificial Intelligence in Smart Energy Systems

P. Suresh Kumar, B. Arundhati, Kishore Yadlapati, and Surender Reddy Salkuti

Abstract In 1834, the first non-rechargeable battery-operated electric vehicle (EV) was built by scientist Devenport. A new revolution in electric vehicles (EVs) has emerged after the invention of lead-acid rechargeable batteries. In India, the first EV was launched in 2002 which is low power and based on separately excited DCM. Globally as the population is increasing the usage of internal combustion (IC) vehicles can result in severe air pollution. So globally there is a revolutionary increase in the electrification of transportation systems and the emerging technologies in EVs. However, this chapter discusses new trends and emerging technologies of EVs integrated with smart energy systems and optimized by artificial intelligence (AI) solutions. This work looks into the issues of integrating EVs with smart energy systems and provides the enhanced applications of integrated into EVs along with smart energy systems.

Keywords Artificial intelligence · Electric vehicles · Machine learning · Hybrid electric vehicles · Plug-in hybrid electric vehicles · Battery electric vehicles

---

P. Suresh Kumar
Department of Electrical and Electronics Engineering, Jawaharlal Nehru Technological University, Kakinada, Andhra Pradesh, India

B. Arundhati
Department of Electrical and Electronics Engineering, Vignan's Institute of Engineering for Women, Visakhapatnam, Andhra Pradesh, India

K. Yadlapati
Department of Electrical and Electronics Engineering, JNTUK University College of Engineering, Narasaraopet, Andhra Pradesh, India
e-mail: kyadlapati@ieee.org

S. R. Salkuti (✉)
Department of Railroad and Electrical Engineering, Woosong University, Daejeon, Republic of Korea
e-mail: surender@wsu.ac.kr

S. R. Salkuti (ed.), *Artificial Intelligence for Integrated Smart Energy Systems in Electric Vehicles*, Lecture Notes in Electrical Engineering 1427,
https://doi.org/10.1007/978-3-031-94276-1_32

Nomenclature

EVs	Electric vehicles
AI	Artificial intelligence
ML	Machine learning
HEV	Hybrid electric vehicles
BEV	Battery electric vehicles
PHEV	Plug-in hybrid electric vehicle
ICE	Internal combustion engine
PCU	Power control unit
B	Battery
G	Generator
M	Motor
F	Fuel
E	Internal combustion engine
P	Power converter
T	Transmission system
DCM	DC motor
REV	Revolutionary electric vehicles
V2G	Vehicle to grid
SOC	State of charge

1 Introduction

Due to the extinction of fossil fuels globally and the adverse effect of fossil fuels on climate change, several countries have initiated plans for reducing CO_2 and carbon emissions [1–3]. The major reason is the emission of these gases by the mode of transportation. Thus, electric cars (EVs) are gaining popularity as a viable way to lessen the impact on environmental problems. EVs are easy to maintain, have less noise, zero carbon emission, etc. However, there are several difficulties while integrating EVs with electrical grid systems such as grid capacity and stability, Peak demand management, infrastructure requirements, data security, and privacy [4, 5]. To minimize the effects a smart grid is to be incorporated for the smooth integration of such sources in our energy system. EVs could potentially store energy from the grid (grid to vehicle) and during surplus deficient the EVs supply back to the grid (vehicle to grid). AI plays a vital role in the integration of EVs into energy systems to overcome the difficulty and can enhance EV infrastructures in several ways. AI Technology makes it possible to optimize the various effects that occur during the integration of EVs to the grid and vice versa. The various AI Techniques such as machine learning (ML) [6–8] data analytics, planners can acquire a significant understanding of various adverse effects of electric vehicle consumption and demand.

Algorithms classified as AI display behaviors that are considered intelligent [9–11]. It takes planning, forecasting, modeling, optimization, and other processes to integrate EVs into the grid efficiently. Numerous AI algorithms are made to handle these kinds of tasks. A substantial body of work has been done in the literature using various AI algorithms to address various facets of the integration of energy and transportation systems. This chapter aims to address the integration technology of EVs and the application of AI Techniques. The following are the contributions made to this chapter.

- Overview of EVs.
- To discuss the present scenario and difficulties faced in developing EVs.
- To outline the difficulties of integrating EVs with conventional grid.
- To list the main areas whereby EV energy system integration can benefit significantly from AI techniques.

2 Electric Vehicles

Based on the type of propulsion, energy carriers, and energy sources electric vehicles (EVs) are classified into various types. Therefore, one way to classify EVs is as pure EVs or HEVs based on their propulsion technologies. An electric motor serves as the sole mode of propulsion in a full EV [12, 13]. A hybrid EV (HEV), on the other hand, is propelled by both an ICE and an electric motor. HEVs are referred to as HVs, whereas popularly known pure EVs are called EVs. Pure EVs can be further divided into battery EVs and fuel cell EVs if EVs are organized according to their energy sources. Therefore, a battery electric car employs a battery as its only energy source, but a hybrid uses both liquid gasoline and a battery. A fuel-cell EV, on the other hand, derives its energy from both a fuel cell and a battery. Energy carriers can also be classified similarly, with electricity serving as the energy carrier for batteries and hydrogen serving as the energy carrier for fuel cells [14, 15].

The family of pure EVs includes four different kinds of EVs. They include ultracapacitors EVs, ultra-flywheel EVs, fuel cell EVs, and battery EVs. The battery serves as a common energy source in each of these cars. A battery serves as the only energy source in a battery EV (BEV), however, batteries also function as a hybrid energy source in fuel cell, ultra-capacitor, and ultra-flywheel EVs [16, 17]. Since a fuel cell cannot store renewable energy, the battery in an EV powered by fuel cells is largely utilized to absorb regenerative power. Batteries are needed to store energy in EVs with ultra-capacitor and ultra-flywheel technology since these components have very low specific energies [18]. As a result, a hybrid electric car gets its power from both an ICE and a motor.

Therefore, it is divided into five categories based on the ratio of hybridization between the IC engine and the electric motor. Therefore, the electric motor rating in Small hybrid vehicles is quite low, but in REVs, it is high. Conversely, the IC engine's rating in a micro-hybrid is higher than that of a REV. Thus, traditional HEVs include micro, mild, and full hybrid models [19, 20]. In V2G HEVs include plug-in hybrids, also known as PHEVs, and range-extended hybrids, or REVs. Therefore, the only way these two types differ is that conventional HEVs can only be refueled at gas stations or pumps. Vehicle-to-Grid (V2G) HEVs, however, can be refueled at gas stations as well as electrically. Thus, in addition to being able to refuel with liquid fuels at a filling station, a V2G HEV allows for direct battery charge through charging ports.

The classification of all types of EVs in terms of propulsion devices, energy carriers, and energy sources is elaborated here. So in terms of propulsion systems, a hybrid EV primarily uses an engine and electric motor while a pure EV uses solely electric motors for propulsion. On the other hand, if EVs are to be categorized according to their energy sources, a pure EV would use hydrogen or batteries, whereas a hybrid vehicle would use either electricity or liquid fuel. If EVs are classified based on energy sources, you will see that hybrid vehicle uses either liquid fuels or batteries as an energy source while pure EV uses batteries and other types of energy sources such as ultracapacitors, ultra-flywheel, and fuel cells as energy sources.

2.1 *Battery Electric Vehicles (BEVs)*

A typical BEV system shown in Fig. 1 consists of a battery pack and a motor. It is powered by a battery pack's stored energy. The range of BEVs depends on the sizing of the battery pack as the vehicle's driving mechanism takes all its power from the battery. BEVs have no exhaust system as a result the emissions of harmful gases are zero. A typical BEV's range on a single charge is between 100 and 250 km, depending on the type. The typical BEV configuration is shown here. An electric drive, which consists of an electrical machine and a power converter, transfers the stored electrical energy to the wheels via transmission gears and a differential. Typically, a charging circuitry is used to directly charge the battery bank of a BEV from the grid. Since this power converter can also be utilized to regenerate wheel power during braking, it must be built with bidirectional power flow in mind. As with conventional IC engine-based vehicles, you can also see that a battery-electric car typically does not require a clutch. However, because of their much lower driving range and much longer charging times, battery electric cars (BEVs) are far less favorable than other forms of EVs [21–23].

2.2 *Hybrid Electric Vehicles (HEVs)*

Any two sources of power or energy can be combined in a hybrid vehicle. Fuel cell (FC)/battery, gasoline/flywheel, and diesel/electric are examples of possible combinations. Usually, there are two types of energy sources: storage and fuel conversion. Two independent propulsion systems could be supported by the combination of two power sources. Thus, the car needs to have at least two modes of propulsion to qualify

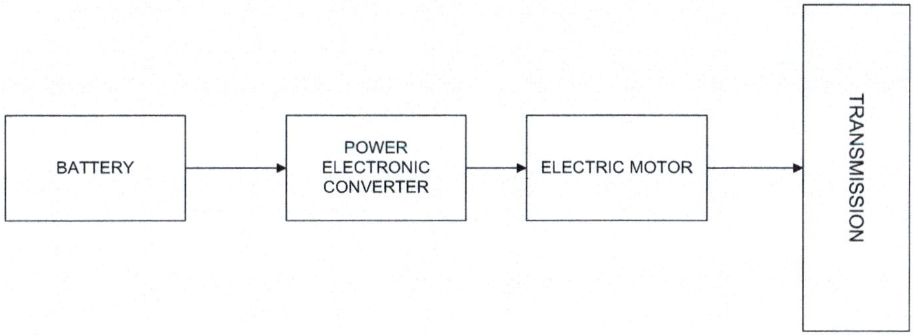

Fig.1 Battery electric vehicle

as a true hybrid [24, 25]. A hybrid electric car combines an electric motor and a gasoline engine, according to the discussion. An electric motor with a diesel engine is an alternative configuration. Parts from pure EVs and pure ICE are combined to create a hybrid electric vehicle (HEV). The regenerative braking feature of an EV is enabled by an M/G placed in the hybrid electric vehicle (HEV). The M/G is situated directly beneath the HEV's engine. In Honda hybrids, the M/G is connected directly to the engine. Next in line is the transmission. The gasoline engine and the M/G in motor mode, M-mode, are the two torque producers in this setup. M/G and the battery are electrically coupled. Based on operation, HEVs are classified as follows:

- Series hybrid electric vehicles
- Parallel hybrid electric vehicles
- Series parallel hybrid electric Vehicles
- Complex hybrid electric vehicles.

2.2.1 Series Hybrid Electric Vehicles

The ICE and generator are coupled in the series HEV observed in Fig. 2 to generate electricity for solely electric propulsion. A generator is used to first transform the mechanical output of a series hybrid system into electricity via the motor and mechanical transmission [26, 27], the converted electricity charges the battery or drives the wheels directly. The benefits of series hybrid drive-trains are:

- The IC engine can operate within a relatively small optimum region when there is mechanical decoupling between the driven wheels and the ICE.
- Multi-gear transmission and electric motors have nearly perfect torque-speed characteristics.

Drawbacks of a series hybrid drive-train:

- The efficiency is decreased since the energy is transformed twice.

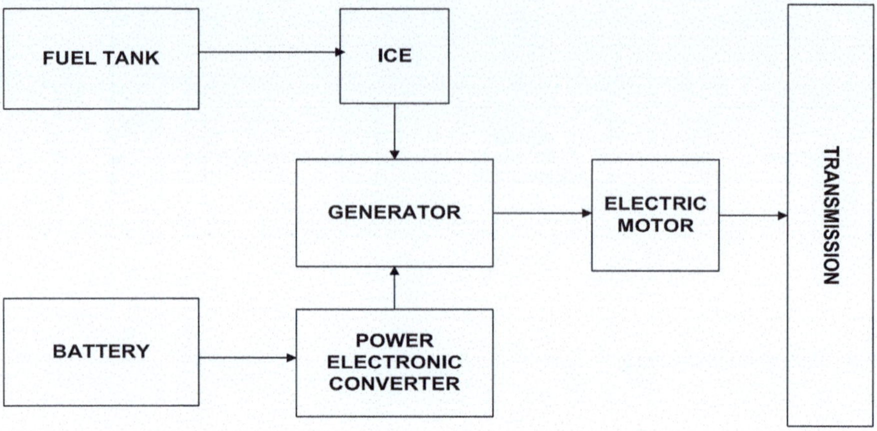

Fig.2 Series hybrid electric vehicle

- It takes two electric machines to move the wheels as the traction motor is the only source of torque for them.

All series hybrid drive-trains are used in heavy-duty trucks. The massive engine/generator system can fit in a larger vehicle, which is the explanation. A power control unit coordinates the power transfer between the engine and the electrical motor in a series hybrid electric vehicle (HEV). The ICE powers a generator in this kind of hybrid architecture, which generates electricity to fuel the electric motor that turns the wheel [28]. Depending on the battery's SOC, the PCU regulates the power flow between the generator, motor, and battery. The drive system is not propelled by the gasoline engine; rather, it produces electricity for battery charging. In conclusion, PCU optimizes the indirect power transfer from engine to wheel in a series HEV.

The functions of a series HEV are as follows:

- The transmission unit and the ICE are not coupled.
- No multi-gear box due to torque–speed characteristics of electric motor (EM).
- Robust speed control technique.

2.2.2 Parallel Hybrid Electric Vehicles

Electric motors (EMs) and internal combustion engines (ICEs) may both supply power to turn the wheels thanks to the parallel HEV depicted in Fig. 3. Because two clutches connect the ICE and EM to the wheel drive shaft, propulsion power can come from either one of the two sources the ICE, the EM, or both [29, 30].

Benefits of parallel hybrid drive-train:

- Since there is no energy from conversion and both the engine and the electric motor directly deliver torques to the driven wheels, there is less energy lost.
- A smaller traction motor and the absence of a generator make it compact.

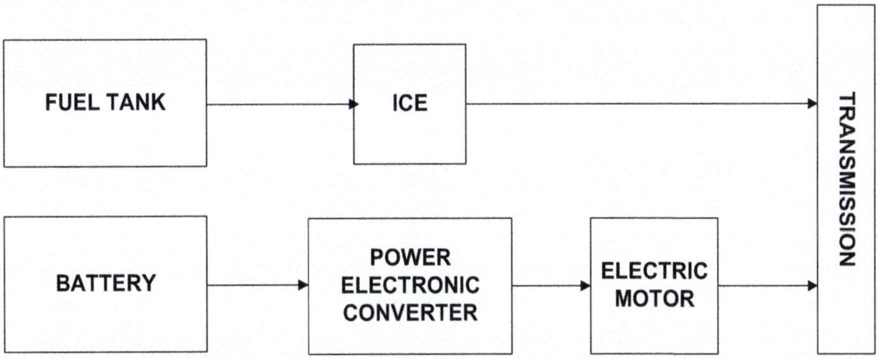

Fig. 3 Parallel hybrid electric vehicles

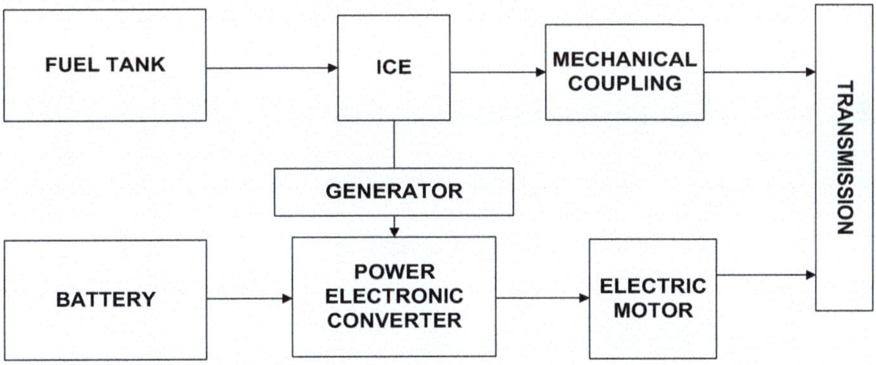

Fig. 4 Series–parallel hybrid electric vehicles

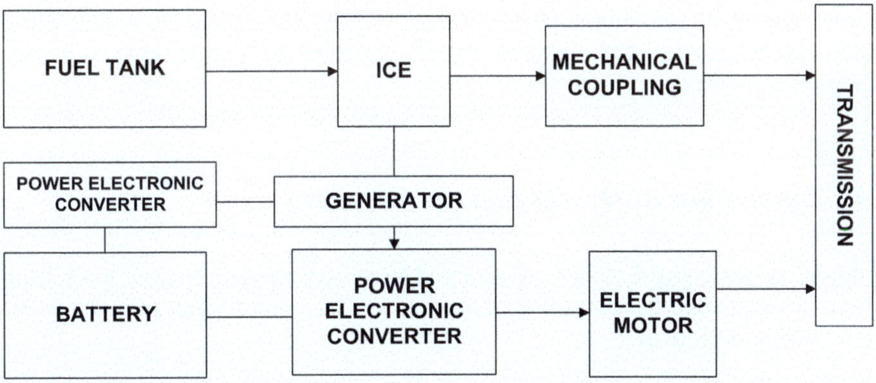

Fig. 5 Complex hybrid electric vehicles

Limitations of parallel hybrid drive-train:

- The mechanical link between the engines and the driven wheels limits the speed range.
- In comparison to a series hybrid drive-train, the mechanical arrangement and the control scheme are more intricate.

Parallel layout is used by little automobiles because of their compact nature. This setup is used by the majority of automobiles for passengers.

A power splitter device (PSD) regulates the power split between the engine and EM in a PHEV. Depending on the drive circumstances, the PSD permits the engine, EM, or both to provide the wheels with power [31]. The majority of the power is sent to the wheels by the EM when the speed and torque are low. The engine provides the majority of the power to the wheels when operating at high speed and low torque. Through the monitoring of many factors like SOC, speed, and load, among others, a hybrid control unit optimizes the power split between the engine and EM.

2.2.3 Series–Parallel Hybrid Electric Vehicles

The series–parallel hybrid configuration combines the characteristics of both series and parallel HEVs. Regretfully, the control of this setup is complicated by the need for a planetary gear unit and an additional electric machine.

2.2.4 Complex Hybrid Electric Vehicles

The configuration of the complicated hybrid system is complex and does not fit into any of the three categories mentioned above. There are parallels between the series–parallel hybrid and the complicated hybrid as both are electric devices—the electric motor and the generator [32, 33]. The primary drawback of complex hybrids is their increased complexity.

2.3 Power Flow in Hybrid Electric Vehicles

Different power control tactics are required to govern the power flow to or from different components because HEV designs differ. The control strategies all achieve the following objectives:

- optimal fuel economy
- low emissions
- low system expenses
- robust driving performance.

Several factors are taken into account when designing power control systems for HEVs, including [34]:

- ***Optimal-ICE operating point:*** The ICE's torque-speed plane's ideal operating point can be determined by maximizing fuel efficiency, minimizing emissions, or striking a balance between the two.
- ***Optimal-ICE operating line:*** An optimal operating line is made up of the appropriate optimal operating points if the ICE is required to supply varying power demands.
- ***Safe battery voltage:*** When regenerative charging, generator charging, or discharging a battery, the voltage may change considerably. Neither the maximum voltage limit nor the minimum voltage limit should be exceeded by this battery voltage.

2.3.1 Power Flow Control in Series Hybrid Vehicles

Based on the power flow depicted in Fig. 6, there are four operating modes for the series hybrid system:

- **Mode 1**: Both the ICE and battery supply electricity to the power converter during regular starting or acceleration of the series HEV. The power converter subsequently powers the electric motor, which in turn powers the wheels through a transmission.

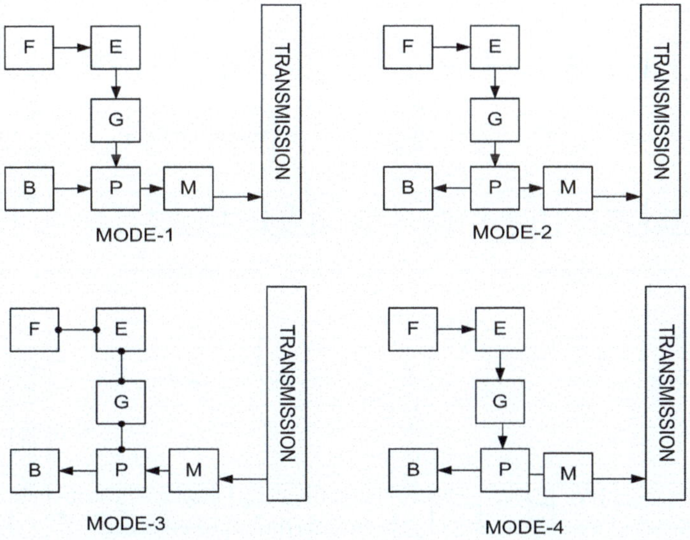

Fig. 6 Power flow control in series hybrid vehicles

- **Mode 2**: The ICE output is more than what is needed to drive the wheels when the load is light. Therefore, to charge the battery, a portion of the electrical energy produced is used. The battery is charged until the battery capacity is at the appropriate level.
- **Mode 3**: The motor works as a generator to power the battery during braking or deceleration.
- **Mode 4**: Even when the car stops completely, the ICE can still charge the battery through the generator.

2.3.2 Power Flow Control in PHEV

There are four ways to operate the hybrid system in parallel. Figure 7 depicts these four ways of action.

- **Mode 1**: The necessary power to move the car forward is shared by the ICE and the EM during startup or full-throttle acceleration. Generally, there is an 80% power distribution on ICE and 20% power distribution on electric motors.
- **Mode 2**: The ICE alone provides the necessary traction power during regular driving, keeping the EM in off mode.
- **Mode 3**: when the vehicle is braking or decelerating. The motor functions as a generator to charge the battery through the power converter.
- **Mode 4**: Under light load conditions the ICE offers traction power and uses the EM to charge the battery.

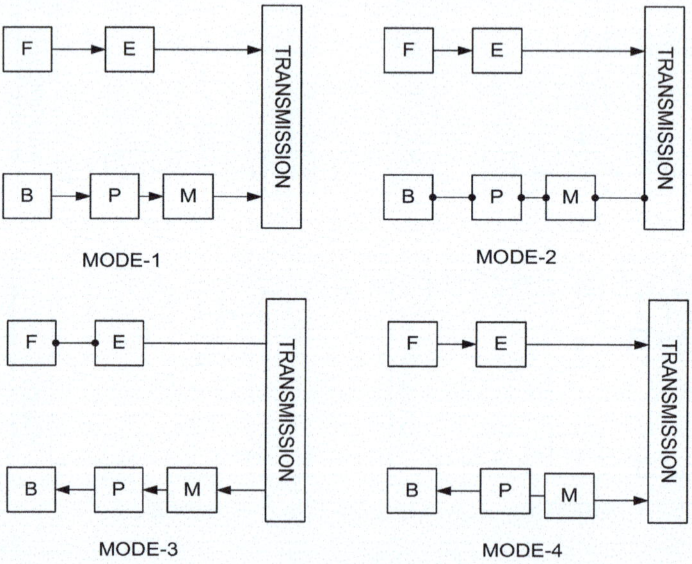

Fig. 7 Power flow control in PHEV

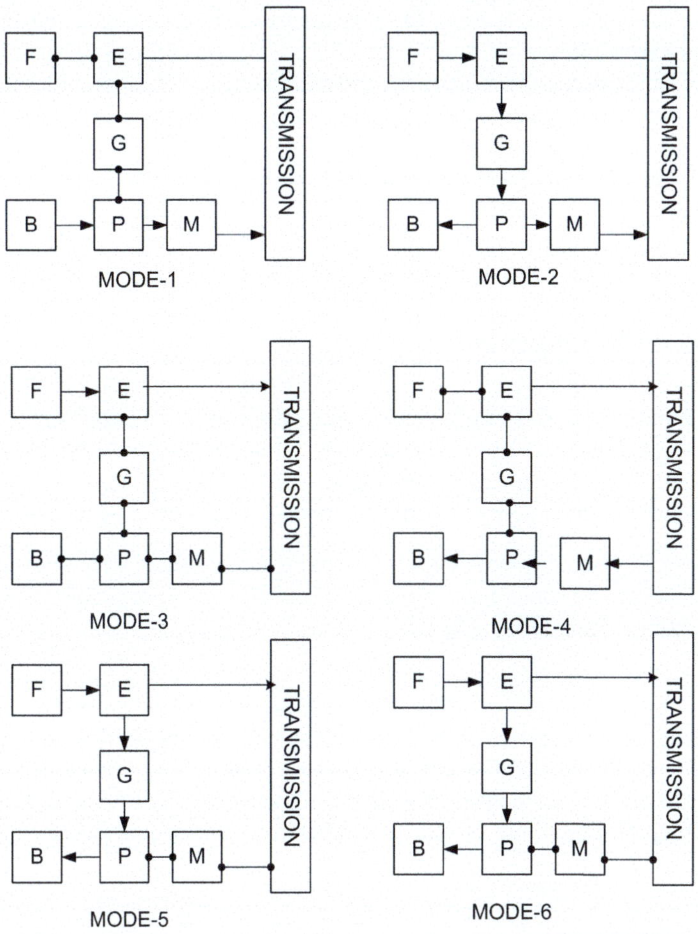

Fig. 8 Power flow control in series–parallel hybrid vehicles

2.3.3 Power Flow Control Series–Parallel Hybrid

The characteristics of both series and parallel hybrid systems are present in the series–parallel hybrid system. The various operating modes of the ICE-dominated system are:

- **Mode 1**: When the automobile first starts up, the ICE is kept in the off mode, and the battery supplies all of the power needed to move the vehicle forward.
- **Mode 2**: The necessary traction power is shared between the ICE and the EM during full-throttle acceleration.
- **Mode 3**: The ICE alone provides the necessary traction power during regular driving, leaving the EM in the off condition.

- **Mode 4**: To charge the battery, the EM functions as a generator during typical braking or deceleration.
- **Mode 5**: When driving, the internal combustion engine (ICE) provides the necessary traction power while simultaneously charging the battery. The EM functions as a generator in this mode.
- **Mode 6**: The ICE can supply power to the battery via the EM while the automobile is at a stop.

3 Artificial Intelligence

Artificial Intelligence (AI) employs diverse algorithms to empower robots to exhibit intelligent behavior, thereby constraining human behaviors. ANNs, Machine Learning, and Deep Learning are some of the subfields of AI. This section focuses on integrating EVs into energy systems and gives an outline of AI approaches.

3.1 Machine Learning (ML)

Machine learning is a branch of artificial intelligence that primarily focuses on simplifying mathematical theoretical models and technical methods so that computers can make decisions automatically according to instructions. The various approaches in ML shown in Fig. 9 are supervised learning, unsupervised learning, and reinforcement learning [24–26]. Table 2 outlines the differences between various approaches in ML.

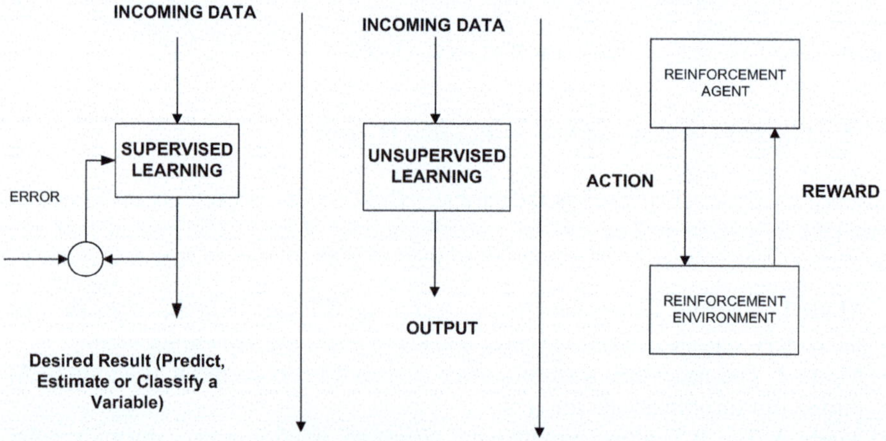

Fig. 9 Various approaches to machine learning

Table 1 Classification of EVs

Type of EVs	Propulsion type	Energy carriers	Energy sources
Hybrid EVs	IC engine and motor	Liquid fuel and electricity	Liquid fuels and battery
PHEVs	IC engine and motor	Liquid fuel and electricity	Liquid fuels and battery
REVs	IC engine and motor	Liquid fuel and electricity	Liquid fuels and battery
Battery EVs	Electric motor	Electricity	Battery
Ultra-flywheel EVs	Electric motor	Electricity	Ultra-flywheel
Ultracapacitor EVs	Electric motor	Electricity	Ultracapacitor
Fuel cell and battery EVs	Electric motor	Hydrogen	Fuel cell

Table 2 Differences between various approaches in ML

Criteria	Supervised ML	Unsupervised ML	Reinforcement ML
Definition	The machine is trained by using labeled data	Without any supervision, the machine is trained using unlabeled data	Agents interact with the environment by performing actions. They are learned by actions and rewards
Type of data	Labeled data	Unlabeled data	No-predefined data
Type of problems	Regression and classification	Association and clustering	Reward and error-based
Supervision	External supervision	No supervision	No supervision
Algorithms	Logistic Regression, NaiveByes, Decision Trees, Linear Regression	K-Mean Clustering, KNN, Principle component analysis, Neural networks	SARSA, Q-learning, and Monte Carlo
Aim	Calculate outcomes	Discover underlying patterns	Acquire knowledge of a sequence of steps
Approach	Maps labeled input to the known output	Understand the pattern and discover the output	The trial and error method is followed
Application	Fault evaluation, Estimation of SOC	Recommendation system, Anomaly Detection	Self-driving cars, Gaming, Healthcare

3.2 Deep Learning

Deep learning is a subset of machine learning that focuses on the algorithms-inspired structure of brain neuron models. The key concepts in Deep learning are ANNs and the various Architecture models in deep learning are CNNs, RNNs, and LSTMs. The adopted methods for training deep learning are backpropagation and stochastic gradient descent [16, 17].

3.2.1 ANN-Based Modeling of EVs

The most promising tools for large-scale, highly uncertain tasks are artificial neural networks (ANNs). For example, ANNs can be used to estimate the traveling behavior of EVs and charge them optimally. Because ANNs learn from historical data they must provide precise input dates to produce the targeted output that is predicted. Historical data is separated into three categories: test, validation, and training. After an ANN has learned the data behavior from training data sets, validation data sets are used to assess the training correctness. Ultimately, the test data set conducts the last evaluation of the ANN's performance following the conclusion of the training and validation phases. By doing this, it can ensure that the ANN performs well with additional data sets [27–30].

The challenge that arises during EV performance informs the design of the ANN structure, therefore to estimate the operational behavior of EVs, it must use a reliable and effective ANN. ANNs with strong forms of neurons, like rough structure-based neurons, which are highly adept at handling input data uncertainty, and Levenberg–Marquardt (LM) neurons, which are generated using the second-order derivative approach. In neural networks, the Rectified Linear unit and Sigmoid functions are frequently used activation functions.

$$f(x) = \frac{1}{1 + e^{-x}} \tag{1}$$

$$f(x) = \max(0, x) \tag{2}$$

4 Role of AI in EV Technology

AI helps EV manufacturers in many aspects, such as manufacturing, cost reduction, performance improvement, battery usage optimization, charging efficiency enhancement, range prediction, and range efficiency improvement. AI helps customers plan their charging activities by analyzing their EV adoption and charging behaviors and helps power system managers monitor load for grid resilience, energy conservation, and supply–demand balance. AI techniques such as deep learning, decision trees, and ML-based clustering algorithms have been utilized in both invasive and nonintrusive load monitoring (ILM and NILM) to identify EV charging profiles and discriminate EV loads from non-EV loads. Proposed NILM algorithms include bounding-box fitting, random forests, support vector machines, and deep generative models [18–20]. AI also helps with grid operations management, such as vehicle-to-grid control, location marginal pricing, optimal EV charging techniques, and microgrid energy management. AI is essential for EV charging infrastructure scheduling and planning, aiming to lower costs, increase use efficiency, and enhance coordination. The quantity and arrangement of charging points, as well as the prediction of infrastructure popularity, have all been accomplished by applying AI algorithms. It has been investigated how to schedule public EV charging infrastructure utilizing techniques including machine learning algorithms, internet of vehicles frameworks, and Markov decision processes. Further research has been done on autonomous AI robots for EV integration, such as robot-based EV test beds and Turtlebot robots for V2G simulation.

5 Conclusions

AI has a significant impact and a crucial role in advancing Integrated smart energy systems for EVs the major contributions of AI in this field are battery energy management, optimization of battery energy, smart grid integration which allows V2G and G2V, Predictive maintenance of EVs, Energy efficiency, safety and security. As AI technology advances, its role in optimizing EV performance and contributing to a cleaner, smarter energy future will only become more significant. AI also plays a transformative role in EVs significantly enhancing various aspects of development in autonomous driving, battery optimization, route optimization, predictive maintenance, market analysis, etc.

Acknowledgements This research work was supported by "Woosong University's Academic Research Funding—2025".

References

1. Wang L, Nian V, Li H, Yuan J (2021) Impacts of electric vehicle deployment on the electricity sector in a highly urbanised environment. J Cleaner Prod 295:456–466. https://doi.org/10.1016/j.jclepro.2021.126386
2. Vaideeswaran V, Ragul M, Keerthana V, Sindhu Priyadharshini C, Prathiksha M (2021) Review on selection of battery packs for pure electric vehicles and technical comparison of battery packs. In: Proceedings of the 2021 innovations in power and advanced computing technologies (i-PACT), vol 1, pp 1–6. https://doi.org/10.1109/i-PACT52855.2021.9696733
3. Ho JC, Huang YHS (2022) Evaluation of electric vehicle power technologies: integration of technological performance and market preference. Cleaner Responsible Consumption 5:1–11. https://doi.org/10.1016/j.clrc.2022.100063
4. Kong Q, Fowler M, Entchev E, Ribberink H, McCallum R (2018) The role of charging infrastructure in electric vehicle implementation within smart grids. Energies 11:1–20. https://doi.org/10.3390/en11123362
5. Ahlqvist V, Holmberg P, Tangerås T (2022) A survey comparing centralized and decentralized electricity markets. Energ Strat Rev 40:1–14. https://doi.org/10.1016/j.esr.2022.100812
6. Ahmed S, Khan ZA, Gul N, Kim J, Kim SM (2021) Machine learning-based clustering of load profiling to study the impact of electric vehicles on smart meter applications. In: 2021 Twelfth international conference on ubiquitous and future networks, vol 1, pp 444–447. https://doi.org/10.1109/ICUFN49451.2021.9528396

7. Bhatt A, Ongsakul W, Madhu N (2020) Machine learning approach to predict the second life capacity of discarded ev batteries for microgrid applications. In: International conference on intelligent computing and optimization, vol 1, pp 633–646. https://doi.org/10.1109/ICUFN49451.2021.9528396
8. Chung YW, Khaki B, Li T, Chu C, Gadh R (2019) Ensemble machine learning-based algorithm for electric vehicle user behavior prediction. Appl Energy 254:1–10. https://doi.org/10.1016/j.apenergy.2019.113732
9. Lei M, Mohammadi M (2021) Hybrid machine learning based energy policy and management in the renewable-based microgrids considering hybrid electric vehicle charging demand. Int J Elec Power 128:1–10. https://doi.org/10.1016/j.ijepes.2020.106702
10. Rigas ES, Ramchurn SD, Bassiliades N (2014) Managing electric vehicles in the smart grid using artificial intelligence: a survey. IEEE Trans Intell Transp Syst 16:1619–1635. https://doi.org/10.1109/TITS.2014.2376873
11. Sanguesa JA, Torres-Sanz V, Garrido P, Martinez JF, Marquez-BarjaMJ (2021) A review on electric vehicles: technologies and challenges. Smart Cities 4:372–404. https://doi.org/10.3390/smartcities4010022
12. Sovacool BK (2010) A transition to plug-in hybrid electric vehicles (phevs): why public health professionals must care 64:185–157. https://doi.org/10.1136/jech.2009.090746
13. Zadeh PT, Joudaki M, Ansari A (2021) A survey on deep learning applications for electric vehicles in micro grids. In: 2021 5th international conference on internet of things and applications (IoT), vol 5, pp 1–6. https://doi.org/10.1109/IoT52625.2021.9469715
14. Gonzalez Vaya M, Andersson G (2013) Integrating renewable energy forecast uncertainty in smart-charging approaches for plug-in electric vehicles. IEEE PowerTech, Grenoble 1:1–6. https://doi.org/10.1109/PTC.2013.6652150
15. Mukherjee JC, Gupta A (2017) Distributed charge scheduling of plug-in electric vehicles using interaggregator collaboration. IEEE Trans Smart Grid 8:331–341. https://doi.org/10.1109/TSG.2016.2515849
16. Hinton GE, Osindero S, The YW (2006) A fast learning algorithm for deep belief nets. Neural Comput 18:1527–1554. https://doi.org/10.1162/neco.2006.18.7.1527
17. Elhoseny M, Nabil A, Hassanien AE, Oliva D (2018) Hybrid rough neural network model for signature recognition. Adv Soft Comput Mach Learn Image Proc 20:295–318. https://doi.org/10.1007/978-3-319-63754-9_14
18. Farhangi H (2010) The path of the smart grid. IEEE Power Energ Mag 8:18–28. https://doi.org/10.1109/MPE.2009.934876
19. Kempton W, Tomic J (2005) Vehicle-to-grid power fundamentals: calculating capacity and net revenue. J Power Sources144:268–279. https://doi.org/ https://doi.org/10.1016/j.jpowsour.2004.12.025
20. Li S, Hu W, Cao D, Dragičević T, Huang Q, Chen Z, Blaabjerg F (2021) Electric vehicle charging management based on deep reinforcement learning. J Mod Power Syst 10:719–730. https://doi.org/10.35833/MPCE.2020.000460
21. Karki A, Phuyal S, Tuladhar D, Basnet S, Shrestha BP (2020) Status of pure electric vehicle power train technology and future prospects. Appl Syst Innovat 3:1–28. https://doi.org/10.3390/asi3030035
22. Abdullah HM, Gastli A, Ben-Brahim L (2021) Reinforcement learning based EV charging management systems–a review. IEEE Access 9:41506–41531. https://doi.org/10.1109/ACCESS.2021.3064354
23. Arfeen ZA, Khairuddin AB, Munir A, Azam MK, Faisal M (2020) En route of electric vehicles with the vehicle to grid technique in distribution networks: status and technological review. Energy Storage 2:105–115. https://doi.org/10.1002/est2.115
24. Mozaffari S, Al-Jarrah OY, Dianati M, Jennings P, Mouzakitis A (2022) Deep learning-based vehicle behavior prediction for autonomous driving applications: a review. IEEE Trans Intell Transp Syst 23:33–47. https://doi.org/10.1109/TITS.2020.3012034
25. Sun X, Li Z, Wang X, Li C (2020) Technology development of electric vehicles: a review. Energies 13:1–29. https://doi.org/10.3390/en13010090

26. Sun S, Yang Q, Yan W (2016) A novel statistical Markov-based approach for modeling charging demand of plug-in electric vehicles. In: 2016 China international conference on electricity distribution (CICED), vol 1, pp 1–6. https://doi.org/10.1109/CICED.2016.7576348
27. Li G, Zhang XP (2012) Modeling of plug-in hybrid electric vehicle charging demand in probabilistic power flow calculations. IEEE Trans Smart Grid 3:492–499. https://doi.org/10.1109/TSG.2011.2172643
28. Salkuti SR (2022) Modeling of various renewable energy resources for smart electrical power systems. In: Salkuti SR, Ray P (eds) Next generation smart grids: modeling, control and optimization. Lecture notes in electrical engineering, vol 824. Springer, Singapore. https://doi.org/10.1007/978-981-16-7794-6_2
29. Salkuti SR, Ray P, Pagidipala S (2022) Overview of next generation smart grids. In: Salkuti SR, Ray P (eds) Next generation smart grids: modeling, control and optimization. Lecture notes in electrical engineering, vol 824. Springer, Singapore. https://doi.org/10.1007/978-981-16-7794-6_1
30. Panahi D, Deilami S, Masoum MAS, Islam SM (2015) Forecasting plug-in electric vehicles load profile using artificial neural networks. Australas Univ Power Eng Conf 1:1–6. https://doi.org/10.1109/AUPEC.2015.7324879
31. Yeole SS, Vasant Kolhe S, Bibave RR, ShivajiChavan V, Kadam BB, SakharamBodhe V (2023) Design of two-wheeler hybrid electric vehicle using series parallel configuration. Third international conference on artificial intelligence and smart energy (ICAIS), Coimbatore, India, vol 1, pp 1595–1598. https://doi.org/10.1109/ICAIS56108.2023.10073870
32. Kumar PS, Arundhati B, Santoshi VVS (2022) Receding horizon h-∞ controller for hypersonic flight. In: First international conference on electrical, electronics, information and communication technologies, vol 1 Trichy, India, pp 1–5. https://doi.org/10.1109/ICEEICT53079.2022.9768491
33. Jung CM, Pagidipala S, Salkuti SR (2024) Emerging technologies for the integration of renewable energy, energy storage and electric vehicles. In: Salkuti SR (eds) Energy and environmental aspects of emerging technologies for smart grid. Green energy and technology. Springer, Cham. https://doi.org/10.1007/978-3-031-18389-8_5
34. Sravanthi P, Sandeep V (2022) Solving realistic reactive power market clearing problem of wind-thermal power system with system security. Int J Emerg Electr Power Syst 23(2):125–144. https://doi.org/10.1515/ijeeps-2021-0060

Design of D-STATCOM for Mitigation of Current Harmonics Using Class Topper Algorithm

T. M. Thamizh Thentral, R. Palanisamy, Rupam Chattopadhyay, Sahil Beniwal, Hrithika Chatterjee, Subhashree Choudhury, Swati Shukla, and Surender Reddy Salkuti

Abstract This work focuses on the design and implementation of a D-STATCOM (Distributed Static Compensator) to mitigate current harmonics in non-linear systems. These harmonics, which are integral multiples of the supply frequency, cause issues such as increased heat generation, reduced power factor, voltage distortion, and decreased system reliability. A novel Class Topper Algorithm is implemented for real-time estimation and control of the current harmonics. This optimization algorithm regulates the DC link capacitor voltage fed to the inverter in the D-STATCOM, minimizing harmonics in the source current and improving system performance compared to traditional methods. This proposed algorithm dynamically adjusts control parameters to maintain a constant DC link voltage, leading to reduced harmonics and enhanced power quality. The proposed system is simulated in MATLAB platform, and different PI controller parameters are analysed to achieve various levels of Total Harmonic Distortion (THD %), ensuring compliance with IEEE519-2014 standards. The effectiveness of the proposed D-STATCOM, utilizing

T. M. Thamizh Thentral · R. Palanisamy · R. Chattopadhyay · S. Beniwal · H. Chatterjee
Department of Electrical and Electronics Engineering, SRM Institute of Science and Technology, Kattankulathur, Chennai, India
e-mail: thamizht@srmist.edu.in

R. Palanisamy
e-mail: palanisr@srmist.edu.in

S. Choudhury
Department of Electrical and Electronics Engineering, ITER, Siksha 'O' Anusandhan Deemed to Be University, Bhubaneswar, India
e-mail: subhashreechoudhury@soa.ac.in

S. Shukla
School of Electronics Engineering, VIT-AP University, Amaravati, Andhra Pradesh, India
e-mail: swati.shukla@vitap.ac.in

S. R. Salkuti (✉)
Department of Railroad and Electrical Engineering, Woosong University, Daejeon, Republic of Korea
e-mail: surender@wsu.ac.kr

S. R. Salkuti (ed.), *Artificial Intelligence for Integrated Smart Energy Systems in Electric Vehicles*, Lecture Notes in Electrical Engineering 1427,
https://doi.org/10.1007/978-3-031-94276-1_33

the Class Topper Algorithm, is evaluated based on its ability to minimize THD % in the source current while maintaining a stable DC link voltage. This approach significantly improves power system efficiency and reliability.

Keywords Class topper optimization · Distributed static compensator · Synchronous reference frame theory · Total harmonic distortion · Pulse width modulation · Voltage source inverter

Nomenclature

VSC	Voltage source converter
SRF	Synchronous reference frame theory
THD	Total harmonic distortion
VSI	Voltage source inverter
DSTATCOM	Distributed static compensator
CTO	Class topper optimization
APF	Active power filter
SAF	Shunt active filter
SMPS	Switched mode power supply
DSO	Digital storage oscilloscope
PCC	Point of common coupling
VFD	Variable frequency drives
FFT	Fast fourier transform
V	Supply voltage
V_{out}	Output voltage
K_p	Proportional gain
K_i	Integral gain
I_{inv}	Inverter output current
V_{inv}	Inverter output voltage
I_s	Source current
I_L	Load current
V_s	Source voltage

1 Introduction

Power quality problems refer to deviations from the ideal characteristics of electrical power, which can cause disruption or damage to electrical equipment [1]. These problems can occur due to various reasons, including load variations, equipment faults, and electrical disturbances. Some common power quality problems include voltage sags or dips, voltage swells, voltage harmonics, current harmonics, voltage fluctuations, and transient voltages [2]. These problems can lead to equipment failure, reduced efficiency, increased maintenance costs, and even safety hazards [3]. Therefore, it is essential to identify and address power quality issues to maintain the stability, efficiency, and longevity of the power system, ensuring that both equipment and infrastructure operate safely and reliably. Harmonics are high-frequency signals that deviate from the fundamental waveform and can cause various issues, such as increased heat generation, a reduced power factor, increased voltage distortion, and decreased system reliability. These harmonics must be removed from the power system for it to be more stable [4–6].

Power filters are electronic devices used to mitigate the effects of current harmonics in electrical systems. Power filters work by using passive or active components to remove or reduce the harmonic content of the current [7]. Passive filters typically use capacitors and inductors to create a resonance circuit that absorbs or filters out certain frequencies, while active filters use controlled semiconductor devices to actively inject the negative harmonic currents to compensate for the existing harmonics present in the source current of the given system. By using power filters, the electrical system can operate more efficiently and reliably, with less risk of damage to equipment and interference with other devices [8, 9].

As discussed earlier, the major power quality problems addressed in a power system are harmonics. To mitigate these harmonics, specifically current harmonics, a tool known as D-STATCOM can be used. D-STATCOM is used to lower the current harmonics in power systems [10–12]. It is a shunt-connected device connected between source and load to balance the current flow in a system by injecting or absorbing reactive power, and thus reduce the current harmonic distortion [13]. The D-STATCOM uses power electronics components, such as MOSFET and capacitors, to dynamically control the amount of reactive power to maintain a stable system voltage and reduce the current harmonics in the system [14, 15].

The design and implementation of a D-STATCOM (distributed static compensator) to reduce current related power quality issues in a power system is the main focus of this work. The mitigation of current harmonics using D-STATCOM can be done using various popular and widely implemented optimization algorithms [16, 17]. This work involves the simulation and design of a D-STATCOM using the Class Topper Algorithm (CTA). The CTA is an optimization algorithm that is used to obtain the optimal value of filter current to minimize the THD%. It is based on a metaphor of a classroom where students (candidate solutions) compete to be the "class topper" or the best solution [18, 19]. The algorithm operates by comparing a population of candidates and then selecting the top-performing candidate to "compete" with each other. The competition involves combining pairs of top-performing solutions and creating a new set of candidate solutions through crossover and mutation operations [20]. The new solutions are then evaluated, and the best-performing ones are selected to compete in the next round. This process continues until a satisfactory solution is found [21–23].

The performance of the designed D-STATCOM is validated in terms of its ability to reduce the THD% and improve the power factor of the system. This paper aims to contribute to the field of power electronics by providing a practical solution for mitigating current harmonics in power systems. The results of this project will have implications for improving the efficiency and reliability of electrical power systems and protecting critical electrical equipment [24]. This project's scope comprises an analysis of the present harmonic problem, the design of a D-STATCOM system, software simulation of the system, the implementation of the Class Topper algorithm for the design of D-STATCOM, and design validation through actual trials. This paper describes the project's findings and offers a thorough analysis of the D-STATCOM system's functionality.

This chapter begins with an introduction that will outline the importance of current harmonics mitigation and introduces the DSTATCOM to be a mitigating device. Further, a literature review presents the current state of research on D-STATCOM and its application in mitigation of current harmonics. The section then explains the methodology and the steps in designing DSTATCOM using the CTO. Next, the result section presents the findings of the studies including the effectiveness of CTO in the mitigation of current harmonics. In conclusion, the chapter discusses the implications of this study and offers recommendations for future research in this field. Overall, it provides a clear and detailed understanding of the DSTATCOM design using the CTO for mitigating current harmonics.

2 Impacts of Harmonics and System Efficiency

Harmonics in distribution systems have been a cause of concern since the early days of alternating current (AC) power transmission [25, 26]. Harmonics are undesirable frequencies generated by nonlinear loads in the power system like electronic devices, variable frequency drives, and devices that use switching technology. In most cases they operate at a fundamental frequency, a power system will serve as a primary means of delivering electricity to residences and commercial establishments [27]. Despite this, the addition of non-linear loads like electronic devices, variable speed drives, etc., to the grid introduces harmonic distortion. Harmonic distortion specifies the existence of frequencies other than the fundamental frequency [28].

As the use of power semiconductor devices became more widespread, engineers began to notice harmonic distortion in power systems [29]. At first, this distortion was characterized by problems with specific pieces of equipment, such as transformers, motors, etc. In the 1930s, researchers began to look at the causes of harmonic distortion exhaustively. They found use of non-linear devices, such as rectifiers, was the major source of harmonics in power systems. Later in the 1960s and 1970s, with the use of power electronics, such as thyristors and transistors, were realized to be highly non-linear and could generate significant amounts of harmonic distortion which led to a new focus on harmonic analysis and mitigation in power systems [30]. Later, since the 1990s, the addition of more and more power semiconductor devices causing non-linearity in the power system has led to harmonics distortion on voltage or current present in the system. Various solutions to mitigate harmonics are implemented, including the usage of harmonic filters, active filters, and phase-shifting transformers. As of today, harmonics in power systems have continued to be a concern, majorly in industrial settings where nonlinear loads are prevalent. Regardless, advances in technology and greater awareness of the issue have led to more effective solutions for reducing harmonics in power systems.

2.1 Sources of Harmonics

In any power distribution system, the nonlinear loads connected to the system causes distortion in current and voltage waveform in the system. Here are some common sources of harmonics in any power system: Power electronics converters, which are essentially rectifiers, inverters, and DC-DC converters, which are majorly used in modern power distribution systems. These are converters that utilize semiconductor devices, like diodes, transistors, and thyristors, which generate harmonics due to their nonlinear characteristics. The harmonics generated by these power electronics converters cause distortion in voltage and current waveform in the system. Arc furnaces are used for melting metals and alloys, and they are an important source of harmonics in power systems. The arc in the furnace is responsible for generating harmonics due to its non-linear characteristics. The harmonics generated by these arc furnaces can cause voltage distortion, which leads to the failure of equipment and power quality issues.

Welding equipment, like arc welders and resistance welders, is another source of harmonics in power systems. The welding process generates voltage and current harmonics due to the nonlinear characteristics of the welding equipment. These also lead to the failure of equipment. Variable Frequency Drives (VFDs) are majorly employed to regulate the AC motors' speed by changing the frequency and voltage in the supply system. VFDs utilize the suitable PWM method to modify the voltage and period of the output waveform. The PWM technique is responsible for generating harmonics due to its nonlinear characteristics. The harmonics generated by VFDs can cause voltage distortion, which can lead to equipment failure and other power quality issues. Lighting loads, like fluorescent lamps and light-emitting diodes (LEDs) are another cause for harmonics in power systems. Ballasts and drivers applied in lighting loads generate harmonics. They have a similar effect like voltage distortion and that leads to equipment failure.

2.2 Effects of Harmonics on Power System Performance and Reliability

The existence of voltage and current harmonics in the system can have several adverse effects. One of the most significant effects is increased current, which can lead to overheated equipment and increased losses. This is because harmonic currents combine with the fundamental current producing a higher total current. This increase in current leads to an increased loss of transformers, and cables, reducing efficiency and increasing the operating costs [30]. One of the impacts that harmonics can have on the power system is voltage distortion. It occurs when the voltage waveform deviates from the sinusoidal waveform. Harmonics cause voltage distortion by distorting the voltage waveform. The voltage distortion leads to several issues, mainly equipment failure, reduction of efficiency, and increased losses. Voltage distortion can also cause flickers, which is the visible effect of voltage variations in lighting systems [31].

Another effect is a distortion of the current. It occurs when the current waveform in the power system deviates from the sinusoidal waveform. Harmonics cause current distortion by distorting the current waveform. This leads to several issues, such as overheating of equipment, increased losses, and reduction of capacity. Current distortion can also cause neutral current overloading, which can damage equipment and lead to safety hazards [32]. The impacts of voltage and current distortion caused by harmonics vary depending on the acuteness and frequency of the harmonics. For example, higher-frequency harmonics can cause issues with electronic equipment, while lower-frequency harmonics can cause issues with motor loads. The impacts of harmonics are based on the type of load and its configuration. Resonance also occurs in the power system due to harmonics. When the normal frequency aligns with the harmonic frequency leading to excessive vibration in the system. This leads to equipment failure. Finally, harmonics cause EMI. It occurs when the harmonic currents create electromagnetic fields that interfere with nearby electronic equipment. This causes operational problems and potential safety hazards [33].

2.3 *Types of Harmonics in Power System*

In power systems, harmonics refer to the additional frequencies present in the electrical power supply that are not part of the supply frequency with 50 or 60 Hz. These harmonics cause various issues in the power system, including equipment damage, reduction in quality of power, and increased power losses. The two major types of harmonics in power systems are even and odd harmonics. Even harmonics are multiples of the supply frequency and are even in numbers. If the 100 Hz frequency is 50 Hz then it is called a second harmonic [34]. Likewise, odd harmonics are multiples of the supply frequency and are odd in numbers. The 150 Hz frequency in 50 Hz is called as third harmonic.

Integer harmonics are the integer multiples of the supply frequency. The fourth harmonic has a frequency of 200 Hz in a 50 Hz power system is called integer harmonics. Non-integer harmonics are not the integer multiples of the supply frequency. These types of harmonics are usually caused by nonlinear loads such as semiconductor devices and can cause significant power quality problems. Resonant harmonics are amplified by resonant circuits called as transformers, capacitors, and inductors. These harmonics can cause equipment damage and power quality problems. It's important to note that while some harmonics are unavoidable, others can be reduced or eliminated by using appropriate power conditioning equipment and avoiding the use of nonlinear loads.

2.4 Measurement and Analysis of Harmonics

Measuring and analyzing the harmonics in power systems are crucial to identify the sources of harmonics and mitigate their effects on power quality and system efficiency. Some tools and techniques used for the measurement and analysis of harmonics in power systems are discussed. Power quality analyzers are portable devices that can measure various parameters of power quality, including voltage and current harmonics. These devices use built-in algorithms to calculate harmonic distortion and display the results in real time. Power quality analyzers can also record the data for further analysis. Spectrum analyzers are used to analyze the frequency spectrum of a signal. In power systems, spectrum analyzers are used to examine the frequency spectrum of the source current and voltage waveforms to categorize the presence of harmonic frequencies [35]. Spectrum analyzers can also provide information on the amplitude and phase of the harmonics.

Oscilloscopes are used to monitor the parameters such as voltage and current waveforms in power systems. Oscilloscopes can also be used to analyze the harmonic content of the waveform by using Fast Fourier Transform (FFT) techniques. Oscilloscopes can display the magnitude and phase angle of different harmonics. Harmonic spectrum analyzers such as fluke analyzers are designed to visualize and monitor the harmonic distortion in power systems. These devices can measure the THD% of the voltage and source current waveforms and provide information on the amplitude and phase of individual harmonics. Simulation software, such as PSCAD and MATLAB, can be used to simulate the behavior of the system with non-linearity under different harmonic orders. Simulation software can also be used to identify the sources of harmonics and evaluate the effectiveness of different mitigation techniques.

2.5 Mitigation Techniques for Harmonics

Harmonics are a common problem in electrical systems, causing overheating, reduced efficiency, and system component failure. To protect the power system from any kind of failure, an appropriate mitigation technique for the harmonics has to be used, which is shown in Fig. 1.

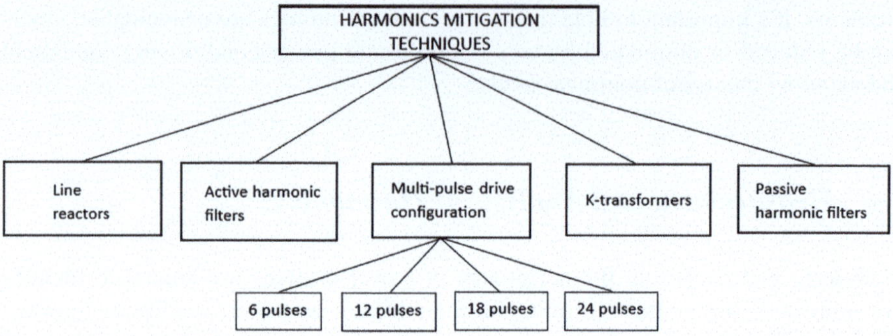

Fig. 1 Harmonics mitigation techniques

3 Design of DSTATCOM

Depending on the needs, the DSTATCOMs (distribution static compensators), which are custom power devices, are used to mitigate some of the issues. DSTATCOMs are primarily used to address current-related power quality problems such as low power factor, poor voltage regulation, unbalanced currents, and elevated neutral current. As a result, the DSTATCOM configuration is chosen based on the issues. In AC distribution networks, the DSTATCOM technology is used to provide compensation for reactive power and current distortion, balancing of load, and neutral current reduction. The key component of a DSTATCOM is the Voltage Source Converter (VSC), which is typically linked to a DC link capacitor. The AC portions of the VSC are usually connected in shunt to the consumer loads or the Point of Common Coupling (PCC). The inverter is considered to enable PWM control, which involves modulating the width of the pulses of the switching signal to achieve the desired voltage or current output.

To mitigate the switching ripples associated with PWM control, small ripple filters are often used [36]. The voltage and current feedback signals to be synthesized are sensed by using Hall Effect sensors. This sensor detects the magnetic field produced by the current flowing through the conductor. The shunt compensator is used to achieve the compensation in neutral current, regulation of voltage, improvement in power factor, reduction in harmonic current, and obtaining the balanced load condition in a three-phase, four-wire system with the distributed loads (combination of linear and non-linear). This compensator uses a capacitor or an inductor as reactive power storage components along with either a VSI or CSI.

The overall control algorithm for DSTATCOM used for the mitigation of current harmonics involves combining the harmonic detection, harmonic compensation, and voltage regulation components into a single control system. The control algorithm for a DSTATCOM must be capable of rapidly and accurately detecting and mitigating harmonics to ensure efficient and effective system operation. There are many control algorithms used for the control of DSTATCOM for the elimination of current harmonics like Synchronous Reference Frame (SRF), PQ Theory, IRPT Theory,

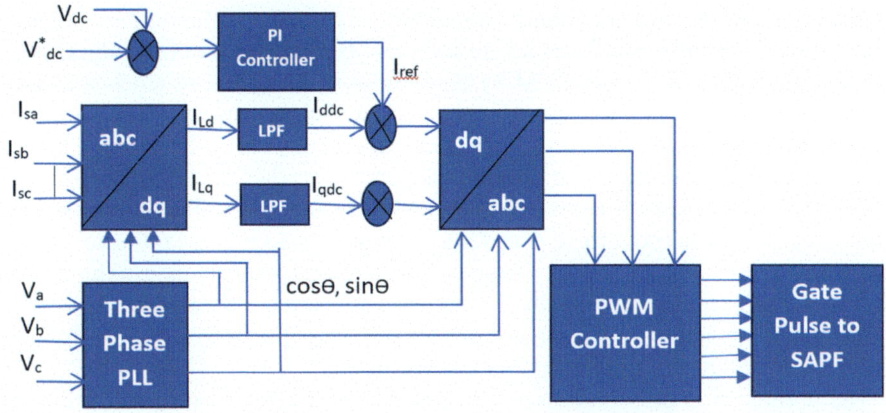

Fig. 2 Synchronous frame theory: control algorithm for DSTATCOM

and many others. In this paper, the SRF method has been applied to control the DSTATCOM, which is displayed in Fig. 2. The SRF theory is a commonly used method for controlling DSTATCOM to mitigate current harmonics in power systems.

SRF theory converts load current's three phases into d-q-0 frame using a mathematical tool named Park's transformation. A three-phase phase-locked loop is considered to synchronize the voltage and current signals at the PCC voltages. Afterwards, the d-q current components go through a low-pass filter (LPF) to isolate the DC components of i_{Ld} and i_{Lq}. Once the system current waveform has been transformed into the synchronous reference frame, the harmonic components of the current can be easily identified and separated from the fundamental component. The harmonic components can then be compensated using a DSTATCOM by generating a compensating current that is 180 degrees out of phase with the harmonic component. In summary, the performance of the DSTATCOM using SRF for harmonic mitigation depends on the accuracy of the synchronization between the reference waveform and the grid voltage. The synchronization can be achieved using various techniques such as PLL or DQ (Clarke and Park) transformation.

4 Class Topper Algorithm (CTA)

The CTA works by iteratively improving the current solution by making small changes to the previously obtained solutions. At each iteration, the algorithm starts evaluating the new solution and compares it to the previous solution that was obtained. If the obtained solution is better than the previous one, it becomes the new current solution, and the same process proceeds to continue. If the newly obtained solution is worse than the previous one, the algorithm must return to the previous solution and try again with a different set of changes and parameters used.

The CTA operates by maintaining a population of candidate solutions, each represented by a different set of variables. In each and every iteration of the algorithm, a new population size is generated by applying a combination of stochastic and deterministic search strategies or methods such as mutation, crossover, and selection processes. The fitness of each and every candidate solution is evaluated based on a particular objective function that is optimized.

In Fig. 3, the maximum number of iterations is fixed to be 300 initially whereas number of classes is taken to be 5 and the population size is fixed to be 15 in the algorithm. The DC link reference voltage is given to be 1000 V in the coding of the algorithm and variation in the value of the reference voltage is considered as 10% which means when the error in the variation is zero, the optimized values of Kp and Ki for the PI controller is obtained by the algorithm.

The optimization algorithm uses a selection scheme that ensures that the best candidate solutions obtained should be preserved from one generation to the next. The top-performing solutions are selected to "compete" with each other so that the best solution can be achieved. The competition also involves creating new solutions through a unique combination of crossover and mutation operations. Crossover operation mainly involves combining different pairs of solutions to create new and modified ones, while the mutation process involves introducing small changes to a solution that already exists to explore new areas of the search space of the given problem. The new solutions can then be evaluated, and the best-performing solutions (or students) are selected to compete in the next rounds to come. This entire flow of various processes like selection, crossover, and mutation shall continue until a satisfactory solution is found, or a termination criterion is particularly met.

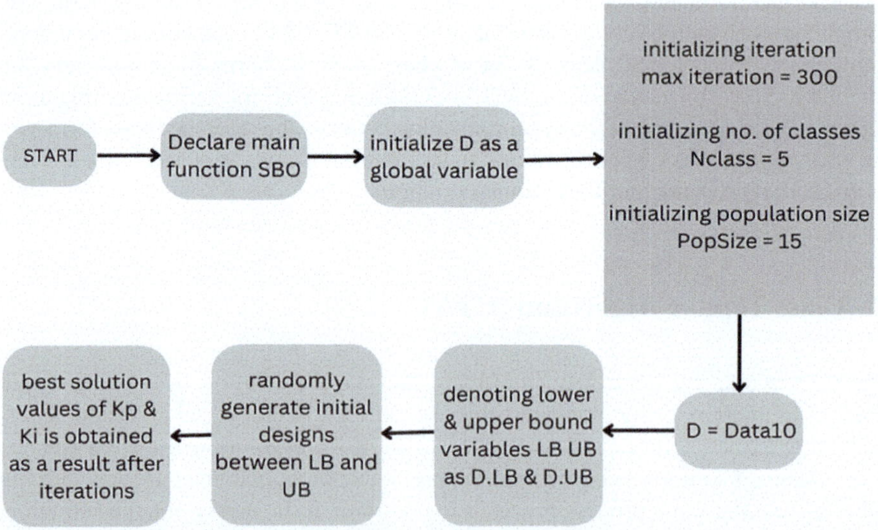

Fig. 3 Flowchart of class topper optimization algorithm

In the CTA, control parameters are adjusted iteratively within specified limits to achieve the desired outcome. Typically, Kp and K_i values are determined through a trial-and-error process, but with the CTA, this process is streamlined. The algorithm iterates multiple times in a short span, making it faster and more efficient to find optimal Kp and K_i values, significantly reducing the time usually required for trial and error.

The CTA that has been used as an optimization algorithm is already a quite popular optimization algorithm that is widely used in machine learning and various other fields. The use of the class topper algorithm in the project was successful as the algorithm was effectively able to give the optimized values of Kp and Ki for the PI controller which play a major role in mitigating the current harmonics that are present in the source current of the standalone system and to get a stabilized and a constant DC link capacitor voltage for smooth functioning of the DSTATCOM system module. The algorithm can also effectively find the best solution out of many possible top-performing best solutions using the combination of stochastic and deterministic search methods. It obtains the best possible by comparing all the solutions and then making some small changes to the previous solutions that were obtained using different sets of parameters or values. The algorithm used is able to run up to the desired number of iterations said to be 300 in this project for obtaining the optimized values of PI parameters. The CTA algorithm has proven to be very efficient, and it can quickly provide us with the optimal solution, making it a very useful and unique tool in various applications.

5 Simulation of DSTATCOM and Results

It presents the outcomes of the simulations carried out for the effectiveness of the proposed system to be analyzed. This chapter holds major importance in this report, as it will provide a detailed analysis and results of the simulation, which will help us confirm the feasibility and practicality of this suggested solution. Here, this chapter starts with a summary of the setup for the proposed simulation that would include the software used for the simulation, the parameters realized, and the test system topology, which is shown in Table 1.

Table 1 Simulation components and their respective ratings

System voltage	415 V
Non-linear loads	Diode bridge rectifier RL types R = 1000 ohms, L = 20 mH
Unbalanced linear load	Ra = 48 ohms, La = 30 mH; Rb = 60 ohms, Lb = 30 mH; Rc = 72 ohms, Lc = 30 mH
DC capacitor voltage	800 V
DC link capacitance	333 μF

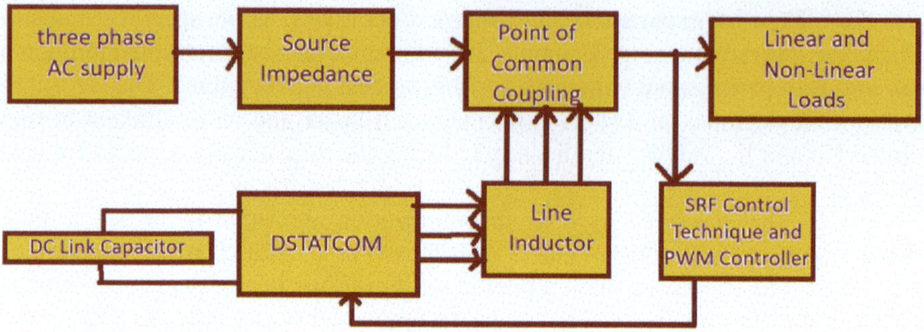

Fig. 4 System topology

The chapter includes a comprehensive study of the simulation results, talking through the impact of the suggested system on the test system's performance. This chapter presents the results in an aligned format and all the specifications that are being used for simulation are mentioned. Ultimately, this section culminates by summarizing the simulation results and their significance in exhibiting the plausibility and usefulness of the suggested solution. The software utilized for the simulation process was MATLAB and it was extensively used to design the DSTATCOM for the mitigation of current harmonics in the power system. The electrical network on which the simulation was conducted is the test system topology. The rudimentary block diagram that was followed for the simulation of this project is displayed in Fig. 4.

To demonstrate that the required Total Harmonic Distortion (THD%) is reduced to below 5%, conforming to the IEEE 519-2014 standard, The MATLAB simulations is utilized. The results show that the THD% was minimized by employing the optimized Kp and Ki gains of the PI controller, achieved through the CTA. These optimized gains allowed the system to meet the desired harmonic reduction effectively. These results have been correlated with the values of THD% acquired utilizing conventional values of Kp and of Ki for the proportional-integral (PI) controller to justify that the harmonics have been diminished in the circuit when optimized values for Kp and Ki were used for the PI controller that controls the DC link capacitor voltage in the DSTATCOM. In spite of the fact that the values of Kp and Ki that were obtained by the Ziegler–Nichols method for fine-tuning the PI controller were used, it is noted higher THD% than the THD% obtained when Kp and Ki values achieved by CTA were used. The table below provides the values of THD% obtained using conventional gain values and THD% obtained using Ziegler–Nichols method for tuning the PI controller.

5.1 *CASE I: Conventional System's Kp and Ki Values and THD%*

Table 2 depicts the system's Kp and Ki values along with their THD%. The given system has two parameters, namely the proportional gain (Kp) and integral gain (Ki). The value of Kp for the system is 0.025, while the value of Ki is 0.14. These two parameters are used in the control algorithm to determine the output signal of the system. The results of FFT analysis for one-dimensional data are a frequency spectrum of the signal. For two-dimensional data, the FFT results in a two-dimensional frequency spectrum. Similarly, for three-dimensional data, the FFT results in a three-dimensional frequency spectrum, which is shown in Fig. 5.

In the first dimension it is noticed that the THD% comes out to be 2.84% which is well under 5% according to the IEEE 519, 2014 standard. Start time is fixed as 0.05 s with fundamental frequency being 50 Hz and maximum frequency measured would be 1000 Hz as given in the figure above. In the second dimension the THD% is a little more and comes out to be 4.37% which is also well under 5% according to the IEEE 519, 2014 standard. In the third dimension the THD% comes out to be 4.44% which is also well under 5% according to the IEEE 519, 2014 standard. The given system has two parameters, namely the proportional gain (Kp) and integral

Table 2 Current THD% values using conventional Kp and Ki values

S. No.	Kp	Ki	THD%		
			Ia	Ib	Ic
1	0.025	0.14	2.84	4.37	4.44
2	0.57	0.0034	14.49	15.5	16.1
3	0.15	0.05	20.31	22.6	23.28

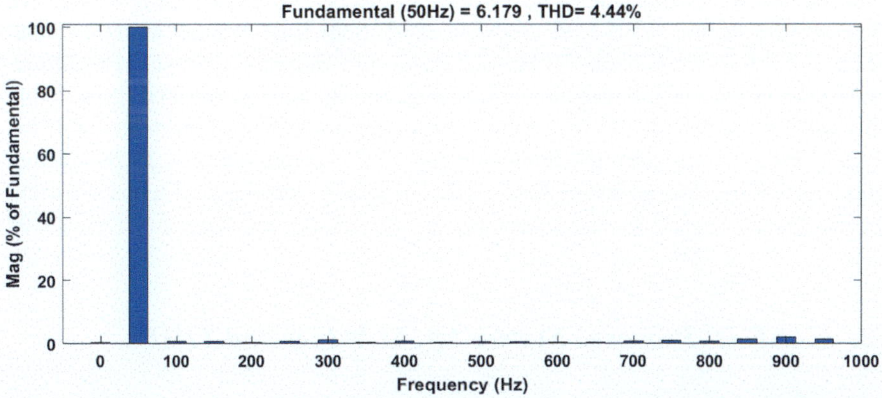

Fig. 5 THD% of the Phase A obtained as 4.44% using 1st set of Conventional Kp and Ki values

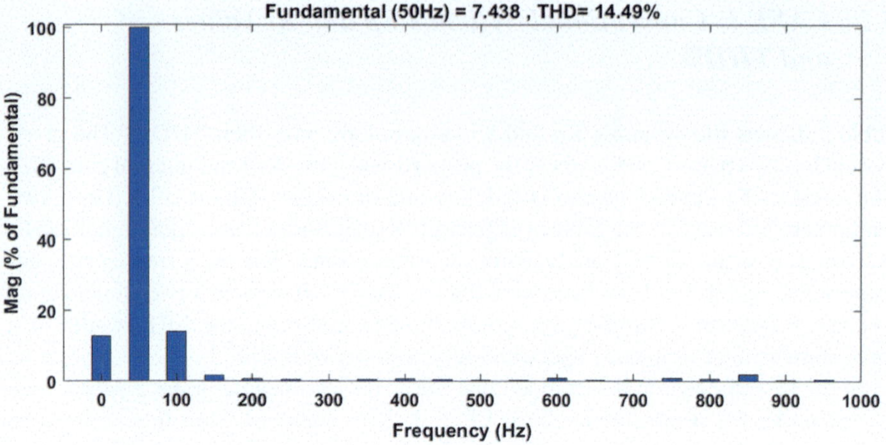

Fig. 6 THD% of the Phase A obtained as 14.49% using 2nd set of conventional Kp and Ki values

gain (Ki). The value of Kp for the system is 0.57, while the value of Ki is 0.0034. For the system FFT analysis for various dimensions are seen to be as follows.

In the first dimension, the THD% comes out to be 14.46% which is large and well over 5% according to the IEEE 519, 2014 standard. Start time is fixed as 0.05 s with fundamental frequency being 50 Hz and maximum frequency measured would be 1000 Hz as given in the figure above. In the second dimension, the THD% is a little more and comes out to be 15.54%. In the third dimension the THD% comes out to be 16.1% which is also well over 5% according to the IEEE 519, 2014 standard. Source side and load side parameters of the above system are given below, which is shown in Fig. 6. The given system has two parameters, namely the proportional gain (Kp) and integral gain (Ki). The value of Kp for the system is 0.15, while the value of Ki is 0.05.

In the first dimension, the THD% comes out to be 20.31% which is well above 5% according to the IEEE 519, 2014 standard. Start time is fixed as 0.05 s with fundamental frequency being 50 Hz and maximum frequency measured would be 1000 Hz as given in the figure above. In the second dimension, the THD% is a little more and comes out to be 22.56% which is also well above 5% according to the IEEE 519, 2014 standard. In the third dimension, the THD% comes out to be 23.28% which is also well above 5% according to the IEEE 519, 2014 standard, which is shown in Fig. 7.

5.2 *CASE II: Proposed System Kp, Ki Values and THD%*

Table 3 depicts the optimized value of Kp and Ki along with the THD%. The given system has two parameters, namely the proportional gain (Kp) and integral gain (Ki)

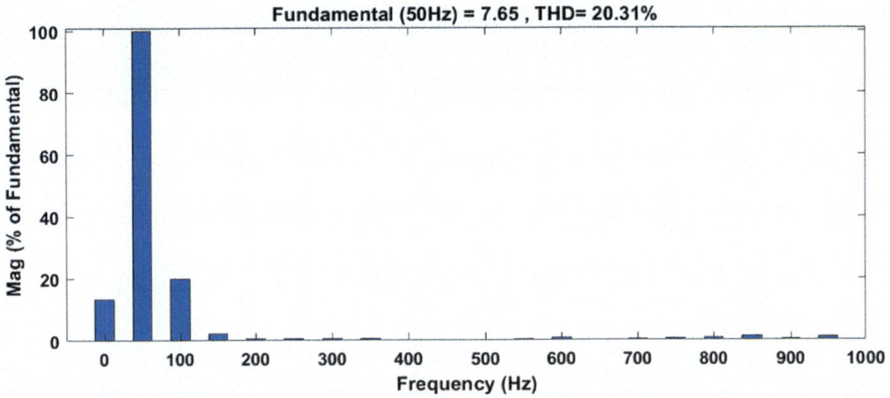

Fig. 7 THD% of the phase A obtained as 20.31% using 3rd set of conventional Kp and Ki values

Table 3 Current THD% values using optimized Kp and Ki values

S. No.	Kp	Ki	THD%		
			Ia	Ib	Ic
1	0.1	0.1	2.69	3.09	3.15

are optimized. The value of Kp for the system is 0.1, while the value of Ki is 0.1. For this system FFT analysis for various dimensions is shown in Table.3. In the first dimension the THD% comes out to be 2.69% which is well under 5% according to the IEEE 519, 2014 standard. Start time is fixed as 0.05 s with fundamental frequency being 50 Hz and maximum frequency measured would be 1000 Hz given in the figure above. In the second dimension the THD% is a little more and comes out to be 3.09% which is also well under 5% according to the IEEE 519, 2014 standard. In the third dimension the THD% comes out to be 3.15% which is also well under 5% according to the IEEE 519, 2014 standard. The THD% as compared to the rest of the combinations is the smallest as desired, which is shown in Fig. 8.

6 Hardware Implementation

To prove the success of the project, different sets of PI controller parameters were provided to the FPGA controller and the changes in the waveforms of different values such as source current (Is), load current (I_L), source voltage (Vs), load voltage (V_L), DC link voltage (V_{dc}), Inverter voltage and current (V_{inv} and I_{inv}), rectifier voltage and current (V_r and I_r) are measured for the different values of Kp and Ki of the PI controller, shown in Fig. 9.

To obtain different kinds of waveforms and verify the success of the parameters of the PI controller used, the value of resistive loads can be varied. The total rating

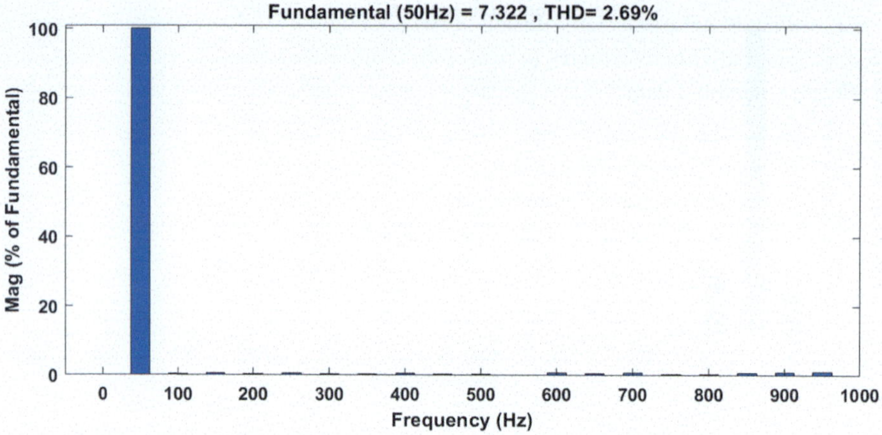

Fig. 8 THD% of the phase A obtained as 2.69% using optimized Kp and Ki values

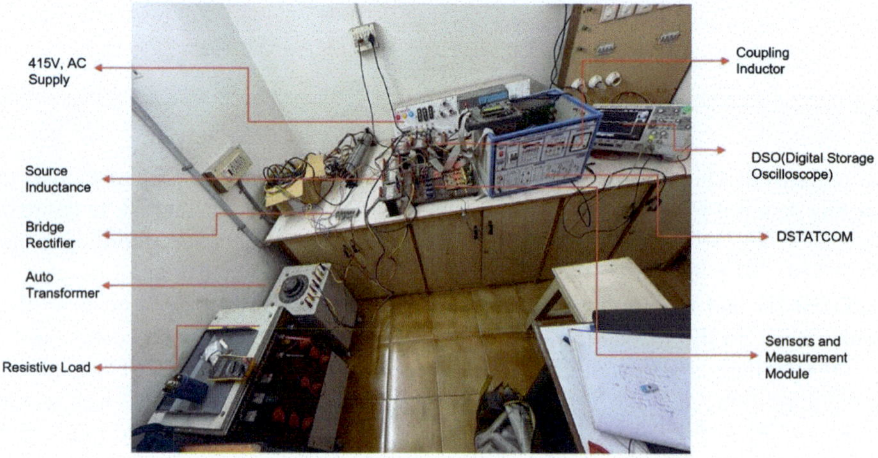

Fig. 9 Experimental setup

of the resistive loading used in the hardware set-up is up to 5 Kw, to take different readings this load can be varied and changed by turning some loads on/off from the resistive load box. Some of the values used of resistive loads and their impact on the hardware set-up and the results obtained are discussed here in this section along with the other variable parameters like source voltage of the hardware set-up.

When the values used of Kp and Ki are obtained by class topper algorithm using MATLAB coding and used for coding in FPGA controller in the hardware set-up, source voltage provided is 100 V and the value of resistive load is 0.25 Kw, the current value measured in the multimeter is 0.7A. The values used for this set of readings are provided in Table 2 along with the current values measured from the

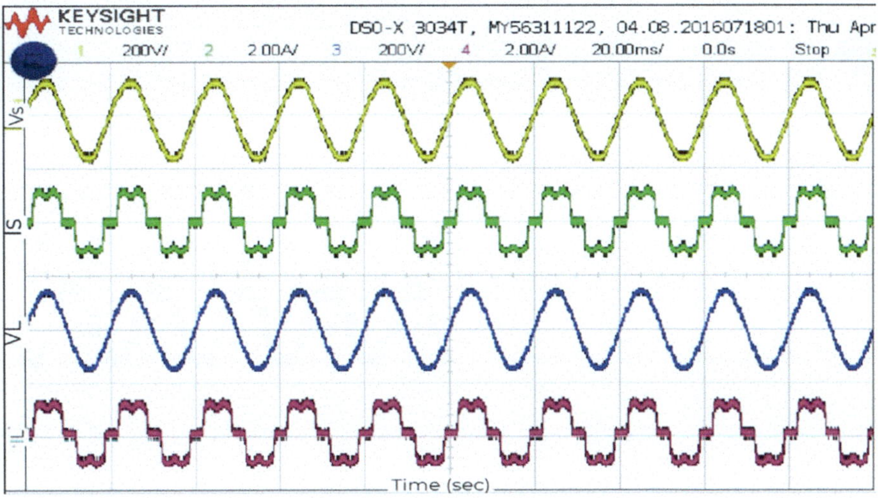

Fig. 10 Output waveforms of Vs, Is, V_L and I_L for optimized PI controller parameters and 100 V and 0.25 kW resistive load before compensation

multimeter. This is before compensation was provided in the power system. The output waveforms of source voltage (Vs), source current (Is), load voltage (V_L) and load current (I_L) obtained from the DSO can be seen in Fig. 10. When the values used of Kp and Ki are optimized ones and source voltage is 100 V and the value of resistive load is 0.25 kW. In this graph taken from DSO, It shows the changes in the waveforms of source voltage (Vs), source current (Is), load voltage (V_L) and load current (I_L) before the compensation and after the compensation was provided in the power system is shown in Fig. 11.

When the values used of Kp and Ki are optimized ones, the source voltage provided is 130 V and the value of resistive load is 0.25 KW. This is before compensation was provided in the power system. The output waveforms of DC link voltage (V_{dc}), source current (Is), Inverter voltage and current (V_{inv} and I_{inv}) obtained from the DSO can be seen in Fig. 11. When the values used of Kp and Ki are optimized ones and source voltage is 130 V and the value of resistive load is 0.25 KW. In this graph taken from DSO, It shows the changes in the waveforms of DC link voltage (V_{dc}), source current (Is), Inverter voltage and current (V_{inv} and I_{inv}) before the compensation and after the compensation was provided in the power system is shown in Fig. 12. When the values used of Kp and Ki are optimized ones and source voltage is 130 V and the value of resistive load is 0.25 KW. This is before the compensation was provided in the power system. The output waveforms of source voltage (Vs), source current (Is) and rectifier voltage (Vr) obtained from the DSO can be seen in Fig. 13.

When the values used of Kp and Ki are optimized ones and source voltage is 130 V and the value of resistive load is 0.25 kW. In this graph taken from DSO, It shows the changes in the waveforms of source voltage (Vs), source current (Is), rectifier voltage (Vr) before the compensation and after the compensation was provided in

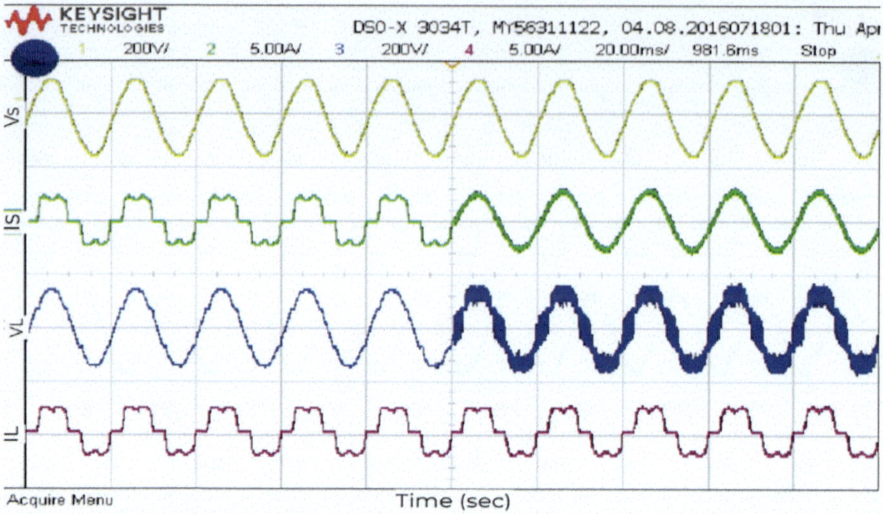

Fig. 11 Output waveforms of Vs, Is, V_L and I_L for optimized PI controller parameters and 100 V, 0.25 KW resistive load before and after compensation

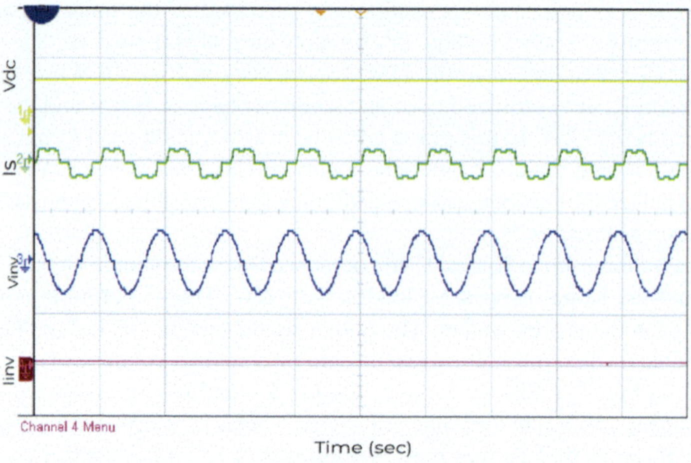

Fig. 12 Output waveforms of V_{dc}, I_s, V_{INV} and I_{INV} for optimized PI controller parameters and 130 V, 0.25 KW resistive load before compensation

the power system is shown in Fig. 14. Instead of using optimized PI parameters (Kp = Ki = 0.1) for all the readings, to prove that the values of Kp and Ki obtained by class topper algorithm are better than the conventional values of Kp and Ki, these conventional values of PI controller were provided to the FPGA controller and results were taken to see the difference, which is shown in Fig. 15.

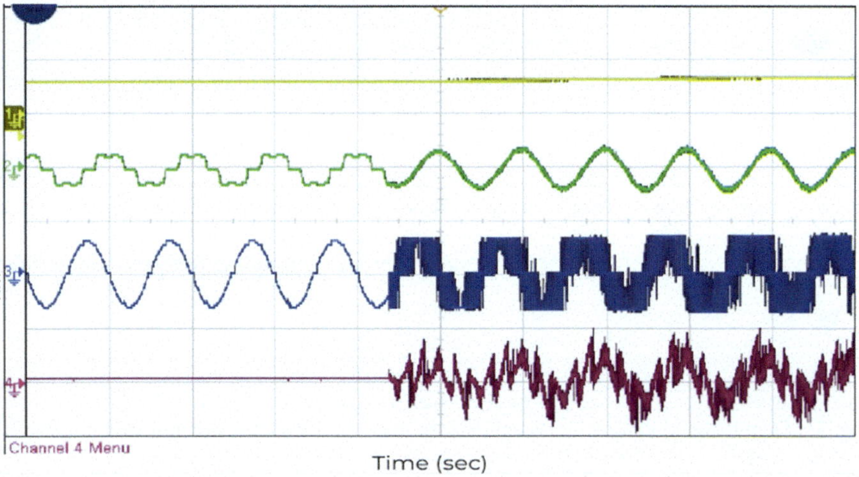

Fig. 13 Output waveforms of V_{dc}, I_s, V_{INV} and I_{INV} for optimized PI controller Parameters and 130 V, 0.25 kW resistive load before and after compensation

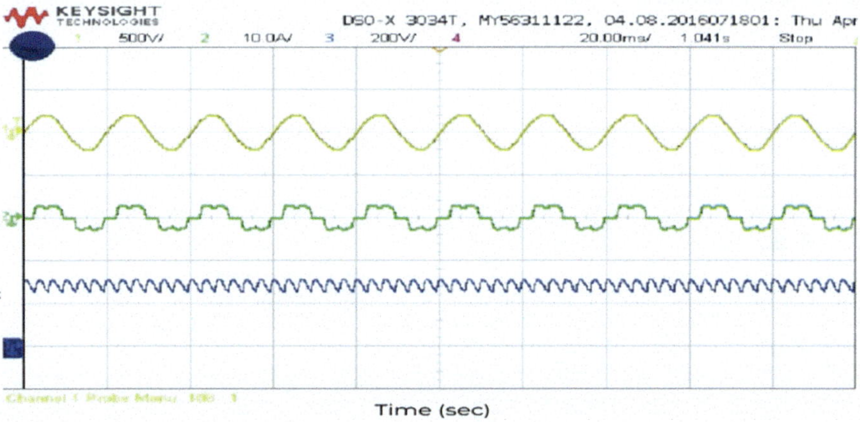

Fig. 14 Output waveforms of Vs, Is and V_R for optimized PI controller parameters and 130 V, 0.25 KW resistive load before compensation

To prove that the values of Kp and Ki obtained by the class topper algorithm are better than the conventional values of Kp and Ki, these conventional values of the PI controller were provided to the FPGA controller and results were taken to see the difference. Output waveforms of I_L, Is, Vs and I_{INV} for conventional PI controller parameters before and after compensation was provided in the power system can be seen in Fig. 16. This can be observed from the above waveform that there are still so many harmonics and PI oscillations present in the output waveform of source current (Is) when the conventional values of Kp and Ki used for the PI controller.

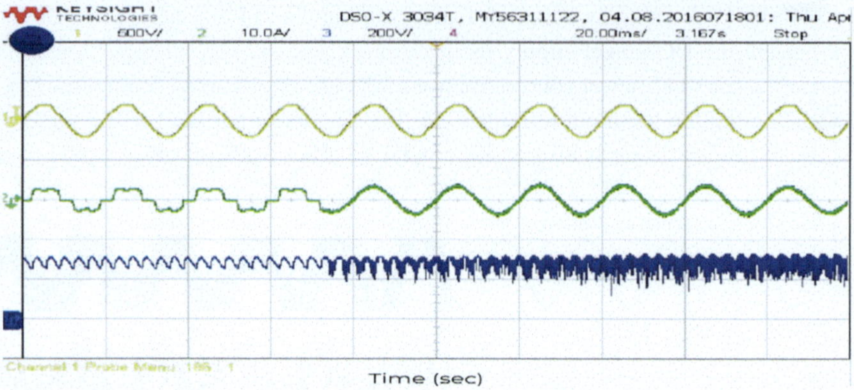

Fig. 15 Output waveforms of Vs, Is and VR for optimized PI controller parameters and 130 V, 0.25 Kw resistive load before and after compensation

The settling time for the waveform in the source current is more too, settling time is important because it affects the ability of a control system to suppress or mitigate harmonics, is shown in Fig. 17.

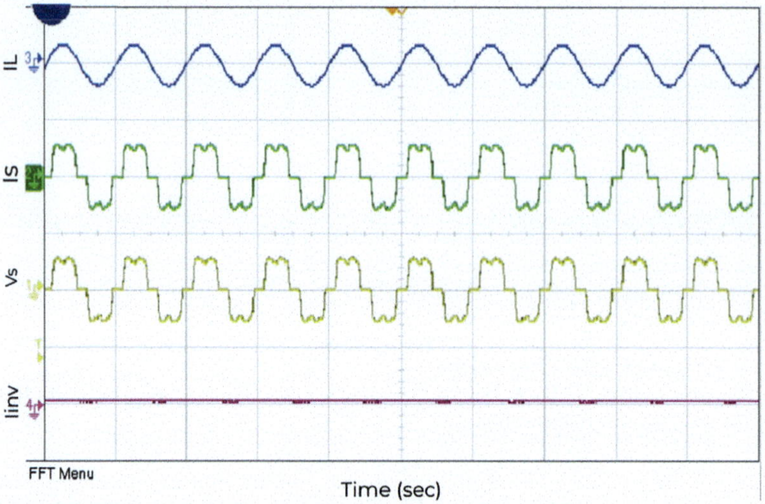

Fig. 16 Output waveforms of I_L, Is, Vs and I_{INV} for conventional PI controller parameters and 130 V, 0.25 KW resistive load before compensation

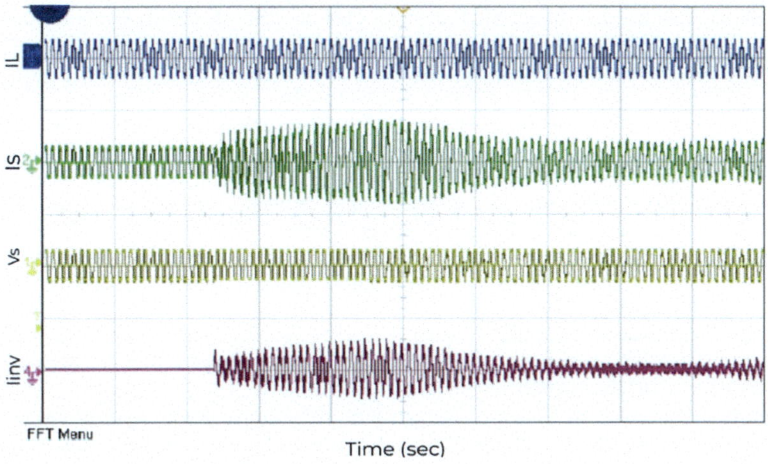

Fig. 17 Output waveforms of I_L, Is, Vs and I_{INV} for conventional PI controller parameters and 130 V, 0.25 Kw resistive load before and after compensation

7 Conclusions

The design of D-STATCOM for mitigation of current harmonics present in source current using the Class Topper Algorithm turns out to be a successful approach for mitigating current harmonics in power systems. The Class Topper Algorithm proves to be an effective optimization algorithm that can efficiently determine the optimized values of Kp and Ki for the PI controller that can be used for the control of the designed D-STATCOM, thereby reducing the total harmonic distortion in the source current of the power system. The simulation results demonstrate that the proposed D-STATCOM with the Class Topper Algorithm can reduce the THD to an acceptable level according to the IEEE 519–2014 standard and improve the power factor of the used standalone system. Additionally, the proposed system is robust and can operate successfully under various load conditions provided by varying the value of the total loads in the resistive load box. Overall, the proposed design can be a very reliable and an efficient solution for mitigating current harmonics in power systems. Future work focused on the very same objectives can explore the implementation of the proposed design in practical systems and its performance under real-world conditions.

References

1. Singh B, Al-haddad K, Chandra A (1999) Power quality improvement. Adv Power Electron 46(5):960–971. https://doi.org/10.1109/ICECENG.2011.6057300
2. Siqueira L, Carneiro AM, De Oliveira M (2000) Assessment of harmonic distortions on power and energy measurements. Proc Int Conf Harmon Qual Power ICHQP 1:105–109. https://doi.org/10.1109/ICHQP.2000.897006

3. Buddhi D, Varghese LJ, Hamid SS, Ramya D, Chakravarthi MK (2022) Harmonic distortion reduction in power system to improve reliability and power quality. In: International conference on innovative computing, intelligent communication and smart electrical systems, vol 12, pp 1–5. https://doi.org/10.1109/ICSES55317.2022.9914129
4. Selvakumar K, Palanisamy R, Vijayakumar K (2017) Hysteresis control 3-Level SI-NPC inverter with wind energy system. Int J Power Electro Drive Syst 8:1764–1770. https://doi.org/10.11591/ijpeds.v8.i4.pp1764-1770
5. Kazem HA (2020) Harmonic mitigation techniques applied to power distribution networks. Adv Power Electron 12(3):2013. https://doi.org/10.1155/2013/591680
6. Fujita H, Akagi H (2001) A practical approach to harmonic compensation in power systems series connection of passive and active filters. IEEE Trans Ind Appl 27(6):1020–1025. https://doi.org/10.1109/28.108451
7. Kumar C, Mishra MK, Mekhilef S (2020) A new voltage control strategy to improve performance of DSTATCOM in electric grid. CES Trans Electr Mach Syst 4(4):295–302. https://doi.org/10.30941/CESTEMS.2020.00036
8. Singh B, Jayaprakash B, Kumar S, Kothari DP (2021) Implementation of neural-network-controlled three-leg VSC and a transformer as three-phase four-wire DSTATCOM. IEEE Trans Ind Appl 47(4):1892–1901. https://doi.org/10.1109/TIA.2011.2153811
9. Singh B, Al-Haddad K, Chandra A (2003) A review of active filters for power quality improvement. IEEE Trans Industr Electron 46(5):960–971. https://doi.org/10.1109/41.793345
10. Labeeb M, Lathika BS (2023) Design and analysis of DSTATCOM using SRFT and ANN-fuzzy based control for power quality improvement. IEEE Recent Adv Intell Computat Syst 1:274–279. https://doi.org/10.1109/RAICS.2011.6069317
11. Singh B, Solanki J (2019) A comparison of control algorithms for DSTATCOM. IEEE Trans Industr Electron 56(7):2738–2745. https://doi.org/10.1109/TIE.2009.2021596
12. Saxena V, Kumar N, Singh B, Panigrahi BK (2021) A rapid circle centre-line concept-based mppt algorithm for solar photovoltaic energy conversion systems. IEEE Trans Circuits Syst I Regul Pap 68(2):940–949. https://doi.org/10.1109/TCSI.2020.3038114
13. Sano K, Takasaki M (2012) A transformerless D-STATCOM based on a multivoltage cascade converter requiring no DC sources. IEEE Trans Power Electron 27:2783–2795. https://doi.org/10.1109/TPEL.2011.2174383
14. Kanjiya P, Singh B, Chandra A, Al-Haddad K (2020) SRF theory revisited to control self-supported dynamic voltage restorer (DVR) for unbalanced and nonlinear loads. IEEE Trans Ind Appl 49(5):2330–2340. https://doi.org/10.1109/TIA.2013.2261273
15. Salmeron P, Litran SP (2010) Improvement of the electric power quality using series active and shunt passive filters. IEEE Trans Power Deliv 25(2):1058–1067. https://doi.org/10.1109/TPWRD.2009.2034902
16. Kumar N, Singh B, Panigrahi BK (2019) Pnklmf-based neural network control and learning-based hc mppt technique for multiobjective grid integrated solar pv based distributed generating system. IEEE Trans Industr Inf 15(6):3732–3742. https://doi.org/10.1109/TII.2019.2901516
17. Sadanand DDK (2021) Modeling of photovoltaic solar cell based on cusbs absorber for the enhancement of performance. IEEE Trans Electron Devices 68(3):1121–1128. https://doi.org/10.1109/TED.2020.3048326
18. Kanjiya P, Singh B, Chandra A, Al-Haddad K (2019) SRF theory revisited to control self-supported dynamic voltage restorer (DVR) for unbalanced and nonlinear loads. IEEE Trans Ind Appl 49(5):2330–2340. https://doi.org/10.1109/TIA.2013.2261273
19. Singh B, Al-Haddad K, Chandra A (2002) A review of active filters for power quality improvement. IEEE Trans Industr Electron 46(5):960–971. https://doi.org/10.1109/41.793345
20. Jabr RA, Dzafic I (2016) A compensation-based conic OPF for weakly meshed networks. IEEE Trans Power Syst 31:4167–4168. https://doi.org/10.1109/tpwrs.2015.2505508
21. Gianto R, Purwoharjono (2022) Trust-region method for load flow solution of three-phase unbalanced electric power distribution system. J Electr Comput Eng 1–17. https://doi.org/10.1155/2022/5415300

22. Kawambwa S, Mwifunyi R, Mnyanghwalo D, et al (2021) An improved backward/forward sweep power flow method based on network tree depth for radial distribution systems. J Electr Syst Inf Technol 8. https://doi.org/10.1186/s43067-021-00031-0
23. Alzaareer K, Saad M, Mehrjerdi H et al (2022) New voltage sensitivity analysis for smart distribution grids using analytical derivation: ABCD model. Int J Electr Power Energy Syst 137:107799. https://doi.org/10.1016/j.ijepes.2021.107799
24. Tyagi A, Kumar K, Ansari MA, Kumar B (2020) An efficient load flow solution for distribution system with addition of distributed generation using improved harmony search algorithms. J Electr Syst Inf Technol 7. https://doi.org/10.1186/s43067-020-00014-7
25. Carreno IL, Scaglione A, Saha SS et al (2023) Log(v) 3LPF: a linear power flow formulation for unbalanced three-phase distribution systems. IEEE Trans Power Syst 38:100–113. https://doi.org/10.1109/tpwrs.2022.3166725
26. Al-Sakkaf A, AlMuhaini M (2018) Power flow analysis of weakly meshed distribution network including DG. Eng Technol Appl Sci Res 8:3398–3404. https://doi.org/10.48084/etasr.2277
27. Martinot E, Kristov L, Erickson JD (2015) Distribution system planning and innovation for distributed energy futures. Curr Sustain/Renew Energy Rep 2:47–54. https://doi.org/10.1007/s40518-015-0027-8
28. Alves G de O, Pereira JLR, Passos Filho JA (2020) A new unbalanced three-phase governor power flow formulation based on the current injections method. Int J Electr Power Energy Syst 123:106184. https://doi.org/10.1016/j.ijepes.2020.106184
29. Bagheri Tolabi H, Ali MH, Rizwan M (2015) Simultaneous reconfiguration, optimal placement of DSTATCOM, and photovoltaic array in a distribution system based on fuzzy-ACO approach. IEEE Trans Sustain Energy 6:210–218. https://doi.org/10.1109/tste.2014.2364230
30. Salkuti SR (2023) Advanced technologies for energy storage and electric vehicles. Energies 16(5):2312. https://doi.org/10.3390/en16052312
31. Tamilselvan V, Jayabarathi T, Raghunathan T, Yang X-S (2018) Optimal capacitor placement in radial distribution systems using flower pollination algorithm. Alex Eng J 57:2775–2786. https://doi.org/10.1016/j.aej.2018.01.004
32. Salkuti SR (2022) Emerging and advanced green energy technologies for sustainable and resilient future grid. Energies 15(18):6667. https://doi.org/10.3390/en15186667
33. Elghaffar NA, Eltamaly AM, Sayed Y, El-Sayed AHM (2018) The effectiveness of dynamic voltage restorer with the distribution networks for voltage sag compensation. Int J Smart Electr Eng 7(02):61–67. https://doi.org/10.1007/S00202-017-0666-4
34. Vural AM (2016) Self-capacitor voltage balancing method for optimally hybrid modulated cascaded H-bridge D-STATCOM. IET Power Electron 9:2731–2740. https://doi.org/10.1049/iet-pel.2016.0256
35. Gawande SP, Ramteke MR (2014) Three-level NPC inverter based new DSTATCOM topologies and their performance evaluation for load compensation. Int J Electr Power Energy Syst 61:576–584. https://doi.org/10.1016/j.ijepes.2014.03.065
36. Salkuti SR, Ray P, Pagidipala S (2022) Overview of next generation smart grids. In: Salkuti SR, Ray P (eds) Next generation smart grids: modeling, control and optimization. lecture notes in electrical engineering, vol 824. Springer, Singapore. https://doi.org/10.1007/978-981-16-7794-6_1

The manufacturer's authorised representative in the EU is Springer Nature Customer Service Centre GmbH, Europaplatz 3, 69115 Heidelberg, Germany. If you have any concerns regarding our products, please contact ProductSafety@springernature.com

Printed and bound by CPI Group (UK) Ltd, Croydon, CR0 4YY

30/09/2026

02225233-0017